VOLUME FIFTY

Annual Reports in
MEDICINAL CHEMISTRY
Platform Technologies in Drug Discovery and Validation

ANNUAL REPORTS IN MEDICINAL CHEMISTRY

VOLUME 50

Serial Editor

STEPHEN NEIDLE

UCL School of Pharmacy,
London, United Kingdom

VOLUME FIFTY

ANNUAL REPORTS IN MEDICINAL CHEMISTRY

Platform Technologies in Drug Discovery and Validation

Volume Editor

ROBERT A. GOODNOW, Jr

Pharmaron Inc., Boston, MA, United States

ACADEMIC PRESS

An imprint of Elsevier

Academic Press is an imprint of Elsevier
50 Hampshire Street, 5th Floor, Cambridge, MA 02139, United States
525 B Street, Suite 1800, San Diego, CA 92101–4495, United States
The Boulevard, Langford Lane, Kidlington, Oxford OX5 1GB, United Kingdom
125 London Wall, London, EC2Y 5AS, United Kingdom

First edition 2017

Notices

Knowledge and best practice in this field are constantly changing. As new research and experience broaden our understanding, changes in research methods, professional practices, or medical treatment may become necessary.

Practitioners and researchers must always rely on their own experience and knowledge in evaluating and using any information, methods, compounds, or experiments described herein. In using such information or methods they should be mindful of their own safety and the safety of others, including parties for whom they have a professional responsibility.

To the fullest extent of the law, neither the Publisher nor the authors, contributors, or editors, assume any liability for any injury and/or damage to persons or property as a matter of products liability, negligence or otherwise, or from any use or operation of any methods, products, instructions, or ideas contained in the material herein.

ISBN: 978-0-12-813069-8
ISSN: 0065-7743

Publisher: Zoe Kruze
Acquisition Editor: Jason Mitchell
Editorial Project Manager: Shellie Bryant
Production Project Manager: Vignesh Tamil
Cover Designer: Alan Studholme

Typeset by SPi Global, India

CONTENTS

CONTRIBUTORS

PREFACE

Modern drug discovery has become ever more interdisciplinary. The advance of technologies that provide new opportunities to a drug discovery scientist often progresses at a pace beyond all but specialist of a particular technology. Strategies to treat unmet medical need now include work on targets that often necessitate multiple approaches throughout the discovery phase of a project. In the past one envisioned a typical medicinal chemist working in a chemistry laboratory, synthesizing compounds of his or her individual imagination and design. Many projects began with a natural product or a small molecule that may have been discovered due to in vivo phenotypic screening. Efficient, manual synthesis of a few hundred compounds were assayed in a limited number of biological and physical characterization assays. That approach was highly successful and productive in many projects and resulted in invention of a plethora of useful drugs.

The medicinal chemist of today must operate in a much different manner. First, there is the flow of much greater quantities of diverse information. There is an increased perception of multiple biological targets, their protein structures as well as the validity of the concept to modulate that target as disease treatment. The likely polypharmacology of many small molecules is now increasingly appreciated and leveraged to the advantage of the project. Medicinal chemists today must confront targets which may not necessarily be tackled with a small molecule and thus, now must be at least conversant in the applications of peptides, antibodies, oligonucleotides, targeted protein degradation, and gene editing. The complexity and challenge of protein–protein interaction targets exemplify much of the complexity that a medicinal chemist must now consider in the prosecution of many discovery campaign. There are also new chemistry technologies such as DNA-encoded chemistry and flow chemistry that provide new opportunities to stream line the efficiency and effective of a medicinal chemist. Despite the power of these approaches, challenges remain for the medicinal chemistry to effectively navigate the many and diverse options.

To this end, this volume is the result of a group of world experts who have come together to write an up-to-date summary of platform technologies for drug discovery and target validation. A common theme has been to provide clear summary and explanation of these technologies to "the busy, perhaps overloaded medicinal chemist." Examples of the applications of these platform technologies provide guidance when and how to include such a strategy in any drug discovery project. Whereas efforts have been made to cover important and new developments, it is not possible within a single volume to include all potential platform technologies. Rather, the intent of this volume is that of a useful, readable guide, not an encyclopedia.

The new format of *Annual Reports in Medicinal Chemistry* befits this common theme. This is the fiftieth volume of the series, but the first of a new, single topic format. *Annual Reports in Medicinal Chemistry* is an Elsevier owned book series that was published previously under the editorship of the ACS Division of Medicinal Chemistry. Now the Elsevier editorial board have changed the series so as to focus on a single topic per volume.

I am grateful to all chapter authors who have taken time from their busy days of drug discovery research and have collaborated in this effort, sharing their expertise and guidance for effective utilization of many different platform technologies for drug discovery and target validation. I am appreciative of Elsevier's staff and web tools that have made convenient this remote collaboration.

Robert A. Goodnow, Jr.
Boston, September 2017

CHAPTER ONE

DNA-Encoded Library Technology: A Brief Guide to Its Evolution and Impact on Drug Discovery

Robert A. Goodnow Jr.*[,1], Christopher P. Davie[†]
*Pharmaron Inc., Boston, MA, United States
[†]GlaxoSmithKline, Waltham, MA, United States
[1]Corresponding author: e-mail address: robert.goodnow@pharmaron.com

Contents

1. INTRODUCTION TO DELT

1.1 The Basic DELT Concept

In creating novel medicines for unmet medical need, many drug discovery scientists look to diversity screening methods (i.e., the assay or screening of many compounds in an unbiased manner for some initial sign of binding affinity and/or pharmacological potency against a biological target of interest). This methodology has been successfully applied, automated, and optimized in multiple high-throughput screening (HTS) formats (see Chapter "High-throughput screening" by Wildey et al. of this volume). The positive impact of HTS on hit findings is inarguable. In HTS, along with the automation technology and assay development for optimizing a process, the value of a quality compound library cannot be overstated. Drug

Annual Reports in Medicinal Chemistry, Volume 50
ISSN 0065-7743
https://doi.org/10.1016/bs.armc.2017.09.002

discovery organizations have spent great sums of money and time building and curating large collections of small molecules, any one of which if identified as an initial hit, would be an attractive starting point for further optimization, ultimately to a clinical candidate. Current methods of automation and cost of reagents often limit the number of compounds that can be screened to not greater than 1–2 million. In fact, many organizations and screening projects for certain target classes (e.g., kinase inhibitors) are conducted using focused screening sets, a subset of the full HTS collection that is intentionally biased according to some ideas about what sorts of compounds ought to be good starting points for subsequent optimization. Organizations lacking large screening collections must seek other strategies than HTS for hit identification.

While HTS has become an efficient and routine process practiced by many organizations, the size of chemical space is vast, much greater than the chemical space covered by a few million compounds.[1,2] As a result, many chemists may often be left with the question, "What high-affinity ligands might I have found for a particular protein target, were I to have used a hit finding method alternative to HTS?" DNA-encoded library technology (DELT) represents exactly a response to that question and need. Because of the split-and-pool nature of DELT synthesis, it is possible to make huge numbers of compounds in a cost- and time-efficient manner (millions to billions). There are reports of trillion compound libraries made possible by DELT, libraries six orders of magnitude greater than a typical HTS collection.

Reviews of various DELT methods and concepts now abound.[3–6] Briefly, a widely practiced format of DELT is a general method in which split-and-pool library synthesis[7] is encoded with strands of oligonucleotides (DNA), thereby tagging each compound with a sequence of DNA that can be linked to the synthetic history that produced such a compound (Fig. 1). The resultant compound mixture is then incubated with a protein target of interest followed by the physical separation through washing away of compounds with lesser binding affinity. Because the incubation of the pooled library with the protein target is done on small scale (μg) and in a time-efficient manner, it is little additional effort to conduct multiple selection campaigns under different conditions (e.g., variable protein concentration, in the presence or absence of a competitor compound, in the presence of different buffers or cofactors). This results in the enrichment of compounds having higher binding affinity; such enrichment is detected based on the sequencing of the PCR products of oligonucleotides that are covalently bound to the high-affinity hits. Important details of this technology include

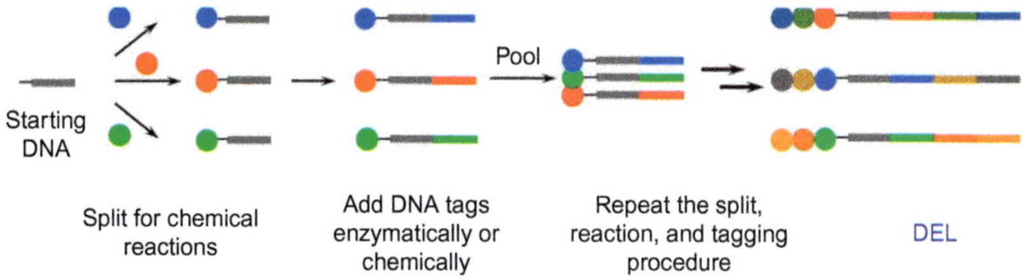

Fig. 1 The basic concept of library synthesis using a DNA-recording method. Each tag sequence represents a different chemical transformation and/or diversity reagent. *Reprinted with permission from Shi, B., et al.* Bioorg. Med. Chem. Lett. ***2017***, 27, *361–369. Copyright 2017 Elsevier.*

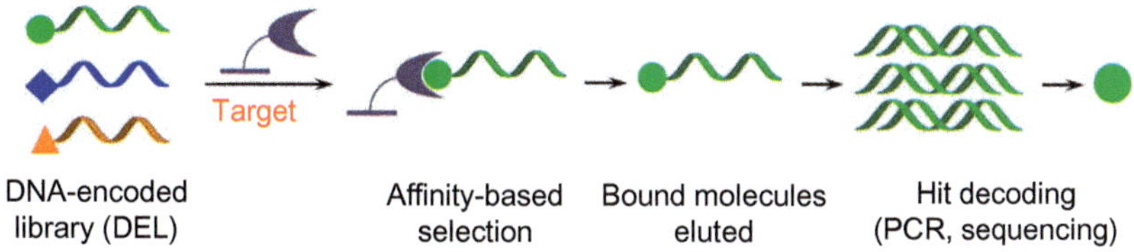

Fig. 2 General method for selection of DNA-encoded small molecule mixtures with subsequent identification of higher affinity binders. *Reprinted with permission from Shi, B., et al.* Bioorg. Med. Chem. Lett. ***2017***, 27, *361–369. Copyright 2017 Elsevier.*

DNA tag design, PCR methods, and next-generation sequencing (NGS). As a result of the combination of these approaches, small quantities of protein target can be exposed to hundreds of millions to billions of compounds simultaneously, producing results in the time it takes to run PCR, sequencing, and data analysis processes (Fig. 2). The high-affinity compounds must then be synthesized "off-DNA," that is without the attachment of a DNA tag. Such compounds are then assayed in order to confirm that they are true hits.

1.2 Evolution of DELT

DNA-encoded library synthesis was originally proposed by Brenner and Lerner in 1992.[8] Phage display of peptides was a well-established technology at that time. These scientists focused on the simplest concept of the phage display technology that could present the synthesis of any small molecule through a DNA sequence; it was an elegant strategy utilizing a phenotype–genotype pairing for a small molecule. The method was shortly exemplified thereafter.[9] Their first proposal was a three-base coding of amino acid sequences, synthesized on solid phase. Over the course of 25 years, many scientists have elaborated the basic methods to create different platforms and formats around the central premise of encoding small molecule libraries with sequences of DNA.[10] Notable highlights in the evolution of this method include disclosures by Clark et al. of the synthesis and selection of an 800 million compound DEL (2009),[11] by Liu and coworkers of a DNA-directing system for the construction of macrocycles (2006),[12] by Neri and colleagues of a fragment-based selection methods (2007),[13] by GSK scientists disclosure of the advance of compounds discovered by DELT into clinical development (2015, vide infra), and by the commercialization of NGS technology by Illumina (2008).[14] Currently there are multiple organizations in pharma, biotech, and academia that have established this technology as a tool for hit identification.

1.3 Different DNA-Encoding and -Directing Formats

As mentioned earlier there are different DELT formats. They include the most commonly practiced DNA recording, in which DNA sequences are attached during the "split" stage along with the reaction of each diversity reagent, thereby forming a unique DNA sequence for each unique combination of building blocks. Such methods have been exemplified by GSK and Nuevolution. There are also strategies in which the formation of sequence-specific DNA duplexes not only report on the combination of particular building blocks but actually guide such building blocks to react with each other (Fig. 3). Such methods are called DNA-directing or DNA-programmed

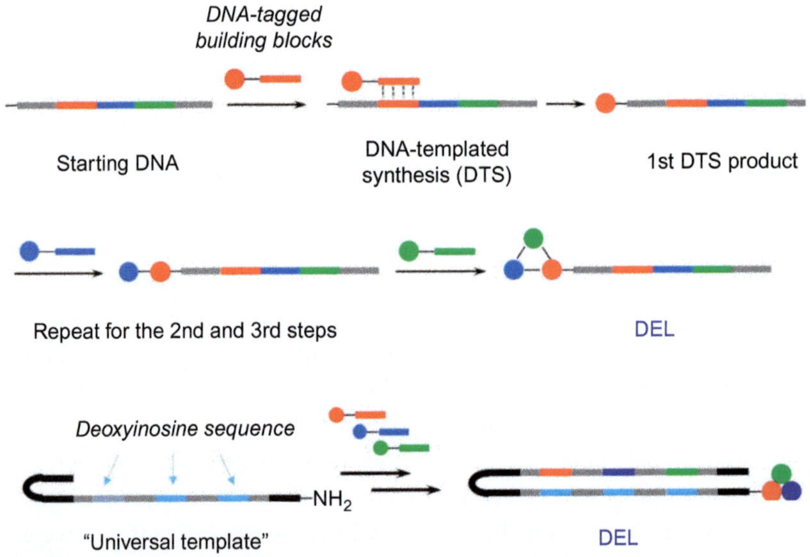

Fig. 3 DNA-directed or DNA-programmed chemistry synthesis methods. DNA tags not only encode the building block identity but guide it to react in an appropriate manner. *Reprinted with permission from Shi, B., et al.* Bioorg. Med. Chem. Lett. ***2017****, 27, 361–369. Copyright 2017 Elsevier.*

chemistry and have been exemplified by Ensemble and Vipergen. Further, there are methods in which DNA duplexes result in the spatial arraying of specific sequences for reaction of the diversity element according to that spatial segregation; such chemistry methods have been reported on by the laboratories of Harbury (Fig. 4). Additionally, DNA sequences may also be used to report on the association of small molecule fragments (Fig. 5) that may combine according to the highest affinity to a protein target by a particular combination of fragments (as published by the laboratories of Neri). Different combinations of these approaches continue to appear in the literature, highlighting the adaptability of DNA as a functional and information-containing medium.

DNA is highly suited to applications for encoded libraries. It has been optimized by nature as a high-density information encoding and recording medium. It is stable under many conditions (e.g., basic, aqueous, thermal),

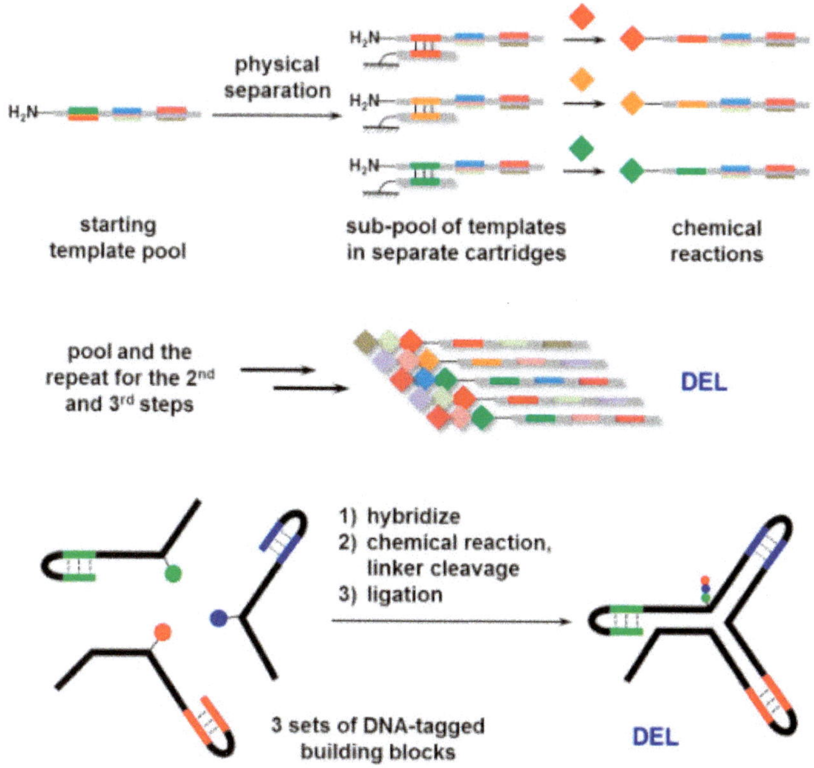

Fig. 4 DNA sequences can also be used in other DNA-directing ways to create sequence-specific reaction environments. *Adapted with permission from Shi, B., et al.* Bioorg. Med. Chem. Lett. ***2017***, *27*, *361–369. Copyright 2017 Elsevier.*

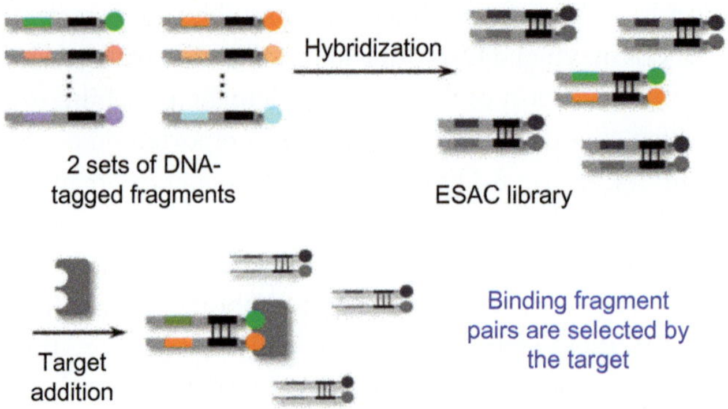

Fig. 5 DNA sequences report on the association of high affinity fragments according to their binding to a protein target. *Adapted with permission from Shi, B., et al.* Bioorg. Med. Chem. Lett. ***2017***, *27, 361–369. Copyright 2017 Elsevier.*

and the growing number of DELT reaction development publications has demonstrated that DNA is also stable upon exposure to a good number of more traditional organic chemistry reagents (e.g., transition metal catalysts and reducing agents). There are many enzymes that operate with great specificity and efficiency on DNA sequences (e.g., T4 ligase and DNA polymerase). These enzymes are important tools in the practice of DNA-encoded chemistry. The ability to amplify ancient sequences of DNA in different biological samples (human, plant, bacterial) is a testament to the robust nature of this encoding medium. There has also been some success to develop other information recording medium such as peptide nucleic acids (PNAs) for encoding and directing the synthesis of chemical libraries.[15]

The basic concept of library synthesis uses a DNA-templated or DNA-directing method. Each tag sequence represents a different chemical transformation and/or diversity reagent.

2. REPORTED HITS FOUND WITH DELT AS EXAMPLE OF THE POTENTIAL TO DELIVER DRUG-LIKE MOLECULES

Scientists who have gone to the trouble to set up and perfect a particular technology are often eager to demonstrate return on investment. One means of demonstration of DELT's potential for return on investment is to point to examples of clinical candidates that have come from the DELT technology. Accordingly, scientists at GlaxoSmithKline (GSK) have disclosed two compounds resulting from DELT that have entered different phases of clinical development. GSK scientist used DELT for discovery of ligands having affinity against RIP1 kinase, a target for interest in immune-mediated inflammatory diseases along with other therapeutic areas. From this selection, GSK 481 was discovered and this led to clinical candidate GSK2982772 (Fig. 6). The initial hit, GSK 481, highlights the importance of diverse building block collections (vide infra).[16] Additionally, using DELT, a hit was discovered which led to GSK2256294A, a potent, reversible, tight-binding inhibitor of isolated recombinant human soluble epoxide hydrolase (sEH, $IC_{50} = 27$ pM).[17]

Another means to demonstrate the utility of the DELT methods is to point to high-affinity hits that have been discovered in DELT selections. Many examples of reported hits resulting from DELs have been tabulated in several manuscripts.[3,18,19] Shown in Fig. 7 are two examples of such DELT hits. These include AZ3451, a potent activator of PAR2 (GPCR) and PAD4 (enzyme) inhibitor by GSK, a compound for interest in oncology research. These molecules contain several features, attractive to medicinal chemists: only one amide bond, stereogenic centers, inclusion of heterocycles and diverse functionality, nonpeptidic, etc. While it is difficult to categorize definitively "drug-likeness," these molecules are good examples of such a terminology. The fact that these molecules and related analogues can be made with DELT and in vast numbers is a convincing indicator of the potential of this technology for drug discovery.

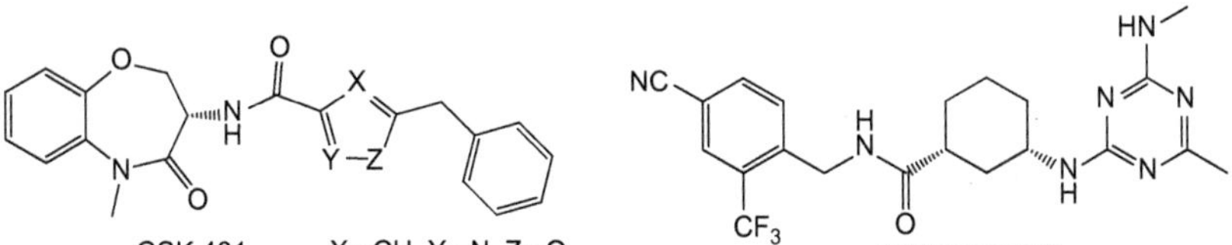

GSK 481 X=CH, Y=N, Z=O
GSK2982772 X=Y=Z=N
GSK2256294A

Fig. 6 Compounds resulting from DELT selections that have advanced to clinical development.

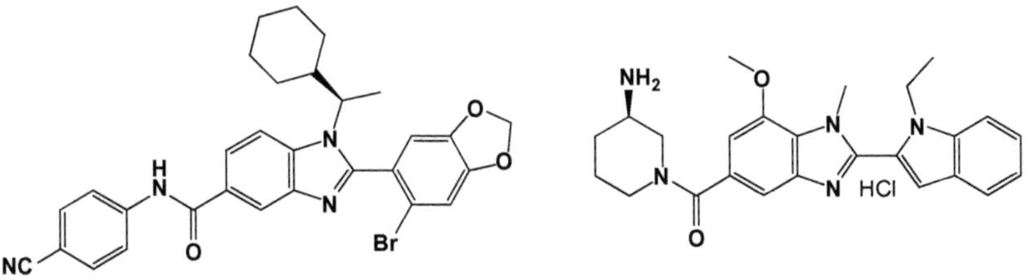

4 AZ3451 **5**

Fig. 7 Compounds reported as hits that have resulted from selections of DELs against PAR2 and PAD4.

These hits have reasonable to good drug-like properties in terms of molecular weight and lipophilicity. This is the result of strategic incorporation of building blocks and functional group motifs for the DNA-compatible chemistry (vide infra) that was used to assemble such compounds. The drug-like properties of molecules made via a split-and-pool synthesis result from the balance of two somewhat opposing motivations. On one hand, a chemist may wish to take advantage of the power of split-and-pool synthesis and make large numbers of compounds, most conveniently possible with three or four cycles of chemistry using as many building blocks as possible at each cycle. However, three or four cycles of chemistry with high molecular weight and/or lipophilic diversity reagents will generate large numbers of molecule well outside the so-called Lipinski space, thus creating potentially active molecules that will then be difficult to optimize into good hits and leads for drug discovery. The solution to this situation is often by consideration of library design with carefully selected building block collections. Organizations that have inventories of many proprietary building blocks are in a particular position of advantage.

In other efforts, scientists have used DELT to build intentionally large, macrocyclic systems to tackle protein–protein interaction (PPI) targets. Such targets are thought to require ligands that span a larger surface area than most small molecules and are at or beyond the limits of small molecule drug-likeness definitions. DELT is a good tool to create such libraries as it can bring together multiple building block inputs in a combinatorial fashion, thereby generating large numbers of complex molecules. It is thought that a large number of building blocks arrayed in complex, dynamic structures is one means to explore the chemical space at the PPI interface. Of course, the challenges to effectively deliver such molecules as therapeutics remain. An example of success here is the IL-17A inhibitor for treatment of inflammatory conditions by Ensemble discovered with the DNA-directing methodology (also called DNA-programmed chemistry) of that company[20] (Fig. 8).

Another excellent example of the diverse output of DEL selection is to be found in recent work by X-Chem scientists who described the discovery of potent Bruton's tyrosine kinase (BTK) inhibitors.[21] BTK is a kinase of interest for its role in B cell development and in anticancer therapeutics. In this report, BTK was the selection target for some 110 million DNA-encoded compounds constructed in a four-cycle library. From that selection, three hit families were discovered; exemplars are shown in Fig. 9. In each of these molecules, the off-DNA analogue was created by substitution of the DNA-attachment point with methylamine at the initial amino

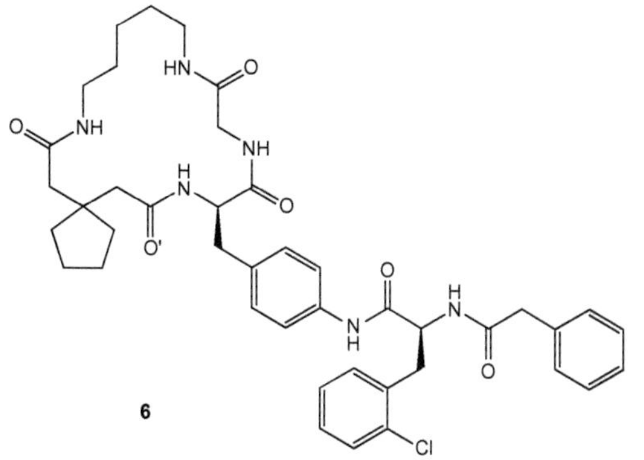

6

Fig. 8 Potent IL-17-modulating macrocycle discovered using DNA-programmed chemistry.

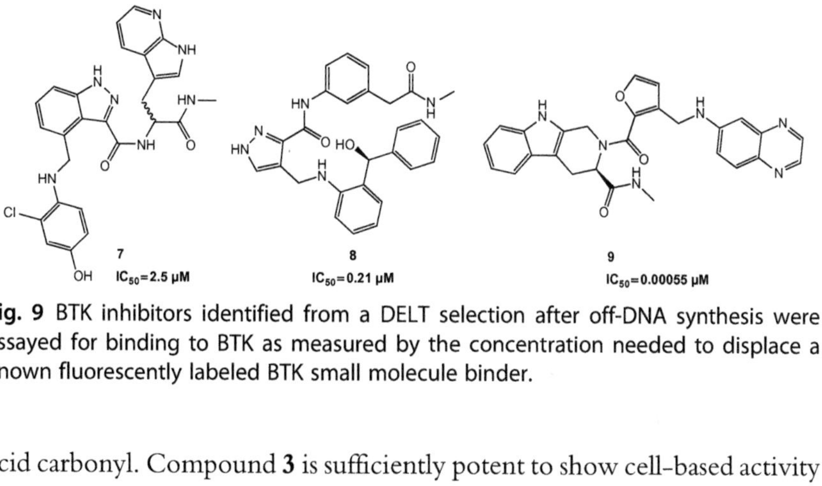

7 IC_{50}=2.5 μM

8 IC_{50}=0.21 μM

9 IC_{50}=0.00055 μM

Fig. 9 BTK inhibitors identified from a DELT selection after off-DNA synthesis were assayed for binding to BTK as measured by the concentration needed to displace a known fluorescently labeled BTK small molecule binder.

acid carbonyl. Compound **3** is sufficiently potent to show cell-based activity in Ramos, Jurkat, and human whole blood assays. The compounds were also characterized in terms of the mechanism of BTK inhibition: whereas Compounds **7** and **8** appear competitive with ATP, Compound **9** is either noncompetitive or binds unusually tightly to BTK. This example is instructive in several ways: (a) identification of high-affinity compounds of a target of high therapeutic interest; (b) the parallel or multiplexed means by which the compounds were identified; (c) the identification of compounds having different modes of target interaction (i.e., competitive, noncompetitive, etc.); (d) the sensitivity of the selection enrichment to protein concentration; and finally (e) the manner in which the DNA-attachment point has been replaced by a small fragment (methylamine) for off-DNA synthesis. Of this latter point, there is little published or generalized, but there are several verbal reports of the sensitivity of the small molecule–protein target interactions at the DNA-attachment point.

2.1 Application of DELT to Cellular Selections

The majority of reported hits for DELT have been with soluble, affinity-tagged proteins. However, there is at least one report of a selection performed using whole cells by GSK scientists in which a cell-active antagonist of tachykinin receptor neurokinin 3 was detected.[22] In this method, high expression of the target protein is important. Limiting the stickiness of the cell surface is another important consideration. Finally, cell surface protein targets seem within scope of this approach. Whereas some form of mediated uptake of a DEL into the cell cannot be discounted, relying on such transport to reach intracellular targets would provide an additional complexity. It is conceivable that transfection agents may be useful in performing selections on intracellular targets followed by isolation of the protein–small molecule complex from lysed cells. These clearly challenging technical details appear to have limited this approach to selection to date but are a likely direction of future development of selection methodology.

3. EVOLUTIONARY DIRECTIONS FOR DNA-ENCODED PLATFORMS

DNA as a platform for encoding and directing binding interactions is highly robust, generally stable, and thus readily adaptable to other applications. This fact is likely one of the root causes by which DNA-encoded methods continue to evolve in divergent and creatively applied ways. A recent innovation of DELT methods has been published in a series of papers by scientists in the Paegel laboratories of Scripps Florida.[23] In these disclosures, DELT chemists have returned to the solid phase by creating libraries that can be guided through cell-sorting systems to accommodate new forms of screening. These scientists have also adapted the DNA-tagging system to incorporate a check of DNA integrity.[24] There is an increasing awareness that running chemistry in a so-called DNA-compatible format does not always spare the DNA of any modification. DELT chemists are increasingly aware of the need to adapt high-yielding reactions to a DNA-compatible format but also to preserve the integrity of the DNA tag such that it is still functional to PCR amplification and sequencing.

DELT has also been creatively used by GSK scientists in the prioritization and small molecule tractability assessment of more than 100 antibacterial targets.[25] In this report, targets from *A. baumannii* and *S. aureus* were selected in parallel with mixtures of hundreds of millions of DNA-encoded compounds in a parallel and platform-automated format. Data analysis in terms of hit rate and assay of specific hits gives an early indication on the tractability of a particular target for small molecule intervention. By comparison with previous HTS data sets, it was understood that DELT results could be used as a reasonable basis for predicting success in an HTS campaign (i.e., 70% concurrence of DELT and HTS success). Given the ease and speed of running a DELT selection relative to an HTS campaign, this illustrates an efficient way to triage and prioritize new biological targets for drug discovery efforts.

In another interesting evolution, the laboratories of Jacob Berlin at City of Hope in Los Angeles have also used solid-phase beads to encode peptide synthesis with DNA.[26] The beads are then arrayed on small plates, and once the DNA has been sequenced, the identity of the compound on the bead and the beads' location are then established. At that point, the DNA sequences are removed and assays that may be sensitive to the presence of a large strand of DNA can be run. Hits are then identified based on their known location.

Scientists at University of Hong Kong have developed a method known as ligate-crosslink-purify that allows for the selection of nonimmobilized protein targets.[27] In this method, binding to the unmodified protein target in solution and irradiation enable the formation of a construct that can be separated from nonbinders by gel. This method expands the proteins that can be incorporated into DELT methods.

Scientist at Peking University have developed dynamic libraries for conducting selection of small fragments against protein targets. Similar to the AESACS technology popularized by the laboratories of Neri at ETH, Li and coworkers use the protein target to bias the pairing of a dynamic collection of DNA tag fragments and lock that pairing together by photochemical reactions. Such "locked" oligo duplexes are then sequenced leading to an identification of the fragment pair that had bound to the target.

DELT chemistry first used amide bond forming chemistry with amino acids. This is a convenient start given the ready availability of appropriately protected reagents and a high-yielding, reliable chemistry that produces often biologically active molecules. However, drug discovery scientists often prefer organic small molecules as more drug-like substances. To this end, there have been efforts to adapt established organic chemistry methods and expand the DELT "toolbox." Applicable reactions have been tabulated,[28,29] tallying to greater than 40 reactions. The report of new methods continues apace.[30,31] Interesting, the laboratories of Sabine Flitsch at University of Manchester have used oligosaccharide forming enzymes to enable synthesis of oligosaccharide-containing libraries tagged with DNA.[32] This is an interesting way to meet the challenge of the DNA-compatible chemistry format: use an aqueous enzyme to affect chemical transformations.

An excellent example of elaboration of DEL hits to potent ligands was recently published by AstraZeneca and X-Chem scientists. A selection against Mcl-1 (a proapoptotic BH3 protein family member) was conducted, resulting in the identification of a linear, amidated structure (Mcl-1 FRET $IC_{50}=2$ μM).[33] The hit was studied in a cocrystal structure with Mcl-1 and shown to bind in the BH3-binding grove of Mcl-1 in a somewhat folded manner. The potency of this molecule was greatly increased by cyclization (Mcl-1 FRET $IC_{50} \leq 3$ nM, Fig. 10). Presumably the DNA-tagged structure was flexible enough to bind the target in a somewhat circular manner. When the system was cyclized, it created a preorganized structure for binding, thereby increasing the compounds' affinity. The elaboration of the hit in this manner illustrates the need to carefully evaluate and, when applicable, reconstruct a DEL hit in order to reach new levels of potency. Some scientists have reported that hits from DEL libraries may represent local minima with respect to the interaction environment of many compounds against a protein surface.[34] When this may indeed be the case, then a detailed analysis, deconstruction, and reconstruction of the hit are warranted.

There is an ongoing debate about the necessary and/or useful size of a library. Some make as many compounds as possible; others will filter the product designs against strict drug-likeness filters; still others report that larger libraries may actually result in high false-negative rates.[35] In any case, the creation of large DELs relies on the ready availability of diverse building block collections in great numbers (>1000 each). There are different estimates of the numbers of building blocks that are available depending on functional class and structural filtration criteria.[36] Nevertheless, organizations wishing

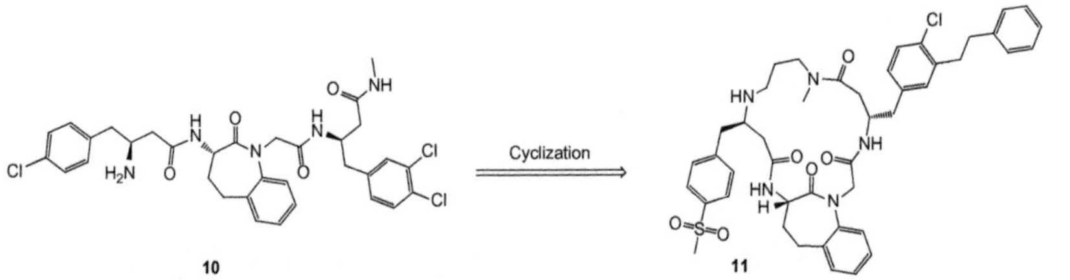

10 11

Fig. 10 Identification of DEL hit against Mcl-1 and its subsequent elaboration to a higher affinity compound.

to make bigger and bigger libraries are stockpiling useful collections of building blocks for DELT reactions. One strategy to incorporate chemistry that is DNA incompatible into a DEL is to do that chemistry on the building block and then use DNA-compatible chemistry for assembly of the library.

Creation of electrophile-containing DELs that may bind covalently to protein targets, either irreversibly or reversibly, is also a new approach to finding high-affinity ligands.[37] The success of covalent kinase inhibitors Afatinib (**12**) and Ibrutinib (**13**) shows the potential of this approach. The Winssinger laboratories have developed a method to synthesize PNA-encoded small molecules that contain cysteine-reactive fragments.[38] In their DNA-array-based detection method, different washing stringencies (PSA vs SDS) will differentiate noncovalent from covalent binders. This platform was used to identify Compound **14** below (Fig. 11) after synthesis of an off-PNA hit. In a 400-kinase panel, Compound **14** at 1 μM bound with selectivity to ERBB2, JAK3, MEK4, and DAP3. This result illustrates an answer to a central question about any method that claims to identify compounds having selective covalent reactivity. Often it is a surprise to find that many covalent systems are quite selective with respect to a broad diversity of reactive nucleophiles within a protein. In hindsight, perhaps this is not so surprising after all. Biological systems require selective reactions against various common substrates according to the use of protein containing nucleophiles. Were such reactive nucleophiles broadly unselective in their reactivity, then biological systems would likely not function properly. Small molecules having functionality that can react with a plethora of biological nucleophiles simply "report" on the inherent selective reactivity of the biological nucleophile.

Finally, an important adaptation of DELT is the discovery of the potent ligands for ubiquitin-proteasome degradation systems (see Chapter "Targeted protein degradation" by Mainolfi and Rasmusson). DELs are well set up to allow the discovery of potent ligands to a particular target. Once a

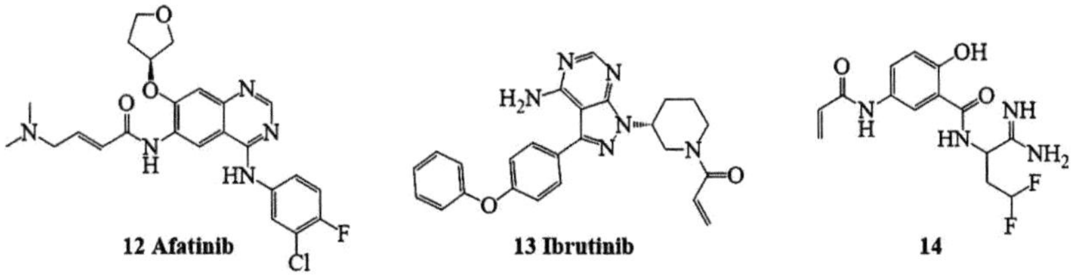

Fig. 11 Identification of a covalently reactive kinase inhibitor using a PNA-encoded library synthesis platform.

high-affinity DNA-encoded hit is found, the DNA portion can be replaced with an E3 ligase-recruiting ligand to recruit and initiate the degradation systems. One can look forward to exciting developments in the combination of these technology approaches.

4. CONCLUSION

In conclusion, DELT has become an established platform technology for hit generation in drug discovery. DELT has provided a rebirth of sorts in the perfection of combinatorial synthesis, which, although popular some 20 years earlier, was viewed to have insufficiently delivered impact and innovation in hit finding. At this time, expectations are growing for DELT to play an increasingly significant role in the discovery of new small molecules for different fields of chemistry. The need to explore new chemical space for novel target biology will continue to propel the growth of DELT and its applications. Given that compound collections have been built in part based on the small molecule designs targeting known biological targets, new biological targets require new designs and new ideas. It is important not to recycle excessively the same chemistry. Thus, DELT has become an attractive means to sample diverse chemistry space in a highly efficient and cost-effective manner.

REFERENCES

1. Bohacek, R. S.; McMartin, C.; Guida, W. C. *Med. Res. Rev.* **1996**, *16*, 3–50.
2. Reymond, J.-L.; van Deursen, R.; Blum, L. C.; Ruddigkeit, L. *Med. Chem. Commun.* **2010**, *1*, 30–38.
3. Goodnow, R. A., Jr.; Dumelin, C. E.; Keef, A. D. *Nat. Rev. Drug Discov.* **2017**, *16*, 131–147.
4. Franzini, R. M.; Randolph, C. *J. Med. Chem.* **2016**, *59*, 6629–6644.
5. Goodnow, R. A. Ed. *A Handbook for DNA-Chemistry*; Wiley: New York, 2014.
6. Shi, B.; et al. *Bioorg. Med. Chem. Lett.* **2017**, *27*, 361–369.
7. Furka, A.; Sebestyen, F.; Asgedom, M., Dibo, G. Proceedings of the 10th International Symposium on Medicinal Chemistry, 1988, Budapest, 288.
8. Brenner, S.; Lerner, R. A. *Proc. Natl. Acad. Sci. U. S. A.* **1992**, *89*, 5381–5383.
9. Nielsen, J.; Brenner, S.; Janda, K. D. *J. Am. Chem. Soc.* **1993**, *115*, 9812–9813.
10. Goodnow, R. A.; Davie, C. P. *Med. Chem. Commun.* **2016**, 7, 1268–1270.
11. Clark, M. A.; Acharya, R. A.; Arico-Muendel, C. C.; et al. *Nat. Chem. Biol.* **2009**, *5*, 647–654.
12. Tse, B. N.; Snyder, T. M.; Shen, Y.; Liu, D. R. *J. Am. Chem. Soc.* **2008**, *130*, 15611–15626.
13. Melkko, S.; Scheuermann, J.; Dumelin, C. E.; Neri, D. *Nat. Biotechnol.* **2004**, *22*, 568–574.
14. Schuster, S. C. *Nat. Methods* **2008**, *5*, 16–18.
15. Zambaldo, C.; Barluenga, S.; Winssinger, N. *Curr. Opin. Chem. Biol.* **2015**, *26*, 8–15.

16. Harris, P. A.; Berger, S. B.; Jeong, J. U.; et al. *J. Med. Chem.* **2017**, *60*, 1247–1261.
17. Belyanskaya, S. L.; Ding, Y.; Callahan, J. F.; et al. *Chembiochem* **2017**, *18*, 837–842.
18. Eidam, O.; Satz, A. L. *Med. Chem. Commun.* **2016**, 7, 1323–1331.
19. Ottl, J. In *A Handbook for DNA-Chemistry*; Goodnow, R. A. Ed.; Wiley: New York, 2014; pp 319–347.
20. Taylor, M; Terrett, N. K.; Connors, W. H. 2013, WO 2013116682, A1.
21. Cuozzo, J. W.; et al. *Chembiochem* **2017**, *18*, 864–871.
22. Wu, Z. N.; et al. *ACS Comb. Sci.* **2015**, *17*, 22–731.
23. MacConnell, A. B.; Price, A. K.; Paegel, B. M. *ACS Comb. Sci.* **2017**, *19*, 181–192.
24. Malone, M. L.; Paegel, B. M. *ACS Comb. Sci.* **2016**, *18*, 182–187.
25. Machutta, C. A.; Kollmann, C. S.; Lind, K. E. *Nat. Commun.* **2017**, *8*, 16081.
26. Berlin, J.; Copeland, G.; Elison, K.; Muradyan, H. 2016, WO 2016138184 A1.
27. Shi, B. B.; Deng, Y.; Zhao, P.; Li, X. *Bioconjug. Chem.* **2017**, *28*, 2293–2301. Article ASAP. https://doi.org/10.1021/acs.bioconjchem.7b00343.
28. Satz, A. L.; Cai, J.; Chen, Y.; et al. *Bioconjug. Chem.* **2015**, *26*, 1623–1632.
29. Satz, A. L.; Luk, K. C. In *A Handbook for DNA-Chemistry*; Goodnow, R. A. Ed.; Wiley: New York, 2014; pp 67–98.
30. Tian, X.; Basarab, G. S.; Selmi, N.; et al. *Bioconjug. Chem.* **2015**, *26*, 1623–1632.
31. Lu, X.; Roberts, S. E.; Franklin, G. J.; Davie, C. P. *Med. Chem. Commun.* **2017**, *8*, 1614–1617.
32. Thomas, B.; Lu, X.; Birmingham, W. R.; Huang, K.; et al. *Chembiochem* **2017**, *18*, 858–863.
33. Johannes, J. W.; Bates, S.; Beigie, C. *ACS Med. Chem. Lett.* **2016**, *8*, 239–244.
34. Deng, H.; et al. *ACS Med. Chem. Lett.* **2015**, *6*, 919–924.
35. Satz, A. L.; Hochstrasser, R.; Petersen, A. C. *ACS Comb. Sci.* **2017**, *19*, 234–238.
36. Kalliokoski, T. *ACS Comb. Sci.* **2015**, *17*, 600–607.
37. De Cesco, S.; Kurian, J.; Dufresne, C.; et al. *Eur. J. Med. Chem.* **2017**, *138*, 96–114.
38. Zambaldo, C.; Daguer, J.-P.; Saarbach, J.; Barluenga, S.; Winssinger, N. *Med. Chem. Commun.* **2017**, 7, 1340–1351.

CHAPTER TWO

The Chemistry of Oligonucleotide Delivery

David B. Rozema[1]
AlexemaBio, LLC, Madison, WI, United States
[1]Corresponding author: e-mail address: dave@alexemabio.com

Contents

Annual Reports in Medicinal Chemistry, Volume 50
ISSN 0065-7743
https://doi.org/10.1016/bs.armc.2017.07.003

1. INTRODUCTION

Due to the central role that nucleic acids play in biology, their potential for use as therapeutics is immense and apparent. In contrast to small-molecule therapeutics that primarily target protein-binding sites to exert their biological effect, oligonucleotides may target the genetic or expression codes for proteins themselves thereby creating new therapeutic platforms that may enable the treatment of previously "undruggable" targets. Unfortunately, oligonucleotides break all the rules of small-molecule drug-likeness, which has slowed the development of oligonucleotide drugs. Indeed—despite the promise and over 30 years of effort to develop oligonucleotide therapeutics—there are a paucity of approved drugs; there are currently five FDA-approved nucleic acids therapeutics. Even though there are only five, it is encouraging to note that two of them have been within the last 2 years, which suggests that the chemistry of oligonucleotides has advanced to make development more facile (Table 1).

Oligonucleotide therapeutics come in a variety of sizes and mechanisms of action. The definition of when an oligonucleotide becomes a polynucleic acid, i.e., a gene, is subjective, but for the purposes of this review we will define oligonucleotide not on any definition of size but its mode of synthesis and whether nucleoside modifications may be incorporated. Similarly,

Table 1 FDA-Approved Nucleic Acid Therapeutics

Name	Mechanism of Action	Administration	Indication
Fomivirsen	Antisense (RNAse H)	Local (intravitreal)	Cytomegalovirus retinitis
Mipomersen	Antisense (RNAse H)	Systemic (iv)	Familial hypercholesterolemia
Pegaptanib	Aptamer	Local (intravitreal)	Neovascular age-related macular degeneration (wet AMD)
Eteplirsen	Splice blocking	Systemic (iv)	Duchenne's muscular dystrophy
Nusinersen	Splice blocking	Local (intrathecal)	Spinal muscular atrophy

nucleic acid delivery may be accomplished using naturally occurring viral vectors, but this review will concentrate solely on nonviral vectors.

2. MECHANISMS OF ACTION

As a broad categorization, oligonucleotide therapeutics can be grouped into four categories based upon mechanism of action: increased gene expression, decreased gene expression, altered (or new) products of expression, and ligands against extracellular receptors. We will discuss the four mechanisms of action along with discussions of the size of the oligonucleotides, and the tolerance vectors have for the incorporation of nonnatural nucleosides (Table 2, Fig. 1).

2.1 Gene Expression Can Be Decreased by Inhibiting Transcription, Increasing mRNA Degradation, or Inhibiting Translation

The original oligonucleotide therapeutic modality was based upon hybridization of DNA oligomers to mRNA resulting in knockdown of expression.[1,2] The mechanism of knockdown for DNA–RNA hybrids is postulated to be mRNA degradation by RNAse H,[3,4] which specifically hydrolyzes chimeric duplexes. Oligonucleotides composed of nucleotides

Table 2 Mechanisms of Action and Size of Oligonucleotides

Mechanism of Action	Examples	Size
Increased transcription/ translation	Antagomirs, lncRNA binding	17–22 nucleotides
Decreased expression	Antisense	15–25 nucleotides
	RNA interference	18–24 base pairs
	Antigene	~16 nucleotides
New gene expression	Splice blocking	15–30 nucleotides
	mRNA	(>1500 nucleotides)
Receptor binding	Aptamers	15–60 nucleotides
	Toll receptor agonists	~20 nucleotides
	Telomere agonists	~15 nucleotides

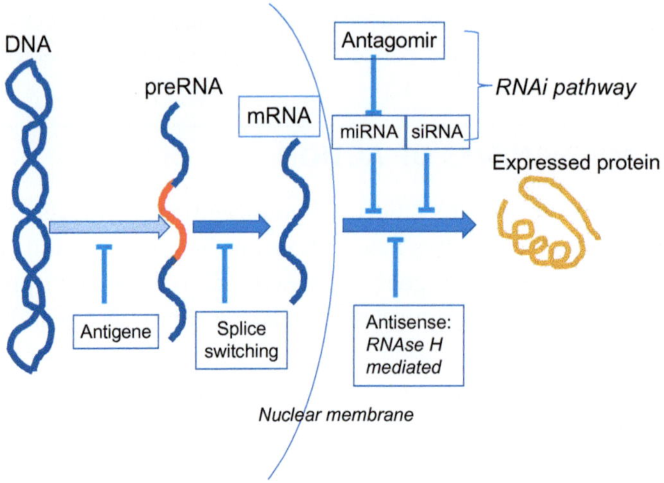

Fig. 1 Mechanisms of action. Mechanisms by which oligonucleotides can alter expression by inhibiting translation, splicing, and mRNA translation. Different oligonucleotide therapeutic modalities are in *boxes*. Not shown are oligonucleotides that bind to lncRNA or those that can act as agonists/antagonists.

with a 3′-endo conformation—like ribose—are not substrates for RNAse H, but bind more tightly to target mRNA and can downregulate expression by a mechanism of steric blocking of ribosomes to access mRNA thereby suppressing translation.[5]

In 2001, the phenomenon of RNA interference (RNAi) was demonstration in mammalian cells,[6] which prompted great interest in this means of mRNA reduction using an endogenous pathway. RNAi is induced by double-stranded RNA where the noncoding strand of the RNA duplex becomes associated with the RNA-induced silencing complex (RISC). MicroRNA's (miR's) largely act by inhibiting mRNA translation and short interfering RNA's (siRNA's) by mRNA cleavage.

The fourth mechanism by which oligonucleotides can inhibit gene expression is through sequence-specific triple helix binding via Hoogsteen base pairing to genomic duplexed DNA. This triplex formation can block transcription factors, thereby reducing mRNA production. In contrast to antisense approaches to decreased expression, antigene oligonucleotides have relatively higher barriers to activity: nuclear transport followed by chromatin insertion to access their sites of action. To date, no antigene oligonucleotides have progressed to the clinic.

2.2 Increased mRNA Translation (miRNA Antagonists, i.e., Antagomirs) and Increased mRNA Transcription (Oligonucleotides That Bind to Long Noncoding RNA)

Oligonucleotides that bind to miR's can inhibit their function and thereby increase the translation of mRNA levels. Oligonucleotides that are active as blocking agents are active in inhibiting miR's, and antagomirs have been designed against several diseases including various cancers (miR 221 and 155).[7,8]

Among the diverse roles of RNA is the activity of long noncoding RNA (lncRNA) to regulate gene expression.[9] A mechanism of increased transcription is to use oligonucleotides to block the binding of lncRNA to chromatin modifiers, which can suppress transcription.[10]

2.3 Use of Oligonucleotides to Create New Products of Translation (Splice Blocking, mRNA, and CRISPR)

In addition to blocking translation of mRNA to inhibit translation, steric-blocking oligonucleotides may be used to block splice sites in pre-mRNA. Splice-switching oligonucleotides (SSOs) can either activate splicing (exon skipping) by binding to an exonic splicing enhancer or inactivate splicing (exon inclusion) by binding to an intronic splicing silencer (see Fig. 2).[11] FDA-approved drugs eteplirsen for Duchene's muscular dystrophy (exon skipping) and nusinersen for spinal muscular atrophy (exon inclusion) are SSOs.

The size, synthesis, and delivery of mRNA, a therapeutic modality of intense current interest, blurs the line between traditional gene therapy and oligonucleotide therapeutics as discussed in this chapter. For the purposes of this chapter, mRNA is considered an oligonucleotide for two reasons: its synthesis may be performed in vitro and nonnatural nucleosides may be incorporated to improve its drug-like properties. Using semisynthetic in vitro transcription, mRNA sequences have been improved to increase translation[12–14] and nonnatural nucleotide mimics have been incorporated to reduce immune stimulation and increase nuclease resistance.[15–18] In addition, mRNA does not require nuclear transport. These positive attributes have made mRNA an attractive means of gene delivery; however, mRNA is a relatively large nucleic acid (>1500 nucleotides), and the machinery of translation limits the ability to incorporate nonnatural nucleotides. For these reasons, mRNA delivery requires either physical methods of delivery such as

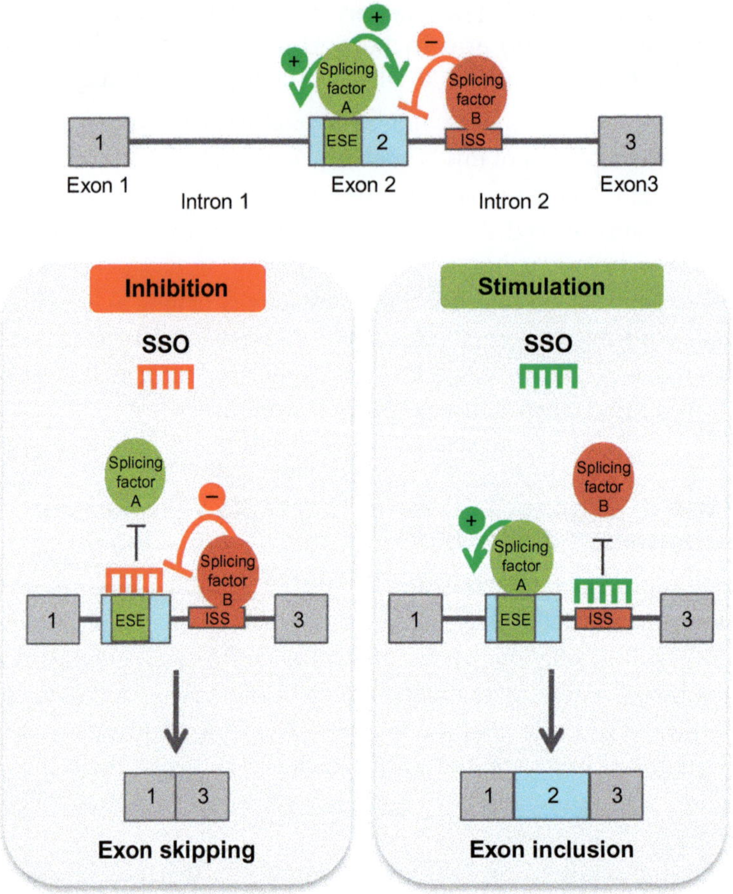

Fig. 2 Splice-blocking mechanisms. Splice-switching oligonucleotides (SSOs) modulate alternative splicing. (*Top*) Diagram of a pre-mRNA transcript with exons depicted as *gray boxes* and introns as *lines*. An intronic splicing silencer (ISS, *red*) and exonic splicing enhancer (ESE, *green*) are shown bound by a transacting inhibitory splicing factor protein (*red oval*) or stimulatory splicing factor (*green oval*). These splicing factor proteins either block (−) or promote (+) splicing at splice sites bordering the surrounding exons. (*Left panel*) An SSO that base pairs to a splicing enhancer sequence creates a steric block to the binding of the stimulatory splicing factor to its cognate enhancer-binding site. This block thereby disrupts splicing and results in exon skipping. (*Right panel*) In contrast, an SSO that base pairs to a splicing silencer sequence element blocks splicing silencer activity by preventing binding of a negatively acting splicing factor. Disruption of the binding of splicing inhibitory proteins to its cognate-binding sequence activates splicing at the splice site that is negatively regulated by the silencer element, resulting in exon inclusion. *From Havens, M.A.; Hastings, M.L. Splice-Switching Antisense Oligonucleotides as Therapeutic Drugs.* Nucleic Acids Res. ***2016,*** 44 *(14), 6549–6563. DOI:10.1093/nar/gkw533. Published by Oxford University Press on behalf of Nucleic Acids Research 2016.*

electroporation, high-pressure injection, or direct tissue injection (e.g., intramuscular) or packaging in particulate delivery vehicles.

Another therapeutic modality that has attracted great interest recently are use of clustered regularly interspaced short palindromic repeats (CRISPR) and its associated endonuclease Cas9 to perform gene editing and transcription reduction. CRISPR technology has two components: a "guide" RNA (gRNA) and Cas9. Gene editing is accomplished using active Cas9 to cut and splice genomic DNA at sites directed using gRNA, while transcription inactivation (CRISPR interference, i.e., CRISPRi) uses a Cas9 that is mutated to inactivate its nuclease activity. There are multiple ways that the Cas9 may be delivered: pDNA, mRNA, or as the protein. The guide strand, which is ~100 nucleotides long, may be generated in situ by transcription of an expression vector for the Cas9 and guide strand, or it may be chemically synthesized and delivered as a separate component from Cas9. In any of these permutations, the problem of delivery is not new—one either needs to deliver a large nucleic acid (i.e., gene delivery) and/or an oligonucleotide.

2.4 Nucleic Acids Can Be Used as Traditional Agonist/Antagonist and Bind Selectively to Receptors (Toll Receptor and Telomere Agonists, Aptamers)

Aptamers are a special class of oligonucleotide therapeutic where their therapeutic activity does not depend on their ability to bind to nucleic acid targets in vivo or be used in the machinery of transcription/translation, but relies on the ability of nucleic acids to form three-dimensional structures that bind to small- and large-molecule targets. Aptamers are generated through rounds of binding selection and amplification to generate sequences based upon RNA or DNA. Aptamers are single stranded of length 15–60 nucleotides. Due to nucleases, unmodified aptamers have half-lives as short as 2 min in vivo.[19] The enzymatic machinery of amplification limits the types of nucleotides analogues that can be incorporated during screening, but optimization to increase potency and nuclease resistance can occur using solid-phase synthesis once a target sequence is identified. In addition to nonnatural nucleosides, conjugation of groups at the 3′ and inversion of nucleotide at the 3′ terminus have shown increased resistance to exonucleases.[20–22] Due to the mechanism of binding to specific ligands by aptamers, they are not packaged into nanoparticles but may be conjugated to groups to increase circulation times or increase nuclease resistance. Due to their lack of membrane permeability—like all oligonucleotides—the use of aptamers therapeutically has primarily been to bind to extracellular targets. The FDA-approved therapeutic pegaptanib for age-related macular degeneration (wet AMD) is an polyethylene glycol (PEG)-conjugated aptamer that binds to and blocks vascular endothelial growth factor. An elegant method of generating nuclease-resistant aptamers is to use the mirror image, nonnatural enantiomer of ligand of choice. The resulting optimized oligonucleotide may then be synthesized as the non-natural, nuclease-resistant D-oligonucleotide "spiegelmers"[23] that now bind the naturally occurring ligand.

In the 1980s, it was observed that bacterial unmethylated CG dinucleotides—in a particular sequence (CpG motifs)—trigger the activation of B-lymphocytes.[24] Subsequently, the CpG receptor was discovered, named Toll-like receptor 9 (TLR9), and cloned.[25] Synthetic oligoDNA containing such CpG motifs mimics bacterial DNA and has been shown to induce a coordinated set of immune responses that comprise innate and acquired immunities.[26] The immune stimulatory effects of CpG DNA are due to the presence of methylated CpG motifs in the genomic DNA of vertebrates and its absence in pathogenic viruses and bacteria.[24] These synthetic oligonucleotides are ~6–26 DNA nucleotides long with the phosphodiesters replaced with phosphorothioate (PS) linkages. Due to the accessibility of TLR's in the lumen of endosomes and the mechanism of TLR agonists, these oligonucleotide therapeutics do not need to reach the cytoplasm to have their pharmacological effect and therefore should not be encapsulated nor do they require active cytoplasmic delivery. CpG oligonucleotides boost the efficacy of vaccines against bacterial, viral, and parasitic pathogens.[27] The recent success of immunooncology therapeutics to use the immune system to attack tumors has renewed the interest in using CpG oligonucleotides as Toll receptor agonists to boost the immune response to tumors.[28]

Although not a mechanism to decrease mRNA levels or translation, another oligonucleotide therapeutic strategy is to provide an oligonucleotide that competitively inhibits telomerase enzymatic activity, thereby reducing the telomere size which allows for apoptosis upon treatment with anticancer therapies.[29,30] Oligonucleotide telomerase antagonists are relatively small (<20 nucleotide) oligonucleotides based upon deoxyribose.[31] Delivery of telomerase antagonists requires access to the nucleus, which is the site of telomerase activity.

3. REQUIREMENTS OF NUCLEIC ACID DELIVERY VEHICLE

It is well known and accepted that there are certain attributes that constitute the properties of a small molecule that make them more likely to be efficacious therapeutics. There are many descriptions, such as Lipinski's rules,[32] of what makes a small-molecule "drug-like" that include limits on molecule weight and solubility. These restrictions on the molecule have been identified empirically and describe attributes of small molecules that are needed for them diffuse through biological membranes. Unfortunately, nucleic acids violate these requirements in molecular weight and solubility. Furthermore, the mechanism of nucleic acid therapeutics that use base pairing between macromolecules may not be mimicked using small molecules. Due to the inability of nucleic acids to become small enough and lipid soluble enough to fit these traditional drug-like properties, let us define a new set of rules that nucleic acids must have for them to be active in vivo.

- Drug-like requirements of oligonucleotide therapeutics
  - Have pharmacological activity
  - Be nuclease resistant
  - Be sufficiently long circulating and targeted to a tissue or cell type (or be administered directly to target cells)
  - Have access to requisite site of activity, e.g., be endosomolytic

3.1 Have Pharmacological Activity

Of course, the requirement that an oligonucleotide has pharmacological activity to be a therapeutic is a tautology. For nucleic acids that target specific sequences, there are a variety of nucleosides derivatives and oligonucleotide sequence patterns that increase base pair affinities and thereby increase potency. An example of using different modification patterns to increase affinity for target site is the use of "Gapmers" (explained later) where the known affinity between RNA–DNA duplexes is used to hybridization and a stretch of deoxy residues allows for RNAse H activity.

3.2 Be Nuclease Resistant

The second requirement for nucleic acid therapeutics is equally obvious in that the nucleic acid must be sufficiently stable to reach its intended target. Nucleic acids are very stable in the absence of nuclease but are readily degraded by nucleases present in vivo. In particular, ribonucleases are ubiquitous, difficult to denature/inhibit, and very efficient in their hydrolysis of RNA. Deoxyribonucleases are not encountered as readily in vivo but are present in the endosomal/lysosomal pathway. Due to the presence of nucleases, any nucleic acid therapeutic needs some mechanism for protection from hydrolysis. There are three strategies for stabilization: (1) incorporation of nucleoside analogues with increased nuclease resistance, (2) blocking of nucleases from their substrates by complexation and/or encapsulation, and (3) neutralization of the phosphodiester charges using cation agents.

3.3 Be Sufficiently Long Circulating

In addition to the relatively straightforward requirements of potency and stability, the task of creating an oligonucleotide drug has the added challenge of delivery to the intended site of activity. For this reason, oligonucleotide therapeutics have additional rules regarding pharmacokinetics and biodistribution that make them drug like. For oligonucleotide therapeutics below the renal threshold, such as siRNAs,[33] the circulation times upon administration can be very short with half-lives of excretion on the order of minutes. The qualification that nucleic acid therapeutics be sufficiently long circulating is contingent upon what nucleic acid is being administered and what tissues needs to be targeted.

To minimize the required circulation times and reduce nonspecific interactions, many applications of oligonucleotide therapeutics use local administration to apply the oligonucleotide directly onto the target cells. There are privileged spaces in the body where there are extracellular or immunological barriers that allow for specific administration, such as intravitreal and intrathecal injection (retinal and neural cells, respectively).

A common application of mRNA and other vectors such as genes that express exogenous gene products is vaccination,[34,35] where local transient delivery of a transgene may be sufficient to produce an immune response. The ability of intramuscular injections to deliver transgenes has been known for many years,[36,37] and intradermal injections of naked and lipoplexed mRNA may elicit immune responses to the delivered transgene.[35] The ability to transfect cells using naked transgenes is surprisingly facile, but the extent of transfection is localized to the site of injection, and the level of transfection has never been sufficient to provide clinical benefit; however, such a level of transfection may be appropriate for vaccinations (Table 3).

Table 3 Examples of Local Routes of Administration and Target Cells

Administration	Target Cells	References	Examples of Indications
Intradermal injection	Dermal and lymphatic cells	37,38	Vaccines
Intravitreal injection	Retinal cells	39,40	Wet AMD, optic neuropathies, retinitis
Intrathecal injection	Neural cells	41,42	Spinal muscular atrophy, neuropathic pain
Enema	Intestinal lamina	43	Crohn's disease, ulcerative colitis
Topical application	Dermal cells	38	Psoriasis
Intratracheal administration	Lung macrophages, endothelium, and epithelium	44,45	Cystic fibrosis, various antivirals such as respiratory syncytial virus
Ex vivo	Leukocytes, stem cells, and dendritic cells	46,47	Oncology including various leukemias

The use of local delivery in clinical testing of oligonucleotides is quite common. Approximately 40% (26 of 62) of current oligonucleotide clinical trials use local injections to administer oligonucleotide therapeutics directly to the intended site of activity.

For indications that require systemic administration, the target tissue must be accessible in a time frame that is achievable by the oligonucleotide therapeutic. The question of how long an oligonucleotide therapeutic needs to circulate depends greatly on the target tissue. Cells and tissues of the reticuloendothelial system (e.g., endothelial cells, liver, spleen, and kidneys) are in rapid contact with blood, and macromolecular therapeutics have ready access to these cells such that targeting may be achieved in a first-pass phenomenon. Less well-vascularized tissues such as tumors, brain, and muscles are less readily accessed and requires longer circulation times. How long is sufficiently long must be answered on a tissue-by-tissue case. The gold standard of long-circulating macromolecular therapeutics is antibodies that have half-lives of circulation on the order of weeks. A more realist pharmacokinetic goal may be given in the enhanced permeability and retention (EPR) literature where a half-life of circulation of >6 h is required for passive tumor targeting.

The pharmacokinetics of systemically administered oligonucleotide therapeutics may be improved by conjugating or associating with agents that effectively increase their molecular weight above the renal threshold. The most well-known agent for increasing molecular weight and decreasing nonspecific interactions is to conjugate PEG either to the oligonucleotide itself or its associated vehicle. A multitude of PEGylated lipids and polymers has been utilized and will be discussed in following sections concerning vehicle designs.

3.4 Be Targeted to a Tissue or Cell Type

The corollary requirement to sufficient circulation is that the oligonucleotide therapeutic has a mechanism for targeting to tissue. Due to the membrane impermeability of oligonucleotide therapeutics, there is no passive diffusion of the therapeutic throughout the body and therefore the therapeutic requires some mechanism by which it reaches its intended target. Again, local delivery applications where the therapeutic is applied to directly to the target cells somewhat remove the targeting requirement.

Systemic administration requires either active or passive targeting. One mechanism of passive targeting is EPR whereby long-circulating macromolecules may disproportionately associate with tumor tissues due to their leaky vasculature. This effect may result in above background levels of material to accumulate in tumor tissues in preclinical models, but its effect has not been demonstrated clinically and one may question the general applicability to all tumors. Another passive mechanism is the association of macromolecules with serum components which facilitate long circulation and some level of targeting to either liver or spleen.

Active targeting may be accomplished by conjugation or association with a targeting group. There are a variety of ligands that have been used to target oligonucleotide therapeutics and their associated vehicles including small molecules, antibodies, aptamers—which are themselves nucleic acids.

3.5 Ability to Access Target

The final requirement of an oligonucleotide therapeutic is that there be some means of cytoplasmic access. Of course, the site of action of some oligonucleotides such as aptamers and Toll receptor agonists is extracellular and hass no need for cytoplasmic delivery; however, most oligonucleotide therapeutics need to reach the cytoplasm or nucleus to exert their pharmacological effect. As mention earlier, oligonucleotides are not membrane permeable and therefore need some means of accessing the cytoplasm. There are a variety of endosomolytic strategies that will be discuss later.

4. TYPES OF DELIVERY VECTORS

Now that we have discussed the mechanism of oligonucleotide therapeutics and the properties of oligonucleotides that make them drug like, let us turn to how these attributes have been combined in the pursuit of clinical efficacy. There are four broad categories of nucleic acid delivery vehicle that we will introduce in levels of increasing complexity: naked, conjugated, liposomal, and polymeric.

4.1 Naked Oligonucleotides

By far, the most clinically pursued and successful delivery vehicle strategy is the use of naked oligonucleotides, which are defined as systems that contain no agents that are associated with the nucleic acid either covalently or noncovalently. The absence of any delivery vehicle requires that the oligonucleotide itself be sufficiently nuclease resistant, sufficiently long circulating, and cell targeted. For small, solid-phase synthesized oligonucleotides such as those used in antisense oligonucleotides, RNAi, and innate immune stimulators, the use of nucleotide mimics provides the required drug-like properties.

4.1.1 Nucleotides That Replace Phosphodiester Group

The substitution of one nonbridging oxygen of a phosphodiester with a sulfur atom creates the PS linkage. A PS bond creates a new stereocenter in the nucleotide and when synthesized under standard achiral conditions creates diastereomeric mixtures of Rp and Sp at the phosphorous atom (Fig. 3).

Sp diastereomer is nuclease resistant, while Rp is relatively susceptible to nucleases.[48] Unfortunately, the Sp diastereomer also has lower affinity for target mRNA.[49] Methods to create stereopure PS oligonucleotides have been subject of much research,[50,51] and distinct diastereomeric oligonucleotides are currently under clinical development at Wave Life Sciences.

Not only does the PS linkage confer oligonucleotides with nuclease resistance, but PS oligonucleotides associate with plasma proteins[52]—albumin[53] in particular—resulting in circulation times that are much longer than one would expect based upon hydrophilicity and molecular weight. Observed plasma pharmacokinetics of PSs had half-lives of several hours and prolonged elimination phases with half-lives of days.[52] The association

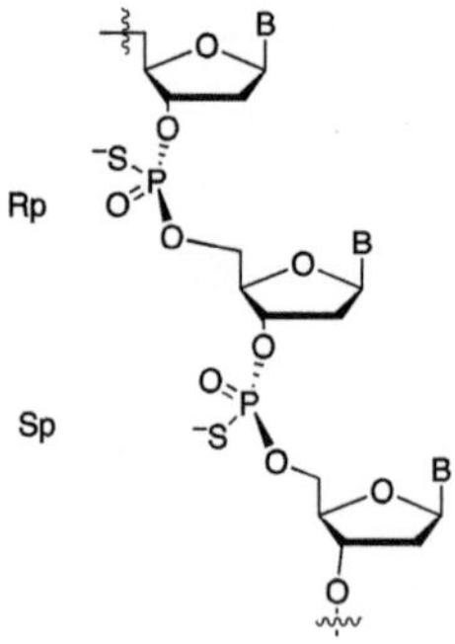

Fig. 3 Structures of Sp and Rp diastereomers of phosphorothioate.

with serum proteins also provides a mechanism of tissue targeting. The biodistribution of PS oligonucleotides is in clearance organs: primarily liver, and kidney followed by bone marrow, adipocytes, and lymph nodes.[41]

There are other functional groups that have been identified as replacements of the phosphodiester group in the oligonucleotide. Like phosphates and PSs, there are a variety of functional groups that are negatively charged such as phosphorodithioate (PS2)[54] and thiophosphoramidates. There are number of analogues that are uncharged such as phosphorodiamidate morpholino oligomer (PMO),[55] peptide nucleic acid (PNA),[56] phosphotriesters,[57] and phosphonates.[58] It has been postulated that the uncharged analogues are not only nuclease resistant but may also be more membrane permeable; however, the size and hydrophilicity of uncharged oligonucleotides still preclude their passive diffusion across membranes.

Although not widely utilized, the PS2 functionality has been the subject of recent reports, suggesting that PS2 functional groups increase the affinity for protein-binding sites on RISC in a RNAi duplex[59] as well as aptamers to increase protein–oligonucleotide binding.[60]

Phosphoramidates are inherently more easily hydrolyzed than naturally occurring phosphodiesters,[61] but N3′→P5′ thiophosphoramidates (Fig. 4) are resistant to nucleases in imetelstat,[62] an inhibitor of telomerase that is in ongoing trials against blood cell cancers myelodysplastic syndrome and myelofibrosis.

Methylphosphonates were among the first neutral phosphate mimics to be evaluated.[58] Like the PS functional group, phosphonates generate a chiral center at phosphorous and—again—the Rp diastereomer of phosphonates has a higher affinity for target DNA sequences.[63,64] Unfortunately, a practical stereoselective method for synthesis of phosphonates has not been developed and their solubility is decreased relative to charge

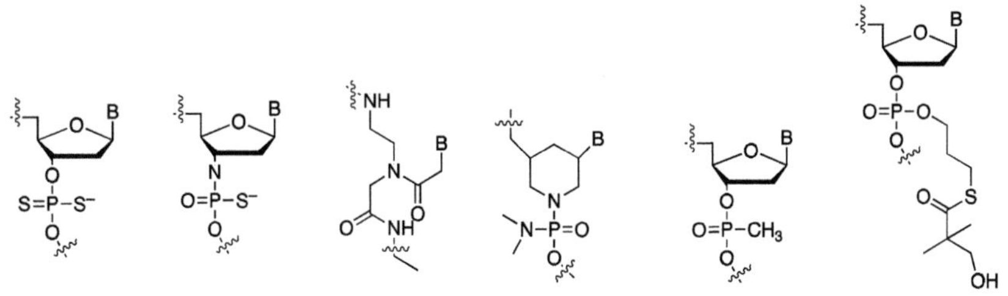

Fig. 4 Phosphodiester analogues.

oligonucleotides. For these reasons, the development of phosphonates has not progressed.

Morpholino oligos (PMOs) use a hydrolytically stable, uncharged phosphordiamidate functional group.[55] They have progressed further than any other nonionic oligonucleotides. They block RNA cellular processes but are not substrates of RNAse H. Although a recent synthetic route has opened the possibility of synthesis using phosphoramidite chemistry that enables the synthesis of PMO-DNA chimeras that are substrates of RNAse H when duplexed to mRNA.[65] PMOs have been used to block translation, splicing, and miRNAs from their targets. PMOs are widely used in zebrafish models of development[66] and are the basis of the approved Duchenne's muscular dystrophy drug eteplirsen.[67]

PNAs are—as their name suggests—based upon the amide functional group.[56] They are the oligonucleotide that are most unlike naturally occurring nucleic acids and have an achiral main chain and no cyclic function group that mimics ribose. Despite the presumed lack of rigidity or pre-organization of the main chain atoms, PNAs hybridize more strongly with DNA than DNA itself. Like PMOs, PNA–mRNA duplexes are not substrates for RNAse H; however, they can be used as steric-blocking agents and triplex forming antigene oligonucleotides.[68] In addition, uncharged oligonucleotides do not associate with serum components and are rapidly excreted upon administration.[69]

Targeting of naked oligonucleotides to tissues outside of the clearance system (liver, kidney, spleen) typically requires the oligonucleotide therapeutic be administered by local injection. Larger oligonucleotides that are synthesized by in vitro transcription, such as mRNA vectors and aptamers that are synthesized enzymatically, have less flexibility with respect to the what nucleotide mimics may be incorporated to increase nuclease resistance. For this reason, all current clinically trials using naked mRNA are administered locally via either intradermal or intramuscular injections.

Enemas and intramuscular, intravitreal, intrathecal injections have been used for the administration of a variety of oligonucleotides with and without PS bonds. In fact, three of the approved oligonucleotide drugs are administered locally: nusinersen is administered intrathecally. Fomivirsen and pegaptanib are injected intravitreally.

4.2 Nucleoside Analogues That Alter the Structure of Ribose

There are a variety of nucleotide mimics wherein the ribose or deoxyribose is modified to increase affinity for target and/or increase nuclease resistance. Modifications to all five positions of the ribose ring have been made; however, the modifications of the 2′ position of ribose have been the most studied and successful.

4.3 1′ Position: The Base

There are a few examples of base modifications that are designed to increase base pairing. The most well known is the G-clamp,[70] which is a cytidine mimic that is designed to have increased affinity for guanosine bases due to hydrogen bonding through an aminoethyl group. C5 propynyl pyrimidines are known to form more stable duplexes; however, they appear to be more toxic as well.[71]

Base modifications have been of great benefit to the use of mRNA as a therapeutic modality. Substitution of uridine and cytidine residues with pseudouridine,[72] *N*-methyladenosine,[16] 2-thiouridine,[16] and 5-methylcytidine[15] reduces innate immune recognition of mRNA and is translated more efficiently. These modifications appear to make mRNA more nuclease resistant; however, mRNA is still large (>500,000 MW) and hydrophilic and therefore requires assistance to reach the cytoplasm of cells (Fig. 5).

4.4 2′ Modifications

Modifications of the hydroxyl group at the 2′ position of ribose have been extensively used to mimic the structure of the ribose ring while inhibiting ribonucleases that require the 2′-OH group for hydrolysis of RNA. 2′-*O*-Methyl ribonucleic acids are naturally occurring nucleosides and have been shown to increase binding affinity to RNA itself[73] while being resistant to ribonuclease.[74] 2′-*O*-Methyl groups can be extensively substituted into

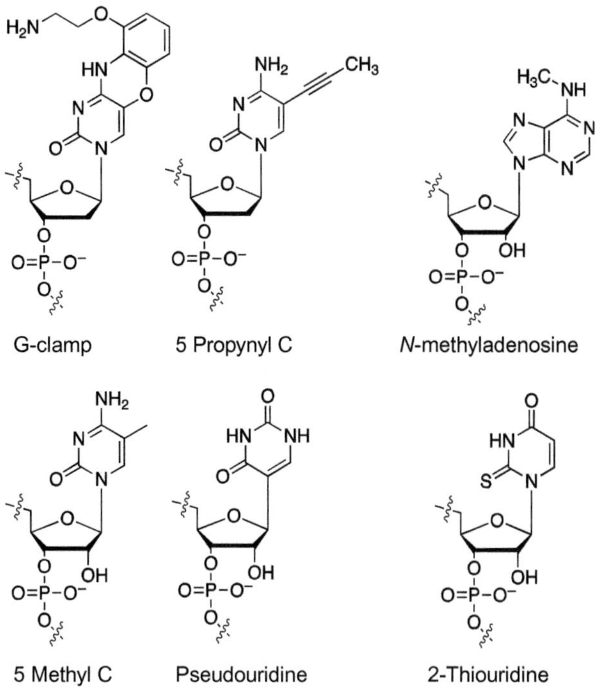

Fig. 5 Base modifications.

RNAi triggers[75] and were the first nucleotide analogues used in "antagomirs."[76] 2′-*O*-Methoxyethyl (MOE) modification was designed to mimic the ribonuclease resistance of *O*-methyl, attenuate protein–oligonucleotide interactions,[77] and have increased affinity for RNA.[78,79] The MOE modification is commonly used in antisense technologies and is incorporated into approved therapeutics nusinersen and mipomersen.

Fluorine is highly electronegative, and 2′-deoxy-2′-fluoro (2′-F) analogues of nucleosides adopt C3′-endo conformations characteristic of the sugars in RNA helices. 2′-Fluoro derivatives were first incorporated into ribozymes[80] and antisense oligonucleotides.[81] The FDA-approved therapeutic aptamer Macugen (pegaptanib) has 2′-F pyrimidines. The machinery of RNAi can tolerate substantial 2′-fluoro modification in both sense and antisense strand (Figs. 6 and 7).[82]

4.5 4′ and 5′ Modifications

Alkoxy substituents at the 4′ position of 2′ deoxyribose mimic the conformation of ribose.[83] It was found that *S* enantiomeric 5′-methyl-substituted

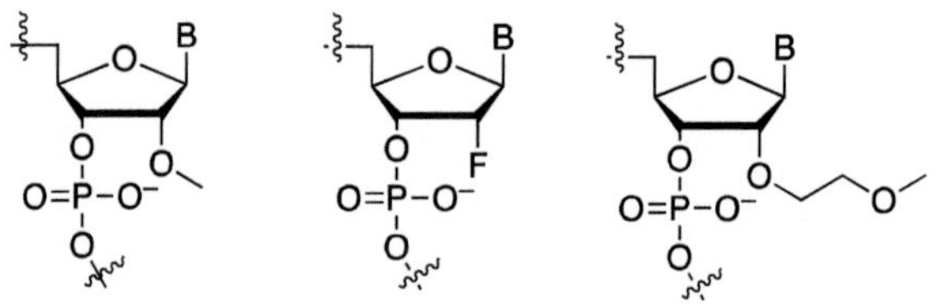

2′-OMe 2′-F 2′-MOE

Fig. 6 Modifications of ribose at 2′ position.

Conformation	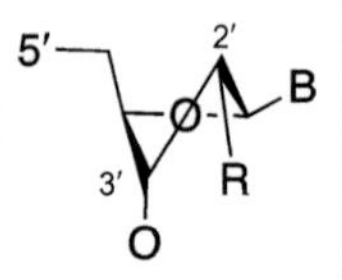	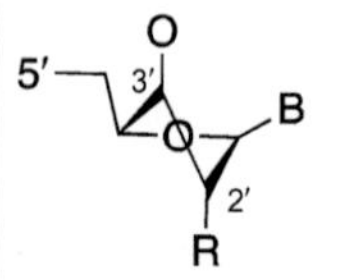
Name	2′-Endo S conformation	3′-Endo N conformation
2′R=	H (DNA)	OH (RNA), OMe, MOE, F, LNA
Uses	RNAse H substrates	RNAi, steric blocking, mRNA

Fig. 7 2′- and 3′-Endo ribose conformations.

nucleotides were in general more RNA like with a C3′-endo conformation. In addition, *S* enantiomer with 2′-fluoro and methoxy substituents was more nuclease resistant than the 5′ *R* enantiomer.[84]

4.6 Bicyclic 2′-4′ Modifications

There are a variety of ribose derivatives that lock the carbohydrate ring into the 3′-endo conformation by the formation of bicyclic structures with a bridge between the 2′ oxygen and the 4′ position. The original bicyclic structure has a methylene bridging group and is termed locked nucleic acids (LNAs).[85,86] The bicyclic structure "locks" the ribose into its preferred 3′-endo conformation and increases base-pairing affinity. It has been shown the that incorporation of LNAs into a DNA duplex can increase melting points up to 8°C per LNA.[87] Subsequently, a variety of bicyclic nucleotides have been developed such as bridged nucleic acids,[88] ethyl-bridged (ENAs),[89] constrained ethyl (cEt)[90] nucleic acids, and tricyclic[91] structures with varying affinity for target sites. LNAs can be incorporated into antagomirs,[92] splice-blocking oligonucleotides, either strand of an RNAi duplex[82,93,94]; however, like other 3′-endo conformers, LNAs are not substrates for RNAse H. On the other end of the stability spectrum from bicyclic nucleic

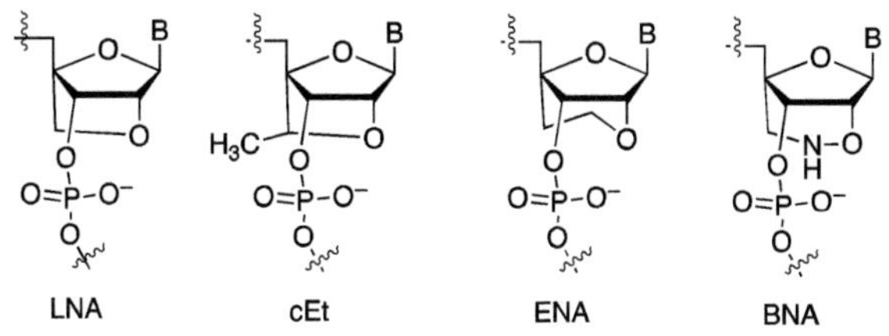

Fig. 8 Bicyclic nucleosides.

acids are acyclic nucleic acids, most commonly called unlocked nucleic acids,[95] which destabilizes hybridization to the target (Fig. 8).

4.7 Modification Patterns: Gapmer Overall Design

The individual nucleotide analogues can be combined in patterns to provide the best combination of efficacy and nuclease resistance. RNAse H-dependent mRNA degradation requires hybridization between DNA and RNA and therefore requires a stretch of nucleotides to be in a 2′-endo conformation; however, the RNA-like 3′-endo conformation of the MOE oligonucleotide has a greater affinity for target mRNA and reduced toxicity. A compromise is the "Gapmer" design, where a central stretch of PS DNA monomers is flanked by modified nucleotides such as 2′-*O*-methyl, MOE, and LNA.[96] The end blocks prevent nucleolytic degradation, increase affinity for target sequence, and decrease toxicity of the oligonucleotide, and the stretch of at least four or five deoxy residues between flanking 2′-*O*-aklyl nucleotides is sufficient for activation of RNAse H.[74] Mipomersen is a Gapmer, with 5 MOE nucleotides on each end and a stretch of 10 deoxy residues in the middle.

For RNAi duplexes, recognition by RISC requires RNA-like 3′-endo nucleotides and some patterns of RNA analogues. It was observed that a pattern of alternating 2′-*O*-methyl groups provides stability against nucleases, but not all permutations of alternating 2′-*O*-methyl are active RNAi agents.[97] The fact that one may remove all 2′-hydroxy groups with alternating 2′-fluoro and 2′-*O*-methyl groups to produce duplexes that are resistant to nucleases and active in RNAi[75] suggests the 2′-hydroxy group is not absolutely required for activity, but that some sites in the RNAi duplex are sensitive to the added steric bulk of the methyl group.

4.8 Conjugated Oligonucleotides

Even if an oligonucleotide therapeutic is to be used in a vehicle—rather than just naked—it is advisable that nucleotide analogues should be screened to increase potency and nuclease stability. The resulting optimized oligonucleotide may then be combined with other delivery components to achieve an optimized combination of cargo and vehicle. The next layer of complexity in oligonucleotide delivery is the conjugation of groups—via covalent bonds—that prolong circulation, provide targeting to tissues, and facilitate intracellular delivery.

The most common strategy to prolong circulation is the conjugation of PEG, which prevents clearance by two mechanisms: the increase in molecular weight above threshold for renal clearance and the prevention of nonspecific interactions with extracellular surfaces and serum components. PEG is most commonly incorporated into nucleic acid delivery vehicles by attachment to components that noncovalently associate with the nucleic acids, e.g., PEGylated lipids and polymers; however, there are examples of direct PEG conjugation to increase nucleic acid circulation times, decrease nonspecific interactions, and alter biodistribution; however, the targeting is passive and the potency of the nucleic may be compromised as PEG MW increases.[33] The FDA-approved aptamer pegaptanib (Macugen) is a conjugate to a PEG with molecular weight of 40,000, which increases the intravitreal residence time. Another class of molecules that can be conjugated in order to increases circulation times is the attachment of lipophilic groups such as cholesterol,[98] which interact with serum components such as albumen and lipoproteins thereby increasing circulation times and passive accumulation in the liver.[99] It should be remembered that extensive PS modification increases circulation times through associations with serum components, with roughly 10 PS groups required for serum binding.[100]

Conjugation of targeting ligands for ligand-mediated tissue-selective delivery has received comparatively much more interest. It is a safe assumption that all known targeting modalities have been conjugated to oligonucleotides from small molecules to antibodies. It is not the intent of this chapter to cover all targeting ligands that have been evaluated, but to discuss the relevant parameters of targeting and the most important targeting ligands. By far, the most studied and developed targeting ligands are based upon *N*-acetylgalactosamine (GalNAc), which targets the asialoglycoprotein receptor (ASGPR) on hepatocytes. The affinity between GalNAc and ASPGR is dependent upon the number of GalNAc residues, with three being the optimal number for targeting and synthetic ease.[101] The precise display of trivalent GalNAc appears to be rather forgiving,[102] but for the purposes of illustration a structure originally conjugated to liposome targeting[103] and is currently used in Alnylam's subcutaneously administered siRNA's is presented in Fig. 9.

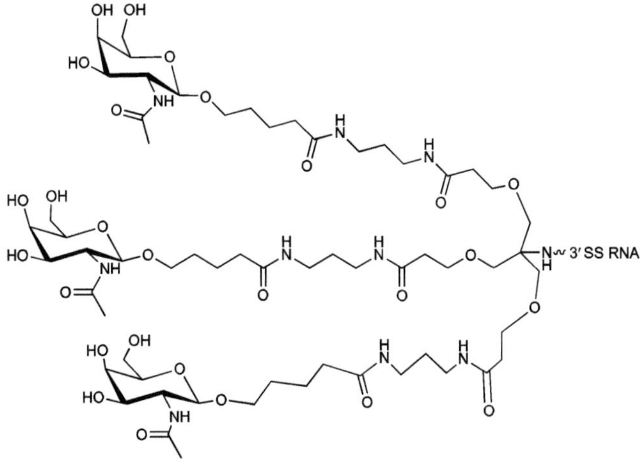

Fig. 9 A triantennary *N*-acetylgalactosamine structure.

Targeting ASGPR has several advantages that make it an attractive receptor for in vivo targeting: it is highly expressed, it is very efficient at internalizing material, and it is on a readily accessible cell type, hepatocytes. For these reason, targeting via the ASGPR is essentially a first-pass phenomenon such that oligonucleotides below the renal threshold do not need an increase in circulation times to reach the target tissue. The half-life of circulation of siRNA duplexes is on the order of 5 min, but the majority of GalNAc-targeted duplexes accumulate in the hepatocytes in less time thereby allowing for hepatocyte targeting.[104] $(GalNAc)_3$ targeting was first applied to targeting of single-stranded antisense oligonucleotides[105] but has become the basis for Alnylam's subcutaneously administered siRNAs.[104] Approximately 70% of injected GalNAc conjugates accumulates in liver in less than 20 min.[104] This ability to accumulate material selectively in hepatocytes, combined with the stability chemistry of using all 2′-modified nucleotides (either fluoro or methoxy), results in cytoplasmic release sufficient to elicit RNAi. From the amount of material associated with RISC, the amount of siRNA that is released into the cytoplasm is estimated to 1%–0.1% of total material[106]; however, the efficiency with which $(GalNAc)_3$ targets material to hepatocytes and the stability of duplexes is sufficient to affect RNAi. Based upon the success of GalNAc-conjugated siRNAs, there has been renewed interest in GalNAc-targeted antisense oligonucleotides and has also been applied to antagomirs.[107]

Although quantitative data are difficult to find for other ligand–receptor pairs, it appears that nothing rivals the ability of ASGPR to target material to a specific tissue. A variety of ligands have been studied preclinically,[108] but none have advanced. Small-molecule ligands such as RGD[109] and folate[110] have been conjugated to oligonucleotides. Therapeutic antibodies are an attractive targeting modality, and antibody–oligonucleotide conjugates have been synthesized and evaluated in vivo[111]; however, the ability to target ~70% of injected dose to target tissue has only been achieved using GalNAc targeting to the ASPGR.

4.8.1 Delivery Vehicles Based Upon Complexation of Nucleic Acid

Oligonucleotides that are too large to efficiently be endocytosed and/or are insufficiently nuclease resistant require the use of vehicles that electrostatically complex with the nucleic acid. The complexation inhibits nuclease from degrading the oligonucleotide by forming a steric barrier and by inhibiting nuclease binding by neutralizing anionic charge. The process of forming compact particles of nucleic acids from their extended chains is called condensation,[112] which may be achieved by the addition of multiply charged cationic species. Multiple positive charges can either be covalently attached to one another in a polycation or noncovalently associated with one another in a complex such as the surface of a cationic liposome. The resulting polycation–polyanion interaction is a colloidal dispersion where the nucleic acid particles vary in size and shape depending on the nucleic acid and the condensing cation. In general, the particles are greater than 20 nm in size,[113] and—in the absence of agents to modulate surface charge such as PEG—have surface charges >20 mV.

For gene delivery that uses large (>1.5 kbp with MW > 1000 KD) nucleic acids, condensation is a necessity for cellular uptake. It is not clear at what point nucleic acids require condensation for cellular entry, but oligonucleotides that are synthesized by transcription (e.g., mRNA vectors, MW > 0.4 M) may not be sufficiently nuclease resistant for systemic administration as naked or conjugated vehicles, and therefore require complexation/condensation to provide nuclease resistance.

One should assess and consider the appropriateness of a formulation's pharmacokinetics when evaluating the applicability of a formulation to a target tissue. Tissues that are readily accessible, such as the liver, have very low requirements for their circulation times (≥~5 min half-life). In contrast, less vascularized tissues such as skeletal muscle and many tumors require hours of circulation to enable extravasation and tissue targeting. The pharmacokinetics and biodistribution of nanoparticles are dependent upon their size and charge. Upon iv administration, large (>200 nm) and/or highly positively charged (surface charge >20 mV) are primarily taken up by macrophages in the liver and spleen and have a half-life of circulation less than 2 h.[114] Reduction in size (<100 nm) and surface charge (~0 mV) results increased circulation times[115,116]; however, negatively charged nanoparticles have decreased circulation times due to association and uptake into macrophages via scavenger receptors.[117]

4.8.2 Strategies for Cytoplasmic Delivery

The in vivo activities of PS-modified antisense oligonucleotides and GalNAc-targeted siRNA's, which distribute primarily to liver tissue and are very resistant to nucleases, suggest that the need for active cytoplasmic delivery is dependent upon the targeting ability of the delivery vehicle and the stability of the oligonucleotides. The ability to transfect cells in tissue culture without the aid of transfection reagents has been termed "gymnotic" delivery.[118] The mechanism of gymnotic delivery is not clear, but it appears to be proceed through the endocytic pathway and be dependent upon enzymes involved in endosome trafficking,[119–121] which are observations consistent with a mechanism of endosomal membrane leakage that occurs during intracellular endosome/lysosome trafficking.[122]

In the absence of sufficient targeting and/or nuclease stability, there are a variety of strategies to facilitate cytoplasmic delivery of oligonucleotides including endosomal buffering (i.e., proton sponge), titratable amphiphiles, cell-penetrating peptides (CPPs), and masked membrane lytic polymers.

The mechanism of endosomal buffering (i.e., proton sponge) to facilitate endosomolysis relies on the ability of agents such as polyamines to buffer endosomal/lysosomal compartments. The resistance to acidification is postulated to result in increased osmotic pressure that results in lysis of the lysosomal compartment.[123] Titratable amphiphiles are polymers/peptides whose structure is pH dependent in such a way that at acidic pH they are hydrophobic and membrane disruptive.[124,125] Typically, titratable amphiphiles are polyanionic polymers or peptides composed of carboxylic acids that become neutral and membrane disruptive upon acidification. CPPs are cationic peptides, with a high propensity of guanidinium groups, that enter cells without any apparent membrane lysis.[126] Masked lytic polymers are membrane-disruptive polymers whose membrane interactivity is attenuated by reversible covalent modification.[127,128] Like titratable amphiphiles, the mechanism of endosomolysis by masked polymers relies on the use of amphipathic polymers whose ability to lyse membranes is controlled such that the activity is only functional in the acidic environment of the endosome/lysosome. In the case of titratable amphiphiles, the mechanism of control is a reversible protonation of carboxylic acids. In the case of masked polymers, the control of membrane activity is the irreversible cleavage of a group that inhibits membrane interactivity of the polymer.

Of these mechanisms, the most commonly published strategy for increasing the cytoplasmic access of oligonucleotides is the use of CPPs. There are a variety of peptides in the literature that have been described as having the ability to diffuse across cell membranes. In general, these peptides are cationic with a propensity for arginine residues. For this reason, their conjugates with anionic oligonucleotides have poor solubility and as a consequence there are many examples of CPPs conjugated to neutral oligonucleotides either PNA[129] or PMO.[130] Although there does seem to be some diffusion of CPPs across cell membranes, the amount of transport is difficult to quantify due to the fact that the majority of CPP conjugate material is endocytosed.

Conjugates of masked lytic peptides and oligonucleotides have been used to aid in cytoplasmic release. The key to delivery is the reversible attachment of groups—typically hydrophilic PEGs—that attenuate the interaction between polymers and membranes. There are several examples of groups that are cleaved in the environment found in the endosome and lysosome. Acid-labile chemistries based upon acetal[125] and maleamate,[127] and protease labile[131] masking groups have been used to reversibly mask membrane lytic polymers

4.9 Liposomal Delivery Systems

Nucleic acids entrapped in lipids (lipoplexes) are attractive vehicles for the delivery of nucleic acids because they provide nuclease resistance and—for larger nucleic acids such mRNA—condensation, which facilitate cellular uptake. Extensive use of PS bonds meets the requirements of nuclease stability, and the size of oligonucleotides less than 100 nucleotides does not need to be condensed. For these reasons, PS-based oligonucleotides are not usually formulated into nanoparticles, and most recent nanoparticle nucleic acid delivery efforts are in the areas of siRNA and mRNA delivery. An exception of note is Toll receptor agonists, which are more immunostimulatory when formulated into nanoparticles.[132]

Entrapment in synthetic liposomes composed of either neutral or negatively charged lipids is exceedingly inefficient due to the hydrophilicity and size of oligonucleotides. For this reason, cationic lipids are required to form electrostatic complexes between nucleic acid and lipids. In addition to the cationic lipids, there are typically neutral or anionic helper lipids which are composed of unsaturated fatty acids and are postulated to assist in fusion between the lipoplex and the cellular membrane, and PEGylated lipids, which prevent aggregation during formulation and storage and nonspecific interactions in vivo.

There are cell-derived secreted vesicles called exosomes that contain a variety of cargo including miRNA and mRNA and are used for extracellular exchange of macromolecules.[133] These natural lipoplexes have obvious therapeutic appeal due to their natural composition—including a lack of cationic lipids—and have attracted attention as therapeutic vehicles. Several significant technological barriers need to be overcome—including scale-up and analytics to enable GMP manufacturing—to make exosomes into drug delivery systems. In addition, it appears that the pharmacokinetics and biodistribution of exosomes are similar to existing lipoplexes with circulation half-lives on the order of minutes and clearance organs liver and spleen being the major sites of deposition.[134] For these reasons, it appears that—for now—the application of exosomes may be restricted to liver delivery and/or local administrations.

Lipids are water insoluble and nucleic acids are organic solvent insoluble. To mix these components in a controlled manner such that formulations are repeatable and relatively homogenous in size, detergents[135] or water-miscible organic solvents such as ethanol[136] are used. After formation of electrostatically associated complexes, the amphipathic detergent or solvent is then removed by dialysis or solvent exchange. The resulting complexes are not typically bilayers with an encapsulated aqueous interior but are multiple alternating layers of nucleic acids and lipids.[137] Depending on the components and the mixing procedure is possible to formulate lipoplexes that are well less than 100 nm, which is a requirement for systemic administration.[138]

Although the transfection efficiencies of lipoplexes are difficult to predict and optimization is empirical, there are a few design features that have been identified to produce optimal transfection efficiency in vivo: pH-sensitive cationic lipids, the use of unsaturation in the lipid chains, and the hydrophobic–hydrophilic balance of PEG lipids to balance circulation times and transfection efficiencies.

There have been several studies that have shown a correlation between the pK_a of the amine groups of the cationic lipid, which is buffer in the range of the endosomal/lysosomal pathway (pH 4–7), and transfection ability.[139,140] A study of 56 different cationic lipids formulated with siRNA showed a narrow range of optimal pK_a from 6.2 to 6.6.[139] To synthesize lipids with such pK_a values, lipids commonly have closely spaced amines or imidazole groups.[141] The effect of these weakly basic amine groups in the lipoplexes produces several attractive attributes that facilitate in vivo transfection: reduced surface charge at neutral pH thereby decreasing non-specific interactions in vivo, increased surface charge in acid environment of endosomes and lysosomes thereby increasing electrostatic interactions with the cellular membrane in these compartments, and providing buffering groups that can provide endosomolytic activity via the proton sponge mechanism.

Another common motif observed in cationic and helper lipids used in lipoplexes is the presence of unsaturation[142] in their component fatty acids with oleic (18 carbon chain with 1 double bond) and linoleic (18 carbons with 2 double bonds) being very common. The incorporation of these groups increases fluidity of membranes, aids in the formation of fusogenic lipid structures, and facilitates the release of cationic lipids from nucleic acids.[143]

PEG-conjugated lipids are incorporated into lipoplexes to aid in the formation of nonaggregating small complexes and for the prevention of non-specific interactions in vivo. Due to the hydrophilicity of PEG, their lipid conjugates are not permanently associated with lipoplexes and diffuse from the complexes with dilution and interaction with amphiphilic components in vivo. This loss of PEG shielding from the surface of the lipoplexes is required for maximal transfection efficiency as PEG also inhibits membrane interactions required for transfection. Due to these mutually exclusive requirements (PEG shielding to aid in circulation and loss of PEG shielding is needed to increase membrane interactions required for transfection) the amount and composition of PEG lipids must be optimized to target desired tissue. In general, longer saturated fatty acid chains increase circulation and extrahepatic targeting with decreased transfection potency, while unsaturation and shorter chains decrease circulation but increase inherent transfection ability.[144]

There are many examples of ligands from small molecules such as GalNAc,[101] folate,[145] and RGD,[146] to large-molecule biologics such as antibodies,[147,148] and transferrin[149] being used to selectively target tissues in vivo. Interestingly, the most effective, "untargeted" lipoplexes for liver delivery associate with endogenous apolipoprotein E (Apo E) thereby becoming targeted to hepatocytes in situ.[150]

A commonly invoked tumor targeting mechanism is the EPR effect, which is when nanoparticles accumulate in tumor tissue much more than they do in normal tissues due to the leaky disorganized vasculature associated with tumor tissues and their lack of lymphatic drainage.[151] EPR-based targeting requires long-circulating particles with maximal tumor accumulation achieved >5 h after injection.[152]

The balance of circulation and transfection abilities of lipoplexes limits the ability to target lipoplexes. It may not be possible with current technologies to find optimal targeting and transection activities for all targeting modalities—especially passive targeting where maximum targeting requires maximum circulation times.[153]

4.9.1 Mechanism of Transfection

Rather than direct fusion with the cellular membrane, the most commonly invoked route of uptake of lipoplexes occurs via endocytosis.[154] Upon endocytosis, pH-titratable cationic lipids become protonated in the acidic environment of the endosome and lysosome. The increase positive charge increases the affinity between the lipoplex and anionic lipids in the endosomal membrane.[155] The cationic lipids can participate in membrane fusion and membrane disruption by the exchange of lipids between the lipoplex and biological membrane. This exchange of lipids facilitates the release of nucleic acids from the lipoplex and its release into the cytoplasm.[156,157]

Liposomes are attractive delivery vehicles for their ability to protect and deliver nucleic acids in vivo. They are also attractive vehicles from a perspective of manufacturing and analytics. The synthesis of the component lipids and their formulation have been shown to be scalable, and the analytics necessary for characterization have been worked out for clinical development. For these reasons, lipoplexes have been the vehicle of choice for larger oligonucleotide therapeutics such as mRNA and CRISPR therapies. The ultimate ability of lipoplexes to target to tissues outside of the liver and local delivery applications is yet to be demonstrated; however, there are many attractive applications of lipoplexes for delivery to relatively accessible tissues.

4.10 Polymer-Based Delivery Vehicles

Like lipoplexes, polymer-based transfection vehicles (polyplexes) provide nuclease protection[158] and condensation of larger nucleic acids. Polyplexes are based upon cationic polymers that form electrostatic complexes with anionic nucleic acids. Typical polyplex endosomal release mechanisms are based upon proton sponge and/or titratable amphiphiles based upon amphipathic polycations.

Polycations form electrostatic complexes with polyanionic nucleic acids. The strength of the association is dependent upon the size of the nucleic acid and the size and charge density of the polycation.[159–161] The presence of salts and polyanionic proteins, extracellular matrices, and cell surfaces can displace nucleic acids from polycations. In particular, complexes between small oligonucleotides and polycations are unstable under physiological conditions.[160] Polyplexes are typically formed in the excess of cationic charge and have a high positive surface charge. In the absence of stabilization and reduction of surface charge, the circulation times of iv administered polyplexes are short with half-lives ~5 min in vivo.[162]

There are three common strategies to improve the stability and surface charge of polyplexes to improve the circulation and targeting of ability of polyplexes: cross-linking of polycation, addition of a synthetic polyanion and conjugation of PEG.

Cross-linking, also called lateral stabilization and caging, is the formation of covalent polyamine–polyamine bonds after complexation/condensation of the nucleic acid.[162,163] The cross-linking is accomplished by the addition of bifunctional, amine-reactive reagents that form a 3-D network of bonds around the nucleic acid, thereby making the polyplex resistant to displacement by salts and polyelectrolytes. The stability of the polyplexes is such that the nucleic acid is no longer active unless a mechanism of reversibility is introduced to allow for release of the nucleic acid. A common way to introduce reversibility is the use of disulfide-containing cross-linking reagent that can be reduced in the cytoplasm allowing release of nucleic acid therapeutic.[163]

Although stabilized, cross-linked polyplexes are still positively charged and have short circulation times in vivo. The half-life of circulation is approximately 3 min with liver being the major site of deposition.[164] A common method to reduce the surface charge of a polyplex is the conjugation of PEG, a method commonly known as steric stabilization. The resulting PEG-modified polyplexes have prolonged circulation in vivo.[164,165] PEG modifications can be added to the side chains of polyamines—either before or after polyplex formation—or at the end of the polymer as a block copolymer of PEG and polycation. However, PEG modification not only inhibits nonspecific interactions but also diminishes the polycation–nucleic acid interactions in the polyplex thereby limiting the extent of PEGylation.

Cross-linking and PEGylation are often combined to make stabilized polyplexes of reduced surface charge for systemic administration that can either be passively or actively targeted. As observed for lipoplexes, a variety of small molecule (such as GalNAc,[127] RGD,[166] and folate[167]) and biologic targeting ligands (such as transferrin[168] and antibodies[169]) have been conjugated to PEG-modified polyplexes for tissues selective targeting.

The most commonly used polymer for polyplexes—and the originator of the proton sponge mechanism—is polyethylenimine (PEI).[170] PEI's high density of amine groups endows it with high charge density and a continuum of amine pK_a's that buffer in the entire pH range of the endosome. The buffering capacity of PEI has been mimicked by the addition of weakly basic imidazole groups.[171–173]

4.10.1 Conclusion: Summary and Future of Delivery Vehicles

To provide a snapshot of the current state of oligonucleotide therapeutics, the oligonucleotides currently under clinical development were categorized based on their mechanism of action and the type of vehicle used in their delivery (Fig. 10A and B). The distribution of mechanisms is somewhat consistent with the history of the field with antisense-based therapies being the most pursued at ~40% of all trials (24 out of 62 total trials). Technologies to knockdown genes (RNAi and antisense) compose more than half of the trials. The renewed interest[174] in mRNA as a therapeutic modality has created a significant (~12%) amount of trials; however, it should be noted that these trials have been to use mRNA as a vaccine. Due to the increased interest, it is expected that the number of trials using mRNA and TLR agonists will increase in the coming years. While it is imprudent to read too much into small data sets, it is interesting to note that the most common mechanism of action for approved oligonucleotide therapeutics (splice blocking with two out of four approved drugs, see Table 1) is a very low percentage of trials currently under development (~3% of trials, Fig. 10A), which implies that this mechanism of action may have few viable clinical targets.

A tabulation of current trials according to the type of vehicle used demonstrates the attractiveness of naked oligonucleotides. The application of know-how to synthesize oligonucleotides that are inherently potent and nuclease resistant such that they need no delivery vehicle has resulted in approved drugs and the large majority (~68%) of current trials. Naked oligonucleotides are attractive from a perspective of manufacturing, analytics, and bioanalytics. In addition, the toxicity of uncomplexed oligonucleotides (naked and conjugated) is generally less than lipoplex- or polyplex-based vehicles. For these reasons, it is the obvious choice to use naked oligonucleotides if one can identify sufficiently potent and stable oligonucleotides that accumulate in the tissue of interest.

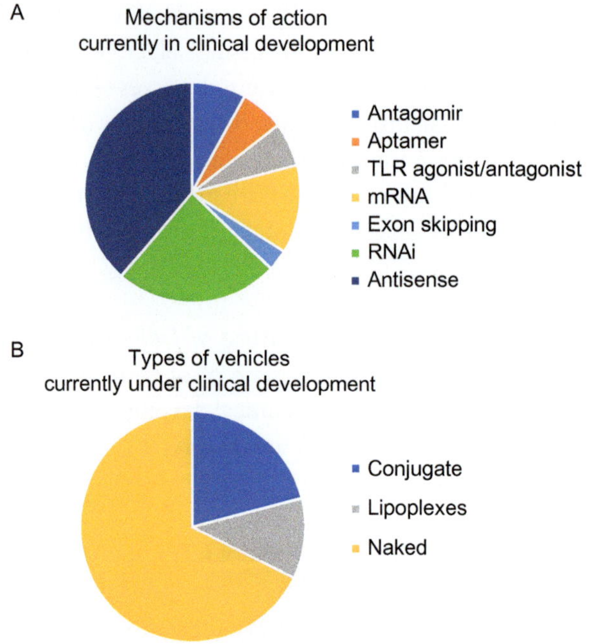

Fig. 10 Current oligonucleotide clinical trials distribution by mechanism of action and type of vehicle.

Oligonucleotides that cannot have extensive PS content—such as siRNAs—do not gain the benefit of the serum binding and pharmacokinetics that enable liver accumulation. For these oligonucleotides, biodistribution may be improved by the simple conjugation of groups that enhance circulation (e.g., PEG, lipids) or target-specific receptors (e.g., GalNAc targeting to ASGPR). These conjugates account for second most common vehicle in clinical development (~20% of trials).

Oligonucleotides that do not have sufficient inherent stability or targeting ability require a more complicated vehicle to package the nucleic acid, protect it from nucleases, and deliver it to the site of interest. The only oligonucleotide "packaging" vehicles that are currently under clinical development are lipoplexes, which have benefited from years of preclinical research and clinical development to not only identify lipoplexes with satisfactory efficacy and toxicity but also the establishment of scalable GMP manufacturing procedures with the attendant analytics and bioanalytics. For these reasons, lipoplexes are in the clinic and will continue to be the subject of development. Despite intense efforts to establish and develop

Table 4 Summary of Delivery Vehicles

Vehicle	Pros	Cons
Naked	• Relatively easy to manufacture and analyze • Small size	• Untargeted • Rapidly excreted • Provides no nuclease protection
Conjugated	• Covalent bond ensures oligo is not displaced from vehicle • Small size	• Rapidly excreted • Not amenable to systemic administration of larger oligonucleotides (e.g., mRNA) • Provides little nuclease protection
Lipoplex	• Relatively more complicated to manufacture and analyze, but regulatory pathway is known • Provides nuclease protection and targetability	• Particulate in nature with inherent limits to pharmacokinetics and biodistribution (PD/BD)
Polyplex	• Provides nuclease protection and targetability	• Particulate in nature with limit to PK/BD • Most difficult to manufacture and analyze • Manufacturing and analytics less established

polyplexes, these vehicles—at the current time—lack the efficacy and/or safety to justify the added complexity of manufacturing and analytics required to develop these complicated formulations.

The field of oligonucleotide therapeutics has benefited from decades of advancements to increase their drug-like properties. The development of nucleoside analogues has produced oligonucleotide therapeutics with sufficient nuclease stability and targetability for liver and local administrations. However, the ability to deliver any oligonucleotides to less accessible tissues such as tumors and the delivery of larger oligonucleotide such as mRNA remains to be demonstrated clinically. For this reason, the future of oligonucleotide therapy rests in advances in these two areas: (1) extrahepatic delivery of systemically administered oligonucleotides that are relatively small (<50 nucleotides) and (2) nonviral delivery of larger oligonucleotides (such as CRISPR vectors and mRNA) that require the formation of colloidal particles to provide sufficient nuclease protection and targeting in vivo (Table 4).

REFERENCES

1. Stephenson, M. L.; Zamecnik, P. C. Inhibition of Rous Sarcoma Viral RNA Translation by a Specific Oligodeoxyribonucleotide. *Proc. Natl. Acad. Sci. U.S.A.* **1978**, *75*(1), 285–288.
2. Zamecnik, P. C.; Stephenson, M. L. Inhibition of Rous Sarcoma Virus Replication and Cell Transformation by a Specific Oligodeoxynucleotide. *Proc. Natl. Acad. Sci. U.S.A.* **1978**, *75*(1), 280–284.
3. Donis-Keller, H. Site Specific Enzymatic Cleavage of RNA. *Nucleic Acids Res.* **1979**, *7*(1), 179–192.
4. Crooke, S. T. Molecular Mechanisms of Antisense Drugs: RNase H. *Antisense Nucleic Acid Drug Dev.* **1998**, *8*(2), 133–134.
5. Kole, R.; Krainer, A. R.; Altman, S. RNA Therapeutics: Beyond RNA Interference and Antisense Oligonucleotides. *Nat. Rev. Drug Discov.* **2012**, *11*(2), 125–140.
6. Elbashir, S. M.; Harborth, J.; Lendeckel, W.; Yalcin, A.; Weber, K.; Tuschl, T. Duplexes of 21-Nucleotide RNAs Mediate RNA Interference in Cultured Mammalian Cells. *Nature* **2001**, *411*(6836), 494–498.
7. Piva, R.; Spandidos, D. A.; Gambari, R. From MicroRNA Functions to MicroRNA Therapeutics: Novel Targets and Novel Drugs in Breast Cancer Research and Treatment (Review). *Int. J. Oncol.* **2013**, *43*(4), 985–994.
8. Poltronieri, P.; D'Urso, P. I.; Mezzolla, V.; D'Urso, O. F. Potential of Anti-Cancer Therapy Based on Anti-miR-155 Oligonucleotides in Glioma and Brain Tumours. *Chem. Biol. Drug Des.* **2013**, *81*(1), 79–84.
9. Prabhakar, B.; Zhong, X. B.; Rasmussen, T. P. Exploiting Long Noncoding RNAs as Pharmacological Targets to Modulate Epigenetic Diseases. *Yale J. Biol. Med.* **2017**, *90*(1), 73–86.
10. Woo, C. J.; Maier, V. K.; Davey, R.; Brennan, J.; Li, G.; Brothers, J.; Schwartz, B.; Gordo, S.; Kasper, A.; Okamoto, T. R.; Johansson, H. E.; Mandefro, B.; Sareen, D.; Bialek, P.; Chau, B. N.; Bhat, B.; Bullough, D.; Barsoum, J. Gene Activation of SMN by Selective Disruption of lncRNA-Mediated Recruitment of PRC2 for the Treatment of Spinal Muscular Atrophy. *Proc. Natl. Acad. Sci. U.S.A.* **2017**, *114*(8), E1509–E1518.
11. Havens, M. A.; Hastings, M. L. Splice-Switching Antisense Oligonucleotides as Therapeutic Drugs. *Nucleic Acids Res.* **2016**, *44*(14), 6549–6563.
12. Holtkamp, S.; Kreiter, S.; Selmi, A.; Simon, P.; Koslowski, M.; Huber, C.; Türeci, O.; Sahin, U. Modification of Antigen-Encoding RNA Increases Stability, Translational Efficacy, and T-Cell Stimulatory Capacity of Dendritic Cells. *Blood* **2006**, *108*(13), 4009–4017.
13. Karikó, K.; Kuo, A.; Barnathan, E. Overexpression of Urokinase Receptor in Mammalian Cells Following Administration of the in vitro Transcribed Encoding mRNA. *Gene Ther.* **1999**, *6*(6), 1092–1100.
14. Kallen, K. J.; Theß, A. A Development That May Evolve Into a Revolution in Medicine: mRNA as the Basis for Novel, Nucleotide-Based Vaccines and Drugs. *Ther. Adv. Vaccines* **2014**, *2*(1), 10–31.
15. Kormann, M. S.; Hasenpusch, G.; Aneja, M. K.; Nica, G.; Flemmer, A. W.; Herber-Jonat, S.; Huppmann, M.; Mays, L. E.; Illenyi, M.; Schams, A.; Griese, M.; Bittmann, I.; Handgretinger, R.; Hartl, D.; Rosenecker, J.; Rudolph, C. Expression of Therapeutic Proteins After Delivery of Chemically Modified mRNA in Mice. *Nat. Biotechnol.* **2011**, *29*(2), 154–157.
16. Anderson, B. R.; Muramatsu, H.; Jha, B. K.; Silverman, R. H.; Weissman, D.; Karikó, K. Nucleoside Modifications in RNA Limit Activation of 2'-5'-Oligoadenylate Synthetase and Increase Resistance to Cleavage by RNase L. *Nucleic Acids Res.* **2011**, *39*(21), 9329–9338.

17. Karikó, K.; Muramatsu, H.; Keller, J. M.; Weissman, D. Increased Erythropoiesis in Mice Injected With Submicrogram Quantities of Pseudouridine-Containing mRNA Encoding Erythropoietin. *Mol. Ther.* **2012**, *20*(5), 948–953.
18. Karikó, K.; Buckstein, M.; Ni, H.; Weissman, D. Suppression of RNA Recognition by Toll-Like Receptors: The Impact of Nucleoside Modification and the Evolutionary Origin of RNA. *Immunity* **2005**, *23*(2), 165–175.
19. Griffin, L. C.; Tidmarsh, G. F.; Bock, L. C.; Toole, J. J.; Leung, L. L. In Vivo Anticoagulant Properties of a Novel Nucleotide-Based Thrombin Inhibitor and Demonstration of Regional Anticoagulation in Extracorporeal Circuits. *Blood* **1993**, *81*(12), 3271–3276.
20. Kasahara, Y.; Kitadume, S.; Morihiro, K.; Kuwahara, M.; Ozaki, H.; Sawai, H.; Imanishi, T.; Obika, S. Effect of 3'-End Capping of Aptamer With Various 2',4'-Bridged Nucleotides: Enzymatic Post-Modification Toward a Practical Use of Polyclonal Aptamers. *Bioorg. Med. Chem. Lett.* **2010**, *20*(5), 1626–1629.
21. Beigelman, L.; Matulic-Adamic, J.; Haeberli, P.; Usman, N.; Dong, B.; Silverman, R. H.; Khamnei, S.; Torrence, P. F. Synthesis and Biological Activities of a Phosphorodithioate Analog of 2',5'-Oligoadenylate. *Nucleic Acids Res.* **1995**, *23*(19), 3989–3994.
22. Keefe, A. D.; Pai, S.; Ellington, A. Aptamers as Therapeutics. *Nat. Rev. Drug Discov.* **2010**, *9*(7), 537–550.
23. Wlotzka, B.; Leva, S.; Eschgfäller, B.; Burmeister, J.; Kleinjung, F.; Kaduk, C.; Muhn, P.; Hess-Stumpp, H.; Klussmann, S. In Vivo Properties of an Anti-GnRH Spiegelmer: An Example of an Oligonucleotide-Based Therapeutic Substance Class. *Proc. Natl. Acad. Sci. U.S.A.* **2002**, *99*(13), 8898–8902.
24. Krieg, A. M.; Yi, A. K.; Matson, S.; Waldschmidt, T. J.; Bishop, G. A.; Teasdale, R.; Koretzky, G. A.; Klinman, D. M. CpG Motifs in Bacterial DNA Trigger Direct B-Cell Activation. *Nature* **1995**, *374*(6522), 546–549.
25. Hemmi, H.; Takeuchi, O.; Kawai, T.; Kaisho, T.; Sato, S.; Sanjo, H.; Matsumoto, M.; Hoshino, K.; Wagner, H.; Takeda, K.; Akira, S. A Toll-Like Receptor Recognizes Bacterial DNA. *Nature* **2000**, *408*(6813), 740–745.
26. Krieg, A. M. CpG Motifs in Bacterial DNA and Their Immune Effects. *Annu. Rev. Immunol.* **2002**, *20*, 709–760.
27. Verthelyi, D.; Kenney, R. T.; Seder, R. A.; Gam, A. A.; Friedag, B.; Klinman, D. M. CpG Oligodeoxynucleotides as Vaccine Adjuvants in Primates. *J. Immunol.* **2002**, *168*(4), 1659–1663.
28. Katsuda, M.; Iwahashi, M.; Matsuda, K.; Miyazawa, M.; Nakamori, M.; Nakamura, M.; Ojima, T.; Iida, T.; Hayata, K.; Yamaue, H. Comparison of Different Classes of CpG-ODN in Augmenting the Generation of Human Epitope Peptide-Specific CTLs. *Int. J. Oncol.* **2011**, *39*(5), 1295–1302.
29. Wang, E. S.; Wu, K.; Chin, A. C.; Chen-Kiang, S.; Pongracz, K.; Gryaznov, S.; Moore, M. A. Telomerase Inhibition With an Oligonucleotide Telomerase Template Antagonist: in vitro and in vivo Studies in Multiple Myeloma and Lymphoma. *Blood* **2004**, *103*(1), 258–266.
30. Wu, X.; Zhang, J.; Yang, S.; Kuang, Z.; Tan, G.; Yang, G.; Wei, Q.; Guo, Z. Telomerase Antagonist Imetelstat Increases Radiation Sensitivity in Esophageal Squamous Cell Carcinoma. *Oncotarget* **2017**, *8*(8), 13600–13619.
31. Asai, A.; Oshima, Y.; Yamamoto, Y.; Uochi, T. A.; Kusaka, H.; Akinaga, S.; Yamashita, Y.; Pongracz, K.; Pruzan, R.; Wunder, E.; Piatyszek, M.; Li, S.; Chin, A. C.; Harley, C. B.; Gryaznov, S. A Novel Telomerase Template Antagonist (GRN163) as a Potential Anticancer Agent. *Cancer Res.* **2003**, *63*(14), 3931–3939.
32. Lipinski, C. A.; Lombardo, F.; Dominy, B. W.; Feeney, P. J. Experimental and Computational Approaches to Estimate Solubility and Permeability in Drug Discovery and Development Settings. *Adv. Drug Deliv. Rev.* **2001**, *46*(1–3), 3–26.

33. Iversen, F.; Yang, C.; Dagnæs-Hansen, F.; Schaffert, D. H.; Kjems, J.; Gao, S. Optimized siRNA-PEG Conjugates for Extended Blood Circulation and Reduced Urine Excretion in Mice. *Theranostics* **2013**, *3*(3), 201–209.
34. Weissman, D. mRNA Transcript Therapy. *Expert Rev. Vaccines* **2015**, *14*(2), 265–281.
35. Pardi, N.; Hogan, M. J.; Pelc, R. S.; Muramatsu, H.; Andersen, H.; DeMaso, C. R.; Dowd, K. A.; Sutherland, L. L.; Scearce, R. M.; Parks, R.; Wagner, W.; Granados, A.; Greenhouse, J.; Walker, M.; Willis, E.; Yu, J. S.; McGee, C. E.; Sempowski, G. D.; Mui, B. L.; Tam, Y. K.; Huang, Y. J.; Vanlandingham, D.; Holmes, V. M.; Balachandran, H.; Sahu, S.; Lifton, M.; Higgs, S.; Hensley, S. E.; Madden, T. D.; Hope, M. J.; Karikó, K.; Santra, S.; Graham, B. S.; Lewis, M. G.; Pierson, T. C.; Haynes, B. F.; Weissman, D. Zika Virus Protection by a Single Low-Dose Nucleoside-Modified mRNA Vaccination. *Nature* **2017**, *543*(7644), 248–251.
36. Wolff, J. A.; Malone, R. W.; Williams, P.; Chong, W.; Acsadi, G.; Jani, A.; Felgner, P. L. Direct Gene Transfer Into Mouse Muscle In Vivo. *Science* **1990**, *247*(4949 Pt. 1), 1465–1468.
37. Probst, J.; Weide, B.; Scheel, B.; Pichler, B. J.; Hoerr, I.; Rammensee, H. G.; Pascolo, S. Spontaneous Cellular Uptake of Exogenous Messenger RNA in vivo Is Nucleic Acid-Specific, Saturable and Ion Dependent. *Gene Ther.* **2007**, *14*(15), 1175–1180.
38. White, P. J.; Atley, L. M.; Wraight, C. J. Antisense Oligonucleotide Treatments for Psoriasis. *Expert Opin. Biol. Ther.* **2004**, *4*(1), 75–81.
39. Leeds, J. M.; Henry, S. P.; Bistner, S.; Scherrill, S.; Williams, K.; Levin, A. A. Pharmacokinetics of an Antisense Oligonucleotide Injected Intravitreally in Monkeys. *Drug Metab. Dispos.* **1998**, *26*(7), 670–675.
40. Bochot, A.; Fattal, E.; Boutet, V.; Deverre, J. R.; Jeanny, J. C.; Chacun, H.; Couvreur, P. Intravitreal Delivery of Oligonucleotides by Sterically Stabilized Liposomes. *Invest. Ophthalmol. Vis. Sci.* **2002**, *43*(1), 253–259.
41. Geary, R. S.; Norris, D.; Yu, R.; Bennett, C. F. Pharmacokinetics, Biodistribution and Cell Uptake of Antisense Oligonucleotides. *Adv. Drug Deliv. Rev.* **2015**, *87*, 46–51.
42. Rigo, F.; Chun, S. J.; Norris, D. A.; Hung, G.; Lee, S.; Matson, J.; Fey, R. A.; Gaus, H.; Hua, Y.; Grundy, J. S.; Krainer, A. R.; Henry, S. P.; Bennett, C. F. Pharmacology of a Central Nervous System Delivered 2'-O-Methoxyethyl-Modified Survival of Motor Neuron Splicing Oligonucleotide in Mice and Nonhuman Primates. *J. Pharmacol. Exp. Ther.* **2014**, *350*(1), 46–55.
43. Marafini, I.; Di Fusco, D.; Calabrese, E.; Sedda, S.; Pallone, F.; Monteleone, G. Antisense Approach to Inflammatory Bowel Disease: Prospects and Challenges. *Drugs* **2015**, *75*(7), 723–730.
44. Uyechi, L. S.; Gagné, L.; Thurston, G.; Szoka, F. C. Mechanism of Lipoplex Gene Delivery in Mouse Lung: Binding and Internalization of Fluorescent Lipid and DNA Components. *Gene Ther.* **2001**, *8*(11), 828–836.
45. Danahay, H.; Giddings, J.; Christian, R. A.; Moser, H. E.; Phillips, J. A. Distribution of a 20-Mer Phosphorothioate Oligonucleotide, CGP69846A (ISIS 5132), Into Airway Leukocytes and Epithelial Cells Following Intratracheal Delivery to Brown-Norway Rats. *Pharm. Res.* **1999**, *16*(10), 1542–1549.
46. Papaioannou, I.; Simons, J. P.; Owen, J. S. Oligonucleotide-Directed Gene-Editing Technology: Mechanisms and Future Prospects. *Expert Opin. Biol. Ther.* **2012**, *12*(3), 329–342.
47. González, F. E.; Gleisner, A.; Falcón-Beas, F.; Osorio, F.; López, M. N.; Salazar-Onfray, F. Tumor cell lysates as immunogenic sources for cancer vaccine design. *Hum. Vaccin. Immunother.* **2014**, *10*(11), 3261–3269.
48. Koziołkiewicz, M.; Wójcik, M.; Kobylańska, A.; Karwowski, B.; Rebowska, B.; Guga, P.; Stec, W. J. Stability of Stereoregular Oligo(Nucleoside Phosphorothioate)s

in Human Plasma: Diastereoselectivity of Plasma 3'-Exonuclease. *Antisense Nucleic Acid Drug Dev.* **1997**, 7(1), 43–48.
49. Koziolkiewicz, M.; Krakowiak, A.; Kwinkowski, M.; Boczkowska, M.; Stec, W. J. Stereodifferentiation—The Effect of P Chirality of Oligo(Nucleoside Phosphorothioates) on the Activity of Bacterial RNase H. *Nucleic Acids Res.* **1995**, *23*(24), 5000–5005.
50. Oka, N.; Yamamoto, M.; Sato, T.; Wada, T. Solid-Phase Synthesis of Stereoregular Oligodeoxyribonucleoside Phosphorothioates Using Bicyclic Oxazaphospholidine Derivatives as Monomer Units. *J. Am. Chem. Soc.* **2008**, *130*(47), 16031–16037.
51. Wan, W. B.; Migawa, M. T.; Vasquez, G.; Murray, H. M.; Nichols, J. G.; Gaus, H.; Berdeja, A.; Lee, S.; Hart, C. E.; Lima, W. F.; Swayze, E. E.; Seth, P. P. Synthesis, Biophysical Properties and Biological Activity of Second Generation Antisense Oligonucleotides Containing Chiral Phosphorothioate Linkages. *Nucleic Acids Res.* **2014**, *42*(22), 13456–13468.
52. Yu, R. Z.; Kim, T. W.; Hong, A.; Watanabe, T. A.; Gaus, H. J.; Geary, R. S. Cross-Species Pharmacokinetic Comparison From Mouse to Man of a Second-Generation Antisense Oligonucleotide, ISIS 301012, Targeting Human Apolipoprotein B-100. *Drug Metab. Dispos.* **2007**, *35*(3), 460–468.
53. Geselowitz, D. A.; Neckers, L. M. Bovine Serum Albumin Is a Major Oligonucleotide-Binding Protein Found on the Surface of Cultured Cells. *Antisense Res. Dev.* **1995**, *5*(3), 213–217.
54. Marshall, W. S.; Caruthers, M. H. Phosphorodithioate DNA as a Potential Therapeutic Drug. *Science* **1993**, *259*(5101), 1564–1570.
55. Summerton, J.; Weller, D. Morpholino Antisense Oligomers: Design, Preparation, and Properties. *Antisense Nucleic Acid Drug Dev.* **1997**, 7(3), 187–195.
56. Nielsen, P. E.; Egholm, M.; Berg, R. H.; Buchardt, O. Sequence-Selective Recognition of DNA by Strand Displacement With a Thymine-Substituted Polyamide. *Science* **1991**, *254*(5037), 1497–1500.
57. Meade, B. R.; Gogoi, K.; Hamil, A. S.; Palm-Apergi, C.; van den Berg, A.; Hagopian, J. C.; Springer, A. D.; Eguchi, A.; Kacsinta, A. D.; Dowdy, C. F.; Presente, A.; Lönn, P.; Kaulich, M.; Yoshioka, N.; Gros, E.; Cui, X. S.; Dowdy, S. F. Efficient Delivery of RNAi Prodrugs Containing Reversible Charge-Neutralizing Phosphotriester Backbone Modifications. *Nat. Biotechnol.* **2014**, *32*(12), 1256–1261.
58. Miller, P. S.; McParland, K. B.; Jayaraman, K.; Ts'o, P. O. Biochemical and Biological Effects of Nonionic Nucleic Acid Methylphosphonates. *Biochemistry* **1981**, *20*(7), 1874–1880.
59. Wu, S. Y.; Yang, X.; Gharpure, K. M.; Hatakeyama, H.; Egli, M.; McGuire, M. H.; Nagaraja, A. S.; Miyake, T. M.; Rupaimoole, R.; Pecot, C. V.; Taylor, M.; Pradeep, S.; Sierant, M.; Rodriguez-Aguayo, C.; Choi, H. J.; Previs, R. A.; Armaiz-Pena, G. N.; Huang, L.; Martinez, C.; Hassell, T.; Ivan, C.; Sehgal, V.; Singhania, R.; Han, H. D.; Su, C.; Kim, J. H.; Dalton, H. J.; Kovvali, C.; Keyomarsi, K.; McMillan, N. A.; Overwijk, W. W.; Liu, J.; Lee, J. S.; Baggerly, K. A.; Lopez-Berestein, G.; Ram, P. T.; Nawrot, B.; Sood, A. K. 2'-OMe-Phosphorodithioate-Modified siRNAs Show Increased Loading Into the RISC Complex and Enhanced Anti-Tumour Activity. *Nat. Commun.* **2014**, *5*, 3459.
60. Abeydeera, N. D.; Egli, M.; Cox, N.; Mercier, K.; Conde, J. N.; Pallan, P. S.; Mizurini, D. M.; Sierant, M.; Hibti, F. E.; Hassell, T.; Wang, T.; Liu, F. W.; Liu, H. M.; Martinez, C.; Sood, A. K.; Lybrand, T. P.; Frydman, C.; Monteiro, R. Q.; Gomer, R. H.; Nawrot, B.; Yang, X. Evoking Picomolar Binding in RNA by a Single Phosphorodithioate Linkage. *Nucleic Acids Res.* **2016**, *44*(17), 8052–8064.

61. Benkovic, S. J.; Sampson, E. J. Structure-Reactivity Correlation for the Hydrolysis of Phosphoramidate Monoanions. *J. Am. Chem. Soc.* **1971**, *93*(16), 4009–4016.
62. Gryaznov, S. M. Oligonucleotide n3'→p5' Phosphoramidates and Thio-Phoshoramidates as Potential Therapeutic Agents. *Chem. Biodivers.* **2010**, 7(3), 477–493.
63. Lesnikowski, Z. J.; Jaworska, M.; Stec, W. J. Octa(Thymidine Methanephosphonates) of Partially Defined Stereochemistry: Synthesis and Effect of Chirality at Phosphorus on Binding to Pentadecadeoxyriboadenylic Acid. *Nucleic Acids Res.* **1990**, *18*(8), 2109–2115.
64. Lesnikowski, Z. J.; Jaworska, M.; Stec, W. J. Stereoselective Synthesis of P-Homochiral Oligo(Thymidine Methanephosphonates). *Nucleic Acids Res.* **1988**, *16*(24), 11675–11689.
65. Paul, S.; Caruthers, M. H. Synthesis of Phosphorodiamidate Morpholino Oligonucleotides and Their Chimeras Using Phosphoramidite Chemistry. *J. Am. Chem. Soc.* **2016**, *138*(48), 15663–15672.
66. Bill, B. R.; Petzold, A. M.; Clark, K. J.; Schimmenti, L. A.; Ekker, S. C. A Primer for Morpholino Use in Zebrafish. *Zebrafish* **2009**, *6*(1), 69–77.
67. Anthony, K.; Feng, L.; Arechavala-Gomeza, V.; Guglieri, M.; Straub, V.; Bushby, K.; Cirak, S.; Morgan, J.; Muntoni, F. Exon Skipping Quantification by Quantitative Reverse-Transcription Polymerase Chain Reaction in Duchenne Muscular Dystrophy Patients Treated With the Antisense Oligomer Eteplirsen. *Hum. Gene Ther. Methods* **2012**, *23*(5), 336–345.
68. Nielsen, P. E. Sequence-Selective Targeting of Duplex DNA by Peptide Nucleic Acids. *Curr. Opin. Mol. Ther.* **2010**, *12*(2), 184–191.
69. Dirin, M.; Winkler, J. Influence of Diverse Chemical Modifications on the ADME Characteristics and Toxicology of Antisense Oligonucleotides. *Expert Opin. Biol. Ther.* **2013**, *13*(6), 875–888.
70. Flanagan, W. M.; Wolf, J. J.; Olson, P.; Grant, D.; Lin, K. Y.; Wagner, R. W.; Matteucci, M. D. A Cytosine Analog That Confers Enhanced Potency to Antisense Oligonucleotides. *Proc. Natl. Acad. Sci. U.S.A.* **1999**, *96*(7), 3513–3518.
71. Shen, L.; Siwkowski, A.; Wancewicz, E. V.; Lesnik, E.; Butler, M.; Witchell, D.; Vasquez, G.; Ross, B.; Acevedo, O.; Inamati, G.; Sasmor, H.; Manoharan, M.; Monia, B. P. Evaluation of C-5 Propynyl Pyrimidine-Containing Oligonucleotides in vitro and In Vivo. *Antisense Nucleic Acid Drug Dev.* **2003**, *13*(3), 129–142.
72. Karikó, K.; Muramatsu, H.; Welsh, F. A.; Ludwig, J.; Kato, H.; Akira, S.; Weissman, D. Incorporation of Pseudouridine Into mRNA Yields Superior Non-immunogenic Vector With Increased Translational Capacity and Biological Stability. *Mol. Ther.* **2008**, *16*(11), 1833–1840.
73. Inoue, H.; Hayase, Y.; Imura, A.; Iwai, S.; Miura, K.; Ohtsuka, E. Synthesis and Hybridization Studies on Two Complementary Nona(2'-O-methyl)ribonucleotides. *Nucleic Acids Res.* **1987**, *15*(15), 6131–6148.
74. Monia, B. P.; Lesnik, E. A.; Gonzalez, C.; Lima, W. F.; McGee, D.; Guinosso, C. J.; Kawasaki, A. M.; Cook, P. D.; Freier, S. M. Evaluation of 2'-Modified Oligonucleotides Containing 2'-Deoxy Gaps as Antisense Inhibitors of Gene Expression. *J. Biol. Chem.* **1993**, *268*(19), 14514–14522.
75. Allerson, C. R.; Sioufi, N.; Jarres, R.; Prakash, T. P.; Naik, N.; Berdeja, A.; Wanders, L.; Griffey, R. H.; Swayze, E. E.; Bhat, B. Fully 2'-Modified Oligonucleotide Duplexes With Improved in vitro Potency and Stability Compared to Unmodified Small Interfering RNA. *J. Med. Chem.* **2005**, *48*(4), 901–904.
76. Krützfeldt, J.; Rajewsky, N.; Braich, R.; Rajeev, K. G.; Tuschl, T.; Manoharan, M.; Stoffel, M. Silencing of MicroRNAs in vivo With 'Antagomirs'. *Nature* **2005**, *438*(7068), 685–689.

77. Baker, B. F.; Lot, S. S.; Condon, T. P.; Cheng-Flournoy, S.; Lesnik, E. A.; Sasmor, H. M.; Bennett, C. F. 2'-O-(2-Methoxy)ethyl-Modified Anti-Intercellular Adhesion Molecule 1 (ICAM-1) Oligonucleotides Selectively Increase the ICAM-1 mRNA Level and Inhibit Formation of the ICAM-1 Translation Initiation Complex in Human Umbilical Vein Endothelial Cells. *J. Biol. Chem.* **1997**, *272*(18), 11994–12000.
78. Freier, S. M.; Altmann, K. H. The Ups and Downs of Nucleic Acid Duplex Stability: Structure-Stability Studies on Chemically-Modified DNA:RNA Duplexes. *Nucleic Acids Res.* **1997**, *25*(22), 4429–4443.
79. Martin, P. A New Access to 2′-O-Alkylated Ribonucleosides and Properties of 2′-O-Alkylated Oligoribonucleotides. *Helv. Chim. Acta* **1995**, *78*, 486.
80. Pieken, W. A.; Olsen, D. B.; Benseler, F.; Aurup, H.; Eckstein, F. Kinetic Characterization of Ribonuclease-Resistant 2'-Modified Hammerhead Ribozymes. *Science* **1991**, *253*(5017), 314–317.
81. Kawasaki, A. M.; Casper, M. D.; Freier, S. M.; Lesnik, E. A.; Zounes, M. C.; Cummins, L. L.; Gonzalez, C.; Cook, P. D. Uniformly Modified 2'-Deoxy-2'-Fluoro Phosphorothioate Oligonucleotides as Nuclease-Resistant Antisense Compounds With High Affinity and Specificity for RNA Targets. *J. Med. Chem.* **1993**, *36*(7), 831–841.
82. Braasch, D. A.; Jensen, S.; Liu, Y.; Kaur, K.; Arar, K.; White, M. A.; Corey, D. R. RNA Interference in Mammalian Cells by Chemically-Modified RNA. *Biochemistry* **2003**, *42*(26), 7967–7975.
83. Liboska, R.; Snášel, J.; Barvík, I.; Buděšínský, M.; Pohl, R.; Točík, Z.; Páv, O.; Rejman, D.; Novák, P.; Rosenberg, I. 4'-Alkoxy Oligodeoxynucleotides: A Novel Class of RNA Mimics. *Org. Biomol. Chem.* **2011**, *9*(24), 8261–8267.
84. Kel'in, A. V.; Zlatev, I.; Harp, J.; Jayaraman, M.; Bisbe, A.; O'Shea, J.; Taneja, N.; Manoharan, R. M.; Khan, S.; Charisse, K.; Maier, M. A.; Egli, M.; Rajeev, K. G.; Manoharan, M. Structural Basis of Duplex Thermodynamic Stability and Enhanced Nuclease Resistance of 5'-C-Methyl Pyrimidine-Modified Oligonucleotides. *J. Org. Chem.* **2016**, *81*(6), 2261–2279.
85. Obika, S.; Nanbu, D.; Hari, Y.; Morio, K. I.; In, Y.; Ishida, T.; Imanishi, T. Synthesis of 2′-O,4′-C-Methyleneuridine and -Cytidine. Novel Bicyclic Nucleosides Having a Fixed C3, -Endo Sugar Puckering. *Tetrahedron Lett.* **1997**, *38*, 8735.
86. Koshkin, A. A.; Singh, S. K.; Nielsen, P.; Rajwanshi, V. K.; Kumar, R.; Meldgaard, M.; Olsen, C. E.; Wengel, J. LNA (Locked Nucleic Acids): Synthesis of the Adenine, Cytosine, Guanine, 5-Methylcytosine, Thymine and Uracil Bicyclonucleoside Monomers, Oligomerisation, and Unprecedented Nucleic Acid Recognition. *Tetrahedron* **1998**, *54*, 3607.
87. Christensen, U.; Jacobsen, N.; Rajwanshi, V. K.; Wengel, J.; Koch, T. Stopped-Flow Kinetics of Locked Nucleic Acid (LNA)-Oligonucleotide Duplex Formation: Studies of LNA-DNA and DNA-DNA Interactions. *Biochem. J.* **2001**, *354*(Pt. 3), 481–484.
88. Rahman, S. M.; Seki, S.; Obika, S.; Haitani, S.; Miyashita, K.; Imanishi, T. Highly Stable Pyrimidine-Motif Triplex Formation at Physiological pH Values by a Bridged Nucleic Acid Analogue. *Angew. Chem. Int. Ed. Engl.* **2007**, *46*(23), 4306–4309.
89. Morita, K.; Hasegawa, C.; Kaneko, M.; Tsutsumi, S.; Sone, J.; Ishikawa, T.; Imanishi, T.; Koizumi, M. 2'-O,4'-C-Ethylene-Bridged Nucleic Acids (ENA) With Nuclease-Resistance and High Affinity for RNA. *Nucleic Acids Symp. Ser.* **2001**, *1*(1), 241–242.
90. Seth, P. P.; Vasquez, G.; Allerson, C. A.; Berdeja, A.; Gaus, H.; Kinberger, G. A.; Prakash, T. P.; Migawa, M. T.; Bhat, B.; Swayze, E. E. Synthesis and Biophysical Evaluation of 2',4'-Constrained 2'-O-Methoxyethyl and 2',4'-Constrained 2'O-Ethyl Nucleic Acid Analogues. *J. Org. Chem.* **2010**, *75*(5), 1569–1581.

91. Steffens, R.; Leumann, C. J. Synthesis and Thermodynamic and Biophysical Properties of Tricyclo-DNA. *J. Am. Chem. Soc.* **1999**, *121*, 3249–3255.
92. Elmén, J.; Lindow, M.; Schütz, S.; Lawrence, M.; Petri, A.; Obad, S.; Lindholm, M.; Hedtjärn, M.; Hansen, H. F.; Berger, U.; Gullans, S.; Kearney, P.; Sarnow, P.; Straarup, E. M.; Kauppinen, S. LNA-Mediated MicroRNA Silencing in Non-Human Primates. *Nature* **2008**, *452*(7189), 896–899.
93. Elmén, J.; Thonberg, H.; Ljungberg, K.; Frieden, M.; Westergaard, M.; Xu, Y.; Wahren, B.; Liang, Z.; Ørum, H.; Koch, T.; Wahlestedt, C. Locked Nucleic Acid (LNA) Mediated Improvements in siRNA Stability and Functionality. *Nucleic Acids Res.* **2005**, *33*(1), 439–447.
94. Lundin, K. E.; Højland, T.; Hansen, B. R.; Persson, R.; Bramsen, J. B.; Kjems, J.; Koch, T.; Wengel, J.; Smith, C. I. Biological Activity and Biotechnological Aspects of Locked Nucleic Acids. *Adv. Genet.* **2013**, *82*, 47–107.
95. Langkjaer, N.; Pasternak, A.; Wengel, J. UNA (Unlocked Nucleic Acid): A Flexible RNA Mimic That Allows Engineering of Nucleic Acid Duplex Stability. *Bioorg. Med. Chem.* **2009**, *17*(15), 5420–5425.
96. Grünweller, A.; Wyszko, E.; Bieber, B.; Jahnel, R.; Erdmann, V. A.; Kurreck, J. Comparison of Different Antisense Strategies in Mammalian Cells Using Locked Nucleic Acids, 2'-O-Methyl RNA, Phosphorothioates and Small Interfering RNA. *Nucleic Acids Res.* **2003**, *31*(12), 3185–3193.
97. Czauderna, F.; Fechtner, M.; Dames, S.; Aygün, H.; Klippel, A.; Pronk, G. J.; Giese, K.; Kaufmann, J. Structural Variations and Stabilising Modifications of Synthetic siRNAs in Mammalian Cells. *Nucleic Acids Res.* **2003**, *31*(11), 2705–2716.
98. Crooke, S. T.; Graham, M. J.; Zuckerman, J. E.; Brooks, D.; Conklin, B. S.; Cummins, L. L.; Greig, M. J.; Guinosso, C. J.; Kornbrust, D.; Manoharan, M.; Sasmor, H. M.; Schleich, T.; Tivel, K. L.; Griffey, R. H. Pharmacokinetic Properties of Several Novel Oligonucleotide Analogs in Mice. *J. Pharmacol. Exp. Ther.* **1996**, *277*(2), 923–937.
99. Wolfrum, C.; Shi, S.; Jayaprakash, K. N.; Jayaraman, M.; Wang, G.; Pandey, R. K.; Rajeev, K. G.; Nakayama, T.; Charrise, K.; Ndungo, E. M.; Zimmermann, T.; Koteliansky, V.; Manoharan, M.; Stoffel, M. Mechanisms and Optimization of in vivo Delivery of Lipophilic siRNAs. *Nat. Biotechnol.* **2007**, *25*(10), 1149–1157.
100. Watanabe, T. A.; Geary, R. S.; Levin, A. A. Plasma Protein Binding of an Antisense Oligonucleotide Targeting Human ICAM-1 (ISIS 2302). *Oligonucleotides* **2006**, *16*(2), 169–180.
101. Rensen, P. C.; Sliedregt, L. A.; Ferns, M.; Kieviet, E.; van Rossenberg, S. M.; van Leeuwen, S. H.; van Berkel, T. J.; Biessen, E. A. Determination of the Upper Size Limit for Uptake and Processing of Ligands by the Asialoglycoprotein Receptor on Hepatocytes in vitro and In Vivo. *J. Biol. Chem.* **2001**, *276*(40), 37577–37584.
102. Prakash, T. P.; Yu, J.; Migawa, M. T.; Kinberger, G. A.; Wan, W. B.; Østergaard, M. E.; Carty, R. L.; Vasquez, G.; Low, A.; Chappell, A.; Schmidt, K.; Aghajan, M.; Crosby, J.; Murray, H. M.; Booten, S. L.; Hsiao, J.; Soriano, A.; Machemer, T.; Cauntay, P.; Burel, S. A.; Murray, S. F.; Gaus, H.; Graham, M. J.; Swayze, E. E.; Seth, P. P. Comprehensive Structure-Activity Relationship of Triantennary N-Acetylgalactosamine Conjugated Antisense Oligonucleotides for Targeted Delivery to Hepatocytes. *J. Med. Chem.* **2016**, *59*(6), 2718–2733.
103. Rensen, P. C.; van Leeuwen, S. H.; Sliedregt, L. A.; van Berkel, T. J.; Biessen, E. A. Design and Synthesis of Novel N-Acetylgalactosamine-Terminated Glycolipids for Targeting of Lipoproteins to the Hepatic Asialoglycoprotein Receptor. *J. Med. Chem.* **2004**, *47*(23), 5798–5808.
104. Nair, J. K.; Willoughby, J. L.; Chan, A.; Charisse, K.; Alam, M. R.; Wang, Q.; Hoekstra, M.; Kandasamy, P.; Kel'in, A. V.; Milstein, S.; Taneja, N.; O'Shea, J.; Shaikh, S.; Zhang, L.; van der Sluis, R. J.; Jung, M. E.; Akinc, A.; Hutabarat, R.;

Kuchimanchi, S.; Fitzgerald, K.; Zimmermann, T.; van Berkel, T. J.; Maier, M. A.; Rajeev, K. G.; Manoharan, M. Multivalent N-Acetylgalactosamine-Conjugated siRNA Localizes in Hepatocytes and Elicits Robust RNAi-Mediated Gene Silencing. *J. Am. Chem. Soc.* **2014**, *136*(49), 16958–16961.

105. Maier, M. A.; Yannopoulos, C. G.; Mohamed, N.; Roland, A.; Fritz, H.; Mohan, V.; Just, G.; Manoharan, M. Synthesis of Antisense Oligonucleotides Conjugated to a Multivalent Carbohydrate Cluster for Cellular Targeting. *Bioconjug. Chem.* **2003**, *14*(1), 18–29.
106. Gupta, A. In *Drug Metabolism and Pharmacokinetic (DMPK) Properties of siRNA-GalNAc Conjugates, DIA/FDA Oligonucleotide Based Therapeutic Conference, Washington, D.C.*; 2015.
107. Huang, Y. Preclinical and Clinical Advances of GalNAc-Decorated Nucleic Acid Therapeutics. *Mol. Ther. Nucleic Acids* **2017**, *6*, 116–132.
108. Gooding, M.; Malhotra, M.; Evans, J. C.; Darcy, R.; O'Driscoll, C. M. Oligonucleotide Conjugates—Candidates for Gene Silencing Therapeutics. *Eur. J. Pharm. Biopharm.* **2016**, *107*, 321–340.
109. Alam, M. R.; Dixit, V.; Kang, H.; Li, Z. B.; Chen, X.; Trejo, J.; Fisher, M.; Juliano, R. L. Intracellular Delivery of an Anionic Antisense Oligonucleotide Via Receptor-Mediated Endocytosis. *Nucleic Acids Res.* **2008**, *36*(8), 2764–2776.
110. Li, S.; Deshmukh, H. M.; Huang, L. Folate-Mediated Targeting of Antisense Oligodeoxynucleotides to Ovarian Cancer Cells. *Pharm. Res.* **1998**, *15*(10), 1540–1545.
111. Tan, M.; Vernes, J. M.; Chan, J.; Cuellar, T. L.; Asundi, A.; Nelson, C.; Yip, V.; Shen, B.; Vandlen, R.; Siebel, C.; Meng, Y. G. Real-Time Quantification of Antibody-Short Interfering RNA Conjugate in Serum by Antigen Capture Reverse Transcription-Polymerase Chain Reaction. *Anal. Biochem.* **2012**, *430*(2), 171–178.
112. Bloomfield, V. A. DNA Condensation. *Curr. Opin. Struct. Biol.* **1996**, *6*(3), 334–341.
113. Vijayanathan, V.; Thomas, T.; Thomas, T. J. DNA Nanoparticles and Development of DNA Delivery Vehicles for Gene Therapy. *Biochemistry* **2002**, *41*(48), 14085–14094.
114. He, C.; Hu, Y.; Yin, L.; Tang, C.; Yin, C. Effects of Particle Size and Surface Charge on Cellular Uptake and Biodistribution of Polymeric Nanoparticles. *Biomaterials* **2010**, *31*(13), 3657–3666.
115. Tang, L.; Yang, X.; Yin, Q.; Cai, K.; Wang, H.; Chaudhury, I.; Yao, C.; Zhou, Q.; Kwon, M.; Hartman, J. A.; Dobrucki, I. T.; Dobrucki, L. W.; Borst, L. B.; Lezmi, S.; Helferich, W. G.; Ferguson, A. L.; Fan, T. M.; Cheng, J. Investigating the Optimal Size of Anticancer Nanomedicine. *Proc. Natl. Acad. Sci. U.S.A.* **2014**, *111*(43), 15344–15349.
116. Cabral, H.; Matsumoto, Y.; Mizuno, K.; Chen, Q.; Murakami, M.; Kimura, M.; Terada, Y.; Kano, M. R.; Miyazono, K.; Uesaka, M.; Nishiyama, N.; Kataoka, K. Accumulation of Sub-100 nm Polymeric Micelles in Poorly Permeable Tumours Depends on Size. *Nat. Nanotechnol.* **2011**, *6*(12), 815–823.
117. Pant, K.; Pufe, J.; Zarschler, K.; Bergmann, R.; Steinbach, J.; Reimann, S.; Haag, R.; Pietzsch, J.; Stephan, H. Surface Charge and Particle Size Determine the Metabolic Fate of Dendritic Polyglycerols. *Nanoscale* **2017**, *9*(25), 8723–8739.
118. Stein, C. A.; Hansen, J. B.; Lai, J.; Wu, S.; Voskresenskiy, A.; Høg, A.; Worm, J.; Hedtjärn, M.; Souleimanian, N.; Miller, P.; Soifer, H. S.; Castanotto, D.; Benimetskaya, L.; Ørum, H.; Koch, T. Efficient Gene Silencing by Delivery of Locked Nucleic Acid Antisense Oligonucleotides, Unassisted by Transfection Reagents. *Nucleic Acids Res.* **2010**, *38*(1), e3.
119. Castanotto, D.; Lin, M.; Kowolik, C.; Koch, T.; Hansen, B. R.; Oerum, H.; Stein, C. A. Protein Kinase C-α Is a Critical Protein for Antisense Oligonucleotide-Mediated Silencing in Mammalian Cells. *Mol. Ther.* **2016**, *24*(6), 1117–1125.

120. Liang, X. H.; Sun, H.; Shen, W.; Crooke, S. T. Identification and Characterization of Intracellular Proteins That Bind Oligonucleotides With Phosphorothioate Linkages. *Nucleic Acids Res.* **2015**, *43*(5), 2927–2945.
121. Crooke, S. T.; Wang, S.; Vickers, T. A.; Shen, W.; Liang, X. H. Cellular Uptake and Trafficking of Antisense Oligonucleotides. *Nat. Biotechnol.* **2017**, *35*(3), 230–237.
122. Juliano, R. L.; Ming, X.; Nakagawa, O. Cellular Uptake and Intracellular Trafficking of Antisense and siRNA Oligonucleotides. *Bioconjug. Chem.* **2012**, *23*(2), 147–157.
123. Akinc, A.; Thomas, M.; Klibanov, A. M.; Langer, R. Exploring Polyethylenimine-Mediated DNA Transfection and the Proton Sponge Hypothesis. *J. Gene Med.* **2005**, 7(5), 657–663.
124. Plank, C.; Zauner, W.; Wagner, E. Application of Membrane-Active Peptides for Drug and Gene Delivery Across Cellular Membranes. *Adv. Drug Deliv. Rev.* **1998**, *34*(1), 21–35.
125. Murthy, N.; Campbell, J.; Fausto, N.; Hoffman, A. S.; Stayton, P. S. Design and Synthesis of pH-Responsive Polymeric Carriers That Target Uptake and Enhance the Intracellular Delivery of Oligonucleotides. *J. Control. Release* **2003**, *89*(3), 365–374.
126. Lehto, T.; Ezzat, K.; Wood, M. J.; El Andaloussi, S. Peptides for Nucleic Acid Delivery. *Adv. Drug Deliv. Rev.* **2016**, *106*(Pt. A), 172–182.
127. Rozema, D. B.; Lewis, D. L.; Wakefield, D. H.; Wong, S. C.; Klein, J. J.; Roesch, P. L.; Bertin, S. L.; Reppen, T. W.; Chu, Q.; Blokhin, A. V.; Hagstrom, J. E.; Wolff, J. A. Dynamic PolyConjugates for Targeted in vivo Delivery of siRNA to Hepatocytes. *Proc. Natl. Acad. Sci. U.S.A.* **2007**, *104*(32), 12982–12987.
128. Murthy, N.; Campbell, J.; Fausto, N.; Hoffman, A. S.; Stayton, P. S. Bioinspired pH-Responsive Polymers for the Intracellular Delivery of Biomolecular Drugs. *Bioconjug. Chem.* **2003**, *14*(2), 412–419.
129. Bendifallah, N.; Rasmussen, F. W.; Zachar, V.; Ebbesen, P.; Nielsen, P. E.; Koppelhus, U. Evaluation of Cell-Penetrating Peptides (CPPs) as Vehicles for Intracellular Delivery of Antisense Peptide Nucleic Acid (PNA). *Bioconjug. Chem.* **2006**, *17*(3), 750–758.
130. Amantana, A.; Moulton, H. M.; Cate, M. L.; Reddy, M. T.; Whitehead, T.; Hassinger, J. N.; Youngblood, D. S.; Iversen, P. L. Pharmacokinetics, Biodistribution, Stability and Toxicity of a Cell-Penetrating Peptide-Morpholino Oligomer Conjugate. *Bioconjug. Chem.* **2007**, *18*(4), 1325–1331.
131. Rozema, D. B.; Blokhin, A. V.; Wakefield, D. H.; Benson, J. D.; Carlson, J. C.; Klein, J. J.; Almeida, L. J.; Nicholas, A. L.; Hamilton, H. L.; Chu, Q.; Hegge, J. O.; Wong, S. C.; Trubetskoy, V. S.; Hagen, C. M.; Kitas, E.; Wolff, J. A.; Lewis, D. L. Protease-Triggered siRNA Delivery Vehicles. *J. Control. Release* **2015**, *209*, 57–66.
132. Wilson, K. D.; Raney, S. G.; Sekirov, L.; Chikh, G.; deJong, S. D.; Cullis, P. R.; Tam, Y. K. Effects of Intravenous and Subcutaneous Administration on the Pharmacokinetics, Biodistribution, Cellular Uptake and Immunostimulatory Activity of CpG ODN Encapsulated in Liposomal Nanoparticles. *Int. Immunopharmacol.* **2007**, 7(8), 1064–1075.
133. Valadi, H.; Ekström, K.; Bossios, A.; Sjöstrand, M.; Lee, J. J.; Lötvall, J. O. Exosome-Mediated Transfer of mRNAs and MicroRNAs Is a Novel Mechanism of Genetic Exchange Between Cells. *Nat. Cell Biol.* **2007**, *9*(6), 654–659.
134. Morishita, M.; Takahashi, Y.; Nishikawa, M.; Takakura, Y. Pharmacokinetics of Exosomes—An Important Factor for Elucidating the Biological Roles of Exosomes and for the Development of Exosome-Based Therapeutics. *J. Pharm. Sci.* **2017**, *106*, 2265–2269.
135. Hofland, H. E.; Shephard, L.; Sullivan, S. M. Formation of Stable Cationic Lipid/DNA Complexes for Gene Transfer. *Proc. Natl. Acad. Sci. U.S.A.* **1996**, *93*(14), 7305–7309.

136. Maurer, N.; Wong, K. F.; Stark, H.; Louie, L.; McIntosh, D.; Wong, T.; Scherrer, P.; Semple, S. C.; Cullis, P. R. Spontaneous Entrapment of Polynucleotides Upon Electrostatic Interaction With Ethanol-Destabilized Cationic Liposomes. *Biophys. J.* **2001**, *80*(5), 2310–2326.
137. Ewert, K.; Evans, H. M.; Ahmad, A.; Slack, N. L.; Lin, A. J.; Martin-Herranz, A.; Safinya, C. R. Lipoplex Structures and Their Distinct Cellular Pathways. *Adv. Genet.* **2005**, *53*, 119–155.
138. Li, W.; Szoka, F. C. Lipid-Based Nanoparticles for Nucleic Acid Delivery. *Pharm. Res.* **2007**, *24*(3), 438–449.
139. Jayaraman, M.; Ansell, S. M.; Mui, B. L.; Tam, Y. K.; Chen, J.; Du, X.; Butler, D.; Eltepu, L.; Matsuda, S.; Narayanannair, J. K.; Rajeev, K. G.; Hafez, I. M.; Akinc, A.; Maier, M. A.; Tracy, M. A.; Cullis, P. R.; Madden, T. D.; Manoharan, M.; Hope, M. J. Maximizing the Potency of siRNA Lipid Nanoparticles for Hepatic Gene Silencing In Vivo. *Angew. Chem. Int. Ed. Engl.* **2012**, *51*(34), 8529–8533.
140. Semple, S. C.; Akinc, A.; Chen, J.; Sandhu, A. P.; Mui, B. L.; Cho, C. K.; Sah, D. W.; Stebbing, D.; Crosley, E. J.; Yaworski, E.; Hafez, I. M.; Dorkin, J. R.; Qin, J.; Lam, K.; Rajeev, K. G.; Wong, K. F.; Jeffs, L. B.; Nechev, L.; Eisenhardt, M. L.; Jayaraman, M.; Kazem, M.; Maier, M. A.; Srinivasulu, M.; Weinstein, M. J.; Chen, Q.; Alvarez, R.; Barros, S. A.; De, S.; Klimuk, S. K.; Borland, T.; Kosovrasti, V.; Cantley, W. L.; Tam, Y. K.; Manoharan, M.; Ciufolini, M. A.; Tracy, M. A.; de Fougerolles, A.; MacLachlan, I.; Cullis, P. R.; Madden, T. D.; Hope, M. J. Rational Design of Cationic Lipids for siRNA Delivery. *Nat. Biotechnol.* **2010**, *28*(2), 172–176.
141. Midoux, P.; Pichon, C.; Yaouanc, J. J.; Jaffrès, P. A. Chemical Vectors for Gene Delivery: A Current Review on Polymers, Peptides and Lipids Containing Histidine or Imidazole as Nucleic Acids Carriers. *Br. J. Pharmacol.* **2009**, *157*(2), 166–178.
142. Koynova, R.; Tenchov, B. Cationic Lipids: Molecular Structure/Transfection Activity Relationships and Interactions With Biomembranes. *Top. Curr. Chem.* **2010**, *296*, 51–93.
143. Wang, L.; Koynova, R.; Parikh, H.; MacDonald, R. C. Transfection Activity of Binary Mixtures of Cationic O-Substituted Phosphatidylcholine Derivatives: The Hydrophobic Core Strongly Modulates Physical Properties and DNA Delivery Efficacy. *Biophys. J.* **2006**, *91*(10), 3692–3706.
144. Mui, B. L.; Tam, Y. K.; Jayaraman, M.; Ansell, S. M.; Du, X.; Tam, Y. Y.; Lin, P. J.; Chen, S.; Narayanannair, J. K.; Rajeev, K. G.; Manoharan, M.; Akinc, A.; Maier, M. A.; Cullis, P.; Madden, T. D.; Hope, M. J. Influence of Polyethylene Glycol Lipid Desorption Rates on Pharmacokinetics and Pharmacodynamics of siRNA Lipid Nanoparticles. *Mol. Ther. Nucleic Acids* **2013**, *2,* e139.
145. Gorle, S.; Ariatti, M.; Singh, M. Novel Serum-Tolerant Lipoplexes Target the Folate Receptor Efficiently. *Eur. J. Pharm. Sci.* **2014**, *59*, 83–93.
146. Juliano, R. L.; Ming, X.; Nakagawa, O.; Xu, R.; Yoo, H. Integrin Targeted Delivery of Gene Therapeutics. *Theranostics* **2011**, *1*, 211–219.
147. Wu, Y.; Ma, J.; Woods, P. S.; Chesarino, N. M.; Liu, C.; Lee, L. J.; Nana-Sinkam, S. P.; Davis, I. C. Selective Targeting of Alveolar Type II Respiratory Epithelial Cells by Anti-Surfactant Protein-C Antibody-Conjugated Lipoplexes. *J. Control. Release* **2015**, *203*, 140–149.
148. van Zanten, J.; Doornbos-Van der Meer, B.; Audouy, S.; Kok, R. J.; de Leij, L. A Nonviral Carrier for Targeted Gene Delivery to Tumor Cells. *Cancer Gene Ther.* **2004**, *11*(2), 156–164.
149. Cinci, M.; Mamidi, S.; Li, W.; Fehring, V.; Kirschfink, M. Targeted Delivery of siRNA Using Transferrin-Coupled Lipoplexes Specifically Sensitizes CD71 High

Expressing Malignant Cells to Antibody-Mediated Complement Attack. *Target Oncol.* **2015**, *10*(3), 405–413.
150. Akinc, A.; Querbes, W.; De, S.; Qin, J.; Frank-Kamenetsky, M.; Jayaprakash, K. N.; Jayaraman, M.; Rajeev, K. G.; Cantley, W. L.; Dorkin, J. R.; Butler, J. S.; Qin, L.; Racie, T.; Sprague, A.; Fava, E.; Zeigerer, A.; Hope, M. J.; Zerial, M.; Sah, D. W.; Fitzgerald, K.; Tracy, M. A.; Manoharan, M.; Koteliansky, V.; Fougerolles, A.; Maier, M. A. Targeted Delivery of RNAi Therapeutics With Endogenous and Exogenous Ligand-Based Mechanisms. *Mol. Ther.* **2010**, *18*(7), 1357–1364.
151. Matsumura, Y.; Maeda, H. A New Concept for Macromolecular Therapeutics in Cancer Chemotherapy: Mechanism of Tumoritropic Accumulation of Proteins and the Antitumor Agent SMANCS. *Cancer Res.* **1986**, *46*(12 Pt. 1), 6387–6392.
152. Wong, A. D.; Ye, M.; Ulmschneider, M. B.; Searson, P. C. Quantitative Analysis of the Enhanced Permeation and Retention (EPR) Effect. *PLoS One* **2015**, *10*(5), e0123461.
153. Kobayashi, H.; Watanabe, R.; Choyke, P. L. Improving Conventional Enhanced Permeability and Retention (EPR) Effects; What Is the Appropriate Target? *Theranostics* **2013**, *4*(1), 81–89.
154. Wrobel, I.; Collins, D. Fusion of Cationic Liposomes With Mammalian Cells Occurs After Endocytosis. *Biochim. Biophys. Acta* **1995**, *1235*(2), 296–304.
155. Hafez, I. M.; Ansell, S.; Cullis, P. R. Tunable pH-Sensitive Liposomes Composed of Mixtures of Cationic and Anionic Lipids. *Biophys. J.* **2000**, *79*(3), 1438–1446.
156. Zelphati, O.; Szoka, F. C. Mechanism of Oligonucleotide Release From Cationic Liposomes. *Proc. Natl. Acad. Sci. U.S.A.* **1996**, *93*(21), 11493–11498.
157. Xu, Y.; Szoka, F. C. Mechanism of DNA Release From Cationic Liposome/DNA Complexes Used in Cell Transfection. *Biochemistry* **1996**, *35*(18), 5616–5623.
158. Chiou, H. C.; Tangco, M. V.; Levine, S. M.; Robertson, D.; Kormis, K.; Wu, C. H.; Wu, G. Y. Enhanced Resistance to Nuclease Degradation of Nucleic Acids Complexed to Asialoglycoprotein-Polylysine Carriers. *Nucleic Acids Res.* **1994**, *22*(24), 5439–5446.
159. Wolfert, M. A.; Dash, P. R.; Nazarova, O.; Oupicky, D.; Seymour, L. W.; Smart, S.; Strohalm, J.; Ulbrich, K. Polyelectrolyte Vectors for Gene Delivery: Influence of Cationic Polymer on Biophysical Properties of Complexes Formed With DNA. *Bioconjug. Chem.* **1999**, *10*(6), 993–1004.
160. Zheng, C.; Niu, L.; Yan, J.; Liu, J.; Luo, Y.; Liang, D. Structure and Stability of the Complex Formed by Oligonucleotides. *Phys. Chem. Chem. Phys.* **2012**, *14*(20), 7352–7359.
161. Kunath, K.; von Harpe, A.; Fischer, D.; Petersen, H.; Bickel, U.; Voigt, K.; Kissel, T. Low-Molecular-Weight Polyethylenimine as a Non-Viral Vector for DNA Delivery: Comparison of Physicochemical Properties, Transfection Efficiency and in vivo Distribution With High-Molecular-Weight Polyethylenimine. *J. Control. Release* **2003**, *89*(1), 113–125.
162. Oupický, D.; Ogris, M.; Seymour, L. W. Development of Long-Circulating Polyelectrolyte Complexes for Systemic Delivery of Genes. *J. Drug Target.* **2002**, *10*(2), 93–98.
163. Trubetskoy, V. S.; Loomis, A.; Slattum, P. M.; Hagstrom, J. E.; Budker, V. G.; Wolff, J. A. Caged DNA Does Not Aggregate in High Ionic Strength Solutions. *Bioconjug. Chem.* **1999**, *10*(4), 624–628.
164. Kunath, K.; von Harpe, A.; Petersen, H.; Fischer, D.; Voigt, K.; Kissel, T.; Bickel, U. The Structure of PEG-Modified Poly(Ethylene Imines) Influences Biodistribution and Pharmacokinetics of Their Complexes With NF-KappaB Decoy in Mice. *Pharm. Res.* **2002**, *19*(6), 810–817.
165. Ogris, M.; Brunner, S.; Schüller, S.; Kircheis, R.; Wagner, E. PEGylated DNA/Transferrin-PEI Complexes: Reduced Interaction With Blood Components, Extended Circulation in Blood and Potential for Systemic Gene Delivery. *Gene Ther.* **1999**, *6*(4), 595–605.

166. Kim, J.; Kim, S. W.; Kim, W. J. PEI-g-PEG-RGD/Small Interference RNA Polyplex-Mediated Silencing of Vascular Endothelial Growth Factor Receptor and Its Potential as an Anti-Angiogenic Tumor Therapeutic Strategy. *Oligonucleotides* **2011**, *21*(2), 101–107.
167. Dohmen, C.; Edinger, D.; Fröhlich, T.; Schreiner, L.; Lächelt, U.; Troiber, C.; Rädler, J.; Hadwiger, P.; Vornlocher, H. P.; Wagner, E. Nanosized Multifunctional Polyplexes for Receptor-Mediated siRNA Delivery. *ACS Nano* **2012**, *6*(6), 5198–5208.
168. Hu-Lieskovan, S.; Heidel, J. D.; Bartlett, D. W.; Davis, M. E.; Triche, T. J. Sequence-Specific Knockdown of EWS-FLI1 by Targeted, Nonviral Delivery of Small Interfering RNA Inhibits Tumor Growth in a Murine Model of Metastatic Ewing's Sarcoma. *Cancer Res.* **2005**, *65*(19), 8984–8992.
169. Shen, M.; Gong, F.; Pang, P.; Zhu, K.; Meng, X.; Wu, C.; Wang, J.; Shan, H.; Shuai, X. An MRI-Visible Non-Viral Vector for Targeted Bcl-2 siRNA Delivery to Neuroblastoma. *Int. J. Nanomedicine* **2012**, 7, 3319–3332.
170. Boussif, O.; Lezoualc'h, F.; Zanta, M. A.; Mergny, M. D.; Scherman, D.; Demeneix, B.; Behr, J. P. A Versatile Vector for Gene and Oligonucleotide Transfer Into Cells in Culture and In Vivo: Polyethylenimine. *Proc. Natl. Acad. Sci. U.S.A.* **1995**, *92*(16), 7297–7301.
171. Midoux, P.; Monsigny, M. Efficient Gene Transfer by Histidylated Polylysine/pDNA Complexes. *Bioconjug. Chem.* **1999**, *10*(3), 406–411.
172. Putnam, D.; Gentry, C. A.; Pack, D. W.; Langer, R. Polymer-Based Gene Delivery With Low Cytotoxicity by a Unique Balance of Side-Chain Termini. *Proc. Natl. Acad. Sci. U.S.A.* **2001**, *98*(3), 1200–1205.
173. Lynn, D. M.; Anderson, D. G.; Putnam, D.; Langer, R. Accelerated Discovery of Synthetic Transfection Vectors: Parallel Synthesis and Screening of a Degradable Polymer Library. *J. Am. Chem. Soc.* **2001**, *123*(33), 8155–8156.
174. Sahin, U.; Karikó, K.; Türeci, Ö. mRNA-Based Therapeutics—Developing a New Class of Drugs. *Nat. Rev. Drug Discov.* **2014**, *13*(10), 759–780.

CHAPTER THREE

CRISPR–Cas9 for Drug Discovery in Oncology

Sylvie M. Guichard[1]
IMED Biotech Unit, AstraZeneca Pharmaceuticals LP, Waltham, MA, United States
[1]Corresponding author: e-mail address: sylvie.guichard@astrazeneca.com

Contents

1. INTRODUCTION

Drug discovery and development is a long and complex process starting with the identification of a medical need and the hypothesis that perturbing a particular target will produce a biological effect that will impact the course of a disease. The proposed target is most often "validated" by modulating its expression and evaluating the consequences in terms of downstream signaling and biological effect. In order to identify small molecules against the target of interest, a screening will be undertaken and sometimes

Annual Reports in Medicinal Chemistry, Volume 50
ISSN 0065-7743
https://doi.org/10.1016/bs.armc.2017.08.006

requires engineering cell lines to detect a signal specific to the modulated target. The project may also require the generation of cell or animal models mimicking the disease to increase confidence that the drug tested will produce the desired outcome. This process represents 3–4 years in the overall development time of a new drug but can significantly impact the overall success of a product.

Indeed, a survey of AstraZeneca's pipeline suggested that failures in PhII for lack of clinical efficacy were due to a lack of genetic target linkage to the disease.[1] In addition, projects were more likely to be successful in phase II if a genetic linkage had been established between the target and the disease indication. More recently, an industry-wide study evaluated whether drug targets with a known genetic association to a disease impact the overall success of a drug project.[2] The team showed that genetic traits in specific indications were enriched fourfold in projects that reached approval compared to projects that stopped in phase I. One of the consequences of these findings is the prioritization of drug targets with genetic linkage to the disease and a project cascade including preclinical models based on that genetic linkage.

Genetic aberrations have long been known to drive tumor development and progression and combine of oncogenes and loss of tumor suppressor genes (TSG).[3] Drug discovery has long tried to identify molecules targeting such aberrations. Indeed, some oncogenes have successfully been targeted leading to significant clinical activity and patient benefits: Imatinib targeting BCR–ABL fusions in CML, Vemurafenib targeting BRAF V600E mutations in melanoma, osimertinib targeting EGFR mutations in nonsmall cell lung cancer to cite a few.[4] However, identifying drug targets associated with loss of TSG has proven more challenging. Inactivating TSG provides a survival advantage during tumor development. However, it may create a vulnerability by increasing the reliance on a "back up" system when the tumor is established. A large number of genetic screens have attempted to identify genes which when inactivated are not detrimental to the cells unless a TSG is missing. An example of such cancer-specific vulnerability is the hypersensitivity of ovarian cancers carrying loss-of-function mutations in BRCA1/2 to the PARP inhibitor olaparib inducing significant benefit in patients.[5]

Molecular biology tools have been developed over the years to evaluate drug targets, to develop assays for compound screening and to create preclinical models. Some of these techniques have limitations, for example, the level of expression of the target of interest in engineered models may be supraphysiological and therefore alter the activity of the compound compared to disease models. RNAi tools may not reduce gene expression sufficiently to induce a biological effect; they can also have off-target activity.[6] Finally, the generation of engineered mouse models can be a lengthy process in the timeline of a drug discovery project. The discovery of clustered regularly interspaced short palindromic repeat (CRISPR) and CRISPR-associated (Cas) protein is the latest technology being deployed in drug discovery.[7]

2. GENE EDITING USING CRISPR–Cas9

2.1 Mechanism of Gene Editing

The modification of genes and their expression is key to our understanding of gene functions. Precise and reproducible modifications of the DNA have been historically carried out by programmable nucleases such as zinc finger nucleases and transcription activator-like effector nucleases.[8] CRISPR–Cas were identified as bacterial defense mechanisms against viruses and phages.[9] Following exposure to viral DNA, bacteria integrate in their genome short fragments derived from the virus among CRISPRs. Bacteria also encode Cas enzymes. Expression of Cas and of CRISPR RNA recognize and cut the foreign DNA, protecting bacteria from further virus infection.[10]

Genetic engineering of CRISPR systems has led to the development of the CRISPR–Cas9 gene-editing tools for mammalian cells. A short complementary RNA associated with a transactivating crRNA (tracrRNA) (called single guide RNA or sgRNA subsequently) recognizes a specific DNA sequence in a target gene. Cas9 binds to the sgRNA and generates a double-strand DNA break (DSB) at the site identified by the sgRNA. The sequence recognition is based on a short motif called PAM (Protospacer adjacent motif) directly located 3′ of the target site. The break can be repaired by an error prone system called nonhomologous end joining (NHEJ), resulting in a short deletion or insertion at the targeted sequence, potentially disrupting the gene sequence and abrogating protein expression. An alternative repair mechanism called homology-directed repair (HDR) can also be used to repair the DSB. In this case, the DNA strand homologous to the cut sequence on the other allele is used for repair. Instead, if an homologous DNA fragment carrying a mutation of interest (donor template) is provided exogenously, this DNA sequence will be incorporated into the genome carrying the gene mutation. So in brief, CRISPR–Cas9 in absence of donor template aims at abrogating protein expression (knockout) and in presence of a donor template, the goal is to modify the genomic sequence to introduce a new variant (knock-in). Cas9 and the sgRNA are usually delivered using expression vectors transfected or virally transduced into cells. However, it is also possible to deliver purified Cas9 protein complexed with sgRNA (Cas9 RNP) into cells by transfection or electroporation.[11] The gene editing is quicker than with a vector-based system since the Cas9 protein is delivered to the cells. The presence of Cas9 and sgRNA in the cells is also transient and may alleviate some risks of off-target activity of Cas9. This is particularly useful for primary cells.[12]

When using CRISPR–Cas9 to knockout a gene, it is worth remembering that each amino acid is encoded by a set of three nucleotides so any insertion or deletion of one or two nucleotides (or multiples of these) will result in a frameshift, more likely to abrogate protein expression. In ~33% of cases, three nucleotides (or multiples of) will be inserted or deleted therefore maintaining the sequence in frame. Since cells contain two alleles of each gene, a null phenotype (frame shift on both alleles) will statistically be observed in ~44% of the cells.[13] To maximize the likelihood that the frameshift occurs early in the gene sequence, sgRNAs are usually designed to target 5′ coding exons. When performing gene editing in presence of a donor template, the frequency of successful recombination is rather low, less than 1%.

Regardless of the methodology adopted to perform the gene editing (vector-based or Cas9 RNP), it is key to enrich as much as possible for the population of gene-edited cells. The expression vectors expressing Cas9 and sgRNA can also encode a gene driving resistance to antibiotics. Cell culture in presence of the antibiotic will therefore enrich the gene-edited cells. Cas9 or the sgRNA can also be tagged with a fluorescent protein, facilitating the selection of the cells likely to have been gene edited by flow cytometry cell sorting or by plating at low density and isolation of fluorescent colonies.

2.2 Off-Target Activity of CRISPR–Cas9

Since CRISPR–Cas9 is based on the recognition of a target sequence by a short oligonucleotide (sgRNA), there is an inherent concern that the sgRNA can bind, albeit with lower affinity, to other unintended target sequences. Detecting and minimizing off-target effects of Cas9 are critical to interpret the results obtained with gene-edited model systems.

The detection and quantification of off-target gene editing is broadly based on the detection of DSB or the alterations of the sequences after gene editing.[14] Digenome-seq is a genome-wide analysis of CRISPR off-target effect in vitro.[15] Genomic DNA is incubated with Cas9 and the sgRNA. Following sequencing, the location of mismatches in the DNA sequence is mapped against the target sequence. Recently reported Circle-Seq and SITE-Seq are recent improvements of Digenome-Seq.[16,17] The BLESS method (direct in situ breaks labeling, enrichment on streptavidin, and next-generation sequencing) assesses the presence of DSB in cells across the genome.[18] GUIDE-seq inserts short sequences (tags) at the cut sites during the repair process. Whole genome sequencing reveals the position of the tags and therefore the off-target sites.[19]

Minimizing off-target effects can be tackled by optimizing the design of the sgRNA, by modifying the way gene editing is performed (Cas9 RNP) or by modifying Cas9 itself. The optimization of sgRNA design has first been established from experimental data,[20] combined with in silico algorithms.[21] Truncated sgRNA have also been shown to reduce off-target effects.[22,23]

2.3 Different CRISPR–Cas9 Systems

Efforts to expand the repertoire of Cas9 have been driven partly by the desire to identify high fidelity endonucleases and to expand the spectrum of functions carried out by modified Cas9. In the following paragraph, we detail three main modifications of Cas9 which emerged over time: variants of Cas9 (nickases high fidelity Cas9); fusions of inactivated Cas9 (dCas9) with various fusion partners allowing the precise targeting of DNA sequence with dCas9 and sgRNA but an array of subsequent functions carried out by the fusion partner (CRISPR inactivation (CRISPRi) and activation (CRISPRa) are based on dCas9 fusions); finally, the development of additional endonucleases such as Cpf1.

Early modifications of Cas9 include mutations in the nuclease domains HNH and RuvC creating Cas9 nickases which induce single-strand DNA breaks rather than DSB.[24,25] More recently, high fidelity Cas9 (SpCas9-HF1) was developed to reduce nonspecific contacts with DNA greatly improves the off-target activity of Cas9 although a slight decrease in on-target activity has been reported for some sgRNA.[26] A number of Cas9 fusions were also created: Fusion of dCas9 with FokI endonuclease has been shown to exhibit an improved selectivity. Fusions of dCas9 with a transcriptional repressor such as KRAB (CRISPRi) or activator such as Vps34 (CRISPRa) can be targeted to promoter regions to effectively silence or induce gene expression.[27–32] Modulation of gene expression can also be achieved by altering methylation and acetylation marks on the chromatin. Fusions of dCas9 with the acetyltransferase domain of p300 or with the demethylase LSD1 alter histone marks at promoters or enhancers thereby modulating gene expression.[33,34] Finally, DNA methylation is a well-known mechanism of gene silencing. DNA methylation is driven by DNA methyltransferases (DNMT1 and 3) while demethylation is initiated by Tet1. Fusion of dCas9 with Dnmt3a or Tet1 modulates expression of specific genes involved in cell differentiation.[35]

Novel nucleases are emerging as complementary tools to the Cas9 arsenal. Cpf1 targets DNA 3′ of T-rich PAM contrary to Cas9 which prefers G-rich PAM.[36] Cpf1 also produces 5′ overhangs rather than blunt end DNA strand breaks. Finally, Cpf1 has less risks for off-target activity compared to Cas9 since it tolerated less mismatches than Cas9.[37]

While the modifications of Cas9 has led to an array of new tools, it is worth mentioning a useful modification to the sgRNA that brings additional functionality to the complex. Adding two MS2 hairpin-binding sites to the sgRNA sequence allows the recruitment of the activation-induced cytidine deaminase (AID) to the target sequence. AID deaminates cytosines in the vicinity of the target site. The cytosines can then be repaired by several repair mechanisms generating a variety of point mutations.[38]

3. USE OF GENE EDITING IN DRUG DISCOVERY

3.1 Target Identification and Validation

Genetic screens in oncology aim at identifying genes whose depletions/deletions impair cell survival (drop out screens) or promote cell survival (enrichment screens). When performed in the presence of small molecules, genome-wide screens can also identify genes which depletion/deletion increases or reduces sensitivity to this particular molecule. RNA interference (RNAi) with siRNA or shRNA have been widely used to carry out genome-wide screens. However, the variability in gene knock down as well as extensive off-target activities have limited their utility for drug target discovery.[6] RNAi screens based on an enrichment strategy (depletion of a gene renders cells resistant to a particular drug) have been more informative. For example, a genome-wide RNAi screen identified the depletion of the NF1 gene as a mechanism of resistance to vemurafenib, a selective BRAF inhibitor active in melanoma cells carrying V600E activating mutation.[39]

CRISPR–Cas9 has been used extensively for target validation by knocking out individual genes and assessing the impact on a signaling pathway or cell survival. However, the knowledge from RNAi to generate libraries of short oligonucleotides to carry out large scale screens allowed the rapid development of CRISPR–Cas9 screens.[20,29,30,40–42] Today, a large number of pooled genome-wide screens have been performed in mammalian cells and several genome-wide libraries exist for CRISPR knockout as well as CRISPRi and CRISPRa. Before progressing further, a practical note on these pooled screens: after expression of statistically one sgRNA per cell, the cell population expressing the entire library grows for several generations. Any detrimental effect of the gene editing on individual cells will reduce their contribution to the overall population. Sequencing the sgRNA library at the start and the end of the experiment and comparing the relative representation of each sgRNA is used as a proxy for gene expression changes and growth inhibition. However, it is important to remember that individual gene expression or growth inhibition is not directly measured in these pooled screens and hit validation activities should aim at determining the amplitude of such changes. Arrayed screens are starting to emerge as complement to pooled screens. They provide a direct individual measurement and can accomodate more complex endpoints such as high content analysis and direct measurement of growth inhibition.[43,44] However, they require significant automation and analysis capabilities depending on the size of the library tested.

The two initial genome-wide knockout screens performed with CRISPR–Cas9 were proof-of-concept that this technology had the breadth of applicability of RNAi screens while providing a much more robust output.[20,40] Essentiality screens carried out subsequently in cell lines with specific revealed cancer-specific vulnerabilities, some of which can be druggable.[45] In some cases, Cancer cell lines may not be representative of the genetic diversity observed in patients so screens in patient-derived cell lines can be an additional tool to identify potential drug targets which can be confirmed with small molecule inhibitors.[46] Some CRISPR screens have also been performed in vivo, for example, to assess the tumor suppressive function of cancer genes in specific genetic context or tissues.[47–52]

Genome-wide screens have successfully been performed but still raise the issue that Cas9 is expressed in cells during the duration of the experiment which increases the risk of off-target gene editing. Fusions of dCas9 with either a transcriptional repressor (CRISPRi) or transcriptional activator (CRISPRa) provide complementary approaches to CRISPR knockout. CRISPRi is highly specific with minimal off-target activity, the efficiency of the sgRNA can lead to differential modulation of gene expression (titratable effect), and the repression or activation is fully reversible.[29,53,54]

However, CRISPRi or CRISPRa requires specific sgRNA designs and libraries.[55,56]

One extension of the essentiality screens is to test the impact of sgRNA pairs on cell growth therefore exploring pairs of aberration/drug target which may not be represented in cancer cell lines.[57,58] This approach could also inform drug combinations.

Another technique worth mentioning is not another variation of Cas9 but nonetheless leads to protein depletion via gene editing. Degron domains target a protein for proteosomal degradation under specific conditions (change in temperature, ligand binding). CRISPR–Cas9 can be used to insert by HDR a degron domain fused at the 3′ end of the sequence of a gene of interest. In presence of the ligand, the protein is expressed normally but upon ligand withdrawal, the protein is rapidly degraded. If the gene targeted is essential to cell survival, reduction in cell viability will be observed shortly after ligand withdrawal.[59] Degron domains can also be used to elucidate the role of mutant proteins or to annotate biomarkers downstream of a drug target.[60]

The approaches described so far deplete or knockout the entire protein. However, in the case of enzymes or multidomain proteins, it is important to distinguish whether the scaffold function or the enzymatic function of the targeted protein drives the observed phenotype. For example, genetic screens had established the synthetic lethal relationship between SMARCA4 and SMARCA2 without clarifying whether the bromodomain or the ATPase domain was responsible for the phenotype observed.[61,62] SMARCA2/4 bromodomain inhibitors subsequently demonstrated that the ATPase domain was responsible for the antiproliferative effect observed.[63,64] Establishing which functional domain drives the biological effect of interest is key before initiating chemistry efforts. The CRISPR–Cas9 system can be coopted to assess the relevance of functional domains based on the fact that a proportion of gene editing will generate in-frame sequences after gene editing. This gain or loss of a few amino acids can inactivate an enzymatic activity or scaffold function within a protein, thereby creating a "dead" drug target. Using this principle, Shi et al. screened a library of sgRNA targeting ~200 chromatin regulatory domains in murine acute myeloid leukemia and showed that sgRNA targeting functional domains were affecting cell survival disproportionally more than sgRNA targeting other areas of the genes.[13] Surveying all exons of SMARCA4, sgRNA targeting the ATPase domain of SMARCA4 overall produced a more severe phenotype. This methodology could help ensure that chemical tractability during the target validation process is determined on the basis of the relevant functional domains.

3.2 CRISPR–Cas9 for Assay Development

3.2.1 Protein Tagging and Protein Detection Systems

Phenotypic screens often rely on the detection and/or tracking of proteins in the cells of interest. A fusion protein containing the protein of interest and a fluorescent probe such as green fluorescent protein (GFP) is usually overexpressed in cells and the impact of the compound is inferred from the changes in the fluorescent protein signal. This system poses several challenges: random integration of the construct encoding the fusion protein may lead to variable expression levels across a cell population; the protein can be inherently expressed at low levels in cells leading to a low fluorescence signal or the fusion protein is expressed at such high levels that the subcellular localization of the endogenous protein is lost or forms aggregates. CRISPR–Cas9 has therefore been used extensively to tag proteins in their endogenous context or to increase the fluorescent signal when proteins are expressed at low level. GFP can also be used to pull-down the protein of interest for purification. To tag endogenous proteins, Cas9 is targeted to the gene of interest either N- or C-terminal with the sgRNA. A donor template containing the sequence of the fluorescent protein is inserted at the cut site by HDR.[65,66] The technical evolution of this approach include the development of a universal donor template,[66] the use of split fluorescent proteins to reduce the size of the integrated sequence,[67] and the scaling up of this technique by using Cas9/sgRNA ribonucleoprotein complexes (RNPs).[68]

To increase the signal from the endogenously tagged protein, a second approach consists of integrating epitope tags such as SunTag in the gene sequence of interest with CRISPR–Cas9.[69] The SunTag epitope can bind several antibody–GFP fusion proteins therefore boosting the signal intensity.

3.2.2 Reporter Systems

Reporter systems are widely used for drug screening to identify modulators of transcription factors. They are usually based on the transient expression of expression vectors carrying response elements for the transcription factors of interest placed in tandem upstream of so-called reporter gene (luciferase, for example). Modulation of transcription factor expression or binding to the response elements elicits a modulation of the luciferase signal. The main limitations of this approach is the absence of the endogenous chromatin context and of cofactors influencing the transcription factor activity. Generation of reporters systems using CRISPR–Cas9 follows the same principles as used for protein tagging. An example of endogenous reporter systems was demonstrated for TGFβ-induced expression of PAI-1.[70]

3.2.3 Chromatin Dynamics and Transcription Factors

The use of CRISPR–Cas9 has increased dramatically the tools available to study and target chromatin modifiers and other epigenetic targets. Fusions of dCas9 with eGFP can be used to visualize promoters or enhancer elements.[71] Since antibodies against Cas9 are readily available, positioning of dCas9 at a specific location on the genome can allow subsequent pull down using an anti-Cas9 antibody and subsequent identification of proteins present at the vicinity of Cas9.[72] Finally, epitope tags can be inserted using CRISPR–Cas9 to create transcription factors fusion proteins.[73]

3.3 Target Deconvolution and Selectivity Profiling

Phenotypic screens are a powerful tool to identify drugs modulating complex biological processes, in particular those engaging different cell types (host-bacteria, for example). However, target deconvolution, a critical component of phenotypic screens, can be time-consuming. CRISPR–Cas9 has recently been added to the array of genetic tools used for target deconvolution. For example, Deans et al. carried out a genome-wide screen to identify genes whose loss was associated with a reduction or an exacerbation of the cytotoxic effect of GSK983.[74] GSK983 was shown to inhibit dihydroorotate dehydrogenase (DHOH) thereby blocking viral RNA replication. However, the reduction in pyrimidine nucleotides following DHOH inhibition was detrimental to human cells, explaining the cytotoxic effect of the drug. In some cases, a drug target is known but the compound is suspected to have off-target activity. Generating KO cell lines for the main target can help evaluate how much of the biological effect is driven by the off-target activity. Recently, MTH1 inhibitors TH588 and TH287 were reported to induce selective cytotoxicity in cancer cells.[75] However, generation of macrocyclic MTH1 inhibitors, chemically distinct from the TH588 and TH287 could not reproduce the claimed biological effect.[76] Using CRISPR/Cas9, MTH1 was knocked out in cancer cells. The biological effect of TH588 and TH287 was similar in parental and MTH1 null cells, confirming that the cytotoxicity induced by these compounds was independent of MTH1 and likely driven by off-target effects.[76] Finally, CRISPR–Cas9 can be used to confirm on-target activity of small molecules as in the recent example of rocaglates and eIF4A or the expotin-1 inhibitor KPT-8602 and XPO1.[77,78]

If a drug target is part of a structurally related family of enzymes or proteins, identifying compounds with good selectivity may be challenging. In vitro selectivity panels have been used extensively but cell-based selectivity is sometimes different due to different concentrations of enzymes or cofactors, specific conformations, or localization in cells. Cell-based selectivity panels have been developed for kinases,[79] but are lacking for other classes of enzymes. Methodologies generating in-frame insertions or deletions in functional domains[13,38] could be used to generate variants unable to bind the compound. The impact on biomarkers or the phenotype could help assess selectivity. These two methodologies generate variants which may or may not encompass all potential amino acid changes altering compound binding. Saturation mutagenesis is a technique in which all possible amino acid changes are generated for a particular or a small number of codons. We have seen that CRISPR–Cas9 can create new gene variants via HDR. Cells targeted by CRISPR–Cas9 at a particular site and exposed to a series of donor templates encompassing all potential amino acid changes will repair the target site by incorporating at random the different templates thereby creating a population of cells carrying all potential amino acid changes for the target of interest.[80] Providing that these changes are not detrimental to the survival of the cells, when exposed to the drug of interest, only cells carrying aa changes altering compound binding will survive therefore providing a detailed map of drug–target interaction.

A specific case concerns the selectivity of covalent inhibitors involving cysteine interaction. Gene editing of the cysteine to a serine and monitoring the biological effect of the inhibitor can provide evidence that the compound acts via its intended target and mechanism.[81]

3.4 Synthesis of Synthetic Drugs and Biopharmaceuticals

Natural products are a significant source of therapeutics and historically have been critical to the development of antiinfectives. However, the chemistry of these "nature-optimized" molecules is complex and optimizing the properties and yield of specific molecules has been challenging. Recently, the use of CRISPR–Cas9 has been suggested to improve the yield of key drugs, paclitaxel and lovastatin, derived from natural products.[82] Biopharmaceuticals is another area where CRISPR–Cas9 is likely to have a significant impact.[83] For example, N-glycosylation in CHO cells can be modified by CRISPR–Cas9 to mirror human glycosylation and improve antibody quality.[84]

3.5 Building Disease Models to Assess Drug Efficacy

Human malignancies usually result from multiple genetic aberrations. Despite the large collection of cancer cell lines available, they do not recapitulate in some cases the combination of aberrations seen in patients. CRISPR–Cas9 has been used extensively to modify cancer cell lines to mimic aberrations seen in patients.[85,86] A few mouse cancer cell lines have also been modified to increase their relevance to the disease. An example of this approach with the ovarian cancer model ID8 showed an increased sensitivity to PARP inhibition when p53 and BRCA2 were knocked out.[87]

CRISPR–Cas9 has also revolutionized the generation of mouse models of cancer. Mouse models of hematological malignancies can be generated using mouse hematopietic stem cells.[88] After a few weeks, the mice developed leukemia associated with detection of blast cells in the blood and bone marrow. Generating models of solid tumors provides an additional challenge in that it develops in adult animals and in a specific tissue. The delivery of CRISPR–Cas9 to carry out the gene editing therefore become crucial. Some organs are more amenable than others to deliver CRISPR–Cas9. An early study, for example, demonstrated that dynamic injection of plasmid DNA expressing Cas9 and sgRNA targeting PTEN and p53 in the liver resulted in induction of liver tumors due to knockout of both genes.[89] Tumorigenesis in solid tumors is often driven by a combination of mutant oncogenes and loss-of-function mutations in tumor suppressors. A mouse model of lung cancer with loss-of-function of p53 and LKB1 as well as KRAS mutation was generated by simultaneous delivery of Cas9, sgRNA, and donor templates to the lung.[90] A similar strategy was used to generate another model of nonsmall cell lung cancer carrying the fusion protein EML4–ALK.[91] Interestingly, when tested in the animals with lung tumors, the ALK inhibitor crizotinib induced complete response in a majority of animals confirming the relevance of the model. Some tumors do not offer an accessible site to deliver CRISPR–Cas9. In this case, the Generation of conditional knockout animals (with gene knockout in a specific tissue) uses a vector construct where the expression of Cas9 and sgRNA is under the control of a tissue-specific promoter. When introduced into zygotes, the gene editing occurs only in the tissue expressing the protein driving the promoter.[92,93]

3.6 Detecting Drug Resistance

Evaluating potential mechanisms of resistance to anticancer agents is an important aspect of drug development. Global mutagenesis using *N*-ethyl-*N*-nitrosourea (ENU) is a powerful tool to generate a large number of mutations across a cell population.[94] When the cells are subsequently exposed to the drug of interest, only cells carrying a mutation impairing the activity of the compound will survive. By analyzing the types of mutations in the surviving cells, it is possible to identify amino acids key to the drug binding. One of the limitations of this approach is that mutations are generated across the genome and can lead to drug resistance without mutating the drug target leading to a significant number of false positives.

We mentioned earlier the use of dCas9 coupled with sgRNA containing MS2-binding hairpins to create an array of point mutations in a specific target site.[38] Expanding this concept using a library of sgRNA covering the entire exonic sequence of the gene of interest, point mutations can be created in the vast majority of amino acids encoded by the gene. When expressed in a population of cells and exposed to the drug of interest, only cells carrying mutations relevant to drug binding will survive. This strategy can also help identify off-target activity suspected to drive growth inhibition: Mutagenesis of the gene suspected to bind the compound and measuring the impact on cell proliferation could help assign its involvement.

CRISPR–Cas9 screens are particularly amenable to detect gene deletion associated with drug resistance. These genes are of special interest if they also happen to carry loss-of-function mutations in cancer. Loss of NF1 is associated with resistance to the BRAF inhibitor vemurafenib.[40] Similarly, deletion of mismatch repair genes is associated with resistance to a nucleotide analogue 6-thioguanine.[20] However, the emergence of resistance to a drug is often associated with the upregulation of specific factors. A technology such as CRISPRa is then a powerful tool to detect such factors. For example, a genome-wide CRISPRa screen showed that upregulation of EGFR as well as several GPCRs were associated to resistance to BRAF inhibition.[30] A recent study also demonstrates the use of CRISPRa in vivo to induce resistance to the alkylating agent temozolomide.[95]

Once the aberrations leading to drug resistance have been detected, it is obviously possible to build cell lines carrying the aberration of interest to confirm the impact on drug sensitivity. For example, introduction of the Y537S mutation in the estrogen receptor using CRISPR–Cas9 and the subsequent generation of xenografts demonstrated the impact of the mutation on the antitumor activity of fulvestrant and the benefit of combining fulvestrant and palbociclib in this model.[96]

3.7 Identifying Patient Selection

We highlighted earlier how the identification of a patient selection for a particular drug is an important factor in the success of an oncology project. A mutation present in tumor cells and not normal cells may create a specific biological context in tumor cells which renders them more sensitive to a drug than normal cells. One of the consequences of this phenomenon is an increase in the therapeutic index for the drug due to the hypersensitivity of the tumor cells. Patient selection hypotheses may have been established before initiation of the drug discovery project. However, they may be tackled much later when small molecules are available. The following few examples show how CRISPR–Cas9 can help identify patient selection hypotheses when small molecules inhibitors are available.

Nutlin3a is a small molecule MDM2 antagonist which results in accumulation of p53 and p53 targets p21, PUMA, Noxa, and MDM2. Knocking out PUMA with CRISPR–Cas9 abrogates Nutlin3a-induced apoptosis suggesting that PUMA is a critical factor for the efficacy of Nutlin3a.[97] PUMA overexpression is observed in a proportion of cancers and could be used to evaluate the efficacy of Nutlin3a. A similar approach generating KO cell lines with CRISPR–Cas9 was used to show that SLFN11 expression was a determinant of sensitivity to PARP inhibition in small cell lung cancer.[98] Finally, a CRISPR screen targeting multiple domains of MLL1 and MLL2 in an NPM1 mutant AML cell line provided a strong rationale for the use of small molecule inhibitor of menin–MLL interaction in these tumors.[99]

An extension of this approach is to identify pairs of genes which when disrupted provide a synergistic biological effect. Reminiscent of the synthetic lethal interactions, it may identify either gene–drug pairs or combinations of drugs if the targets are druggable.[57]

3.8 Assessment of Drug Disposition

The evaluation of drug disposition is an important component of drug discovery as drug efflux or drug metabolism by cytochrome P450s (CYP) are key features impacting the likelihood of progression to the clinic. Madin–Darby canine kidney II (MDCK) cells ectopically expressing one or several human transporters are widely used to assess whether a drug is a substrate or an inhibitor of specific transporters. However, MDCK express canine transporters which may also influence drug efflux. Ablation of canine MDR1 using CRISPR–Cas9 to create a KO cells followed by ectopic expression of human MDR1 constitutes the first model of humanized MDCK cells improving the assessment of drug transporters and efflux.[100]

Rat is a key rodent system to assess drug disposition and drug–drug interactions. However, generation of knockout models has been challenging due to the lack of robust methods to establish rat embryonic stem cells.[101] Cytochrome P450 enzymes play a key role in metabolism of endogenous products as well as xenobiotics. A proof-of-concept of the utility of CRISPR models in rats emerged from the generation of CYP2E1 KO rats using CRISPR–Cas9. In vivo, the clearance of chlorzoxazone, a CYP2E1 substrate, in CYP2E1 KO rats was reduced by 81% compared to WT rats. A CYP3A1/2 double KO rat model subsequently created similarly showed significant alteration in drug disposition of inhibitor substrates of the CYP isoform.[102] In the future, these rat models may provide an optimized way to predict drug disposition in man.

3.9 Safety Assessment

Success of drug discovery projects is largely influenced by the safety profile of the molecule. However, animal models do not necessarily represent the behavior of the drug in patients. In vitro microphysiological systems (MPS) are being deployed in drug discovery projects, bridging selectivity screening, and animal studies. With their high throughput and human genetic background they offer an additional tool to predict drug-induced toxicity.[103] MPS are based on human-induced pluripotent stem cells (HiPSCs) differentiated in the cells from the tissue of interest and organized in a 3D structure in a controlled microenvironment including mechanical and physiological stimuli of the functioning organ. Cardiac, liver, and lung epithelium are some of the MPS in use today. HiPSCs can be gene edited using CRISPR–Cas9 providing a way to evaluate the impact of key genes on drug-induced toxicity in MPS.[104] It can also recapitulate some key single nucleotide polymorphisms observed in patients which can be associated with drug-induced toxicity. Some cancer cell lines like HepG2 cells derived from a liver hepatocellular carcinoma have been used to assess drug-induced liver injuries. A recent genome-wide CRISPR screen in HeG2 cells identified several genes associated with the cytotoxic effect of triclosan.[105]

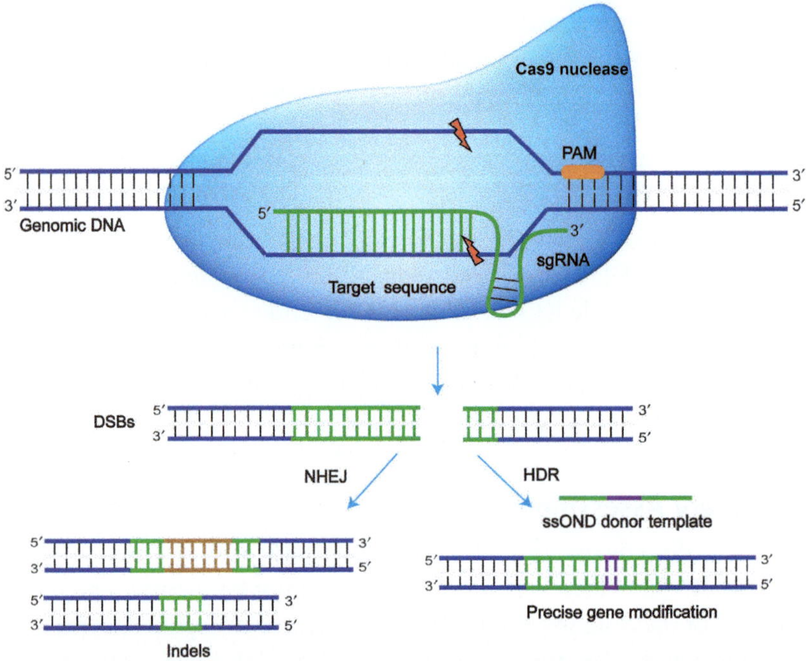

Fig. 1 Schematic illustration of genome editing by CRISPR–Cas9 system. Cas9 endonuclease is guided by a sgRNA that base pairs with a specific target sequence followed by PAM sequence. Cas9-mediated induction of DSBs are repaired either via NHEJ which give rise to indel mutations, and consequently, lead to gene knock-in and knockout, or via HDR with a donor DNA template that generate precise gene correction or replacement. *CRISPR*, clustered regularly interspaced short palindromic repeat; *DSBs*, double-strand breaks; *HDR*, homology directed repair; *indels*, insertions or deletions; *NHEJ*, non-homologous end joining; *PAM*, protospacer adjacent motif; *sgRNA*, single guide RNA; *ssODNs*, single-stranded DNA oligonucleotides. *From L. Yi and J. Li, CRISPR-Cas9 Therapeutics in Cancer: Promising Strategies and Present Challenges, Biochim. Biophys. Acta 1866 (2) (2016) 197–207.*

4. CONCLUSIONS

CRISPR–Cas9 is rapidly becoming an essential component of the drug discovery process in oncology. The versatility of the technology and the broad engagement of the resource community have provided a fertile ground for innovation in the pharmaceutical industry. Over the next few years, there is no doubt that CRISPR–Cas will have revolutionized the way we approach drug discovery projects today, from a better understanding of the drug target, to a streamlined screening process, and a better characterization of the drug candidates (Fig. 1).

REFERENCES

1. Cook, D.; Brown, D.; Alexander, R.; March, R.; Morgan, P.; Satterthwaite, G.; Pangalos, M. N. Lessons Learned From the Fate of AstraZeneca's Drug Pipeline: A Five-Dimensional Framework. *Nat. Rev. Drug Discov.* **2014**, *13*(6), 419–431.
2. Nelson, M. R.; Tipney, H.; Painter, J. L.; Shen, J.; Nicoletti, P.; Shen, Y.; Floratos, A.; Sham, P. C.; Li, M. J.; Wang, J.; Cardon, L. R.; Whittaker, J. C.; Sanseau, P. The Support of Human Genetic Evidence for Approved Drug Indications. *Nat. Genet.* **2015**, *47*(8), 856–860.
3. Vogelstein, B.; Papadopoulos, N.; Velculescu, V. E.; Zhou, S.; Diaz, L. A., Jr.; Kinzler, K. W. Cancer Genome Landscapes. *Science* **2013**, *339*(6127), 1546–1558.
4. Fleuren, E. D.; Zhang, L.; Wu, J.; Daly, R. J. The Kinome 'at Large' in Cancer. *Nat. Rev. Cancer* **2016**, *16*(2), 83–98.
5. Ledermann, J. A.; Harter, P.; Gourley, C.; Friedlander, M.; Vergote, I.; Rustin, G.; Scott, C.; Meier, W.; Shapira-Frommer, R.; Safra, T.; Matei, D.; Fielding, A.; Spencer, S.; Rowe, P.; Lowe, E.; Hodgson, D.; Sovak, M. A.; Matulonis, U. Overall Survival in Patients With Platinum-Sensitive Recurrent Serous Ovarian Cancer Receiving Olaparib Maintenance Monotherapy: An Updated Analysis From a Randomised, Placebo-Controlled, Double-Blind, Phase 2 Trial. *Lancet Oncol.* **2016**, *17*(11), 1579–1589.
6. Echeverri, C. J.; Beachy, P. A.; Baum, B.; Boutros, M.; Buchholz, F.; Chanda, S. K.; Downward, J.; Ellenberg, J.; Fraser, A. G.; Hacohen, N.; Hahn, W. C.; Jackson, A. L.; Kiger, A.; Linsley, P. S.; Lum, L.; Ma, Y.; Mathey-Prevot, B.; Root, D. E.; Sabatini, D. M.; Taipale, J.; Perrimon, N.; Bernards, R. Minimizing the Risk of Reporting False Positives in Large-Scale RNAi Screens. *Nat. Methods* **2006**, *3*(10), 777–779.
7. Fellmann, C.; Gowen, B. G.; Lin, P. C.; Doudna, J. A.; Corn, J. E. Cornerstones of CRISPR-Cas in Drug Discovery and Therapy. *Nat. Rev. Drug Discov.* **2017**, *16*(2), 89–100.
8. Hu, J. H.; Davis, K. M.; Liu, D. R. Chemical Biology Approaches to Genome Editing: Understanding, Controlling, and Delivering Programmable Nucleases. *Cell Chem. Biol.* **2016**, *23*(1), 57–73.
9. Doudna, J. A.; Charpentier, E. Genome Editing. The New Frontier of Genome Engineering With CRISPR-Cas9. *Science* **2014**, *346*(6213), 1258096.
10. Rath, D.; Amlinger, L.; Rath, A.; Lundgren, M. The CRISPR-Cas Immune System: Biology, Mechanisms and Applications. *Biochimie* **2015**, *117*, 119–128.

11. Kim, S.; Kim, D.; Cho, S. W.; Kim, J.; Kim, J. S. Highly Efficient RNA-Guided Genome Editing in Human Cells Via Delivery of Purified Cas9 Ribonucleoproteins. *Genome Res.* **2014**, *24*(6), 1012–1019.
12. Schumann, K.; Lin, S.; Boyer, E.; Simeonov, D. R.; Subramaniam, M.; Gate, R. E.; Haliburton, G. E.; Ye, C. J.; Bluestone, J. A.; Doudna, J. A.; Marson, A. Generation of Knock-in Primary Human T Cells Using Cas9 Ribonucleoproteins. *Proc. Natl. Acad. Sci. U. S. A.* **2015**, *112*(33), 10437–10442.
13. Shi, J.; Wang, E.; Milazzo, J. P.; Wang, Z.; Kinney, J. B.; Vakoc, C. R. Discovery of Cancer Drug Targets by CRISPR-Cas9 Screening of Protein Domains. *Nat. Biotechnol.* **2015**, *33*(6), 661–667.
14. Zischewski, J.; Fischer, R.; Bortesi, L. Detection of on-Target and Off-Target Mutations Generated by CRISPR/Cas9 and Other Sequence-Specific Nucleases. *Biotechnol. Adv.* **2017**, *35*(1), 95–104.
15. Kim, D.; Bae, S.; Park, J.; Kim, E.; Kim, S.; Yu, H. R.; Hwang, J.; Kim, J. I.; Kim, J. S. Digenome-Seq: Genome-Wide Profiling of CRISPR-Cas9 Off-Target Effects in Human Cells. *Nat. Methods* **2015**, *12*(3), 237–243. 1 p following 243.
16. Tsai, S. Q.; Nguyen, N. T.; Malagon-Lopez, J.; Topkar, V. V.; Aryee, M. J.; Joung, J. K. CIRCLE-Seq: A Highly Sensitive in vitro Screen for Genome-Wide CRISPR-Cas9 Nuclease Off-Targets. *Nat. Methods* **2017**, *14*(6), 607–614.
17. Cameron, P.; Fuller, C. K.; Donohoue, P. D.; Jones, B. N.; Thompson, M. S.; Carter, M. M.; Gradia, S.; Vidal, B.; Garner, E.; Slorach, E. M.; Lau, E.; Banh, L. M.; Lied, A. M.; Edwards, L. S.; Settle, A. H.; Capurso, D.; Llaca, V.; Deschamps, S.; Cigan, M.; Young, J. K.; May, A. P. Mapping the Genomic Landscape of CRISPR-Cas9 Cleavage. *Nat. Methods* **2017**, *14*(6), 600–606.
18. Crosetto, N.; Mitra, A.; Silva, M. J.; Bienko, M.; Dojer, N.; Wang, Q.; Karaca, E.; Chiarle, R.; Skrzypczak, M.; Ginalski, K.; Pasero, P.; Rowicka, M.; Dikic, I. Nucleotide-Resolution DNA Double-Strand Break Mapping by Next-Generation Sequencing. *Nat. Methods* **2013**, *10*(4), 361–365.
19. Tsai, S. Q.; Zheng, Z.; Nguyen, N. T.; Liebers, M.; Topkar, V. V.; Thapar, V.; Wyvekens, N.; Khayter, C.; Iafrate, A. J.; Le, L. P.; Aryee, M. J.; Joung, J. K. GUIDE-Seq Enables Genome-Wide Profiling of Off-Target Cleavage by CRISPR-Cas Nucleases. *Nat. Biotechnol.* **2015**, *33*(2), 187–197.
20. Wang, T.; Wei, J. J.; Sabatini, D. M.; Lander, E. S. Genetic Screens in Human Cells Using the CRISPR-Cas9 System. *Science* **2014**, *343*(6166), 80–84.
21. Chuai, G. H.; Wang, Q. L.; Liu, Q. In Silico Meets In Vivo: Towards Computational CRISPR-Based sgRNA Design. *Trends Biotechnol.* **2017**, *35*(1), 12–21.
22. Fu, Y.; Sander, J. D.; Reyon, D.; Cascio, V. M.; Joung, J. K. Improving CRISPR-Cas Nuclease Specificity Using Truncated Guide RNAs. *Nat. Biotechnol.* **2014**, *32*(3), 279–284.
23. Morgens, D. W.; Wainberg, M.; Boyle, E. A.; Ursu, O.; Araya, C. L.; Tsui, C. K.; Haney, M. S.; Hess, G. T.; Han, K.; Jeng, E. E.; Li, A.; Snyder, M. P.; Greenleaf, W. J.; Kundaje, A.; Bassik, M. C. Genome-Scale Measurement of Off-Target Activity Using Cas9 Toxicity in High-Throughput Screens. *Nat. Commun.* **2017**, *8*, 15178.
24. Mali, P.; Aach, J.; Stranges, P. B.; Esvelt, K. M.; Moosburner, M.; Kosuri, S.; Yang, L.; Church, G. M. CAS9 Transcriptional Activators for Target Specificity Screening and Paired Nickases for Cooperative Genome Engineering. *Nat. Biotechnol.* **2013**, *31*(9), 833–838.
25. Ran, F. A.; Hsu, P. D.; Lin, C. Y.; Gootenberg, J. S.; Konermann, S.; Trevino, A. E.; Scott, D. A.; Inoue, A.; Matoba, S.; Zhang, Y.; Zhang, F. Double Nicking by RNA-Guided CRISPR Cas9 for Enhanced Genome Editing Specificity. *Cell* **2013**, *154*(6), 1380–1389.

26. Kleinstiver, B. P.; Pattanayak, V.; Prew, M. S.; Tsai, S. Q.; Nguyen, N. T.; Zheng, Z.; Joung, J. K. High-Fidelity CRISPR-Cas9 Nucleases With No Detectable Genome-Wide Off-Target Effects. *Nature* **2016**, *529*(7587), 490–495.
27. Cheng, A. W.; Wang, H.; Yang, H.; Shi, L.; Katz, Y.; Theunissen, T. W.; Rangarajan, S.; Shivalila, C. S.; Dadon, D. B.; Jaenisch, R. Multiplexed Activation of Endogenous Genes by CRISPR-on, an RNA-Guided Transcriptional Activator System. *Cell Res.* **2013**, *23*(10), 1163–1171.
28. Guilinger, J. P.; Thompson, D. B.; Liu, D. R. Fusion of Catalytically Inactive Cas9 to FokI Nuclease Improves the Specificity of Genome Modification. *Nat. Biotechnol.* **2014**, *32*(6), 577–582.
29. Gilbert, L. A.; Larson, M. H.; Morsut, L.; Liu, Z.; Brar, G. A.; Torres, S. E.; Stern-Ginossar, N.; Brandman, O.; Whitehead, E. H.; Doudna, J. A.; Lim, W. A.; Weissman, J. S.; Qi, L. S. CRISPR-Mediated Modular RNA-Guided Regulation of Transcription in Eukaryotes. *Cell* **2013**, *154*(2), 442–451.
30. Konermann, S.; Brigham, M. D.; Trevino, A. E.; Joung, J.; Abudayyeh, O. O.; Barcena, C.; Hsu, P. D.; Habib, N.; Gootenberg, J. S.; Nishimasu, H.; Nureki, O.; Zhang, F. Genome-Scale Transcriptional Activation by an Engineered CRISPR-Cas9 Complex. *Nature* **2015**, *517*(7536), 583–588.
31. Perez-Pinera, P.; Kocak, D. D.; Vockley, C. M.; Adler, A. F.; Kabadi, A. M.; Polstein, L. R.; Thakore, P. I.; Glass, K. A.; Ousterout, D. G.; Leong, K. W.; Guilak, F.; Crawford, G. E.; Reddy, T. E.; Gersbach, C. A. RNA-Guided Gene Activation by CRISPR-Cas9-Based Transcription Factors. *Nat. Methods* **2013**, *10*(10), 973–976.
32. Qi, L. S.; Larson, M. H.; Gilbert, L. A.; Doudna, J. A.; Weissman, J. S.; Arkin, A. P.; Lim, W. A. Repurposing CRISPR as an RNA-Guided Platform for Sequence-Specific Control of Gene Expression. *Cell* **2013**, *152*(5), 1173–1183.
33. Hilton, I. B.; D'Ippolito, A. M.; Vockley, C. M.; Thakore, P. I.; Crawford, G. E.; Reddy, T. E.; Gersbach, C. A. Epigenome Editing by a CRISPR-Cas9-Based Acetyltransferase Activates Genes From Promoters and Enhancers. *Nat. Biotechnol.* **2015**, *33*(5), 510–517.
34. Kearns, N. A.; Pham, H.; Tabak, B.; Genga, R. M.; Silverstein, N. J.; Garber, M.; Maehr, R. Functional Annotation of Native Enhancers With a Cas9-Histone Demethylase Fusion. *Nat. Methods* **2015**, *12*(5), 401–403.
35. Liu, X. S.; Wu, H.; Ji, X.; Stelzer, Y.; Wu, X.; Czauderna, S.; Shu, J.; Dadon, D.; Young, R. A.; Jaenisch, R. Editing DNA Methylation in the Mammalian Genome. *Cell* **2016**, *167*(1), 233–247. e17.
36. Toth, E.; Weinhardt, N.; Bencsura, P.; Huszar, K.; Kulcsar, P. I.; Talas, A.; Fodor, E.; Welker, E. Cpf1 Nucleases Demonstrate Robust Activity to Induce DNA Modification by Exploiting Homology Directed Repair Pathways in Mammalian Cells. *Biol. Direct* **2016**, *11*, 46.
37. Kim, D.; Kim, J.; Hur, J. K.; Been, K. W.; Yoon, S. H.; Kim, J. S. Genome-Wide Analysis Reveals Specificities of Cpf1 Endonucleases in Human Cells. *Nat. Biotechnol.* **2016**, *34*(8), 863–868.
38. Hess, G. T.; Fresard, L.; Han, K.; Lee, C. H.; Li, A.; Cimprich, K. A.; Montgomery, S. B.; Bassik, M. C. Directed Evolution Using dCas9-Targeted Somatic Hypermutation in Mammalian Cells. *Nat. Methods* **2016**, *13*(12), 1036–1042.
39. Whittaker, S. R.; Theurillat, J. P.; Van Allen, E.; Wagle, N.; Hsiao, J.; Cowley, G. S.; Schadendorf, D.; Root, D. E.; Garraway, L. A. A Genome-Scale RNA Interference Screen Implicates NF1 Loss in Resistance to RAF Inhibition. *Cancer Discov.* **2013**, *3*(3), 350–362.
40. Shalem, O.; Sanjana, N. E.; Hartenian, E.; Shi, X.; Scott, D. A.; Mikkelsen, T. S.; Heckl, D.; Ebert, B. L.; Root, D. E.; Doench, J. G.; Zhang, F. Genome-Scale CRISPR-Cas9 Knockout Screening in Human Cells. *Science* **2014**, *343*(6166), 84–87.

41. Koike-Yusa, H.; Li, Y.; Tan, E. P.; Velasco-Herrera Mdel, C.; Yusa, K. Genome-Wide Recessive Genetic Screening in Mammalian Cells With a Lentiviral CRISPR-Guide RNA Library. *Nat. Biotechnol.* **2014**, *32*(3), 267–273.
42. Zhou, Y.; Zhu, S.; Cai, C.; Yuan, P.; Li, C.; Huang, Y.; Wei, W. High-Throughput Screening of a CRISPR/Cas9 Library for Functional Genomics in Human Cells. *Nature* **2014**, *509*(7501), 487–491.
43. Strezoska, Z.; Perkett, M. R.; Chou, E. T.; Maksimova, E.; Anderson, E. M.; McClelland, S.; Kelley, M. L.; Vermeulen, A.; Smith, A. V. B. High-Content Analysis Screening for Cell Cycle Regulators Using Arrayed Synthetic crRNA Libraries. *J. Biotechnol.* **2017**, *251*, 189–200.
44. Tan, J.; Martin, S. E. Validation of Synthetic CRISPR Reagents as a Tool for Arrayed Functional Genomic Screening. *PLoS One* **2016**, *11*(12), e0168968.
45. Hart, T.; Chandrashekhar, M.; Aregger, M.; Steinhart, Z.; Brown, K. R.; MacLeod, G.; Mis, M.; Zimmermann, M.; Fradet-Turcotte, A.; Sun, S.; Mero, P.; Dirks, P.; Sidhu, S.; Roth, F. P.; Rissland, O. S.; Durocher, D.; Angers, S.; Moffat, J. High-Resolution CRISPR Screens Reveal Fitness Genes and Genotype-Specific Cancer Liabilities. *Cell* **2015**, *163*(6), 1515–1526.
46. Hong, A. L.; Tseng, Y. Y.; Cowley, G. S.; Jonas, O.; Cheah, J. H.; Kynnap, B. D.; Doshi, M. B.; Oh, C.; Meyer, S. C.; Church, A. J.; Gill, S.; Bielski, C. M.; Keskula, P.; Imamovic, A.; Howell, S.; Kryukov, G. V.; Clemons, P. A.; Tsherniak, A.; Vazquez, F.; Crompton, B. D.; Shamji, A. F.; Rodriguez-Galindo, C.; Janeway, K. A.; Roberts, C. W.; Stegmaier, K.; van Hummelen, P.; Cima, M. J.; Langer, R. S.; Garraway, L. A.; Schreiber, S. L.; Root, D. E.; Hahn, W. C.; Boehm, J. S. Integrated Genetic and Pharmacologic Interrogation of Rare Cancers. *Nat. Commun.* **2016**, 7, 11987.
47. Sanchez-Rivera, F. J.; Papagiannakopoulos, T.; Romero, R.; Tammela, T.; Bauer, M. R.; Bhutkar, A.; Joshi, N. S.; Subbaraj, L.; Bronson, R. T.; Xue, W.; Jacks, T. Rapid Modelling of Cooperating Genetic Events in Cancer Through Somatic Genome Editing. *Nature* **2014**, *516*(7531), 428–431.
48. Zuckermann, M.; Hovestadt, V.; Knobbe-Thomsen, C. B.; Zapatka, M.; Northcott, P. A.; Schramm, K.; Belic, J.; Jones, D. T.; Tschida, B.; Moriarity, B.; Largaespada, D.; Roussel, M. F.; Korshunov, A.; Reifenberger, G.; Pfister, S. M.; Lichter, P.; Kawauchi, D.; Gronych, J. Somatic CRISPR/Cas9-Mediated Tumour Suppressor Disruption Enables Versatile Brain Tumour Modelling. *Nat. Commun.* **2015**, *6*, 7391.
49. Chiou, S. H.; Winters, I. P.; Wang, J.; Naranjo, S.; Dudgeon, C.; Tamburini, F. B.; Brady, J. J.; Yang, D.; Gruner, B. M.; Chuang, C. H.; Caswell, D. R.; Zeng, H.; Chu, P.; Kim, G. E.; Carpizo, D. R.; Kim, S. K.; Winslow, M. M. Pancreatic Cancer Modeling Using Retrograde Viral Vector Delivery and in vivo CRISPR/Cas9-Mediated Somatic Genome Editing. *Genes Dev.* **2015**, *29*(14), 1576–1585.
50. Weber, J.; Ollinger, R.; Friedrich, M.; Ehmer, U.; Barenboim, M.; Steiger, K.; Heid, I.; Mueller, S.; Maresch, R.; Engleitner, T.; Gross, N.; Geumann, U.; Fu, B.; Segler, A.; Yuan, D.; Lange, S.; Strong, A.; de la Rosa, J.; Esposito, I.; Liu, P.; Cadinanos, J.; Vassiliou, G. S.; Schmid, R. M.; Schneider, G.; Unger, K.; Yang, F.; Braren, R.; Heikenwalder, M.; Varela, I.; Saur, D.; Bradley, A.; Rad, R. CRISPR/Cas9 Somatic Multiplex-Mutagenesis for High-Throughput Functional Cancer Genomics in Mice. *Proc. Natl. Acad. Sci. U. S. A.* **2015**, *112*(45), 13982–13987.
51. Katigbak, A.; Cencic, R.; Robert, F.; Senecha, P.; Scuoppo, C.; Pelletier, J. A CRISPR/Cas9 Functional Screen Identifies Rare Tumor Suppressors. *Sci. Rep.* **2016**, *6*, 38968.
52. Rogers, Z. N.; McFarland, C. D.; Winters, I. P.; Naranjo, S.; Chuang, C. H.; Petrov, D.; Winslow, M. M. A Quantitative and Multiplexed Approach to Uncover

the Fitness Landscape of Tumor Suppression In Vivo. *Nat. Methods* **2017**, *14*(7), 737–742.
53. Zhao, Y.; Liu, Q.; Acharya, P.; Stengel, K. R.; Sheng, Q.; Zhou, X.; Kwak, H.; Fischer, M. A.; Bradner, J. E.; Strickland, S. A.; Mohan, S. R.; Savona, M. R.; Venters, B. J.; Zhou, M. M.; Lis, J. T.; Hiebert, S. W. High-Resolution Mapping of RNA Polymerases Identifies Mechanisms of Sensitivity and Resistance to BET Inhibitors in t(8;21) AML. *Cell Rep.* **2016**, *16*(7), 2003–2016.
54. Fulco, C. P.; Munschauer, M.; Anyoha, R.; Munson, G.; Grossman, S. R.; Perez, E. M.; Kane, M.; Cleary, B.; Lander, E. S.; Engreitz, J. M. Systematic Mapping of Functional Enhancer-Promoter Connections With CRISPR Interference. *Science* **2016**, *354*(6313), 769–773.
55. Kampmann, M.; Horlbeck, M. A.; Chen, Y.; Tsai, J. C.; Bassik, M. C.; Gilbert, L. A.; Villalta, J. E.; Kwon, S. C.; Chang, H.; Kim, V. N.; Weissman, J. S. Next-Generation Libraries for Robust RNA Interference-Based Genome-Wide Screens. *Proc. Natl. Acad. Sci. U. S. A.* **2015**, *112*(26), E3384–91.
56. Radzisheuskaya, A.; Shlyueva, D.; Muller, I.; Helin, K. Optimizing sgRNA Position Markedly Improves the Efficiency of CRISPR/dCas9-Mediated Transcriptional Repression. *Nucleic Acids Res.* **2016**, *44*(18), e141.
57. Shen, J. P.; Zhao, D.; Sasik, R.; Luebeck, J.; Birmingham, A.; Bojorquez-Gomez, A.; Licon, K.; Klepper, K.; Pekin, D.; Beckett, A. N.; Sanchez, K. S.; Thomas, A.; Kuo, C. C.; Du, D.; Roguev, A.; Lewis, N. E.; Chang, A. N.; Kreisberg, J. F.; Krogan, N.; Qi, L.; Ideker, T.; Mali, P. Combinatorial CRISPR-Cas9 Screens for De Novo Mapping of Genetic Interactions. *Nat. Methods* **2017**, *14*(6), 573–576.
58. Du, D.; Roguev, A.; Gordon, D. E.; Chen, M.; Chen, S. H.; Shales, M.; Shen, J. P.; Ideker, T.; Mali, P.; Qi, L. S.; Krogan, N. J. Genetic Interaction Mapping in Mammalian Cells Using CRISPR Interference. *Nat. Methods* **2017**, *14*(6), 577–580.
59. Sheridan, R. M.; Bentley, D. L. Selectable One-Step PCR-Mediated Integration of a Degron for Rapid Depletion of Endogenous Human Proteins. *Biotechniques* **2016**, *60*(2), 69–74.
60. Zhou, Q.; Derti, A.; Ruddy, D.; Rakiec, D.; Kao, I.; Lira, M.; Gibaja, V.; Chan, H.; Yang, Y.; Min, J.; Schlabach, M. R.; Stegmeier, F. A Chemical Genetics Approach for the Functional Assessment of Novel Cancer Genes. *Cancer Res.* **2015**, *75*(10), 1949–1958.
61. Oike, T.; Ogiwara, H.; Tominaga, Y.; Ito, K.; Ando, O.; Tsuta, K.; Mizukami, T.; Shimada, Y.; Isomura, H.; Komachi, M.; Furuta, K.; Watanabe, S.; Nakano, T.; Yokota, J.; Kohno, T. A Synthetic Lethality-Based Strategy to Treat Cancers Harboring a Genetic Deficiency in the Chromatin Remodeling Factor BRG1. *Cancer Res.* **2013**, *73*(17), 5508–5518.
62. Hoffman, G. R.; Rahal, R.; Buxton, F.; Xiang, K.; McAllister, G.; Frias, E.; Bagdasarian, L.; Huber, J.; Lindeman, A.; Chen, D.; Romero, R.; Ramadan, N.; Phadke, T.; Haas, K.; Jaskelioff, M.; Wilson, B. G.; Meyer, M. J.; Saenz-Vash, V.; Zhai, H.; Myer, V. E.; Porter, J. A.; Keen, N.; McLaughlin, M. E.; Mickanin, C.; Roberts, C. W.; Stegmeier, F.; Jagani, Z. Functional Epigenetics Approach Identifies BRM/SMARCA2 as a Critical Synthetic Lethal Target in BRG1-Deficient Cancers. *Proc. Natl. Acad. Sci. U. S. A.* **2014**, *111*(8), 3128–3133.
63. Romero, F. A.; Taylor, A. M.; Crawford, T. D.; Tsui, V.; Cote, A.; Magnuson, S. Disrupting Acetyl-Lysine Recognition: Progress in the Development of Bromodomain Inhibitors. *J. Med. Chem.* **2016**, *59*(4), 1271–1298.
64. Vangamudi, B.; Paul, T. A.; Shah, P. K.; Kost-Alimova, M.; Nottebaum, L.; Shi, X.; Zhan, Y.; Leo, E.; Mahadeshwar, H. S.; Protopopov, A.; Futreal, A.; Tieu, T. N.; Peoples, M.; Heffernan, T. P.; Marszalek, J. R.; Toniatti, C.; Petrocchi, A.;

Verhelle, D.; Owen, D. R.; Draetta, G.; Jones, P.; Palmer, W. S.; Sharma, S.; Andersen, J. N. The SMARCA2/4 ATPase Domain Surpasses the Bromodomain as a Drug Target in SWI/SNF-Mutant Cancers: Insights From cDNA Rescue and PFI-3 Inhibitor Studies. *Cancer Res.* **2015**, *75*(18), 3865–3878.

65. Ratz, M.; Testa, I.; Hell, S. W.; Jakobs, S. CRISPR/Cas9-Mediated Endogenous Protein Tagging for RESOLFT Super-Resolution Microscopy of Living Human Cells. *Sci. Rep.* **2015**, *5*, 9592.
66. Lackner, D. H.; Carre, A.; Guzzardo, P. M.; Banning, C.; Mangena, R.; Henley, T.; Oberndorfer, S.; Gapp, B. V.; Nijman, S. M.; Brummelkamp, T. R.; Burckstummer, T. A Generic Strategy for CRISPR-Cas9-Mediated Gene Tagging. *Nat. Commun.* **2015**, *6*, 10237.
67. Kamiyama, D.; Sekine, S.; Barsi-Rhyne, B.; Hu, J.; Chen, B.; Gilbert, L. A.; Ishikawa, H.; Leonetti, M. D.; Marshall, W. F.; Weissman, J. S.; Huang, B. Versatile Protein Tagging in Cells With Split Fluorescent Protein. *Nat. Commun.* **2016**, *7*, 11046.
68. Leonetti, M. D.; Sekine, S.; Kamiyama, D.; Weissman, J. S.; Huang, B. A Scalable Strategy for High-Throughput GFP Tagging of Endogenous Human Proteins. *Proc. Natl. Acad. Sci. U. S. A.* **2016**, *113*(25), E3501–8.
69. Tanenbaum, M. E.; Gilbert, L. A.; Qi, L. S.; Weissman, J. S.; Vale, R. D. A Protein-Tagging System for Signal Amplification in Gene Expression and Fluorescence Imaging. *Cell* **2014**, *159*(3), 635–646.
70. Rojas-Fernandez, A.; Herhaus, L.; Macartney, T.; Lachaud, C.; Hay, R. T.; Sapkota, G. P. Rapid Generation of Endogenously Driven Transcriptional Reporters in Cells Through CRISPR/Cas9. *Sci. Rep.* **2015**, *5*, 9811.
71. Engel, K. L.; Mackiewicz, M.; Hardigan, A. A.; Myers, R. M.; Savic, D. Decoding Transcriptional Enhancers: Evolving From Annotation to Functional Interpretation. *Semin. Cell Dev. Biol.* **2016**, *57*, 40–50.
72. Fujii, H.; Fujita, T. Isolation of Specific Genomic Regions and Identification of Their Associated Molecules by Engineered DNA-Binding Molecule-Mediated Chromatin Immunoprecipitation (enChIP) Using the CRISPR System and TAL Proteins. *Int. J. Mol. Sci.* **2015**, *16*(9), 21802–21812.
73. Xiong, X.; Zhang, Y.; Yan, J.; Jain, S.; Chee, S.; Ren, B.; Zhao, H. A Scalable Epitope Tagging Approach for High Throughput ChIP-Seq Analysis. *ACS Synth. Biol.* **2017**, *6*(6), 1034–1042.
74. Deans, R. M.; Morgens, D. W.; Okesli, A.; Pillay, S.; Horlbeck, M. A.; Kampmann, M.; Gilbert, L. A.; Li, A.; Mateo, R.; Smith, M.; Glenn, J. S.; Carette, J. E.; Khosla, C.; Bassik, M. C. Parallel shRNA and CRISPR-Cas9 Screens Enable Antiviral Drug Target Identification. *Nat. Chem. Biol.* **2016**, *12*(5), 361–366.
75. Gad, H.; Koolmeister, T.; Jemth, A. S.; Eshtad, S.; Jacques, S. A.; Strom, C. E.; Svensson, L. M.; Schultz, N.; Lundback, T.; Einarsdottir, B. O.; Saleh, A.; Gokturk, C.; Baranczewski, P.; Svensson, R.; Berntsson, R. P.; Gustafsson, R.; Stromberg, K.; Sanjiv, K.; Jacques-Cordonnier, M. C.; Desroses, M.; Gustavsson, A. L.; Olofsson, R.; Johansson, F.; Homan, E. J.; Loseva, O.; Brautigam, L.; Johansson, L.; Hoglund, A.; Hagenkort, A.; Pham, T.; Altun, M.; Gaugaz, F. Z.; Vikingsson, S.; Evers, B.; Henriksson, M.; Vallin, K. S.; Wallner, O. A.; Hammarstrom, L. G.; Wiita, E.; Almlof, I.; Kalderen, C.; Axelsson, H.; Djureinovic, T.; Puigvert, J. C.; Haggblad, M.; Jeppsson, F.; Martens, U.; Lundin, C.; Lundgren, B.; Granelli, I.; Jensen, A. J.; Artursson, P.; Nilsson, J. A.; Stenmark, P.; Scobie, M.; Berglund, U. W.; Helleday, T. MTH1 Inhibition Eradicates Cancer by Preventing Sanitation of the dNTP Pool. *Nature* **2014**, *508*(7495), 215–221.

76. Kettle, J. G.; Alwan, H.; Bista, M.; Breed, J.; Davies, N. L.; Eckersley, K.; Fillery, S.; Foote, K. M.; Goodwin, L.; Jones, D. R.; Kack, H.; Lau, A.; Nissink, J. W.; Read, J.; Scott, J. S.; Taylor, B.; Walker, G.; Wissler, L.; Wylot, M. Potent and Selective Inhibitors of MTH1 Probe Its Role in Cancer Cell Survival. *J. Med. Chem.* **2016**, *59*(6), 2346–2361.
77. Chu, J.; Galicia-Vazquez, G.; Cencic, R.; Mills, J. R.; Katigbak, A.; Porco, J. A., Jr.; Pelletier, J. CRISPR-Mediated Drug-Target Validation Reveals Selective Pharmacological Inhibition of the RNA Helicase, eIF4A. *Cell Rep.* **2016**, *15*(11), 2340–2347.
78. Vercruysse, T.; De Bie, J.; Neggers, J. E.; Jacquemyn, M.; Vanstreels, E.; Schmid-Burgk, J. L.; Hornung, V.; Baloglu, E.; Landesman, Y.; Senapedis, W.; Shacham, S.; Dagklis, A.; Cools, J.; Daelemans, D. The Second-Generation Exportin-1 Inhibitor KPT-8602 Demonstrates Potent Activity Against Acute Lymphoblastic Leukemia. *Clin. Cancer Res.* **2017**, *23*(10), 2528–2541.
79. Warmuth, M.; Kim, S.; Gu, X. J.; Xia, G.; Adrian, F. Ba/F3 Cells and their use in Kinase Drug Discovery. *Curr. Opin. Oncol.* **2007**, *19*(1), 55–60.
80. Findlay, G. M.; Boyle, E. A.; Hause, R. J.; Klein, J. C.; Shendure, J. Saturation Editing of Genomic Regions by Multiplex Homology-Directed Repair. *Nature* **2014**, *513*(7516), 120–123.
81. Neggers, J. E.; Vercruysse, T.; Jacquemyn, M.; Vanstreels, E.; Baloglu, E.; Shacham, S.; Crochiere, M.; Landesman, Y.; Daelemans, D. Identifying Drug-Target Selectivity of Small-Molecule CRM1/XPO1 Inhibitors by CRISPR/Cas9 Genome Editing. *Chem. Biol.* **2015**, *22*(1), 107–116.
82. El-Sayed, A. S. A.; Abdel-Ghany, S. E.; Ali, G. S. Genome Editing Approaches: Manipulating of Lovastatin and Taxol Synthesis of Filamentous Fungi by CRISPR/Cas9 System. *Appl. Microbiol. Biotechnol.* **2017**, *101*(10), 3953–3976.
83. Lee, J. S.; Grav, L. M.; Lewis, N. E.; Faustrup Kildegaard, H. CRISPR/Cas9-Mediated Genome Engineering of CHO Cell Factories: Application and Perspectives. *Biotechnol. J.* **2015**, *10*(7), 979–994.
84. Yang, Z.; Wang, S.; Halim, A.; Schulz, M. A.; Frodin, M.; Rahman, S. H.; Vester-Christensen, M. B.; Behrens, C.; Kristensen, C.; Vakhrushev, S. Y.; Bennett, E. P.; Wandall, H. H.; Clausen, H. Engineered CHO Cells for Production of Diverse, Homogeneous Glycoproteins. *Nat. Biotechnol.* **2015**, *33*(8), 842–844.
85. Park, M. Y.; Jung, M. H.; Eo, E. Y.; Kim, S.; Lee, S. H.; Lee, Y. J.; Park, J. S.; Cho, Y. J.; Chung, J. H.; Kim, C. H.; Yoon, H. I.; Lee, J. H.; Lee, C. T. Generation of Lung Cancer Cell Lines Harboring EGFR T790M Mutation by CRISPR/Cas9-Mediated Genome Editing. *Oncotarget* **2017**, *8*(22), 36331–36338.
86. Vanoli, F.; Tomishima, M.; Feng, W.; Lamribet, K.; Babin, L.; Brunet, E.; Jasin, M. CRISPR-Cas9-Guided Oncogenic Chromosomal Translocations With Conditional Fusion Protein Expression in Human Mesenchymal Cells. *Proc. Natl. Acad. Sci. U. S. A.* **2017**, *114*(14), 3696–3701.
87. Walton, J.; Blagih, J.; Ennis, D.; Leung, E.; Dowson, S.; Farquharson, M.; Tookman, L. A.; Orange, C.; Athineos, D.; Mason, S.; Stevenson, D.; Blyth, K.; Strathdee, D.; Balkwill, F. R.; Vousden, K.; Lockley, M.; McNeish, I. A. CRISPR/Cas9-Mediated Trp53 and Brca2 Knockout to Generate Improved Murine Models of Ovarian High-Grade Serous Carcinoma. *Cancer Res.* **2016**, *76*(20), 6118–6129.
88. Heckl, D.; Kowalczyk, M. S.; Yudovich, D.; Belizaire, R.; Puram, R. V.; McConkey, M. E.; Thielke, A.; Aster, J. C.; Regev, A.; Ebert, B. L. Generation of Mouse Models of Myeloid Malignancy With Combinatorial Genetic Lesions Using CRISPR-Cas9 Genome Editing. *Nat. Biotechnol.* **2014**, *32*(9), 941–946.

89. Xue, W.; Chen, S.; Yin, H.; Tammela, T.; Papagiannakopoulos, T.; Joshi, N. S.; Cai, W.; Yang, G.; Bronson, R.; Crowley, D. G.; Zhang, F.; Anderson, D. G.; Sharp, P. A.; Jacks, T. CRISPR-Mediated Direct Mutation of Cancer Genes in the Mouse Liver. *Nature* **2014**, *514*(7522), 380–384.
90. Platt, R. J.; Chen, S.; Zhou, Y.; Yim, M. J.; Swiech, L.; Kempton, H. R.; Dahlman, J. E.; Parnas, O.; Eisenhaure, T. M.; Jovanovic, M.; Graham, D. B.; Jhunjhunwala, S.; Heidenreich, M.; Xavier, R. J.; Langer, R.; Anderson, D. G.; Hacohen, N.; Regev, A.; Feng, G.; Sharp, P. A.; Zhang, F. CRISPR-Cas9 Knockin Mice for Genome Editing and Cancer Modeling. *Cell* **2014**, *159*(2), 440–455.
91. Maddalo, D.; Manchado, E.; Concepcion, C. P.; Bonetti, C.; Vidigal, J. A.; Han, Y. C.; Ogrodowski, P.; Crippa, A.; Rekhtman, N.; de Stanchina, E.; Lowe, S. W.; Ventura, A. In Vivo Engineering of Oncogenic Chromosomal Rearrangements With the CRISPR/Cas9 System. *Nature* **2014**, *516*(7531), 423–427.
92. Mandasari, M.; Sawangarun, W.; Katsube, K.; Kayamori, K.; Yamaguchi, A.; Sakamoto, K. A Facile One-Step Strategy for the Generation of Conditional Knockout Mice to Explore the Role of Notch1 in Oroesophageal Tumorigenesis. *Biochem. Biophys. Res. Commun.* **2016**, *469*(3), 761–767.
93. Annunziato, S.; Kas, S. M.; Nethe, M.; Yucel, H.; Del Bravo, J.; Pritchard, C.; Bin Ali, R.; van Gerwen, B.; Siteur, B.; Drenth, A. P.; Schut, E.; van de Ven, M.; Boelens, M. C.; Klarenbeek, S.; Huijbers, I. J.; van Miltenburg, M. H.; Jonkers, J. Modeling Invasive Lobular Breast Carcinoma by CRISPR/Cas9-Mediated Somatic Genome Editing of the Mammary Gland. *Genes Dev.* **2016**, *30*(12), 1470–1480.
94. Acevedo-Arozena, A.; Wells, S.; Potter, P.; Kelly, M.; Cox, R. D.; Brown, S. D. ENU Mutagenesis, a Way Forward to Understand Gene Function. *Annu. Rev. Genomics Hum. Genet.* **2008**, *9*, 49–69.
95. Braun, C. J.; Bruno, P. M.; Horlbeck, M. A.; Gilbert, L. A.; Weissman, J. S.; Hemann, M. T. Versatile In Vivo Regulation of Tumor Phenotypes by dCas9-Mediated Transcriptional Perturbation. *Proc. Natl. Acad. Sci. U. S. A.* **2016**, *113*(27), E3892–900.
96. Ladd, B.; Mazzola, A. M.; Bihani, T.; Lai, Z.; Bradford, J.; Collins, M.; Barry, E.; Goeppert, A. U.; Weir, H. M.; Hearne, K.; Renshaw, J. G.; Mohseni, M.; Hurt, E.; Jalla, S.; Bao, H.; Hollingsworth, R.; Reimer, C.; Zinda, M.; Fawell, S.; D'Cruz, C. M. Effective Combination Therapies in Preclinical Endocrine Resistant Breast Cancer Models Harboring ER Mutations. *Oncotarget* **2016**, 7(34), 54120–54136.
97. Valente, L. J.; Aubrey, B. J.; Herold, M. J.; Kelly, G. L.; Happo, L.; Scott, C. L.; Newbold, A.; Johnstone, R. W.; Huang, D. C.; Vassilev, L. T.; Strasser, A. Therapeutic Response to Non-genotoxic Activation of p53 by Nutlin3a Is Driven by PUMA-Mediated Apoptosis in Lymphoma Cells. *Cell Rep.* **2016**, *14*(8), 1858–1866.
98. Lok, B. H.; Gardner, E. E.; Schneeberger, V. E.; Ni, A.; Desmeules, P.; Rekhtman, N.; de Stanchina, E.; Teicher, B. A.; Riaz, N.; Powell, S. N.; Poirier, J. T.; Rudin, C. M. PARP Inhibitor Activity Correlates With SLFN11 Expression and Demonstrates Synergy With Temozolomide in Small Cell Lung Cancer. *Clin. Cancer Res.* **2017**, *23*(2), 523–535.
99. Kuhn, M. W.; Song, E.; Feng, Z.; Sinha, A.; Chen, C. W.; Deshpande, A. J.; Cusan, M.; Farnoud, N.; Mupo, A.; Grove, C.; Koche, R.; Bradner, J. E.; de Stanchina, E.; Vassiliou, G. S.; Hoshii, T.; Armstrong, S. A. Targeting Chromatin Regulators Inhibits Leukemogenic Gene Expression in NPM1 Mutant Leukemia. *Cancer Discov.* **2016**, *6*(10), 1166–1181.
100. Karlgren, M.; Simoff, I.; Backlund, M.; Wegler, C.; Keiser, M.; Handin, N.; Muller, J.; Lundquist, P.; Jareborg, A. C.; Oswald, S.; Artursson, P. A CRISPR-Cas9 Generated Madin-Darby Canine Kidney Expressing Human MDR1 Without Endogenous

Canine MDR1 (cABCB1): An Improved Tool for Drug Efflux Studies. *J. Pharm. Sci.* **2017**, *106*(9), 2909–2913.

101. Kawaharada, K.; Kawamata, M.; Ochiya, T. Rat Embryonic Stem Cells Create New Era in Development of Genetically Manipulated Rat Models. *World J. Stem Cells* **2015**, 7(7), 1054–1063.
102. Lu, J.; Shao, Y.; Qin, X.; Liu, D.; Chen, A.; Li, D.; Liu, M.; Wang, X. CRISPR Knockout Rat Cytochrome P450 3A1/2 Model for Advancing Drug Metabolism and Pharmacokinetics Research. *Sci. Rep.* **2017**, 7, 42922.
103. Dehne, E. M.; Hasenberg, T.; Marx, U. The Ascendance of Microphysiological Systems to Solve the Drug Testing Dilemma. *Future Sci. OA* **2017**, *3*(2), FSO185.
104. Yumlu, S.; Stumm, J.; Bashir, S.; Dreyer, A. K.; Lisowski, P.; Danner, E.; Kuhn, R. Gene Editing and Clonal Isolation of Human Induced Pluripotent Stem Cells Using CRISPR/Cas9. *Methods* **2017**, *121-122*, 29–44.
105. Xia, P.; Zhang, X.; Xie, Y.; Guan, M.; Villeneuve, D. L.; Yu, H. Functional Toxicogenomic Assessment of Triclosan in Human HepG2 Cells Using Genome-Wide CRISPR-Cas9 Screening. *Environ. Sci. Technol.* **2016**, *50*(19), 10682–10692.

CHAPTER FOUR

Recent Advances of Microfluidics Technologies in the Field of Medicinal Chemistry

László Ürge*[,1], Jesus Alcazar[†], Lena Huck[†], György Dormán[‡]
*DBH Group, Budapest, Hungary
[†]Janssen Research and Development, Toledo, Spain
[‡]Innostudio Inc., Budapest, Hungary
[1]Corresponding author: e-mail address: laszlo.urge@dbh-group.com

Contents

Annual Reports in Medicinal Chemistry, Volume 50
ISSN 0065-7743
https://doi.org/10.1016/bs.armc.2017.09.001

1. INTRODUCTION

Microfluidics, mesofluidics, and lab-on-a-chip technologies have the potential to expedite a wide variety of scientific discoveries, e.g., synthesis, analytics, biological screening, and formulation development.[1] In the last two decades, the development of commercially available tools utilizing these principles was expedited, and as a result cost effective commercially available bench-top products came out that have enabled scientist to explore previously unexplored territories in many different areas.[1,2] Key advantages of microfluidic and mesofluidic technologies include the operation with small volumes in a well-controlled environment, flexible technical setup, and parameter control to the extent that is not available for batch-based methods. Their technical setup typically consists of a pumping unit, a mixer, reactor chamber, detectors, and separating or receiving units.

The significantly increased surface-to-volume ratio and short distances in microfluidic channels translate to more efficient mixing and enable the controlled and rapid heat and mass transfer that changes reaction characteristics compared to traditional batch approaches. This often results in fast reaction times, increased throughput, better and reproducible product profile with fewer impurities, and increased yields.[3–5] The most important process parameters such as mixing, temperature, pressure, flow rate, and residence time can be controlled to the extent that is not available in traditional batch-based methods, allowing fast parameter screening and process optimization.

The modular mixing and reactor units may have special geometries that offer highly specific operation opportunities such as separating and splitting liquid flow, eliminating desired or undesired products immediately after reaction, or efficient mixing of heterogeneous phases such as gas, liquid, solid, or catalytic surfaces.[6]

The wide variety of materials available to fabricate micro- and mesofluidics reactors, including glass, silicon, polymers, teflon, other plastics, metal, and ceramics, allow the use of a wide range of parameters, pressure, temperature, residence time, solvents, reactants, corrosive reagents, etc. Furthermore, unavailable until now, special reactor geometry may also be achieved with recent advances in 3D printing technologies.[7] The above summarized features are especially useful in handling dangerous or runaway, highly exothermic reactions (e.g., hydrogenation, oxidation, nitration), or reactions that require hazardous or unstable materials (e.g., halogens, cyanides, carbon monoxide). In addition individual microreactors or just reaction chambers can be tethered and sequenced so that a microfluidic-based chemical plant is obtained.

Due to this wide flexibility, microfluidics and mesofluidics technologies have been successfully applied in many different scientific areas including synthetic chemistry, analytical chemistry, mass spectrometry,[8] protein crystallization,[9] high-throughput screening,[10] flow cytometry and single-cell analysis,[11,12] drug delivery,[13] API (active pharmaceutical ingredient) production,[14] final drug production, and several other areas.[15–17]

1.1 Application Areas in the Pharmaceutical Industry

Microfluidics, mesofluidics, and lab-on-a-chip technologies started to penetrate the pharmaceutical, biotechnology, and life science-related industries about two decades ago and entered different stages of the discovery and development process. Today, the industry is using these technologies in many different aspects, and in certain cases they became integrated into the standard discovery and development process. The application areas are the following:

Manufacturing

- continuous production of APIs[18]
- continuous combined production of APIs and final drug product end-to-end pharmaceutical manufacturing
- on-demand drug synthesis[14]
- distributed (local) manufacturing[19]
- portable mini factories in flow[20,21]

Development

- total synthesis of approved APIs and APIs in clinical development
- final formulation
- combination of API synthesis with final product formulation[22]

Early development

- quick scale up
- process development and optimization
- regulatory synthesis
- formulation development
- sustainable medicinal chemistry in flow[23]

Early discovery

- synthesis of building blocks
- hit-to-lead/lead optimization[24]

- new chemical entities[6]
- early formulations
- screening
- combination of synthesis and screening, and automated lead optimization[25]
- metabolism studies

1.2 Key Drivers for Adoption

The development of continuous manufacturing of APIs is proactively supported not only by key industry players but also by regulatory agencies.[26,27] The first large-scale initiative was announced by Novartis in collaboration with MIT in 2007 to develop and implement technologies to transform traditional batch processes to continuous mode.[28]

Other pharmaceutical companies such as GlaxoSmithKline and Eli Lilly have announced plans to build continuous manufacturing plants in Singapore and Ireland, respectively. Novartis, Merck, Bayer, AstraZeneca, and Richter Gedeon have been doing R&D on implementing continuous manufacturing for years. In addition some of the contract manufacturing organizations have also implemented continuous-flow technologies and continuous manufacturing in order to stay competitive and to provide better support to their customer base.[29,30]

Two milestone examples of industrial-scale continuous API production have been approved by the FDA recently, indicating a significant paradigm shift in the industry. Vertex's cystic fibrosis product Orkambi (Lumaca/Ivaca) in 2015 was granted approval as the first New Drug Application (NDA) for using a continuous manufacturing technology for production, and a 4000-square-foot continuous manufacturing facility has been built for production.[31] In 2016 Johnson & Johnson's Prezista for HIV (Darunavir) was the first NDA supplement approval for switching from batch manufacturing to continuous manufacturing.[32] J&J also aims to produce 70% of its highest-volume products through continuous manufacturing within 8 years.

Switching to continuous processes in API production is calculated to be economically viable and beneficial.[33–35] However, the full industrial implementation of these novel processes is highly knowledge intensive[36] and the lack of experienced and trained scientist is still a major barrier. The implementation of these approaches is still more difficult than in the case of batch processes.[37–40]

The FDA is also proactively trying to support the adaptation of the technologies since the launch of the Pharmaceutical cGMP initiative in 2002. The Agency also claims that no regulatory barriers would hinder implementation of continuous manufacturing and encourages the players to consult regularly and frequently before and during any implementation.

In addition to these large-scale initiatives other drivers are also playing a significant role in certain areas.

Traditional medicinal chemistry is complemented with microfluidics-based chemistry technologies for safety reasons. Dangerous transformations can utilize dedicated commercially available reactors for particular reactions, e.g., hydrogenation, ozonolysis—that make them safe, easy to use and easy and effective to optimize.

Automated and autonomous devices may expedite optimization processes, shorten the SAR cycle time, and lower the cost of drug discovery by integrating flow chemistry devices and screening methodologies. This is particularly applicable for lead optimization, quick analogue, and SAR generation.

It is also important to note that there are disadvantages of microfluidics technologies that may be the reasons for hindering the adaptation. In general reactions must stay in solution in order to avoid blocking of the channels, pumping slurries is technically quite challenging, and precipitation of products or side products is a major risk for disturbing operations. On top of the technical barriers, the human factors also need to be considered. The lack of training of chemists in continuous technologies and the lack of key expertise of engineering-minded chemists quite often contribute to the hindering of the adaptation.[30,36,37,41–44]

1.3 Scientific Activity on Microfluidics-Driven Synthetic Chemistry

Table 1 summarizes our literature analysis based on synthetic chemistry applications of the commercial and in-house-made microfluidics- and mesofluidics-based chemistry technologies.

Over the last 20 years, about 1000 papers have already been published on utilizing microfluidics technologies for organic synthesis, while at the same time the application areas are limited to certain key aspects. The vast majority fall into the experimentation with novel chemistries, novel compound synthesis and library synthesis,[45] suggesting that chemists are still searching for the direction for applications. In addition there is a significant number of papers focusing on process optimization and API synthesis. At the same time

Table 1 Number of Scientific Publications Relating to Different Synthetic Chemistry Applications With Commercial and Noncommercial Bench Top Microfluidics Technologies

Bench to Microfluidics Reactor Technology Publications	Total Number of Peer-Reviewed Publications	Total Number of Papers in Hit-to-Lead and Lead Optimization	Total Number of Process Chemistry (API)	Other (Integrated Discovery/ Metabolism Model/ PET Tracer)	Experimentation With Novel Chemistries and Novel Compounds in General
Thales H-Cube	391	89	No	3	290
Thales other	94	No	No	No	94
Vapourtec	250	5	14	3	220
Uniqsis	58	1	3	1	51
Chemtrix	108	No	No	1	102
Syrris	67	No	1	2	61
Other commercial	6	No	No	No	5
Noncommercial instrument	>60	No	17	12	9

in the hit-to-lead and lead optimization phase of the discovery process, the flow hydrogenation technology seems to be the only technology that has established standard position, with the Vapourtec technology providing a few more examples. The rest of the technologies are not present or results are not being published within the lead optimization phase at all. Therefore, the flow hydrogenation technology is the only technology that seems to be routinely applied today in the medicinal chemistry toolbox. Some of the aspects of flow hydrogenation to support medicinal chemistry have also been reviewed earlier.[46]

This may be explained by the fact that medicinal chemists are working on small scale and typically not restricted with safety, yield, or reproducibility issues, while these are some of the major drivers for adaptation. Therefore, microfluidics and mesofluidics chemistries have made less of an impact on their workflow.

2. FLOW CHEMISTRY IN THE PATH FOR CONTINUOUS MANUFACTURING

Pharmaceutical production has not changed in the last 50 years. However, comparing overall processes with other industries, the pharma sector, which is largely batch in nature, remains relatively inefficient and less understood.[47] For this reason, the FDA and other regulatory agencies are encouraging companies to consider continuous manufacturing to overcome current limitations.[48] Continuous manufacturing involves not only chemical transformations, but other processes, such as quench, work up, isolation, purification, crystallization, and drying, are also part of the production process.[49] However, this chapter will be mainly focused on the synthesis part as it is the main piece of the puzzle. It will be divided per technology used to cover many different situations that can be found in the development of pharmaceutical products.

2.1 Solution-Phase Chemistry

One of the key requirements for flow chemistry is that all reagents must be in solution. Then the mixture is matured in microreactors or coils until reaction is completed. In the following examples, several flow reactions will be presented and their motivation explained.

The key step for the synthesis of $S1P_1$ receptor agonist **3**, a potential antiinflammatory agent, was the preparation of amidoxime **2**. This intermediate was obtained by reaction of nitrile **1** with hydroxylamine. To avoid the issues associated with the handling of this reagent in large scale, the procedure was translated to flow. Running the reaction in a Vapourtec continuous-flow equipment in *n*-butanol as solvent, at 120°C, $t_R = 1$ h, and 8–9 bar of pressure the conversion remains stable over hours allowing a safer scale up of amidoxime intermediate **2** (Scheme 1).[50]

The combination of the principle of design experiments with automated flow synthesizers offers a rationalized approach to improve reaction outcomes. In this way, the Gioiello's group presented an alternative route to the batch procedure for the synthesis of thieno(2,3-*c*)isoquinolin-5(4*H*)-one core **7**, a valuable source of PARP-1 inhibitors.[51] In this case, they show the optimization of the Suzuki coupling reaction and the thermal Curtius rearrangement, evaluating the effect of flow rate and temperature. Finally, with the optimized conditions, the core was obtained in 50% overall yield starting from boronic acid **5** and thiophene **4** (Scheme 2).

An application of continuous-flow micromixing reactor technology for the synthesis of benzimidazole proton pump inhibitors (**10**) was presented in 2013.[52] The first step is a condensation of pyridine compounds (**8**) with

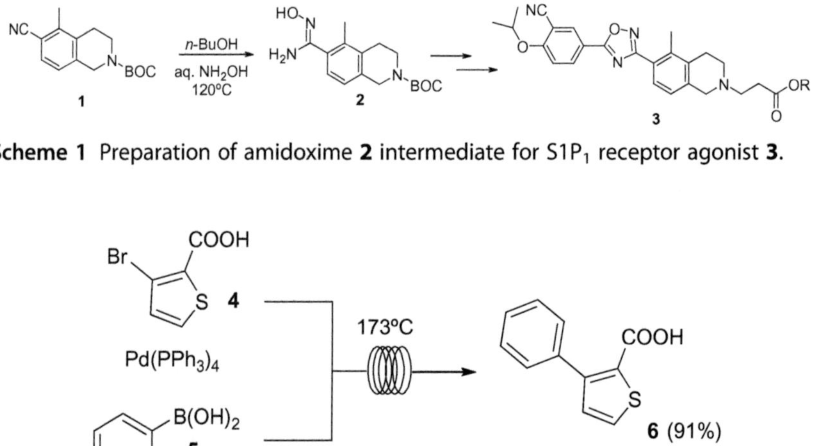

Scheme 1 Preparation of amidoxime **2** intermediate for S1P$_1$ receptor agonist **3**.

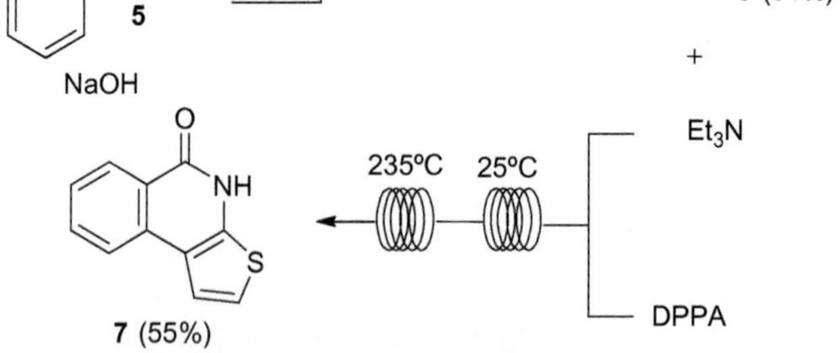

Scheme 2 Preparation of thieno(2,3-*c*)isoquinolin-5(4*H*)-one core **7**.

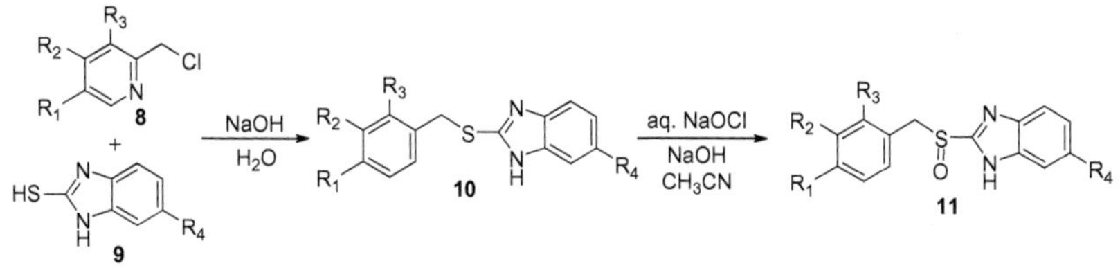

Scheme 3 Preparation of benzimidazole analogues **11**.

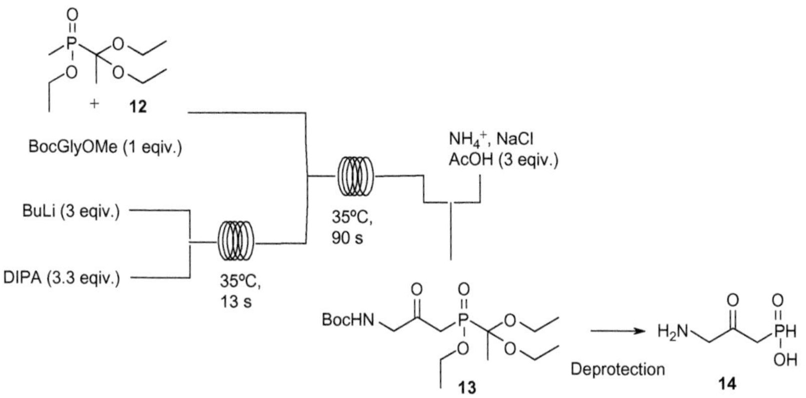

Scheme 4 Preparation of AZD6906 **14**.

thiols (**9**) followed by their oxidation (Scheme 3). Aqueous NaOCl is used instead of expensive *m*-CPBA, improving yield and quality of the final product. The residence time was decreased from 3 h to 1 s and the sulfone side product minimized.

Continuing with applications to gastrointestinal drugs, a flow approach was presented for the synthesis of reflux inhibitor AZD6906 **14**.[53] The synthesis consisted in three steps: formation of protected phosphinate, acylation with *N*-Boc-glycine methyl ester, and finally deprotection and salt removal (Scheme 4). The first two steps are known to be highly exothermic which could be controlled under flow conditions. LDA was prepared in situ to reduce variability between LDA batches and the stoichiometry of glycine: phosphinate was reduced to 1:1.

Spirocyclic compounds are important motifs in drug discovery. In this regard, AstraZeneca scientists have developed a large-scale route to an MCH_1 receptor antagonist.[54] For the synthesis of a spiroazetidine, β-lactam **18** was prepared following Staudinger reaction of imine **15** with an in situ generated ketene at room temperature (Scheme 5). The reaction was run on 50 g scale with an isolated yield of 56%. To follow the progress of the

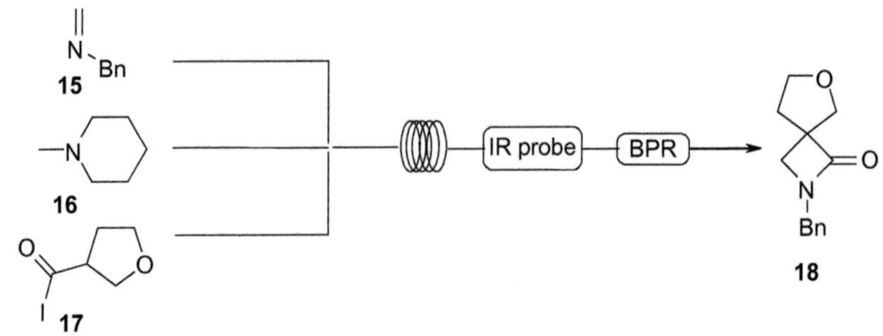

Scheme 5 Synthesis of intermediate **18**.

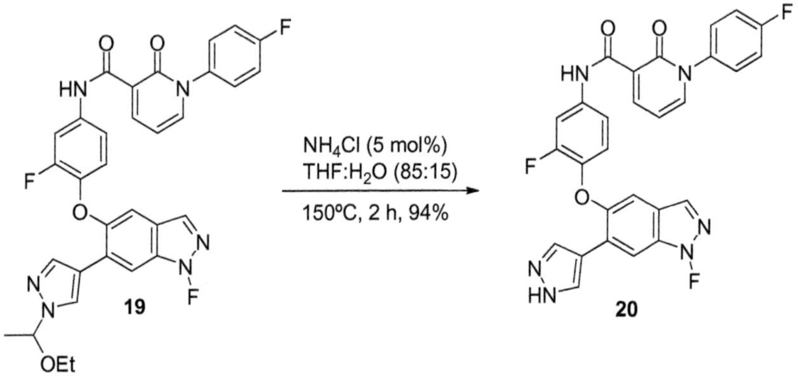

Scheme 6 Synthesis of Merestinib **20**.

reaction an in-line IR was used. Further reduction and deprotection was performed in batch.

For the synthesis of antitumoral drug Merestinib **20** an NH_4Cl-catalyzed ethoxyethyl deprotection in flow was developed (Scheme 6).[55] The batch deprotection of the intermediate **19** under acid conditions cleaved also the amide group. To control the deprotection flow conditions were examined using a flow reactor and different additives (protic and aprotic acids). NH_4Cl gave the best results. Reaction scale up was performed by introducing separate feed solutions for every component, then heating the mixture in a coil at 150°C for 2 h of residence time. Reaction monitoring was done by online HPLC. In total, over 100 kg was prepared with a total yield of 94%.

For the synthesis of Vildagliptin **27**, an oral antidiabetic drug, Sedelmeier and Pellegatti described the preparation and in situ utilization of the Vilsmeier reagent (VR) **24**.[56] This transformation is ideal to be done in flow because the VR is irritant and has a high thermal energy of decomposition (Scheme 7). *N*-acylated proline amide **23** is directly converted to

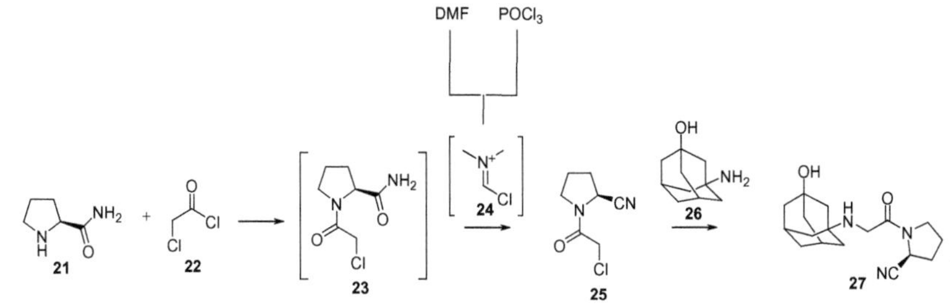

Scheme 7 Synthesis of Vildagliptin **27**.

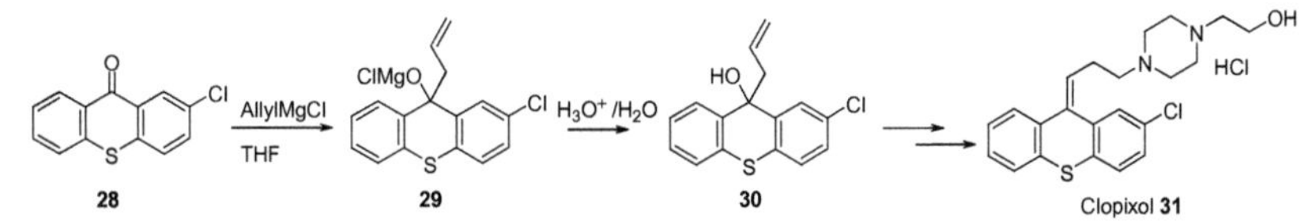

Scheme 8 Synthesis of Clopixol **31**.

cyanopyrrolidine **25** by in-line quenching with freshly prepared VR **77**. Synthesis of **23** was performed in batch, using the crude solution in stream 1 to join stream 2, providing the VR, in a T-piece and further reactor. The residence time is 90 s at 20°C, and the conversion to cyanopyrrolidine was quantified by HPLC analysis.

Organometallic reagents have been a key reagent for API synthesis. However, a safer and more controllable way to handle them has been identified as a key element for their use in production. Flow chemistry has appeared as a technology able to improve their chemo- and regioselectivity. Moreover, the translation to flow makes these reagents easier to be handle, decreasing the presence of side products.

One of the earliest precedents in the use of organometallic chemistry in flow for APIs is the synthesis of Clopixol **31**, an antipsychotic drug, performed in a setup consisting of two reactors in series, combining continuous stirred tank reactor and plug flow reactor.[57,58] The beneficial part of such setups is the potential to operate at charges far greater than those that allow complete dissolution of the reactants. One of the key steps in the synthesis of Clopixol is a classic Grignard alkylation of ketone **28** and further hydrolysis (Scheme 8). The heterogeneous setup is needed because of the low solubility of ketone **28** in THF. The use of the reactor train in combination with in-line NIR analysis supposed a reduction of the solvent volume, and the formation of an impurity was suppressed.

One of the key issues handling organometallic reagents is pumping. In 2013, the Ley group presented a newly developed, chemically resistant, peristaltic pumping system.[59] Their potential value was evaluated in different chemical transformations such the use of *n*-butyllithium, Grignard reagents, and DIBAL-H were reported. Finally, they present the synthesis of (*E*/*Z*)-Tamoxifen using continuous-flow organometallic reagent-mediated transformation (Fig. 1). Running the system continuously over 80 min provided 12.43 g of pure (*E*/*Z*)-Tamoxifen.

For the synthesis of Verubecestat **35**, a candidate for treatment of Alzheimer's disease, a diastereoselective Mannich-type addition of methyl sulfonamide **32** to chiral Ellman sulfinyl ketamine **33** to generate **34** was required.[60] Subsequent coupling with 5-fluoropicolinamide, global deprotection, and guanidinylation provided the API **35** in 44% yield over four steps. The Mannich-type addition of lithiated to the ketimine is energy intensive because it must be conducted under cryogenic conditions to avoid α-deprotonation of **34** and the subsequent side products (Scheme 9). For this reason, flow technology was ideal for this transformation not only controlling the organometallic reagent but also improving the yields obtained in batch.

A kilogram synthesis of Akt kinase inhibitor **40** is presented in 17 total steps.[61] One of the key transformations is a formylation of a dianion **38** derived by deprotonation and subsequent lithium–halogen exchange from a 2-bromo-3-aminopyridine precursor **36**. When the reaction was performed in a 100-L vessel at −60°C, the solution of the second batch turned into a thick gel at the dianion stage, thereby hindering stirring. For practical reasons the step was translated to a flow procedure. The anion **37** derived from 1 kg of bromide **36** was processed through the flow reactor in 1 h with efficient cooling and residence time of seconds (Scheme 10). The overall yield of the run was higher than the one obtained in batch, and less debrominated impurity was obtained.

In the synthesis of H_3 antagonist **43**, an agent for neurological disorders, a one-pot addition of a magnesium complex **41** to ketone **42** was required (Scheme 11).[62] Studying the optimal conditions in batch for the addition, the group found different problems to address: formation of side products, dehalogenation of starting material, and difficult to maintain necessary cryogenic conditions. The best conditions were found by adding both components of the reactions simultaneously into a batch reactor; this showed that a flow process would be ideal. The addition reaction was run in a kilo lab flow chemistry rig with a residence time of 6–7 min at 0°C. The outcome was

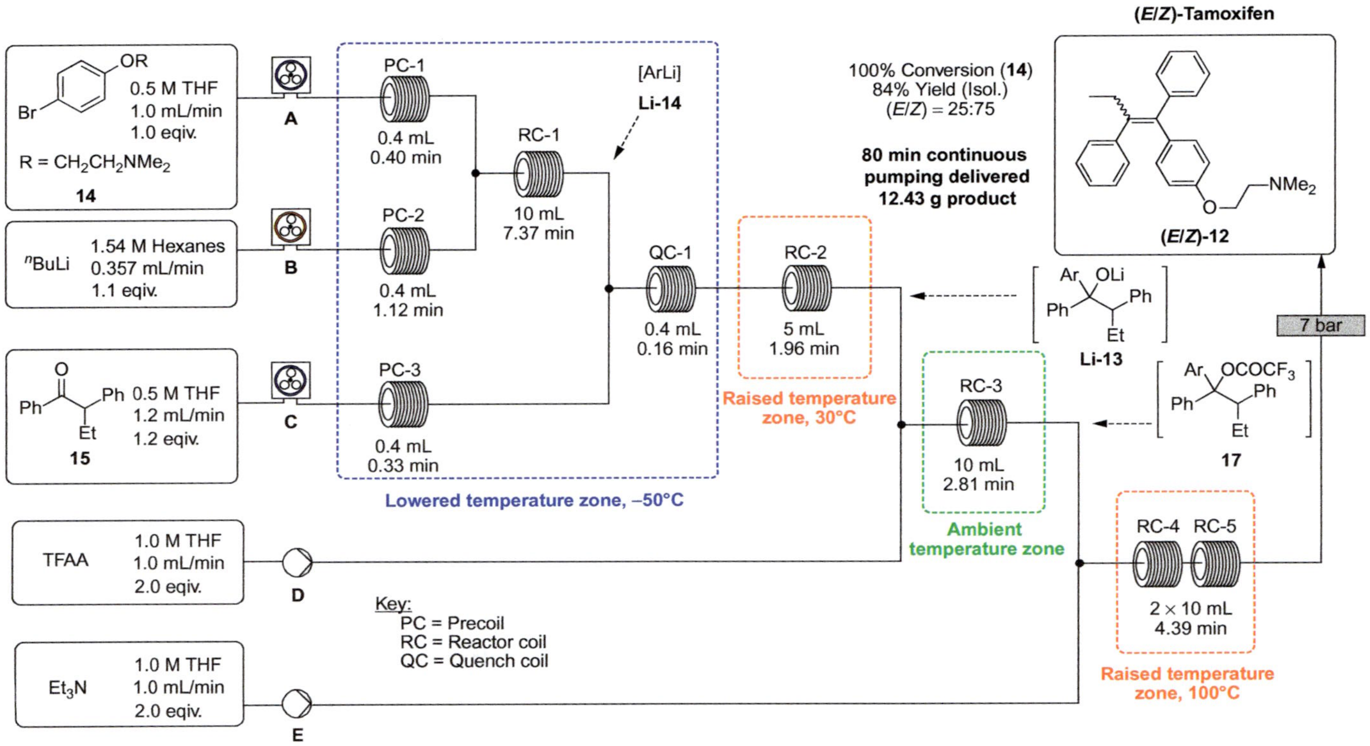

Fig. 1 Synthesis of (*E*/*Z*)-Tamoxifen. New peristaltic pumps A, B, and C. *Reproduced from Murray, P. R. D.; Browne, D. L.; Pastre, J. C.; Butters, C.; Guthrie, D.; Ley, S.V.* Org. Process Res. Dev. ***2013***, 17, *1192–1208. Copyright 2013 American Chemical Society.*

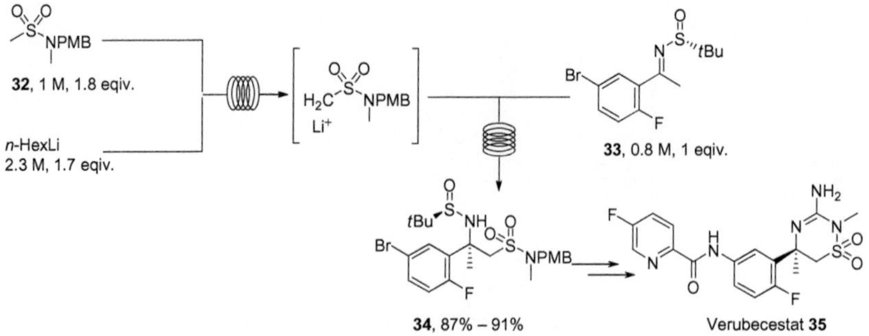

Scheme 9 Application to the synthesis of Verubecestat **35**.

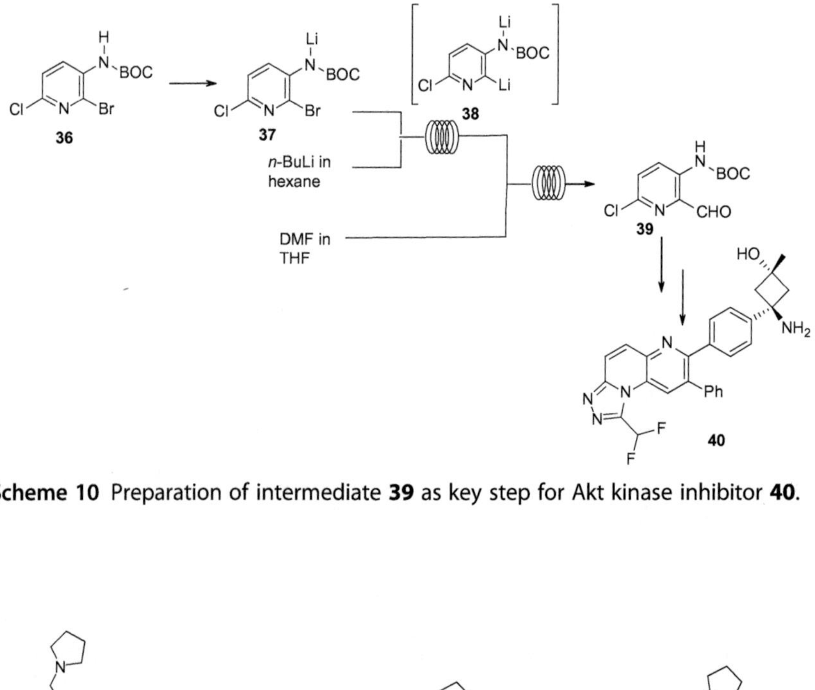

Scheme 10 Preparation of intermediate **39** as key step for Akt kinase inhibitor **40**.

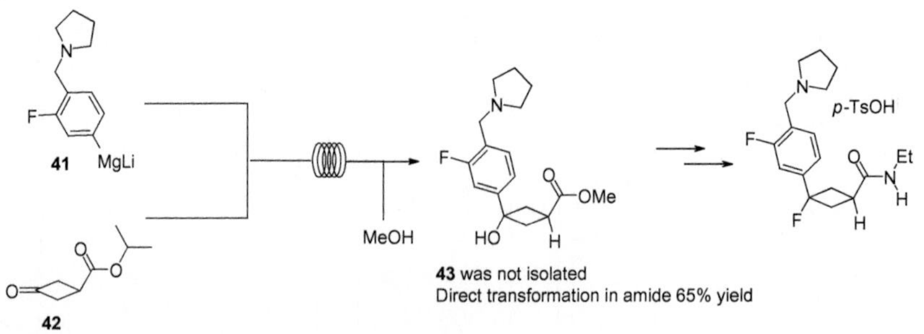

Scheme 11 Preparation of intermediate **43**.

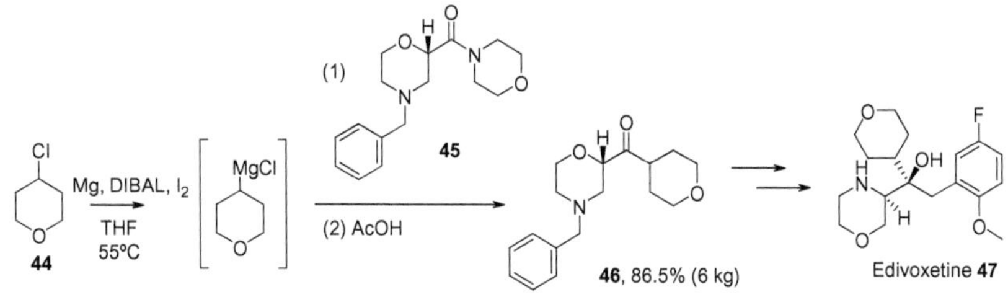

Scheme 12 Preparation of Edivoxetine **47**.

collected in a vessel with methanol for transesterification and ultimate amination with ethylamine. The work up of this step was streamlined to minimize the number of unit operation.

A key intermediate in the synthesis of Edivoxetine **47** was prepared by a flow Barbier process (Scheme 12).[63] In this case a continuous stirred tank reactor was used minimizing hazards by operating at a small reaction volume and performing metal activation only once per campaign. First a solvent study was performed to avoid secondary products due to the use of toluene. A mixture of THF/2-Me THF was found as best option. A viable process was developed to obtain around 6 kg of **46** in 86.5% overall yield.

2.2 Applying Supported Catalysts in API Synthesis

Metal-catalyzed cross-coupling chemistry has played a major role in the synthesis of natural products and other biologically active compounds.[64] However, these metal catalysts are not soluble in the solvent mixture and become a problem when flowing through the system. Supported catalysis has appeared as a very efficient solution for this issue as it allows to recycle and reuse the catalyst and reduce its leaching off the solid support, increasing its turnover number.

One example was the synthesis of a key intermediate of Valsartan **52** accomplished by using five parallel monolithic microreactors connected to a split-and-recombine-type flow reactor.[65] The first step was the synthesis of the organoboron component **49** starting from cyanobromobenzene **48**. The resulting solution was collected and *p*-iodobenzaldehyde **50** was added. The mixture was introduced to a reactor charged with palladium on monolith at 120°C and a residence time of 4 min to get the final intermediate **51** with high yield and good purity (Scheme 13).

Lapkin's group presented a Buchwald–Hartwig amination using a Pd-NHC-based supported catalyst for the synthesis of an important pharmaceutical intermediate **55** (Scheme 14).[66] A comparison between batch,

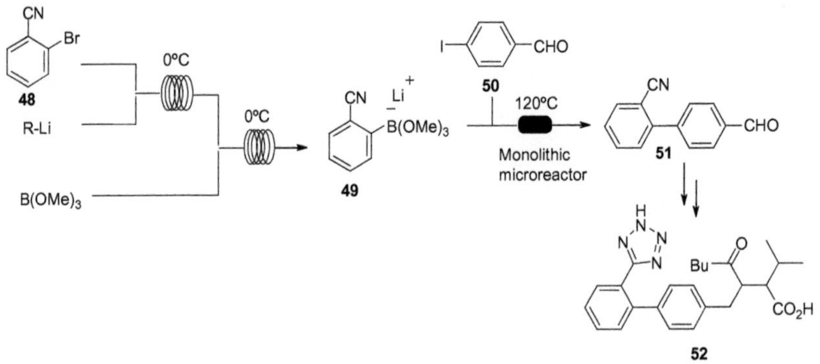

Scheme 13 Synthesis of Valsartan intermediate **51**.

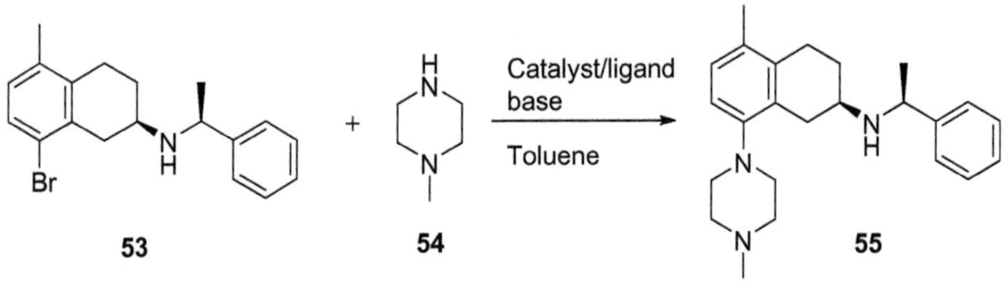

Scheme 14 Synthesis of intermediate **55**.

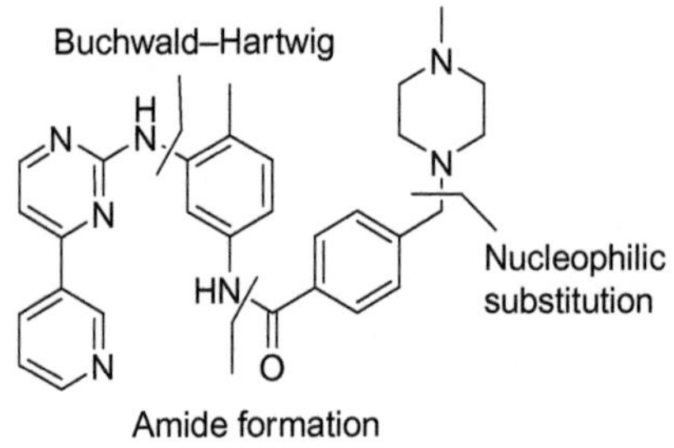

Imatinib **56**

Scheme 15 Structure and disconnection approach for Imatinib **56**.

lab-scale mini plant and pilot-scale plant was made looking at different parameters.

Another example of a Buchwald–Hartwig reaction in flow was the synthesis of Imatinib **56** (Scheme 15), a tyrosine kinase inhibitor.[67] The synthesis was realized as a continuous process featuring an amide formation, a nucleophilic substitution, and a Buchwald–Hartwig coupling. Scavenger resins were used to purify the intermediates which limited the uninterrupted run of the synthesis and the quantity of the final products. Solvent switching operations and in-line UV-monitoring made this approach completely automated.

In 2015, a convergent strategy for the synthesis of Telmisartan **60**,[68] an important antihypertensive drug, with no intermediate purifications or solvent exchanges was described by Prof. Gupton. The key step was a Suzuki–Miyaura cross-coupling between two functionalized benzimidazoles catalyzed by a silica-supported Pd catalyst (Scheme 16). A tubular reactor system coupled with a plug flow packed bed cartridge unit is used. An overall yield of 81% was obtained in a milligram scale.

Fukuyama and Ryu presented a 100-g scale of a key intermediate of matrix metalloproteinase **63** using a novel approach combining supported and solution catalysis. In their approach, the reaction took place in solution, but having the catalyst and the reactants dissolved in immiscible solvents (Scheme 17). Thus, the catalyst is separated from the crude product by an in-line separation device and fed back into the reaction system.[69] The process was run for 5.5 h, and 103 g of pure product was obtained in 86% yield.

Catalytic hydrogenation is another key process for the synthesis of APIs. For instance, an economic and scalable process for glucosylceramide synthase inhibitor **66** has been developed by reductive amination of aldehyde **64** with deoxynojirimycin **65** using an H-Cube MIDI (Scheme 18).[70] The biggest challenge was to find a solvent for the solubility of the starting material and easy extraction of the final product; different catalysts were also tested. The best conditions were found using THF/MeOH/H_2O (2:2:1) as solvent system and $Pd(OH)_2$ on charcoal as catalyst. Hydrogenation was performed at 100 bar, 150°C, and a flow rate of 15 mL/min to obtain compound **78** in 99% purity and 61% yield.

2.3 Applying Flow Photochemistry in API Synthesis

Photochemistry is a known technology able to perform complex organic reaction efficiently. However, often it has been highly disregarded due to the limitation of light absorption as described in the Bouguer–Lambert–Beer law. To overcome this limitation, flow chemistry offers a clear advantage over batch protocols as it ensures a uniform irradiation of the narrow channels. In this way, all the reaction mixtures will be uniformly irradiated, meanwhile in batch, only the outer part of the container will receive enough number of photons to make the reaction work.[71]

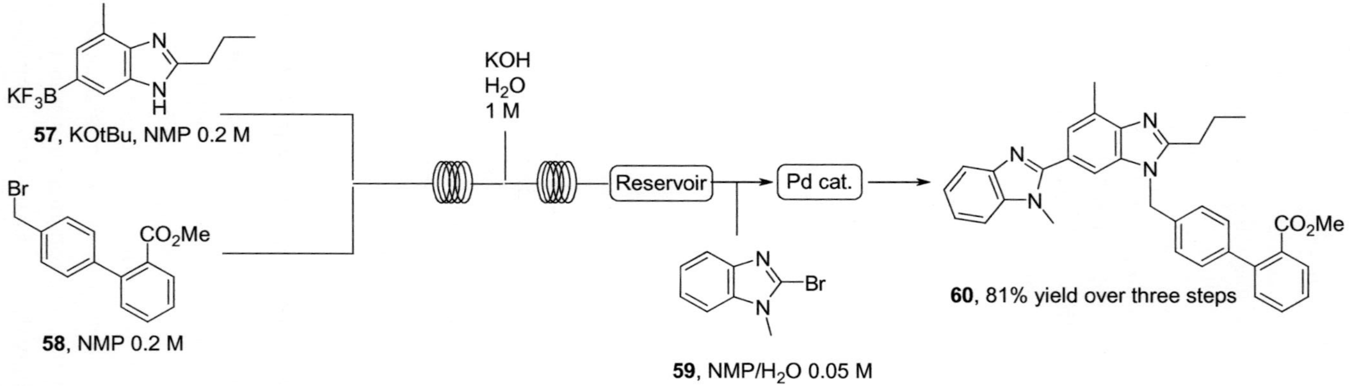

Scheme 16 Suzuki–Miyaura approach to Telmisartan **60**.

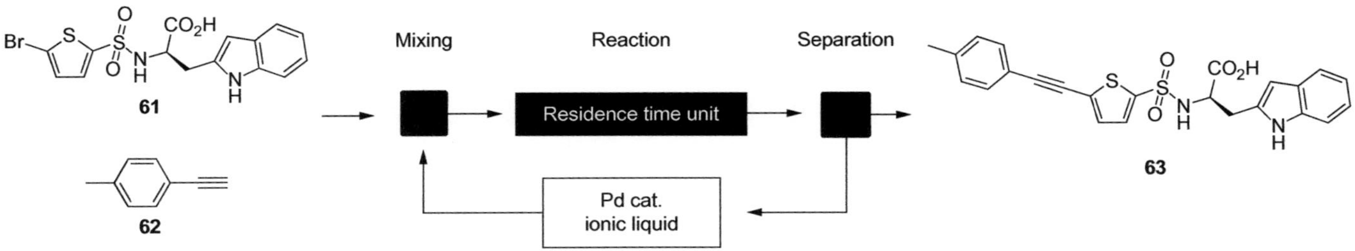

Scheme 17 Ionic liquid approach to matrix metalloproteinase inhibitor **63**.

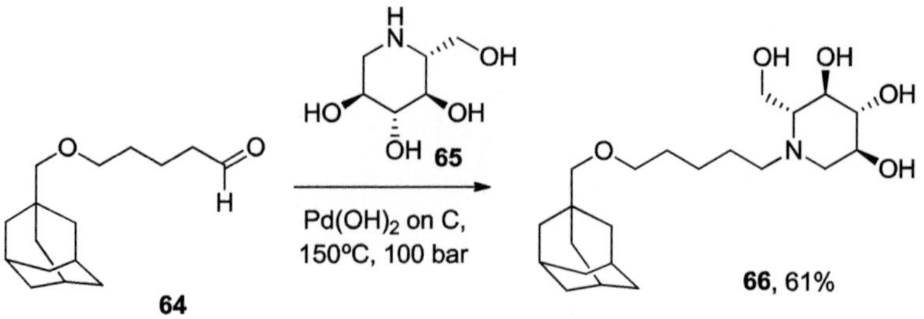

Scheme 18 Reductive amination using catalytic hydrogenation conditions.

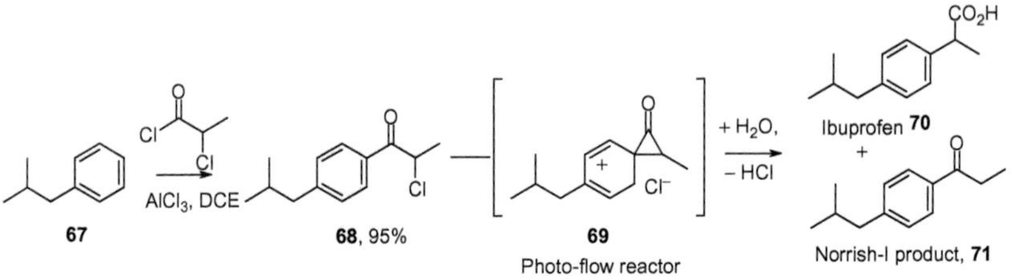

Scheme 19 Photo-Favorskii approach toward Ibuprofen **70**.

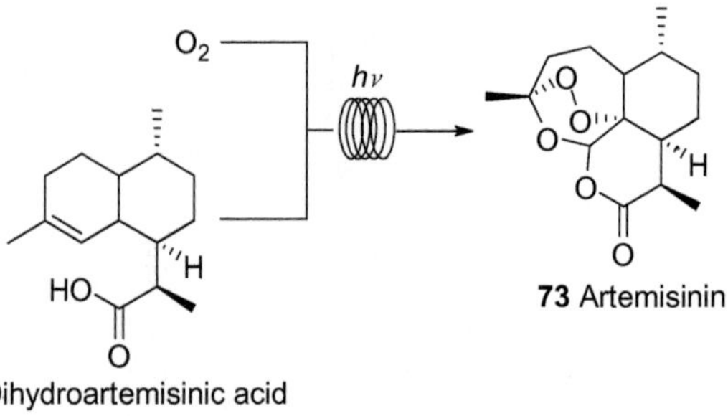

Scheme 20 Preparation of Artemisinin using photochemistry.

A new approach to reach Ibuprofen **70** in an efficient manner has been presented by Baxendale using photochemistry in flow (Scheme 19).[72] The continuous synthesis is based on a photo-Favorskii rearrangement reaction of a readily available α-chloropropiophenone precursor **68**. Real-time analysis through a small footprint photospectrometer permits to get the best conditions in short time and perform a multigram synthesis in a single working day.

Another example using photochemistry to synthetize a pharmaceutical was performed by the Seeberger group for the preparation of Artemisinin **73**.[73] Dihydroartemisinic acid **72** is used as a starting point (Scheme 20). The key transformations demanded a reaction cascade including a singlet oxygen-mediated ene reaction, a Hock cleavage followed by oxidation, and final peracetalization. More recently, they presented a strategy to obtain different derivatives of artemisinin.[28,74] The key step is a $NaBH_4$-based flow reduction to obtain dihydroartemisinin, and the combination of different reaction modules allows the production of a variety of APIs. All monitorized with in-line flow IR and in-line purifications.

This reduction step was improved by Lapkin's group.[75] They presented a facile stoichiometric reduction of artemisinin to dihydroartemisinin using $LiBHEt_3$ instead of $NaBH_4$ after testing different reduction agents. The excellent reducing power of $LiBHEt_3$ decreases the residence time to 30 s in THF and at room temperature. Finally, the use of 2-methyl THF as solvent made the procedure even greener.

2.4 Integrated Approaches in API Synthesis

Novel platforms with integrated flow synthesis are appearing in the literature combining the different features presented so far in this chapter. Herein, some examples of this new trend will be disclosed.

Nevirapine is an important drug for the HIV treatment. Prof. Gupton and coworkers have presented the full synthesis of this drug in flow.[76] First they reported the continuous synthesis of a key nicotinonitrile precursor.[31] This was done by a three-step synthesis: formation of enamine, Knoevenagel condensation, and final cyclization. From this precursor, final synthesis of the API was achieved by process intensification in flow increasing the isolated yield from 63% to 91% while reducing the PMI value from 46 to 11.[77]

Researchers at the MIT presented in 2013 the first end-to-end integrated continuous manufacturing plant for a pharmaceutical product, in this case, Aliskiren hemifumarate.[22] The plant performs intermediate reactions, separations, crystallizations, drying, and formulation to get a formed final tablet (Fig. 2). Starting materials were pumped into a tubular reactor, where they are mixed with an acid catalyst, extraction is performed in-line to continue with deprotection, and finally a reactive crystallization is performed to create and purify the final salt.

Seeberger and coworkers have developed a modular platform to access different chemical families, depending on the module combination (Fig. 3).[20] Each single module represented a widely used medicinal chemistry transformation. For instance, Module 1 allows the oxidation of primary and secondary alcohols, Module 2 allows olefination of carbonyl compounds, Module 3 is used for Michael addition reactions, Module 4 is the

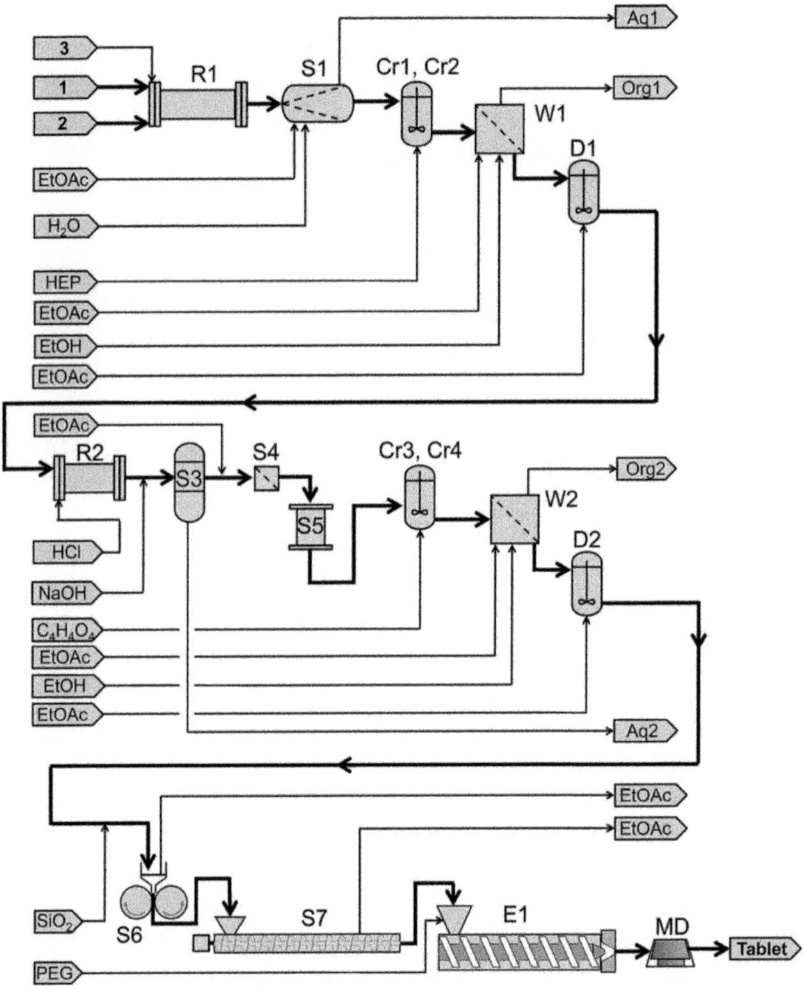

Fig. 2 Process flow diagram for the preparation of Aliskiren hemifumarate. R, reactor; S, separation; Cr, crystallization; W, filter/wash; D, dilution tank; E, extruder; MD, mold. *Reproduced with permission from Mascia, S.; Heider, P. L.; Zhang, H.; Lakerveld, R.; Benyahia, B.; Barton, P. I.; Braatz, R. D.; Cooney, C. L.; Evans, J. M. B.; Jamison, T. F.; Jensen, K. F.; Myerson, A. F.; Trout, B. L.* Angew. Chem. Int. Ed. ***2013***, 52, *12359–12363. Copyright 2013 Wiley-VCH.*

hydrogenation step, and Module 5 is the saponification reaction. This example represents the first proof of concept demonstrating how the production of small-molecule drugs may be customized in the future covering a variety of molecular frameworks. For instance, combining modules 1, 2, 4, and 5 β-amino acids were obtained; the combination 1, 2, 3, and 4 provided γ-lactams; and the combination 1, 2, 3, and 5 provided γ-amino acids.

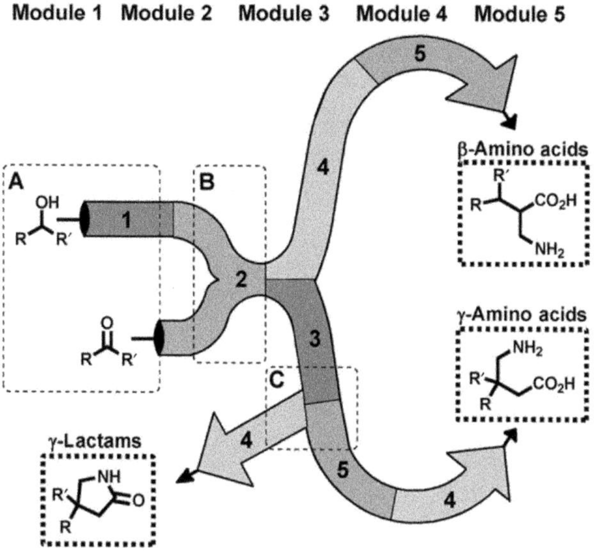

Fig. 3 Modular assembly system for the preparation of different APIs. *Reproduced with permission from Ghislieri, D.; Gilmore, K.; Seeberger, P. H.* Angew. Chem. Int. Ed. ***2015***, 54, *678–682. Copyright 2015 Wiley-VCH.*

3. APPLICATION OF MICROFLUIDICS TECHNOLOGY TO FORMULATE LEADS AND DEVELOPMENT CANDIDATES WITH POOR SOLUBILITY AND BIOAVAILABILITY

The solubility of APIs, development candidates and lead compounds is an important factor in the PK profiling of any leads. At the same time despite the widespread acceptance of guidelines related to desirable physicochemical properties of potential small-molecule drugs, key properties of advanced leads have been shown to have poorer PK profiles such as higher logP and logD, and poorer water solubility.[78] The low solubility of the compound is generally accepted to be one of the rate-limiting factors of oral absorption.[79]

To address the solubility problems of APIs, development candidates and lead compounds, scientists at Nangenex have developed and validated a proprietary, scalable, and fully integrated continuous-flow instrument technology based on microfluidics principles.[80] The so-called Super-API platform technology deploys analytics, instrumentation, and know-how to design, create, develop, and manufacture proprietary Super-API drugs (Fig. 4). The system provides the means to efficiently select potential pipeline

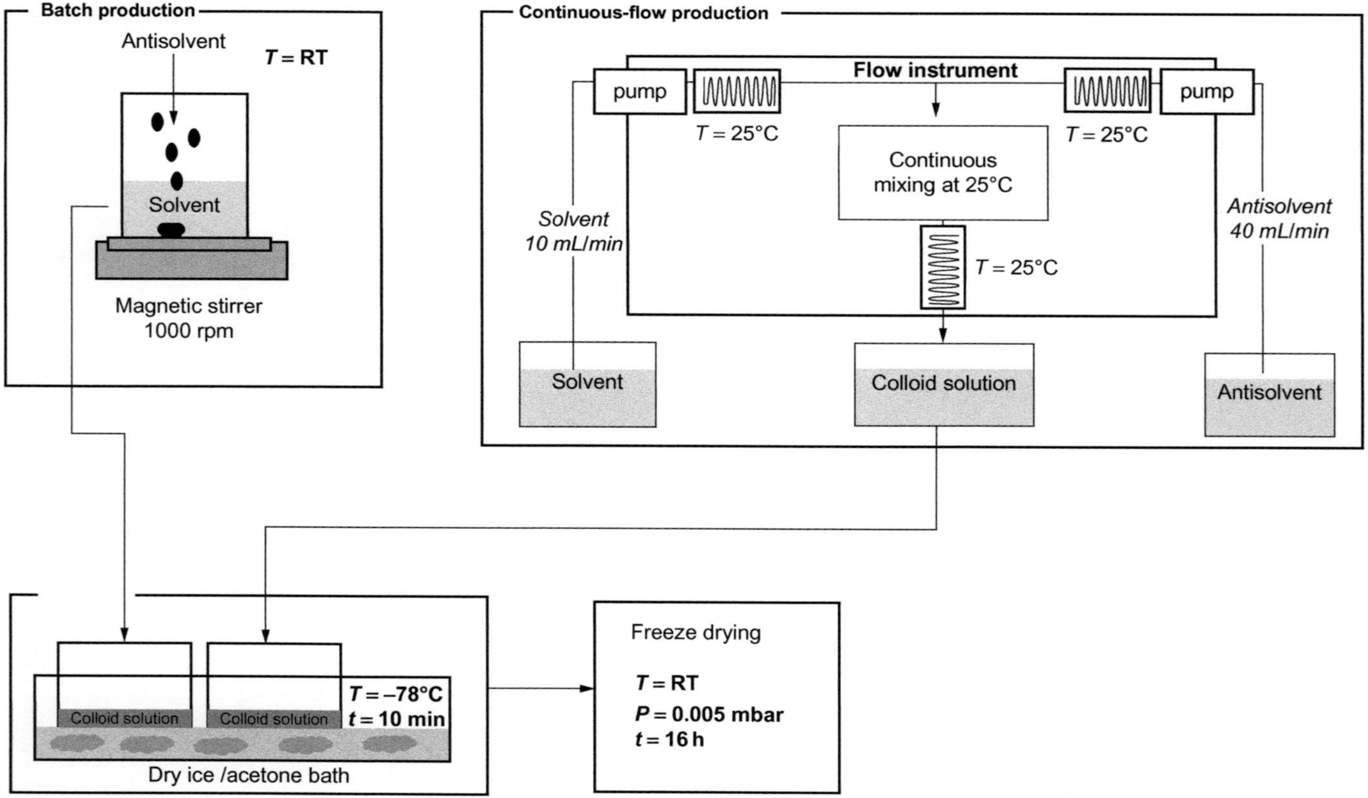

Fig. 4 The microfluidics-based Super-API platform developed and validated by Nangenex. *Reproduced with permission from Solymosi, T.; Angi, R.; Basa-Dénes, O.; Ránky, S.; Ötvös, Z.; Glavinas, H.; Filipcsei, G.; Heltovics, G.* Eur. J. Pharm. Biopharm. ***2015****, 94, 135–140. Copyright 2015 Elsevier.*

candidates. The developed HT screening and in vitro assays enable the screening of thousands of parameter combinations and predict their in vivo performance.

The Super-API drugs have unique supramolecular structures with improved clinical utility such as increased exposure, reduced or eliminated food effect, and changed administration route. The novel supramolecular structures deliver substantial improvements in material properties and biological performances: The microfluidics-based technology has been used successfully in medicinal chemistry programs to formulate otherwise active but poorly soluble leads and development candidates.[81] This methodology provides an advantage for medicinal chemists converting active and promising but otherwise poorly bioavailable lead compounds into soluble and bioavailable drug candidates.

Validation programs have been reported on different APIs. Nanostructuring of Sirolimus (Rapamune) resulted in faster absorption, higher exposure, and higher trough concentrations when compared to the marketed form. These advantageous properties could allow the development of solid oral *Sirolimus formulae* with lower strength and gel-based topical delivery systems.[82]

In vivo beagle dog pharmacokinetic studies with nanostructured Aprepitant (Emend) showed that the novel formula developed by microfluidics techniques exhibited greatly improved pharmacokinetic characteristics when compared to the reference compound.[83] The marked food effect observed for the reference compound was practically eliminated by this formulation method.

These results indicate that the novel continuous-flow precipitation technology is suitable to improve the PK properties especially oral bioavailability and solubility of otherwise poorly soluble compounds.

The high-throughput microfluidics technology may improve productivity significantly as the number of advanced leads to be dropped for poor PK properties may be reduced, and leads that otherwise would be eliminated can be advanced safely in the pipeline, reducing the cost and resources for the medicinal chemistry programs.

4. MICROFLUIDICS TECHNIQUES TO SUPPORT ADME STUDIES

One of the main reasons for failure of drug candidates is poor pharmacokinetics and drug metabolism (DMPK) characteristics.[84] So today these studies are crucial part of the medicinal chemistry lead optimization and selection programs. A good development candidate requires a well-thought balance of potency, safety, and PK; therefore, techniques that contribute to understanding these characteristics are important to enable researchers to design more robust candidates.

Significant research effort has been devoted to develop synthetic methodologies to model the metabolism of drug candidates. Besides well-established in vitro and in vivo methods using biological matrices, other biomimetic models have been developed. The methods applied today are (a) microsomal or tissue incubation, (b) porphyrin-catalyzed chemical oxidation, (c) Fenton-type reactions, and (d) electrochemical oxidations.[85]

Microfluidics-based methods have been researched and validated in a wide range of DMPK and metabolomics applications. The application of microchips to improve and increase the sensitivity of the analytical methods has been the focus of several academic and industrial groups a decade ago, and these applications have extensively been reviewed by others earlier.[86] Our review efforts will only focus on the latest practical developments on this field.

4.1 Microfluidics in Biological Media

Microfluidics-based continuous-flow enzyme reactors applying immobilized enzymes have been developed as practical tools to expedite and study biotransformations.[87] These methods have also been successfully adopted and validated to microsomal enzymes such as cytochrome P450. Nicoli et al. applied biotinylated recombinant CYP2D6 or CYP3A4. The reconstituted systems were anchored to the surface of two monolithic mini-columns (2 mm × 6 mm I.D.), which had been covalently grafted with NeutrAvidin.[88]

van Midwoud developed a microdevice to perfuse precision-cut liver slices (PCLS) under flow conditions.[89] PCLS were embedded on a poly(dimethylsiloxane) device containing 25-μL microchambers that was coupled to a perfusion system, to enable a constant delivery of nutrients and oxygen and a continuous removal of waste products (Fig. 5). High metabolite detection sensitivity could be achieved both for Phase I and Phase II metabolism for 7-ethoxycoumarin (7-EC) as test compound. The results correlated well with earlier reported well plate-based results.[90]

The applicability of this system was also validated for precision-cut intestinal slices, and for the sequential perifusion of rat intestinal and liver slices to

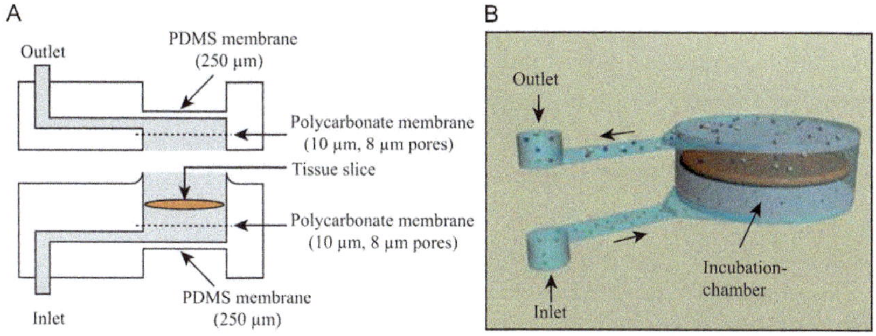

Fig. 5 Microfluidic system developed by van Midwoud for in vitro assessment of interorgan interactions in drug metabolism using intestinal and liver slices. *Reproduced with permission from van Midwoud, P. M.; Merema, M. T.; Verpoorte, E.; Groothuis, G. M.* Lab Chip ***2010***, 10 *(20), 2778–2786. Copyright 2010 Royal Society of Chemistry.*

mimic the in vivo first pass situation.[91] It has been shown that this system was very powerful to study the interorgan interactions, and this microfluidics system holds the potential to elucidate as yet unknown mechanisms involved in toxicity, gene regulation, and drug–drug interactions. The technology was later validated for human tissue slices and holds great potential to support the ADME studies of advanced leads.[92]

4.2 Electrosynthetic Methods

To overcome the disadvantages of using biological matrices, other biomimetic methods have been developed to model the enzymatic drug metabolism. These nonenzymatic strategies include metalloporphyrins as surrogates of the active center of cytochrome P450, Fenton's reagent, and the electrochemical oxidation of drug compounds. Obviously the systems cannot model the whole range of cytochrome P450-catalyzed reactions adequately; however, in certain aspects they show some advantages over standard in vitro methods.[93] The electrochemical-based metabolite generation technique tethered to liquid chromatography/mass spectrometry has been applied successfully for studying reactive metabolites and can be amenable in automated high-throughput screening approaches. Detailed comparisons with cytochrome P450 catalysis have been published, and the advantages and disadvantages of the respective methods were analyzed by Lohmann and Karst elsewhere.[93]

While electrosynthetic oxidations have been extensively performed in batch mode,[94,95] microfluidics methods provide further advantages such as increased electrode surface-to-volume ratios and shortened distance between the working electrode and the counter electrode.[12] This can help overcome conversion limitations observed in batch mode due to solution resistivity. It can even offer electrolyte-free reactions.[96] Microfluidics techniques have also been adopted to drug metabolism studies by electrosynthetic oxidations and have also been coupled in line with electrochemical/mass spectrometry (EC-MS) systems to generate Phase I and Phase II metabolites.[97,98]

These combinations result in powerful analytical methods to model and study drug metabolism and proteomics. The advantages of these methods for the metabolomics, proteomics, and biomarker discovery as well as their practical limitations have also been reviewed earlier by Permentier et al.[98]

At the same time the analytical nature of this technology provides limitations for the structure elucidation of the metabolites, since typically the product scale is too low to purify and isolate products for NMR characterization. Preparative amounts of fully characterized pure drug metabolites, however, can play a significant role in assays and improve the drug development and candidate selection process.

Stalder and Roth have further developed this method and reported the validation of a preparative scale continuous-flow electrosynthesis technique for the production of Phase I metabolites of several commercial drugs including diclofenac, tolbutamide, primidone, albendazole, and chlorpromazine.[99] The method was adopted to a commercially available continuous-flow electrochemical system[100] (Fig. 6), and a wide range of chemical reactivity, functional groups, and solubility were investigated. Several oxidative chemical transformations were validated including aliphatic oxidation, aromatic hydroxylation, S-oxidation, N-oxidation, or dehydrogenation. They successfully synthesized such metabolites at a preparative scale and demonstrated applicability of this flow electrolysis method to make and isolate 10–100 mg compounds. The purified product structures were fully elucidated by NMR and other analytical methods.

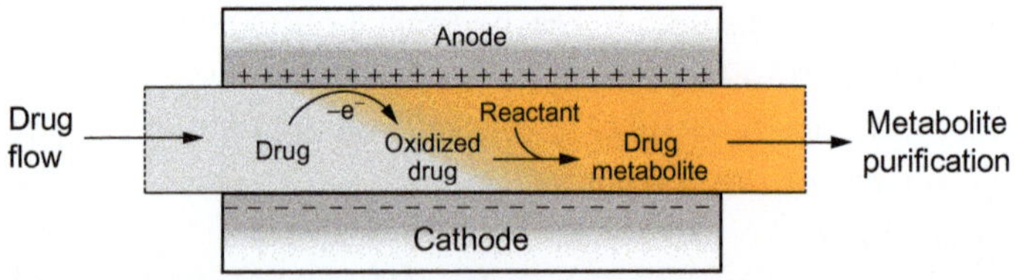

Fig. 6 The continuous process electrochemical synthesis system developed by Stalder and Roth. *Reproduced with permission from Stalder, R.; Roth, G. P.* ACS Med. Chem. Lett. ***2013****, 4 (11), 1119–1123. Copyright 2013 American Chemical Society.*

In addition to Phase I oxidation Phase II glutathione (GSH) adducts were also shown to be synthetically accessible and isolable. The reaction rate and the capacity were also in line with synthetic methods and were between 33 and 100 mg/h. This was several orders of magnitude higher than that of earlier reported electroanalytical techniques.

Electrosynthesis will not replace analytical biosynthetic techniques such as microsomal incubations for complete in vitro drug metabolism pathway studies. However, medicinal chemistry programs can benefit from the ability to create potential oxidative metabolites at a preparative scale that can be deciphered in further SAR as well as liability studies of expected metabolic transformations. The adaptation of the microfluidics methodology made this technology widely available for medicinal chemistry by overcoming limitations in batch-based methods. As a result, these technologies are expected to be more widely adopted in the lead optimization and selection of phase of pharma discovery and development programs.

5. INTEGRATED MICROFLUIDICS-BASED SYNTHESIS AND SCREENING

The pharmaceutical industry's research groups in early drug discovery are facing a challenging task of delivering new development candidates at a reasonable cost, while the complexity of optimizing leads increases significantly. Medicinal chemists need to identify new patentable chemical entities, with high activity, selectivity, increased drug likeness, and good PK profile, while they constantly need to eliminate liability factors as early as possible. Medicinal chemistry programs consist of SAR cycle that requires compound design, synthesis, purification, chemical structure confirmation, biological evaluation, and data analysis and feedback. This complex workflow typically distributed across highly specialized facilities must contain time-consuming compound logistics/management, and transfer steps. Typically biological data on a new chemical structure are generated after more than a week quite often several weeks after the completion of the synthesis and identification. The development of parallel techniques provided some advantage, but it is still cost intensive and still cannot reach real-time feedback loop. In addition it results in a consequential generation of a large number of irrelevant chemical examples. Time is also of essence; since parallel discovery programs get reprioritized regularly in big pharma companies, new programs evolve and ongoing programs get lower priority or eliminated regularly. Therefore, the industry is constantly looking for novel solutions to boost productivity in delivering viable candidates, with reduced resources and reduced cost and more importantly reduced time. Microfluidics solutions especially the combination of them with screening methods, to reduce optimization cycles, and thus improving throughput has been in the forefront to achieve these objectives. The potential for using a fraction of the reagents and chemicals, the short reaction time, the fast optimization cycle time, the potential for complete automation of the process, and the avoidance of storage materials made this technology attractive to revolutionize drug discovery in several companies. In most cases the crucial element of intelligent computational methods to feedback and design the next round of SAR loop was also developed and implemented in the feedback loop to further reduce human interaction.

This automated concept was first reported by colleagues at GSK who used a single channel microchip-based technology. By utilizing hydrodynamic flow control they created a 7×3 library of pyrazoles. Chemistry was adopted from published batch-based Knorr chemistry procedures.[101] This work demonstrated the capabilities of an automated microfluidic microreactor-based system to synthesize and analyze online multiple analogue reactions. The group set the future objectives of the project to develop multistep synthesis on microchannels, to link a microreactor system to a screening device as well. In addition the incorporation of polymer-supported reagents was also forecasted.

This integration of synthesis purification and screening was also conceptually predicted by other authors.[102] Biological screening of small molecules was foreseen to dramatically reduce the cycle time in the SAR loop between synthesis and assay, resulting in "real-time" design and/or optimization of the next generation of chemical candidates.

The "real-time" biological data generation concept to drive the chemical design and optimization program was also on the focus of many other pharma and academic research groups and has been researched extensively.[103–109]

The experimental bottlenecks such as rapid synthesis under controlled conditions followed by almost immediate measurement of biological response may be solved by a microfluidic-based chemistry and biology platform.[110] It can also offer the autonomous operation that eliminates all the bottlenecks and gets close to the ideal real-time biology-driven optimization. However, even if the experimental setup optimized, it is essential that an appropriate software solution to provide automatic feedback on the SAR loops is developed in order to reach the autonomous 24/7 operation.[25]

Scientists at GSK have successfully implemented the individual components of such microfluidics systems.[111,112] Their pioneering work has resulted in several patents around the functional components of the microfluidic platform for drug discovery, constantly taking advantage of reducing the gap between screening and synthesis.[113] The result of their approach was a platform capable of the high-speed synthesis of diverse single compounds coupled to a microfluidic bioassay which provides biological information in "real time" and allows intelligent "on-the-fly" decision making.[114,115] GSK has completed the platform development by integrating all components into their proprietary Automated Lead Optimization Equipment (ALOE) Platform (see Fig. 7).[116] It comprises three key components: the microfluidic chemistry platform to optimize and perform reactions in microfluidic system, the bioassay microfluidic platform to screen the newly synthesized compounds, and the intelligent algorithm (for instance genetic algorithm, GA) that builds the model from the bioassay data and guides the operator in the choice of the next set of reagents to use to refine the model. This technically integrated and iterative process can accelerate lead optimization. In typical medicinal chemistry project setups and in conventional drug discovery these three components are usually separate processes.

GSK validated this system on three different biological targets, by producing amides with known biological activities (Table 2). The targets were: apoptosis inducement in U-937 (lymphoma) cancer cells for compounds **74, 75** (Table 2); stimulatory effects on insulin secretion in vitro for compounds **76, 77**; and compound **78**, capsaicin (Table 2), has proven a potent radical scavenging activity.

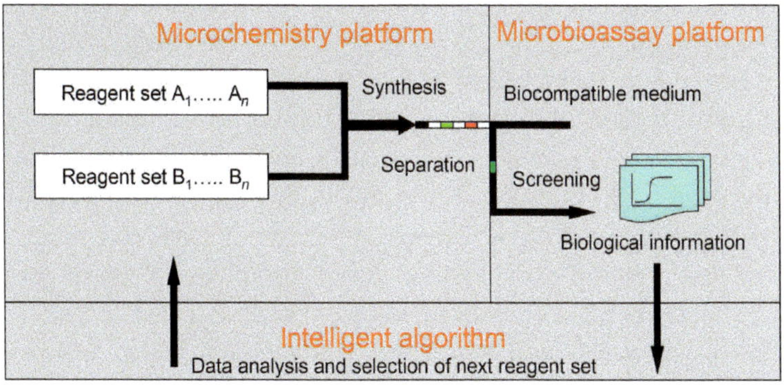

Fig. 7 The components of the ALOE platform developed by GSK. *Reproduced with permission from Fernandez Suarez, M.; Garcia-Egido, E.; Montembault, M.; Chapela, M. J.; Wong-Hawkes, S. Y. F.* Proc. ICNMM **2006**, *997–1002. Copyright 2006 ASME.*

Table 2 Compounds Generated to Validate the ALOE Platform

Acid	Amine	Product	Conversion (%)
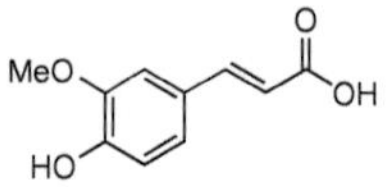	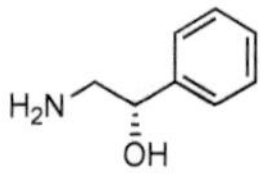	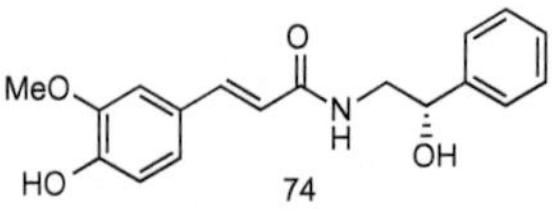 74	72
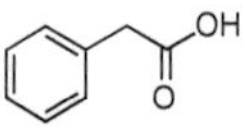	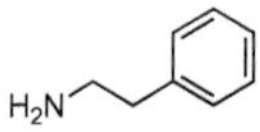	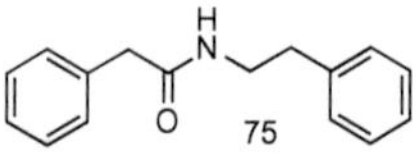 75	94
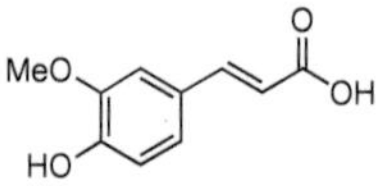	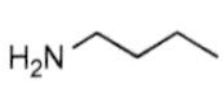	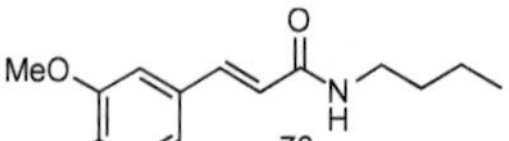 76	56
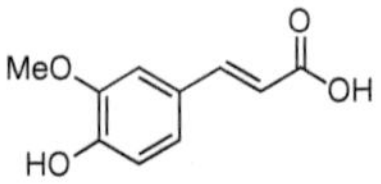	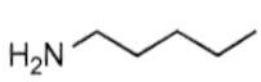	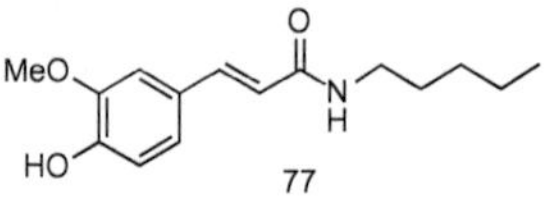 77	68
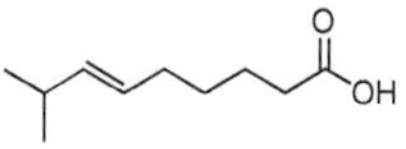	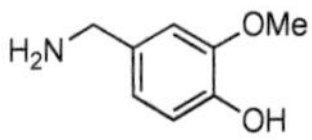	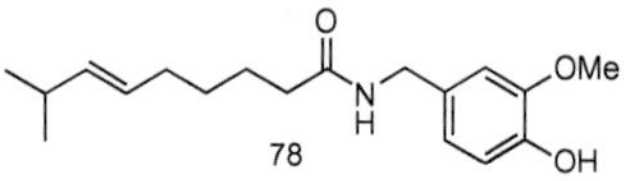 78	80

After successfully implementing the ALOE platform, scientists at GSK have made several additional attempts to drive traditional lead generation or lead optimization projects toward full automation in order to reduce make/test cycle for chemistries that are more complex and non-straightforward as well. The key component that needed to be optimized was the guiding software to provide the autonomous intelligent operations. Additional GAs were implemented, and the applicability of this integrated approach on different targets such as MMP-12[25] and a T-cell tyrosine phosphatase was demonstrated. Chemistries were performed on NS110 Caliper chip. Target sulfonamides were synthesized by standard coupling, while methyl esters were achieved by SN_2 transesterification in methanol.

It is important to note, however, that this optimization experiment was conducted on a single parameter such as the activity against the selected target, the MMP-12 inhibition. It remains to be proven that the underlying methods may be used more generally in the multiparameter optimization strategy that is typical of today's integrated medicinal chemistry programs.

Cresset Therapeutics has invented an automated iterative drug discovery process that is also building on these guiding principles. Their de novo design method is uniquely able to cope with multiple (and biologically active) products arising from the same reaction and is able to process data when compounds of unknown structure are generated by relating the reagents and reaction conditions to the pharmacophore.[117]

Another interesting chemistry-focused approach for integrated discovery was presented by researchers at UCLA in 2006.[118] They have designed a flow reactor that was capable of automating in situ click chemistry. The reactor was coupled with a parallel microfluidic reactor screening device and using bovine carbonic anhydrase II as target. Thirty-two compounds were synthesized and screened in 0.5 h.

The technology was further developed, multiplied, and scaled up later to be able to synthesize 1024 compounds.[119] Their device also integrated a solid-phase extraction step for purification and an electrospray ionization mass spectrometry device for analysis. As this technology was focusing on one chemistry, their wider application in industrial medicinal chemistry was limited.

Other online synthesis and screening of drug-like compounds were presented by Kool et al.[120] They used an online bioaffinity analysis to screen a library of fragments for activity against acetylcholine binding protein. Guetzoyan et al. applied frontal affinity chromatography to integrate biological screening and synthesis. A series of imidazo[1,2-*a*]pyridines, including zolpidem and alpidem, was synthesized in flow mode, and connected to a frontal affinity chromatography screening assay to investigate their interaction with human serum albumin.[121]

Developing on their experience from GSK, Pfizer, and other pharma and academic research, scientists at Cyclofluidic have developed and validated a microfluidics-based integrated drug discovery platform CyclOps™ in order to revolutionize hit and lead optimization.[122] The CyclOps™ platform integrates flow chemistry, purification, screening, and drug design with complex algorithms. The outcome provides a unique performance in many parameters important in lead optimization. Drug lead molecules are assayed minutes, rather than weeks, after they are designed and synthesized. By intelligently applying the power of microfluidics chemistry technologies that was developed by Steve Ley at Cambridge University, the CyclOps™ platform puts synthesis, screening, and molecular design "under one roof" in a single laboratory. An integral part of the platform is the activity prediction using random-forest regression with chemical space sampling algorithms. This provides a self-refining SAR model that adjusts itself after every synthetic iteration and biological assay cycle.

This allows fast iterative exploration of vast areas of diverse chemical space. The platform may revolutionize hit-to-lead discovery since this can optimize hits in real time, making and screening only those key compounds that progress a hypothesis toward the best possible lead molecules. It is important to note, however, that the platform uses well plate-based bioassays, so only the synthesis is based on microfluidics technologies. The validity of the platform was demonstrated on two different targets in two reports, an Ab1 kinase inhibitor[123] and Xanthine-derived dipeptidyl peptidase 4 (DPP4).[124]

The SAR modeling for the discovery of Ab1 kinase was demonstrated on compounds showed in Fig. 8. After 21 iterative screening cycles in 24 h, and using automated Sonogashira chemistry the platform identified novel templates and hinge-binding motif with pIC_{50} >8 against both wild-type and clinically relevant mutants of Abl kinase.

The Xanthine-derived DPP4 antagonists program was another milestone validation for the platform as it used purely flow-based biological screening for determining IC_{50} values. The results obtained by the platform was in full correlation with reference data provided by using traditional medicinal chemistry approach. The integrated biological data can also be obtained from after low-yielding reactions, so the optimization time can be minimized. This is an additional advantage as the flow synthesis method does not need extensive optimization for less reactive reagents. Each compound was synthesized and tested in less than 2 h and the total time to make and test the 29 compounds took less than 3 days.

Recently, researchers at F. Hoffmann–La Roche reported the design and validation of a system (Fig. 9) that addresses all the key experimental bottlenecks of medicinal chemistry and lead optimization/generation. Those areas are compound logistics and management, transfer times between synthesis, and biological testing.[125] Amidation reactions conducted on a Vapourtec R4 flow synthesizer subsequently coupled to in-line preparative HPLC purification platform, compound identification platform by LC–MS, and quantification using a calibrated evaporative light-scattering detector. Beta secretase 1 (BACE1) was utilized as the target and the bioassay setup was subsequently coupled to the platform resulting in a seamless integration with a screening assay.

The complete synthesis–purification–assay cycle took only 60 min per compound, with the biochemical assay taking over 30 min on a glass chip, thus proceeding at the same timescale. As far as the scientific literature scanned, so far we have not found any other platform that could generate SAR data for identification and optimization of lead compounds faster than this platform. In addition the bioassay was capable of measuring both very potent and inactive compounds accurately.

The next generation of fully integrated screening and chemistry platform was very recently revealed by scientists at Bactevo (www.bactevo.com).

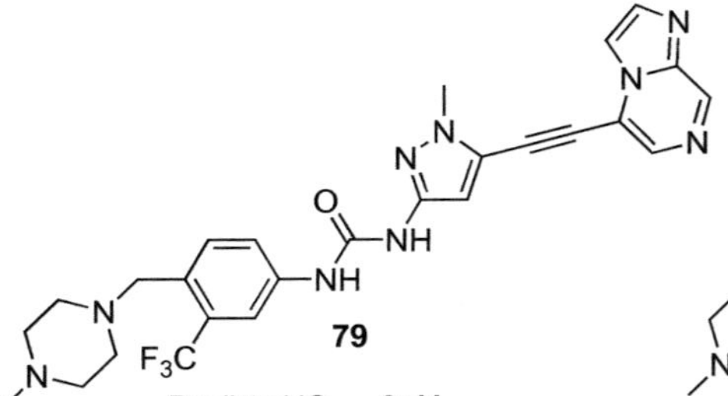

79

Predicted $IC_{50} = 3\,nM$
Experimental $IC_{50} = 2\,nM$

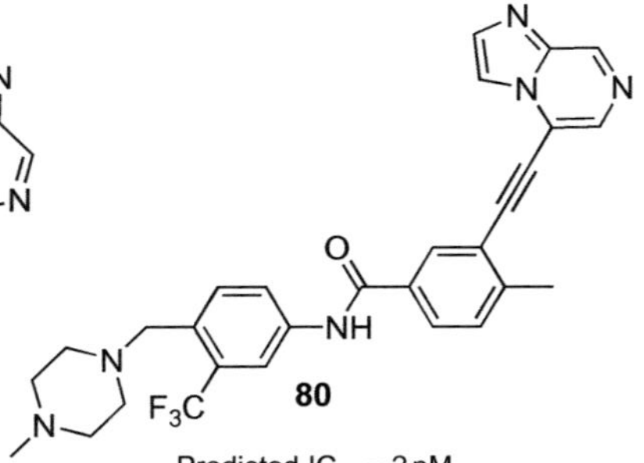

80

Predicted $IC_{50} = 2\,nM$
Experimental $IC_{50} = 0.2\,nM$

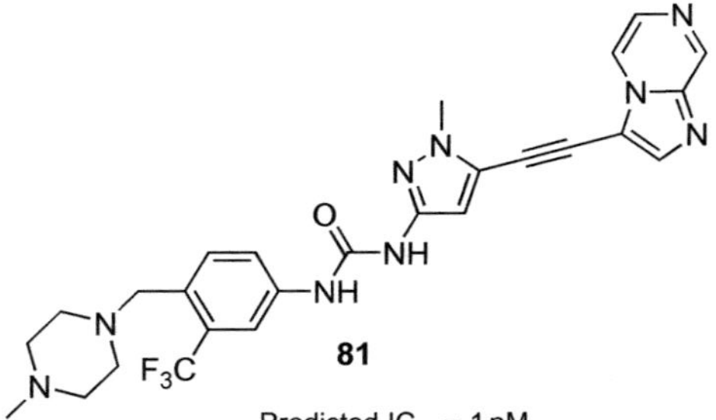

81

Predicted $IC_{50} = 1\,nM$
Experimental $IC_{50} = 0.4\,nM$

Fig. 8 The most active compounds after SAR modeling for the discovery of Ab1 kinase.[123]

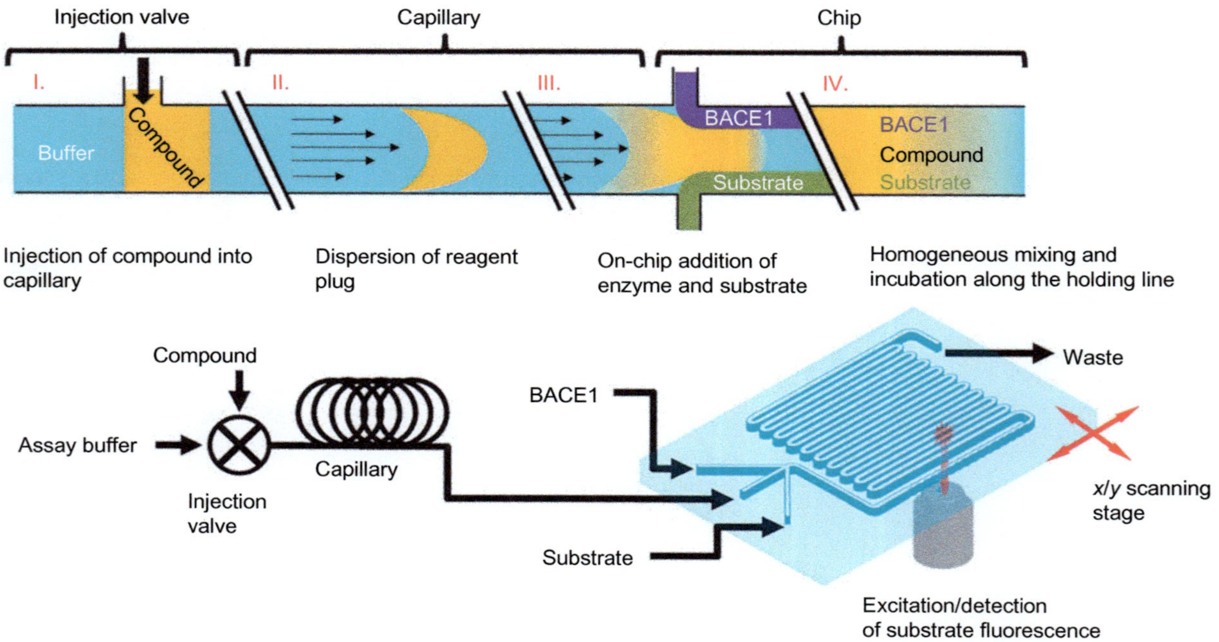

Fig. 9 Schematic of the flow-based assay procedure and outline of the employed microfluidic setup. *Reproduced with permission from Werner, M.; Kuratli, C.; Martin, R. E.; Hochstrasser, R.; Wechsler, D.; Enderle, T.; Alanine, A. I.; Vogel, H.* Angew. Chem., Int. Ed. ***2014***, 53 *(6), 1704–1708. Copyright 2014 Wiley-VCH.*

Bactevo's Totally Integrated Medicines Engine (TIME) platform based on microfluidics droplet technology can synthesize and assay 100 million novel compounds per hour against relevant patient cells, iPSC's as well as molecular target assays. By multiplexing phenotypic assays such as activity, in vitro safety and ADMET can be achieved simultaneously. The huge data volume generated for each compound enables a massive acceleration in medicinal chemistry leading to therapeutic compounds in less than a year. Results of platform validation experiments were not yet disclosed at the time of writing.

Although several attempts have been made at different pharmaceutical companies, and several integrated microfluidics-based synthesis, purification, and bioassay platforms have been validated that hold the potential for significant increase in productivity improvement, it is important to note that so far we were not able to identify in the literature any clinical candidate that was the direct or indirect results of the implementation of these integrated platforms. These platforms although hold lots of potential, still have limitations that needs to be resolved. For chemical synthesis only a handful of reactions have been optimized for microreactor chemistry and there are still some reagents that are not compatible with microfluidics-based setup. As far as the biological assay concerned, only a limited number of target-based bioassays have been adopted. Today's medicinal chemistry programs require multiparameter optimizations, and the application of the widest arsenal of chemical transformations, that is accessible so far for only with traditional batch-based chemistry. In addition there are only a limited number of experts in the flow-based chemistry since the training of chemists is based on traditional batch-based methods. Until these issues are sorted out, medicinal chemistry programs will very likely rely on traditional methods of compound synthesis, compound logistics, and transfer between synthesis and assay, resulting in a long feedback loop and inefficient optimization cycles.

6. MICROFLUIDICS CHEMISTRY IN HIT-TO-LEAD AND LEAD OPTIMIZATION

Hit-to-lead process and lead optimization aim to refine each hit series to obtain confirmed SAR while producing more potent and selective compounds with improved PK properties. This effort requires to generate and test many series of compounds, including novel scaffolds and compounds with defined substitution patterns. The commercial availability of flow reactors particularly with supported reagents or catalysts opened novel avenues for medicinal chemists; therefore, flow chemistry approaches were gradually implemented to the traditional medicinal chemistry practices. The vast majority of the applications in this field are heterogeneous flow hydrogenation.[46] The other applications are other heterogeneous catalytic reactions (such as cross-coupling reactions) and other dangerous transformations (ozonolysis, oxidation, Curtius reaction, etc.). In general, flow chemistry allows fast generation of analogues libraries in small quantity sufficient for in vitro biological screening. Further advantage is the use of increased parameter space which allows to carry out reactions, which require harsh conditions.

6.1 Continuous-Flow Hydrogenation

The advantages of flow hydrogenation compared with the batch process include: increased safety (particularly if hydrogen is in situ generated); improved yield and selectivity; easy control of the reaction parameters; and flow hydrogenation often leads to less side products.[126] Furthermore, flow hydrogenation at high pressure and temperature allows full saturation of (hetero)aromatic ring.

One of the commercial reactors (H-Cube®; ThalesNano)[127] gained much popularity due to the safe, in situ generation of H_2 gas through water electrolysis sparing the use of gas cylinder and placing the supported catalyst into a disposable cartridge.[128]

Flow hydrogenation could be applied in many different reaction pathways and other transformation could also be practically replaced by hydrogenation. Often multiple hydrogenations could be carried out in a one-flow process (e.g., nitro group reduction plus double bond saturation or aromatic ring saturation and dehalogenation[129]). Furthermore, nearly one fourth of the registered drugs require at least one hydrogenation step in their manufacturing process.

Nearly 100 papers refer flow hydrogenation applications that complement the traditional batch medicinal chemistry efforts. The applications could be divided into four major reaction classes[130]:

1. Functional group transformations leading to amines from azides, nitriles, and nitro groups. This is a particularly attractive approach since many nitro and nitrile precursors are commercially available.
2. Benzyl type of protecting group manipulation (*O*-benzyl, *N*-benzyl, Cbz removal). Flow hydrogenation also allows selective or sequential deprotection.[131]

3. Reduction of unsaturated bonds (from olefin, alkynes to olefines and alkanes) including reductive aminations. This group is particularly useful in many synthetic strategies since numerous cross-coupling (Suzuki, Sonogashira, etc.), condensations (aldol, Knoevenagel, etc.), chain elongation (e.g., Wittig), and metathesis reactions lead to double or triple bonds that can be saturated to double or single bonds. An important subclass of this section is the full saturation of aromatic or partially saturated rings toward sp^3-rich systems, which is a popular design trend in the last decade.[132]
4. Dehalogenation reactions from aromatic rings, in some cases halogens block or activate certain positions; therefore, temporary introduction would be beneficial in many synthetic pathways.

The following examples demonstrate the diverse utility of flow hydrogenation in the hit-to-lead and lead optimization efforts. There are various scenarios of flow hydrogenations:

1. It plays a key role in the construction of the active moieties or scaffolds.
2. It takes part in the intermediate synthesis including benzyl type of protecting group manipulations or functional group transformations (e.g., nitro group reduction).
3. It is involved in the synthesis of one particular series of compounds during H2L or LO. These distinct series of compounds could either lead to more potent and bioavailable compounds or to lesser active or inactive compounds; however, the biological data are highly important for SAR determination.

At least one flow synthesis step is involved in the synthesis of the skeleton or important structural motifs of the biological active chemical entities. Darout from Pfizer[133] described the synthesis of a series of GPR119 (G protein-coupled receptor 119) agonists based on a diazatricyclodecane ring system. During the assembly of the ring system one-flow benzyl group removal and Boc protection was carried out in quantitative yield (Scheme 21) using

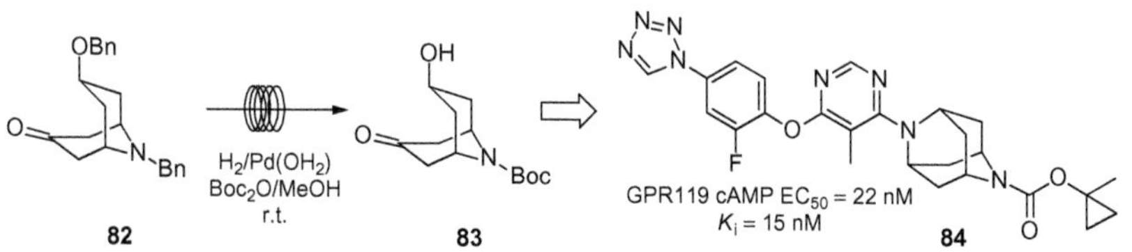

Scheme 21 One-flow benzyl group removal and Boc protection.

Boc_2O in MeOH and catalytic hydrogenation through a $Pd(OH)_2$ cartridge at 70 bar and 70°C for 1 h.

Optimization of the carbamate analogues of the diazatricylic compounds led to the identification of a potent agonist of the GPR119 receptor (**84**, GPCR119 cAMP $EC_{50}=22$ nM; $K_i=15$ nM).

Janssen scientists[134] synthesized several 7-(phenylpiperidinyl)-1,-2,4-triazolo[4,3-*a*]pyridines as potential positive allosteric modulators (PAMs) of the mGlu receptor 2 (mGluR2). These novel compounds were radiolabeled with carbon-11 and evaluated as potential PET radioligands for in vivo imaging of the mGluR2 allosteric binding site. Based on its potency and selectivity they selected the most promising PET radioligand candidate (**89**) (Scheme 22, right).

The arylpiperidine motif (**88**) of the active compound was synthesized through a Suzuki coupling/double bond saturation sequence. The reduction of the double bond was carried out under flow conditions using $Pd(OH)_2/C$ at 80°C and 1 mL/min flow rate.

In a medicinal chemistry program developing novel protein phosphatase 1 and 2A and dynamin GTPase inhibitors Tarleton and McCluskey[135] synthesized a bis-aromatic amine library (**93** deriv.) as precursors to attach to dehydronorcantharidin (**94**), which is a natural product-related antitumor drug inhibiting protein phosphatase type-2A. The amine library was generated in two steps: first, Knoevenagel condensation between pyrrole-2-carboxaldehyde (**90**) and substituted benzyl nitriles (**91** deriv.) yielded α,β-unsaturated nitriles (**92** deriv.), and flow hydrogenation converted the nitrile to amine and saturated the double bond in a "one-flow" process (Scheme 23). The saturated amines were obtained in >99% yield using Raney Nickel catalyst cartridge at 70°C, 70 bar H_2 pressure, and 0.5–1.0 mL/min flow rates.

The bis-aromatic amine library members (**93** deriv.) were linked to 5,6-dehydronorcantharidin (**94**) under flow hydrogenation conditions (50°C, 50 bar H_2, 1.0 mL/min) including generation of the amide bond and double bond saturation (Scheme 24).

One of the ring-opened norcantharidin hybrid was found as a potent protein phosphatase 1 and 2A inhibitor (**95**).

Day et al.[136] investigated the phenomenon that some of the PDE4 inhibitors trigger intracellular aggregation of PDE4A4 into accretion foci through the association with ubiquitin-binding scaffold protein p62 (SQSTM1).

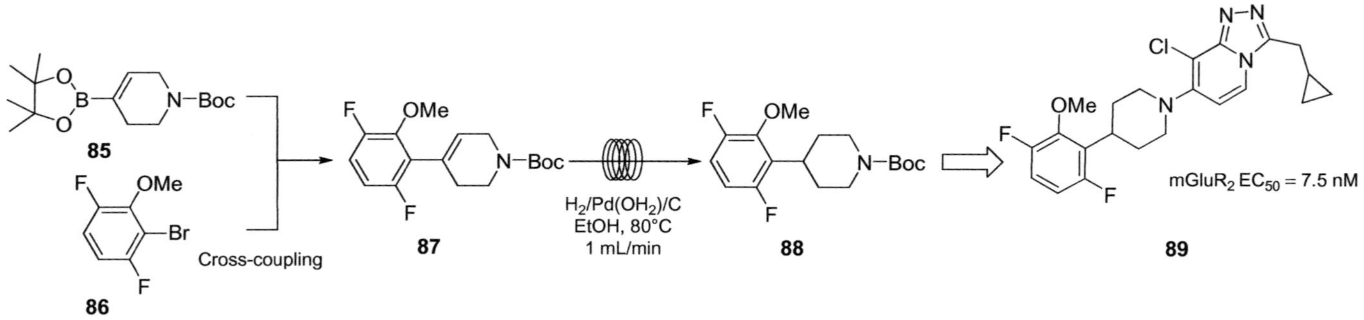

Scheme 22 Suzuki coupling/flow double bond saturation sequence toward a potential PET radioligand for in vivo imaging of the mGluR2 allosteric binding site.

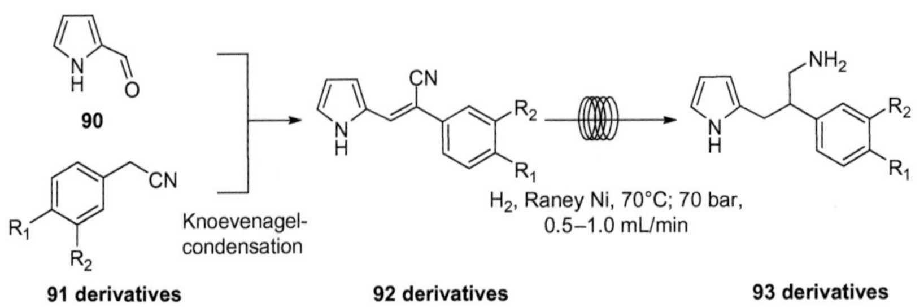

Scheme 23 "One-flow" process to convert nitriles to amines and saturate the double bond in the synthesis of a bis-aromatic amine library.

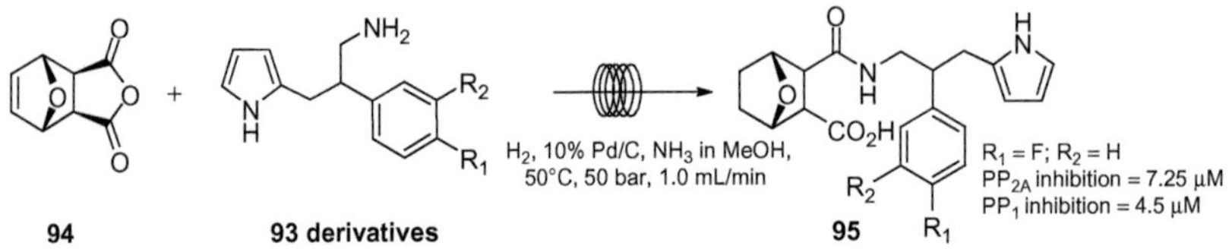

Scheme 24 "One-flow" process for amide bond formation and double bond saturation leading to norcantharidin hybrides.

During the study several rolipram analogues were synthesized. In one series the rolipram cyclopentyloxy group was replaced with simple 1–3 C alkyl groups.

The compounds were prepared in an optically pure form from diastereomeric amides derived from (*S*)-1-phenylethylamine. These separated compounds were hydrogenated under flow condition (H_2, 10% Pd/C, EtOH, 90 bar, 90°C) to convert the nitro group to amines together with the removal of the *O*-benzyl group, and the resulting intermediate was cyclized enantioselectively to the rolipram analogue precursor phenolic species (**97S**/**97R**, Scheme 25). Finally, the phenol hydroxyl group was alkylated with alkyl iodides leading the desired analogues (not shown).

The methoxy (*R*/*S*) group reduced the rolipram's PDE4-inhibitory activity by an order of magnitude, and none of the enantiomers had foci-inducing activity at 100 μM concentrations. Increasing the alkyl chain length from methoxy to ethoxy or propyloxy PDE4 inhibitory was increased as expected.

Researchers at Novartis[137] synthesized 8-hydroxyquinolinone 2-aminoindane compound series as β_2-adrenoceptor agonists (β_2-Adc) and evaluated their long-acting bronchodilator potency. The authors investigated the effect of lipophilicity by applying alkyl groups with increasing chain length linked to the aromatic ring of the indane core.

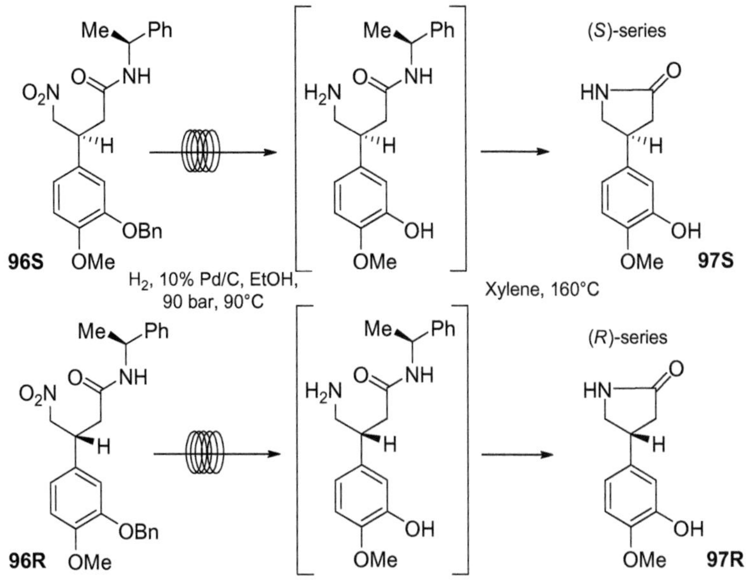

Scheme 25 "One-flow" nitro group reduction and removal of the *O*-benzyl group toward the synthesis of rolipram analogue precursors.

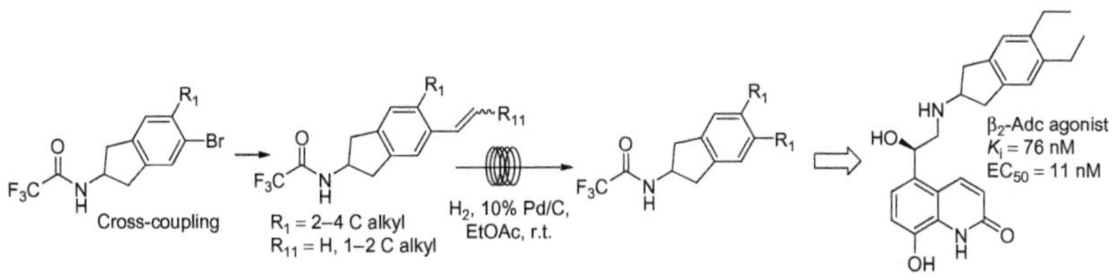

Scheme 26 Introduction of an alkyl substituent to the aromatic ring by using palladium-catalyzed cross-coupling reaction and double bond hydrogenation.

The sequential introduction of the 5,6-dialkyl groups was carried out in two steps: palladium-catalyzed cross-coupling reaction with the appropriate 1-alkenyl boronic acid was followed by double bond reduction (Scheme 26). Flow hydrogenation (1 atm H_2, 10% Pd/C in EtOAc at r.t.) resulted in the 5,6-dialkyl intermediates (**100** derivatives) in 90%–95% yield.

Based on the biological evaluation the 5,6-diethyl-substituted indan analogue (**101**) was found as the most active β_2-adrenoceptor agonist and exhibited a long intrinsic duration of its action profile.

Jerabek et al.[138] synthesized tacrine-resveratrol fused hybrids and investigated their multitargeted effects (human acetylcholinesterase;

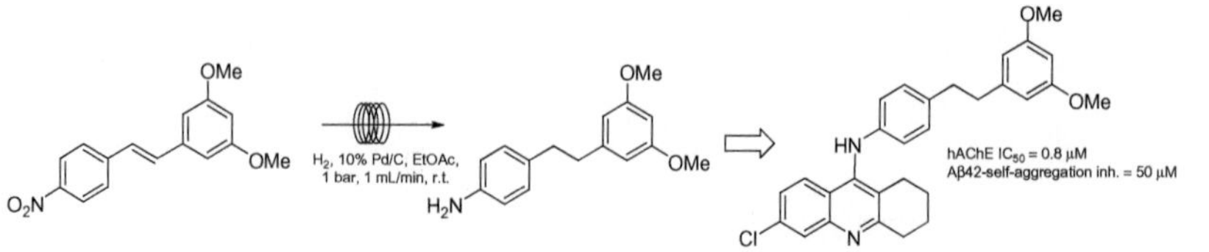

Scheme 27 "One-flow" hydrogenation process including nitro group reduction and double bond saturation leading to a resveratrol–tacrine hybrid precursor.

butyrylcholinesterase activity) against Alzheimer's disease. A small hybrid library was generated by varying the substituents at the tacrine unit together with saturation of the bridging double bond in resveratrol. The modified resveratrol precursor unit (**103**) was prepared in high yield by a "one-flow" hydrogenation process (10% Pd/C; 1 bar, r.t.) including nitro group reduction and double bond saturation (Scheme 27).

The highest activity hybrid compound (**104**) inhibited human acetylcholinesterase at low micromolar concentrations and effectively modulated Aβ self-aggregation in vitro.

AstraZeneca researchers[139] reported the discovery of a series of substituted 4H-chromen-4-one derivatives as PI3Kβ/δ inhibitors for the treatment of PTEN-deficient tumors. The authors intended to explore the chemical space around their Phase I clinical candidate AZD8186 (**105**), and based on molecular modeling they postulated that cyclization (preferably a five-membered ring) at the aniline moiety may restrict the conformational freedom and lock the molecule into the active conformation. The authors followed various synthetic strategies to realize the molecular targets. In one approach (Scheme 28, upper route) *N*-Boc-pyrrole was introduced by Suzuki coupling and selective flow hydrogenation (5% Rh/Al_2O_3, MeOH, H_2, 10 atm, 50°C, 3.5 h, then 10% Pd/C, MeOH, H_2, 60 atm, 60°C, 6 h) led to Boc-protected pyrrolidine (**107**). In an alternative procedure (Scheme 28, lower route) an *N*-Boc-2,3-dihydropyrrole derivative (**108**) was obtained by using Heck coupling and again flow hydrogenation (5% Pd/C, MeOH, H_2, 5 atm, 45°C, 3 h) resulted in the corresponding pyrrolidine (**109**).

The most active analogue (**110**) was found potent and selective: PI3Kβ cell IC_{50} 11 nM in PTEN null MDA-MB-468 cell and PI3Kδ cell IC_{50} 14 nM in Jeko-1 B-cells. This compound also showed in vivo excellent tumor growth inhibition in a mouse PTEN-null PC3 prostate tumor xenograft model.

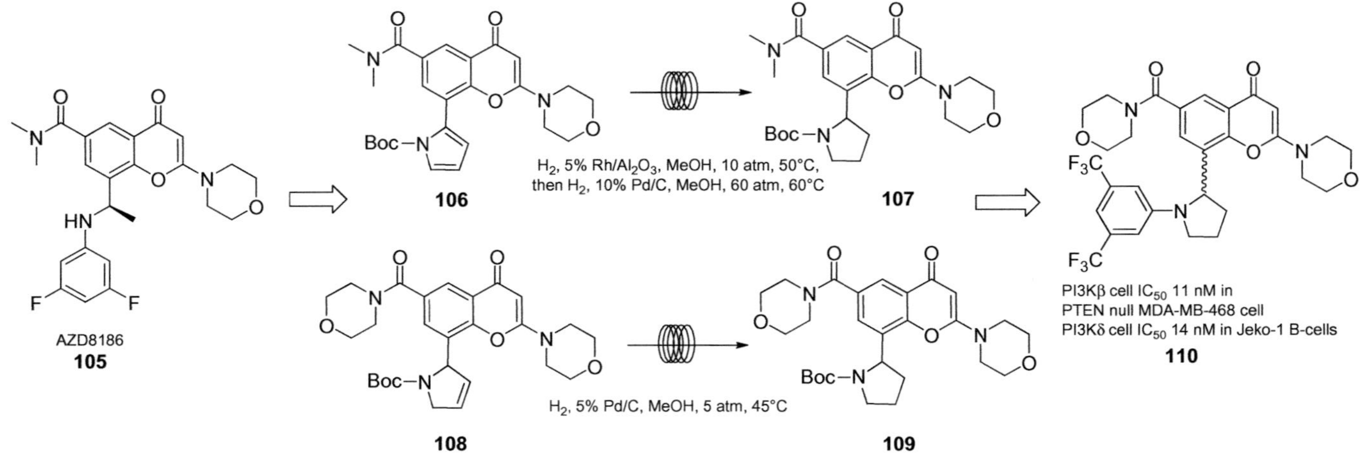

Scheme 28 Cross-coupling/flow hydrogenation routes the aromatic ring coupled pyrrolidines.

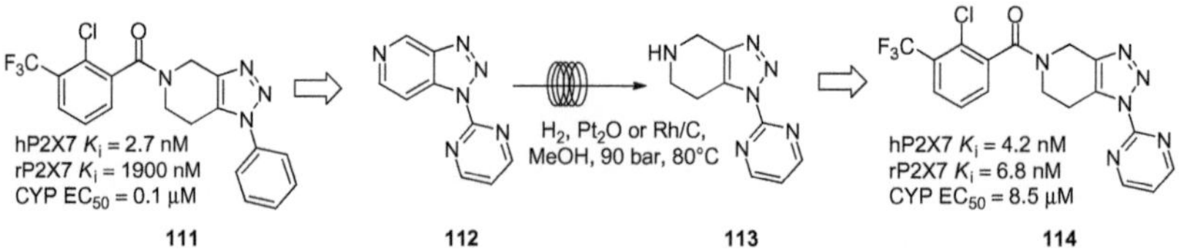

Scheme 29 Selective flow hydrogenation of heteroaromatic-substituted triazolopyridine ring system to the corresponding triazolopiperidines in a P2X7R antagonist lead optimization study.

Janssen scientists conducted[140] a lead optimization study of 1,2,3-triazolopiperidines as a novel series of potent purinoceptor 7 (P2X7) antagonists. The initial lead compound (**111**) had potent human P2X7R antagonists with weak rodent efficacy, CYP inhibition, poor metabolic stability, and low solubility. In order to improve the above parameters the researchers intended to explore other heterocyclic cores connected to the triazole ring. Flow hydrogenation allowed to selectively saturate the heteroaromatic-substituted triazolopyridine ring system (**112**) to the corresponding triazolopiperidines (**113**) (Scheme 29). The different heteroaromatic rings attached to the triazole moiety required extensive optimization of the catalysts and flow parameters, therefore, Pt_2O or Rh/C, 70–90 bar conditions were used at 80°C with careful reaction monitoring and product recycling.

As a result of the LO study by replacing the phenyl group with various heterocycles a novel compound (**114**) was identified that exhibited human and rat P2X7R antagonist activity with acceptable physicochemical properties.

Often during the hit-to-lead or lead optimization efforts flow transformations Patel et al. (Janssen)[141] reported the optimization and SAR determination efforts on an early dual leucine zipper kinase (DLK, MAP3K12) inhibitor compound (**115**).

The four key structural elements were successively replaced and optimized including the heteroarylamine moiety, the 2- and 6-pyrimidine core substituents, and the central core (changing the 2,4-diaminopyrimidine to a 2,6-diaminopyridine scaffold). Pyridine was used as the favored heteroarylamine moiety and its substitution was also explored. During the synthesis the piperidine precursor tetrahydropyridine ring was introduced by Suzuki–Miyaura cross-coupling, and the resulting double bond (in **116** derivatives) was saturated by flow hydrogenation at a later phase after the full decoration of the pyridine core (Scheme 30, leading to **117** derivatives). Further studies revealed

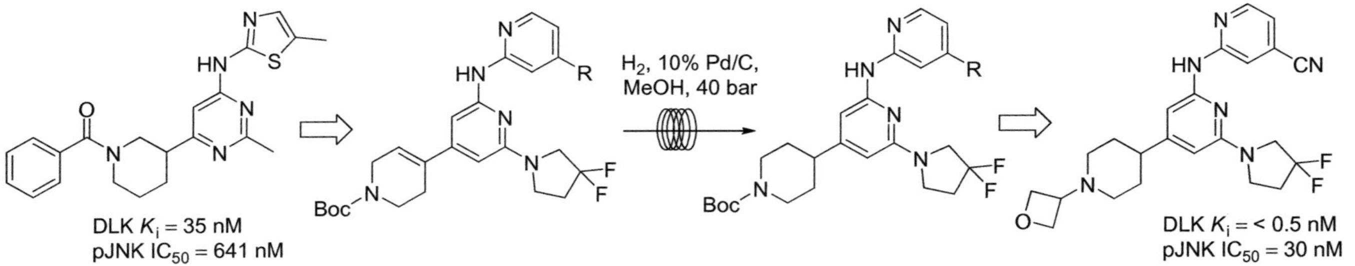

115 **116 derivatives** **117 derivatives** **118**

Scheme 30 Cross-coupling/flow hydrogenation sequence leading to aromatic ring coupled piperidines in a dual leucine zipper kinase (DLK, MAP3K12) inhibitor optimization study.

that cyano group is the most beneficial at the R position of the aminopyridine ring (**118**).

Martin-Martin[142] et al. (Janssen) reported the discovery and SAR analysis of novel series of imidazopyrimidinones and dihydroimidazopyrimidinones as positive allosteric modulators (PAMs) of metabotropic glutamate receptor 5 (mGlu5).

During this lead optimization effort based on the structure of an initial hit other 5,6-bicyclic systems (**119**) together with various types of substituents were explored. Dihydroimidazopyrimidinones were particularly promising targets for further investigations. The partially saturated bicyclic core system was prepared from the imidazopyrimidinone precursor by selective flow hydrogenation (H_2, Raney-Ni, 80°C, DMF/MeOH, Scheme 31).

The best compound (**122**) showed good in vitro potency and efficacy, acceptable drug metabolism, and pharmacokinetic properties as well as in vivo efficacy.

Flow hydrogenation is a particularly useful tool for saturating the double bond after peptide stapling by metathesis. Liu et al.[143] reported the syntheses and pharmacological evaluation of dimeric peptides as potent and selective neuropeptide Y Y4 receptor agonists. The alkyl bridge between the monomers was extended by 2-carbones using a metathesis, flow hydrogenation sequence.

The cross-metathesis reaction was carried out on peptide resins; Fmoc deprotection followed by cleavage from the resin yielded the alkenyl peptides that were hydrogenated in flow (10% Pd/C cartridge, H_2 (50 psi), EtOAc, 50°C, 1 h).

Often the synthesis of a particular, attractive side chain or substituent requires multistep, de novo synthesis. Bazin et al.[144] (GlaxoSmithKline) reported the synthesis and biological evaluation of a new series of 8-oxoadenines substituted at the 9-position with a 4-piperidinylalkyl moiety in toll-like receptor 7/8 (TLR7/8) activation. During the hit-to-lead exploration study a series of 4-piperidinylalkyl derivatives of 8-oxoadenines was

119 **120 derivatives** **121 derivatives** **122**

Scheme 31 Preparation of dihydroimidazopyrimidinones by partial saturation of the imidazopyrimidinone precursor by selective flow hydrogenation.

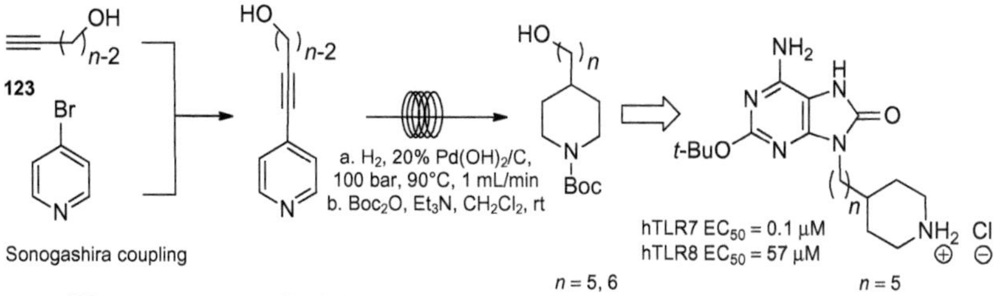

124 **125 derivatives** **126 derivatives** **127**

Scheme 32 Sonogashira coupling/flow hydrogenation sequence toward protected piperidinyl alkanols ($n=5$, 6) as precursors in the synthesis of toll-like receptor 7/8 (TLR7/8) activating 9-substituted 8-oxoadenines.

synthesized, where the length of the N9 alkyl chain was systematically varied. Since the two Boc-protected piperidinyl alcohols were not commercially available these intermediates were prepared in three steps from 4-bromopyridine (**124**) by Sonogashira coupling with acetylenic alcohols ($n=3$, 4, **123**) followed by flow hydrogenation of the alkynyl pyridines under relatively harsh conditions ($Pd(OH)_2/C$, 100 bar, 90°C) and *N*-Boc protection resulted in the protected piperidinyl alkanols ($n=5$, 6, **126**) (Scheme 32).

Among the synthesized 9-substituted 8-oxoadenines the 5-carbon-linked piperidinyl derivative (**127**) was found as the most active in TLR7/8 activation as determined by a reporter gene assay using HEK293 cells stably transfected with either hTLR7 or hTLR8 and the NFrB SEAP (secreted embryonic alkaline phosphatase) reporter.

In a similar approach Fiasella et al.[145] synthesized a series of aminoazetidin-2-one derivatives having long and flexible alkyl chains as potent as *N*-acylethanolamine acid amidase (NAAA) inhibitors. The long-chain alkanoic acid precursor having a terminal cyclohexyl moiety of one of the most active compounds ($IC_{50}=280$ nM) was synthesized in a Wittig reaction, flow hydrogenation (double bond saturation) sequence.

6.2 Multistep and Library Synthesis Under Flow Condition

There are few medicinal chemistry examples for multistep flow synthesis of analogue libraries included in biological evaluation.[146]

Ley et al.[147] developed a modified telescoped machine-assisted flow procedure (involving Uniqsis FlowSyn platform) to generate a small adamantane-based library as potential analgesic agents through the inhibition of P2X7-induced glutamate release.

The flow sequence started from an exocyclic methylene oxazoline derivative (**128**), which was cleaved by ozonolysis (Fig. 10). The excess ozone and the ozonides were quenched with immobilized thiourea. The resulting oxazolinone (**129**) was reacted with different nitrogen nucleophiles at high pressure and temperature to afford the azalactones. In vivo animal experiments (mouse abdominal constriction test) revealed that one compound (**130**, R = cyclopropyl) showed excellent antinociceptive effect at low dose of 3 mg/kg.

Petersen et al.[148] reported a three-step continuous-flow synthesis system to generate a new series of chemokine CCR8 receptor ligands from commercial building blocks in a hit-to-lead effort. The library was inspired by previously identified potent CCR8 ligands (**131** and **132**, Scheme 33).

In the first step 0.1 M solution of an amine and isocyanate in DMF was mixed at 0.15 mL/min flow rate through a 2-mL stainless steel tube reactor at 75°C. The excess isocyanate was scavenged by supported amine resins. The resulting Cbz-protected piperazine amide was deprotected by flow hydrogenation H-Cube (full H_2 mode, 80°C, 10% Pd/C cartridge). After removal of the excess gases the stream containing the deprotected piperazine was mixed with a solution of the alkylating agent (both at 0.30 mL/min) at 75°C. After *N*-methyl morpholin and trisamine scavenging the corresponding substituted piperazines were isolated in 80%–95% yield (Fig. 11).

This system was modified in a later phase which allowed to remove the scavenging steps without causing a decrease of the yield. Altogether 33 compounds were synthesized; the best CCR8 agonist (**133**) had an EC_{50} of 3 nM.

Rosatelli[149] et al. reported the synthesis of a secondary and tertiary aryl sulfonamide library under flow conditions (Fig. 12), and the resulting compounds were tested to selectively inhibit tumor-associated carbonic anhydrase isoforms IX and XII.

The library members were generated in a single step using a 10-mL PTFE coil reactor. The sulfonyl chloride solution (0.20 mmol) in acetone (2 mL), and the appropriate primary or secondary amine (0.22 mmol) solution together with $NaHCO_3$ (0.40 mmol) in water/PEG-400 (2 mL, 1:1, v/v) were mixed and pumped through the coil reactor in 0.5 mL/min low rate at 25°C and 100 psi (residence time: 15 min). The sulfonamides were obtained in 82%–97% yield after standard extractive workup. The most active compound (4-methyl-*N*-phenyl-benzenesulfonamide) inhibited CA IX with an IC_{50} of 90 nM.

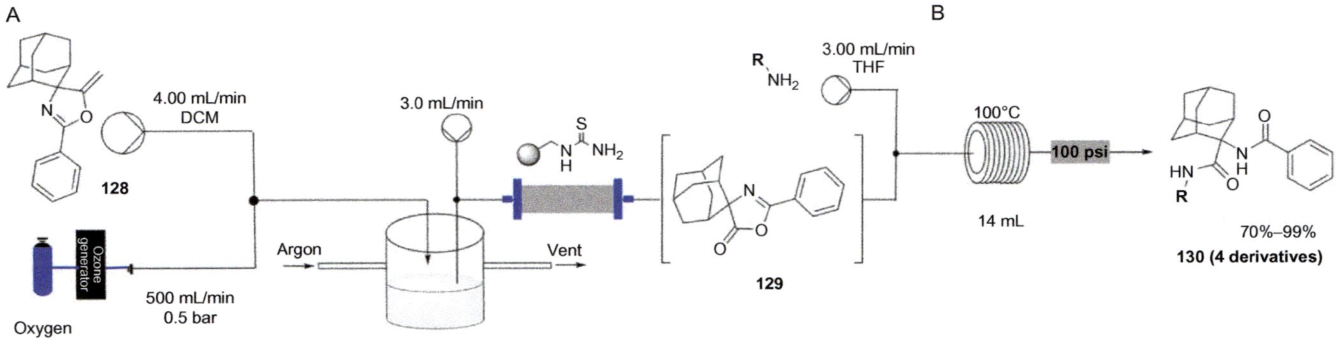

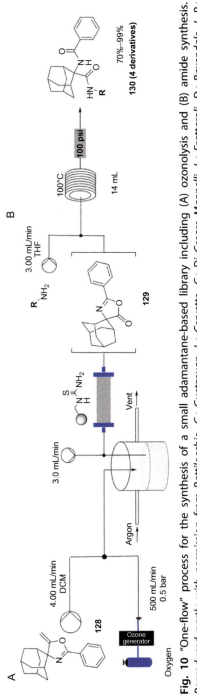

Fig. 10 "One-flow" process for the synthesis of a small adamantane-based library including (A) ozonolysis and (B) amide synthesis. *Reproduced partly with permission from Battilocchio, C.; Guetzoyan, L.; Cervetto, C.; Di Cesare Mannelli, L.; Frattaroli, D.; Baxendale, I. R.; Maura, G.; Rossi, A.; Sautebin, L.; Biava, M.; Ghelardini, C.; Marcoli, M.; Ley, S. V.* ACS Med. Chem. Lett. ***2013****, 4 (8), 704–9. Copyright 2013 American Chemical Society.*

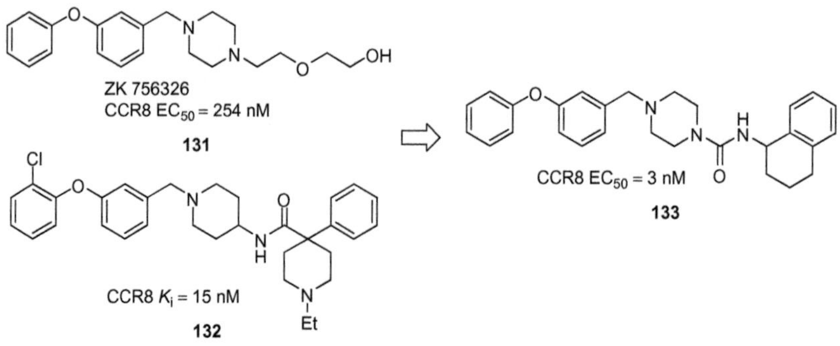

Scheme 33 Hit-to-lead strategy to improve the CCR8 receptor ligand binding activity based on early hits.

Guetzoyan et al.[150] investigated the synthesis of modulators of the histone reader BRD9 using flow methods of chemistry and frontal affinity chromatography for biological evaluation. Bromosporine (**134**) was used as a starting point for the library design (Scheme 34).

For preparing the key intermediate in large scale (40 mmol scale) an automated (machine-assisted) flow setup (including a Vapourtec R4 reactor) was developed with remote monitoring by MS for the Curtius rearrangement using **135** and diphenylphosphoryl azide (DPPA) (Fig. 13).

The obtained bromosporine analogues were evaluated with a BRD9 bromodomain frontal affinity chromatography column.

7. OTHER APPLICATIONS OF MICROFLUIDICS TECHNOLOGIES TO SUPPORT DRUG DISCOVERY

7.1 Microfluidics in DNA-Encoded Library Technologies

Over the last few years DNA-encoded library technologies penetrated the pharma and life science industry and is getting wider use in hit and lead generation.[151] The technology enables scientists to tap into and explore a 4–5 orders of magnitude bigger chemical space than traditional HTS-based hit and lead generation methods.[152] Microfluidics technologies have been successfully utilized to expedite the application of DNA-encoded chemistry technologies in small molecule as well as biological discovery. Those applications have been reviewed earlier by others,[151] while the newest developments are reviewed under the DNA-encoded libraries chapter of this review.

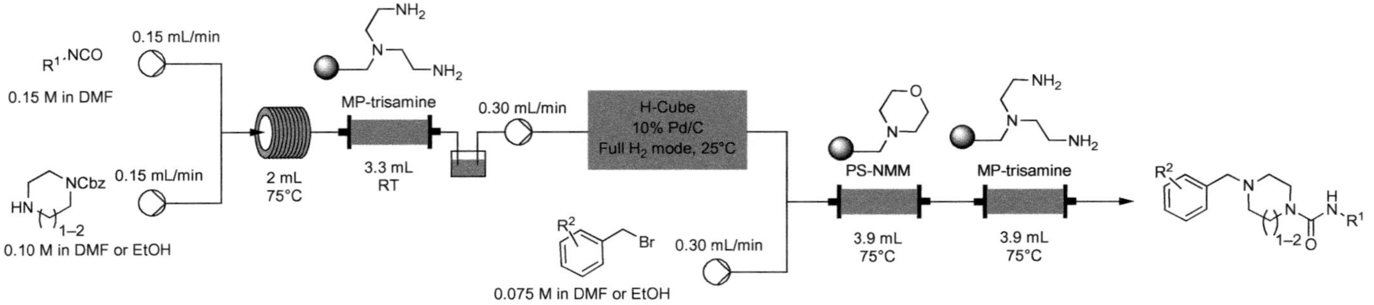

Fig. 11 First generation multistep flow procedure to generate a substituted piperazine library. *Reproduced with permission from Petersen, T. P.; Mirsharghi, S.; Rummel, P. C.; Thiele, S.; Rosenkilde, M. M.; Ritzen, A.; Ulven, T.* Chemistry ***2013***, 19 *(28), 9343–9350. Copyright 2013 Wiley-VCH.*

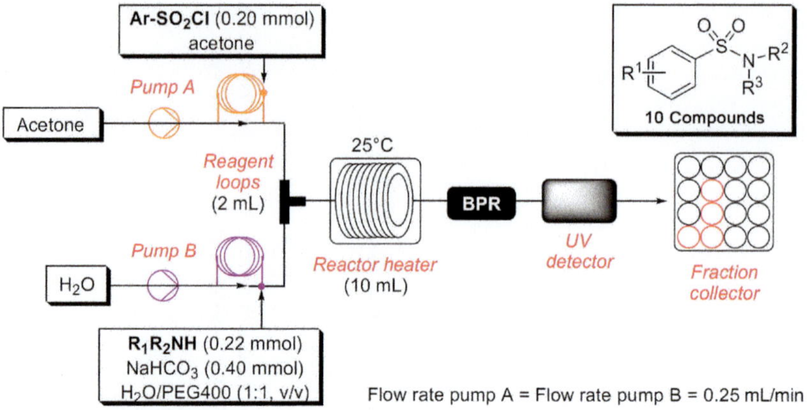

Fig. 12 Flow synthesis of a secondary and tertiary aryl sulfonamide library. *Reproduced in part with permission from Rosatelli, E.; Carotti, A.; Ceruso, M.; Supuran, C. T.; Gioiello, A.* Bioorg. Med. Chem. Lett. ***2014****, 24 (15), 3422–3425. Copyright 2014 Elsevier.*

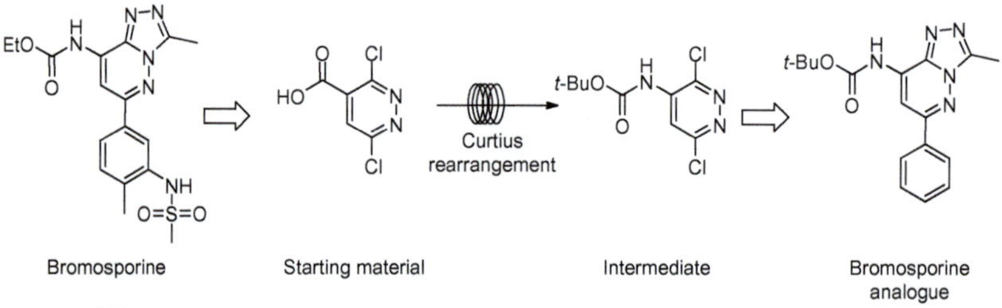

134 **135** **136** **137**

Scheme 34 Flow synthesis of histone reader BRD9 modulators via Curtius rearrangement.

7.2 New Building Block, Reagent, and Natural Product Synthesis

The application of microfluidics reactors,[153] and coil-based reactors for the synthesis of new building blocks, reagents, heterocycles, and natural products, has been reviewed by several research groups (e.g., Ley and colleagues[154]) extensively. Although medicinal chemistry programs may use those effectively, it is not the focus of this review material, so those summaries can be read elsewhere.[153,155–157]

8. CONCLUSION AND FUTURE DIRECTIONS

Microfluidics, mesofluidics, and lab-on-a-chip technologies have the potential to revolutionize pharma R&D in many areas. In certain cases, such

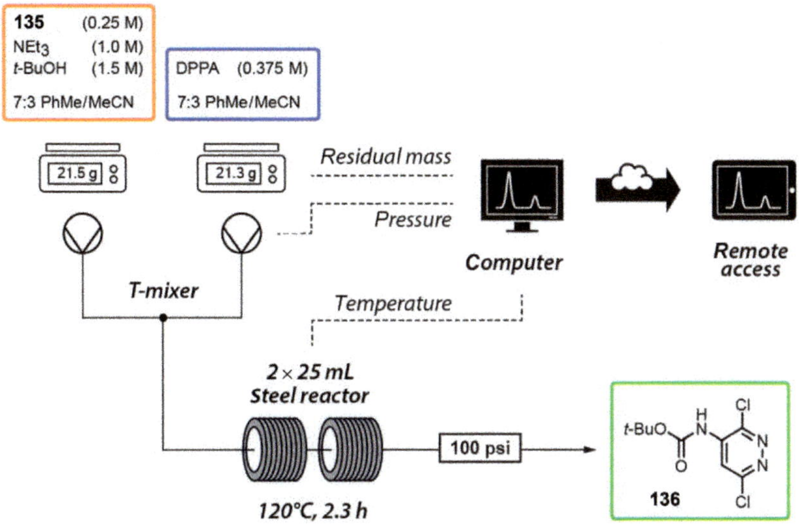

Fig. 13 Machine-assisted flow synthesis of the key intermediate for bromosporine analogues. *Reproduced with permission from Guetzoyan, L.; Ingham, R. J.; Nikbin, N.; Rossignol, J.; Wolling, M.; Baumert, M.; Burgess-Brown, N. A.; Strain-Damerell, C. M.; Shrestha, L.; Brennan, P. E.; Fedorov, O.; Knapp, S.; Ley, S. V.* MedChemComm ***2014***, 5 *(4), 540. Copyright 2014 Royal Society of Chemistry.*

as API manufacturing (see the very recent report on kg-scale prexasertib monolactate monohydrate synthesis under continuous-flow CGMP[158]), end-to-end pharma production, and on-demand drug production, they have made significant progress[159] and traditional processes are being replaced with more flexible methodology. At the same time, their penetration into the pharma research, medicinal chemistry, and lead optimization area is still hindered.[160] Further refinement of the technology and additional applications needs to be developed in order to getting them more widely adopted.[161]

We expect that future technological advancements[162] in this field will result in a paradigm shift. However until then they will be utilized as part of the special toolbox kit of the pharma R&D process.

REFERENCES

1. Whitesides, G. M. *Nature* **2006**, *442*(7101), 368.
2. Elvira, K. S.; Solvas, X. C.; Wootton, R. C. *Nat. Chem.* **2013**, *5*(11), 905–915.
3. Hartman, R. L.; McMullen, J. P.; Jensen, K. F. *Angew. Chem.* **2011**, *123*(33), 7642–7661.
4. Schwalbe, T.; Autze, V.; Wille, G. *CHIMIA Int. J. Chem.* **2002**, *56*(11), 636–646.
5. Reyes, D. R.; Iossifidis, D.; Auroux, P.-A.; Manz, A. *Anal. Chem.* **2002**, *74*(12), 2623–2636.

6. Rodrigues, T.; Schneider, P.; Schneider, G. *Angew. Chem. Int. Ed.* **2014**, *53*(23), 5750–5758.
7. Wirth, T. *Microreactors in Organic Chemistry and Catalysis*. John Wiley & Sons: Hoboken, NJ, 2013.
8. Ramsey, R.; Ramsey, J. *Anal. Chem.* **1997**, *69*(6), 1174–1178.
9. Hansen, C. L.; Skordalakes, E.; Berger, J. M.; Quake, S. R. *Proc. Natl. Acad. Sci. U. S. A.* **2002**, *99*(26), 16531–16536.
10. Johnson, M.; Li, C.; Rasnow, B.; Grandsard, P.; Xing, H.; Fields, A. *J. Assoc. Lab. Aut.* **2002**, 7(4), 62–68.
11. Yin, H.; Marshall, D. *Curr. Opin. Biotechnol.* **2012**, *23*(1), 110–119.
12. Joensson, H. N.; Andersson Svahn, H. *Angew. Chem.* **2012**, *124*(49), 12342–12359.
13. Steinhilber, D.; Rossow, T.; Wedepohl, S.; Paulus, F.; Seiffert, S.; Haag, R. *Angew. Chem. Int. Ed.* **2013**, *52*(51), 13538–13543.
14. Adamo, A.; Beingessner, R. L.; Behnam, M.; Chen, J.; Jamison, T. F.; Jensen, K. F.; Monbaliu, J.-C. M.; Myerson, A. S.; Revalor, E. M.; Snead, D. R. *Science* **2016**, *352*(6281), 61–67.
15. Kang, L.; Chung, B. G.; Langer, R.; Khademhosseini, A. *Drug Discov. Today* **2008**, *13*(1), 1–13.
16. Neuzi, P.; Giselbrecht, S.; Länge, K.; Huang, T. J.; Manz, A. *Nat. Rev. Drug Discov.* **2012**, *11*(8), 620.
17. Kabeshov, M. A.; Musio, B.; Murray, P. R.; Browne, D. L.; Ley, S. V. *Org. Lett.* **2014**, *16*(17), 4618–4621.
18. Baumann, M.; Baxendale, I. R. *Beilstein J. Org. Chem.* **2015**, *11*, 1194.
19. Howard, J. L.; Schotten, C.; Browne, D. L. *React. Chem. Eng.* **2017**, *2*(3), 281–287.
20. Ghislieri, D.; Gilmore, K.; Seeberger, P. H. *Angew. Chem. Int. Ed.* **2015**, *54*(2), 678–682.
21. Borukhova, S.; Hessel, V. Continuous Manufacturing of Pharmaceuticals. In: Vaccaro, L. Ed.; John Wiley & Sons: Hoboken, NJ, 2017, pp 127–168.
22. Mascia, S.; Heider, P. L.; Zhang, H.; Lakerveld, R.; Benyahia, B.; Barton, P. I.; Braatz, R. D.; Cooney, C. L.; Evans, J.; Jamison, T. F. *Angew. Chem. Int. Ed.* **2013**, *52*(47), 12359–12363.
23. Bryan, M. C.; Dillon, B.; Hamann, L. G.; Hughes, G. J.; Kopach, M. E.; Peterson, E. A.; Pourashraf, M.; Raheem, I.; Richardson, P.; Richter, D. *J. Med. Chem.* **2013**, *56*(15), 6007–6021.
24. Wong Hawkes, S. Y.; Chapela, M. J.; Montembault, M. *QSAR Comb. Sci.* **2005**, *24*(6), 712–721.
25. Pickett, S. D.; Green, D. V.; Hunt, D. L.; Pardoe, D. A.; Hughes, I. *ACS Med. Chem. Lett.* **2010**, *2*(1), 28–33.
26. Brennan, Z. FDA calls on manufacturers to begin switch from batch to continuous production. In *In-Pharma Technologist*, 2015.
27. http://www.in-pharmatechnologist.com/content/search?SearchText=FDA+continuous +manufacturing, Accessed July 21, 2017.
28. https://novartis-mit.mit.edu/ http://www.pharmafile.com/news/514335/continuous-manufacturing-novartis-experiment-future-drug-production.
29. Alcázar, J. Sustainable Flow Chemistry: Methods and Applications. In: Vaccaro, L.Ed.; John Wiley & Sons: Hoboken, NJ, 2017, pp 135–164.
30. Malet-Sanz, L.; Susanne, F. *Med. Chem.* **2012**, *55*(9), 4062–4098.
31. https://iscmp2016.mit.edu/sites/default/files/documents/FDA%20MIT-CMAC%20for %20CM%202016%20Ver6.pdf.
32. Lee, S. MIT-CMAC 2nd International Symposium on Continuous Manufacturing of Pharmaceuticals, September 26–27, 2016.
33. Sundaramoorthy, A.; Evans, J. M.; Barton, P. I. *Ind. Eng. Chem. Res.* **2012**, *51*(42), 13692–13702.

34. Gerogiorgis, D. I.; Jolliffe, H. G. *Chim. Oggi* **2015**, *33*, 6.
35. Bana, P.; Örkényi, R.; Lövei, K.; Lakó, Á.; Túrós, G. I.; Éles, J.; Faigl, F.; Greiner, I. *Bioorg. Med. Chem.* **2016**, (in press). https://doi.org/10.1016/j.bmc.2016.12.046.
36. Stanton, D. *Lack of Talent will Hamper Continuous Manufacturing Adoption, Says MIT Prof.* http://www.in-pharmatechnologist.com/Processing/Lack-of-talent-will-hamper-continuous-manufacturing-uptake-MIT-Prof; 2015. Accessed 20 July 2017.
37. Kobayashi, S. *Chem. Asian J.* **2016**, *11*(4), 425–436.
38. Wegner, J.; Ceylan, S.; Kirschning, A. *Chem. Comm.* **2011**, *47*(16), 4583–4592.
39. Valera, F. E.; Quaranta, M.; Moran, A.; Blacker, J.; Armstrong, A.; Cabral, J. T.; Blackmond, D. G. *Angew. Chem. Int. Ed.* **2010**, *49*(14), 2478–2485.
40. Rossetti, I.; Compagnoni, M. *Chem. Eng. J.* **2016**, *296*, 56–70.
41. Brennan, Z. *FDA Calls on Manufacturers to Begin Switch From Batch to Continuous Production.* http://www.in-pharmatechnologist.com/Processing/, FDA-calls-on-manufacturers-to-begin-switch-from-batch-to-continuousproduction; 2015. Accessed 9 April 2016.
42. Yu, L. *Continuous Manufacturing Has a Strong Impact on Drug Quality.* http://blogs.fda.gov/fdavoice/index.php/2016/04/continuousmanufacturing-has-a-strong-impact-on-drug-quality>; 2016. Accessed 9 April 2016.
43. Schaber, S. D.; Gerogiorgis, D. I.; Ramachandran, R.; Evans, J. M.; Barton, P. I.; Trout, B. L. *Ind. Eng. Chem. Res.* **2011**, *50*(17), 10083–10092.
44. Jolliffe, H. G.; Gerogiorgis, D. I. *Chem. Eng. Res. Des.* **2015**, *97*, 175–191.
45. Baxendale, I. R. *J. Chem. Technol. Biotechnol.* **2013**, *88*(4), 519–552.
46. Russell, C. C.; Baker, J. R.; Cossar, P. J.; McCluskey, A. New Advances in Hydrogenation Processes—Fundamentals and Applications. In: Ravanchi, M. T. Ed.; InTech: Rijeka, Croatia, 2017, pp 269–288.
47. Myerson, A. S.; Krumme, M.; Nasr, M.; Thomas, H.; Braatz, R. D. *J. Pharm. Sci.* **2015**, *104*, 832–839.
48. Lee, S. L.; O'Connor, T. F.; Yang, X.; Cruz, C. N.; Chatterjee, S.; Madurawe, R. D.; Moore, C. M. V.; Yu, L. X.; Woodcock, J. *J. Pharm. Innov.* **2015**, *10*, 191–199.
49. Porta, R.; Benaglia, M.; Puglisi, A. *Org. Process Res. Dev.* **2016**, *20*, 2–25.
50. Harris, R. M.; Andrews, B. I.; Clark, S.; Cooke, J. W. B.; Gray, J. C. S.; Ng, S. Q. Q. *Org. Process Res. Dev.* **2013**, *17*, 1239–1246.
51. Filipponi, P.; Ostacolo, C.; Novellino, E.; Pellicciari, R.; Gioiello, A. *Org. Process Res. Dev.* **2014**, *18*, 1345–1353.
52. Reddy, G. S.; Reddy, N. S.; Manudhane, K.; Krishna, M. V. R.; Ramachandra, K. J. S.; Gangula, S. *Org. Process Res. Dev.* **2013**, *17*, 1272–1276.
53. Gustafsson, T.; Sörensen, H.; Pontén, F. *Org. Process Res. Dev.* **2012**, *16*, 925–929.
54. Karlsson, S.; Bergman, R.; Löfberg, C.; Moore, P. R.; Pontén, F.; Tholander, J.; Sörensen, H. *Org. Process Res. Dev.* **2015**, *19*, 2067–2074.
55. Frederick, M. O.; Calvin, J. R.; Cope, R. F.; LeTourneau, M. E.; Lorenz, K. T.; Johnson, M. D.; Maloney, T. D.; Pu, Y. J.; Miller, R. D.; Cziesla, L. E. *Org. Process Res. Dev.* **2015**, *19*, 1411–1417.
56. Pellegatti, L.; Sedelmeier, J. *Org. Process Res. Dev.* **2015**, *19*, 551–554.
57. Christensen, K. M.; Pedersen, M. J.; Dam-Johansen, K.; Holm, T. L.; Skovby, T.; Kiil, S. *Chem. Eng. Sci.* **2012**, *71*, 111–117.
58. Pedersen, M. J.; Dam-Johansen, K.; Holm, T. L.; Rahbek, J. P.; Skovby, T.; Mealy, M. J.; Kiil, S. *Org. Process Res. Dev.* **2013**, *17*, 1141–1148.
59. Murray, P. R. D.; Browne, D. L.; Pastre, J. C.; Butters, C.; Guthrie, D.; Ley, S. V. *Org. Process Res. Dev.* **2013**, *17*, 1192–1208.
60. Thaisrivongs, D. A.; Naber, J. R.; McMullen, J. P. *Org. Process Res. Dev.* **2016**, *20*, 1997–2004.
61. Grongsaard, P.; Bulger, P. G.; Wallace, D. J.; Tan, L.; Chen, Q.; Dolman, S. J.; Nyrop, J.; Hoerrner, R. S.; Weisel, M.; Arredondo, J.; Itoh, T.; Xie, C.; Wen, X.;

Zhao, D.; Muzzio, D. J.; Bassan, E. M.; Schultz, C. S. *Org. Process Res. Dev.* **2012**, *16*, 1069–1081.

62. Hawkins, J. M.; Dubé, P.; Maloney, M. T.; Wei, L.; Ewing, M.; Chesnut, S. M.; Denette, J. R.; Lillie, B. M.; Vaidyanathan, R. *Org. Process Res. Dev.* **2012**, *16*, 1393–1403.
63. Braden, T. M.; Johnson, M. D.; Kopach, M. E.; Groh, J. M.; Spencer, R. D.; Lewis, J.; Heller, M. R.; Schafer, J. P.; Adler, J. J. *Org. Process Res. Dev.* **2017**, *21*, 317–326.
64. Nicolaou, K. C.; Bulger, P. G.; Sarlah, D. *Angew. Chem. Int. Ed.* **2005**, *44*, 4442–4489.
65. Nagaki, A.; Hirose, K.; Tonomura, O.; Taniguchi, S.; Taga, T.; Hasebe, S.; Ishizuka, N.; Yoshida, J. *Org. Process Res. Dev.* **2016**, *20*, 687–691.
66. Yaseneva, P.; Hodgson, P.; Zakrzewski, J.; Falss, S.; Meadows, R. E.; Lapkin, A. A. *React. Chem. Eng.* **2016**, *1*, 229–238.
67. Hopkin, M. D.; Baxendale, I. R.; Ley, S. V. *Chem. Commun.* **2010**, *46*, 2450–2452.
68. Martin, A. D.; Siamaki, A. R.; Belecki, K.; Gupton, B. F. *J. Flow Chem.* **2015**, *5*, 145–147.
69. Fukuyama, T.; Rahman, M. T.; Sumino, Y.; Ryu, I. *Synlett* **2012**, *23*, 2279–2283.
70. Cooper, C. G. F.; Lee, E. R.; Silva, R. A.; Bourque, A. J.; Clark, S.; Katti, S.; Nivoroshkin, V. *Org. Process Res. Dev.* **2012**, *16*, 1090–1097.
71. Cambié, D.; Bottecchia, C.; Straathof, N. J. W.; Hessel, V.; Noël, T. *Chem. Rev.* **2016**, *116*, 10276–10341.
72. Baumann, M.; Baxendale, I. *React. Chem. Eng.* **2016**, *1*, 147–150.
73. Lévesque, F.; Seeberger, P. H. *Angew. Chem. Int. Ed.* **2012**, *51*, 1706–1709.
74. Gilmore, K.; Kopetzki, D.; Lee, J. W.; Horváth, Z.; McQuade, D. T.; Seidel-Morgenstern, A.; Seeberger, P. H. *Chem. Commun.* **2014**, *50*, 12652–12655.
75. Fan, X.; Sans, V.; Yaseneva, P.; Plaza, D. D.; Williams, J.; Lapkin, A. *Org. Process Res. Dev.* **2012**, *16*, 1039–1042.
76. Verghese, J.; Kong, C. J.; Rivalti, D.; Yu, E. C.; Krack, R.; Alcázar, J.; Manley, J. B.; McQuade, D. T.; Ahmad, S.; Belecki, K.; Gupton, B. F. *Green Chem.* **2017**, *19*, 2986–2991.
77. Longstreet, A. R.; Opalka, S. M.; Campbell, B. S.; Gupton, B. F.; Mcquade, D. T. *Beilstein J. Org. Chem.* **2013**, *9*, 2570–2578.
78. Keserü, G. M.; Makara, G. M. *Nat. Rev. Drug Discov.* **2009**, *8*(3), 203.
79. Luo, P.-C.; Cheng, Y.; Jin, Y.; Yang, W.-H.; Ding, J.-S. *Chem. Eng. Sci.* **2007**, *62*(22), 6178–6190.
80. www.nangenex.com.
81. http://www.contractpharma.com/contents/view_breaking-news/2013-12-05/bayer-drgt-in-strategic-formulation-pact/9778.
82. Solymosi, T.; Angi, R.; Basa-Dénes, O.; Ránky, S.; Ötvös, Z.; Glavinas, H.; Filipcsei, G.; Heltovics, G. *Eur. J. Pharm. Biopharm.* **2015**, *94*, 135–140.
83. Angi, R.; Solymosi, T.; Ötvös, Z.; Ordasi, B.; Glavinas, H.; Filipcsei, G.; Heltovics, G.; Darvas, F. *Eur. J. Pharm. Biopharm.* **2014**, *86*(3), 361–368.
84. Roberts, S. Drug Metabolism and Pharmacokinetics in Drug Discovery. *Curr. Opin. Drug Discov. Devel.* **2003**, *6*(1), 66–80.
85. Johansson, T.; Weidolf, L.; Jurva, U. *Rapid Commun. Mass Spectrom.* **2007**, *21*(14), 2323–2331.
86. Kraly, J. R.; Holcomb, R. E.; Guan, Q.; Henry, C. S. *Anal. Chim. Acta* **2009**, *653*(1), 23–35.
87. Tomin, A.; Dorkó, Z.; Hornyánszky, G.; Weiser, D.; Darvas, F.; Ürge, L.; Poppe, L. *BioTrans* **2009**, *2009*, 5–9.
88. Nicoli, R.; Bartolini, M.; Rudaz, S.; Andrisano, V.; Veuthey, J.-L. *J. Chromatogr. A* **2008**, *1206*(1), 2–10.

89. van Midwoud, P. M.; Groothuis, G. M.; Merema, M. T.; Verpoorte, E. *Biotechnol. Bioeng.* **2010**, *105*(1), 184–194.
90. De Kanter, R.; Monshouwer, M.; Draaisma, A.; De Jager, M.; De Graaf, I.; Proost, J.; Meijer, D.; Groothuis, G. *Xenobiotica* **2004**, *34*(3), 229–241.
91. van Midwoud, P. M.; Merema, M. T.; Verpoorte, E.; Groothuis, G. M. *Lab Chip* **2010**, *10*(20), 2778–2786.
92. van Midwoud, P. M.; Merema, M. T.; Verpoorte, E.; Groothuis, G. M. *J. Assoc. Lab. Aut.* **2011**, *16*(6), 468–476.
93. Lohmann, W.; Karst, U. *Anal. Bioanal. Chem.* **2008**, *391*(1), 79–96.
94. Moeller, K. D. *Tetrahedron* **2000**, *56*(49), 9527–9554.
95. Yoshida, J.-I.; Kataoka, K.; Horcajada, R.; Nagaki, A. *Chem. Rev.* **2008**, *108*(7), 2265–2299.
96. Horcajada, R.; Okajima, M.; Suga, S.; Yoshida, J.-I. *Chem. Commun.* **2005**, *10*, 1303–1305.
97. Karst, U. *Angew. Chem. Int. Ed.* **2004**, *43*(19), 2476–2478.
98. Permentier, H. P.; Bruins, A. P.; Bischoff, R. *Mini-Rev. Med. Chem.* **2008**, *8*(1), 46–56.
99. Stalder, R.; Roth, G. P. *ACS Med. Chem. Lett.* **2013**, *4*(11), 1119–1123.
100. Roth, G. P.; Stalder, R.; Long, T. R.; Sauer, D. R.; Djuric, S. W. *J. Flow Chem.* **2013**, *3*, 34–40.
101. Garcia-Egido, E.; Spikmans, V.; Wong, S. Y.; Warrington, B. H. *Lab Chip* **2003**, *3*(2), 73–76.
102. Hessel, V.; Löwe, H. *Chem. Eng. Technol.* **2005**, *28*(3), 267–284.
103. Weber, L.; Wallbaum, S.; Broger, C.; Gubernator, K. *Angew. Chem. Int. Ed.* **1995**, *34*(20), 2280–2282.
104. Singh, J.; Ator, M. A.; Jaeger, E. P.; Allen, M. P.; Whipple, D. A.; Soloweij, J. E.; Chowdhary, S.; Treasurywala, A. M. *J. Am. Chem. Soc.* **1996**, *118*(7), 1669–1676.
105. Yokobayashi, Y.; Ikebukuro, K.; McNiven, S.; Karube, I. *J. Chem. Soc. Perkin Trans.* **1996**, *1*(20), 2435–2437.
106. Illgen, K.; Enderle, T.; Broger, C.; Weber, L. *Chem. Biol.* **2000**, 7(6), 433–441.
107. Kamphausen, S.; Höltge, N.; Wirsching, F.; Morys-Wortmann, C.; Riester, D.; Goetz, R.; Thürk, M.; Schwienhorst, A. *J. Comput. Aided Mol. Des.* **2002**, *16*(8), 551–567.
108. Weber, L. *Drug Discov. Today* **1998**, *3*(8), 379–385.
109. Bräuer, S.; Almstetter, M.; Antuch, W.; Behnke, D.; Taube, R.; Furer, P.; Hess, S. *J. Comb. Chem.* **2005**, 7(2), 218–226.
110. Hong, J.; Edel, J. B. *Drug Discov. Today* **2009**, *14*(3), 134–146.
111. Wong Hawkes, S. Y.; Chapela, M. J.; Montembault, M. *Mol. Inform.* **2005**, *24*(6), 712–721.
112. Hughes, I.; Warrington, B. H.; Wong, Y. F. WO-2004089533 A1, 2004.
113. Warrington, B. H.; Hoyle, C. K.; Pell, T.; Pardoe, D. A. WO-2007021813 A2, 2007.
114. Hughes, I.; Warrington, B. H.; Wong, Y. F. WO-2006038014 A1, 2006.
115. Hoyle, C. K.; Pell, T.; Hawkes, S. Y. F. W.; Warrington, B. H. WO-2007021815 A2, 2007.
116. Fernandez Suarez, M.; Garcia-Egido, E.; Montembault, M.; Chapela, M. J.; Wong-Hawkes, S. Y. F., Proc. ICNMM 2006, 997–1002.
117. Warrington, B.; Vinter, J.; Mackay M. WO 2007/148130, 2007.
118. Wang, J.; Sui, G.; Mocharla, V. P.; Lin, R. J.; Phelps, M. E.; Kolb, H. C.; Tseng, H. R. *Angew. Chem.* **2006**, *118*(32), 5402–5407.
119. Wang, Y.; Lin, W.-Y.; Liu, K.; Lin, R. J.; Selke, M.; Kolb, H. C.; Zhang, N.; Zhao, X.-Z.; Phelps, M. E.; Shen, C. K. *Lab Chip* **2009**, *9*(16), 2281–2285.
120. Kool, J.; de Kloe, G. E.; Bruyneel, B.; de Vlieger, J. S.; Retra, K.; Wijtmans, M.; van Elk, R.; Smit, A. B.; Leurs, R.; Lingeman, H. *J. Med. Chem.* **2010**, *53*(12), 4720–4730.

121. Guetzoyan, L.; Nikbin, N.; Baxendale, I. R.; Ley, S. V. *Chem. Sci.* **2013**, *4*(2), 764–769.
122. http://cyclofluidic.co.uk.
123. Desai, B.; Dixon, K.; Farrant, E.; Feng, Q.; Gibson, K. R.; van Hoorn, W. P.; Mills, J.; Morgan, T.; Parry, D. M.; Ramjee, M. K.; Selway, C. N.; Tarver, G. J.; Whitlock, G.; Wright, A. G. *J. Med. Chem.* **2013**, *56*(7), 3033–3047.
124. Czechtizky, W.; Dedio, J.; Desai, B.; Dixon, K.; Farrant, E.; Feng, Q.; Morgan, T.; Parry, D. M.; Ramjee, M. K.; Selway, C. N.; Schmidt, T.; Tarver, G. J.; Wright, A. G. *ACS Med. Chem. Lett.* **2013**, *4*(8), 768–772.
125. Werner, M.; Kuratli, C.; Martin, R. E.; Hochstrasser, R.; Wechsler, D.; Enderle, T.; Alanine, A. I.; Vogel, H. *Angew. Chem. Int. Ed.* **2014**, *53*(6), 1704–1708.
126. Dormán, G.; Kocsis, L.; Jones, R.; Darvas, F. *J. Chem. Health Saf.* **2013**, *20*(4), 3–8.
127. www.thalesnano.com.
128. Jones, R. V.; Godorhazy, L.; Varga, N.; Szalay, D.; Urge, L.; Darvas, F. *J. Comb. Chem.* **2006**, *8*(1), 110–116.
129. Whelligan, D. K.; Solanki, S.; Taylor, D.; Thomson, D. W.; Cheung, K. M.; Boxall, K.; Mas-Droux, C.; Barillari, C.; Burns, S.; Grummitt, C. G.; Collins, I.; van Montfort, R. L.; Aherne, G. W.; Bayliss, R.; Hoelder, S. *J. Med. Chem.* **2010**, *53*(21), 7682–7698.
130. Cossar, P. J.; Hizartzidis, L.; Simone, M. I.; McCluskey, A.; Gordon, C. P. *Org. Biomol. Chem.* **2015**, *13*(26), 7119–7130.
131. Borcard, F.; Baud, M.; Bello, C.; Dal Bello, G.; Grossi, F.; Pronzato, P.; Cea, M.; Nencioni, A.; Vogel, P. *Bioorg. Med. Chem. Lett.* **2010**, *20*(17), 5353–5356.
132. Lovering, F.; Bikker, J.; Humblet, C. *J. Med. Chem.* **2009**, *52*(21), 6752–6756.
133. Darout, E.; Robinson, R. P.; McClure, K. F.; Corbett, M.; Li, B.; Shavnya, A.; Andrews, M. P.; Jones, C. S.; Li, Q.; Minich, M. L. *J. Med. Chem.* **2012**, *56*(1), 301–319.
134. Andres, J. I.; Alcazar, J.; Cid, J. M.; De Angelis, M.; Iturrino, L.; Langlois, X.; Lavreysen, H.; Trabanco, A. A.; Celen, S.; Bormans, G. *J. Med. Chem.* **2012**, *55*(20), 8685–8699.
135. Tarleton, M.; McCluskey, A. *Tetrahedron Lett.* **2011**, *52*(14), 1583–1586.
136. Day, J. P.; Lindsay, B.; Riddell, T.; Jiang, Z.; Allcock, R. W.; Abraham, A.; Sookup, S.; Christian, F.; Bogum, J.; Martin, E. K.; Rae, R. L.; Anthony, D.; Rosair, G. M.; Houslay, D. M.; Huston, E.; Baillie, G. S.; Klussmann, E.; Houslay, M. D.; Adams, D. R. *J. Med. Chem.* **2011**, *54*(9), 3331–3347.
137. Baur, F.; Beattie, D.; Beer, D.; Bentley, D.; Bradley, M.; Bruce, I.; Charlton, S. J.; Cuenoud, B.; Ernst, R.; Fairhurst, R. A.; Faller, B.; Farr, D.; Keller, T.; Fozard, J. R.; Fullerton, J.; Garman, S.; Hatto, J.; Hayden, C.; He, H.; Howes, C.; Janus, D.; Jiang, Z.; Lewis, C.; Loeuillet-Ritzler, F.; Moser, H.; Reilly, J.; Steward, A.; Sykes, D.; Tedaldi, L.; Trifilieff, A.; Tweed, M.; Watson, S.; Wissler, E.; Wyss, D. *J. Med. Chem.* **2010**, *53*(9), 3675–3684.
138. Jerabek, J.; Uliassi, E.; Guidotti, L.; Korabecny, J.; Soukup, O.; Sepsova, V.; Hrabinova, M.; Kuca, K.; Bartolini, M.; Pena-Altamira, L. E.; Petralla, S.; Monti, B.; Roberti, M.; Bolognesi, M. L. *Eur. J. Med. Chem.* **2017**, *127*, 250–262.
139. Barlaam, B.; Cosulich, S.; Degorce, S.; Ellston, R.; Fitzek, M.; Green, S.; Hancox, U.; Lambert-van der Brempt, C.; Lohmann, J. J.; Maudet, M.; Morgentin, R.; Ple, P.; Ward, L.; Warin, N. *Bioorg. Med. Chem. Lett.* **2017**, *27*(9), 1949–1954.
140. Savall, B. M.; Wu, D.; De Angelis, M.; Carruthers, N. I.; Ao, H.; Wang, Q.; Lord, B.; Bhattacharya, A.; Letavic, M. A. *ACS Med. Chem. Lett.* **2015**, *6*(6), 671–676.
141. Patel, S.; Cohen, F.; Dean, B. J.; De La Torre, K.; Deshmukh, G.; Estrada, A. A.; Ghosh, A. S.; Gibbons, P.; Gustafson, A.; Huestis, M. P.; Le Pichon, C. E.; Lin, H.; Liu, W.; Liu, X.; Liu, Y.; Ly, C. Q.; Lyssikatos, J. P.; Ma, C.; Scearce-Levie, K.;

Shin, Y. G.; Solanoy, H.; Stark, K. L.; Wang, J.; Wang, B.; Zhao, X.; Lewcock, J. W.; Siu, M. *J. Med. Chem.* **2015**, *58*(1), 401–418.

142. Martin-Martin, M. L.; Bartolome-Nebreda, J. M.; Conde-Ceide, S.; Alonso de Diego, S. A.; Lopez, S.; Martinez-Viturro, C. M.; Tong, H. M.; Lavreysen, H.; Macdonald, G. J.; Steckler, T.; Mackie, C.; Bridges, T. M.; Daniels, J. S.; Niswender, C. M.; Noetzel, M. J.; Jones, C. K.; Conn, P. J.; Lindsley, C. W.; Stauffer, S. R. *Bioorg. Med. Chem. Lett.* **2015**, *25*(6), 1310–1317.
143. Liu, M.; Mountford, S. J.; Richardson, R. R.; Groenen, M.; Holliday, N. D.; Thompson, P. E. *J. Med. Chem.* **2016**, *59*(13), 6059–6069.
144. Bazin, H. G.; Li, Y.; Khalaf, J. K.; Mwakwari, S.; Livesay, M. T.; Evans, J. T.; Johnson, D. A. *Bioorg. Med. Chem. Lett.* **2015**, *25*(6), 1318–1323.
145. Fiasella, A.; Nuzzi, A.; Summa, M.; Armirotti, A.; Tarozzo, G.; Tarzia, G.; Mor, M.; Bertozzi, F.; Bandiera, T.; Piomelli, D. *ChemMedChem* **2014**, *9*(7), 1602–1614.
146. Vasudevan, A.; Bogdan, A. R.; Koolman, H. F.; Wang, Y.; Djuric, S. W. *Prog. Med. Chem.* **2017**, *56*, 1–35.
147. Battilocchio, C.; Guetzoyan, L.; Cervetto, C.; Di Cesare Mannelli, L.; Frattaroli, D.; Baxendale, I. R.; Maura, G.; Rossi, A.; Sautebin, L.; Biava, M.; Ghelardini, C.; Marcoli, M.; Ley, S. V. *ACS Med. Chem. Lett.* **2013**, *4*(8), 704–709.
148. Petersen, T. P.; Mirsharghi, S.; Rummel, P. C.; Thiele, S.; Rosenkilde, M. M.; Ritzen, A.; Ulven, T. *Chemistry* **2013**, *19*(28), 9343–9350.
149. Rosatelli, E.; Carotti, A.; Ceruso, M.; Supuran, C. T.; Gioiello, A. *Bioorg. Med. Chem. Lett.* **2014**, *24*(15), 3422–3425.
150. Guetzoyan, L.; Ingham, R. J.; Nikbin, N.; Rossignol, J.; Wolling, M.; Baumert, M.; Burgess-Brown, N. A.; Strain-Damerell, C. M.; Shrestha, L.; Brennan, P. E.; Fedorov, O.; Knapp, S.; Ley, S. V. *MedChemComm* **2014**, *5*(4), 540.
151. Goodnow, R. A., Jr.; *A Handbook for DNA-Encoded Chemistry: Theory and Applications for Exploring Chemical Space and Drug Discovery*; John Wiley & Sons: Hoboken, NJ, 2014.
152. Goodnow, R. A., Jr.; Dumelin, C. E.; Keefe, A. D. *Nat. Rev. Drug Discov.* **2017**, *16*(2), 131–147.
153. Darvas, F., Dorman, G., Hessel, V., Eds. *Flow Chemistry*, Vols. 1–2; De Gruyter GmbH: Berlin/Boston, 2014.
154. Baumann, M.; Baxendale, I. R.; Ley, S. V. *Mol. Divers.* **2011**, *15*(3), 613–630.
155. McQuade, D. T.; Seeberger, P. H. J. *Org. Chem.* **2013**, *78*(13), 6384–6389.
156. Pastre, J. C.; Browne, D. L.; Ley, S. V. *Chem. Soc. Rev.* **2013**, *42*(23), 8849–8869.
157. Webb, D.; Jamison, T. F. *Chem. Sci.* **2010**, *1*(6), 675–680.
158. Cole, K. P.; Groh, J. M.; Johnson, M. D.; Burcham, C. L.; Campbell, B. M.; Diseroad, W. D.; Heller, M. R.; Howell, J. R.; Kallman, N. J.; Koenig, T. M. *Science* **2017**, *356*(6343), 1144–1150.
159. Britton, J.; Raston, C. L. *Chem. Soc. Rev.* **2017**, *46*(5), 1250–1271.
160. Djuric, S. W.; Hutchins, C. W.; Talaty, N. N. *F1000Res* **2016**, *5*, 2426.
161. Wiles, C.; Watts, P. *Micro Reaction Technology in Organic Synthesis*. CRC Press: Boca Raton, FL, 2016.
162. Jensen, K. F.; Reizman, B. J.; Newman, S. G. *Lab Chip* **2014**, *14*(17), 3206–3212.

CHAPTER FIVE

High-Throughput Screening

Mary Jo Wildey*, Anders Haunso*, Matthew Tudor*, Maria Webb*, Jonathan H. Connick†,1
*Merck & Co., Inc., Kenilworth, NJ, United States
†Independent Consultant, Glasgow, Lanarkshire, United Kingdom
1Corresponding author: e-mail address: connickj@gmail.com

Contents

Annual Reports in Medicinal Chemistry, Volume 50
ISSN 0065-7743
https://doi.org/10.1016/bs.armc.2017.08.004

1. INTRODUCTION—A BRIEF HISTORY OF HIGH-THROUGHPUT SCREENING

The concept of high-throughput screening (HTS) first appeared in the mid-1980s and has evolved over the past 25 years to serve the changing needs of pharmaceutical research. Sometimes unjustly derided as "antiintellectual," HTS now forms one of the cornerstones of modern small-molecule drug discovery sitting at the interface between pharmacology, computational chemistry, and medicinal chemistry.

Prior to the advent of HTS, the starting point for drug discovery would evolve around a medicinal chemists' modification of known biologically active compounds such as endogenous ligands, natural products, or even cytotoxic agents. For example, modification of morphine, isolated from the opium poppy, yielded drugs with improved drug metabolism characteristics such as Oxycodone.[1] Compounds were synthesised in milligram to gram quantities and would often be tested on whole cells, tissue preparations, or directly in animal models. Structure-based drug design (SBDD) was largely unknown and indeed, it was not until 1990 that the first examples of SBDD as applied to HIV drug discovery emerged in the literature.[2] Even the concept of archived compound libraries was largely unknown. Compounds (from tens to a few thousand) were typically kept by individual chemists or were grouped by project until a rare lab clean out led to deposit in a stock room and the creation of an archive.

Each individual pharmaceutical company or institution has a different history and reasons for developing HTS capabilities. For many companies this was to serve the demands of using natural products to identify starting points for drug discovery,[3] servicing the requirement to screen multiple fermentation broths and extracts containing multiple compounds. It is not coincidental, however, that HTS emerged at the same time as the convergence of many scientific and technical advances, the emergence of molecular biology, protein crystallography, combinatorial chemistry, as well as laboratory instrumentation and computational science. In particular, early adopters of HTS took advantage of the new availability of automated liquid handling equipment as well as the first microcomputers to enter the laboratory.

The 1990s were a revolutionary decade in the pharmaceutical industry. The new science of molecular biology generated a myriad of new potential targets, as multiple isoforms of known enzymes and receptors were identified, the pharmacology of which was only previously addressable using tissue extracts or cell lines expressing multiple endogenous proteins. The decade from 1991 to 2001 (when the human genome project first published a 90% complete sequence of all 3 billion base pairs in the human genome)[4] witnessed a technological transformation in pharmaceutical research. A race was initiated between the world's Pharma companies to match tool validation compounds and new drugs, to the anticipated hundreds of thousands of new drug targets. In common with the human genome project, HTS and the related developments in combinatorial chemistry shared a need to work faster, cheaper, and with increasing quality in order to meet the demands of the industry. In retrospect, the number of human genes was surprisingly smaller than anticipated (at around 30,000) and progressable drug targets only a subset of these.

1.1 HTS: Process, Timelines, Expectations, and Terminology

HTS is an intensive but time-limited activity which may occur one or more times during the process of drug discovery. While screening is sometimes likened to looking for a needle in a haystack (some relate the needle to a new drug), it is important to understand that HTS almost *never* identifies a new drug but rather a chemical starting point or cluster of chemical analogues around which a hit to lead chemistry process can be initiated. Indeed, the phrase "drug discovery" is a misnomer as drugs are not discovered, they are invented after years of iterative synthesis and testing in vitro and in vivo. Although each company may have variants in the terminology and milestones involved in HTS and at each side of it in the drug discovery trajectory (see examples in Table 1), it is important to understand the process, organizational requirements, and critical factors for success.

The most critical elements for success in HTS are the assay and the quality of the compound collection to be screened.

1.1.1 Assays

The development of the appropriate assay or collection of assays is fundamental to executing a successful HTS campaign. One of the challenges in the evolution of HTS has been to develop assays which can be performed at a throughput, statistical robustness, and reproducibility which is consistent

Table 1 Definitions

Assay: Precisely defined and efficiently designed experiment measuring the effect of a substance on a biochemical or cellular process of interest.
High-throughput screen (HTS): Iterative testing of different substances in a common assay. Screen is generally considered high throughput for >10,000 wells per day. Ultra HTS (uHTS) is reserved for >100,000 wells per day.
Active: Biochemical activity at 1 concentration, 1 well.
Confirmed active: Retest of active in replicate.
Hit: Artifacts removed by deselection assays, typically single point.
Confirmed Hit: Dose–response curve, basic structure confirmation, and purity tested by LCMS.
Lead: Member of a series of compounds for which a chemical optimization plan can be foreseen.
False Positive: HTS "active" that is not active at the target.
False Negative: A compound with activity toward the target biology that is not identified in HTS.

with the budgetary constraints of any organization. It is important to note that assay formats and detection methods all bias the results in one form or another and are also prone to artifacts. A comprehensive assay guidance manual is available via the National Institute of Health and is highly recommended.[5]

In general, the assay formats best suited for HTS are homogeneous or "mix and measure" (Table 2). These greatly simplify the automation requirements (e.g., no wash steps) and tend to provide more reliable results. In particular, the development of several highly sensitive homogeneous fluorescent technologies in the mid-1990s also enabled the reduction of assay volume and thus a higher density, up to 1536 or even 3456 wells per microtitre plate. Together, these have resulted in significant cost and time reductions, importantly consuming less reagents and requiring a much smaller volume of test compound.

The technology developments in detection, high-density formats, and automation, together with the competition between large Pharma to exploit the output from the Human Genome Project, resulted in the establishment of several "factory-like" HTS centers, based around platform technologies marketed by specialist companies such as the Automation Partnership and Aurora.[6] The very significant investments needed to establish such centers tended to favor centralized HTS groups. As these technologies have matured, these constraints on organizational design have been somewhat reduced.

Table 2 Assay Formats and Detection Methods in HTS

Ligand binding – Competition	Absorbance Radioactivity – Scintillation proximity assay (SPA)[a]
Enzymatic activity – Biochemical – Cellular	Fluorescence[a] – Intensity – Fluorescent resonance energy transfer (FRET)
Ion or ligand transport – Ion-sensitive dyes – Membrane potential dyes	– Time resolved FRET (TR-FRET) – Polarization – Fluorescent confocal spectroscopy (FCS)
Protein–protein interactions – Biochemical – Cellular	Luminescence[a] – Chemiluminescence – Bioluminescence – Amplified luminescent proximity homogeneous assay (ALPHA)
Cellular signal transduction – Reporter gene – Second messenger	ELISA (wash steps very challenging in 1536-well format
Phenotypic – Protein redistribution – Cell viability	

[a]Preferred formats for HTS in higher density.
All formats are prone to artifacts.

1.1.2 Compound Libraries

The management of compound collections and the science of curation of libraries within a company or institution have developed into a discipline in its own-right, advancing in parallel with the experiences and lessons learned from HTS. Over the past decades, many sets of guidelines and recommendations for the selection of compounds in any given library have been developed. It is outside the scope of this review to delve deeply into the structural design characteristics of HTS libraries. In brief, experience has shown that the extensive involvement of cheminformatics tools and the leverage of as much data on a compounds chemical (e.g., molecular weight, complexity, diversity, reactivity uniqueness, etc.), physicochemical (e.g., solubility, aggregation, etc.), and predicted characteristics (e.g., ADME, toxicology, etc.) are key to the assembly of a high quality HTS library[7]. A particular landmark in this area was the concept of the "rule of five" guidelines published by Lipinski et al.[8] This, together with analysis of as much information as possible of previous pharmacology experience with the compound or close analogues, enables a partnership with the HTS screen execution to maximize productivity and minimize cost.

A long-running debate within the HTS community has been the optimal size of the compound collection to enable a maximally efficient HTS campaign—in this case defined as a screen which yields multiple confirmed hit compounds, ideally also providing information on the SAR of a hit class for the target for minimal time and expense. Table 3 illustrates historical trends in the growth of Pharma screening collections. Pragmatic cost considerations have, however, constrained the execution of full deck screens of an ever-growing number of compounds. Attention has generally switched toward improving quality and the execution of focused or iterative screens. Again, the clustering and diversity methodologies needed to assemble the collection are outside the scope of this review as is the various types of compound in the library; small molecules, fragments, peptides, natural products, etc.

Three considerations of the compound library which are most relevant to the execution of a successful HTS campaign are discussed in more detail later; storage format, purity/integrity, and retrieval flexibility.

From the early days of HTS, dimethyl sulfoxide (DMSO) has been adopted as the solvent of choice for most libraries. This is due to its ability to solubilize most small molecules, generally at millimolar concentrations. This enables ease of storage (frozen at −20°C to −80°C), and when diluted in assay systems to the usual starting concentration of 10 μM and subsequent dilution to <1% in aqueous medium, the solvent will have minimal interference in the test well. Fragment libraries may require solubilization at higher concentrations of DMSO and other specific classes of compounds, e.g., peptides may require alternative solvents. DMSO is, however, a hygroscopic compound and attention must be given to minimizing water content during storage.

Table 3 Screening Collections

- Before HTS (pre-1990s) most pharma archived a few thousand compounds from historical programs
- Mid-1990s 50–100,000 compound collections
- End-1990s >500,000
- Mid-2000s >1.5 million and growing
  - Capture of all medicinal chemistry compounds (and intermediates)
  - Combinatorial chemistry
  - Purchase of commercially available collections (academia, former Soviet Union, library vendors)

Coupled to this, the number of freeze–thaw cycles also needs to be kept to a minimum, in-order to prevent precipitation of the small molecule.

With the increasing attention to compound quality, the ability to select and remove unsuitable compounds has become necessary. While compounds were originally stored as liquids in 96-well blocks, the current practice is to store in 96 or 384 well, sealed microtubes. Periodic quality control is often performed by liquid chromatography-mass spectrometry to monitor both purity of the sample (at least by presence of the expected mass) and to ensure compounds have not degraded upon storage.

Although the most common approach is to screen a single compound in one well, the concept of pooling or testing mixtures of compounds has frequently featured in screening strategies over many decades. Many natural product screens followed the concept of screening mixtures and the pooling of individual compounds has been advocated on the basis of efficiency gains.[9] Debate will no doubt continue regarding the virtues of testing single or pooled libraries. For certain collection types, e.g., DNA-encoded libraries and very diverse combinatorial collections, pooling is a necessary approach.

Altogether, applying computational filters and compound integrity processes has resulted in the removal of as many as 50% of compounds from some libraries.[10,11]

As costs of chemical synthesis and compound acquisition are high, continued attention is required to minimize consumption of compounds in libraries. Compounds are expensive to synthesize and are often only made in milligram quantities. If the collection is to be effectively maintained for a period of many years, miniaturization of storage format, and minimal waste during pipetting is necessary. Developments in acoustic dispensing have greatly helped in this respect.

1.1.3 Process

Once a suitable assay is validated with respect to pharmacology and the number of false positive and negative compounds has been minimized, the HTS campaign is executed with the appropriate compound library. Usually with a pilot assay of several thousand compounds to estimate the expected hit rate, followed either with a full deck screen (the complete compound collection of up to a few million compounds) or in a more iterative, focused approach. Advances in computational chemistry and informatics have greatly influenced the strategy to be adopted for a new screening campaign (see further discussion in Section 4.1).

The output of the screen is active wells. The compounds demonstrating activity may then be cherry-picked from duplicate liquid samples of compound library and retested in the same assay to deliver confirmed actives. When available, it is often useful to also test these compounds using another assay format (often referred to as an orthogonal assay which may not be compatible with HTS but which provides additional information in low throughput; e.g., if HTS is in reporter gene format, the orthogonal assay may be ligand binding). Subsequently, the confirmed actives are purified or resynthesized to deliver, after confirmation of the biological activity, confirmed hits; structurally identified molecular entities of which a few are selected for a hit optimization project to deliver a lead compound which meets predetermined selection criteria. While potency in the primary assay is important, many other characteristics of the hit will often be examined at this stage to determine the compound most likely to progress to the clinic. The use of computational chemistry tools is critical at this stage. Selectivity, solubility, physiochemical characteristics, and assessment of other adverse protein interactions may be determined for many of the most interesting compounds. The lead compound or series with SAR resulting from several rounds of medicinal chemistry (Lead Identification) will then typically undergo further medicinal chemistry optimization (Lead Optimization) to address deficiencies in metabolism, toxicology, and hopefully the identification of a suitable candidate to advance to the clinic.

2. HTS PLATFORMS AND TECHNOLOGIES: AUTOMATION, LIQUID HANDLING, DETECTION

2.1 Automation

Advances in the automation and miniaturization of in vitro assays to 384-well and 1536-well microtiter formats has been enabling in the development of reproducible and sustainable HTS processes capable of testing hundreds of thousands of compounds daily. These automated HTS campaigns have resulted in increased data quality and consistency and have catalyzed advances in data analytics, visualization, and informatics, enabling a more holistic view of a potential hit before additional drug discovery scientific resources are engaged.[12] In this section, we will review several automation, liquid handling, and detection platforms that can be found in screening laboratories.

Assay miniaturization is one of the key components enabling HTS automation. As stated earlier, HTS operates predominantly in two plate formats: 384 well and 1536 well, relying on assay miniaturization to drive reduction in biological reagent and compound use and to enable testing of large compound libraries in short timeframes. Depending on the success of miniaturization efforts, assays may be limited to a 96-well format (e.g., some filter binding assays) but the benefits mentioned earlier are usually put at risk.

Table 4 summarizes typical working volumes for each plate density format.

A comparison outlined by Boettxher and Mayr shows the impact of density for a 1 million compound library screening campaign with a commercially available protease assay.[13] In a 96-well assay format, with assay volumes of 150 μL/well, over 150 L of reaction mix was required with a corresponding substrate cost of $1.5 million, and requiring 11,364 assay plates. Increasing density to a 384-well assay reduced reagents and cost about threefold. Further increase in density, from a 384-well format to a 1536-well format, decreased reagent volumes almost sixfold, resulted in substrate cost savings of $416,000, and reduced the number of assay plates from 2841 in the 384-well density to 711 for a 1536-well density. These data reinforce the value and necessity of miniaturization when the biology and quality metrics of the assay supports it.

In general, there are three types of automation "modes" used in HTS laboratories (1) batch, (2) semi-automated, and (3) integrated. Table 5 highlights some of the key characteristics of each of these automation "modes."

Batch mode uses plate stackers for each peripheral and is in general "manned" by scientists, who move stacks of plates from one peripheral to another for processing each screening assay step. Traditional workstations such as the Thermo Combi, CyBio SELMA, and Perkin Elmer Envision are examples of batch mode processing. Typically, plate-to-plate and run-to-run consistency can be an issue when running in batch mode as differences in the timing of individual assay steps are inherent. Depending on the kinetics of the assay, these differences can result in an increase in overall variability and reduce the effectiveness of tools that can correct for systematic trends in the screen. Since batch mode devices rely on scientists, daily

Table 4 Typical Working Volumes for 96-Well, 384-Well, and 1536-Well Screening Modes

Well Format	Total Assay Volume (μL)	Discrete Addition Volumes (μL)
96	100–200	1–100
384 standard volume	10–100	1–50
384 low volume	5–15	0.01–5
1536	2–10	0.01–2

Table 5 Characteristics of the Three Typical Screening Automation Modes: Batch, Semiautomated, and Integrated

Characteristic	Batch Mode	Semiautomated Mode	Integrated Mode
Flexibility	Limited	Moderate	High
Scheduling software	None	Limited depending on the configuration	Dedicated and extensive Required for operation
System size/utilities	Benchtop Usually operates on house power and utilities	Benchtop or small custom tables (e.g., 12 × 12 sq. ft.) Usually operates on house power and utilities	Can be as large as a room (e.g., 20 × 30 sq. ft.) Requires dedicated power and utilities, UPS, and often HVAC
Walk-away processing capacity for microplates	96- or 384-W: ~25	96- or 384-W: ~75	384- or 1536-W: ~1000
Plate movement logic	Manually loaded attached plate stackers	Limited axis "pick and place" arm, SCARA, cylindrical, Cartesian gantry arms	Fully articulated robot arm, typically on a track, conveyer, or pedestal
Number/complexity of tasks performed	Limited number of tasks and complexity	Moderate range of complexity and tasks, depending on configuration	Wide number of tasks and range of complexities supported, depending on peripherals
Automation skills needed to support	Limited skills needed, reasonable to self-teach	Moderate skills required, vendor training needed	Considerable basic automation skills and vendor training required
Primary use environment	<25 plates for 384-W or 96-W assays	Focused library screening Low-medium HTS <50,000 compounds	High volume HTS and uHTS >50,000 compounds
Price range	\$25,000–\$300,000	\$100,000–\$400,000	\$750,000–≫\$1 million
Examples	Biotek MultiFlo CyBio SELMA Molecular Devices FLIPR Perkin Elmer Envision Perkin Elmer MicroBeta 2 Thermo Combi	Beckman Biomek i-Series CyBio FeliX Hamilton VANTAGE Perkin Elmer Janus Tecan Freedom EVO	CyBio Screen-machine HighRes Biosystems Kalypsys Systems MicroStar ThermoFisher Dimension4

throughput is limited by the capacity of the stackers and scientists in the laboratory.

The semiautomated mode is defined by the use of small systems that carry out some of the assay steps, for example, a benchtop liquid handler system such as a Hamilton VANTAGE, Beckman Biomek i-Series, or a Tecan Freedom EVO. Semiautomated systems reduce some of the process error and variability that is observed in batch mode processes and are enabling for focused library screening and low-medium throughput HTS (<50,000 compounds).

The integrated mode of screening is usually defined as a collection of diverse peripherals managed by an articulating arm or conveyer system and the system is usually controlled by scheduling software. Figs. 1 and 2 show examples of integrated systems and a scheduled assay workflow.

These systems can perform similar or diverse assays, singly or interleaved, depending on the biology and run parameters of the assays being programmed. Integrated mode screening, which can be defined as a robotic system that is capable of carrying out an in vitro assay in its entirety from compound addition to detection, reduces plate-to-plate and run-to-run inconsistencies and as a result typically provides higher quality data and enables the use of automated quality control (QC) tools to address systematic errors. Another advantage of an integrated system is the ability to screen continuously, supporting after hours unmanned work. One disadvantage is that these systems usually require operators with an automation engineering background in addition to specialized system-specific training and are capital intensive and often not practical for use outside the HTS environment, for example, in general pharmacology profiling.

Typical components of an HTS system include plate storage, robotic arms, liquid handlers, centrifuges, plate washers, and detectors.

2.2 Plate Storage

There are two general types of plate storage systems: shelf-based systems and stackers. Within the shelf-based storage systems there are (1) open shelved racks (hotels), which provide ambient temperature and humidity conditions and (2) incubators with controlled temperature, humidity, and gas environments. Each of these types of storage systems holds a microplate on an individual shelf and each microplate is exposed to the same temperature, humidity, and CO_2 environment on the top and bottom of the plate being stored. This can be a critical component in reducing assay variability when

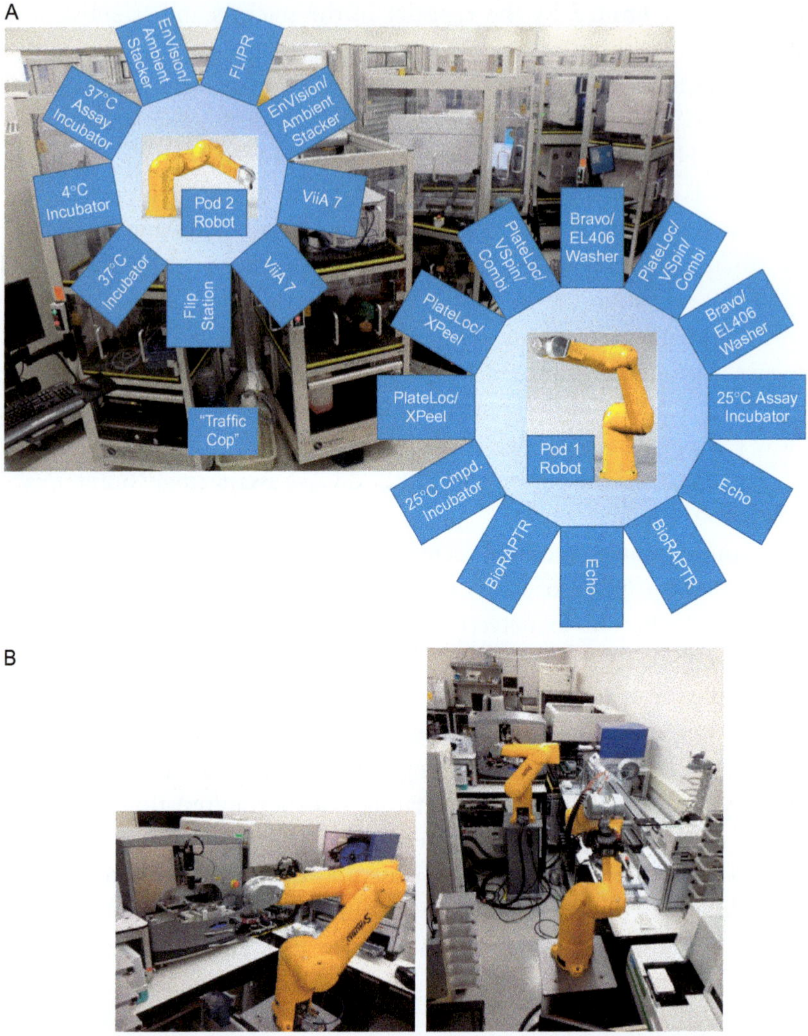

Fig. 1 Examples of integrated robotic systems. (A) HighRes Biosolutions Modular System. (B) Telios-based custom integrated platform supporting radioactive assays.

incubation times are short and temperatures are not ambient. Capacity of robotic incubators can range from 20 plates to hundreds of plates, depending on the size of the incubator being used. Some of the robotic incubators are also capable of shaking the microplate as it incubates (e.g., ThermoFisher Cytomat 2). In the second type of plate storage, the plate stacker, microplates rest on top of each other, in a cassette holder which is then linked to another device such as a plate washer or detection device. Stackers can hold up to

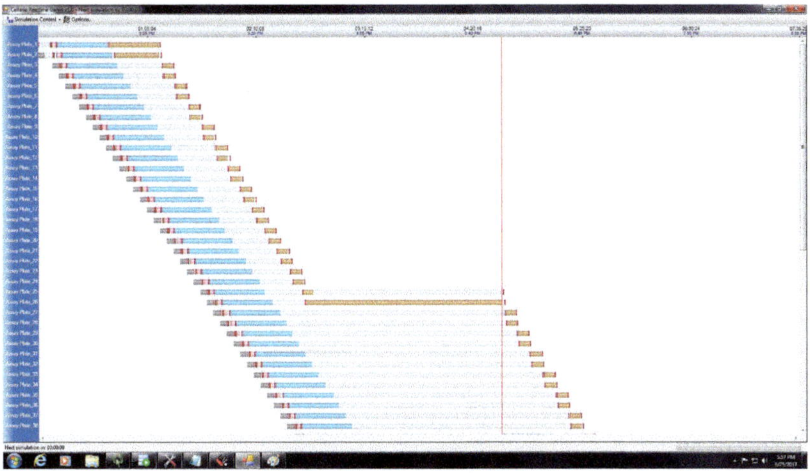

Fig. 2 Example of scheduling software from an integrated robotic system.

50 microplates and are usually kept at ambient conditions. Because each microplate sits on top of another, temperature gradients can form between the first and last plates and gas exchange varies within a plate and from plate to plate, leading to increased assay variability. Stackers are the typical plate storage systems for batch mode workflows.

2.3 Robotic Arms

The primary purpose of robotic arms and their grippers in screening systems is to move consumables and reagents from one assay step to another, with the end goal of maintaining consistency between each assay plate for the entire assay run. There are several types of arms typically found in screening applications: simple plate movers, limited access arms, and fully articulated arms.

Simple plate movers transfer a microplate from a plate stacker to an associated device and have built-in controllers and condensed command sets, generally under the control software of the associated device. Examples of simple plate movers are shown in Fig. 3.

Limited access arms are more complex, can be circular or linear, and usually have 2–4 degrees of freedom. The arms can address stackers, peripherals that have the ability to robotically present consumables, and hotels, depending on the gripper capabilities. Traditional examples of limited access arms are Hudson's PlateCrane and Perkin Elmer's Twister II shown in Fig. 4.

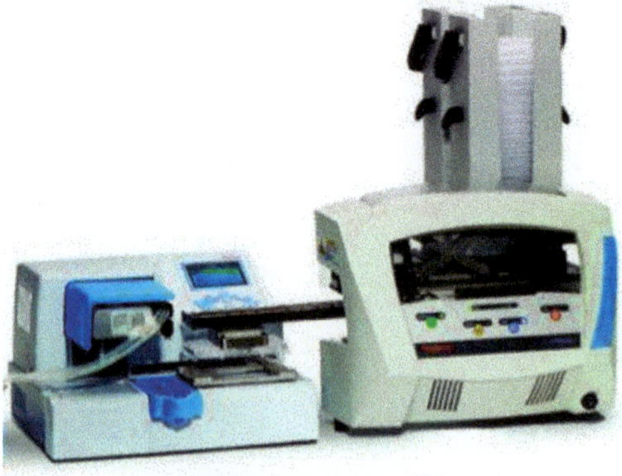

Thermo RapidStak™

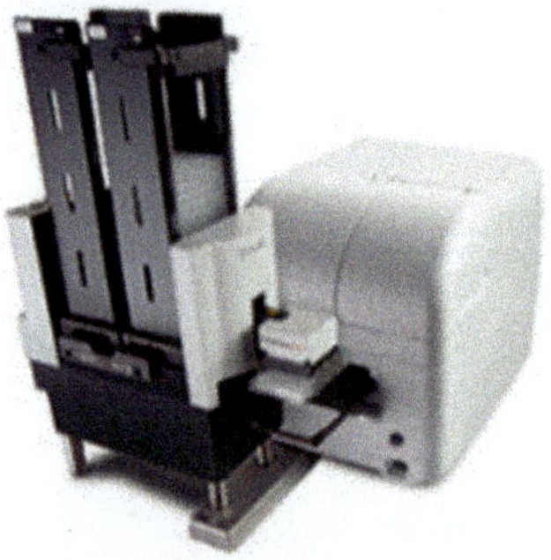

BioTek BioStack™ Microplate Stacker.
Image courtesy of BioTek Instruments, Inc.

Fig. 3 Examples of simple plate mover robotic arms.

Fully articulated robotic arms are found at the heart of many integrated robotic HTS systems. They usually have 5–6 degrees of freedom, use rotary joints to access the peripherals, and are able to support a wider array of applications and peripherals compared to less flexible robotic arms. The arms are usually located in a central fixed position or on a conveyer. Viewed as "industrial robots," these arms and their systems are often guarded with physical barriers for the safety of scientists and have fairly complex scheduling software and programming to control them. Examples of articulated arms are ThermoFisher's F5 Robot, Staubli's TX and TX2 series, Denso Robotic's VS series (Fig. 5).

Historically, there are several inherent challenges in the routine use of robotic arms regardless of whether they are simple plate movers, limited access arms, or fully articulated robots. Robotic arms need to be "taught" and "retaught" positions on a regular basis to maintain microplate and

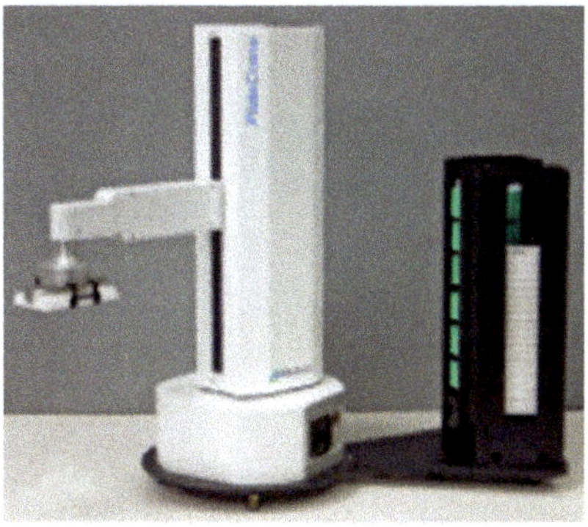

Hudson Robotics PlateCrane

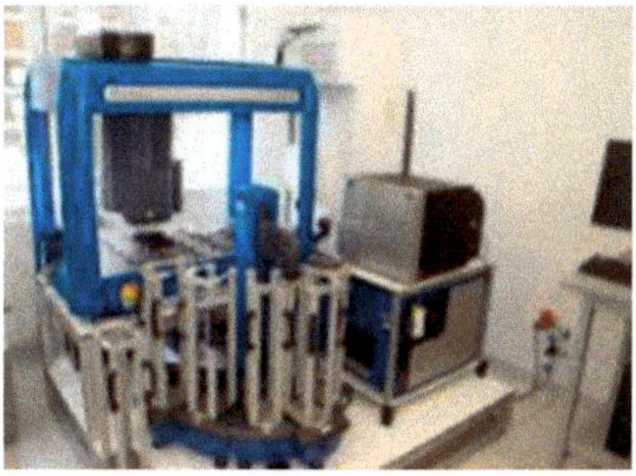

PE Twister II

Fig. 4 Examples of limited access robotic arms (2–4 degrees of freedom).

peripheral alignment during a screen. Teaching can be time-consuming and tedious and often requires special training, depending on the robotic arm and its controlling software. Additionally, providing a safe working environment between humans and robots can be expensive and requires in-depth effort. Machine guarding barriers and protocols can make interacting with the HTS system to refresh reagents or address errors difficult and as a result the guarding is often circumvented, putting operators at risk. Recent technological advances have started to address these challenges by developing arms with integrated sensors that can recognize external forces (e.g., detection of overcurrents when a collision occurs) and respond before injury to the scientist or system is incurred. To overcome the tedious teach task, newer arms can be taught by demonstration with a simple "teach" command, not programming. This method requires no in-depth operator training or expertise. The HighRes Biosolutions ACell is one example of enhanced teaching capabilities and tactile sensing.[14] Other arms, such as

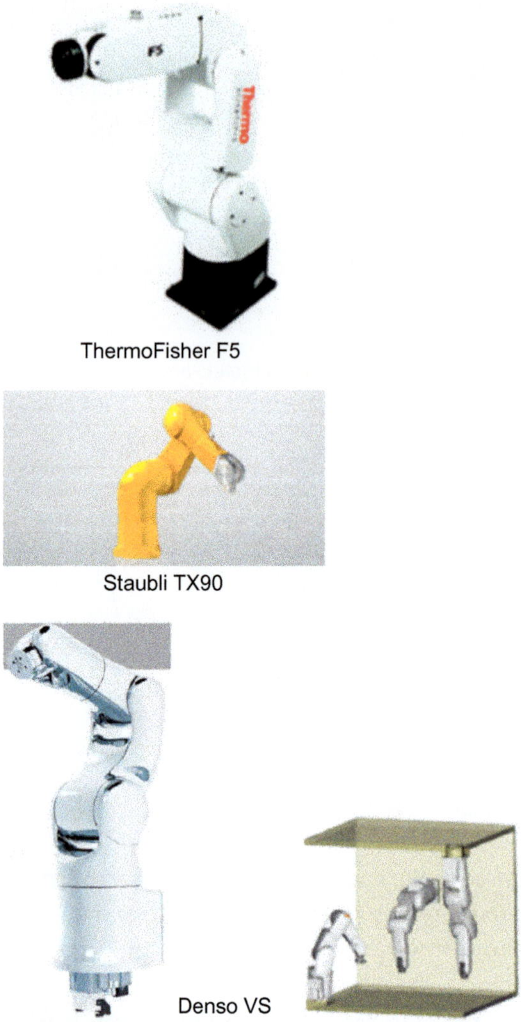

Fig. 5 Examples of fully articulating robotic arms.

the Thermo Spinnaker, have integrated vision-assisted teaching and barcode-reading capabilities in addition to the ability to self-correct for instrument drift.[15] BLUECAT BIO has introduced a collaborative robot to work as a simple plate mover.[16] Illustrations of these systems can be seen in Fig. 6. These innovations have already shown their value toward reducing time required to program assays, in maintaining a high quality robotic system, reducing overall cost of the system, and enabling scientists and robots to work without the need for extensive machine guarding.[17–19]

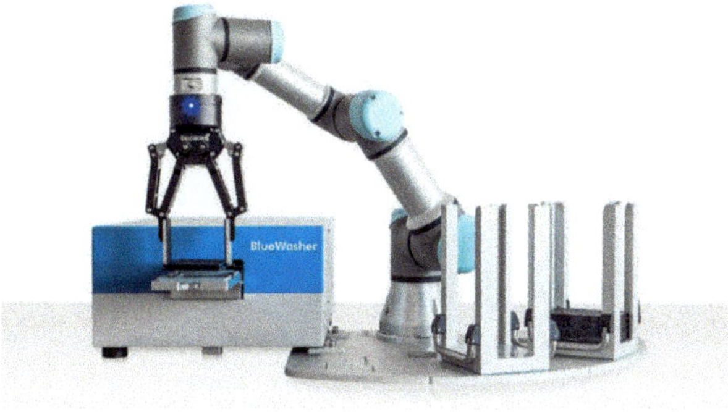

BLUECAT BIO BlueBench

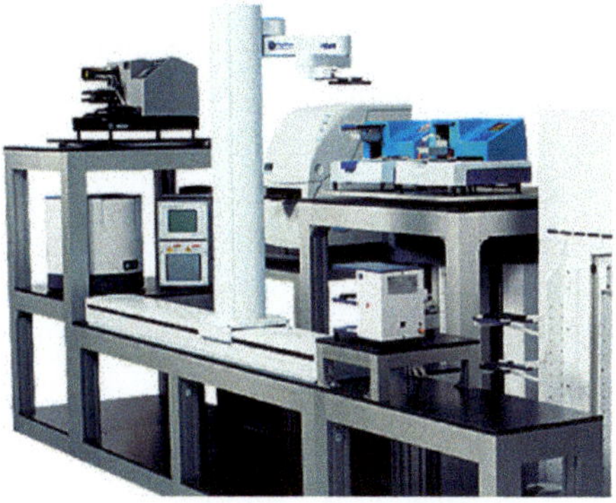

HighRes Biosolution Acell

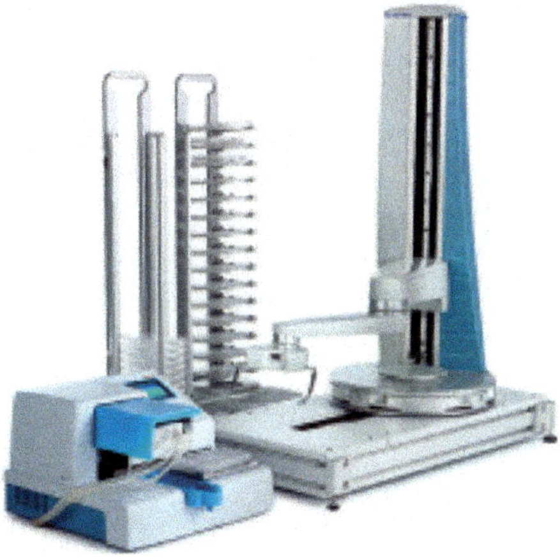

ThermoFisher's Spinnaker

Fig. 6 Examples of recent robotic arm designs aimed at reducing and simplifying teach time, increasing reliability, and enabling a collaborative human work environment.

Robotic arm technology is a rapidly changing field, leading to novel ways of enabling HTS to focus on less traditional detection platforms such as FLIPR and High Content Screening. Recent publications describe a dual gripper on a Universal Robot collaborative arm as shown in Fig. 7.[20] It is not difficult to imagine the throughput impacts of a dual gripper on a screening campaign!

2.4 Liquid Handling

Within screening workflows, liquid transfers are a critical component, often being one of the main contributors to assay variability. There are several frequently used dispenser types representing a variety of dispensing mechanisms. Traditionally, tip-based dispenser types with air and positive displacement dispensing mechanisms have been most commonly used in screening. In more recent years, nontouch dispensing types have gained in popularity, especially acoustic dispensing. Table 6 shows examples of common liquid handling types and mechanisms found in screening labs.[21,22]

2.5 Air and Positive Displacement

These systems use plungers working within some type of cylinder or dispense block. The action of the plunger establishes an aspiration or dispense step. In air displacement dispensers, there is a small air gap between the plunger and the liquid being aspirated, with the aim of separating the two. To maximize the effectiveness of air displacement dispensers, attention must be given to minimizing the air gap to reduce pipetting variability. In

Fig. 7 Universal Robotics dual gripper collaborative robot.

Table 6 Dispense Mechanisms and Types Found in HTS Labs

Dispensing Mechanism	Dispenser Type	Dispensing Range
Air displacement	Disposable/ changeable tip	0.25 μL and higher
Positive displacement	Fixed tip	Traditional: 25 nL–1.2 μL Dragonfly 2: 200 nL–4 mL
Direct transfer	Pin tool	2 nL–5 μL
Acoustic transducer	Acoustic	LabCyte: 2.5 nL droplet; 2.5 nL–10 μL EDC Biosystems: 1–20 nL drop size; 1 nL–100 μL
Peristaltic pump	Mechanical force	Multidrop Combi: 0.5–2500 μL Multidrop Combi nL: 50 nL–50 μL Biotek Multiflo FX: 500 nL–3000 μL
Solenoid syringe and bottle	Valve	Tecan D300e: 11 pL–10 μL Formulatrix Tempest: 0.2–1 μL
Piezo stack actuators	Piezo stack actuators	200 pL–50 μL

addition, the fit of the plunger and cylinder and the cylinder and tip must be monitored to ensure there are no seal breaks, which would lead to increased variability. Disposable tip-based pipette systems are typically air displacement systems with pipetting ranges as low as 0.25 μL and compatibility with 96-, 384-, and 1536-well microplates. Examples are Beckman, CyBio, Hamilton, Perkin Elmer, and Tecan systems.[23–27]

Positive displacement systems still use the plunger and cylinder concept, however, there is no air gap between the plunger and liquid being dispensed. Historically, most positive displacement systems use fixed tips, made from stainless steel and possibly coated to reduce compound adsorption. These types of fixed tip systems must use wash steps in between pipetting steps, introducing potential carryover concerns. Most liquid handling companies offer a fixed tip option with their systems, and although they are typically compatible with 96- and 384-well microplates, they are not 1536-well compatible due to physical spacing constraints with the cannulas. However, the Mosquito (TTP Labtech) is an example of a disposable positive displacement pipetter, with pipetting ranges in the 25 nL to 1.2 μL range, dead volumes under 0.3 μL, and compatibility with 96-, 384-, and 1536-well assay

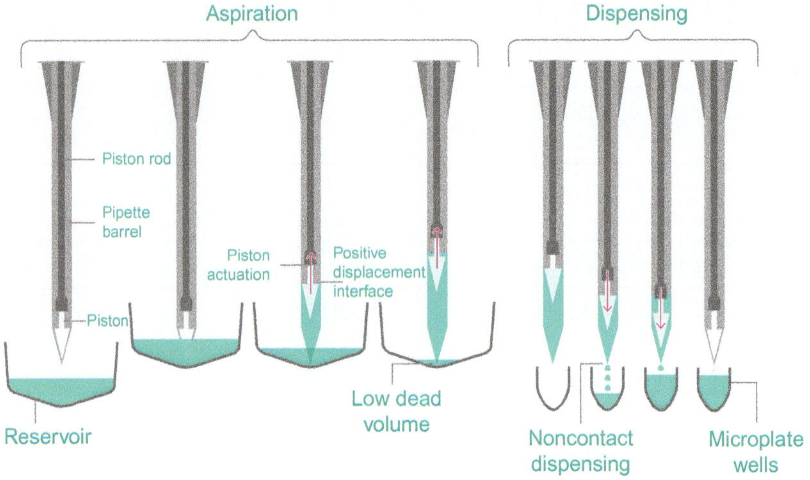

Fig. 8 TTP Labtech Dragonfly 2 aspirate and dispense logic.

formats.[28] This is possible because the Mosquito tips are presented on a continuous reel with pitches of either 4.5 or 9 mm. More recently, a disposable positive displacement tip-based system, the Dragonfly 2, was introduced by TTP Labtech.[29] The Dragonfly 2 addresses many of the concerns of traditional fixed tip and positive displacement systems, being compatible with 96-, 384-, and 1536-well plates. It can be configured for up to 10 channels, all independently controllable and has a fill time of less than 1 min for a 384-well plate an <3 min for a 1536-well plate. Fig. 8 illustrates the logic of the Dragonfly 2 tip.

2.6 Pin Tools

With pin tools, the liquid being dispensed sticks to the end of the pin and then transfers to the destination plate using a touch-off (contact dispense) to remove the drop from the pin. In HTS, pin tools are typically used to transfer test compounds from a source plate to the assay plate.[30] There are several factors that affect the transfer volume, including the pin shape (e.g., slotted, smooth, grooved, hollow), pin diameter, the depth that the pin is moved into the source liquid, surface tension of the involved liquids, and speed of the "dip and touch." There are varying reports in the literature for pin tool accuracy, with some reports at <5% when manufacturing and QC process for the pin tools were improved.[31] In addition, more stringent robotic control of the speed and heights in the dispense process have helped decrease variability. Pin tools are not disposable and must therefore be washed between transfers, introducing the potential for cross contamination. Some of the advantages of pin tools are reduced cost, the ability to support 96-, 384-, 1536-, and even 3456-well formats, and compatibility with a variety of automated dispensing systems.

2.7 Acoustic Transfer

Acoustic transfer systems are based on transducers sending sound waves through a liquid, to dispense specific sized droplets.[32,33] These are noncontact dispenses and some of the advantages are no cross contamination, minimized waste, the creation of compound concentration curves on the fly, support of 96-, 384-, and 1536-well plate densities, and monitoring of water uptake in DMSO solutions. Two of the disadvantages are the need for specific source plates that are compatible with the transducer system and the relatively high cost of the instrument. Acoustic transfer systems are typically used for compound addition steps, but more recently they have been used in the addition of other assay reagents.[34]

2.8 Peristaltic Pumps

Systems using peristaltic pumps are noncontact and use flexible tubing that is compressed to move liquid from a reservoir through the tubing and into a receiving microplate through a series of tips. The Thermo Multidrop Combi, Combi nL, and the Biotek Multiflo are three examples of this type of a system.[35,36] The Combi has a different liquid path for each of 8 or 16 channels, depending on the cassette type used. The cassette is resistant to many solvents and can be calibrated to maintain precision and accuracy. Dispense speeds can be adjusted to account for varying reagent properties and for dispensing cells.

2.9 Solenoid Syringe and Solenoid Pressure Bottle Systems

The solenoid syringe system uses a syringe to aspirate the reagent to be dispensed and supplies a pressure source against a closed microsolenoid valve. A tip is used to regulate nanoliter droplet sizes, with working ranges regulated by the syringes and tip, but typically in the 5 nL to 50 μL range.

The pressure bottle system replaces the syringe in the above system with a pressurized bottle. Examples are the Perkin Elmer FlexDrop, Tecan D300, Certus Nano, Formulatrix Tempest, and Mantis.[37–39] These systems are capable of running at high rates and can have the ability to dispense multiple reagents simultaneously, taking advantage of separate valves and fluid paths. They are compatible with 96-, 384-, and 1536-well microplates.

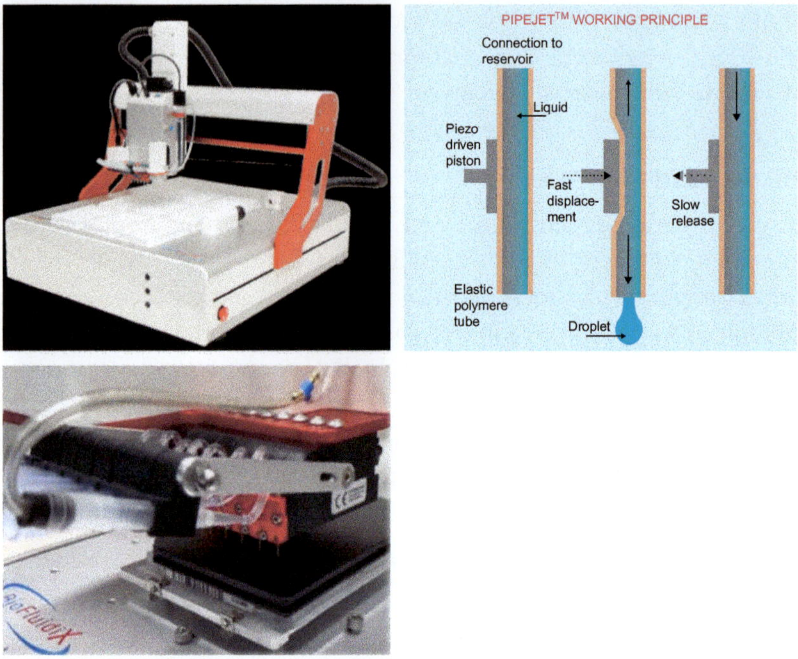

Fig. 9 Tekmatic BioSpot dispenses aqueous liquids from 200 pL to 50 μL.

2.10 Piezo-Actuator-Based Liquid Handling

Piezo stack actuators make use of the deformation of electroactive lead/Zirconia/titanate ceramics caused by exposure to an electrical field. The deformation is used to produce a force or motion.[40] Tekmatic has combined an elastic micro pipe with a piezo stack actuator resulting in the "Biospot," a high speed reagent and cell dispensing system (Fig. 9). The BioSpot has dead volumes of only a few microliters with accuracy and reproducibility of <3% for typical aqueous liquids. Dispensing ranges are from 200 pL to 50 μL.[41]

3. DETECTION TECHNOLOGIES

There are several types of detection modalities used in screening applications, each designed to detect and quantitate a biological, chemical, or physical phenomenon. Examples of more widely used modalities are absorbance, fluorescence, luminesence and radiometric; most are available in single and multimode readers and examples of each are outlined in Table 7.

Table 7 Detection Modalities Used in HTS

Modality	Detection Technology
Absorbance	Photometric
	Colorimetric
Fluorescence	Fluorescent intensity (FI)
	Time-resolved fluorescence (TRF)
	Fluorescence resonance energy transfer (FRET)
	Time-resolved FRET (TR_FRET)
	Homogenous time-resolved FRET (HTRF)
	Fluorescence polarization (FP)
	Fluorescence lifetime (FLT)
	Fluorescence correlation spectroscopy (FCS)
Luminescence	Flash
	Glow
	Amplified luminescent proximity homogenous assay (Alpha) Technology
	Bioluminescence resonance energy transfer (BRET)
	Electrochemiluminescence (ECL)
Radiometric	Filter binding
	Scintillation proximity assay (SPA): Flash plate
	Scintillation proximity assay (SPA): Bead based

3.1 Absorbance

Absorbance measures the amount of light at a selected wavelength that is absorbed as it passes through the microplate well contents. The detector measures the amount of light from the opposing side of the well and light source.

3.2 Fluorescence

Fluorescence is one of the more widely used modalities and there are several variations as outlined in Table 7. In a basic FI system, a light source with a specific wavelength illuminates the sample well containing fluorescent molecules. At the same time, light is emitted from the sample well where it can be filtered from the light source with an emission wavelength filter and then measured or detected. Detection is usually a photomultiplier tube (PMT).

HTRF measures analytes in a homogenous format and is a combination of FRET with time-resolved measurement. In this technology, there is a donor and a receptor fluorophore and the donor is excited by an energy source such as a laser or flash lamp. This energy is transferred to the acceptor fluorophore if the two are in close enough proximity to each other and the acceptor emits light at its characteristic wavelength. HTRF is sensitive, can be miniaturized to 1536-well format, is robust and has been applied to many different assay systems.[42]

Another example is FP where the excitation and emission filters are polarized and the readout intensity is measured in parallel and perpendicular orientations, relative to the excitation plane. The Brownian tumbling of the fluorescent molecule is measured. Larger molecules rotate slower and retain a greater fraction of incident polarization than do those that tumble rapidly.[43]

3.3 Luminescence

Luminescence does not require a light source and systems usually consist of a lightproof chamber and PMT detector. Variations in the type of PMT detector selected provide opportunities to select specific wavelengths or ranges, to multiplex assay systems, or to optimize signal detection.

Electrochemiluminescent labels generate light when stimulated by electricity in the microplate well. ECL is sensitive and specific and typically has a low background.[44]

Alpha Technology is bead based, based on an oxygen channeling technology, and measures the interaction of two molecules that are conjugated to donor and acceptor beads. The technology is represented by two assay types: AlphaScreen and AlphaLISA. Fig. 10 illustrates the principle of the technology.[45]

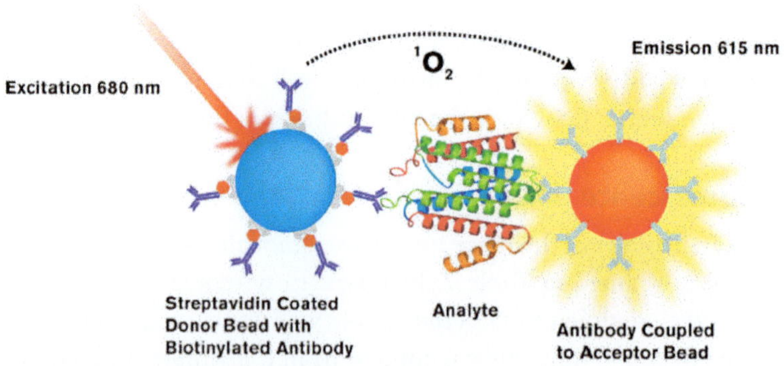

Fig. 10 AlphaScreen/AlphaLISA assay principle.[45]

3.4 Radiometric

Radiometric assays use radioisotopes to monitor the activity and/or kinetics of a specific receptor or enzyme assay.

Scintillation Proximity Assay (SPA): SPA is a bead-based assay technique that has been applied to radioimmunoassays, receptor-binding assays and enzyme assays. It has also been validated in the evaluation of protein–peptide interactions, protein–DNA interactions, and cellular adhesion molecule binding. It is a homogenous assay format and therefore does not require the classical physical separation step or the need for scintillation cocktails. It is compatible with ^{3}H, ^{14}C, ^{33}P, ^{35}S, and ^{125}I-based assays where the beads contain an embedded scintillant that converts the energy from radioactive decay to light when the radionuclide and bead are in close proximity. The blue light emission from the SPA scintillation bead is then detected in a PMT-based scintillation counter. SPA can also be used in imaging detection systems, where the bead emits a red light that can be detected in a charge-coupled device (CCD) camera detector. Radioactive decay that occurs in solution at a distance greater than the decay path length of the B-particle in the reaction mixture will not stimulate the scintillant bead. Fig. 11 illustrates the principle.[46]

SPA: FlashPlate is a plate-based version of SPA. Each well of the microplate is coated with a thin layer of polystyrene-based scintillant which provides the platform for the nonseparation assay. Similar to the bead-based SPA, no scintillation cocktail is required. Flashplates are available in 96-well and 384-well format. Fig. 12 illustrates the design of the well interior of a FlashPlate.

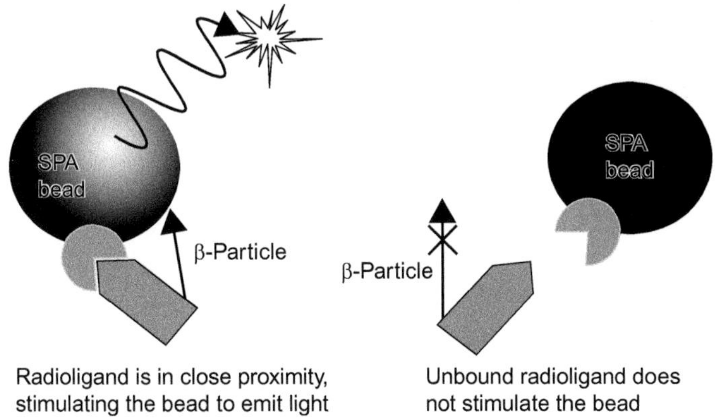

Fig. 11 Principle of the SPA technology.

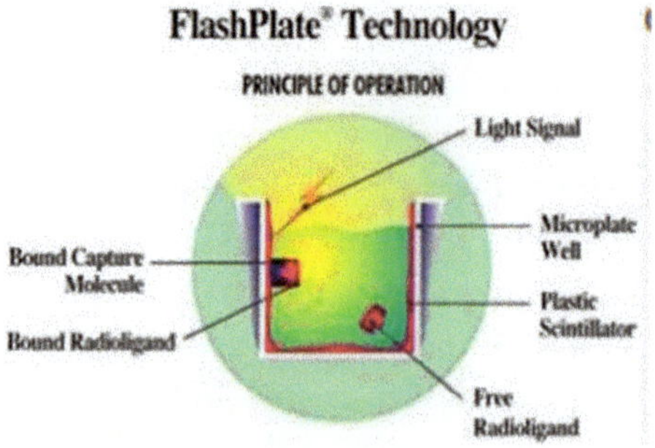

Fig. 12 FlashPlate technology.[46]

3.5 Plate Readers

In HTS, the majority of readers use well-based detection systems where the signal is measured from the entire microplate well and the reader is multimodal to enable support of the different assay technologies used in a typical screening lab. Most of these readers rely on PMTs where one of several types of light sources are combined with specific excitation and emission filters to manage the wavelengths required for a specific assay technology. A second type of reader is CCD based and records the image of an entire plate in one read. These CCD-based readers are enabling for high-density assay plates because of the fast read detection times and lower well-to-well variability.[47]

3.6 PMT-Based Detectors

Most PMT-based readers use a white light source such as a tungsten lamp, a xenon flash lamp, or a laser (providing additional sensitivity). More recently, LEDs have been used as a light source at a specific wavelength. PMT readers can again be divided based on how excitation and emission wavelengths are determined; one is filter based and the other is monochromator based.[48] Table 8 shows a comparison of the two options.

Filter-based detectors are more typical in screening labs due to improved efficiency of light transmission, increased sensitivity, lower overall cost, and faster ability to alternate between two wavelengths. Monochromator-based systems are typically found in assay development and mechanism-of-action labs where the ability to scan a spectrum of wavelengths is an important capability.

Examples of different types of PMT-based microplate readers and their capabilities are shown in Table 9.

Table 8 Filter and Monochromator-Based Wavelength Selection

	Filter Based	Monochromator Based
Definition	Optical filters with specific wavelengths and bandwidths are added to the excitation and emission light paths	Diffraction grating(s) are used to separate white light into the desired excitation and emission wavelengths. The wavelengths are "selectable" using the instrument software
Convenience	Multiple filters and filter sets must be managed and properly stored. Filters may need to be changed out by the operator before use	Flexible and convenient does not require filter inventories
Breadth of applicability	Cannot do spectral scans and breadth of use is dependent on available filters	Broad applications from performing a spectral scan to supporting almost any fluor in an assay
Sensitivity	Signal and sensitivity are high due to specific separation of excitation and emission wavelengths	Signal and sensitivity can be compromised

3.7 CCD-Based Detectors

CCD-based readers are enabling for high-density plate format screening, such as 1536 well, because of the fast detection speeds and reduced well-to-well variability. Depending on the specific imager, fluorescence, luminescence, absorbance, and radioactivity are supported (examples are PerkinElmer's ViewLux and MesoScale Discovery's SECTOR Imager 6000). Some of the characteristics of these systems are cooled CCDs to enhance sensitivity, coupled telecentric lenses to minimize parallax, longer exposure times in low light detection assays due to result integration on the CCD chip before read-out, ability to read format-free, and presentation of results in both numerical representation and as a visual image. Some of the disadvantages are the need for multiple raw data corrections (e.g., parallax, flatfield, shading, pixel binning, vignetting), dust interference, and maintenance of ultra-low-temperature cameras.

3.8 Radiometric Detectors

HTS radiometric detection is typically supported using either PMT-based systems such as the PerkinElmer TopCount NXT or MicroBeta2 or with

Table 9 Examples of PMT-Based Microplate Readers and the Capabilities Supported

	Detection Modality						
Reader Example	**ABS**	**FI**	**TRF**	**FP**	**Lum**	**ALPHA Techn.**	**Additional Information**
Filter based							
Infinite F Nano Plus	+	+	+	–	–	–	http://lifesciences.tecan.com/products/reader_and_washer/microplate_readers/reader_comparison?p=%20Multimode%20Reader%20Guide
PHERAstar FSX	+	+	+	+	–	+	https://cms-bmglabtech.viomassl.com/condeon/cdata/media/52886/1629817.pdf
FilterMax F5	+	+	+	+	+	–	https://www.moleculardevices.com/systems/microplate-readers/multi-mode-readers/multi-mode-microplate-reader-comparison-table
VICTOR X5	+	+	+	+	+	–	http://www.perkinelmer.com/product/victor-x5-for-fl-lum-uv-trf-fp-2030-0050
Monochromater based							
Infinite M Nano Plus	+	+	+	–	–	–	http://lifesciences.tecan.com/products/reader_and_washer/microplate_readers/reader_comparison?p=%20Multimode%20Reader%20Guide
CLARIOstar	+	+	+	+	–	–	https://cms-bmglabtech.viomassl.com/condeon/cdata/media/52886/1629817.pdf

SpectraMax i3x	+	+	+	+	+	+	https://www.moleculardevices.com/systems/microplate-readers/multi-mode-readers/multi-mode-microplate-reader-comparison-table
Filter and Monochromater based							
SynergyNeo2	+	+	+	+	+	Laser	https://www.biotek.com/products/microplate_detection/compare_multimode.html
Synergy2	+	Filter	+	+	+	Standard	https://www.biotek.com/products/microplate_detection/compare_multimode.html
Spark	+	+	+	+	+	+	http://lifesciences.tecan.com/products/reader_and_washer/microplate_readers/reader_comparison?p=%20Multimode%20Reader%20Guide
EnVision	Filter mono	Filter mono	Filter	Filter	Filter	Filter	http://www.perkinelmer.com/product/envision-multilabel-reader-2104-0010a
Varioskan LUX	Mono	Mono	Filter	–	Filter mono	Filter	https://www.thermofisher.com/order/catalog/product/VLBL0TD2?ICID=search-product

Filter, filter based; *mono*, monochromater based.

Table 10 Radiometric Detectors for High-Throughput Screening

Reader	Detector Type	Number of Detectors	Microplate Format Compatibility	Additional Information
MicroBeta2	Dual PMT	1, 2, 3, 6, or 12	96/384	http://www.perkinelmer.com/product/microbeta2-with-12-detectors-16-shelf-2450-0120?searchTerm=&pushBackUrl=
TopCount	PMT	2, 4, 6, or 12	24, 96, 384	www.perkinelmer.com/content/relatedmaterials/brochures/bro_microbetaandtopcountscint.pdf
ViewLux	CCD	Not applicable	Format independent	http://www.perkinelmer.com/product/viewlux-all-technology-1430-0010a

a CCD-based system like the PerkinElmer ViewLux. Table 10 summarizes their characteristics.

3.9 Whole Plate Kinetic Imaging

Plate readers such as the Molecular Devices FLIPR Tetra and the Hamamatsu FDSS7000EX™ enable fluorescence and luminescent-based kinetic measurements in a 96-, 384-, and 1536-well format.[49,50] These readers use cooled CCD detectors and typical applications measure intracellular calcium, support membrane potential assays, enable transporter assays, and facilitate cardiotoxicity assays that require repeated measurements.

4. ANALYSIS AND QUALITY CONTROL

The quality of screening data, as with all data, has a significant influence in the probability of success in the drug discovery process. Regardless of the difficulty of the target, improving reproducibility and the usefulness of HTS data, the very early stage of this process, is being approached at Merck & Co., Inc., Kenilworth, NJ, USA by consistent use of statistics, common data repositories across the network of research sites, and standardized reporting of data so that screening and project teams, as well as modeling and informatics groups have real-time and transparent access to the data.

Assay technology, replication of the primary screen, and QC parameters influence the degree of confidence in screening results. Some technologies (e.g., reporter gene, fluorescence intensity assays) are known to have a high degree of detection artifacts and appropriate follow-up must be done to ensure that reported activity data relate to the desired biology rather than the detection method. Independent replicates of the primary screening assay can reduce false positives and negatives due to stochastic sources such as liquid handling or reader errors. This level of redundancy may be necessary if an assay has a small "window" (difference between negative and positive control), or in the extreme case that a good positive control does not exist and thus the window is unknown. QC parameters (signal/noise, Z', repeatability) can inform acceptance/rejection of assay plates during screening and can give a sense of screen "health" throughout the campaign.

Assays with high variability or small windows are candidates for replicated primary screening, but should also be analyzed accordingly for hit picking purposes. An assay expected to have high assay-dependent false positives can be compared to historical screens of related assays to identify artifacts, while an assay expected to have nonnegligible false negative rate can be analyzed in conjunction with compound structure and bioactivity profile information to rescue missed hits.

The notion that screening data is of poor quality is only correct if one chooses not to exert the same experimental controls in screening as in any other experimental assay.

4.1 Screening Informatics

High-throughput screens generate a continuous range of assay activity values. In the case of single dose, single readout primary screens, the activity value will normally be a point estimate of percent/fractional activation/inhibition. Richer readouts, e.g., high content imaging, can yield dozens to hundreds of parameters measured for each data point, and these must be filtered or processed to reduce to a small number of metrics (e.g., activity and toxicity) that can be used to select compounds for follow-up. Regardless of the assay readout, the next stages of hit triage typically have reduced throughput, and thus a prioritized selection must be made for subsequent characterization. Typically, prioritization is made on the basis of activity in the primary assay, though specificity can be used in the initial screening if relevant measures are available. In addition to selecting the highest activity measurements, other criteria can be considered such as chemical diversity (if many compounds from the same class are active, it may not be necessary to pursue all), potential assay artifacts (does a compound frequently show activity in a given assay readout, e.g., fluorescence), and potential assay interference (e.g., does a compound with documented toxicity show activity in a loss-of-signal cell-based screen). (As a convention, we refer here to "higher" activity as the desired activity being screened for, thus more assay activity in an agonist assay and more inhibition in an antagonist assay.)

Stochastic effects can lead to both false positives and negatives. Bubbles or liquid handling errors can lead to both inactive compounds seeming active and vice versa. Such random errors are well addressed by repetition since it is unlikely that an independent experiment will suffer the same random error. Replication can be used up front (i.e., conducting a screening assay in duplicate or triplicate) or can be used in follow-up to a $N=1$ primary screen. The advantage of performing primary screening in replicate is the decrease in false negatives afforded by the ability to negate stochastic assay failures. The disadvantage is that, for a given screening capacity, this approach permits a smaller/sparser chemical space to be screened. Performing a primary screen as a single measurement followed by replicate confirmation reduces false positives but does not rescue false negatives. On balance, it would seem that resources are better spent on screening more compounds rather than compound replicates in primary screens, though assay-dependent considerations should be weighed in determining a screening strategy.[51]

Assay performance is important to optimize as much as possible.[52] Standard guidelines for assay quality (e.g., $Z' > 0.5$), assume $N=1$ primary screening, but smaller assay windows can be adequate if replication is used. Hit thresholds can be determined statistically or practically. An example of a statistical hit threshold is mean plus three times standard deviation (mean $+3\sigma$). This guideline assumes normally distributed errors and permits ~0.1% false-positive rate (i.e., 0.1% of screened compounds will pass this threshold by chance in the absence of any activity on the biology of interest). Another approach is to set a limit on number of compounds to be progressed based on resources/capacity and take that number of highest-scoring actives forward. These approaches, applied naively, may undersample actives in assays with a high rate of activity and oversample inactives in the low hit-rate case. However, both can adaptively consider chemical diversity and selectivity to downsample large hit lists and phenotypic profiling to expand small hit lists.

Systematic artifacts can affect the measurement of the activity of interest. These include assay readout artifacts (e.g., fluorescent compounds) and errors introduced by the experimental platform (plate based processing). These types of errors will not be ameliorated by repetition and are dependent on the assay type and detailed conditions.

Assay artifacts such as interference with a fluorescent readout by fluorescent compounds or quenchers, interference with a reporter assay by inhibitors of the reporter enzyme (e.g., luciferase) and interference with a metalloenzyme assay by nonspecific metal chelators can all lead to false positives and negatives. Historical data can be explored, conditional on assay readout, to identify potential problem compounds, for example, an assay screening for an increase in fluorescence can be compared to historical fluorescence assays to identify frequent hitters. Such historical bad actors might be downweighted when selecting compounds for follow-up. False negatives (e.g., a compound active in the biology of interest that also quenches the fluorescent signal) are more challenging to overcome, requiring testing in orthogonal assays.

Automated processing of plates with liquid handlers and robotics is designed to minimize variability but assay artifacts are always a possibility. Uneven heating, gas exchange, or evaporation can lead to plate effects where the edges of the plates behave in a reproducibly different manner than the center. In addition, liquid handlers and readers that scan rows/columns of a plate can lead to row/column effects. These systematic differences in measured activity, if sufficiently large, can introduce false positives and false negatives. Realizing that reproducible bias can be modelled and reduced, one potential approach is to "subtract" the position effects using, e.g., the B score.[53] The temptation to overprocess data can, however, lead to introducing noise and such approaches should be used conservatively. An example of a reasonable approach of modeling primary data is to perform analyses both with and without modeling of artifacts and to pursue the union of resulting hit lists. However, this approach minimizes false negatives at the expense of false positives and must be considered in the context of secondary assay capacity.

Active compounds can be missed due to stochastic or systematic errors associated with the assay, or because they are not tested (in the case of subset screening). In order to recover potentially interesting compounds, a number of informatics approaches can be deployed. In addition, systematic effects such as assay interference must be addressed using orthogonal readouts in the follow-up stage (e.g., an ELISA assay to follow-up hits from an HTRF primary screen). In order to identify compounds that are potentially of interest to a project team, there are at least two approaches that can be used, one based on chemical structure and the second based on phenotype. Actives from the primary screen can be used to estimate the chemical space of all possible actives. Untested compounds that fall in the same space are candidates for other potential actives. Chemical similarity in 2D (fingerprint Tanimoto index) and 3D (conformer "fuzzy" matching), can be used to search chemical space in the neighborhood of known actives and the candidates tested to ascertain their activity. Moreover, compounds with similar activity profiles across assays can be identified to search the phenotypic space in the vicinity of the observed actives. For example, HTS fingerprints of active compounds can be compared to the rest of available compounds and those with a sufficiently similar profile across assays can be nominated for additional characterization.[54] Such an approach is chemotype independent, though it can be susceptible to assay artifacts (e.g., fluorescent compounds will have similar profiles across assays).

An important consideration with regard to expanding from the set of observed activities to other compounds, either real or virtual, is the best stage to deploy such an approach. Expanding hits after the primary screen has the potential advantage of being able to seamlessly integrate the model predictions with the observed hits in the assay triage funnel. The disadvantage is that the systematic and stochastic false positives and negatives in the primary assay have the potential to pollute the modeling effort. Since there is limited capacity for follow-up, this may lead to missed opportunities if some compounds suggested by biologically interesting hits are not followed up in favor of testing compounds of similar structure or "stronger" assay actives that are in fact artifacts. As the primary hit list is triaged, the activity data increase in quality and thus modeling based on chemical structure or phenotypic profiles can better prioritize compounds for expanded testing.

After confirmation of desired activity and removal of artifacts, hit lists are typically reduced in size but can still be too large to permit detailed mechanistic and pharmacological studies. At this point, hits may be prioritized for follow-up based on available structure–activity relationships, synthetic tractability, and intellectual property (IP). We include empirical triage of hits to identify compounds with promising activity profiles that might not be the most potent or chemically attractive. Empirical triage of hit lists can benefit from phenotypic profiling, whereby compounds are tested in broad/generic assays for biological function to identify on- and off-target effects. Approaches such as gene expression profiling or cell painting can be used to categorize compounds into phenotypic classes, to estimate specificity/pleiotropy of the compounds, to predict potential liabilities and to propose molecular mode of action for phenotypic screening hits.[55,56] The output of such methods is typically of high dimensionality and may be difficult to interpret, requiring significant investment in bio/cheminformatic analysis to convert measurements into insights. Nevertheless, such broad and unbiased approaches have the potential to reveal unexpected connections which can aid in the selection of the most promising candidates for follow-up.

Finally, quantitative structure–activity relationship modeling can be deployed to predict properties of molecules that may influence their progressability. Modeling absorption, distribution, metabolism, excretion, and toxicity (ADMET) properties along with high risk off-target activities (e.g., family members, hERG) can highlight liabilities of compounds/series that need to be tested and overcome in subsequent medicinal chemistry optimization. Conversely, identifying compounds without predicted bad marks is not a guarantee of a hurdle-free progression but can be an indication of lower risk and thus a component of prioritization of classes for downstream efforts.

5. CURRENT AND FUTURE TRENDS

The drive toward treating unmet medical needs with increasingly complex pathobiology has pushed screening science toward unprecedented targets and increased the complexity of the screening operation to integrated campaigns that use multiple modalities. In this section, we will address (1) the changing landscape of screening in pharma, (2) some current integrated screening strategies, (3) screening at academic labs and contract research organizations (CROs) and (4) future directions.

6. CHANGING LANDSCAPE OF SCREENING IN BIG PHARMA

How do we measure success in HTS? As mentioned earlier, HTS began in the late to mid-1980s and HTS publications started to appear in the early 1990s. The Society for Biomolecular Screening, now the Society for Lab Automation and Screening, was founded in 1994. With HTS solidly in its third decade and knowing that target identification to FDA approval averages 13.5 years, there is now sufficient time and track record to evaluate the impact HTS has had to small-molecule drug discovery.[57]

It is generally accepted that the best indication of HTS success is the identification of compounds that can be advanced to success in the clinic. This success tends to correlate to sufficiently diverse and "lead-like" chemical series discovered in HTS, such that frequently, several diverse classes of chemical matter are required to reach this successful endpoint. Use of simple "hit rate," i.e., % of confirmed hits (confirmed in a concentration–response curve (CRC) and frequently also in a subsequent orthogonal assay), is often a misleading metric as significant redundancy and conversely insufficient chemical diversity, exists in some large pharma compound collections. Thus, having a high "hit rate" may not sustain a successful medical chemistry effort. Similarly, targets with low hit rates can have successful endpoints especially if different structural series identify pharmacophores that modulate target activity and do not carry off-target liabilities. Therefore, it is the ability to sample broad chemical diversity that is more valuable than high hit rates per se. Attempts by drug discovery scientists to maximize sampling diversity has led to the genesis of screening campaigns, i.e., deploying several concurrent approaches (functional, affinity based, fragment, virtual) at multiple nonoverlapping collections. The concept of a screening campaign allows one to mine the chemical matter with different technologies as opposed to a single-pronged approach to lead identification at a target. How does this translate to success?

Macarron et al. reported that HTS campaigns have a 48%–84% rate of success in finding chemical matter to start a chemical optimization process with 36%–38% of programs advancing to candidate selection.[12] If one assesses the number of drugs derived from starting points identified in a screening campaign, the "screen to drug success rate" is 33%. An analysis by Perola found that of 58 drugs derived from well-documented leads, 19 of these came from HTS.[58] Some examples are (1) Merck Sharp and Dohme's (MSDs) HIV integrase inhibitor raltegravir (Isentress) and sitagliptin (Januvia); (2) Boehringer Ingelheim's HIV protease inhibitor, tipranavier (Aptivus); (3) Bayer's Factor Xa inhibitor rivaroxaban (Xarelto) for thromboembolic disorders; (4) Pfizer's HIV entry inhibitor, maraviroc, a CCR5 antagonist (Selzentry); and (5) GSK-Ligand's TPO mimetic eltrombopag (Promacta) for short-term idiopathic thrombocytopenic purpura (ITP).[59–62] This 33% "screen to drug" rate must be looked at from the perspective of the vagaries of target validation and clinical development, the low diversity and quality of early compound collections, screening technologies and the target to approval timeline in drug discovery. One can optimistically say the screen to drug success rate will increase given the many improvements in screening science, however, today's targets have much less precedent and will therefore likely require new strategies for success.

7. CURRENT HTS STRATEGIES

Screening large numbers of compounds vs screening chemical diversity. Macarron et al. reported that screening of a 2–3 million diverse compound library from big pharma was sufficient to find leads for ~60% of targets. As this represented the targets of the previous two decades, it is unclear how these collections will fare against unprecedented targets of today.[12] Today's large protein targets, often with large molecule binding pockets, or membrane proteins that are hard to solubilize represent challenges when targeting a small molecule intervention.

As mentioned earlier, it is generally accepted that it is not how many compounds screened that is the most important factor for lead identification, but the ability to screen a diverse collection of compounds using varied technologies. However, the temptation to screen every compound is great in order to "not miss anything." This leads to numerous discussions among scientists on project teams regarding whether screening a representative set of the "parent collection" is sufficient for the particular target, or whether screening of the parental collection is warranted. Several recent studies of small, selected compound clusters have shown that the total number of wells screened can be reduced, while capturing 75%–80% of the true actives. This was achieved by screening a subset of the parental set.[63–66] Karnachi and Brown[63] used compound clustering and iterative screening rounds to identify 97% of the structural classes while screening ~25% of their compound collection. Screening a well-chosen representative set of compounds that captures ~80% of the diversity of the parent collection is a frequently used approach that has been especially successful when statistical and iterative screening, i.e., combine screening at $N=3$ to reduce false positives and negatives (false positives are problematic for model building), with mathematical model building and informatics driven similarity searches of the parent collection, Fig. 13. If the hit rate is too low, one can test another subset of the parent deck and if the deck is plated "progressively" this provides a rapid means to screen in a step-wise fashion. Thus, iterative focused screening (IFS) of a well-chosen representative set of the parent collection is a reasonable alternative to "full deck screens" and provides the opportunity to screen more targets by virtue of its improved efficiencies in costs and compounds.

Another current advance in screening is the recognition that the compound collections need not be all drug-like small molecules (MW < 500), but can and do include larger molecules (MW 500–1000), fragments (MW 250–350), macrocycles, peptides, and cyclic-peptides. Protein: protein interactions represent many current targets, and peptides are viewed as attractive candidates for interrupting these interactions.[67] Again, druggability comes into play but with most large pharma's primary small molecule libraries averaging 2–3 million compounds, one must ask, what

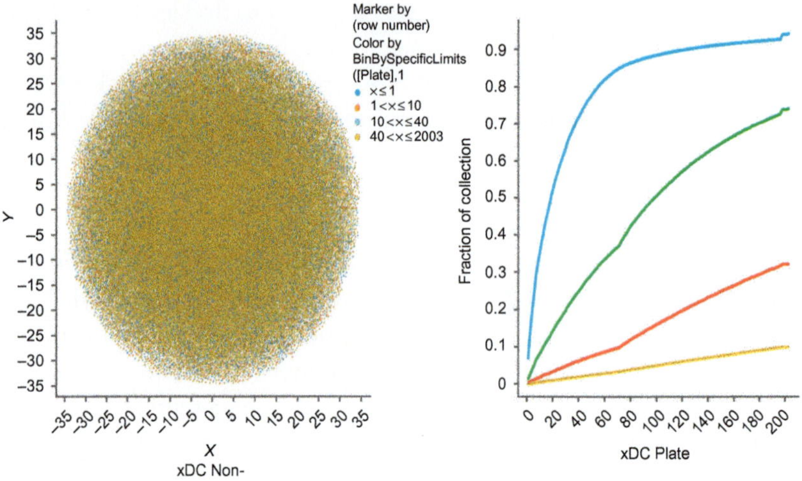

Fig. 13 Representation of a parent collection with a model and informatics-driven compound subset.

are the best strategies to increase probability of success at today's hard targets? Choosing a strategy that combines sampling a representative set of the company's primary collection in a functional screen with other approaches such as affinity based and fragment screens adds to that target's probability of success. If one adds in the practical issue of budget constraints, the issue can be framed as one of opportunity cost and diminishing returns vs quality sampling of chemical diversity with multiple approaches.

While doing the right science should always be the main consideration of any drug discovery strategy, budgets often have to be considered. The mean annual capital budgets for 10 large pharma respondents in the 2014 HTStec survey was \$3.5 M and the reagents/consumables budget \$3.9 M. This is a fraction of the R&D expenditures for 13 major pharma which ranged from \$22B to \$72B over the 8-year period from 2006 to 2014. Given the number of new molecular entities filed over the same time period, these major companies had an R&D efficiency of \$3-32B/NME.[68] This points to the need to consider costs in all drug discovery innovation stages including in doing purposeful screening. To this end, many companies have turned to modeling and informatics assessments of their large compound collections to determine if and how a representative subset(s) can be well-chosen to cover the chemical diversity of the parent collection(s). Further, as discussed earlier, the use of "statistical-based screening" or "IFS" improves data quality and facilitates the rapid follow-up by choosing similar to further mine the parent collection. It builds a knowledge environment vs a data point one.

8. INTEGRATED SCREENING APPROACHES

In large pharma, integrated and multimodality approaches to screening are commonly used to maximize success. Current screening campaigns are usually a combination of (1) a cell-free or cell-based "functional" screen directed at a target, a pathway or a phenotype; (2) a fragment screen; and (3) an affinity-based technology. Structure-enabled and virtual screening are additional technologies that complement a fully integrated screening approach but are beyond the scope of this review. It should be noted that rarely are all modalities used at a single target but often an integration of several of these approaches is necessary to increase the chemical diversity and probability of screen to drug success.

8.1 Fragment-Based Lead Discovery

In contrast to screening millions of compounds in functional methods that read-out as an activity increase or decrease, fragment-based lead discovery (FBLD) screens a few thousand compounds of a much reduced molecular mass of approximately <300 Da that reads-out as a biophysical event. The contrasts with functional screening do not end there; the screening methodology has a strong dependence on structural, conformational, and computational methods and a tolerance for weak potency in initial stages of a medical chemistry effort to build a small molecule out to a larger and more potent one. The availability of a crystal structure (or other structural information) of the protein target is of great importance for FBLD. There are now numerous successes from FBLD most notably the approval of PLX4032 (Zelboraf).[69] The rule of three in FBLD (<300 Da, up to 3 H bond donors, up to 3 H bond acceptors and clogP <3) reduces the number of possible molecules and improves the qualitative interactions with high ligand efficiency.[70] The low potency in fragment hits necessitates a sensitive and robust assay capable of detecting weak interactions. NMR spectroscopy and surface plasmon resonance and notable FBLD methods for screening and thermal shift methods are among newer methods with application to fragment screening. One can see that the structural-based nature of FBLD is an alternative method to the activity read-out of functional screening and therefore is complementary part of an integrated screening strategy.

8.2 Affinity-Based Technologies

Affinity selection mass spectroscopy (ASMS) is another complement to function-based screening, and one with great potential. Since the mid-

1990s, ASMS methods have been employed to screen mixtures of large numbers of compounds with the readout being a simple binding event to the target of interest. Affinity methods employ mass spectrometric detection of compound/target binding, as opposed to substrate turnover or probe competition. Three different ASMS approaches from Abbott, Novartis, and MSD have been reviewed recently by O'Connell et al.[71] These methods involve preincubation of target with a mixture of compounds, isolating the target with the compound(s) bound to it, and analysis of the bound compounds. This solution-based ASMS affords good throughput (1 million compounds per day) and isolates compounds based on binding to various allosteric as well as orthosteric sites in a target. Combined with an orthogonal functional assay, detection of new classes of molecules is possible. A limitation of the ASMS approach is the lack of a unified commercial solution. Rather, one must build and integrate a system from commercially available components and integrate with a software solution to deconvolute the compound identity.

9. PHYSIOLOGICALLY RELEVANT CELLS

In addition to the above-mentioned components of an integrated screen, one of the most significant short-comings in current screening is the (in) ability to use rare or physiologically relevant cells (PRC) at scale for a primary screen. Despite many advances, this is an issue in low-throughput assays as well.[72] HTS has long tried to use more PRC in orthogonal assays in a screening funnel and even here the reviews are mixed with oncology groups using PRC more than other disease areas. The question should be asked: what is a "physiologically relevant cell"? The answer is a relative answer in that a PRC is more relevant to the target than an engineered cell line or a transformed or immortalized cell line but it may not necessarily be a primary cell, a stem cell, patient-derived cells, or other 3D or organotypic cell types that closely represents the physiological or pathological host cell. For instance, a THP-1 cell is more relevant than an engineered cell for some targets if, for instance, it is in the same cell lineage. However, are data derived from them more translatable and therefore more relevant? A renal carcinoma cell line likely yields more translatable data than an engineered CHO line but that cell type is as good or better than podocytes? The challenges that hold this field back include artificial cell culture environments, reproducibility and scalability. For instance, the microenvironments that healthy or diseased cells normally grow in are not optimized for the rapid growth that culture conditions typically engender. High serum and nutrient environments required for fast growth dedifferentiate cells or cause drift in their genetic and epigenetic profiles.[73] Today, coculture conditions are being engineered with more appropriate substrates than the plastic-ware of cell culture flasks in a 2D environment. Substrates such as matrigel and collagen type I may also not be ideal mimetics of the complex extracellular environment in the human body. In addition, growth of a single cell type that is neither contacting nor communicating with other cells is also artificial. Deriving coculture conditions that address these limitations in a scalable and reproducible way is a significant challenge that must be overcome to address "relevance." Assays built on 3D cell cultures better reflect the architecture of tissues and organs are a compromise to support throughput and relevance.[72] With current technology limitations, such assays may be better as orthogonal assays in a screening funnel to build confidence that the hits from a screen are more translatable.

10. SCREENING AT ACADEMIC INSTITUTIONS AND CROs

Though HTS started in the late 1980s to early 1990s at pharmaceutical companies, over the years medium and small biotech companies, academic, governmental, and not-for-profit screening sites as well as CROs have also built up HTS capabilities. The consolidation within the pharmaceutical industry over the last decades has reduced the number of industrial screening sites. Consolidation has also been observed in the nonindustry HTS sites as illustrated by the Molecular Libraries Screening Center Network being replaced by Molecular Libraries Production Center Network to consolidate identification of screening, SAR and chemical probes for chemical biology and help to generate probes to dissect ever more complex biology. The number of academic screening groups has (reportedly) decreased based on mixed reviews, low success, and often the need to share data publicly.[74] However, some large academic groups continue.

Over the last decades, there has been an increase in capable CROs providing HTS options for large and small pharmaceutical companies, as well as biotech and academics to conduct screening of either the CRO's or the client's compound collections, with the client retaining IP rights. To some extent, this growth has been driven by the pharmaceutical industry simplifying to core competencies, reducing fixed costs and consolidating to larger vendors, the CRO community's strategic desire to provide end-to-end early drug discovery support for integrated programs and increased funding for small or virtual biotech organizations focusing on early discovery with a need for new chemical matter. According to HTStec survey data of 10 global pharmaceutical company participants in 2014 their interest to outsource screening declined from ~22% in 2011 to 10% in 2013.[74] Reasons to outsource were primarily to manage capacity restraints but also to access complementary capabilities or instrumentation that may not be available at the pharmaceutical company such as electrophysiology, higher biosafety level facilities, high content screening, etc.

The ability to aggregate the screening operations into fewer screening sites either through consolidation within the pharmaceutical industry, academia, or governmental screening centers or CROs supporting multiple clients could provide economy of scale and cost benefits beyond what any single organization can manage. This is perhaps best illustrated when one considers the direct costs of any screening approach, comprising the cost of people and overhead, the capital depreciation as well as laboratory supplies, all of which can have cost efficiencies when performed at scales greater than any individual organization may need or be able to do.

Irrespective of the approach to identify new active compounds, the underlying need to identify the best starting points the quickest and at the lowest cost transcends all screening modalities described in this review (classical HTS, screening mixtures vs singletons, DNA-encoded libraries/binding affinity screens, fragment screens, virtual screens, etc.). The different screening modalities described in this review have their inherent strengths, weaknesses and cost structures and like-for-like cost comparisons are not necessarily easy. The overarching goal of any screening campaign should be to understand the strengths and weaknesses of the different screening technologies available and thus to adopt a flexible approach to the screening campaign and tactically deploy the right screen(s)/screening modality, at the right time for the right target. In reality, this will be driven by an organization's existing capabilities or capabilities they can source elsewhere but stressing one modality's strengths over another such as DNA-encoded libraries vs classical libraries likely misses the point described earlier as more integrated screening campaigns that use multiple screening modalities may well be the best approach to identify new chemical matter.

11. CONCLUSION AND FUTURE DIRECTIONS

The sources of new chemotypes for current targets being prosecuted are viewed as coming from screening internal compound collections, with other sources being fragment screening, ligand-based design, in silico, and licensing (HTStec 2014). The need for new chemical technologies such as DNA-encoded libraries[75] or mRNA-encoded peptide libraries,[67] where compound numbers are in the 10^{10} or 10^{13}, respectively, are also viewed as essential needs for today's targets. New investments in screening will also include new assay technologies, more and smarter robotics, and training staff to fully utilize the flexibility and power of robotic technologies.

"HTS" is likely to remain the main route to lead identification, however, it is likely to transform from screening small molecules in a single activity, i.e., a "functional" assay to an integrated set of modalities that employs more modeling informatics, mechanistic diversification of chemical matter. No strategy should be unchallenged, or unchanged for long. Clearly, screening strategies in combination with various new data and technologies can be more successful as they adapt regularly to changing demands of drug discovery. The search for new chemical entities for novel drug targets will typically now involve several HTS campaigns conducted during the lifetime of the project. Each screen may differ in the way libraries are chosen and using a variety of assay formats to bias the screen as the requirements of the project become apparent. Today, HTS is a mature technology, the effectiveness of which is maximized when used in combination with complementary technologies and the leverage of emerging knowledge to identify the starting points for the medicines of tomorrow.

ACKNOWLEDGMENTS

The authors thank our colleagues in the HTS field both within and outside of MSD, who have shared their data, insight, and passion for doing good science with us over the years. We especially recognize the automation expertise of Jason Cassaday and Brian Squadroni for development of the Telios-based systems shown in Fig. 1B. We also humbly dedicate this work to Dr. Frank Brown who was a visionary, pioneer, and advocate for using statistically based screening data in modeling and informatics. Your voice is missed Frank, but we still hear you.

REFERENCES

1. Kalso, E. Oxycodone. *J. Pain Symptom Manage.* **2005**, *29*(5S), S47–S56.
2. Roberts, N.; Martin, J.; Kinchington, D.; Broadhurst, A.; Craig, J.; Duncan, I.; Galpin, S.; Handa, B.; Kay, J.; Krohn, A.; et al. Rational Design of Peptide-Based HIV Proteinase Inhibitors. *Science* **1990**, *248*, 358–361.
3. Pereira, D. A.; Williams, J. A. Origin and Evolution of High Throughput Screening. *Br. J. Pharmacol.* **2007**, *152*(1), 53–61.
4. Lander, E.; et al. Initial Sequencing and Analysis of the Human Genome. *Nature* **2001**, *409*, 860–921.
5. https://www.ncbi.nlm.nih.gov/books/NBK53196/.

6. Dove, A. Drug Screening—Beyond the Bottleneck. *Nat. Biotechnol.* **1999**, *17*, 859–863.
7. Cumming, J. G. Chemical Predictive Modelling to Improve Compound Quality. *Nat. Rev. Drug Discov.* **2013**, *12*, 948–962.
8. Lipinski, C. A.; Lombardo, F.; Dominy, B.W.; Feeney, P.J. Experimental and Computational Approaches to Estimate Solubility and Permeability in Drug Discovery and Development Settings. Adv. Drug Deliv. Rev. Vol. 23, Issues 1–3 January 1997, pp. 3–25.
9. Raghunandan, M. K.; Woolf, P. J. Pooling in High-Throughput Drug Screening. *Curr. Opin. Drug Discov. Devel.* **2009**, *12*(3), 339–350.
10. Jacoby, E.; Schuffenhauer, A.; Popov, M.; Azzaoui, K.; Havill, B.; Rigollier, P.; Stoll, F.; Koch, G.; Meier, P.; Orain, D.; Giger, R.; Hinrichs, J.; Malagu, K.; Zimmermann, J.; Rioth, H.-J. Key Aspects of the Novartis Compound Collection Enhancement Project for the Compilation of a Comprehensive Chemogenomics Drug Discovery Screening Collection. *Curr. Top. Med. Chem.* **2005**, *5*, 397–411.
11. Lane, S. J.; Eggleston, D. S.; Brinded, K. A.; Hollerton, J. C.; Taylor, N. L.; Readshaw, S. A. Defining and Maintaining a High-Quality Screening Collection: The GSK Experience. *Drug Discov. Today* **2006**, *11*, 267–272.
12. Macarron, R.; Banks, M. N.; Bojanic, D.; Burns, D. J.; Cirovic, D. A.; Garyantes, T.; Green, D. V. S.; Hertzberg, R. P.; Janzen, W. P.; Paslay, J. W.; Schopfer, U.; Sittampalam, G. S. Impact of High-Throughput Screening in Biomedical Research. *Nat. Rev. Drug Discov.* **2011**, *10*, 188–195.
13. Boettxher, A.; Mayr, L. Miniaturisation of Assay Development and Screening. Drug Discov. World. 2006, 2006, 17–27 Summer. http://www.ddw-online.com/screening/p97061-miniaturisation-of-assay-development-and-screening-summer-2006.html (accessed Feb 22, 2017).
14. HighRes Biosolutions *ACell.* http://highresbio.com/systems/acell.php (accessed Feb 22, 2017).
15. Thermo Spinnaker. https://tools.thermofisher.com/content/sfs/brochures/Spinnaker-Robot-specsheet.pdf (accessed Feb 22, 2017).
16. BLUECAT BIO *Bluebench.* http://www.bluecatbio.com/bluebot.html (accessed Feb 22, 2017).
17. Cobots. http://www.perosh.eu/safe-co-operation-between-human-beings-and-robots-cobots (accessed Feb 22, 2017).
18. Cobots. http://www.engineering.com/AdvancedManufacturing/ArticleID/12169 (accessed Feb 22, 2017).
19. Cobots. http://www.engineering.com/AdvancedManufacturing/ArticleID/13540/A-History-of-Collaborative-Robots-From-Intelligent-Lift-Assists-to-Cobots.aspx (accessed Feb 22, 2017).
20. Universal Robots Collaborative dual gripper. https://blog.universal-robots.com/launching-at-automate-2017-is-the-new-dual-gripper-urcap (accessed Feb 22, 2017).
21. Zheng, W.; Chen, C. Screening Automation. In: *A Practical Guide to Assay Development and High-Throughput Screening in Drug Discovery*; Czarnik, T., Yan, A. W., Chen, B., Eds.; CRC Press: Boco Raton, FL, USA, 2010, pp 184–185.
22. Jones, E.; Michael, S.; Sittampalam, G. S. Basics of Assay Equipment and Instrumentation for High Throughput Screening. In: *NIH Assay Guidance Manual*; Sittanpalam, G. S., Coussens, N. P., Brimacombe, K., Eds.; Eli Lilly & Company/National Center for Advancing Translational Sciences: Bethesda, MD, USA, 2016. https://www.ncbi.nlm.nih.gov/books/NBK92014/.
23. Beckman Coulter Life Sciences Liquid Handling Home Page. http://www.beckman.com/liquid-handling-and-robotics (accessed Mar 30, 2017).
24. Hamilton Company Home Page. https://www.hamiltoncompany.com/products/automated-liquid-handling (accessed Mar 30, 2017).

25. Analytil-Jena Home Page. https://www.analytik-jena.de/en/lab-automation/products-lab-automation/liquid-handling.html (accessed Mar 30, 2017).
26. Perkin Elmer Home Page. http://www.perkinelmer.com/category/automation-liquid-handling-instruments (accessed Mar 30, 2017).
27. Tecan Life Sciences Home Page. http://lifesciences.tecan.com/products/liquid_handling_and_automation (accessed Mar 30, 2017).
28. TTP LabTech Liquid Handling Mosquito. http://ttplabtech.com/liquid-handling/mosquito_hts (accessed Mar 28, 2017).
29. TTP LabTech Liquid Handling DragonFly. http://ttplabtech.com/liquid-handling/dragonfly_screen_optimisation (accessed Mar 30, 2017).
30. VP Scientific Pin Tools. http://www.vp-scientific.com/prod_gr_robot_pin_tools.htm (accessed Mar 28, 2017).
31. Cleveland, P. H.; Koutz, P. J. Nanoliter Dispensing for uHTS Using Pin Tools. *Assay Drug Dev. Technol.* **2005**, *3*(2), 213–225.
32. Labcyte Home Page. http://www.labcyte.com/products/liquidhandling/echo-555-liquid-handler (accessed Mar 28, 2017).
33. EDC Biosystems Home Page. http://www.edcbiosystems.com (accessed Mar 28, 2017).
34. Agrawal, S.; Cifelli, S.; Johnstone, R.; Pechter, D.; Barbey, D. A.; Lin, K.; Allison, T.; Agrawal, S.; Rivera-Gines, A.; Milligan, J. A.; Schneeweis, J.; Houle, K.; Struck, A. J.; Visconti, R.; Sills, M.; Wildey, M. J. Utilizing Low-Volume Aqueous Acoustic Transfer With the Echo 525 to Enable Miniaturization of qRT-PCR Assay. *J. Lab. Autom.* **2016**, *21*(1), 57–63.
35. Thermo Fisher Combi. https://tools.thermofisher.com/content/sfs/brochures/D11002~.pdf (accessed Mar 28, 2017).
36. Biotek Multifo. http://www.biotek.com/products/liquid_handling/multiflo_microplate_dispenser.html (accessed Mar 28, 2017).
37. Formulatrix Home Page. http://www.formulatrix.com/liquid-handling/index.html (accessed Mar 28, 2017).
38. Tecan Digital Dispenser. http://lifesciences.tecan.com/products/liquid_handling_and_automation/tecan_d300e_digital_dispenser (accessed Mar 28, 2017).
39. PerkinElmer FlexDrop. https://shop.perkinelmer.com/Content/RelatedMaterials/SpecificationSheets/spc_flexdrop.pdf (accessed Mar 28, 2017).
40. APC International, Ltd. Piezo-Mechanics: An Introduction. https://www.americanpiezo.com/images/stories/content_images/pdf/apc_stack_principles.pdf; 2015 (accessed Mar 28, 2017).
41. Tekmatic Home Page. http://tekmatic.com (accessed Mar 28, 2017).
42. Degorce, F.; Card, A.; Soh, S.; Trinquet, E.; Knapik, G. P.; Xie, B. HTRF: A Technology Tailored for Drug Discovery—A Review of Theoretical Aspects and Recent Applications. *Curr. Chem. Genomics* **2009**, *3*, 22–32.
43. Hall, M. D.; Yasgar, A.; Peryea, T.; Braisted, J. C.; Jadhav, A.; Simeonov, A.; Coussens, N. P. Fluorescent Probes Sensitive to Changes in the Cholesterol-to-Phospholipids Molar Ratio in Human Platelet Membranes During Atherosclerosis. *Methods Appl. Fluoresc.* **2016**, *4*(2), 034013.
44. Miao, W. Electrogenerated Chemiluminescence and Its Biorelated Applications. *Chem. Rev.* **2008**, *108*, 2506–2553.
45. Eglin, R. M.; Reisine, T.; Roby, P.; Rouleau, N.; Illy, C.; Bosse, R.; Bielefeld, M. The Use of AlphaScreen Technology in HTS: Current Status. *Curr. Chem. Genomics* **2008**, *1*, 2–10.
46. PerkinElmer Scintillation Proximity. http://www.perkinelmer.com/lab-products-and-services/application-support-knowledgebase/radiometric/spa-ligand-binding.html (accessed May 1, 2017).

47. Chen, T. *A Practical Guide to Assay Development and High-Throughput Screening in Drug Discovery*. CRC Press: Boca Raton, FL, USA, 2009; Print ISBN: 978-1-4200-7050-7, eBook ISBN: 978-1-4200-7051-4.
48. Comley, J. Monochromator vs Filter-based Plate Readers; Horses for Courses, or a Winning Combination? *Drug Discov. World* Fall **2007**, 34–51.
49. Molecular Devices FLIPR Tetra. https://www.moleculardevices.com/systems/flipr-tetra-high-throughput-cellular-screening-system (accessed May 1, 2017).
50. Hamamatsu *FDSS 7000*. http://www.hamamatsu.com/us/en/product/category/5002/5021/FDSS7000EX/index.html (accessed May 1, 2017).
51. Pertusi D.A., et al.; Prospective Assessment of Virtual Screening Heuristics Derived Using a Novel Fusion Score, SLAS Discov., 2017. https://doi.org/10.1177/2472555217706058.
52. Zhang, J.-H.; Chung, T. D.; Oldenburg, K. R. A Simple Statistical Parameter for Use in Evaluation and Validation of High Throughput Screening Assays. *J. Biomol. Screen.* **1999**, *4*(2), 67–73.
53. Brideau, C.; et al. Improved Statistical Methods for Hit Selection in High-Throughput Screening. *J. Biomol. Screen.* **2003**, *8*(6), 634–647.
54. Petrone, P. M.; et al. Biodiversity of Small Molecules—A New Perspective in Screening Set Selection. *Drug Discov. Today* **2013**, *18*(13), 674–680.
55. Subramanian A.; et al., A Next Generation Connectivity Map: L1000 Platform and the First 1,000,000 Profiles, bioRxiv, 2017, 136168.
56. Wawer, M. J.; et al. Toward Performance-Diverse Small-Molecule Libraries for Cell-Based Phenotypic Screening Using Multiplexed High-Dimensional Profiling. *Proc. Natl. Acad. Sci.* **2014**, *111*(30), 10911–10916.
57. Paul, S. M.; Mytelka, D. S.; Dunwiddie, C. T.; Persinger, C. C.; Munos, B. H.; Lindborg, S. R.; Schacht, A. L. How to Improve R&D Productivity: The Pharmaceutical Industry's Grand Challenge. *Nat. Rev. Drug Discov.* **2010**, *9*, 203–214.
58. Perola, E. An Analysis of the Binding Efficiencies of Drugs and Their Leads in Successful Drug Discovery Programs. *J. Med. Chem.* **2010**, *53*, 2986–2997.
59. Hazuda, D. J.; Felock, P.; Witmer, M.; Wolfe, A.; Stillmock, K.; Grobler, J. A.; Espeseth, A.; Gabryelski, L.; Schleif, W.; Blau, C.; Miller, M. D. Inhibitors of Strand Transfer That Prevent Integration and Inhibit HIV-1 Replication in Cells. *Science* **2000**, *287*, 646–650.
60. Brockunier, L. L.; He, J.; Colwell, L. F., Jr.; Habulihaz, B.; He, H.; Leiting, B.; Lyons, K. A.; Marsilio, F.; Patel, R. A.; Teffera, Y.; Wu, J. K.; Thornberry, N. A.; Weber, A. E.; Parmee, E. R. Substituted Piperazines as Novel Dipeptidyl Peptidase IV Inhibitors. *Bioorg. Med. Chem. Lett.* **2004**, *14*, 4763–4766.
61. Xu, J.; Ok, H. O.; Gonzalez, E. J.; Colwell, L. F., Jr.; Habulihaz, B.; He, H.; Leiting, B.; Lyons, K. A.; Marsilio, F.; Patel, R. A.; Wu, J. K.; Thornberry, N. A.; Weber, A. E.; Parmee, E. R. Discovery of Potent and Selective Beta-Homophenylalanine Based Dipeptidyl Peptidase IV Inhibitors. *Bioorg. Med. Chem. Lett.* **2004**, *14*, 4759–4762.
62. Thaisrivongs, S.; Tomich, P. K.; Watenpaugh, K. D.; Chong, K. T.; Howe, W. J.; Yang, C. P.; Strohbach, J. W.; Turner, S. R.; McGrath, J. P.; Bohanon, M. J.; et al. Structure-Based Design of HIV Protease Inhibitors: 4-Hydroxycoumarins and 4-Hydroxy-2-Pyrones as Non-Peptidic Inhibitors. *J. Med. Chem.* **1994**, *37*, 3200–3204.
63. Karnachi, P. S.; Brown, F. Practical Approaches to Efficient Screening: Information-Rich Screening Protocol. *J. Biomol. Screen.* **2004**, *9*, 678–686.
64. van Rhee, A. M.; Stocker, J.; Printzenhoff, D.; Creech, C.; Wagoner, P. K.; Spear, K. L. Retrospective Analysis of An Experimental High-Throughput Screening Data Set by Recursive Partitioning. *J. Comb. Chem.* **2001**, *3*, 267–277.
65. Blower, P. E.; Cross, K. P.; Eichler, G. S.; Myatt, G. J.; Weinstein, J. N.; Yang, C. Comparison of Methods for Sequential Screening of Large Compound Sets. *Comb. Chem. High Throughput Screen.* **2006**, *9*, 115–122.

66. Sun, D.; Jung, J.; Rush, T. S.; Xu, Z.; Weber, M. J.; Bobkova, E.; Northrup, A.; Kariv, I. Efficient Identification of Novel Leads by Dynamic Focused Screening: PDK1 Case Stud. *Comb. Chem. High Throughput Screen.* **2010**, *13*, 16–26.
67. Hacker, D. E.; Hoinka, J.; Iqbal, E. S.; Przytycka, T. M.; Hartman, M. C. T. Highly Constrained Bicyclic Scaffolds for the Discovery of Protease-Stable Peptides via mRNA Display. *ACS Chem. Biol.* **2017**, *12*, 795–804.
68. Schuhmacher, A.; Gassmann, O.; Hinder, M. J. Highly Constrained Bicyclic Scaffolds for the Discovery of Protease-Stable Peptides via mRNA Display. *J. Transl. Med.* **2016**, *14*, 1–11.
69. Tsai, J.; Lee, J. T.; Wang, W.; Zhang, J.; Cho, H.; Mamo, S.; Bremer, R.; Gillette, S.; Kong, J.; Haass, N. K.; Sproesser, K.; Ki, L.; Smalley, K. S.; Fong, D.; Zhu, Y. L.; Marimuthu, A.; Nguyen, H.; Lam, B.; Liu, J.; Cheung, I.; et al. Discovery of a Selective Inhibitor of Oncogenic B-Raf Kinase With Potent Antimelanoma Activity. *Proc. Natl. Acad. Sci. U. S. A.* **2008**, *105*, 3041–3046.
70. Congreve, M.; Carr, R.; Murray, C.; Jhoti, H. A 'Rule of Three' for Fragment-Based Lead Discovery?*Drug Discov. Today* **2003**, *8*, 876–877.
71. O'Connell, T. N.; Ramsay, J.; Rieth, S. F.; Shapiro, M. J.; Stroh, J. G. Solution-Based Indirect Affinity Selection Mass Spectrometry—A General Tool for High-Throughput Screening of Pharmaceutical Compound Libraries. *Anal. Chem.* **2014**, *86*, 7413–7420.
72. Horvath, P.; Aulner, N.; Bickle, M.; Davies, A. M.; Nery, E. D.; Ebner, D.; Montoya, M. C.; Ostling, P.; Pietiainen, V.; Price, L. S.; Shorte, S. L.; Turcatti, G.; von Schantz, C.; Carragher, N. O. Screening Out Irrelevant Cell-Based Models of Disease. *Nat. Rev. Drug Discov.* **2016**, *15*, 751–769.
73. Nestor, C. E.; Ottaviano, R.; Reinhardt, D.; Cruickshanks, H. A.; Mjoseng, H. K.; McPherson, R. C.; Lentini, A.; Thomson, J. P.; Dunican, D. S.; Pennings, S.; Anderton, S. M.; Benson, M.; Meehan, R. R. Rapid Reprogramming of Epigenetic and Transcriptional Profiles in Mammalian Culture Systems. *Genome Biol.* **2015**, *16*, 11.
74. Comley, J. HTS Metrics and Future Directions Trends 2014, **2014**. http://www.htstec.com/consultancyitem.aspx?Item=408.
75. Goodnow, R. A.; Dumelin, C. E.; Keefe, A. D. DNA-Encoded Chemistry: Enabling the Deeper Sampling of Chemical Space. *Nat. Rev. Drug Discov.* **2017**, 131–147.

FURTHER READING

76. Brown, N. *Bioisosteres in Medicinal Chemistry*. Wiley-VCH: Godalming, Surrey, United Kingdom, 2012;237.
77. Law, J.; Zsoldos, Z.; Simon, A.; Reid, D.; Liu, Y.; Khew, S. Y.; Johnson, A. P.; Major, S.; Wade, R. A.; Ando, H. Y. *Route Designer: A Retrosynthetic Analysis Tool Utilizing Automated Retrosynthetic Rule Generation. J. Chem. Inf. Model.* **2009**, *49*(3), 593–602. https://doi.org/10.1021/ci800228y. http://pubs.acs.org/doi/abs/10.1021/ci800228y.

CHAPTER SIX

Kinase-Centric Computational Drug Development

Albert J. Kooistra*,†,2, Andrea Volkamer‡,1,2
*Centre for Molecular and Biomolecular Informatics (CMBI), Radboud University Medical Center, Nijmegen, The Netherlands
†Amsterdam Institute for Molecules, Medicines and Systems (AIMMS), Vrije Universiteit Amsterdam, Amsterdam, The Netherlands
‡Charité—Universitätsmedizin Berlin, Institute of Physiology, In-silico Toxicology Group, Berlin, Germany
[1]Corresponding author: e-mail address: andrea.volkamer@charite.de

Contents

1. INTRODUCTION: KINASES IN A NUTSHELL

Protein kinases are involved in regulation of a majority of cellular processes. In their role as enzymes, kinases modify or activate proteins by phosphorylation, thereby controlling signal transduction in the cell. Dysregulation of kinases, e.g., induced by mutations or overexpression, has been detected

[2] Both authors contributed equally to this work.

Annual Reports in Medicinal Chemistry, Volume 50
ISSN 0065-7743
https://doi.org/10.1016/bs.armc.2017.08.001

in a variety of diseases and turned kinases into well-studied drug targets for cancer, inflammation, and autoimmune disorders.[1,2] While kinases were considered "undruggable" in earlier days, the clinical success of Gleevec (Imatinib, approved by the FDA for the treatment of chronic myelogenous leukemia in 2001)[3] and many subsequent drugs turned protein kinases into one of the most pursued drug target families.

The human kinome consists of ~540 kinase-encoding genes (2% of the human genome), which were clustered by Manning and coworkers based on their sequence identity into eight main groups of eukaryotic protein kinases (ACG, CAMK, CK1, CMGC, STE, TK, TKL, other) and the atypical protein kinase families.[4] All regular protein kinases bind adenosine triphosphate (ATP) and catalyze the transfer of the phosphate group to serine, threonine, or tyrosine residues of themselves (autophosphorylation) or of other proteins. About 10% of the kinases within the human kinome lack one or more conserved residues required for catalytic activity and are therefore classified as pseudokinases.[4]

1.1 The Structure of Protein Kinases

Crystal structures of protein kinases in complex with their natural substrate or designed small-molecule inhibitors have been solved intensively. Nowadays, more than 3600 crystal structures of human kinases are available in the Protein Data Bank (PDB),[5] not including the massive amount of proprietary structures of pharmaceutical companies. An organized and structurally annotated overview of the available kinase structures can be found in the KLIFS database.[6] Many kinases are only represented by one or few crystal structures, whereas others were crystallized up to 100 times (e.g., EGFR with 127 structures, see Fig. 1A).[11] In total, more than half of the kinome is covered by the available crystal structures.

Most structures have been resolved in the active form, but only 8% in the inactive form. Kinase structures undergo large conformational changes upon activation with the most prominent states being the active (DFG-in) and inactive (DFG-out) form.[12,13] The states are named after the conformation of the DFG-motif, as the orientation of the aspartic acid (D) and phenylalanine (F) side-chains flip. This flip allows (DFG-in) or blocks (DFG-out) ATP binding and is accompanied by a large movement of the activation loop, as well as an outward movement of the G-loop and αC-helix (see Fig. 1B).

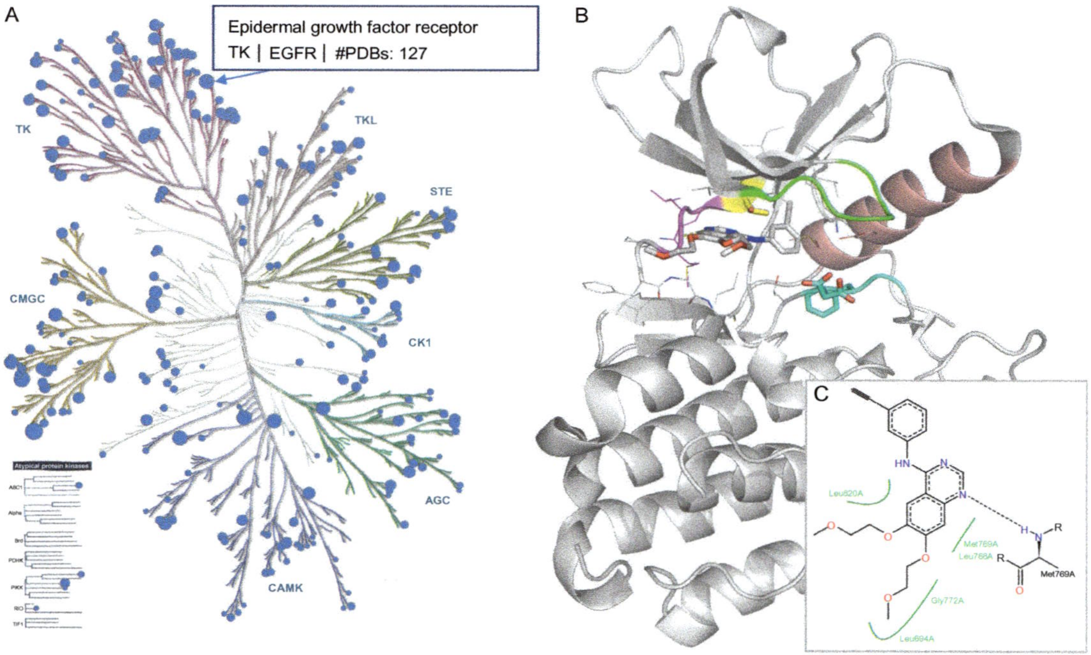

Fig. 1 (A) Number of PDBs available for human kinases (log-scale, *larger circle* indicates higher number) collected from KLIFS[6] and plotted to the kinome tree. Figure created using KinMap server.[7] (B) A representative structure of EGFR kinase (PDB code: 1m17, *cartoon* representation) bound to Erlotinib (*stick* representation),[8] highlighted are hinge region (*pink*), G-loop (*green*), αC-helix (*red*), DFG-motif (*blue*), and gatekeeper (*yellow*). Figure created using PyMol.[9] (C) Close-up of interactions of Erlotinib with a hydrogen bond to the hinge region and hydrophobic interactions in the sugar pocket (figure created using PoseView functionality in Proteins*Plus* Server[10]). *Panel (A): Illustration reproduced courtesy of Cell Signaling Technology, Inc. (www.cellsignal.com).*

1.2 FDA-Approved Kinase Inhibitors

The broad therapeutic relevance of kinases already resulted in over 34 FDA-approved small-molecule inhibitors on the market, see Fig. 2 (as of May 2017, excluding the three larger immunosuppressive drugs Everolimus, Sirolimus, and Temsirolimus).[14,15] Most of the approved drugs were designed as ATP-competitive inhibitors (Type I and II). They occupy part of the adenine pocket and form hydrogen bonds with the backbone of the hinge region (see Fig. 1B). Type I inhibitors bind to the active form of kinases (DFG-in) and block ATP binding, e.g., Erlotinib. DFG-out binding inhibitors (Type II) stabilize the enzymatically inactive kinase conformation, e.g., Imatinib. Type III and IV inhibitors are both allosteric in nature. Type III inhibitors occupy the close-by allosteric or back pocket, e.g., Trametinib, while Type IV allosteric inhibitors bind to remote sites and stabilize the inactive conformation by an allosteric regulatory mechanism, e.g., Everolimus.[16,17] Type V inhibitors, e.g., Lenvatinib, are bivalent and can bind to two different regions of the catalytic kinase domain.[18] While all these inhibitors bind noncovalently, lately interest has shifted to covalently binding inhibitors (Type VI). Covalent inhibitors mostly target noncatalytic cysteines in the ATP-binding site (front pocket) and generally bind with higher affinities due to their unlimited (irreversible) or long (reversible) residence times, e.g., Afatinib, Ibrutinib, and Osimertinib. Reactive cysteines at specific locations in the front pocket were recently investigated in a structure-based study as valuable anchor points for covalent inhibition.[19] Note that the described kinase inhibitor type nomenclature may differ throughout literature, we refer here to the classification as introduced by Roskoski.[14]

1.3 Obstacles in Kinase Research

While kinases have been well studied over the last decades (Fig. 2), there are still major hurdles in kinase research, namely (I) the need for more selective inhibitors, (II) the need for drug design campaigns focusing on novel, less explored, kinases, and (III) the need to cope with drug resistance due to mutations.

Obtaining selectivity: The identification of selective kinase inhibitors is hindered by the high conservation of the ATP-binding site of protein kinases. Many kinase inhibitors were found to be unintentionally promiscuous and to bind other kinases in addition to their aimed key target. For example, profiling studies revealed that Sunitinib inhibits more than 50% of a tested kinase panel of 290 kinases and Dasatinib showed a broad

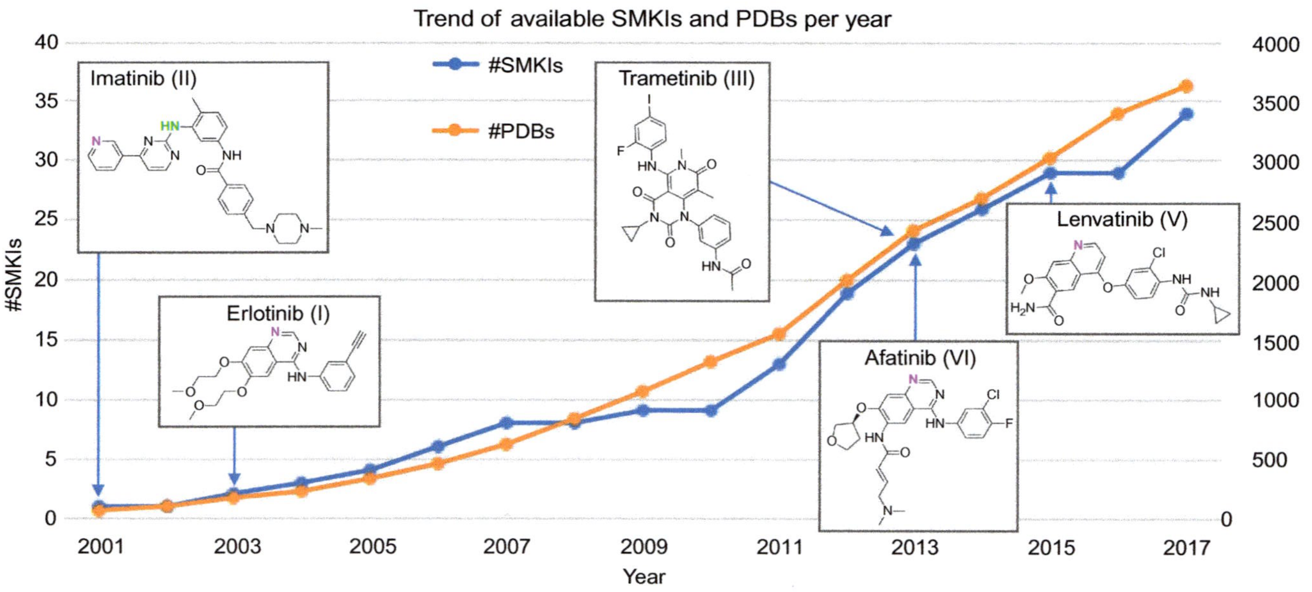

Fig. 2 Overview of the total number of FDA-approved small-molecule kinase inhibitors (# SMKIs, *blue line*), as well as the total number of available structures of human kinases in the PDB (# PDBs, *orange line*) since 2001. Additionally, the structures of five inhibitors of the different classes are shown: Type I: Erlotinib, Type II: Imatinib, Type III: Trametinib, Type V: Lenvatinib, and Type VI: Afatinib.[14] Atoms involved in hydrogen bonding with the hinge region and gatekeeper residue are colored *pink* and *green*, respectively. *Note*: PDB structures resolved before 2001 were included in the 2001 data point, numbers were extracted from KLIFS.[6]

inhibitory profile focused on several tyrosine kinases (TK).[20] In contrast, Erlotinib binds fewer kinases and only four with high affinity (EGFR, GAK, LOK, and SLK). Lapatinib shows a very selective spectrum (five kinases in total, three of which with high affinity: EGFR, ErbB2, and ErbB4; see profiles in Fig. 3A). Polypharmacology can have a synergistic effect when the drug interacts with multiple targets on a particular pathway or by covering alternative pathways to overcome resistance mechanisms.[23] In contrast, unintended drug–target interactions can cause severe side effects[24,25] and are among the most prominent causes for project termination.[26,27] Thus, tackling the challenge to selectively target a well-defined set of kinases to obtain the desired (poly)pharmacological effect remains a highly sought-after goal in drug discovery studies.[28]

Bias toward TK kinases: Historically, most FDA-approved kinase inhibitors were designed to target tyrosine kinases (TK), thus, revealing a strong bias toward clinically validated targets (depicted as green triangles in Fig. 3A).[29,30] Nevertheless, mutation studies have uncovered a different trend, indicating somatic mutations across the whole kinome tree.[31] A recent study showed that cancer driver genes are enriched for kinases distributed over several families.[32] Thus, comparing the fraction of established kinase drug targets and the amount of disease-relevant and potentially druggable targets fortifies the need to investigate novel kinase targets.[33–35]

Drug resistance: Although kinase inhibitors have been effective in the treatment of different cancer types, a major setback has been the occurrence of drug resistance. Frequently, patients initially respond well to the treatment with kinase inhibitors; however, within a year many patients develop drug resistance.[36] This is often the result of specific mutations within protein kinases that impair drug binding. Treatment of non-small cell lung cancer with Erlotinib or Gefitinib, for example, is often hampered by the mutation of the gatekeeper residue of EGFR (depicted with yellow sticks in Fig. 1B) from threonine to methionine (T790M). The T790M mutation sterically hinders the binding of these drugs and therefore the successful continued treatment of the patients.[36] This was countered by the development of Osimertinib, a precision medicine, which is able to inhibit this mutated form of EGFR.[37] As cancerous tissues generally evolve quickly (witnessed by the many annotated oncogenetic mutations annotated in the COSMIC database[38]), there is a need for the development of mutation-resistant inhibitors and for the identification of synergetic drug combination treatments[39,40].

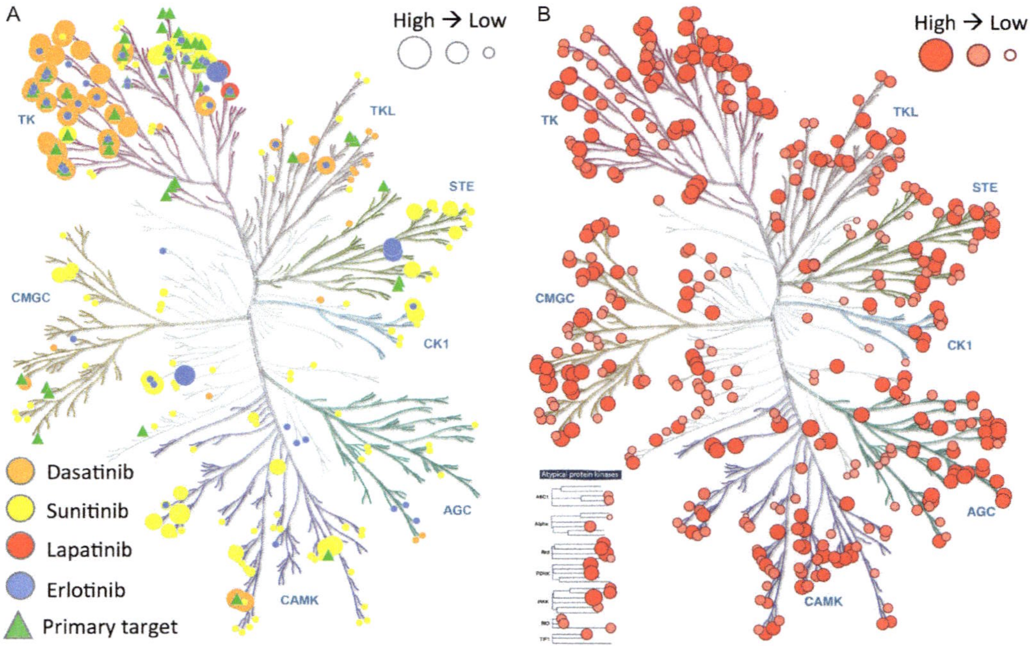

Fig. 3 (A) Binding profiles of four FDA-approved drugs tested against 290 wild-type kinases[20]: Dasatinib (*orange*), Sunitinib (*yellow*), Lapatinib (*red*), and Erlotinib (*blue*) combined with the primary targets (*green triangles*) of FDA-approved drugs as annotated in the DrugBank.[21] (B) The number of unique bioactive ($-\log K_i/K_D/AC_{50}/EC_{50}/IC_{50}$ value ≥ 5) kinase inhibitors for each human protein kinase extracted from ChEMBL.[22] Both figures were created with KinMap.[7] A larger dot size represents a higher affinity (A) or a higher number of unique inhibitors (B) on a logarithmic scale. *Panels (A and B): Illustration reproduced courtesy of Cell Signaling Technology, Inc. (www.cellsignal.com).*

1.4 Outlook of the Chapter

With a growing amount of available sequence, biochemical, and structural data comes the need and the chance for computational methods to collect, structure, and analyze this pool of information. Computational methods can offer insights into the structural determinants of kinase–ligand interactions and selectivity. In this chapter, we will present kinase-centric tools, resources, and databases as well as computational platforms and methods that can assist in kinase inhibitor design and development.

2. DATABASES, RESOURCES, AND TOOLS

Due to the pivotal role of protein kinases, they have been the topic of many research projects in both academia as well as pharmaceutical industry. This has resulted in a wealth of experimental data, which has been collected, annotated, curated, and made available via several (kinase-specific) resources that are highlighted in Tables 1 and 2. These resources have been categorized into (I) bioactivity resources (Table 1), and (II) disease, (III) sequence, and (IV) structure resources (Table 2) dependent on their primary data content. In addition, several kinase-specific computational tools (V) are also listed. For a recent and comprehensive overview of generic chemogenomics resources, we would like to refer the reader to the work of Kanev et al.[74]

Below we discuss the resources with a primary focus on human kinases in more detail and provide several examples in which the respective sources or tools have been applied. Where applicable, we will refer to EGFR as case study.

2.1 Bioactivity and Profiling Resources

A summary of the contents of the data sources discussed in the following paragraph is provided in Table 1.

2.1.1 Databases

ChEMBL—https://www.ebi.ac.uk/chembl/

A primary resource for biochemical data is the freely available ChEMBL database[22] which contains data for 11K protein targets, 1.7M unique compounds, and 14M (bio)activities (ChEMBL v23). A relatively large portion of the data within ChEMBL covers protein kinases, as it comprises more than 739K bioactivity data points for 659 kinases from 20 different species for over 132K inhibitors. For example, solely for EGFR there are 19,166

Table 1 A Comprehensive Overview of Kinase Inhibitor Bioactivity Resources From Literature and Databases

Resource	Name	URL	Species	Kinases	Inhibitors	Bioactivities	Diseases
Databases							
ChEMBL[22]	EMBL-EBI ChEMBL bioactivity database (v23)	https://www.ebi.ac.uk/chembl/	20	659	~132K	739K	—
KKB	Eidogen Kinase Knowledgebase (Q2_2017)	http://www.eidogen.com/kinasekb.php	n.a.	>500	~696K	1.85M	—
PKIS	Published Kinase Inhibitor Set	https://www.ebi.ac.uk/chembldb/extra/PKIS/	1	454	376	168K	—
LINCS[41]	Library of Integrated Network-Based Cellular Signatures	http://www.lincsproject.org	3	597 WT 56 Mut 12 Var	153	78.6K (%)	54 Mut 1126 cell lines
Literature							
2007[42]	Fedorov et al.—Inhibitor profiling	https://doi.org/10.1073/pnas.0708800104	1	60 WT	156	9183 T_m	—
2007[43]	McDermott et al.—Oncolines profiling	https://doi.org/10.1073/pnas.0707498104	1	—	14	6641 sensitivities	500 oncolines
2008[20]	Karaman et al.—Inhibitor profiling	https://doi.org/10.1038/nbt1358	1	290 WT 27 Mut	38	8964 (>10 μM) 3100 IC_{50}	27 Mut
2011[44]	Davis et al.—Inhibitor profiling	https://doi.org/10.1038/nbt.1990	3	385 WT 51 Mut 6 Var	72	22.4K inactives 9424 K_D	—
2011[45]	Metz et al.—Comparative study	https://doi.org/10.1038/nchembio.530	1	172 WT	3858	258K	—

Continued

Table 1 A Comprehensive Overview of Kinase Inhibitor Bioactivity Resources From Literature and Databases—cont'd

Resource	Name	URL	Species	Kinases	Inhibitors	Bioactivities	Diseases
2011[46]	Anastassiadis et al.—Inhibitor profiling	https://doi.org/10.1038/nbt.2017	1	300 WT	178	52.8K (%)	—
2012[47]	Kitagawa et al.—Approved TK inhibitors profiling	https://doi.org/10.1111/gtc.12022	1	281 WT 29 Mut	9	3410 (%) 624 IC_{50}	29 Mut
2014[48]	Tang et al.—Comparative study	https://doi.org/10.1021/ci400709d	1	467 mixed	52.5K	246K	—
2014[49]	Uitdehaag et al.—Approved drugs profiling	https://doi.org/10.1371/journal.pone.0092146	1	283 WT 30 Mut	25	1496 IC_{50} 7723 (%)	44 oncolines 30 Mut
2014[50]	Huber et al.—Profiling stereospecificity of Crizotinib	https://doi.org/10.1038/nature13194	3	393 WT 55 Mut 8 Var	2	912 (%)	55 Mut
2016[51]	Christmann-Franck et al.—Comparative study	https://doi.org/10.1021/acs.jcim.6b00122	1	466 WT 135 Mut 34 Var	2106	357K	—
2016[52]	Duong-Ly et al.—Inhibitor profiling mutant kinases	https://doi.org/10.1016/j.celrep.2015.12.080	1	21 WT 76 Mut	183	17.7K (%)	76 Mut
2016[53]	Uitdehaag et al.—Anticancer agents profiling	https://doi.org/10.1158/1535-7163.MCT-16-0403	1	284 WT 27 Mut	68 kinase (122 total)	6251 IC_{50} 311 (%)	66 oncolines
2017[54]	Willemsen-Seegers et al.—Kinetic profiling	https://doi.org/10.1016/j.jmb.2016.12.019	1	46 WT	18	80 kinetics	—

%, percentage inhibition or percentage remaining activity; *Mut*, mutated kinases; T_m, thermal shift; *Var*, kinase variants; *WT*, wild-type kinases.

bioactivities annotated for 9917 unique compounds. The coverage of the human kinome with data points of unique active kinase inhibitors is depicted per kinase in Fig. 3B.

LINCS—http://www.lincsproject.org
The LINCS library[41] is a resource of large-scale biochemical and cell-based assays for the identification of network-based cellular signatures and contains, among others, the inhibition profiles of 153 compounds on 665 kinases including 56 mutants, 12 phosphorylated kinases, and 52 combinations (June 2017). Moreover, it contains 1190 different disease-related cell lines with associated assay data, including gene expression data, small-molecule inhibition profiles, and epigenetic profiles.

KKB—http://eidogen-sertanty.com/kinasekb.php
The Eidogen–Sertanty Kinase Knowledgebase (KKB) is a commercial database of biochemical data and chemical synthesis data on kinase inhibitors. The data have been obtained from literature and patents and curated with a proprietary web-based technology. It currently forms the largest biochemical resource on kinase inhibitors with 1.8M bioactivities for 355K unique inhibitors covering >500 kinases (release Q2_2017). Recently, a subset of the KKB was released to the public comprising 258K bioactivity data points for 76K kinase inhibitors on eight different protein kinases.[75] Interestingly, more than 20% (54K) of the data points cover the EGFR kinase.

2.1.2 Profiling Datasets in Literature

The need to investigate the selectivity of kinase inhibitors has resulted in the creation of kinome profiling platforms. In the past decade, several of the resulting profiling panels have been published. In the same time, also large-scale inhibition profiles were performed for kinase inhibitors on series of cancer cell lines (oncolines) providing valuable data to move further toward precision medicine.

In 2007, Fedorov et al. published the first large-scale kinase inhibitor profiling study in which 156 kinase inhibitors were systematically screened against 60 kinases,[42] and McDermott et al. published 14 kinase inhibitor profiles against 500 oncolines.[43] The year after, Karaman et al. published a screening of 38 inhibitors against a panel of 317 kinases (including 27 therapeutically relevant mutated kinases).[20] In 2011, two kinase inhibitors profile studies were simultaneously presented: (i) Anastassiadis et al. screened 178 inhibitors against 300 kinases,[46] and (ii) Davis et al. screened 72 inhibitors against 442 kinases (including 51 mutants and 6 phosphorylation-state variants).[44] Kitagawa and colleagues released a dataset in 2012 in which they focused on approved TK inhibitors (nine in total) that were profiled against 310 kinases (including 29 mutants).[47] In 2014, Huber et al. investigated the impact of the stereochemistry of Crizotinib on kinase inhibition and screened both the *R* and *S* enantiomer against 454 kinases (including 55 kinase mutants and 8 phosphorylation-state variants).[50] In the same year Uitdehaag and colleagues published a cancer-focused screening of 25 approved kinase inhibitors (and 7 other compounds) against 44 oncolines together with the kinome profiles for the 25 inhibitors.[49] In 2016, Duong-Ly et al. reported the profile of 183 inhibitors against 76 mutated and 21 WT kinases[52] and Uitdehaag et al. published an extended version of their previous panel now including 66 oncolines, 68 kinase inhibitors (122 compounds in total), and inhibition profiles against 311 kinases (including 27 mutants).[53] One year later, the latter group also published, to the best of our knowledge, the first kinetic profiling of kinase inhibitors against selected targets involving 18 inhibitors and 46 kinases.[54]

During this period GSK made a Published Kinase Inhibitor Set (PKIS) comprising 376 compounds available for screening.[76] This joint effort by pharmaceutical industry and academia resulted in the profiling of this set against 454 kinases and other targets, including but not limited to 24 GPCRs.[77,78]

These kinome profiles have been the subject of comparative and integrative research in multiple studies. In these studies (e.g., by Metz et al.,[45] Tang et al.,[48] Christmann-Franck et al.,[51] and Hu et al.[79]), the different bioactivity datasets were collected, curated, combined, and analyzed with respect to, e.g., kinome coverage, reproducibility, selectivity/promiscuity, structure–activity relationships, and scaffold diversity.

2.2 Disease-Associated Kinase Resources

Open Targets Platform (formerly CTTV)—https://targetvalidation.org
Over 2.8M associations between 26K protein targets and 9150 diseases are annotated in the Open Targets Platform (accessed June 2017).[70] Already for the EGFR kinase there are 834 disease associations annotated within the platform.

KIDFamMap—http://gemdock.life.nctu.edu.tw/kidfammap/
KIDFamMap[55] (kinase–inhibitor-disease family maps) contains sequence, structure, protein–ligand interaction, and disease information on 399 protein kinases, covering 36K kinase inhibitors, 187K kinase inhibitor assays, 339 diseases, and 962 disease associations. KIDFamMap was built based on site-moiety maps generated with the SiMMap tool.[80] The maps include

annotation of the conserved anchors within the binding pocket, the inhibitor, and pocket-moiety interaction types. These maps were subsequently classified into kinase inhibitor families (KIF) and kinase-inhibitor-disease (KID) relationships based on their similarity. Starting from a kinase (name, sequence, or by browsing), a small molecule (name, 3-letter HET/PDB code), or disease (name or by browsing) one can go through the identified associations. For each kinase in KIDFamMap, the results are clustered into one or more KIF templates that are again interlinked with other kinases, inhibitors, and diseases. For each template, a 3D view is provided in which the kinase pocket is shown with an inhibitor and the location of five sub-pockets: ATP, N-lobe, head of activation loop, C-lobe, and substrate pocket.

KinMutBase—http://structure.bmc.lu.se/idbase/KinMutBase/

The KinMutBase is a collection of disease-causing variations in protein kinase domains. The first version of the KinMutBase[56] was initially released in 2005 with the latest update (version 4.0) in 2015. The KinMutBase provides an overview of 1414 variations in the human kinome (catalytic kinase domains) and the disease association of these variations. This dataset is only available in a flat file format that can be easily parsed and is enriched with cross references to literature and other external resources. Thirty-three variations have been annotated for the EGFR receptor.

MoKCa—Mutations of Kinases in Cancer—http://strubiol.icr.ac.uk/extra/mokca/

MoKCa[57,81] is an extensive resource comprising mutations of kinases and their association with cancer. The initial database was collected from the genome of 210 human cancer cell lines (i.e., oncolines) and complemented with mutants annotated in the COSMIC database[38] (the COSMIC database from the Sanger Institute is a catalog of over 4.5M somatic mutations from 1020 cancer cell lines and literature[38]). Via the MoKCa web interface, expert annotations are presented concerning the functional implications of the annotated mutations. The location of the mutations can also be shown on the kinase structure (when available), and using the CanPredict[82] algorithm, the oncogenetic contribution of the mutant is predicted. Moreover, for each mutation it is annotated in which tissue it was observed and whether there are any posttranslational modifications known for this residue position. This resource is interlinked with many external resources and lists all GO annotations for each protein. For EGFR, 565 distinct mutations have been annotated.

A summary of the contents of the data sources discussed in this and following paragraphs is provided in Table 2.

Table 2 A Comprehensive Overview of Kinase-Focused Databases, Datasets, and Tools

Resource	Name	URL	Version	Open Access	Species	Kinases	Sequences	Structures	Inhibitors	Bioactivities	Diseases
Disease											
KIDFamMap[55]	Kinase-inhibitor-disease family map	http://gemdock.life.nctu.edu.tw/kidfammap/	Aug 2012	Y	1	399	399	1208	36K	187K	962 associations
KinMutBase[56]	Disease-causing variations in protein kinase domains	http://structure.bmc.lu.se/idbase/KinMutBase	4.0	Y	1	52	—	—	—	—	1414 Var/Mut
MoKCa[57]	Mutations, oncogenes, knowledge, and cancer	http://strubiol.icr.ac.uk/extra/mokca/	—	Y	1	423	—	Visualization	—	—	1406 Mut
Sequence											
BYKdb[58]	Bacterial tYrosine-kinase database	https://bykdb.ibcp.fr/BYKdb/	53.0	Y	53	9095	9647	—	—	—	—
EKPD[59]	Eukaryotic kinase and phosphatase database	http://ekpd.biocuckoo.org	1.1	Y	84	50K	50K	—	—	—	—
Kinase.com[4]	Kinase.com/KinBase/WiKinome	http://kinase.com	—	Y	15 kinomes 56 BSK	7597 318 BSK	7597 318 BSK	—	—	—	—

KinG[60]	Kinases encoded in genomes	http://king.mbu.iisc.ernet.in	July 2014	Y		49K	49K	—	—	—	—
KinWeb[61]	Protein kinases in human genome	http://www.itb.cnr.it/kinweb/	—	Y	1	518	518	394 models	—	—	—
KSD	Kinase sequence database	http://sequoia.ucsf.edu/ksd/	—	Y	948	7128	7128	—	—	—	—
MAPK resource[62]	MAP kinase resource	http://www.mapkinases.eu	—	Y	24	99	99	—	—	—	—
ProKinO[63]	Protein kinase ontology browser	http://vulcan.cs.uga.edu/prokino/	2.0	Y	15 (KinBase)	7597	7597	—	—	—	COSMIC[38]
RKD[64]	Rice kinase database	http://ricephylogenomics.ucdavis.edu/kinase	2015	Y	1	1467	1467	—	—	—	—
Structure											
Kinase SARfari	EMBL-EBI Kinase SARfari	https://www.ebi.ac.uk/chembl/sarfari/kinasesarfari	6.00	Y	12	989	989	~823	54K	532K	—
KLIFS[6,65]	KLIFS	http://klifs.vu-compmedchem.nl	v2.3	Y	2	269	1087	3832	2458	ChEMBL[22]	—
MOE kinase database	Protein kinase and PI3K databases	https://www.chemcomp.com	2016.0802	N	46	412	412	3962	2540	—	—

Continued

Table 2 A Comprehensive Overview of Kinase-Focused Databases, Datasets, and Tools—cont'd

Resource	Name	URL	Version	Open Access	Species	Kinases	Sequences	Structures	Inhibitors	Bioactivities	Diseases
Tools											
K-map[66]	Kinase—inhibitor mapper	http://tanlab.ucdenver.edu/kMap/	v1.0	Y	1	519	—	—	241	85K	48 Mut
KAR[67]	Kinase addiction ranker	http://tanlab.ucdenver.edu/KAR/	2015	Y	User input	User input	—	—	User input	User input	User input
Kinannote[68]	Kinase identification and classification	https://sourceforge.net/projects/kinannote/	1.0	Y	User input	User input	User input	—	—	—	—
KinConform[69]	Kinase conformation classification	https://github.com/esbg/kinconform	v1.0	Y	—	—	—	User input	—	—	—
KinMap[7]	Kinome mapper	http://kinhub.org/kinmap/	Beta	Y	1	496	—	PDB	ChEMBL[22]/DrugBank[21]/Karaman[20]/Davis[44]		CTTV[70]
Kinome Render[71]	Kinome mapper	http://bcb.med.usherbrooke.ca/kinomerender	—	Y	1	518	—	—	—	—	—
Kinomer[72]	HMM for kinase identification	http://www.compbio.dundee.ac.uk/kinomer/	v1.0	Y	52	14K	User input	—	—	—	—

BSK, bacterial spore kinases.[73]

2.3 Sequence-Based Kinase Resources

UniProt—http://www.uniprot.org

The UniProt database contains high-quality data on protein sequences with currently 555K manually annotated and reviewed entries (Swiss-Prot) and 87M automatically annotated (TrEMBL) sequences[83] (accessed June 2017). Within this wealth of protein data are 236K sequences that have been classified as protein kinases according to the PFAM protein kinase motif (PF00069).

Kinase.com (including KinBase, WiKinome)—http://kinase.com

The mapping of the human kinome by Manning et al. in 2002 was a large step in the field of kinase research.[4] Meanwhile, the article has been cited over 6200 times and over 100K posters of the accompanying dendrogram of the human kinome have been distributed and decorated lab interiors. KinBase currently contains the kinomes of 15 species including the classification of all kinases into their respective phylogenetic kinase groups and (sub)families (in-line with the Hanks and Hunter classification), the annotation of the different domains in the kinase sequences, annotation of gene aliases, references to external databases, and sequence alignments of the catalytic kinase domains. Apart from these full kinomes also a set of 318 unique bacterial spore kinases[73] has been fully characterized for 56 bacteria (107 variants). The WiKinome is another resource on kinase.com and is a wiki with additional information on each of the kinase groups and families and discusses different topics including protein kinase evolution and kinase classification. Kinase.com is one of the primary resources in the field of kinase research, and most (if not all) other resources mentioned here are in part based on this work.

KinWeb—http://www.itb.cnr.it/kinweb/

KinWeb is a kinase database in which curated information on the human kinome is collected. Apart from the classification and gene annotation of human kinases, it also provides an interface for (graphical) gene analysis, predicted membrane regions, and disulfide bridges, as well as 3D (homology) models. The data is searchable via gene name, kinase group, domain name, and a graphical genome browser.

ProKinO—Protein Kinase Ontology Browser—http://vulcan.cs.uga.edu/prokino

The ProKinO[63] is an ontology browser build on top of the KinBase (see Kinase.com) and is an integration of kinase-related data from COSMIC,[38] Reactome,[84] UniProt,[83] and subdomain data.[85] It allows the user to browse and search the resource by gene name, disease, pathways, and has additional options to browse by organism, kinase, and functional domains. For advanced data retrieval, it allows the user to write SPARQL queries based on the ProKinO database scheme that is provided.

It also contains a visualization tool that allows the selection and comparison of subsets of kinases. This tool plots the sequence logo based on natural sequence variations, highlights the secondary structure, and shows both cancer-associated variants from the COSMIC database and experimentally validated posttranslational modifications.

2.4 Structure-Based Kinase Resources

2.4.1 Databases and Collections

RCSB Protein Data Bank (PDB)—http://www.rcsb.org
The PDB[5] is the primary resource for (crystal) structures of macromolecules, currently containing over 130K structures (June 2017), and forms the basis for all structure-based kinase resources listed here.

Kinase SARfari—https://www.ebi.ac.uk/chembl/sarfari/kinasesarfari
The Kinase SARfari is an integrated chemogenomics kinase resource covering 989 protein kinase domains comprising 989 sequences, 823 structures, 54,189 inhibitors, and 532,000 bioactivity data points. The database consists of five main elements: protein families, binding sites, 3D structures, bioactivities, and inhibitors. The Kinase SARfari can be searched by kinase sequence, name, family, and group, as well as, small-molecule name, synonyms, (sub)structure, and structural similarity. Besides browsing through the data it also allows filtering and extracting the data, and every entry is further linked to kinase, inhibitor, structure, and bioactivity data. The binding site option allows users to compare the similarity of binding sites from different kinases based on their pairwise distance. The 3D structure allows for the superposition and comparison of protein kinase structures in an interactive 3D viewer. The viewer provides the options to display, label, or restrict one of the eight kinase motifs/residues for easy comparisons.

KLIFS—http://klifs.vu-compmedchem.nl
The Kinase–Ligand Interaction Fingerprints and Structures database (KLIFS)[6,65] contains sequence, structure, and protein–ligand interaction information on 269 protein kinases (human and mouse) covering 3832 PDB structures and 2458 unique ligands (June 2017, v2.3, updated weekly). The database can be searched by PDB code, kinase name, family and group, small-molecule structure, protein conformation (DFG, G-loop, and αC-helix), interaction pattern, sequence composition, ligand and structural properties, subpockets, and any combination of conserved pocket waters.

Every kinase structure obtained from the PDB is split into separate chains and subsequently aligned, curated, and annotated with relevant kinase data, interaction fingerprints, and processed according to a predefined pocket definition and nomenclature. Every ligand is matched against the ChEMBL database, and all kinase bioactivities are presented. The interaction fingerprints can be used to search through all kinase structures to identify kinase–ligand complexes with similar interaction profiles. Additionally, KLIFS allows inspecting the entries via an interactive web-based 3D viewer in which the user can perform a 3D comparison of ligand binding modes from different structures. Each entry is linked to external databases, including the PDB,[5] sc-PDB,[86] KIDFamMap,[55] PubMed, and Guide To Pharmacology/IUPHARdb. Recent updates of KLIFS also included the open source development of nodes[87] for the open analytics platform KNIME[88] (version 2.1), a virtual reality viewer (version 2.2), and the incorporation of atypical protein kinases (version 2.3).

MOE kinase suite—https://www.chemcomp.com

The commercially available Molecular Operating Environment (MOE) from the Chemical Computing Group is a computational drug discovery platform. The MOE suite comes with several databases, including an annotated structural protein kinase database. The MOE protein kinase database contains a clustered and aligned collection of currently 3962 structures comprising a total of 412 kinases from 46 different species (including 3223 structures from human kinases). The database provides annotations of the kinase as well as the structural conformation of the kinase (DFG and αC-helix) and highlights the known motifs (e.g., hinge, G-loop, DFG, and αC-helix). All small molecules in complex with the kinases are extracted from the complexes and classified according to the core scaffold scheme introduced by Vieth et al.[89] and De Moliner et al.[90]

2.4.2 Kinase-Specific Visualization Resources

Together with the large amount of available kinase data comes the need for intuitive ways to visualize, analyze, and share this data. A common way to do so is by using kinome tree viewers. The basis for these viewers is the phylogenetic kinase tree established by Manning and coworkers.[4] In the human kinome tree, kinases are clustered by sequence similarity, single branches describe the different kinase groups and families, and data can be annotated accordingly (see Figs. 1, 3, and 4). Several kinome tree viewers have been developed over the years including Kinome Render and KinMap tools. We will concentrate on the latter two, which are freely available tools that provide ways to visualize kinase-related data, such as bioactivity data, structural data, key targets, disease relevance, and functional relationships. This visualization can help to identify relationships between different kinases and to analyze, for example, kinase inhibitor promiscuity.

Kinome Render—http://bcb.med.usherbrooke.ca/kinomerender.php

Kinome Render[71] is available as a stand-alone program and as a web-based tool. The viewer allows the user to map either binding data onto the kinome or custom data and labels using the more advanced Kinome Render syntax. It also supports batch processing of binding data and maps the binding data from multiple inhibitors onto the same or separate kinome trees. The user data are plotted onto the kinome according to an adjusted form of the Manning[4] names which are provided as a lookup table.

KinMap—http://www.kinhub.org/kinmap/

KinMap[7] is a web-based kinase tool that facilitates interactive navigation and exploration of the kinome as well as mapping of complex data onto the kinome. In addition to plotting user-defined data, KinMap uniquely integrates preprocessed kinase-related data from different sources, i.e., from ChEMBL,[22] PDB,[5] and CTTV[70] (now Open Targets). Available preprocessed structure and compound data include the number of available PDB structures per kinase (2639 structures for 233 kinases), the number of tested compounds, and the number of assays in ChEMBL associated with a particular kinase (covering 435 kinases), as well as knowledge about key targets of approved drug (51 kinases). Furthermore, the published inhibitor profiling datasets from Karaman et al.[20] and Davis et al.[44] are integrated as well as kinase disease associations obtained from CTTV.[70] This combination of data can help to uncover new therapeutic indications of known inhibitors or to prioritize novel kinases for drug development. KinMap furthermore allows for browsing, sharing of data, and individual styling. It provides support for various input and output formats, and features the generation of high-resolution images.

3. STRUCTURE- AND LIGAND-BASED APPLICATIONS FOR KINASE DRUG DESIGN

Computers have long entered the early drug development pipeline to reduce the expenses in terms of R&D time and costs. Structure-based concepts frequently applied are virtual screening to find new drugs,[92] target prioritization, druggability predictions, and target comparison (e.g., for polypharmacology and side effect predictions). A detailed explanation on structure-based target assessment, such as function prediction, disease relevance, or target modulation based on similarities to known targets, can be found in recent reviews.[33,93,94] Selected examples of concepts and successful applications are provided here, showcasing how to use the available platforms and tools for specific kinase research tasks.

3.1 Structure-Based Target Assessment

3.1.1 Protein Structure Preprocessing

Vital for application of structure-based computational methods is the selection and processing of high-quality structural data. If no structure is available, homology models can be generated—given the structure of a closely homologous protein has been solved—and used for subsequent computational steps.[95] Before most computational methods can be safely applied, the input structures should be preprocessed and further analyzed, including the addition of hydrogen atoms, the identification of potential binding sites, or the assembly of alternative conformations.[96] The Proteins*Plus* Server, for example, provides several tools for structure analysis of macromolecules in an easy to use manner.[10] Proteins*Plus* features the search for alternative protein conformations (SIENA[91]) and the visualization of protein–ligand interactions in an intuitive 2D diagram (PoseView,[97] as shown in Fig. 1C). Furthermore, it also incorporates pocket prediction and druggability assessment tools (DoGSiteScorer[98,99]), as described in the following paragraphs.

3.1.2 Binding Site Detection

The active site of a protein is the key to its function. Given a protein structure, an integral step is the identification of potential binding sites. If a structure with bound ligand is available, the cocrystallized ligand defines the binding site. Nevertheless, additional sites may be of interest. Furthermore, ligand-free structures need to be considered in prospective analyses. For this purpose, several automatic methods to predict cavities[100] have been developed, e.g., FPocket,[101] SiteMap,[102] Volsite,[103] and DoGSiteScorer.[98] Such binding site detection methods rely solely on the 3D structure of the protein and use their geometric and/or energetic composition to detect cavities. In DoGSiteScorer, the protein is assigned to a 3D Cartesian grid. Grid points are labeled as occupied if they overlay with a protein atom and as free otherwise. Subsequently, a difference of Gaussian (DoG) filter is invoked to identify those grid points lying in cavities. These favorable grid points are clustered to subpockets, and, subsequently, neighboring subpockets are merged to pockets. Since binding site detection methods predict several potential cavities on the protein surface, the cavities can be ranked simply by their volume or, more elaborately, by their druggability.

It is important to note that generally proteins are treated as static snapshots, in contrast some pockets might be induced upon ligand binding (induced-fit), and will thus be overseen. Special methods such as TRAPP[104,105] try to address this issue by introducing protein flexibility by molecular dynamics simulations to analyze transient pockets.

3.1.3 Structure-Based Druggability

Prioritizing a suitable target for the drug development pipeline, or a specific protein structure for docking studies, is essential for the success of the drug design project.[106]

In this context, the term druggability has been coined and describes the ability of a (disease modifying) target to bind a drug-like molecule.[106] Structure-based in silico druggability predictions rely on the 3D structure of the protein binding site only and identify discriminative features from known drug targets that imply druggability and transfer this knowledge to novel targets. Current descriptor-based methods consist of three steps: Detection and encoding of the discriminant features such as geometric and physicochemical properties, a classification algorithm, e.g., exponential functions,[107] linear combinations,[102] or machine learning,[99] and a labeled dataset for model training.[107–110] DoGSiteScorer, for example, uses a small set of geometric and physicochemical descriptors combined with a support vector machine (SVM). As a rule of thumb, sufficient volume, depth, and hydrophobicity were found to be important contributors to a pocket's druggability. Thus, computational druggability predictions may point to novel promising target structures a priori without the need to undergo expensive experimental screening efforts.

3.1.4 Application: Identifying Novel and Druggable Pockets

One of the grand challenges in (kinase) ligand discovery remains the identification of allosteric modulators. In order to apply structure-based methods (e.g., docking) to identify new compounds, the respective pocket has to be defined beforehand. As a model scenario, we focus here on PDPK1, an AGC kinase, which is an important anticancer target and known to inhere an allosteric peptide-binding site (PIF-pocket). We used PDPK1 kinase structure 1uu3,[111] which is only bound to a nanomolar protein kinase C inhibitor (ATP-pocket) and unoccupied in the allosteric PIF-pocket, as query for the DoGSiteScorer[99] binding site and druggability prediction with the Proteins*Plus*[10] Server. The algorithm returned 13 pockets of varying size and druggability. The ATP-pocket is detected as largest pocket and ranked with a high druggability score of 0.84 (score range: 0 (undruggable) to 1 (highly druggable), see Fig. 4A). Additionally, the

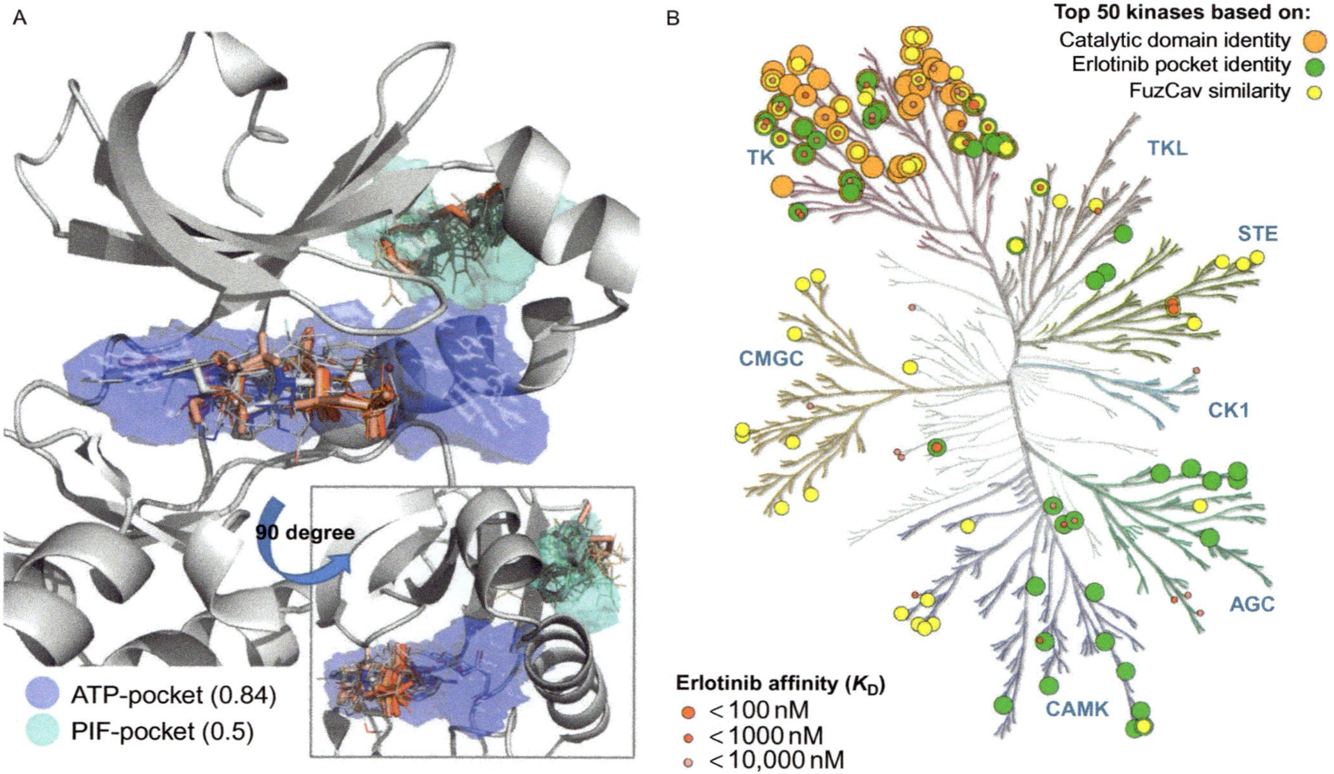

Fig. 4 (A) Structure of PDPK1 kinase (PDB code: 1uu3, cartoon representation, *gray*) with bound inhibitor (stick representation), highlighted are two predicted pocket volumes (ATP: *blue*, PIF: *cyan*). From the 36 aligned structures collected from SIENA,[91] ATP- and PIF-bound ligands are shown as *gray lines*. Ligand structures from 3hrf (*red*) and 4xx9 (*salmon*) are shown as *sticks*. Figure created using PyMol.[9] (B) The application example on the investigation of the bioactivity profile of Erlotinib: the KinMap[7] kinome decorated with the top 50 kinases according to (I) the full catalytic kinase domain sequence identity to EGFR (*orange dots*), (II) the Erlotinib pocket sequence identity to EGFR (*green dots*), (III) the FuzCav binding site similarity compared to the Erlotinib–EGFR complex (*yellow dots*), and (IV) the Erlotinib bioactivity profile. *Panel (B): Illustration reproduced courtesy of Cell Signaling Technology, Inc. (www.cellsignal.com).*

PIF-pocket is identified as third largest pocket with a lower druggability score of 0.5. Note that allosteric pockets are generally smaller and shallower than the orthosteric pocket and do not necessarily fulfill the same criteria as the catalytic pockets. To verify the identified allosteric PIF-pocket, we used SIENA[91] from the same server to generate an ensemble of PDPK1 structures (restricted to crystal structures with a resolution below 2 Å and no mutations), yielding 36 aligned structures. Several structures with PIF modulators (activators and inhibitors, e.g., 3hrf[112] and 4xx9[113]) are contained in the ensemble. Strikingly, the modulators nicely overlay with the detected allosteric pocket (see Fig. 4A). Once the binding site is known, virtual screening can be applied, as exemplified in the study by Rettenmaier et al.[113] to identify novel PDPK1 modulators.

3.2 Protein–Ligand Interactions and Binding Site Comparisons

3.2.1 Interaction Fingerprints

To enable quick and insightful comparisons of the interactions that occur between a protein and a small molecule so-called molecular interaction fingerprinting methods have been developed.[114] These fingerprinting methods encode the 3D information of a protein–ligand complex into a binary 1D representation. The frequently used fingerprinting method IFP was initially created by Marcou and Rognan to optimize the predicted binding mode of fragments and scaffolds.[115] For this, they developed an interaction fingerprint in which seven interaction types (apolar, aromatic face-to-face, aromatic face-to-edge, H-bond donor, H-bond acceptor, ionic negative, ionic positive) between each pocket residue and the small molecule are encoded in a bit string. By comparing the IFP of the docked fragments/scaffolds with the IFP of a crystallographic pose they increased the accuracy in retrospective structure-based virtual screening against multiple protein targets. This IFP method has since its introduction been successfully applied for the filtering and selection of hits in virtual screening campaigns, for the creation and application of predictive models regarding ligand effect, and for the comparison of ligand-binding modes of different crystal structures.[6,116] In the past decade, the IFP method has been further developed by Rognan et al. yielding among others the graph-based Triplets of Interaction Fingerprints (TIFP) technique.[117] Moreover, the same group also demonstrated that they could predict the ligand-binding mode based on IFP-trained neural networks with high efficiency.[118]

The original IFP approach is integrated into the KLIFS database to annotate interactions between kinases and ligands for each structure. The IFPs in KLIFS can, for example, be used to identify structures with specific interaction patterns, to compare the interaction profile of a single structure with all available structures, and to identify interesting chemical moieties with highly similar interaction patterns that can be used for scaffold hopping.[6,87]

3.2.2 Binding Site Comparison

As indicated by the conserved ATP-binding site of kinases, structurally similar sites can bind similar molecules. Although protein sequence comparisons are a good indication for functional similarities and relations, examples show that the structure of the binding site, i.e., the relative spatial arrangement of specific residues in the binding site, can be more conserved than sequences.[119] Thus, computational annotations of structure-based similarities between active sites can be used to identify on- or off-targets and are good indicators for potential drug promiscuity, polypharmacology, or adverse effects. Several tools for structure-based binding site comparison have been developed over the past years and can mainly be grouped in alignment-based and alignment-free methods.[120,121] Both groups of methods detect similar arrangements of physicochemical features between the respective sites. Alignment-based methods superimpose the binding sites first based on, for example, geometric hashing or clique detection algorithms, and the resulting alignment is scored based on the match of physicochemical features (e.g., Cavbase,[122] ProBis[123]). Alignment-free methods encode the features and distances in a one-dimensional fingerprint, which can then be matched efficiently (e.g., FLAP,[124] SiteAlign,[125] FuzCav[126]). Lately, combined or enhanced methods have been developed that unite the advantages from both groups, the interpretability from alignment-based and the speed from fingerprint-based methods (e.g., BSAlign,[127] TrixP[128]). Generally, pairwise structure comparisons can be performed in milliseconds.

3.2.3 Application: Explaining Bioactivity Profiles

To investigate the bioactivity profile of Erlotinib a combined sequence and structure-based analysis was performed. Using the multiple sequence alignment of the catalytic kinase domain by Manning et al.[4] the top 50 kinases with the highest sequence identity to EGFR were identified (orange dots in Fig. 4B). Not surprisingly, all top 50 kinases are tyrosine kinases that are clustered closely together within the kinome tree. Within this top 50, 16 kinases show activity for Erlotinib based on the bioactivity profiling by Karaman et al.[20]

Starting from the crystal structure of Erlotinib in complex with EGFR[8] (PDB code: 1m17, Fig. 1), residues in contact with Erlotinib were identified in KLIFS[6] and a multiple sequence alignment of these residues was created using the KLIFS KNIME nodes.[87] Subsequently, the sequence identity of this alignment was calculated and again the 50 closest kinases to EGFR were selected (green dots in Fig. 4B). Interestingly, these top 50 kinases were not solely found within the TK group, but more dispersed across the kinome. Of these 50 kinases, 18 showed affinity for Erlotinib. With a FuzCav comparison, the structure-based pocket similarity was assessed, comparing the Erlotinib–EGFR structure to all structures of human kinases within KLIFS with a quality score of 8 or higher. The top 50 highest ranked kinases, among which 12 had affinity for Erlotinib, were again mapped onto the kinome using KinMap (yellow dots in Fig. 4B).

This application example clearly shows that different techniques can provide different explanations for the observed bioactivity profile of Erlotinib. Where the domain-wide sequence identity comparison only highlighted off-targets within the TK family, the interaction-focused Erlotinib pocket sequence identity analysis highlighted off-targets such as RIPK2 and the Aurora family. Moreover, this focused pocket approach was the only approach that overlapped with the high-affinity off-target GAK ($K_D = 3.1$ nM). On the other hand, the structure-based FuzCav approach was the only approach that identified the remaining two high-affinity off-targets, namely SLK ($K_D = 26$ nM) and LOK ($K_D = 19$ nM).

3.3 Computational Approaches to Tackle Obstacles in Kinase Research

3.3.1 The Pocketome of Human Kinases

As discussed in Section 1.3, most FDA-approved small-molecule kinase inhibitors have been designed to address TK kinases as primary targets, which contradicts with the number of kinases found to be (potentially) therapeutically relevant and druggable.[30,34] To overcome this hurdle and to prioritize novel kinases as potential drug targets, the ATP-binding pockets of all human kinases (defined as pocketome) have been investigated.[11] All available human kinase X-ray structures were collected from the PDB, covering almost half of the human kinome. As the number of structures differs per kinase, a structural representative was manually selected preferring structures with high resolution, structural completeness, and presence of a cocrystallized ligand. This yielded structures for roughly 200 kinases in the DFG-in conformation (due to the low coverage of kinases in DFG-out state, the study focused on the DFG-in state). Homology models were built for the remaining part of the kinome using Yasara.[129] For quality assessment of the models redocking studies using ATP were performed as well as analyses of the internal energies and stereochemistry. DoGSiteScorer[99] (as described in the previous paragraph) was invoked for all structures (apo and holo) to detect, describe, and rate the pockets to prioritize novel kinases for drug development. As expected, the features for most kinase structures were found to be in the range of druggable proteins. 75% of the kinome has ATP-binding pocket structures that are rated as druggable (score > 0.5, scale: undruggable 0-highly druggable 1). Most interesting were the 25% of kinase structures showing druggability estimated higher than 0.7, which related to the mean value of a control set of kinase structures cocrystallized with approved drugs. The top-scoring kinases included primary targets of approved drug targets such as ABL1 (a key target for Imatinib, Dasatinib, and Bosutinib) and EGFR (key target for six approved drugs and well studied[130]). Other, less explored, potentially druggable kinases with indications of being disease relevant were identified, such as MLK1 (involved in MAPK and JNK pathways), CK1g2 (involved in Wnt pathway), and CLKs (regulate mRNA splicing). The pocketome of human kinases allowed the prioritization of the ATP-pockets of the entire kinome with respect to structural druggability estimates, and puts the spotlight to less explored kinases. An illustration about the most druggable kinases can be found in the original pocketome of human kinases paper.[11]

3.3.2 Detection of Selectivity Pockets

One of the main hurdles in kinase-focused drug design is the development of selective protein kinase inhibitors to reduce unwanted side effects, as well as to provide tool compounds for validation studies. The same feature, in favor when building homology models for unresolved kinases, complicates the design of selective inhibitors: The high conservation of the overall-fold of the ATP-binding site of protein kinases.

While binding site comparison methods focus on the identification of similarities for functional classification, selectivity determination methods focus on the small but important differences between closely related kinases. To identify the selectivity determining features often physicochemical hot spots are identified. Such methods are energy-grid based and calculate the molecular interaction fields within the respective binding sites.[131–133] BioGPS,[133] for example, calculates pharmacophore interaction points for structural ensembles, thereby taking protein flexibility into account. The calculated interaction points highlight hydrophilic and lipophilic parts, and can be used to manually compare and identify the differences between targets and off-targets. Furthermore, in proteochemometric models ligand and receptor information can be combined for the prediction of ligand/protein target combinations.[134,135] Another novel shape-based approach aims at the identification of specificity determining subpockets, i.e., small differences, between otherwise very similar overall pockets.[136] To account for the intrinsic flexibility, multiple crystal structures per key target and off-target are investigated at once. The volume-based method employs the idea of selectivity grids using structural information only. First, to consider protein flexibility upon ligand binding, a combined pocket is calculated for a set of on-target and off-target structures, respectively (see Fig. 5A and B). After all structures have been aligned, the individual pockets are detected and represented as grids.[99] Next, these pocket grids per protein set are fused to a combined pocket by retaining the frequency on each grid point. The frequency describes how often the grid point was present in the ensemble to capture the core regions as well as the more mobile parts of the pocket (see Fig. 5C). In a second step, the combined pockets from the two grids are subtracted from each other to reveal the difference pocket which contains three exclusive areas in a user-friendly visualization scheme: The target-specific points

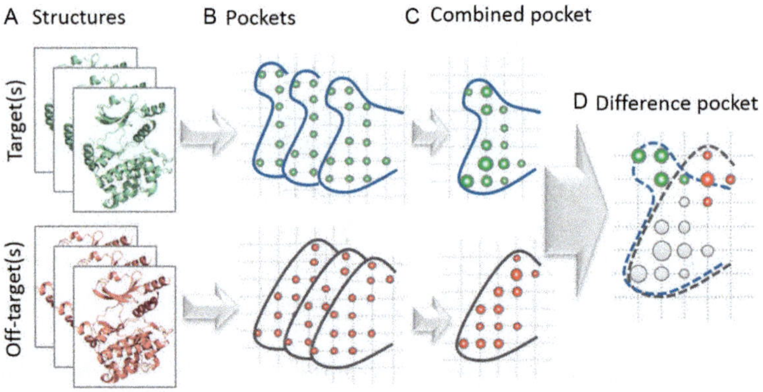

Fig. 5 Procedure to calculate the combined and difference pockets. (A and B) Pockets are calculated for each target (*green*) and off-target (*red*) structure. (C) For each set, a combined pocket is derived containing information about the frequency of each grid point. (D) The final difference pocket represents commonalities (*gray points*) and differences between the two combined pockets (target-specific, *green*; off-target-specific, *red*). *Reprinted with permission from Volkamer, A.; Eid, S.; Turk, S.; Rippmann, F.; Fulle, S. Identification and Visualization of Kinase-Specific Subpockets.* J. Chem. Inf. Model. ***2016,*** 56 *(2), 335–346. Copyright 2016 American Chemical Society.*

(green), the common core (gray), and the off-target-specific points (red), see Fig. 5D. The method showed good results in several retrospective evaluation studies, including MAP kinases and p21-activated kinases. The method can be applied to rationally guide the design of more selective inhibitors, as also exemplified in an analysis of 37 ITK and 45 AurA structures. An ITK selective subpocket above the ligand plane became apparent throughout the 82 structures, which results from differences in the gatekeeper residues. This selective subpocket was previously also manually identified in a rational design study from Genentech and by targeting the pocket with a lipophilic substituent of a selective indazole series yielded a 660-fold activity difference.[137] Small differences between multiple structures and subpockets can be highlighted which are unique for specific kinases, and thus, can be used to guide the design of more selective inhibitors.

3.3.3 Profiling Data for Activity/Selectivity Prediction

The incrementing amount of available bioactivity data in the kinase field enables promising developments such as standard virtual screening pipelines to identify novel binders against a particular kinase as well as cross-reactivity estimation throughout a kinase panel (see Table 1). Several success stories using multidimensional structure–activity relationships have been published.[138,139] In virtual screening applications, searching for compounds similar to known active compounds (ligand-based) or docking of compounds (structure-based) from large datasets of up to several millions of virtual compounds is currently feasible for drug discovery and design.[140–142]

Profiling data allow the evaluation of structure–activity relationships against hundreds of targets simultaneously. This is, for example, used in the similarity ensemble (SEA) and Optimized Cross rEActivity estimation (OCEAN) approaches where the similarity of a query ligand to known inhibitors is utilized for the prediction of its molecular targets.[143–145] Furthermore, machine learning methods are increasingly being used, such as SVMs, random forest, naïve Bayesian, neural networks, and lately deep learning as tool for ligand-centric target activity and selectivity predictions.[146–153] SVMs have successfully been utilized to prospectively identify novel inhibitors of GPCRs and kinases.[153] Naïve Bayesian classification and regression also showed high predictive power for kinase activity.[147,151] Neural networks trained on 240 active kinase compounds and 240 random compounds achieved 79% correct classifications on an external test set of 120 kinase inhibitors[148] and deep learning outperformed other prediction methods with an area under the curve (AUC) of 0.83 on ChEMBL data (not focused on kinases).[149]

Recently, Merget and coworkers used the available open data (Tang,[48] ChEMBL[22]) as well as in-house data (Merck kinase panel) to train several machine learning models for cross-reactivity estimation.[139] They analyzed the impact of the dataset composition, the utilization of different fingerprints, and evaluated different under- and oversampling techniques for data balancing. Kinase activity classification with random forest in a fivefold cross-validation experiment yielded the highest quality models. The models trained on the combined data (open + in-house) outperformed other models regarding average model quality metrics and robustness (i.e., mean AUC of 0.76, sensitivity 0.63, and specificity of 0.78) on a panel of 291 kinases and a dataset of over 44K compounds.

Christmann-Franck et al. compiled a large dataset of 356K data points for 482 kinases covering 2106 unique inhibitors and utilized this dataset for the creation of proteochemometric[134] models (combining both ligand-based and protein-based information) using random forest for the prediction of kinase activity.[51] They explored the contribution of each descriptor to the final model, the accuracy of the models in relation to the amount of data used for training, the overall predictive value per target and per compound, as well as for different inhibitor types.

4. SUMMARY AND OUTLOOK

Due to the scientific interest in protein kinases as drug targets for cancer and other therapeutic indications, an increasing amount of disease-associated, sequence, biochemical, and structural data has become available and will continuously increase. With this accumulation of data comes the need and the chance for computational methods to collect, structure, analyze, and utilize this extraordinary pool of information. In this chapter, we supplied an overview of kinase-centric tools, platforms, resources, and databases, and provided examples of computational kinase-centric drug development. Ultimately, the aim of computational chemistry is to enable accurate *in silico* predictions for the rational design of novel kinase inhibitors, i.e., more selective inhibitors as well as optimization of existing hits against established as well as novel targets.

One drawback in many structure-based approaches is the limited consideration of protein flexibility. Crystal structures only represent a snapshot of otherwise flexible molecules. Therefore, intrinsic consideration of protein motion and, in particular, conformational changes upon ligand binding is essential. Indeed, already from the combination and comparison of the many available crystal structures of kinases these motions become apparent. To obtain more detailed insights into the motions of kinases molecular dynamics simulations are nowadays integrated into the drug design process, albeit still limited to short time frames.[154,155]

Another drawback is the quality of the experimental data, as good computational models can only be developed based on good experimental input data. For example, the ChEMBL database contains several cases in which the recorded inhibition data seems to provide contradictory information, i.e., a specific compound being measured as highly active by one research group was found to be inactive by another.[48] Another pitfall for the bioactivity prediction for kinases is that due to the availability of more inactive than active bioactivity data machine learning models often need to be trained on unbalanced data. To cope with this bias, over- or undersampling algorithms should be used to balance the training set in order to prevent overfitting.[139] Good data curation, as well as the availability of reproducible and consistent data is of great importance for the advancement of computational predictions.

Although kinase inhibitors are primarily evaluated based on their affinity for a protein target, it seems that the residence time of a compound is more significantly correlated to the therapeutic efficacy than its binding affinity.[156,157] Therefore, the importance of ligand-binding kinetics for drug development is slowly being recognized.[156] The first kinetics profiling of kinase inhibitors by Willemsen-Seegers et al.[54] emphasized the importance of binding kinetics profiling.[54] For example, although the affinity of Ponatinib for ABL and TIE2 is similar ($pIC_{50} = 8.4$–8.5), Ponatinib shows a more than 1500-fold kinetic selectivity for ABL over TIE2 with a residence time of 180 h for ABL and only 6 min for TIE2. The same group utilized this binding kinetics knowledge to drive the design of TTK inhibitors with optimized antiproliferative activity.[158]

With increasing experimental data, knowledge, and computational power, methods like free energy calculations and molecular dynamics simulations become feasible on larger scales. Nevertheless, these methods still only cover the interaction between few molecules and neglect the systematic level, up- or downstream effects, and the routes the molecules might take in the human body (ADME). Therefore, understanding the impact of the interactions in the cell and the kinetics will engage the life sciences field for a long time to come.

REFERENCES

1. Klebl, B.; Muller, G.; Hamacher, M.; Mannhold, R.; Kubinyi, H.; Folkers, G. Protein Kinases as Drug Targets. In: *Methods and Principles in Medicinal Chemistry*, 49th ed.; Klebl, B., Muller, G., Hamacher, M., Eds.; Wiley-VCH Verlag GmbH & Co. KGaA: Weinheim, Germany, 2011.
2. Wood, L. D.; Parsons, D. W.; Jones, S.; Lin, J.; Sjoblom, T.; Leary, R. J.; Shen, D.; Boca, S. M.; Barber, T.; Ptak, J.; et al. The Genomic Landscapes of Human Breast and Colorectal Cancers. *Science* **2007**, *318*(5853), 1108–1113.
3. Pray, L. A. Gleevec: The Breakthrough in Cancer Treatment. *Nat. Educ.* **2008**, *1*(1), 37.
4. Manning, G.; Whyte, D. B.; Martinez, R.; Hunter, T.; Sudarsanam, S. The Protein Kinase Complement of the Human Genome. *Science* **2002**, *298*(5600), 1912–1934.
5. Berman, H. M.; Westbrook, J.; Feng, Z.; Gilliland, G.; Bhat, T. N.; Weissig, H.; Shindyalov, I. N.; Bourne, P. E. The Protein Data Bank. *Nucleic Acids Res.* **2000**, *28*(1), 235–242.
6. Kooistra, A. J.; Kanev, G. K.; van Linden, O. P. J. J.; Leurs, R.; de Esch, I. J. P. P.; de Graaf, C. KLIFS: A Structural Kinase-Ligand Interaction Database. *Nucleic Acids Res.* **2016**, *44*(D1), D365–D371.
7. Eid, S.; Turk, S.; Volkamer, A.; Rippmann, F.; Fulle, S. KinMap: A Web-Based Tool for Interactive Navigation through Human Kinome Data. *BMC Bioinformatics* **2017**, *18*(1), 16.
8. Stamos, J.; Sliwkowski, M. X.; Eigenbrot, C. Structure of the Epidermal Growth Factor Receptor Kinase Domain Alone and in Complex With a 4-Anilinoquinazoline Inhibitor. *J. Biol. Chem.* **2002**, *277*(48), 46265–46272.
9. DeLano, W. Pymol: An Open-Source Molecular Graphics Tool. *CCP4 Newsl. Protein Crystallogr.* **2002**, *700*(11), 44–53.
10. Fährrolfes, R.; Bietz, S.; Flachsenberg, F.; Meyder, A.; Nittinger, E.; Otto, T.; Volkamer, A.; Rarey, M. ProteinsPlus: A Web Portal for Structure Analysis of Macromolecules. *Nucleic Acids Res.* **2017**, *45*, 1–7.
11. Volkamer, A.; Eid, S.; Turk, S.; Jaeger, S.; Rippmann, F.; Fulle, S. Pocketome of Human Kinases: Prioritizing the ATP Binding Sites of (Yet) Untapped Protein Kinases for Drug Discovery. *J. Chem. Inf. Model.* **2015**, *55*(3), 538–549.
12. Brooijmans, N.; Chang, Y. W.; Mobilio, D.; Denny, R. A.; Humblet, C. An Enriched Structural Kinase Database to Enable Kinome-Wide Structure-Based Analyses and Drug Discovery. *Protein Sci.* **2010**, *19*(4), 763–774.
13. Möbitz, H. The ABC of Protein Kinase Conformations. *Biochim. Biophys. Acta* **2015**, *1854*(10), 1555–1566.
14. Roskoski, R. Classification of Small Molecule Protein Kinase Inhibitors Based Upon the Structures of Their Drug-Enzyme Complexes. *Pharmacol. Res.* **2016**, *103*, 26–48.
15. Wu, P.; Nielsen, T. E.; Clausen, M. H. Small-Molecule Kinase Inhibitors: An Analysis of FDA-Approved Drugs. *Drug Discov. Today* **2016**, *21*(1), 5–10.
16. Zhang, J.; Yang, P.; Gray, N. Targeting Cancer With Small Molecule Kinase Inhibitors. *Nat. Rev. Cancer* **2009**, *9*(1), 28–39.
17. Fang, Z.; Grütter, C.; Rauh, D. Strategies for the Selective Regulation of Kinases With Allosteric Modulators: Exploiting Exclusive Structural Features. *ACS Chem. Biol.* **2013**, *8*(1), 58–70.
18. Okamoto, K.; Ikemori-Kawada, M.; Jestel, A.; Von König, K.; Funahashi, Y.; Matsushima, T.; Tsuruoka, A.; Inoue, A.; Matsui, J. Distinct Binding Mode of Multikinase Inhibitor Lenvatinib Revealed by Biochemical Characterization. *ACS Med. Chem. Lett.* **2015**, *6*(1), 89–94.
19. Zhao, Z.; Liu, Q.; Bliven, S.; Xie, L.; Bourne, P. E. Determining Cysteines Available for Covalent Inhibition Across the Human Kinome. *J. Med. Chem.* **2017**, *60*(7), 2879–2889.

20. Karaman, M. W.; Herrgard, S.; Treiber, D. K.; Gallant, P.; Atteridge, C. E.; Campbell, B. T.; Chan, K. W.; Ciceri, P.; Davis, M. I.; Edeen, P. T.; et al. A Quantitative Analysis of Kinase Inhibitor Selectivity. *Nat. Biotechnol.* **2008**, *26*(1), 127–132.
21. Law, V.; Knox, C.; Djoumbou, Y.; Jewison, T.; Guo, A. C.; Liu, Y.; Maclejewski, A.; Arndt, D.; Wilson, M.; Neveu, V.; et al. DrugBank 4.0: Shedding New Light on Drug Metabolism. *Nucleic Acids Res.* **2014**, *42*(D1), D1091–D1097.
22. Bento, A. P.; Gaulton, A.; Hersey, A.; Bellis, L. J.; Chambers, J.; Davies, M.; Krüger, F. A.; Light, Y.; Mak, L.; McGlinchey, S.; et al. The ChEMBL Bioactivity Database: An Update. *Nucleic Acids Res.* **2014**, *42*(D1), D1083–D1090.
23. Morphy, R. Selectively Nonselective Kinase Inhibition: Striking the Right Balance. *J. Med. Chem.* **2010**, *53*(4), 1413–1437.
24. Bamborough, P. System-Based Drug Discovery Within the Human Kinome. *Expert Opin. Drug Discov.* **2012**, 7(11), 1053–1070.
25. Scapin, G. Protein Kinase Inhibition: Different Approaches to Selective Inhibitor Design. *Curr. Drug Targets* **2006**, 7(11), 1443–1454.
26. Azzaoui, K.; Hamon, J.; Faller, B.; Whitebread, S.; Jacoby, E.; Bender, A.; Jenkins, J. L.; Urban, L. Modeling Promiscuity Based on in vitro Safety Pharmacology Profiling Data. *ChemMedChem* **2007**, *2*(6), 874–880.
27. Uitdehaag, J. C. M.; Verkaar, F.; Alwan, H.; De Man, J.; Buijsman, R. C.; Zaman, G. J. R. A Guide to Picking the Most Selective Kinase Inhibitor Tool Compounds for Pharmacological Validation of Drug Targets. *Br. J. Pharmacol.* **2012**, *166*(3), 858–876.
28. Arrowsmith, C. H.; Audia, J. E.; Austin, C.; Baell, J.; Bennett, J.; Blagg, J.; Bountra, C.; Brennan, P. E.; Brown, P. J.; Bunnage, M. E.; et al. The Promise and Peril of Chemical Probes. *Nat. Chem. Biol.* **2015**, *11*(8), 536–541.
29. Zhang, L.; Daly, R. J. Targeting the Human Kinome for Cancer Therapy: Current Perspectives. *Crit. Rev. Oncog.* **2012**, *17*(2), 233–246.
30. Fedorov, O.; Müller, S.; Knapp, S. The (Un)targeted Cancer Kinome. *Nat. Chem. Biol.* **2010**, *6*(3), 166–169.
31. Greenman, C.; Stephens, P. R.; Bignell, G.; Birney, E.; Stratton, M. R.; Smith, R. M.; Dalgliesh, G.; Hunter, C.; Davies, H.; Teague, J.; et al. Patterns of Somatic Mutation in Human Cancer Genomes. *Nature* **2007**, *446*(7132), 153–158.
32. Fleuren, E. D. G.; Zhang, L.; Wu, J.; Daly, R. J. The Kinome "at Large" in Cancer. *Nat. Rev. Cancer* **2016**, *16*(2), 83–98.
33. Patel, M. N.; Halling-Brown, M. D.; Tym, J. E.; Workman, P.; Al-Lazikani, B. Objective Assessment of Cancer Genes for Drug Discovery. *Nat. Rev. Drug Discov.* **2012**, *12*(1), 35–50.
34. Manning, B. D. Challenges and Opportunities in Defining the Essential Cancer Kinome. *Sci. Signal.* **2009**, *2*(63), pe15.
35. Workman, P.; Al-Lazikani, B. Drugging Cancer Genomes. *Nat. Rev. Drug Discov.* **2013**, *12*(12), 889–890.
36. Holohan, C.; Van Schaeybroeck, S.; Longley, D. B.; Johnston, P. G. Cancer Drug Resistance: An Evolving Paradigm. *Nat. Rev. Cancer* **2013**, *13*(10), 714–726.
37. Mok, T. S.; Wu, Y.-L.; Ahn, M.-J.; Garassino, M. C.; Kim, H. R.; Ramalingam, S. S.; Shepherd, F. A.; He, Y.; Akamatsu, H.; Theelen, W. S. M. E.; et al. Osimertinib or Platinum–Pemetrexed in *EGFR* T790M–Positive Lung Cancer. *N. Engl. J. Med.* **2017**, *376*(7), 629–640.
38. Forbes, S. A.; Tang, G.; Bindal, N.; Bamford, S.; Dawson, E.; Cole, C.; Kok, C. Y.; Jia, M.; Ewing, R.; Menzies, A.; et al. COSMIC (the Catalogue of Somatic Mutations In Cancer): A Resource to Investigate Acquired Mutations in Human Cancer. *Nucleic Acids Res.* **2009**, *38*(Suppl. 1), D652–7.
39. Tang, Z.; Du, R.; Jiang, S.; Wu, C.; Barkauskas, D. S.; Richey, J.; Molter, J.; Lam, M.; Flask, C.; Gerson, S.; et al. Dual MET-EGFR Combinatorial Inhibition Against

T790M-EGFR-Mediated Erlotinib-Resistant Lung Cancer. *Br. J. Cancer* **2008**, *99*(6), 911–922.

40. Mendoza, M. C.; Er, E. E.; Blenis, J. The Ras-ERK and PI3K-mTOR Pathways: Cross-Talk and Compensation. *Trends Biochem. Sci.* **2011**, *36*, 320–328.
41. Vidović, D.; Koleti, A.; Schürer, S. C. Large-Scale Integration of Small Molecule-Induced Genome-Wide Transcriptional Responses, Kinome-Wide Binding Affinities and Cell-Growth Inhibition Profiles Reveal Global Trends Characterizing Systems-Level Drug Action. *Front. Genet.* **2014**, *5*, 342.
42. Fedorov, O.; Marsden, B.; Pogacic, V.; Rellos, P.; Müller, S.; Bullock, A. N.; Schwaller, J.; Sundström, M.; Knapp, S. A Systematic Interaction Map of Validated Kinase Inhibitors With Ser/Thr Kinases. *Proc. Natl. Acad. Sci. U.S.A.* **2007**, *104*(51), 20523–20528.
43. McDermott, U.; Sharma, S. V.; Dowell, L.; Greninger, P.; Montagut, C.; Lamb, J.; Archibald, H.; Raudales, R.; Tam, A.; Lee, D.; et al. Identification of Genotype-Correlated Sensitivity to Selective Kinase Inhibitors by Using High-Throughput Tumor Cell Line Profiling. *Proc. Natl. Acad. Sci. U.S.A.* **2007**, *104*(50), 19936–19941.
44. Davis, M. I.; Hunt, J. P.; Herrgard, S.; Ciceri, P.; Wodicka, L. M.; Pallares, G.; Hocker, M.; Treiber, D. K.; Zarrinkar, P. P. Comprehensive Analysis of Kinase Inhibitor Selectivity. *Nat. Biotechnol.* **2011**, *29*(11), 1046–1051.
45. Metz, J. T.; Johnson, E. F.; Soni, N. B.; Merta, P. J.; Kifle, L.; Hajduk, P. J. Navigating the Kinome. *Nat. Chem. Biol.* **2011**, 7(4), 200–202.
46. Anastassiadis, T.; Deacon, S. W.; Devarajan, K.; Ma, H.; Peterson, J. R. Comprehensive Assay of Kinase Catalytic Activity Reveals Features of Kinase Inhibitor Selectivity. *Nat. Biotechnol.* **2011**, *29*(11), 1039–1045.
47. Kitagawa, D.; Yokota, K.; Gouda, M.; Narumi, Y.; Ohmoto, H.; Nishiwaki, E.; Akita, K.; Kirii, Y. Activity-Based Kinase Profiling of Approved Tyrosine Kinase Inhibitors. *Genes Cells* **2013**, *18*(2), 110–122.
48. Tang, J.; Szwajda, A.; Shakyawar, S.; Xu, T.; Hintsanen, P.; Wennerberg, K.; Aittokallio, T. Making Sense of Large-Scale Kinase Inhibitor Bioactivity Data Sets: A Comparative and Integrative Analysis. *J. Chem. Inf. Model.* **2014**, *54*(3), 735–743.
49. Uitdehaag, J. C. M.; De Roos, J. A. D. M.; Van Doornmalen, A. M.; Prinsen, M. B. W.; De Man, J.; Tanizawa, Y.; Kawase, Y.; Yoshino, K.; Buijsman, R. C.; Zaman, G. J. R. Comparison of the Cancer Gene Targeting and Biochemical Selectivities of All Targeted Kinase Inhibitors Approved for Clinical Use. *PLoS One* **2014**, *9*(3), e92146.
50. Huber, K. V. M.; Salah, E.; Radic, B.; Gridling, M.; Elkins, J. M.; Stukalov, A.; Jemth, A.-S.; Göktürk, C.; Sanjiv, K.; Strömberg, K.; et al. Stereospecific Targeting of MTH1 by (S)-Crizotinib as an Anticancer Strategy. *Nature* **2014**, *508*(7495), 222–227.
51. Christmann-Franck, S.; Van Westen, G. J. P.; Papadatos, G.; Beltran Escudie, F.; Roberts, A.; Overington, J. P.; Domine, D. Unprecedently Large-Scale Kinase Inhibitor Set Enabling the Accurate Prediction of Compound-Kinase Activities: A Way Toward Selective Promiscuity by Design?*J. Chem. Inf. Model.* **2016**, *56*(9), 1654–1675.
52. Duong-Ly, K. C.; Devarajan, K.; Liang, S.; Horiuchi, K. Y.; Wang, Y.; Ma, H.; Peterson, J. R. Kinase Inhibitor Profiling Reveals Unexpected Opportunities to Inhibit Disease-Associated Mutant Kinases. *Cell Rep.* **2016**, *14*(4), 772–781.
53. Uitdehaag, J. C. M.; de Roos, J. A. D. M.; Prinsen, M. B. W.; Willemsen-Seegers, N.; de Vetter, J. R. F.; Dylus, J.; van Doornmalen, A. M.; Kooijman, J.; Sawa, M.; van Gerwen, S. J. C.; et al. Cell Panel Profiling Reveals Conserved Therapeutic Clusters and Differentiates the Mechanism of Action of Different PI3K/mTOR, Aurora Kinase and EZH2 Inhibitors. *Mol. Cancer Ther.* **2016**, *15*(12), 3097–3109.
54. Willemsen-Seegers, N.; Uitdehaag, J. C. M.; Prinsen, M. B. W.; de Vetter, J. R. F.; de Man, J.; Sawa, M.; Kawase, Y.; Buijsman, R. C.; Zaman, G. J. R. Compound

Selectivity and Target Residence Time of Kinase Inhibitors Studied With Surface Plasmon Resonance. *J. Mol. Biol.* **2017**, *429*(4), 574–586.
55. Chiu, Y.-Y. Y.; Lin, C.-T. T.; Huang, J.-W. W.; Hsu, K.-C. C.; Tseng, J.-H. H.; You, S.-R. R.; Yang, J.-M. M. KIDFamMap: A Database of Kinase-Inhibitor-Disease Family Maps for Kinase Inhibitor Selectivity and Binding Mechanisms. *Nucleic Acids Res.* **2013**, *41*(D1), D430–40.
56. Ortutay, C.; Väliaho, J.; Stenberg, K.; Vihinen, M. KinMutBase: A Registry of Disease-Causing Mutations in Protein Kinase Domains. *Hum. Mutat.* **2005**, *25*(5), 435–442.
57. Richardson, C. J.; Gao, Q.; Mitsopoulous, C.; Zvelebil, M.; Pearl, L. H.; Pearl, F. M. G. MoKCa Database—Mutations of Kinases in Cancer. *Nucleic Acids Res.* **2009**, *37*(Database issue), D824–31.
58. Jadeau, F.; Grangeasse, C.; Shi, L.; Mijakovic, I.; Deleage, G.; Combet, C. BYKdb: The Bacterial Protein tYrosine Kinase Database. *Nucleic Acids Res.* **2012**, *40*(Database issue), D321–4.
59. Wang, Y.; Liu, Z.; Cheng, H.; Gao, T.; Pan, Z.; Yang, Q.; Guo, A.; Xue, Y. EKPD: A Hierarchical Database of Eukaryotic Protein Kinases and Protein Phosphatases. *Nucleic Acids Res.* **2014**, *42*(D1), D496–D502.
60. Krupa, A.; Abhinandan, K. R.; Srinivasan, N. KinG: A Database of Protein Kinases in Genomes. *Nucleic Acids Res.* **2004**, *32*(Database issue), D153–5.
61. Milanesi, L.; Petrillo, M.; Sepe, L.; Boccia, A.; D'Agostino, N.; Passamano, M.; Di Nardo, S.; Tasco, G.; Casadio, R.; Paolella, G. Systematic Analysis of Human Kinase Genes: A Large Number of Genes and Alternative Splicing Events Result in Functional and Structural Diversity. *BMC Bioinformatics* **2005**, *6*(Suppl. 4), S20.
62. Cicenas, J. Welcome to the Incredible World of MAP Kinases. *MAP Kinase* **2013**, *2*(1), 1.
63. Mcskimming, D. I.; Dastgheib, S.; Talevich, E.; Narayanan, A.; Katiyar, S.; Taylor, S. S.; Kochut, K.; Kannan, N. ProKinO: A Unified Resource for Mining the Cancer Kinome. *Hum. Mutat.* **2015**, *36*(2), 175–186.
64. Dardick, C.; Chen, J.; Richter, T.; Ouyang, S.; Ronald, P. The Rice Kinase Database. A Phylogenomic Database for the Rice Kinome. *Plant Physiol.* **2007**, *143*(2), 579–586.
65. Van Linden, O. P. J.; Kooistra, A. J.; Leurs, R.; De Esch, I. J. P.; De Graaf, C. KLIFS: A Knowledge-Based Structural Database to Navigate Kinase-Ligand Interaction Space. *J. Med. Chem.* **2014**, *57*(2), 249–277.
66. Kim, J.; Yoo, M.; Kang, J.; Tan, A. C. K-Map: Connecting Kinases With Therapeutics for Drug Repurposing and Development. *Hum. Genomics* **2013**, 7(1), 20.
67. Ryall, K. A.; Shin, J.; Yoo, M.; Hinz, T. K.; Kim, J.; Kang, J.; Heasley, L. E.; Tan, A. C. Identifying Kinase Dependency in Cancer Cells by Integrating High-Throughput Drug Screening and Kinase Inhibition Data. *Bioinformatics* **2015**, *31*(23), 3799–3806.
68. Goldberg, J. M.; Griggs, A. D.; Smith, J. L.; Haas, B. J.; Wortman, J. R.; Zeng, Q. Kinannote, a Computer Program to Identify and Classify Members of the Eukaryotic Protein Kinase Superfamily. *Bioinformatics* **2013**, *29*(19), 2387–2394.
69. McSkimming, D. I.; Rasheed, K.; Kannan, N. Classifying Kinase Conformations Using a Machine Learning Approach. *BMC Bioinformatics* **2017**, *18*(1), 86.
70. Koscielny, G.; An, P.; Carvalho-Silva, D.; Cham, J. A.; Fumis, L.; Gasparyan, R.; Hasan, S.; Karamanis, N.; Maguire, M.; Papa, E.; et al. Open Targets: A Platform for Therapeutic Target Identification and Validation. *Nucleic Acids Res.* **2017**, *45*(D1), D985–D994.
71. Chartier, M.; Chénard, T.; Barker, J.; Najmanovich, R. Kinome Render: A Stand-Alone and Web-Accessible Tool to Annotate the Human Protein Kinome Tree. *PeerJ* **2013**, *1, e126.*

72. Martin, D. M. A.; Miranda-Saavedra, D.; Barton, G. J. Kinomer v. 1.0: A Database of Systematically Classified Eukaryotic Protein Kinases. *Nucleic Acids Res.* **2009**, *37*(Suppl. 1), D244–D250.
73. Scheeff, E. D.; Axelrod, H. L.; Miller, M. D.; Chiu, H. J.; Deacon, A. M.; Wilson, I. A.; Manning, G. Genomics, Evolution, and Crystal Structure of a New Family of Bacterial Spore Kinases. *Proteins Struct. Funct. Bioinf.* **2010**, *78*(6), 1470–1482.
74. Kanev, G. K.; Kooistra, A. J.; de Esch, I. J. P.; de Graaf, C. Structural Chemogenomics Databases to Navigate Protein–Ligand Interaction Space. In: *Comprehensive Medicinal Chemistry III*; Chackalamannil, S., Rotella, D., Ward, S., Eds.; Vol. 1, Elsevier: Amsterdam, Netherlands; Oxford, United Kingdom; Cambridge, United States, 2017; pp 444–471.
75. Sharma, R.; Schürer, S. C.; Muskal, S. M. High Quality, Small Molecule-Activity Datasets for Kinase Research. *F1000Res.* **2016**, *5*, 1366.
76. Drewry, D. H.; Willson, T. M.; Zuercher, W. J. Seeding Collaborations to Advance Kinase Science With the GSK Published Kinase Inhibitor Set (PKIS). *Curr. Top. Med. Chem.* **2014**, *14*(3), 340–342.
77. Elkins, J. M.; Fedele, V.; Szklarz, M.; Abdul Azeez, K. R.; Salah, E.; Mikolajczyk, J.; Romanov, S.; Sepetov, N.; Huang, X.-P.; Roth, B. L.; et al. Comprehensive Characterization of the Published Kinase Inhibitor Set. *Nat. Biotechnol.* **2015**, *34*(1), 95–103.
78. Dranchak, P.; MacArthur, R.; Guha, R.; Zuercher, W. J.; Drewry, D. H.; Auld, D. S.; Inglese, J. Profile of the GSK Published Protein Kinase Inhibitor Set Across ATP-Dependent and -Independent Luciferases: Implications for Reporter-Gene Assays. *PLoS One* **2013**, *8*(3), e57888.
79. Hu, Y.; Furtmann, N.; Bajorath, J. Current Compound Coverage of the Kinome. *J. Med. Chem.* **2015**, *58*(1), 30–40.
80. Chen, Y. F.; Hsu, K. C.; Lin, S. R.; Wang, W. C.; Huang, Y. C.; Yang, J. M. SiMMap: A Web Server for Inferring Site-Moiety Map to Recognize Interaction Preferences Between Protein Pockets and Compound Moieties. *Nucleic Acids Res.* **2010**, *38*(Suppl. 2), W424–W430.
81. Baeissa, H. M.; Benstead-Hume, G.; Richardson, C. J.; Pearl, F. M. G. Mutational Patterns in Oncogenes and Tumour Suppressors. *Biochem. Soc. Trans.* **2016**, *44*(3), 925–931.
82. Kaminker, J. S.; Zhang, Y.; Watanabe, C.; Zhang, Z. CanPredict: A Computational Tool for Predicting Cancer-Associated Missense Mutations. *Nucleic Acids Res.* **2007**, *35*(Suppl. 2), W595–W598.
83. The UniProt Consortium UniProt: The Universal Protein Knowledgebase. *Nucleic Acids Res.* **2017**, *45*(D1), D158–D169.
84. Fabregat, A.; Sidiropoulos, K.; Garapati, P.; Gillespie, M.; Hausmann, K.; Haw, R.; Jassal, B.; Jupe, S.; Korninger, F.; McKay, S.; et al. The Reactome Pathway Knowledgebase. *Nucleic Acids Res.* **2016**, *44*(D1), D481–D487.
85. Kannan, N.; Neuwald, A. F. Did Protein Kinase Regulatory Mechanisms Evolve Through Elaboration of a Simple Structural Component?*J. Mol. Biol.* **2005**, *351*(5), 956–972.
86. Meslamani, J.; Rognan, D.; Kellenberger, E. Sc-PDB: A Database for Identifying Variations and Multiplicity of "Druggable" Binding Sites in Proteins. *Bioinformatics* **2011**, *27*(9), 1324–1326.
87. McGuire, R.; Verhoeven, S.; Vass, M.; Vriend, G.; De Esch, I. J. P.; Lusher, S. J.; Leurs, R.; Ridder, L.; Kooistra, A. J.; Ritschel, T.; et al. 3D-E-Chem-VM: Structural Cheminformatics Research Infrastructure in a Freely Available Virtual Machine. *J. Chem. Inf. Model.* **2017**, *57*(2), 115–121.
88. Berthold, M. R.; Cebron, N.; Dill, F.; Gabriel, T. R.; Kötter, T.; Meinl, T.; Ohl, P.; Sieb, C.; Thiel, K.; Wiswedel, B. KNIME: The Konstanz Information Miner. In: *Data*

Analysis, Machine Learning and Applications; Preisach, C., Burkhardt, H., Schmidt-Thieme, L., Decker, R., Eds.; Springer Berlin Heidelberg: Berlin, Heidelberg, 2007; pp 319–326.

89. Vieth, M.; Higgs, R. E.; Robertson, D. H.; Shapiro, M.; Gragg, E. A.; Hemmerle, H. Kinomics—Structural Biology and Chemogenomics of Kinase Inhibitors and Targets. *Biochim. Biophys. Acta—Proteins Proteomics* **2004**, *1697*(1–2), 243–257.
90. De Moliner, E.; Moro, S.; Sarno, S.; Zagotto, G.; Zanotti, G.; Pinna, L. A.; Battistutta, R. Inhibition of Protein Kinase CK2 by Anthraquinone-Related Compounds. A Structural Insight. *J. Biol. Chem.* **2003**, *278*(3), 1831–1836.
91. Bietz, S.; Rarey, M. SIENA: Efficient Compilation of Selective Protein Binding Site Ensembles. *J. Chem. Inf. Model.* **2016**, *56*(1), 248–259.
92. Sotriffer, C. Virtual Screening : Principles, Challenges, and Practical Guidelines. *Curr. Opin. Drug Discov. Devel.* **2009**, *12*(3), 519.
93. Volkamer, A.; Rarey, M. Exploiting Structural Information for Drug-Target Assessment. *Futur. Med. Chem.* **2014**, *6*(3), 319–331.
94. Villoutreix, B. O.; Lagorce, D.; Labbé, C. M.; Sperandio, O.; Miteva, M. A. One Hundred Thousand Mouse Clicks Down the Road: Selected Online Resources Supporting Drug Discovery Collected Over a Decade. *Drug Discov. Today* **2013**, *18*(21–22), 1081–1089.
95. Liu, T.; Tang, G. W.; Capriotti, E. Comparative Modeling: The State of the Art and Protein Drug Target Structure Prediction. *Comb. Chem. High Throughput Screen.* **2011**, *14*(6), 532–547.
96. Gore, S.; Velankar, S.; Kleywegt, G. J. Implementing an X-Ray Validation Pipeline for the Protein Data Bank. *Acta Crystallogr. Sect. D Biol. Crystallogr.* **2012**, *68*(4), 478–483.
97. Stierand, K.; Rarey, M. From Modeling to Medicinal Chemistry: Automatic Generation of Two-Dimensional Complex Diagrams. *ChemMedChem* **2007**, *2*(6), 853–860.
98. Volkamer, A.; Griewel, A.; Grombacher, T.; Rarey, M. Analyzing the Topology of Active Sites: On the Prediction of Pockets and Subpockets. *J. Chem. Inf. Model.* **2010**, *50*(11), 2041–2052.
99. Volkamer, A.; Kuhn, D.; Grombacher, T.; Rippmann, F.; Rarey, M. Combining Global and Local Measures for Structure-Based Druggability Predictions. *J. Chem. Inf. Model.* **2012**, *52*(2), 360–372.
100. Henrich, S.; Salo-Ahen, O. M. H.; Huang, B.; Rippmann, F.; Cruciani, G.; Wade, R. C. Computational Approaches to Identifying and Characterizing Protein Binding Sites for Ligand Design. *J. Mol. Recognit.* **2010**, *23*(2), 209–219.
101. Le Guilloux, V.; Schmidtke, P.; Tuffery, P. Fpocket: An Open Source Platform for Ligand Pocket Detection. *BMC Bioinformatics* **2009**, *10*, 168.
102. Halgren, T. A. Identifying and Characterizing Binding Sites and Assessing Druggability. *J. Chem. Inf. Model.* **2009**, *49*(2), 377–389.
103. Desaphy, J.; Azdimousa, K.; Kellenberger, E.; Rognan, D. Comparison and Druggability Prediction of Protein-Ligand Binding Sites From Pharmacophore-Annotated Cavity Shapes. *J. Chem. Inf. Model.* **2012**, *52*(8), 2287–2299.
104. Kokh, D. B.; Richter, S.; Henrich, S.; Czodrowski, P.; Rippmann, F.; Wade, R. C. TRAPP: A Tool for Analysis of Transient Binding Pockets in Proteins. *J. Chem. Inf. Model.* **2013**, *53*(5), 1235–1252.
105. Stank, A.; Kokh, D. B.; Horn, M.; Sizikova, E.; Neil, R.; Panecka, J.; Richter, S.; Wade, R. C. TRAPP Webserver: Predicting Protein Binding Site Flexibility and Detecting Transient Binding Pockets. *Nucleic Acids Res.* **2017**, *W1*, W325–W330. gkx277.
106. Egner, U.; Hillig, R. C. A Structural Biology View of Target Drugability. *Expert Opin. Drug Discov.* **2008**, *3*(4), 391–401.

107. Schmidtke, P.; Barril, X. Understanding and Predicting Druggability. A High-Throughput Method for Detection of Drug Binding Sites. *J. Med. Chem.* **2010**, *53*(15), 5858–5867.
108. Krasowski, A.; Muthas, D.; Sarkar, A.; Schmitt, S.; Brenk, R. DrugPred: A Structure-Based Approach to Predict Protein Druggability Developed Using an Extensive Non-redundant Data Set. *J. Chem. Inf. Model.* **2011**, *51*(11), 2829–2842.
109. Sheridan, R. P.; Maiorov, V. N.; Holloway, M. K.; Cornell, W. D.; Gao, Y. D. Drug-Like Density: A Method of Quantifying The "Bindability" of a Protein Target Based on a Very Large Set of Pockets and Drug-Like Ligands From the Protein Data Bank. *J. Chem. Inf. Model.* **2010**, *50*(11), 2029–2040.
110. Hajduk, P. J.; Huth, J. R.; Fesik, S. W. Druggability Indices for Protein Targets Derived From NMr-Based Screening Data. *J. Med. Chem.* **2005**, *48*(7), 2518–2525.
111. Komander, D.; Kular, G. S.; Schüttelkopf, A. W.; Deak, M.; Prakash, K. R. C.; Bain, J.; Elliott, M.; Garrido-Franco, M.; Kozikowski, A. P.; Alessi, D. R.; et al. Interactions of LY333531 and Other Bisindolyl Maleimide Inhibitors With PDK1. *Structure* **2004**, *12*(2), 215–226.
112. Hindie, V.; Stroba, A.; Zhang, H.; Lopez-Garcia, L. A.; Idrissova, L.; Zeuzem, S.; Hirschberg, D.; Schaeffer, F.; Jorgensen, T. J.; Engel, M.; et al. Structure and Allosteric Effects of Low-Molecular-Weight Activators on the Protein Kinase PDK1. *Nat. Chem. Biol* **2009**, *5*(10), 758–764.
113. Rettenmaier, T. J.; Fan, H.; Karpiak, J.; Doak, A.; Sali, A.; Shoichet, B. K.; Wells, J. A. Small-Molecule Allosteric Modulators of the Protein Kinase PDK1 From Structure-Based Docking. *J. Med. Chem.* **2015**, *58*(20), 8285–8291.
114. Vass, M.; Kooistra, A. J.; Ritschel, T.; Leurs, R.; de Esch, I. J.; de Graaf, C. Molecular Interaction Fingerprint Approaches for GPCR Drug Discovery. *Curr. Opin. Pharmacol.* **2016**, *30*, 59–68.
115. Marcou, G.; Rognan, D. Optimizing Fragment and Scaffold Docking by Use of Molecular Interaction Fingerprints. *J. Chem. Inf. Model.* **2007**, *47*(1), 195–207.
116. Jansen, C.; Kooistra, A. J.; Kanev, G. K.; Leurs, R.; De Esch, I. J. P.; De Graaf, C. PDEStrIAn: A Phosphodiesterase Structure and Ligand Interaction Annotated Database as a Tool for Structure-Based Drug Design. *J. Med. Chem.* **2016**, *59*(15), 7029–7065.
117. Desaphy, J.; Raimbaud, E.; Ducrot, P.; Rognan, D. Encoding Protein-Ligand Interaction Patterns in Fingerprints and Graphs. *J. Chem. Inf. Model.* **2013**, *53*(3), 623–637.
118. Chupakhin, V.; Marcou, G.; Baskin, I.; Varnek, A.; Rognan, D. Predicting Ligand Binding Modes From Neural Networks Trained on Protein-Ligand Interaction Fingerprints. *J. Chem. Inf. Model.* **2013**, *53*(4), 763–772.
119. Illergård, K.; Ardell, D. H.; Elofsson, A. Structure Is Three to Ten Times More Conserved Than Sequence—A Study of Structural Response in Protein Cores. *Proteins Struct. Funct. Bioinf.* **2009**, 77(3), 499–508.
120. Kellenberger, E.; Schalon, C.; Rognan, D. How to Measure the Similarity Between Protein Ligand-Binding Sites?*Curr. Comput.—Aided Drug Des.* **2008**, *4*(3), 209–220.
121. Nisius, B.; Sha, F.; Gohlke, H. Structure-Based Computational Analysis of Protein Binding Sites for Function and Druggability Prediction. *J. Biotechnol.* **2012**, *159*(3), 123–134.
122. Kuhn, D.; Weskamp, N.; Schmitt, S.; Hüllermeier, E.; Klebe, G. From the Similarity Analysis of Protein Cavities to the Functional Classification of Protein Families Using Cavbase. *J. Mol. Biol.* **2006**, *359*(4), 1023–1044.
123. Konc, J.; Janežič, D. ProBiS Algorithm for Detection of Structurally Similar Protein Binding Sites by Local Structural Alignment. *Bioinformatics* **2010**, *26*(9), 1160–1168.

124. Baroni, M.; Cruciani, G.; Sciabola, S.; Perruccio, F.; Mason, J. S. A Common Reference Framework for Analyzing/Comparing Proteins and Ligands. Fingerprints for Ligands and Proteins (FLAP): Theory and Application. *J. Chem. Inf. Model.* **2007**, *47*(2), 279–294.
125. Schalon, C.; Surgand, J. S.; Kellenberger, E.; Rognan, D. A Simple and Fuzzy Method to Align and Compare Druggable Ligand-Binding Sites. *Proteins Struct. Funct. Genet.* **2008**, *71*(4), 1755–1778.
126. Weill, N.; Rognan, D. Alignment-Free Ultra-High-Throughput Comparison of Druggable Protein-Ligand Binding Sites. *J. Chem. Inf. Model.* **2010**, *50*(1), 123–135.
127. Aung, Z.; Tong, J. C. BSAlign: A Rapid Graph-Based Algorithm for Detecting Ligand-Binding Sites in Protein Structures. *Genome Inform.* **2008**, *21*, 65–76.
128. Von Behren, M. M.; Volkamer, A.; Henzler, A. M.; Schomburg, K. T.; Urbaczek, S.; Rarey, M. Fast Protein Binding Site Comparison via an Index-Based Screening Technology. *J. Chem. Inf. Model.* **2013**, *53*(2), 411–422.
129. Krieger, E.; Vriend, G. *YASARA—Yet Another Scientific Artificial Reality Application.* http://www.yasara.org/. Accessed 27 August 2014.
130. Roskoski, R. The ErbB/HER Family of Protein-Tyrosine Kinases and Cancer. *Pharmacol. Res.* **2014**, *79*, 34–74.
131. Goodford, P. J. A Computational Procedure for Determining Energetically Favorable Binding Sites on Biologically Important Macromolecules. *J. Med. Chem.* **1985**, *28*(7), 849–857.
132. Cross, S.; Baroni, M.; Carosati, E.; Benedetti, P.; Clementi, S. FLAP: GRID Molecular Interaction Fields in Virtual Screening. Validation Using the DUD Data Set. *J. Chem. Inf. Model.* **2010**, *50*(8), 1442–1450.
133. Ferrario, V.; Siragusa, L.; Ebert, C.; Baroni, M.; Foscato, M.; Cruciani, G.; Gardossi, L. BioGPS Descriptors for Rational Engineering of Enzyme Promiscuity and Structure Based Bioinformatic Analysis. *PLoS One* **2014**, *9*(10), e109354.
134. Cortés-Ciriano, I.; Ain, Q. U.; Subramanian, V.; Lenselink, E. B.; Méndez-Lucio, O.; IJzerman, A. P.; Wohlfahrt, G.; Prusis, P.; Malliavin, T. E.; van Westen, G. J. P. Polypharmacology Modelling Using Proteochemometrics (PCM): Recent Methodological Developments, Applications to Target Families, and Future Prospects. *MedChemComm* **2015**, *6*(1), 24–50.
135. Subramanian, V.; Prusis, P.; Pietilä, L. O.; Xhaard, H.; Wohlfahrt, G. Visually Interpretable Models of Kinase Selectivity Related Features Derived From Field-Based Proteochemometrics. *J. Chem. Inf. Model.* **2013**, *53*(11), 3021–3030.
136. Volkamer, A.; Eid, S.; Turk, S.; Rippmann, F.; Fulle, S. Identification and Visualization of Kinase-Specific Subpockets. *J. Chem. Inf. Model.* **2016**, *56*(2), 335–346.
137. Burch, J. D.; Lau, K.; Barker, J. J.; Brookfield, F.; Chen, Y.; Chen, Y.; Eigenbrot, C.; Ellebrandt, C.; Ismaili, M. H. A.; Johnson, A.; et al. Property- and Structure-Guided Discovery of a Tetrahydroindazole Series of Interleukin-2 Inducible T-Cell Kinase Inhibitors. *J. Med. Chem.* **2014**, *57*(13), 5714–5727.
138. Goldstein, D. M.; Gray, N. S.; Zarrinkar, P. P. High-Throughput Kinase Profiling as a Platform for Drug Discovery. *Nat. Rev. Drug Discov.* **2008**, 7(5), 391–397.
139. Merget, B.; Turk, S.; Eid, S.; Rippmann, F.; Fulle, S. Profiling Prediction of Kinase Inhibitors: Toward the Virtual Assay. *J. Med. Chem.* **2017**, *60*(1), 474–485.
140. Ruddigkeit, L.; Van Deursen, R.; Blum, L. C.; Reymond, J. L. Enumeration of 166 Billion Organic Small Molecules in the Chemical Universe Database GDB-17. *J. Chem. Inf. Model.* **2012**, *52*(11), 2864–2875.
141. Ripphausen, P.; Nisius, B.; Bajorath, J. State-of-the-Art in Ligand-Based Virtual Screening. *Drug Discov. Today* **2011**, *16*(9–10), 372–376.
142. Schneider, G. Virtual Screening: An Endless Staircase?*Nat. Rev. Drug Discov.* **2010**, *9*(4), 273–276.

143. Keiser, M. J.; Roth, B. L.; Armbruster, B. N.; Ernsberger, P.; Irwin, J. J.; Shoichet, B. K. Relating Protein Pharmacology by Ligand Chemistry. *Nat. Biotechnol.* **2007**, *25*(2), 197–206.
144. Keiser, M. J.; Setola, V.; Irwin, J. J.; Laggner, C.; Abbas, A. I.; Hufeisen, S. J.; Jensen, N. H.; Kuijer, M. B.; Matos, R. C.; Tran, T. B.; et al. Predicting New Molecular Targets for Known Drugs. *Nature* **2009**, *462*(7270), 175–181.
145. Czodrowski, P.; Bolick, W. G. OCEAN: Optimized Cross rEActivity estimatioN. *J. Chem. Inf. Model.* **2016**, *56*(10), 2013–2023.
146. Ferrè, F.; Palmeri, A.; Helmer-Citterich, M. Computational Methods for Analysis and Inference of Kinase/Inhibitor Relationships. *Front. Genet.* **2014**, *5*, 196.
147. Martin, E.; Mukherjee, P.; Sullivan, D.; Jansen, J. Profile-QSAR: A Novel Meta-QSAR Method That Combines Activities Across the Kinase Family to Accurately Predict Affinity, Selectivity, and Cellular Activity. *J. Chem. Inf. Model.* **2011**, *51*(8), 1942–1956.
148. Manallack, D. T.; Pitt, W. R.; Gancia, E.; Montana, J. G.; Livingstone, D. J.; Ford, M. G.; Whitley, D. C. Selecting Screening Candidates for Kinase and G Protein-Coupled Receptor Targets Using Neural Networks. *J. Chem. Inf. Comput. Sci.* **2002**, *42*(5), 1256–1262.
149. Unterthiner, T.; Mayr, A.; Klambauer, G.; Steijaert, M.; Wegner, J. K.; Ceulemans, H. Deep Learning as an Opportunity in Virtual Screening. In: *Proceedings of the Deep Learning Workshop at NIPS, 2014,* 2014; pp 1–9.
150. Schürer, S. C.; Muskal, S. M. Kinome-Wide Activity Modeling From Diverse Public High-Quality Data Sets. *J. Chem. Inf. Model.* **2013**, *53*(1), 27–38.
151. Chen, B.; Sheridan, R. P.; Hornak, V.; Voigt, J. H. Comparison of Random Forest and Pipeline Pilot Naïve Bayes in Prospective QSAR Predictions. *J. Chem. Inf. Model.* **2012**, *52*(3), 792–803.
152. Xia, X.; Maliski, E. G.; Gallant, P.; Rogers, D. Classification of Kinase Inhibitors Using a Bayesian Model. *J. Med. Chem.* **2004**, *47*(18), 4463–4470.
153. Yabuuchi, H.; Niijima, S.; Takematsu, H.; Ida, T.; Hirokawa, T.; Hara, T.; Ogawa, T. Analysis of Multiple Compound—Protein Interactions Reveals Novel Bioactive Molecules. *Mol. Syst. Biol.* **2011**, 7(472), 1–12.
154. Muzzioli, E.; Del Rio, A.; Rastelli, G. Assessing Protein Kinase Selectivity With Molecular Dynamics and MM-PBSA Binding Free Energy Calculations. *Chem. Biol. Drug Des.* **2011**, *78*(2), 252–259.
155. Shukla, D.; Meng, Y.; Roux, B.; Pande, V. S. Activation Pathway of Src Kinase Reveals Intermediate States as Targets for Drug Design. *Nat. Commun.* **2014**, *5*, 3397.
156. Copeland, R. A. The Drug-Target Residence Time Model: A 10-Year Retrospective. *Nat. Rev. Drug Discov.* **2016**, *15*(2), 87–95.
157. Lu, H.; England, K.; Ende, C. A.; Truglio, J. J.; Luckner, S.; Reddy, B. G.; Marlenee, N. L.; Knudson, S. E.; Knudson, D. L.; Bowen, R. A.; et al. Slow-Onset Inhibition of the FabI Enoyl Reductase From Francisella tularensis: Residence Time and in vivo Activity. *ACS Chem. Biol.* **2009**, *4*(3), 221–231.
158. Uitdehaag, J. C. M.; de Man, J.; Willemsen-Seegers, N.; Prinsen, M. B. W.; Libouban, M. A. A.; Sterrenburg, J. G.; de Wit, J. J. P.; de Vetter, J. R. F.; de Roos, J. A. D. M.; Buijsman, R. C.; et al. Target Residence Time-Guided Optimization on TTK Kinase Results in Inhibitors With Potent Anti-Proliferative Activity. *J. Mol. Biol.* **2017**, *429*(14), 2211–2230.

CHAPTER SEVEN

Free Energy Calculation Guided Virtual Screening of Synthetically Feasible Ligand R-Group and Scaffold Modifications: An Emerging Paradigm for Lead Optimization

Robert Abel[1], Sathesh Bhat
Schrödinger, Inc., New York, NY, United States
[1]Corresponding author: e-mail address: robert.abel@schrodinger.com

Contents

1. INTRODUCTION TO VIRTUAL SCREENING

Virtual screening of commercially purchasable and in-house small-molecule collections has become a widespread practice both in academic laboratories and in pharmaceutical companies to rapidly and inexpensively identify small molecules which bind to the intended target with sufficient affinity to facilitate initiation of a drug discovery project.[1–4] These identified small molecules are generally referred to as "hits," and the most promising of these may become a "lead molecule" for the project. A histogram analysis of the number of papers reporting virtual screening results for a wide variety of targets in both retrospective and prospective applications shows monotonic growth, with approximately 600 such studies now published per year (Fig. 1). Initial small-molecule hits identified by virtual screening typically

Annual Reports in Medicinal Chemistry, Volume 50
ISSN 0065-7743
https://doi.org/10.1016/bs.armc.2017.08.007

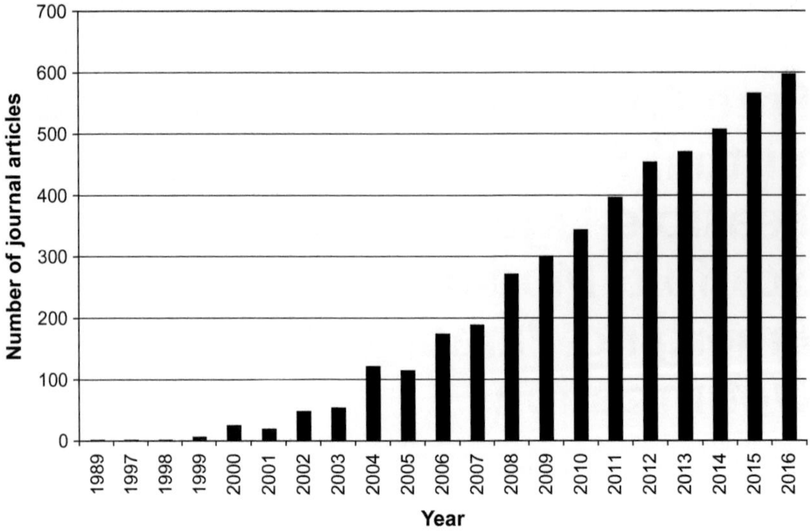

Fig. 1 We here depict a histogram analysis of the PubMed search "virtual screening" by year. Ever wider adoption is suggested as the number of journal articles reporting on the topic continues to grow.

bind with micromolar affinities and must be modified over the course of the lead optimization phase of a drug discovery project to bind with sufficient potency to achieve efficacy against the identified biological target, as well as exhibit an ADMET profile consistent with small-molecule drug therapies.[5] Given a traditionally structured high-throughput experimental screen may cost >$1M USD per target, the advantage of being able to obtain initial hit molecules required to initiate a drug discovery project via a computational approach should be self-evident.[5,6]

Interestingly, although this approach of computationally evaluating very large numbers of small molecules, and then selecting a small subset of these for additional experimental testing is widespread in the lead discovery phase of a drug discovery projects, the equivalent paradigm is seldom employed in the lead optimization phase. In the lead optimization phase of a drug discovery project, the lead molecule must be appropriately synthetically modified to exhibit high affinity for the target protein, selectivity against homologous antitarget proteins, and exhibit a drug-like ADMET profile.[5] Important work by Paul et al. has clearly established lead optimization to be the single most expensive phase of drug discovery, often accounting for more than 20% of the total capitalized costs to bring a drug to market, and further accounting for more than half of all preclinical drug discovery project capitalized costs.[5,7] As such, any approach that may help to reduce the significant capitalized costs associated with lead optimization is expected to greatly improve the productivity of pharmaceutical drug discovery efforts to bring novel and more effective small-molecule drug therapies to the clinic.

Given the nature of the lead optimization problem, i.e., identifying a small molecule within the vast synthetically tractable drug-like chemical space around the lead molecule with an appropriate property profile for the small molecule to be sent into the clinic, it is perhaps surprising this computational screening paradigm has not historically made greater contributions to the lead optimization phase of drug discovery projects. We believe this to be primarily due to two key challenges:

1. Given an identified lead molecule, how can one enumerate some reasonable approximation of the full set of all synthetically tractable near neighbor compounds to the identified lead molecule?
2. Given the enumerated set of all synthetically tractable near neighbor compounds, how does one computationally identify the most promising molecules for synthesis?

Both these problems must be addressed in order for large-scale computational evaluation of idea molecules to play a greater role in lead optimization. Excitingly, great progress has recently been made toward providing solutions to both of these challenges which might be amenable to large-scale use in the pharmaceutical industry.

The first requirement for programmatic enumeration of synthetically feasible idea compounds is a translation of chemical reactions into their respective computational transforms. A common approach to converting chemical reactions into transforms is utilizing languages for interpreting chemical patterns, such as reaction SMARTS.[8] The versatility of the SMARTS-based approach allows for encoding of a wide variety of chemistries including coupling, condensation, and elimination reactions (Fig. 2). Many of the most common organic medicinal chemistry reactions have

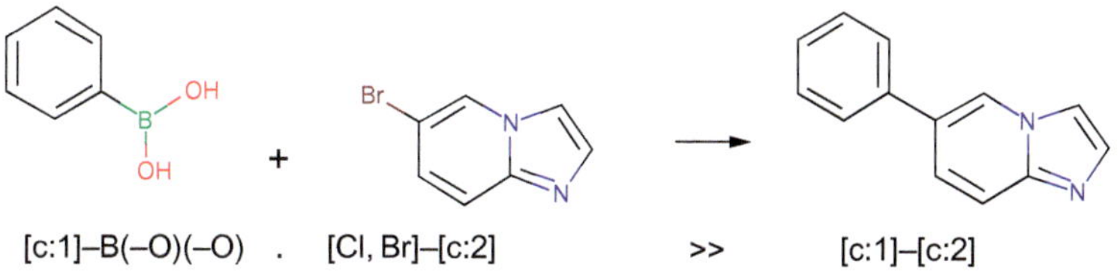

Fig. 2 Example of Suzuki coupling reaction and the conversion to reaction SMARTS.

already been encoded and made available to the public.[9] It is important to note that for purposes of ideation, these transforms can be greatly simplified relative to their wet lab counterparts. For example, aspects such as protecting groups, catalysts, and yields which can be vital for wet lab chemistry are less integral for virtual idea generation.

The next step involves curating a set of virtual reagents that can participate in the reactions, which are readily available from a variety of vendors.[10,11] It is also possible to associate each building block with information such as cost, purity, and availability, which can be useful for filtering downstream virtual idea molecules.

Once a set of reactions and building blocks have been compiled, the next step is choosing a route to ideate on. The route is essentially a subset of the reactions operating on relevant building blocks to generate the initial lead compound we are interested in enumerating around (Fig. 3). There are two approaches to generate such routes: (a) selecting routes based on known wet lab chemistry and (b) generating virtual routes using computational tools. A variety of computational approaches have been devised to assess synthetic tractability and generate routes, ranging from simple rule-based methods to full retrosynthetic analysis.[12–15]

The fourth step is to decide on the type of enumeration to be carried out, which can be considered to fall into two broad categories: (a) R-group enumeration and (b) core hopping. Operationally, the two categories are differentiated by which portions of the route are kept fixed and which are varied when the forward reaction is carried out. For example, keeping all reagents fixed except for the secondary amines in Fig. 3 would results in an R-group enumeration around the amide portion of the lead compound. Conversely, fixing all reagents except the acid chloride would result in core hopping ideas (Fig. 4). Depending on the number of routes generated from the lead compound as well as the size of the building block collection, this approach has the potential to generate millions of synthetically tractable idea molecules that are readily amenable to computational profiling.

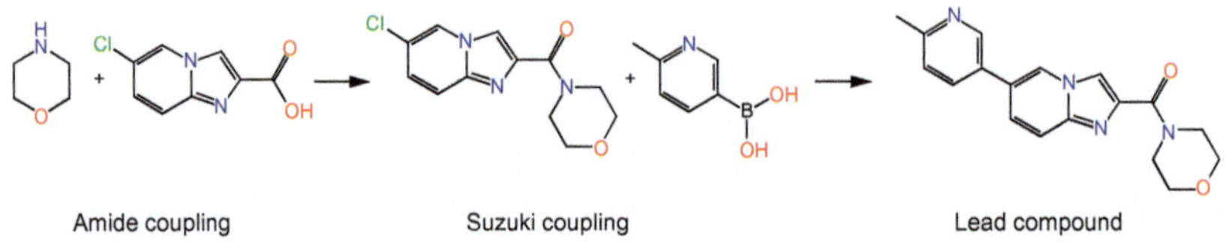

Fig. 3 Example of a potential route to arrive at a lead compound.

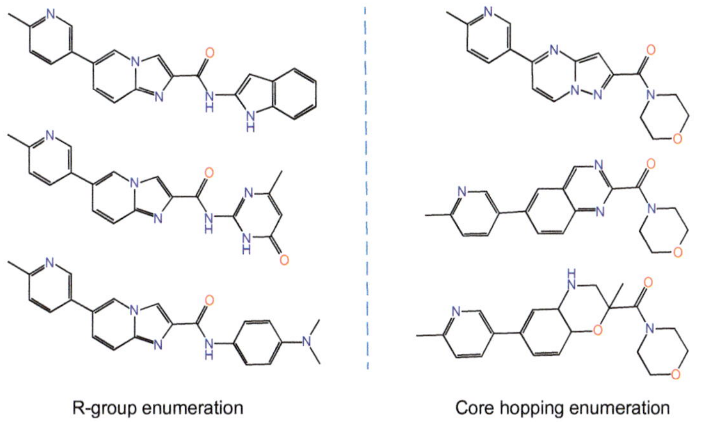

Fig. 4 Examples of different types of enumeration results based on which portions of the route are varied and fixed.

In addition to the programmatic enumeration of the synthetically tractable idea space around the most promising lead molecules, in order for this paradigm to make great contributions to lead optimization, it must also be feasible to identify the most promising subset of molecules for synthesis from this quite large set of idea molecules. In order to ensure this to be tractable, we anticipate a computational scoring funnel will need to be constructed, similar in spirit to routinely utilized experimental screening funnels or screening cascades, where only the most promising compounds are gradually advanced to the most expensive and time-consuming experimental assays (Fig. 5).[2,5] In such a computational screening funnel, filtering the idea molecules based on their physical chemical and anticipated medicinal chemical properties, such as TPSA, PAINS filtering, clogP, molecular weight, and other such common filtering criteria, can help to ensure the resulting molecules will be of interest to the project team. Subsequent to this first-pass filtering, fast empirical docking calculations can be used to quickly reject molecules with very severe steric clashes and other gross incompatibilities with the binding site.[16–23] Next a highly accurate scoring method must be used to identify the subset of molecules most likely to maintain or improve the affinity of the lead molecule to the target protein. And, finally in the fifth stage, those few molecules surviving the stages of this computational screening funnel would be recommended for experimental synthesis and assay. Such a screening funnel would be architected to ensure only the most computationally inexpensive calculations would be run over all the enumerated molecules, with the compute investment per molecule, and

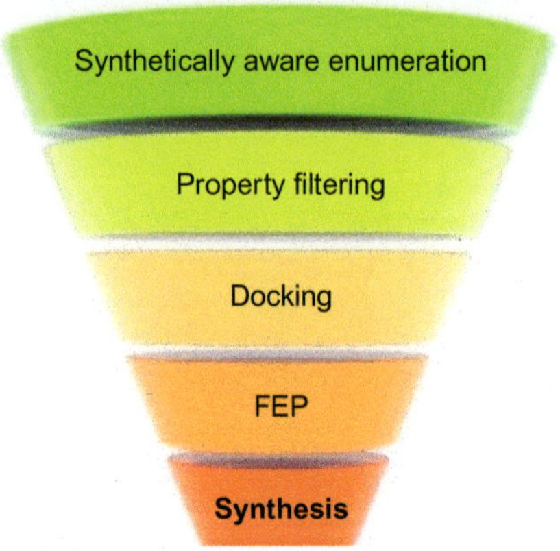

Fig. 5 We here depict a proposed computational screening funnel for lead optimization. In the first stage, the set of likely synthetically tractable molecules similar to the lead molecule would be enumerated. In the second stage, physical chemistry and medicinal chemistry property filtering can significantly reduce the number of molecules required to be considered. In the third stage fast empirical docking calculations can be used to quickly reject molecules with very severe steric clashes and other gross incompatibilities with the binding site. In the fourth stage free energy calculations (FEP) can be used to identify the subset of molecules most likely to maintain or improve their affinity to the target protein. And, finally in the fifth stage those few molecules surviving the stages of this computational screening funnel would be recommended for experimental synthesis and assay.

presumably also the accuracy of the scoring, increasing as the molecules were progressed to the subsequent stages of the screening funnel.

Although such a computational screening funnel can be configured to ensure computational tractability of the evaluation of such large sets of molecules is feasible, this leaves open the question of whether or not the scoring itself is sufficiently accurate, at any feasible level of computational investment, to make the exercise worthwhile to pursue in the first place. Further, it is in its own right an interesting question what scoring accuracy itself is needed for such an activity to be justified.

This topic of what scoring accuracy might be needed to address lead optimization questions has been addressed by several authors. Brown et al. have suggested that in a lead optimization setting, scoring accuracy must have a root-mean-square error of no more than 1.4 kcal/mol to achieve any significant improvement to the productivity of a lead optimization campaign.[24] Shirts et al., using a slightly different framing of the problem, have argued a scoring accuracy of less than 2 kcal/mol should be able to improve the efficiency of lead optimization efforts.[25] Wang et al. have also advanced a more stringent criteria of 1 kcal/mol.[26] One common theme in all these analyses has been the assumption of the error of the scoring method to be Gaussian distributed and invariant over the experimental activity range of the compounds. Although a Gaussian error distribution is convenient for numerical modeling of the required scoring accuracy to positively impact project chemistry, it should be noted observing such an error distribution is not guaranteed, and it remains unsettled how the how these estimates of the scoring accuracy should be interpreted should the calculation error distribution not necessarily follow a Gaussian distribution.

Irrespective of whether or not the required scoring accuracy to materially impact lead optimization might actually be a root-mean-square error of 1.0, 1.4, or 2.0 kcal/mol, it has been a source of great excitement to the field of computational chemistry that recently introduced free energy calculation methods appear to be routinely meeting or exceeding these accuracy requirements under realistic conditions.[26–29] Further, multiple biotechnology and pharmaceutical companies are now reporting successful prospective application of these techniques in their own discovery efforts.[30–34]

Given these recent advances, it perhaps should not be surprising successful application of virtual screening of synthetically feasible compounds in the context of lead optimization, either to enhance the binding potency of the lead matter or to maintain potency while improving other ADMET properties, is now just beginning to be reported.[35,36] The most extensive of these may be the work recently reported by Nimbus Pharmaceuticals, where 4000 synthetically feasible enumerated idea molecules were computationally screened using free energy calculations for Tyk2 binding potency, binding selectivity, and solubility.[35] Of these 4000 computationally screened molecules, only 46 were prioritized for synthesis on the basis of the computational scoring, and 9 of the 46 synthesized molecules were found to meet the targeted experimental potency, selectivity, and solubility property profile. Likewise, Roche Pharmaceuticals has recently reported successful application of free energy calculation guided R-group virtual screening for the optimization of the binding potency of 200 nM inhibitor of Cathepsin L.[36] The Roche study is particularly notable for disclosing the full structures and calculations results for all the studied molecules, which in turn enables independent statistical analysis of the quality of the results, which is pursued in the next section of this chapter.

2. DETAILED STATISTICAL ANALYSIS OF A RECENT R-GROUP VIRTUAL SCREENING CAMPAIGN

In recent work reported by Roche Pharmaceuticals, a 200-nM inhibitor of Cathepsin L was selected for further potency optimization by replacement of a cyclohexyl R-group binding the S2 pocket with a more appropriate substituent in a round of chemistry (Fig. 6).[36] Seeking to optimize the potency of an inhibitor through variation modification of its R-groups is a very common lead optimization activity and should be expected to manifest many of the challenges associated with that phase of drug discovery. To facilitate the identification of the desired potency enhancing R-groups, a 3325 R-group library was enumerated where all R-groups in the library could be easily coupled to the inhibitor through the appropriate reaction chemistry. Only limited S2 R-group SAR was known at the initiation of the exercise, as is depicted in Fig. 7. This earlier SAR data is quite notable in that, despite the SAR being merely related to the filing of a hydrophobic groove by a small hydrophobic moiety, it is difficult to rationalize various activity cliffs that are present. A particularly notable pair is the isopropyl exhibiting a 30-nM affinity, whereas the quite similar cyclopropyl exhibits only a 1600-nM affinity. Correct modeling of such nonobvious activity cliffs is a key test of any scoring method, especially so in a lead optimization context, and it was expected the identification of new tight binding S2 pocket R-groups would manifest many of the challenges routinely encountered in lead optimization.

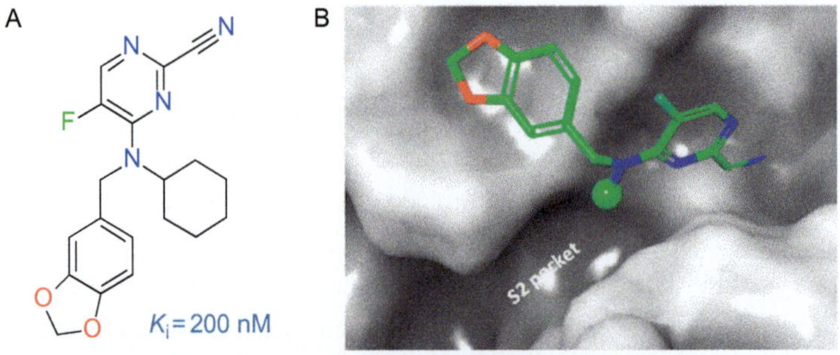

Fig. 6 (A) We depict the 200 nM Cathepsin L inhibitor selected for further optimization. (B) We depict the expected crystallographic binding mode of the inhibitor as threaded onto the closely related inhibitor found in PDB crystal structure 4AXM.[36]

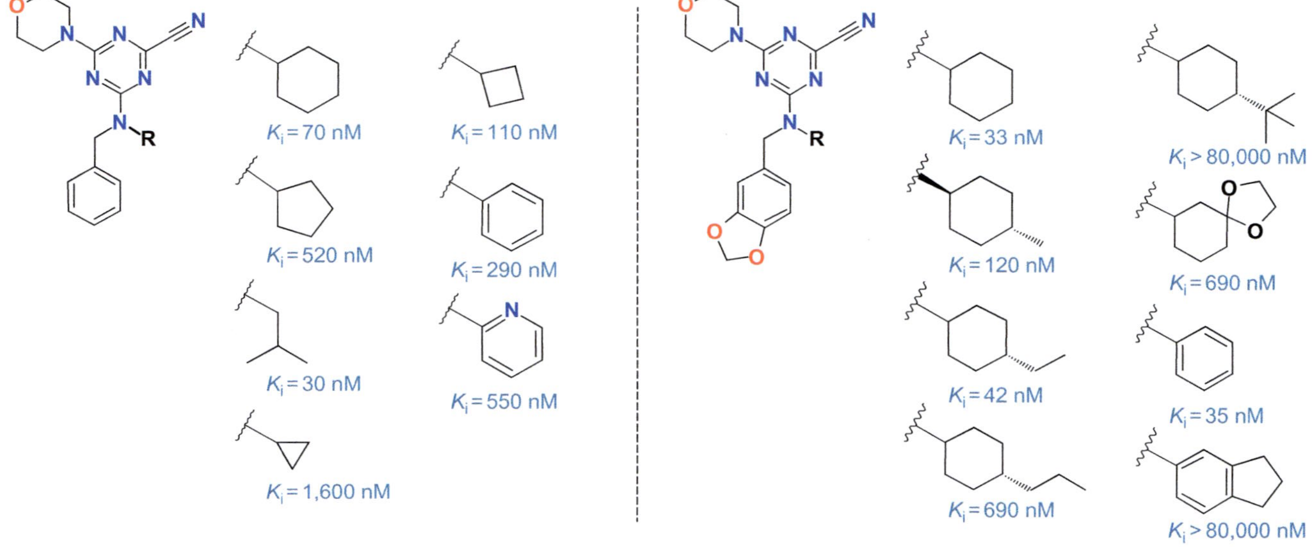

Fig. 7 The previously known Cathepsin L S2 pocket SAR reported in Ref. 36 is here depicted.

Four different compound selection methods were systematically compared regarding their respective abilities to identify novel Cathepsin L S2 pocket R-groups that might lead to more potent inhibition:

1. Selection by an experienced medicinal chemist;
2. Selection by an experienced structure-based modeler using any technique of their choosing other than free energy calculations;
3. Docking and filtering; and
4. Free energy calculation-based scoring, where a diverse subset of 92 molecules obtained by way of a computational screening funnel were subjected to free energy calculation-based scoring.

Each selection method was used in an entirely prospective manner to prioritize 10 R-groups from synthesis from the full library. The R-groups selected for synthesis by the various scoring methods are depicted in Fig. 8.

The synthesis and assay of selected compounds lead to a striking finding: the use of free energy calculations to prioritize the synthesis of the molecules leads to eightfold enrichment of molecules tighter binding than the lead molecule compared with any other tested approach (Table 1). The numerical accuracy of the free energy calculations also appears to have been quite good, the results of which are tabulated in Table 2. Interestingly, in agreement with the probability analysis introduced in Ref. 37, the free energy calculation results appear to manifest a significant selection bias effect, and much more accurate predictions were obtained by prospectively applying selection bias corrections suggested in that work ahead of the synthesis of the compounds.

In addition to providing enrichment of tight binding molecules, the reported calculations also identified a novel extended S2 pocket binding topology manifest by molecules 30, 31, 33, 34, 35, and 38 which to the knowledge of the authors had no earlier precedent in Cathepsin L literature. Thus, in addition to enriching the number of tight binding compounds observed in the round of chemistry, the calculations were also able to identify novel productive R-groups which might steer a drug discovery project in new directions.

However, despite the success reported in this study, in some ways, it raises more questions than it may answer, in particular, the following questions come to mind:

1. Were the strikingly good results of the free energy calculations (8 of 10 molecules tighter binding than the lead molecule) a matter of random chance, or should similar positive results be expected in future applications?

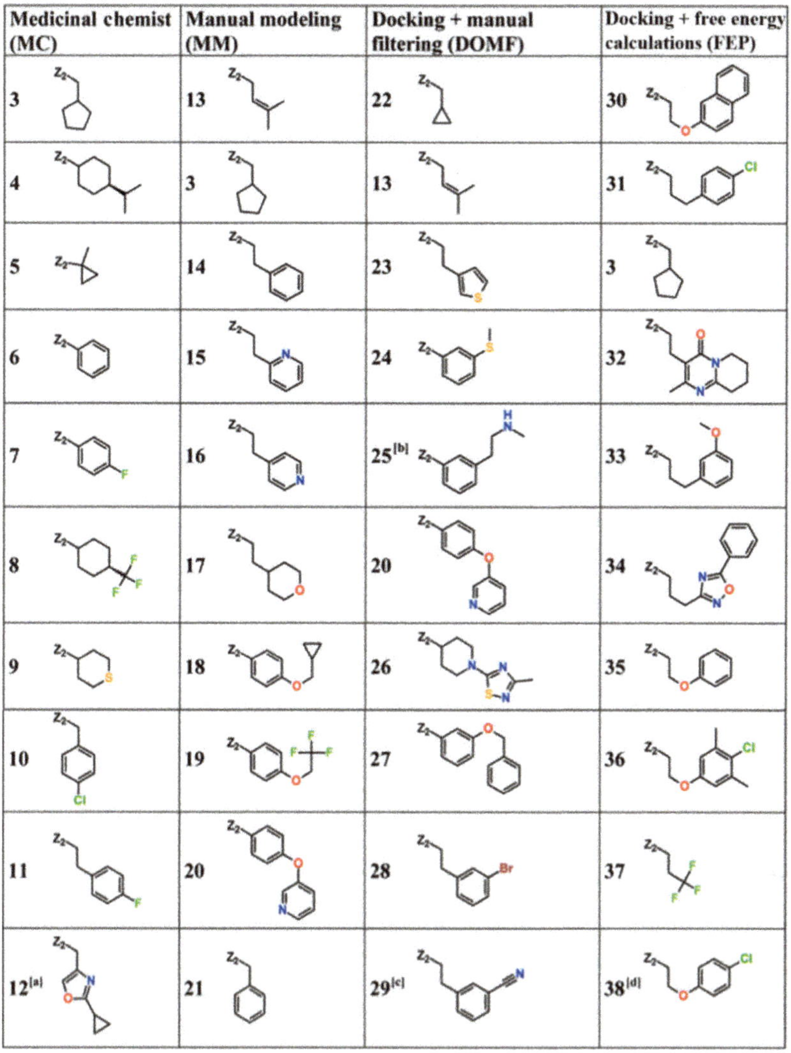
Medicinal chemist (MC)
Manual modeling (MM)
Docking + manual filtering (DOMF)
Docking + free energy calculations (FEP)
3
13
22
30
4
3
13
31
5
14
23
3
6
15
24
32
7
16
25[b]
33
8
17
20
34
9
18
26
35
10
19
27
36
11
20
28
37
12[a]
21
29[c]
38[d]

Fig. 8 We here depict the molecules selected for synthesis by the various tested approaches. [a] Compound **12** could not be synthesized. [b] Compound **25** is missing a methyl on the nitrogen compared to the intended molecule, still maintaining the positive charge. [c] Compound **29** is an unintended product but was approved before compounds were tested. It shows high similarity to compound **28**. [d] The original submission for compound **38** had one more carbon atom in the linker. Three building blocks (3, 13, 20) were prioritized by more than one approach. *Adapted from Kuhn, B.; Tichý, M.; Wang, L.; Robinson, S.; Martin, R.E.; Kuglstatter, A.; Benz, J.; Giroud, M.; Schirmeister, T.; Abel, R.; Diederich, F.; Hert, J.* J. Med. Chem. ***2017***, 60 *(6), 2485.*

Table 1 Cathepsin L Inhibition Constants for the Synthesized Molecules Reported in Ref. 36

Free Energy Calculations		Medicinal Chemist Selection		Structure-Based Drug Design		Docking and Filtering	
Cmpd. ID	Expt. K_i (nM)	Cmpd. ID	Expt. K_i (nM)	Cmpd. ID	Expt. K_i (nM)	Cmpd. ID	Expt. K_i (nM)
3	12	3	12	3	12	22	77
37	25	11	279	14	217	29	304
31	27	9	515	16	505	28	358
33	30	7	952	13	1010	26	411
35	77	6	1800	20	1020	23	671
34	91	4	3020	17	2790	13	1010
30	123	8 (*cis*)	3500	15	3860	20	1020
38	167	5	>5100	18	>5100	24	5100
36	1430	8 (*trans*)	>5100	19	>5100	25	>5100
32	1750	10	>5100	21	>5100	27	>5100

Table 2 Free Energy Calculation Results and Experimental Binding Free Energies for the Inhibitors Selected for Synthesis Are Here Tabulated

Cmpd. ID	Expt. ΔG (kcal/mol)	Calculated ΔG (kcal/mol)	Abs. Error	Selection Bias Corrected ΔG (kcal/mol)	Abs. Error
3	−10.77	−11.12	0.35	−10.47	0.3
33	−10.23	−10.92	0.69	−10.31	0.08
35	−9.67	−11.13	1.46	−10.48	0.81
37	−10.34	−11.35	1.01	−10.65	0.31
31	−10.29	−11.29	1.00	−10.60	0.31
30	−9.40	−11.49	2.09	−10.76	1.36
36	−7.95	−11.37	3.42	−10.66	2.71
34	−9.58	−9.80	0.22	−9.44	0.14
32	−7.83	−11.25	3.42	−10.57	2.74
38	−9.22	−11.89	2.67	−11.07	1.85
Lead Mol.	−9.11	N/A		N/A	
	MUE:	1.63	MUE:	1.06	
	R^2:	0.03	R^2:	0.03	

The selection bias corrected free energies were found to be in much better agreement with the experimental data. Prior to the synthesis of the molecules selection bias correction parameters $\langle pK_i \rangle = 6$ and $stdev(pK_i) = 1.5$ had been assumed as per the recommendations of Ref. 37.

2. Was the performance of the medicinal chemist, structure-based modeler, and docking approaches (each only identifying 1 of 10 molecules tighter binding than the lead molecule) unusually poor in this application?
3. Was it coincidence that the medicinal chemist, structure-based modeler, and docking approaches should each find only one molecule tighter binding than the lead molecule?

In the analysis that follows, we conjectured the performance of the medicinal chemist, structure-based modeler, and docking was comparable to random selection. Although, this assertion would be difficult to prove without additional chemistry resources to further substantiate it, it is likely the most logical assumption. First, all three of these scoring methods are quite different from each other; and second, if these methods provided any enrichment of tight binding molecules at all, then that enrichment would have had to be in very similar direction and magnitude for the scoring performance of the methods to appear comparable. Thus, given the similarity of performance of these scoring methods, but the very different nature of the scoring methods, it seems likely their performance was no better than random selection.

A consequence of this is the expectation that the potency distribution of the 3325 R-group library might be well modeled as a Gaussian distribution with $\langle pK_i \rangle = 5.76$ and $\mathrm{stdev}(pK_i) = 1.03$, which has been determined from the appropriate averaging and numerical analysis of the affinities of the assay results of the molecules prioritized by these scoring methods. The characterization of this affinity distribution of the 3325 library compounds is important because it now in turn allows us to quantitatively consider if the strikingly good performance of the free energy calculations was fortuitous or to be expected.

Utilizing the Monte Carlo approach introduced in Ref. 37, we may now quantitatively investigate if the good performance of the free energy calculations was fortuitous or within the expected operating characteristics of the method. The Monte Carlo algorithm used to investigate the expected performance of the FEP+ scoring was the following:

1. Generate a fictitious 3325 compound library where the affinities of all the compounds are known and are drawn from an assumed $\langle pK_i \rangle = 5.76$ and $\mathrm{stdev}(pK_i) = 1.03$ Gaussian distribution;
2. Randomly select a 92 molecule subset from this full library;
3. Compute fictitious free energy calculation determined affinities by applying a $\sigma = 1$ kcal/mol Gaussian distributed random error to the compound affinities, in line with earlier characterization of the error associated with free energy calculation scoring;

4. Rank-order the 92 molecule subset by their free energy calculation determined affinities;
5. Determine how many of the top 10 ranked molecules have an originally assigned affinity value tighter than 200 nM;
6. Repeat this procedure many times to obtain good statistics regarding the most likely outcomes.

Application of this procedure finds that the expected outcome of the FEP+ scoring was the identification of 6.3 ± 1.7 molecules tighter binding than the initial 200 nM lead compound. This in turn suggests the results of the free energy calculations identifying 8 out of the 10 prioritized molecules to be tighter binding than the 200-nM lead compound were perhaps very slightly fortuitous, but well within the expected operating characteristics of the scoring method, where anywhere to five to eight such molecules would be expected to be identified in similar exercises.

Interestingly, this result independently supports the performance of the medicinal chemist, structure-based modeler, and docking was comparable to random selection. If these methods had provided significant enrichment of tight binding molecules, then the actual affinity distribution of the 3325 R-group library should be shifted to a $<\mathrm{p}K_\mathrm{i}>$ less than 5.76, and the expected performance of the free energy calculation-based scoring should have been much worse than what was observed in the exercise.

A related interesting question is whether or not the earlier formulated Monte Carlo model can be used to infer an effective root-mean-square error for the medicinal chemist, structure-based modeler, and docking-based scoring approaches. Since these scoring methods were applied to all 3325 library compounds, the Monte Carlo algorithm used to determine the outcomes was as follows:

1. Generate a fictitious 3325 compound library where the affinities of all the compounds are known and are drawn from an assumed $<\mathrm{p}K_\mathrm{i}> = 5.76$ and $\mathrm{stdev}(\mathrm{p}K_\mathrm{i}) = 1.03$ Gaussian distribution;
2. Compute fictitious medicinal chemist, structure-based modeler, and docking-based estimated affinities by applying a $\sigma = X$ log unit Gaussian distributed random error to the compound affinities;
3. Rank-order the set of molecules by their medicinal chemist, structure-based modeler, and docking-based estimated affinities;
4. Determine how many of the top 10 ranked molecules have an originally assigned affinity value tighter than 200 nM;
5. Repeat this procedure many times to obtain good statistics regarding the most likely outcomes for multiple reasonable choices of the scoring error parameter X.

The outcomes of this Monte Carlo analysis are tabulated in Table 3. The results may be quite surprising. Even provided a very high root-mean-square scoring error of 6 log units, the Monte Carlo analysis suggests the medicinal chemist, structure-based modeler, and docking should have been better than random, with at least one of the scoring methods expected to identify at least three molecules tighter binding than the 200 nM lead molecule. A 6-log unit RMSE is an enormous error, and an error reliably much better than 6 log unit RMSE could easily be obtained by simply predicting a 1-μM affinity for any presented compound. How then could the medicinal chemist, structure-based modeler, and docking-based selection methods applied in good faith by experienced practitioners perform worse than this?

The solution to this paradox may be easily obtained by considering the expected performance obtained by simply predicting a 1-μM affinity for any presented compound. Although the RMSE of such a scoring method will certainly be less than 6 log units since very few molecules are tighter binding than 1 nM, and the detection limit of most assays is less than 1 mM. However, the performance of such a trivial scoring method in prioritizing compounds for synthesis will clearly be strictly equivalent to random selection. The fundamental reason for this is the error distribution of this scoring method will be highly nonuniform over the activity range of the compounds, whereas the preceding Monte Carlo analysis assumes the scoring error to be uniform over the entire range. This insight in turn leads us to consider what performance of medicinal chemist, structure-based modeler, and docking-based compound selection algorithms might be expected by

Table 3 Results of the Monte Carlo Analysis of the Expected Number of Molecules Tighter Binding Than the Lead Molecule in Top 10, Given 3325 Molecules Scored Assuming a Gaussian Error Distribution

RMSE/σ (Log Units)	Expected Number of Identified Tight Binding Molecules (<200 nM)
0.8	10.0 ± 0.1
1.0	9.9 ± 0.4
2.0	7.6 ± 1.3
3.0	5.6 ± 1.6
4.0	4.2 ± 1.5
5.0	3.5 ± 1.5
6.0	3.0 ± 1.5

the Monte Carlo if those compound selection methods might also have a biased error distribution or a non-Gaussian error distribution.

To investigate the possible effects of non-Gaussian error distributions on performance of various scoring methods one might consider the Cauchy distribution an illustrative case. The Cauchy distribution has a simple analytical form of

$$P(\text{Error}) = \frac{1}{\gamma\pi\left[1 + \left(\frac{\text{Error}}{\gamma}\right)^2\right]}$$

where *P*(Error) is the probability of observing any particular value of the scoring error, and spread parameter γ controls the width of the distribution. Although, at first glance the Cauchy distribution might appear quite similar in to a Gaussian distribution, the Cauchy distribution has "fat tails" leading to the mean and standard deviation of the distribution being formally undefined, and the sample estimates of their values will diverge with growing sample size (Fig. 9).

We now can repeat the above Monte Carlo procedure where rather than drawing our computed scores from a Gaussian distribution, we will draw the computed scores from a Cauchy distribution. The results of the Monte

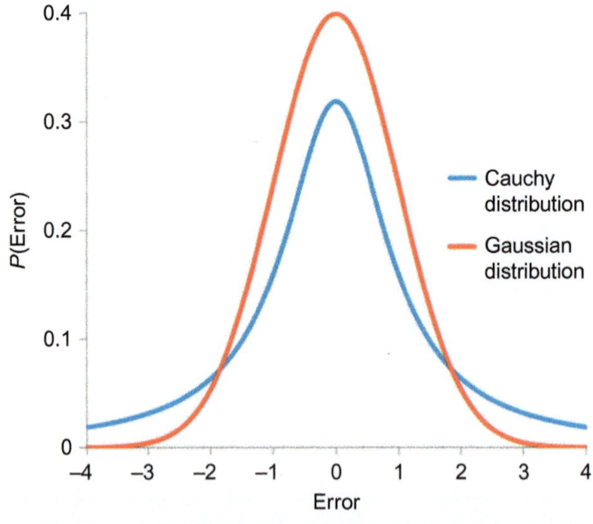

Fig. 9 We here depict a Cauchy distribution and Gaussian distribution. The Cauchy distribution visually appears quite similar to the Gaussian distribution, but the tails of the distribution contain much more probability density than the Gaussian distribution.

Table 4 Results of the Monte Carlo Analysis of the Expected Number of Molecules Tighter Binding Than the Lead Molecule in Top 10, Given 3325 Molecules Scored Assuming a Cauchy Error Distribution

Spread/γ (Log Units)	Expected Number of Identified Tight Binding Molecules (<200 nM)
0.8	1.1 ± 1.0
1.0	1.0 ± 0.9
2.0	1.0 ± 0.9
3.0	1.0 ± 0.9
4.0	1.0 ± 0.9
5.0	1.0 ± 0.9
6.0	1.0 ± 0.9

Carlo analysis are presented in Table 4. Interestingly, a Cauchy error distribution, even for quite small values of the spread parameter γ, results in scoring performance statistically indistinguishable from random selection. The reason for this is when scoring large numbers of molecules, a fat tail error distribution will lead to the large outliers crowding out all well-scored molecules from the small subset of molecules selected for synthesis.

In hindsight, it appears highly probable the molecules flagged for synthesis by the medicinal chemist might be drawn from a fat tail error distribution. An intuition regarding why this might be the case may be gained from careful inspection of Fig. 10. In Case 1, the molecule to be scored is an interpolation between two close analogues, and the prediction of the med chemist is likely to be very highly accurate. However, in Case 2, the medicinal chemist is forced to extrapolate out to a new compound with little obvious information to inform his or her thinking. As such, the error associated with the medicinal chemist in cerebro scoring should be expected to have fat tails as the molecules scored deviate with earlier known SAR. Further, the size of the chemical space accessible to the project team is vast, and the pressures to optimize ADMET, potency, and selectivity profiles will repeatedly push the medicinal chemist into ever more Case 2 situations.

Similar problems, albethey more subtle, are likely to be manifest by the structure-based modeling and docking-based approaches. These approaches make gross approximations neglecting explicit solvent effects, protein motion, configurational entropy, as well as other effects crucial to the thermodynamics of protein–ligand binding. Thus, very large errors are to be

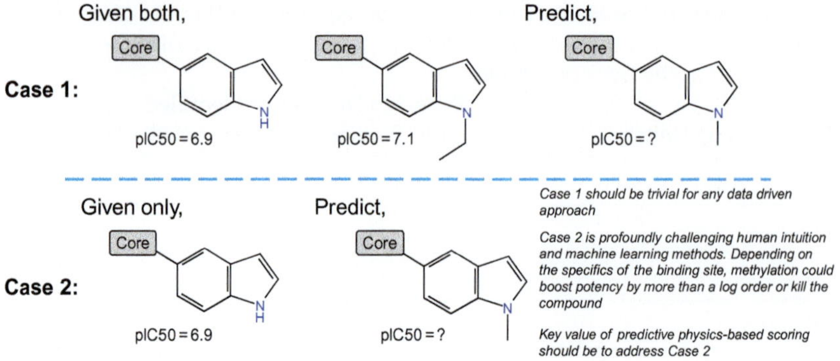

Given both,
Predict,
Case 1:
Core
pIC50 = 6.9
pIC50 = 7.1
pIC50 = ?
Given only,
Predict,
Case 2:
pIC50 = 6.9
pIC50 = ?
Case 1 should be trivial for any data driven approach
Case 2 is profoundly challenging human intuition and machine learning methods. Depending on the specifics of the binding site, methylation could boost potency by more than a log order or kill the compound
Key value of predictive physics-based scoring should be to address Case 2

Fig. 10 We here depict two limiting cases of SAR-informed compound scoring. In Case 1, the earlier known SAR data make prediction of the affinity of the methyl-indole variant a trivial task. However, in Case 2, the prediction is far more challenging. Depending on the particulars of the binding site, the methylation of the indole moiety might increase the binding affinity of the compound by more than a log unit or kill the compound.

expected when these methods are used to evaluate compounds where the primary physical driving force of their binding is neglected or only very grossly approximated in the scoring method.

As such, rather than the root-mean-square error of a scoring method being the essential quantity to optimize in the methods development process, it may be more valuable to verify the error distribution of the scoring method is Gaussian when applied to large sets of compounds. For example, in Refs. 26 and 27, an approximately Gaussian shape of the error distribution was explicitly confirmed out to 2-sigma, which may be related to why the free energy calculation-based scoring was so successful in this particular application. However, it should be noted that this introduces an unavoidable requirement for very large test sets, where hundreds of compounds would need to be scored to confirm the error distribution is approximately Gaussian out to 2-sigma and thousands of compounds would need to be scored to confirm the error distribution is approximately Gaussian out to 3-sigma. Should one also need to check the scoring method error distribution is unbiased across the range of the experimental binding affinity of the compounds, even large test sets would be required.

3. CONSIDERATION OF AN APPROPRIATE NULL MODEL FOR THE EVALUATION OF AFFINITY SCORING

In the preceding section, four different scoring methods were compared regarding their abilities to effectively select molecules for synthesis in a round of chemistry. However, given the expense associated with the synthesis of small molecules in lead optimization, where the full FTE costs of the chemistry team are included, has been estimated to be as high as ~$5000 per molecule, it is unlikely for data of this type to be routinely generated. Rather, in most industrial drug discovery projects, we anticipate only a single scoring approach will be used, and the project team will only later ask in retrospect was the scoring effective? Posing this question in the absence of the appropriate control data (i.e., the results for the med chemist, modeler, and docking in the earlier section) might lead the project team to attempt to formulate an appropriate "null model" to assess the statistical quality of the scoring results, where the null model would ideally be utilized to assess the quality of the scoring results without any actual compounds being flagged for synthesis on basis of the null model.

Several such ex post facto null models for the evaluation of the quality of scoring results have been suggested including:

1. Is the mean-unsigned error of the predicted compound affinities more accurate than simply guessing the average affinity of the compounds for all the synthesized compounds (i.e., the guess the average null model)?;
2. Is the mean-unsigned error of the predicted compound affinities more accurate than simply guessing the affinity of the original lead compound for all the synthesized compounds (i.e., the uniformly guess zero $\Delta\Delta G$ null model)?[38]; and
3. Is the correlation coefficient of the predicted affinities with the experimental affinities significantly different than the correlation coefficient of molecular weight with the experimental affinities (i.e., the molecular weight null model)?[39]

Although the desire to assess the quality of the modeling results without needing to incur the additional costs of conducting a proper control experiment (i.e., synthesizing additional molecules prioritized with another technique or a random selection of molecules from the original enumerated set) is quite understandable; all these forementioned ex post facto null models yield highly misleading results when applied to the previously described Roche study, which should make their inherent pathologies more clear. In what follows, we imagine the situation where only the 10 compounds selected for synthesis by the free energy calculations were synthesized, and then with only that much more limited information at hand, the project team then attempts to make a judgment regarding the quality of the scoring based on the various proposed null model analyses.

First, let us consider the guess the average affinity null model. The average binding free energy of the 10 compounds selected for synthesis by FEP+ scoring was −9.53 kcal/mol. If that affinity was to be "predicted" for all 10 molecules, then the MUE of that set of predictions would be 0.74 kcal/mol, significantly more accurate than the MUE of the free energy calculation selection bias corrected affinity estimates themselves. However, it is important to then ask, does this exercise suggest anything informative about the quality of the FEP+ scoring? We believe the answer to be clearly not. Rather than being informative, the guess the average null model is actually formalized cheating—i.e., prior to the synthesis and assay of the compounds, the average affinity of the compounds cannot be exactly known, and therefore cannot be used to take the place of a predicted value. Further, all this exercise is actually probing is the dynamic range of the experimental affinities of the synthesized compounds, rather than actually interrogating any feature of the prospective scoring. If compounds selected for synthesis happen to have a small dynamic range of experimental affinities, the guess the average affinity null model will have an exceptionally small MUE. If the dynamic range of experimental affinities of the synthesized molecules is small, we see no clear basis to conclude the round of chemistry should be considered of poor quality. As was the case with the preceding described Roche study, if many of the synthesized molecules are tightly clustered in activity, but generally as potent or more so than the lead molecule, the round of chemistry under consideration should be considered in our view quite productive. Thus, rather than simply being uninformative, this null model is actively misleading and should be avoided.

Next, we may consider the uniformly guess zero $\Delta\Delta G$ null model. In the Roche FEP-guided R-group optimization, the binding free energy of the lead compound was −9.11 kcal/mol. If we were to assign this value as the predicted binding free energy for all 10 molecules selected by the free energy calculations, then the MUE of uniformly guess zero $\Delta\Delta G$ null model would be 0.91 kcal/mol, again more accurate than the 1.06 kcal/mol MUE of the free energy calculations. Again though, we must ask ourselves is this exercise determining anything informative regarding the quality, utility, or statistical significance of the FEP-computed binding free energy calculations, which provided an eightfold enrichment of tight binders over industry standard techniques? Once again, it appears clear the answer is no. In realistic applications, it is very unusual to gain more than a log unit or two of potency in a single round of chemistry.[25] So, a highly successful scoring-guided round of chemistry should be expected to generate compounds that will be anywhere from roughly equipotent to the lead to up to perhaps a log order or two tighter binding, which in turn leads to the spurious apparent good performance of the uniformly guess zero $\Delta\Delta G$ null model. Conversely, had the uniformly guess zero $\Delta\Delta G$ null model been used to prospectively *select and score* 10 compounds, rather than simply retrospectively rescore 10 compounds selected by another approach, it's MUE would very likely have been significantly worse. Assuming the med chemist, modeler, or docking selections reported in the Roche study were at least as good as random selection, we would anticipate roughly a third of the uniformly guess zero $\Delta\Delta G$ null model selected compounds would have had no measurable binding affinity, despite all being predicted by the uniformly guess zero $\Delta\Delta G$ null model to equipotent to the lead molecule. Lastly, one of the key anticipated use cases of free energy calculation guided screening of synthetically tractable molecules is to maintain the affinity of the lead molecule while appropriately modifying the ligand R-groups to address some other ADMET liability, such as membrane permeability, solubility, selectivity, metabolism, or other properties. Yet paradoxically, the better the performance of the free energy calculations to achieve this very challenging lead optimization activity, the worse the results of the calculations will appear when compared to the uniformly guess zero $\Delta\Delta G$ null model. So, similar to the guess the average affinity null model, the uniformly guess zero $\Delta\Delta G$ null model, is actively misleading and should be avoided.

Finally we turn our attention to consider the molecular weight null model. In FEP-guided R-group screen reported by Roche Pharmaceuticals the R^2 value of the FEP calculated ligand affinities and the experimental affinities was only 0.03, a very small value generally considered to be indicative of poor performance. This may naturally lead one to wonder, how then did FEP-guided scoring succeeded to suggest 8 of 10 molecules, if the scoring was seemingly so uncorrelated with the experimental affinities? A completely satisfactory solution to this apparent paradox has been fully elaborated in Ref. 37. Therein, the authors develop a Monte Carlo model of the scoring a one-thousand compound R-group library, where only the small subset predicted to bind more tightly than some cutoff value are selected for synthesis. The outcome of that selection step, where only a small subset are selected synthesis greatly compresses the dynamic range of the predicted affinities such that achieving a good R^2 value becomes very highly improbable. As noted by the authors, if the full one-thousand idea molecules were synthesized and assayed, the R^2 value would be outstanding, and it is only a selection bias effect that makes the R^2 value of the synthesized subset appear to be poor.

Clear manifestations of this selection bias effect can be observed in the experimental and predicted affinities for the compounds selected by FEP-guided scoring for synthesis in the Roche study here presented in Table 2. The dynamic range of the selection bias corrected predicted affinities, due to the biased synthesis of only the best-scoring compounds, is reduced to only 1.63 kcal/mol despite the experimental affinities having a dynamic range of 2.94 kcal/mol. Given such a small predicted affinity range, obtaining a good R^2 value becomes highly statistically improbable, even in the limit of very accurate scoring.[24,37]

With a better understanding of the problematic nature of an R^2 value in this context in-hand, we can now turn our attention more directly toward the molecular weight null model. The R^2 value of the correlation of the molecular weight of the FEP-selected compounds with their experimental affinities is a seemingly quite respectable 0.57, albeit with the sign of the correlation indicating *reductions* in molecular weight should lead to tighter binding. This is clearly a nonsense result which could not be used to inform synthesis decisions. Would the advocates of this null model actually conclude the project team should attempt to reduce the mass of the compounds to attempt to improve the binding affinities of the compounds? Would one actually expect the smallest compounds from the original R-group library would all uniformly be tight binding? Even worse, would the advocates of this null hypothesis argue the results of the FEP-guided R-group virtual screening were not "statistically significant" due to the very high ligand efficiency of the two tightest binding molecules identified in FEP+ scoring? The ex post facto identification of a quantity that fortuitously correlates with the experimental affinities of the compounds selected for synthesis has no logical relevance to question of whether or not the original scoring method performed well. In order to answer that question, a proper control experiment must be conducted where randomly selected compounds, or compounds selected by other industry standard techniques, must also be synthesized and assayed to generate the appropriate comparison data to judge the value of the scoring method under consideration.

The above highlights a much broader problem regarding how null models are commonly suggested to be used in the field of molecular modeling, especially in the context of methods development. Typically, when developing a scoring method, one assembles a large number of data sets to attempt to retrospectively profile the performance characteristics of the scoring method. Recently, if any group suggests a positive retrospective result with any scoring method for any assembled data set, it has become fashionable to attempt to identify any trivial descriptor or set of descriptors which might also correlate with the affinities of the compounds in the data set, under some manner of misguided belief that if any trivial correlation can be found, then the scoring method must be succeeding only due to the presence of the trivial correlation.[40] This is often asserted without even attempting to check if the scoring method in question is more correlated with the trivial descriptor than it was with the original compound affinities, which is the only situation in which the existence of the trivial correlation might be expected to be indicative of a problem.

In analogy with type 1 and type 2 statistical errors, we would propose the existence of type 1 and type 2 statistical deceptions:

- *Type 1 deception*: The misleading presentation or incorrect analysis of data to motivate acceptance of an illegitimate result (e.g., *P*-value hacking)[41]
- *Type 2 deception*: The contriving of intellectually spurious hurdles, possibly adorned with a false pretense of statistical rigor, to prevent the acceptance of a legitimate result.

It is rather ironic one of the most notable type 2 deceptions in history was perpetrated by the father of modern frequentist statistics, R. A. Fisher, who was an avid smoker, and fought for years to prevent acceptance of the hypothesis smoking tobacco caused lung cancer.[42] His arguments, while rhetorically brilliant, were personally motivated and deeply flawed. In a similar spirit, the molecular modeling community as a whole would likely benefit from a more critical review of poorly formulated null models. Our own advocacy, in line with recent guidelines released by the American Statistical Association, is that "no single [statistical test] should substitute for scientific reasoning."[41]

4. CONCLUSIONS AND FUTURE PERSPECTIVE

The development of more efficient and effective strategies to successfully optimize the potency, selectivity, and ADMET profile of a lead molecule is one of the key challenges facing the pharmaceutical industry. Excitingly, an emerging paradigm involving the virtual screening of an approximation to the full enumerated set of all synthetically tractable near neighbor compounds to the lead molecule through an appropriately constructed screening funnel appears poised to make dramatic contributions to improve the productivity of lead optimization phase drug discovery projects. In a typical lead optimization project, the synthesis of more than 1000 molecules in a full year of project chemistry is quite rare. Yet, through application of this virtual screening approach, thousands to perhaps hundreds of thousands of additional idea molecules can now be explicitly considered by project teams during these lead optimization efforts. We anticipate broader adoption of this approach should improve the efficiency of lead optimization, as well improve the likelihood lead optimization efforts against highly challenging targets should succeed.

A key enabling technology making this approach to lead optimization feasible has been the development of accurate and reliable free energy calculation methods for the determination of ligand binding potency and selectivity. A wide variety of recent advances related to improvements in molecular mechanics force fields, enhanced sampling methods, GPU computing, cloud computing, and automated calculation setup have made it feasible to apply these methods on a very large scale without compromising prediction accuracy, as is needed to address the very large synthetically feasible idea space available to the project team.

Excitingly, application of these techniques to real-world industrial lead optimization problems is just now beginning to be reported. Where possible, we strongly advocate conducting the proper control experiments when utilizing these techniques so that the entire field of molecular modeling may over time develop a better appreciation of the strengths and weaknesses of these approaches. However, if the proper control experiments are out of reach due to cost or labor considerations, we would caution users of these techniques away from the application of intellectually ill-posed null models. Said more forcefully, no suggested statistical test should ever be allowed to supercede sound scientific reasoning; and if the information content and relevance of a particular null model is not obvious, then the burden of proof for establishing its relevance for the problem at hand should rest with the individual suggesting the use of the null model, rather than the individual reporting the results for a particular type of scoring method.

REFERENCES

1. Lyne, P. D. *Drug Discov. Today* **2002**, 7(20), 1047.
2. Lionta, E.; Spyrou, G.; Vassilatis, D. K.; Cournia, Z. *Curr. Top. Med. Chem.* **2014**, *14*(16), 1923.
3. Cheng, T.; Li, Q.; Zhou, Z.; Wang, Y.; Bryant, S. H. *AAPS J.* **2012**, *14*(1), 133.
4. Kitchen, D. B.; Decornez, H.; Furr, J. R.; Bajorath, J. *Nat. Rev. Drug Discov.* **2004**, *3*(11), 935.
5. Hughes, J. P.; Rees, S.; Kalindjian, S. B.; Philpott, K. L. *Br. J. Pharmacol.* **2011**, *162*(6), 1239.

6. Agresti, J. J.; Antipov, E.; Abate, A. R.; Ahn, K.; Rowat, A. C.; Baret, J.-C.; Marquez, M.; Klibanov, A. M.; Griffiths, A. D.; Weitz, D. A. *Proc. Natl. Acad. Sci. U.S.A.* **2010**, *107*(9), 4004.
7. Paul, S. M.; Mytelka, D. S.; Dunwiddie, C. T.; Persinger, C. C.; Munos, B. H.; Lindborg, S. R.; Schacht, A. L. *Nat. Rev. Drug Discov.* **2010**, *9*, 203–214.
8. Y. Taitz, D. Weininger and J.J. Delany, Daylight Theory: SMARTS—A Language for Describing Molecular Patterns. http://www.daylight.com/dayhtml/doc/theory/theory.smarts.html. Accessed 30 July 2017.
9. Hartenfeller, M.; Eberle, M.; Meier, P.; Nieto-Oberhuber, C.; Altmann, K.-H.; Schneider, G.; Jacoby, E.; Renner, S. *J. Chem. Inf. Model.* **2011**, *51*(12), 3093.
10. eMolecules http://www.emolecules.com (accessed Jul 30, 2017).
11. *Sigma-Aldrich: Analytical, Biology, Chemistry & Materials Science Products and Services.* http://www.sigmaaldrich.com/. Accessed 30 July 2017.
12. Baber, J. C.; Feher, M. *Mini Rev. Med. Chem.* **2004**, *4*(6), 681.
13. Huang, Q.; Li, L.-L.; Yang, S.-Y. *J. Chem. Inf. Model.* **2011**, *51*(10), 2768.
14. Law, J.; Zsoldos, Z.; Simon, A.; Reid, D.; Liu, Y.; Khew, S. Y.; Johnson, A. P.; Major, S.; Wade, R. A.; Ando, H. Y. *J. Chem. Inf. Model.* **2009**, *49*(3), 593.
15. Corey, E.; Long, A.; Rubenstein, S. *Science* **1985**, *228*(4698), 408.
16. Murphy, R. B.; Repasky, M. P.; Greenwood, J. R.; Tubert-Brohman, I.; Jerome, S.; Annabhimoju, R.; Boyles, N. A.; Schmitz, C. D.; Abel, R.; Farid, R.; Friesner, R. A. *J. Med. Chem.* **2016**, *59*(9), 4364.
17. Friesner, R. A.; Murphy, R. B.; Repasky, M. P.; Frye, L. L.; Greenwood, J. R.; Halgren, T. A.; Sanschagrin, P. C.; Mainz, D. T. *J. Med. Chem.* **2006**, *49*(21), 6177.
18. Friesner, R. A.; Banks, J. L.; Murphy, R. B.; Halgren, T. A.; Klicic, J. J.; Mainz, D. T.; Repasky, M. P.; Knoll, E. H.; Shelley, M.; Perry, J. K.; Shaw, D. E.; Francis, P.; Shenkin, P. S. *J. Med. Chem.* **2004**, *47*(7), 1739.
19. Halgren, T. A.; Murphy, R. B.; Friesner, R. A.; Beard, H. S.; Frye, L. L.; Pollard, W. T.; Banks, J. L. *J. Med. Chem.* **2004**, *47*(7), 1750.
20. Verdonk, M. L.; Cole, J. C.; Hartshorn, M. J.; Murray, C. W.; Taylor, R. D. *Proteins* **2003**, *52*(4), 609.
21. McGann, M. *J. Chem. Inf. Model.* **2011**, *51*(3), 578.
22. Rarey, M.; Kramer, B.; Lengauer, T.; Klebe, G. *J. Mol. Biol.* **1996**, *261*(3), 470.
23. Jain, A. N. *J. Med. Chem.* **2003**, *46*(4), 499.
24. Brown, S. P.; Muchmore, S. W.; Hajduk, P. J. *Drug Discov. Today* **2009**, *14*(7–8), 420.
25. Shirts, M. R.; Mobley, D. L.; Brown, S. P. In: *Drug Design: Structure- and Ligand-Based Approaches*; Merz, K. M., Ringe, D., Reynold, C. H., Eds.; Cambridge University Press: Cambridge, 2010, pp 61–86.
26. Wang, L.; Wu, Y.; Deng, Y.; Kim, B.; Pierce, L.; Krilov, G.; Lupyan, D.; Robinson, S.; Dahlgren, M. K.; Greenwood, J.; Romero, D. L.; Masse, C.; Knight, J. L.; Steinbrecher, T.; Beuming, T.; Damm, W.; Harder, E.; Sherman, W.; Brewer, M.; Wester, R.; Murcko, M.; Frye, L.; Farid, R.; Lin, T.; Mobley, D. L.; Jorgensen, W. L.; Berne, B. J.; Friesner, R. A.; Abel, R. *J. Am. Chem. Soc.* **2015**, *137*(7), 2695.
27. Harder, E.; Damm, W.; Maple, J.; Wu, C.; Reboul, M.; Xiang, J. Y.; Wang, L.; Lupyan, D.; Dahlgren, M. K.; Knight, J. L.; Kaus, J. W.; Cerutti, D. S.; Krilov, G.; Jorgensen, W. L.; Abel, R.; Friesner, R. A. *J. Chem. Theory Comput.* **2016**, *12*(1), 281.
28. Aldeghi, M.; Heifetz, A.; Bodkin, M. J.; Knapp, S.; Biggin, P. C. *Chem. Sci.* **2016**, 7(1), 207.
29. Bhati, A. P.; Wan, S.; Wright, D. W.; Coveney, P. V. *J. Chem. Theory Comput.* **2017**, *13*(1), 210.
30. Abel, R.; Wang, L.; Harder, E. D.; Berne, B. J.; Friesner, R. A. *Acc. Chem. Res.* **2017**, *50*(7), 1625.

31. Rombouts, F. J. R.; Tresadern, G.; Buijnsters, P.; Langlois, X.; Tovar, F.; Steinbrecher, T. B.; Vanhoof, G.; Somers, M.; Andrés, J.-I.; Trabanco, A. A. *ACS Med. Chem. Lett.* **2015**, *6*(3), 282.
32. Lovering, F.; Aevazelis, C.; Chang, J.; Dehnhardt, C.; Fitz, L.; Han, S.; Janz, K.; Lee, J.; Kaila, N.; McDonald, J.; Moore, W.; Moretto, A.; Papaioannou, N.; Richard, D.; Ryan, M. S.; Wan, Z.-K.; Thorarensen, A. *ChemMedChem* **2016**, *11*(2), 217.
33. Jorgensen, W. L. *Bioorg. Med. Chem.* **2016**, *24*(20), 4768.
34. Lenselink, E. B.; Louvel, J.; Forti, A. F.; van Veldhoven, J. P. D.; de Vries, H.; Mulder-Krieger, T.; McRobb, F. M.; Negri, A.; Goose, J.; Abel, R.; van Vlijmen, H. W. T.; Wang, L.; Harder, E.; Sherman, W.; IJzerman, A. P.; Beuming, T. *ACS Omega* **2016**, *1*(2), 293.
35. Abel, R.; Mondal, S.; Masse, C.; Greenwood, J.; Harriman, G.; Ashwell, M. A.; Bhat, S.; Wester, R.; Frye, L.; Kapeller, R.; Friesner, R. A. *Curr. Opin. Struct. Biol.* **2017**, *43*, 38.
36. Kuhn, B.; Tichý, M.; Wang, L.; Robinson, S.; Martin, R. E.; Kuglstatter, A.; Benz, J.; Giroud, M.; Schirmeister, T.; Abel, R.; Diederich, F.; Hert, J. *J. Med. Chem.* **2017**, *60*(6), 2485.
37. Abel, R.; Wang, L.; Mobley, D. L.; Friesner, R. A. *Curr. Top. Med. Chem.* **2017**, *17*, 2577–2585.
38. Nicholls, A. 2017, personal communication.
39. Nicholls, A. *A Different Conference.* https://www.eyesopen.com/ants-rants/different-conference. Accessed 30 July 2017.
40. Website https://openeye.app.box.com/s/een1zjxli5yebo002jgl (accessed Jul 30, 2017).
41. Memory Madondo, S.; Madondo, S. M. *Sci. J. Appl. Math. Stat.* **2017**, *5*(1), 41.
42. Stolley, P. D. *Am. J. Epidemiol.* **1991**, *133*(5), 416.

CHAPTER EIGHT

Phenotypic Screening

Alleyn T. Plowright*[,1], Lauren Drowley[†]
*Integrated Drug Discovery, Sanofi-Aventis Deutschland GmbH, Industriepark Höchst, Frankfurt am Main, Germany
[†]Global Exploratory Development, UCB Pharma, Braine-l'Alleud, Belgium
[1]Corresponding author: e-mail address: alleyn.plowright@sanofi.com

Contents

1. INTRODUCTION

Phenotypic screening—defined herein as a target-agnostic approach for the identification of molecules and targets resulting in a desired phenotype putatively associated with a therapeutic effect—is in many ways a return to the drug discovery methods used in the past when molecular target information was limited or unknown. Phenotypic screening is now reestablished in academia and industry as an approach to discover novel drug targets, pathways, and high-quality hit and lead molecules. The increased access to patient-derived cells and tissues as well as developments in induced

Annual Reports in Medicinal Chemistry, Volume 50
ISSN 0065-7743
https://doi.org/10.1016/bs.armc.2017.07.001

pluripotent stem (iPS) cell technology coupled to more complex cellular systems such as three-dimensional cultures or organs on a chip is providing opportunities for more disease-relevant screens. Incorporating advances in gene editing and screening technologies such as single-cell imaging as well as developments in target identification and mode of action elucidation is further enhancing the opportunities to discover human disease relevant biological targets and hit molecules.

The most significant reason for failure of candidate drugs in the clinic is reported to be lack of efficacy.[1] This is partly due to adverse events elicited by the compounds preventing them being dosed high enough to achieve sufficient target engagement. However, in other cases, sufficient target engagement is achieved but elicits insufficient clinical efficacy, often due to suboptimal target selection and screening cascade design and lack of translation from animal models to human patients.[2,3] Utilizing phenotypic screening with cellular models with strong translational links to human pathophysiology provides an opportunity to identify human disease-relevant biological targets and molecules with the potential to enhance efficacy and differentiate over standard of care to treat patients.

Often cited review articles have highlighted the value and impact of phenotypic screening.[4,5] Swinney and Anthony published an analysis in 2011 highlighting that between 1999 and 2008, 75 new clinically approved first-in-class molecular entities with new molecular modes of action were approved by the US FDA. Of the 75, 50 were small molecules and 25 were biologics. Delving into the discovery of the 50 small molecules, 28 (56%) were discovered through phenotypic screening. This compared to 17 (34%) which were discovered through target-based approaches. A second analysis by Eder et al. in 2014 analyzed the origins of all 113 first-in-class drugs approved by the FDA between 1999 and 2013. In this case the updated analysis showed that 78 of the 113 were discovered through a target-based approach where 45 were small-molecule drugs and 33 were biologics. Of the 33 molecules which were identified in the absence of a target hypothesis, the authors classified 25 into a set discovered through a chemocentric approach leaving 8 coming from the authors' definition of phenotypic screening. While this is a lower percentage than the 2011 disclosure and the numbers vary based on the specific definitions of phenotypic screening (Fig. 1), both analyses highlight that phenotypic screening and optimizing compounds in a target-agnostic manner are complementary to target-based discovery and an impactful approach to drug discovery with

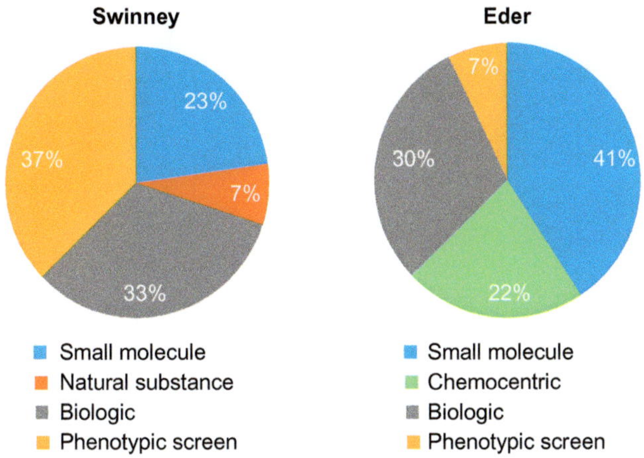

Fig. 1 Comparison between origin of drugs between the two analyses of Swinney et al. and Eder et al. In the Eder et al. analysis the category chemocentric is defined as an approach based around a specific compound or a compound class.

the ability to deliver new first-in-class medicines to patients. A few examples of drugs discovered by phenotypic screening are shown in Table 1.

A successful phenotypic screening campaign requires significant investments from different disciplines including disease biology, screening capabilities, bioinformatics, chemical biology, and computational and medicinal chemistry. This also highlights the diverse skills which are required and the necessity for cross-disciplinary collaboration to make phenotypic discovery projects successful.

As academic and industrial research moves into emerging areas of biology in the search for new and highly efficacious biological targets as well as pharmacological tools to probe their function, phenotypic screening provides an opportunity to discover the unknown. In addition, this can provide the potential to treat currently untreatable diseases as well as addressing multiple aspects of complex diseases simultaneously through polypharmacology.[6] This chapter will describe some key aspects of phenotypic screening including screening technologies and libraries of agents to screen with recent examples. Initial analyses of the screening outcome and early follow-up will be described herein including target inference methods, although more detailed descriptions of target deconvolution and molecular mechanism of action (MMoA) strategies will be described in chapter "Chemical biology in drug discovery" by Castaldi et al.

Table 1 Examples of Drugs Discovered or Optimized Based on the Phenotype With the Chemical Structures and Indications

Name	Chemical Structure	Indication	Discovery/Optimization
Daptomycin	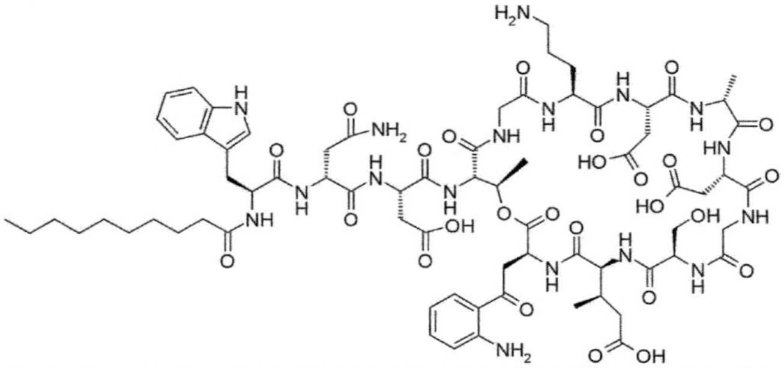	Infection	Screen of bacterial fermentation extracts for antibiotic activity
Ezetimibe	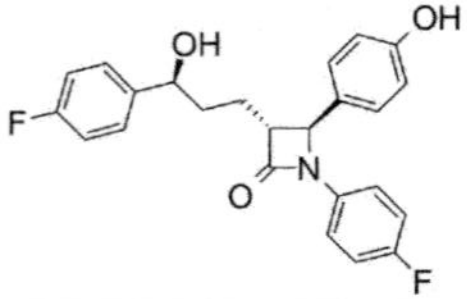	Hyperlipidemia	In vivo screen (rat/rhesus monkey) for agents that specifically blocked intestinal cholesterol absorption
Linezolid	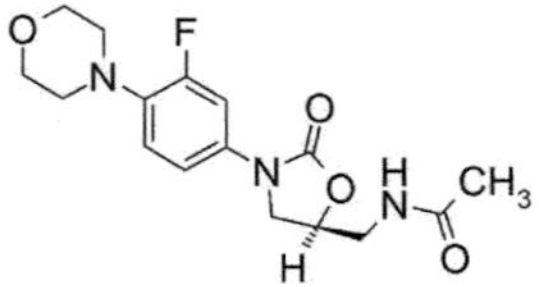	Infection	SAR on oxazolidinone chemistry and biology in bacterial strains
Lacosamide	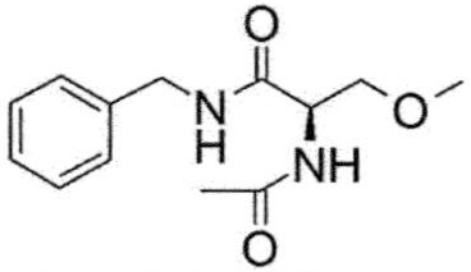	Epilepsy	SAR on 2-substituted *N*-benzyl-2-acetamidoacetamides in rat epilepsy model

Fingolimod	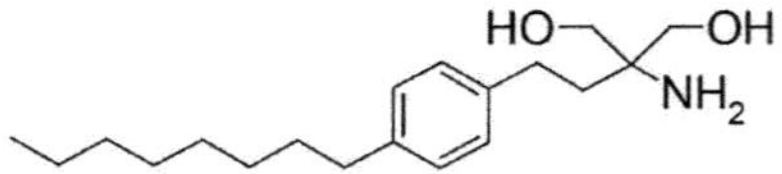	Immunology	SAR from natural substance on in vitro/in vivo immune suppression
Eribulin	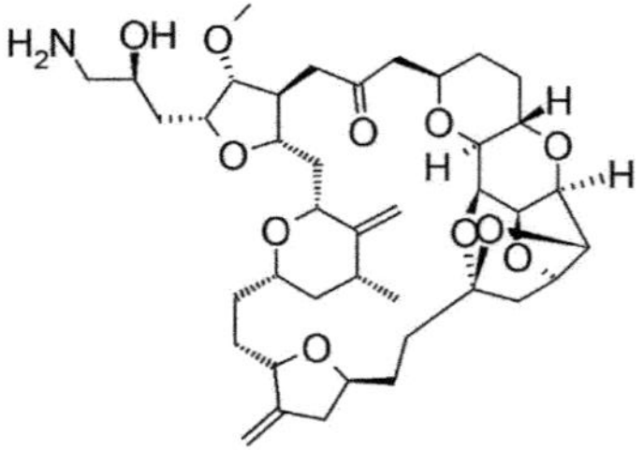	Oncology	SAR on natural substance in cancer cell line screen
Bedaquiline	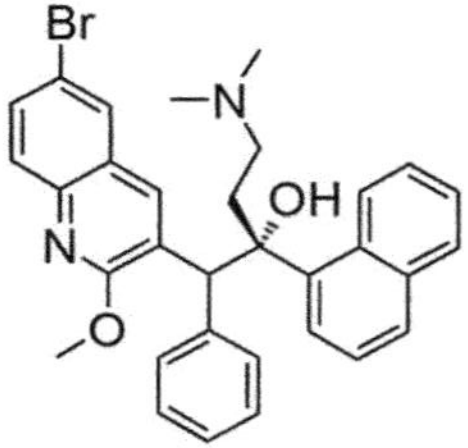	Infection	Screen of more than 70,000 compounds with activity against the saprophytic *Mycobacterium smegmatis*
Trametinib	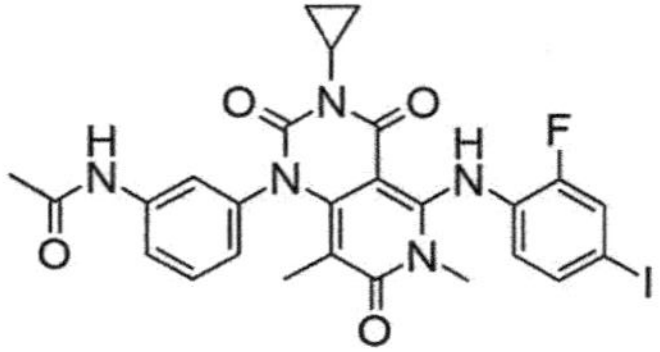	Oncology	Proliferation of panel of cell lines
Alemtuzumab	Antibody	Oncology, multiple sclerosis	Raised antibodies specific for human lymphocyte proteins, tested for T-cell depletion

2. SCREENING TECHNOLOGIES

Screening technology is developing rapidly and has enabled the use of more physiologically relevant cell systems. This includes access to a range of iPS-derived cell types, including cardiomyocytes and cardiac progenitor cells for cardiac diseases,[7,8] iPS-derived beta cells for diabetes,[9] and iPS-derived neuron and glial cells for neurological diseases such as Parkinson's disease and amyotrophic lateral sclerosis.[10,11] In addition, access to human patient-derived cells and tissues is becoming more widespread,[12] for example, small airway epithelial cells and T cells, providing options to screen directly on primary patient cells.[13,14] One such example is the recently reported use of patient epicardium-derived cells for cardiac disease.[15] These technologies are being extended to three-dimensional or coculture systems where different cell types are mixed to provide a more physiologically relevant assay system (Table 2). A recent example disclosed that a multilayered culture

Table 2 Comparison of Types of Screening Models, the Throughput, Benefits, and Drawbacks

Model	Throughput	Benefit	Drawback
Cell free	High	Speed, throughput, cost	Relevance
Cell line	High	Speed, throughput, cost	Relevance
Primary cells	Low	Increased relevance	Lower throughput
iPS/ES-derived cells	Medium	Ability to screen in difficult cell types	Question about translation
Coculture	Low to medium	Increased relevance	Low throughput
3D	Low to medium	Increased relevance	Low throughput
Biomechanical forces	Low	Increased relevance	Low throughput
Fly	Low	Living organism	Question about translation
Zebrafish	Low	Living organism	Question about translation
Mouse	Very low	Living organism	Cost, throughput

containing primary human fibroblasts, mesothelial cells, and extracellular matrix could recapitulate the human ovarian cancer metastatic microenvironment and was adapted to 384- and 1536-multiwell HTS (high-throughput screen) assays.[16] Technology will continue to develop and provide more relevant options for screening to discover novel modulators of specific phenotypes and targets to modulate disease.

3. SCREENING ENDPOINTS

As translation, or rather the lack of, has come to the forefront after the high failure rate of new drugs from industry in the past decades, it has become clear that investing time and effort at the beginning of the drug discovery process is key to improve the probability of success. The attraction for modern phenotypic screens is increased biological relevance and the ability to be hypothesis-free in the screening process, which is feasible due to the advances in assay technologies such as high-content imaging that make it amenable to drug discovery, combining automated imaging and analysis with techniques to better characterize cellular phenotypes in response to treatment.[17,18] These advancements in technology will improve drug discovery as the data provided from a phenotypic screen will allow identification of molecules that would be missed in a traditional target-based screen as well as removing compounds with unwanted biological effects earlier in the process. Previously, compounds with toxicity may not have been identified until they had progressed to animal models, but now some potential safety issues can be detected earlier which will save both resources and time. Phenotypic screening can also identify compounds that have beneficial polypharmacology, which would have been missed in a traditional target-based assay.

High-content screening (HCS) allows for multiplexing of multiple endpoints to gain resolution on how compounds affect several biological aspects at the same time. Cellular phenotypes can be characterized by image readouts including shape, intensity, and texture, in addition, to specific markers. The corresponding feature vectors can be combined into a phenotypic fingerprint that can serve as a compound descriptor.[19] One illustration by Hakkim et al. measured the effects of small molecules on four defined nuclear morphologies relevant to the formation of neutrophil extracellular traps in a screen on primary human neutrophils.[20] A further excellent example demonstrating the power of this type of analysis is demonstrated by a group at Novartis who used HCS fingerprints to link cellular phenotypes to cellular processes and targets.[19] They measured phenotypic effects in six different cellular compartments, including nucleus, cytoskeleton, the Golgi apparatus, the cytoplasm, and the endoplasmic reticulum (ER). To set up the fingerprints, they profiled a set of almost 3000 well-annotated compounds and clustered them based on their phenotypes across all 6 cellular compartments. After clustering, they enriched for gene sets within the clusters and then looked at the compounds to determine if cellular phenotypes could be linked to the modulation of specific targets, target classes, or biological processes. One phenotype that clustered together had cells with an elongated shape and nuclei coupled with a reduction in cell number. All 15 compounds in that cluster were annotated as HDAC inhibitors, and a closely related phenotypic cluster also showed strong enrichment for HDAC inhibitors demonstrating the ability to identify related mechanisms based on phenotype. They continued on to analyze if novel compound–target associations could be predicted based on the phenotypic fingerprint clustering and chose compounds that clustered with PI3K, mTOR, and PDPK1 but had not previously been associated with modulation of those genes. Doing this, they were able to identify Silmitasertib, an ATP-competitive protein kinase inhibitor of casein kinase II, which they tested and found to have a strong inhibitory effect on both mTOR and PI3Kα. This demonstrated that they were successfully able to link cellular phenotype to pathways. Measuring multiple endpoints can provide beneficial information and a more complete picture of the cellular effects. However, while it is easy to measure a vast number of endpoints, it is critical to ensure that the information is relevant. Such things are highly dependent on the cells, model system, and assay endpoints used in the screen.

The assays in which compounds are initially screened are critical to long-term success. If the cells, culture conditions, or assay endpoints do not reflect the patient environment, resultant data become significantly less useful and can lead to large amounts of time wasted on compounds that will not translate.[21] Indeed, with high-content biology, it is easy to generate data, but ensuring that it is useful and translatable information is more difficult. One of the first decisions when setting up a screen is the cell source. Traditionally cell-free assays or cell lines have been used, but although easier and cheaper, these models have often proven to be too simplistic. Both industry and academia have moved toward primary human and iPS-derived cells, both of which are a step closer in terms of biological relevance, though they each come with their own set of challenges.[22]

Primary cells are ideal when feasible as they can be isolated from the patient and tissue of interest and are therefore closer to the desired biology.

It is important to keep in mind that using primary human cells will require additional work in advance to examine several different donors to ensure that the assay and endpoints are not specific to a single donor and that the cells do not alter their phenotype after removal from their tissue microenvironment or during passage. The paper by Paunovic et al. demonstrates how this can be done, validating the effects on proliferation and differentiation of identified compounds across four different epicardium-derived cell (EPDC) donors after the primary screen was run on a single donor.[15] Overall similar effects were seen across donors, but with one compound up to a twofold difference in proliferation was seen between the tested donors. Despite the benefits of using primary cells, they are not always an option due to issues with isolating or expanding sufficient quantities of cells for screening, and thus other sources then need to be examined. iPS-derived cells have become increasingly popular as screening and model systems because they are easier to derive and can provide a more reproducible assay as large quantities of cells can be made and frozen at the same time, obviating the need for frequent batch testing. However, these cells do not always accurately mimic the adult human situation as demonstrated by iPS-derived cardiomyocytes, which have a fetal phenotype. This fact must be kept in mind when deciding if they are appropriate to use.[23] The work performed to develop a screen for diabetic cardiomyopathy at Roche highlights the work needed to validate the cells and assay for use in drug discovery.[7] In this example, iPS-derived cardiomyocytes were the cell source, but due to their fetal phenotype, they did not accurately reflect the more mature cardiomyocytes found in adults. To address this, the authors found a method to promote the adult patterns of metabolic activity in the cardiomyocytes, which increased the maturity of the cells and gave a more appropriate baseline for the screen. They went further and reintroduced glucose into the model after the maturation step to mimic the conditions found in diabetes, where glucose excess is seen along with addition of known hormonal mediators of diabetes. The work undertaken to ensure that the assay was as physiologically relevant as possible was worthwhile as the cardiomyocytes after treatment displayed hallmarks of diabetic cardiomyocytes, including lipid accumulation and oxidative stress (Fig. 2).

After deciding on the optimal cell population(s) to use, another aspect to consider is the level of complexity required in the assay to ensure interrogation of relevant biology. For some assays, typical cell culture may be enough, whereas others require additional parameters to truly mimic the biology of interest. An example of this is the microenvironment, including 3D aspects, which can be key for many cell types and is clearly shown in lung

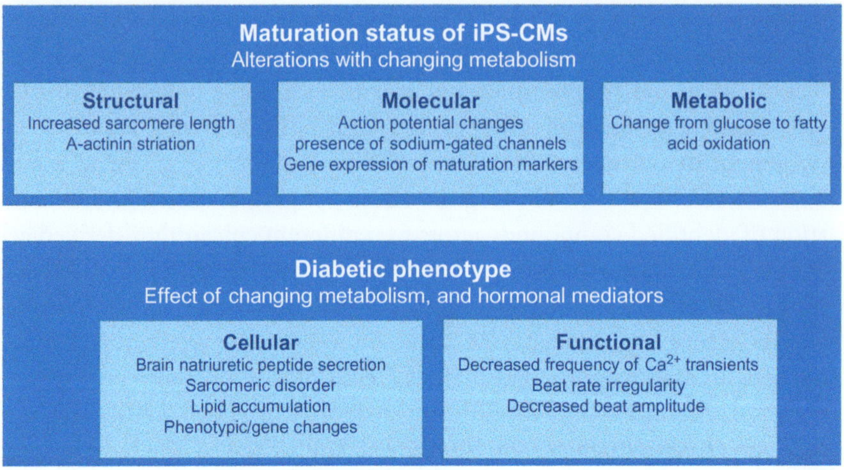

Fig. 2 The Roche group[7] determined suitability of iPS-derived cardiomyocytes prior to use in screening to ensure that the phenotypic readout was biologically relevant. Focus was on two key questions: the maturation status of the cardiomyocytes and the induction of a diabetic phenotype, each with distinct characteristics that needed to be met prior to screening.

progenitors.[24] After isolation from patients these cells lose their phenotype when cultured in 2D, but can maintain their plasticity, response to stimuli, and paracrine factor secretion profile to a much greater degree when cultured in 3D. These characteristics included key cytokines such as IGF-1, HGF, and GDNF, which are known to play roles in stem cell biology. In addition, the 3D cells had very low expression of TGFβ1 as compared to the cells cultured in 2D, which secreted higher levels and were more mesenchymal in phenotype. This clearly highlights the importance of a physiologically relevant environment to maintain the appropriate cell phenotype of interest in order to screen compounds that will have a better chance of translating into humans. Further increasing the complexity, for certain systems the interactions between several cell types are key and coculture models are becoming increasingly popular. In the cardiac area, researchers at AstraZeneca have demonstrated the importance of these multicellular interactions by developing a human microtissue model that contains human iPS-derived cardiomyocytes, primary cardiac endothelial cells, and primary cardiac fibroblasts.[25] The cardiac microtissues are uniform in size and beat spontaneously, and each cell type can maintain key markers during culture. Using this system, the group tested compounds known to affect cardiomyocyte contractility in vivo and were able to detect the compounds at a high sensitivity and specificity at therapeutically relevant concentrations, a significant improvement over using cardiomyocytes alone and much less expensive than in vivo testing.

There have also been some recent advances in microphysiological systems that can now include biomechanical stimuli to better simulate the in vivo environment. An example of how this addition can increase the physiological relevance is demonstrated in bone tumors, where the Mikos group included shear stress in their model system to reflect the host microenvironment including biomechanical stimuli which can be exploited by the tumor.[26] They were able to demonstrate perfusion-dependent effects on key signaling pathways, including IGF, and saw protein expression levels of oncoproteins c-KIT and HER2 at levels that were more consistent with in vivo values rather than what is seen in 2D. In addition, they were able to show effects of dalotuzumab in vitro, which even though the clinical efficacy has been demonstrated in several studies, little to no in vitro activity has been seen previously, highlighting the physiological relevance of recapitulating biochemical stimuli in the in vitro model system. Throughput is increasing for these more complicated systems such as the one utilized by Kim et al. which connects chambers via microchannels and allows for different tissues to interact.[27] In the proof-of-concept study, they connected rat liver and colorectal tumor tissues and cultured them in the presence of cyclophosphamide, which is known to impact tumor growth but requires activation in the liver. Effects were seen on tumor growth only when cocultured in connection with liver tissue, which again highlights the importance of having translatable systems in which to test compounds.

The advances in cellular systems, microtissues, and organs on a chip are progressing rapidly and new technologies are increasing throughput.[28] However, in most cases, throughput is either still too limited or the cost is prohibitive to run a large primary screen. Hence, further technological advances will be required to miniaturize the experiments and lower the cost, while still maintaining the relevance to human patients. As the complexity increases, it may also become more challenging to derive structure–activity relationships (SARs) of the tested molecules in these assays due to the number of effects a compound may be causing. However, this needs to be tested and it will be exciting to see examples appearing which showcase these technologies.

One of the key issues with translating compounds is understanding how they will work in more complex systems, as many compounds fail when taken in vivo. Zebrafish and flies have been used to screen compounds as these models include a greater degree of biological complexity and have some physiological similarities to human. In addition, they have a throughput that is greater than in mammalian in vivo models, as well as being significantly less expensive.[29,30] An example would be the screen that identified dorsomorphin, which was set up to identify compounds that affect body organization.[31] This screen would not have been possible in cell culture alone and would have been lower throughput in other commonly used animal models such as mouse or rat. However, using zebrafish a few thousand compounds were screened. For certain areas, such as neuropsychiatric disease, cell-based screens may not provide a clear picture due to their inherent simplicity. Many current neuropsychiatric drugs were originally discovered via clinical observations on behavior from drugs that were originally intended for other indications. In this area, zebrafish can be a good model system since behavior can be studied. Rihel and colleagues used this system to screen for compounds that altered zebrafish behavior focusing on modulation of restfulness and wakefulness.[32] Using this, they identified novel neuroactive substances as well as known modulators of major neurotransmitters such as clonidine, which is a medication for the treatment of ADHD but was starting to gain use as a sedative and was found to have sedating effect in the zebrafish assay, providing additional validation. Although there are many promising reports demonstrating the use of flies and zebrafish for drug discovery, it remains to be seen how the findings in these types of screen will translate to humans.

4. OPTIONS OF MOLECULES AND MODALITIES TO SCREEN

Different modalities have been used for phenotypic screening ranging from small molecules to antibodies and proteins to RNA reagents including RNAi and microRNAs and most recently to whole genome clustered regularly interspaced short palindromic repeats (CRISPR) screens.[33] Different modalities have a variety of advantages and disadvantages and provide diverse screening outcomes as well as alternative options for following up in a project setting. For example, screening of small molecules provides the opportunity to rapidly discover compounds which provide the required phenotype in a relevant cellular setting including potential polypharmacology which can enhance the observed efficacy of the biological effect. Small molecules also have the advantage over other modalities such as antibodies or siRNA reagents in that they can often penetrate cells to target intracellular targets in the absence of a delivery agent. However, downstream target deconvolution, as will be discussed more in the chemical biology chapter, is a significant challenge. In comparison, performing a screen using RNAi or CRISPR reagents can more rapidly highlight a biological target of interest, the modulation of which can provide the required phenotype. In this case a project can move forward using the tested modality itself with the aim to develop RNA- or CRISPR-based tool molecules and potential therapeutics. Alternatively, it is possible to switch to, or follow up, with other modalities such as small molecules which are known to modulate the given target. If no alternative modalities (e.g., small molecules or proteins) exist, then a target-based screen can be performed against the specific target to identify novel molecules.

A broad diversity of libraries of molecules now exist which can be exploited in phenotypic screens. The most common are small-molecule libraries that exist across many academic and industrial institutions as well as commercial suppliers and range in size from hundreds up to a few million molecules. One popular category of small molecules for phenotypic screening are natural products due to their inherent and evolved activity in biological systems[34] or compounds elaborated from the core scaffolds of natural products themselves.[35] Natural products are attractive in that they have evolved with proteins and have a biological purpose and hence activity. They are therefore an important class of molecules to profile as they are considered biologically relevant and can provide fast access to tool molecules. However, they can also be highly challenging to follow up due to their lack of biological target information and lengthy chemical syntheses. Examples of natural product sources include venoms and naturally occurring peptides where drug leads have previously been discovered.[36] These include angiotensin-converting enzyme inhibitors for the treatment of hypertension and exenatide, a synthetic analogue of exendin-4, which is a venom peptide of the Gila Monster lizard. Libraries of antibodies can also be screened using phage display technologies such as svFc and designed ankyrin repeat protein (DARPin) formats in complex cellular models to discover antibody-tractable targets.[37] This approach is complementary to other approaches to find novel targets amenable to modulation with antibodies such as utilizing expression data from patient tissue or disease models. The size of the library tested depends on the availability of the cells, the cost of the screen, and the specific question being asked. If a large number of cells are available which behave in a robust and reproducible way in the screening setting, then a HTS can be performed.

5. RECENT EXAMPLE OF A HIGH-THROUGHPUT PHENOTYPIC SCREEN

A recent example of a high-throughput phenotypic screen was disclosed by Pfizer where they searched for molecules inhibiting the secretion of PCSK9.[38] In this case a HTS was performed using a library of 2.55 million small molecules against a CHO-K1 cell line overexpressing recombinant ProLabel-tagged PCSK9 to identify compounds that inhibited PCSK9 secretion as assessed by the reduced amount of PCSK9 in the media. One compound, termed *R*-IMPP, was identified which showed PCSK9 antisecretagogue activity (IC_{50} = 4.8 μM) but did not decrease the secretion of control proteins such as alkaline phosphatase or show cytotoxicity as measured by the effect on intracellular ATP levels, cell viability, or impairment of mitochondria (Fig. 3). Additional cellular experiments in Huh7 cells showed that treatment with R-IMPP led to an increase in LDL-R on the cell surface and enhanced LDL uptake, as hypothesized for an inhibitor of PCSK9 secretion. Interestingly the enantiomer of *R*-IMPP, *S*-IMPP, did not decrease the secretion of PCSK9, revealing a stereospecific response and providing an excellent negative control in subsequent experiments. Additional work to elucidate the mechanism of action through ruling out possibilities such as decreasing PCSK9 transcription or enhancing degradation of PCSK9 revealed that the compound blocked the synthesis of PCSK9,

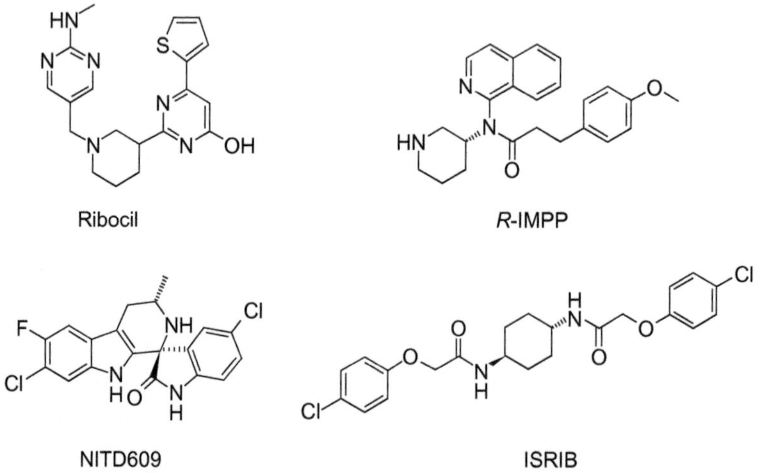

Fig. 3 Structures of molecules identified by the outlined phenotypic approaches using different compound libraries.

more specifically via modulation of PCSK9 protein translation through binding to the ribosome. The identification of a single compound with the required characteristics out of a phenotypic screen of greater than 2.5 million compounds operating through a new mechanism for inhibiting the secretion of PCSK9 is a great example of how phenotypic screening can identify novel mechanisms of action and, in this case, potentially provide a new mechanistic approach for targeting challenging biological targets.

6. REDUCING LIBRARY SIZE BY PRIORITIZING KNOWN PHENOTYPIC RESPONSES

In cases where running a HTS is not possible or the cost of doing so is prohibitive, a range of methods have been used to reduce the number of compounds including random compound selection, maximizing chemical or biological diversity, or selecting compounds based on broad phenotypic changes or a known phenotype where greater specificity is required. A recent example utilizing a small-molecule compound set which had previously been shown to elicit a gross phenotype, in this case antibacterial activity, was reported by Howe et al..[39] This example also highlights the impact of phenotypic screening for the discovery of novel molecules against an emerging target class known as riboswitches, which are noncoding RNA structures which regulate gene expression following the binding of endogenous ligands. However, although different hit finding strategies and screening approaches have been employed to identify mimics of riboswitch ligands, very little success has been shown. In this example, the riboflavin biosynthesis pathway was the focus as riboflavin is essential for microbial growth. Riboflavin is the precursor of flavin mononucleotide (FMN), a key ligand for the FMN riboswitches which regulate the gene expression of enzymes and transporters of riboflavin biosynthesis and uptake. Here 57,000 small-molecule compounds with previously shown antibacterial activity were screened against a specific *Escherichia coli* strain to identify compounds which inhibited the viability of *E. coli* in a manner which was dependent on riboflavin. The screen revealed the hit molecule, termed ribocil, whose antibacterial activity could be fully suppressed with the addition of riboflavin (Fig. 3). Ribocil-treated cells gave the same phenotype as *E. coli* mutants where genes involved in the early steps of riboflavin biosynthesis had been knocked out. In addition, cells treated with ribocil had reduced levels of riboflavin and the downstream metabolite FMN, highlighting that the molecule targets the riboflavin biosynthetic pathway.

In identifying the target of ribocil, 19 ribocil-resistant *E. coli* mutants were isolated and whole genome sequenced to identify the mutations. All mutations were found in the FMN riboswitch within the ribB gene, strongly suggesting that mutations in the FMN riboswitch lead to resistance to ribocil. A suite of analyses subsequently confirmed the direct interaction of ribocil with the FMN riboswitch including a range of cellular experiments and biophysical measurements including binding to the synthesized FMN aptamer. In addition, an X-ray crystal structure of ribocil bound to the FMN riboswitch of *Fusobacterium nucleatum* was solved. Finally, ribocil showed in vivo activity in a murine *E. coli* septicemia model of infection. This excellent example demonstrates that riboswitches that regulate gene expression can be druggable with small synthetic molecules and can expand accessible target space.

7. UTILIZING NATURAL PRODUCTS/NATURAL PRODUCT-LIKE LIBRARIES

An area where phenotypic screening has proven to be rewarding is to discover leads for antimalarial agents. Recent examples have screened diverse chemical libraries to evaluate *Plasmodium* cell proliferation in human erythrocytes.[40] A recent disclosure utilized a chemical library consisting of 12,000 compounds with a mixture of natural products and compounds containing structural features found in natural products and identified 275 hits against *Plasmodium falciparum*. Further screening against multidrug-resistant parasites and for cytotoxicity against mammalian cells left 17 compounds which were then evaluated for pharmacokinetic and physical properties. This left one compound, structurally related to the spiroazepineindole compound class, as the lead compound which underwent medicinal chemistry optimization. Some 200 derivatives of this lead compound were synthesized and evaluated, resulting in the spiroindolone compound NITD609 (Fig. 3). NITD609 remained active against a range of drug-resistant strains and in ex vivo assays using fresh isolates of *P. falciparum* and *Plasmodium vivax* collected from malaria patients where chloroquine resistance has been widely reported. The compound had excellent pharmacokinetic properties allowing a predicted once daily dose in humans and could also clear infection of malaria in an in vivo mouse model. Performing in vitro selection of resistance by applying drug pressure and the subsequent sequencing of the resistant clones suggested that treatment with NITD609 selects for mutations in the gene encoding cation transporting P-type adenosine triphosphatase

(PfATP4), a different mechanism of action compared with current antimalarial drugs. NITD609 is now undergoing clinical trials for the treatment of malaria.

8. LIBRARIES BASED ON BIOLOGICAL MEASUREMENTS OF DIVERSITY

Drug-like space is vast with an estimation of greater than 10^{60} molecules. Hence, any screening library can only cover a small fraction of compounds that are theoretically accessible.[41] This has the consequence that how a screening library is composed is likely to be biased and only cover a certain chemical space, thereby limiting the possibilities achievable within the screen.

It is an easy assumption that choosing a library based on the chemical structure diversity would lead to diverse biological performance, but this is not always the case. Instead, some groups have proposed to construct libraries based on biological measurements of diversity that have been derived from compound profiling in cells. If biological diversity in the chosen compound set is low, a few clusters will be highly overrepresented and can easily dominate the top of the screening hit lists, especially if they are associated with relatively nonspecific biological effects. Random selection of library compounds would maintain the relative representation of each cluster, but would reduce the number of small clusters that are included. Selecting compounds based on the biological profile avoids this by retaining unique biological effects and would not lead to a loss of small clusters. Work by Kauvar and others has suggested that selecting compounds with distinct in vitro biochemical profiles would avoid overrepresentation or clustering in certain parts of bioactivity space.[42] Retrospective analysis from Novartis supported this premise, showing that library sets selected for high-performance diversity instead of high chemical diversity resulted in hits in more assays.[43] This type of analysis is possible when there are previous biological data that can be used to reduce in size the well-tested libraries, but there is still a gap in how to use this type of data to create novel libraries or to expand current ones. The approach demonstrated by Wawer et al. to address this question used profiling technologies, one gene expression based and one for cell morphology, to select a diverse library of compounds.[44] They compared the hits and diversity from these libraries to libraries that had instead been selected either randomly or for chemical diversity, and both of the diverse biological sets led to higher hit rates. In order to expand hits, a newer strategy is moving the concept of SAR from chemical to biological space. Rather than focusing on chemical structure similarity, grouping compounds by mechanism of action based on the biological profiling can aid in the creation of a biologically diverse compound library. This can provide different hits as chemical starting points that lead to the phenotype of interest as well as providing more confidence in mechanism of action as structurally diverse compounds are less likely to share the same screening artifacts.

A similar approach was taken to select a biologically active compound library in the phenotypic screen on EPDCs.[15] The challenge with this screen was the limited throughput due to the use of primary cells, originally isolated from patients undergoing cardiac surgery, which meant that only 7400 compounds were tested in the primary assay. Each compound was tested at three concentrations to avoid dismissing compounds as false negatives due to bell-shaped concentration–response curves, i.e., if just a single high dose had been screened or a lack of activity had been obtained if a single lower dose had been used. The selected library included two compound sets, one that was known to act on certain biological targets (defined as having an annotated mechanism of action); this was an attempt to make subsequent target deconvolution a more straightforward process through the generation of target hypotheses at an early stage. The second compound set demonstrates another way to enrich potential hits, which was to select compounds that have been reported to have effects in similar cell types or on the phenotype of interest. These two approaches combined resulted in a hit rate of 3% from the primary screen, of which over 90% were then confirmed in full concentration–response experiments. In this example the annotated targets of the hit compounds were not the key targets responsible for driving the phenotypic effect (proliferation of EPDCs), demonstrating that annotated targets do not necessarily mean that those targets are responsible for the phenotype and that alternative targets or polypharmacology may be playing a role (Fig. 4).

9. TARGETING SPECIFIC PATHWAYS

An alternative approach to avoid the complexity of running completely unbiased phenotypic screens and allowing a more focused approach is to target specific pathways that are known or hypothesized to be important in disease. A recent example focused on the integrated stress response, or more specifically on the phosphorylation of the α-subunit of initiation factor 2 (eIF2a).[45] Phosphorylation of eIF2a reduces most protein

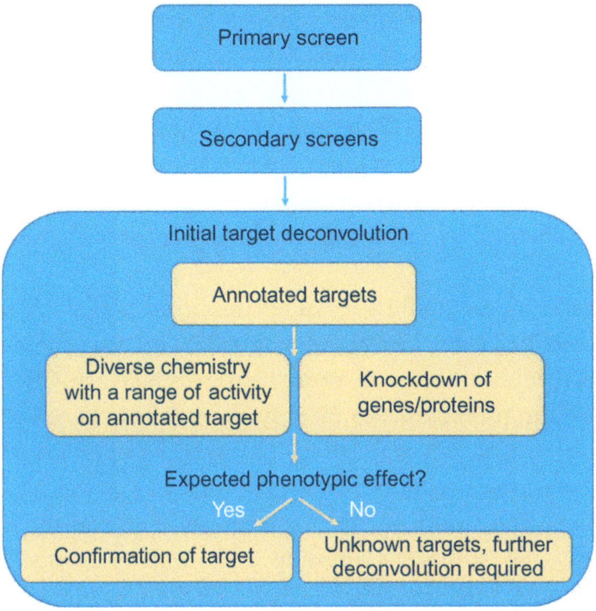

Fig. 4 Screening cascade highlighting how the initial screen can be used to aid in target deconvolution through target inference and initial testing of target hypotheses.

synthesis while upregulating the production of specific regulatory proteins, which allows cells to adjust to different stresses. eIF2a can be activated through phosphorylation by four protein kinases, one of which, PERK, is activated by the accumulation of unfolded proteins in the ER. In this case, a reporter-based screen was engineered where the 5′ UTR of ATF4 was fused to luciferase to monitor the activation of PERK signaling in HEK293T cells. A known stressor of the ER, thapsigargin, was used to activate PERK signaling and induce eIF2a phosphorylation providing a 4.9-fold induction in the reporter system and a window to show effective and robust inhibition. Greater than 100,000 compounds based on a broad chemical space were screened at 10 μM and furnished 460 hit molecules that reduced the luciferase signal greater than 3 standard deviations from the thapsigargin control. The hit molecules were progressed through counter screens to identify molecules which specifically inhibited the unfolded protein response through PERK signaling and secondary assays using a different reporter system, cell type and readout (microscopy-based), and finally the ability to inhibit the induction of the endogenous downstream marker ATF4. Of the 28 compounds which successfully met the criteria of these assays, one containing a bis-glycolamide had particularly high potency in cells ($IC_{50}=40$ nM). The disubstituted cyclohexane-containing compound comprises two diastereoisomers, *cis* and *trans*. Both of these were synthesized and tested in the reporter gene assay and the *trans*'isomer was 100-fold more potent ($IC_{50}=5$ nM). This molecule was termed ISRIB—integrated stress response (ISR) inhibitor—and the first reported inhibitor of the ISR (Fig. 3). Finally, it was shown that reducing eIF2a phosphorylation through ISRIB treatment led to an enhancement in memory consolidation and learning in rodents.

The specific target of ISRIB remained unknown until a subsequent publication revealed the mechanism of action.[46] This was achieved using a genetic screen with a library of short hairpin RNAs (shRNAs) for 2933 genes involved in proteostasis to identify genes whose knockdown modulated the sensitivity of cells to the drug. This shRNA library was designed to target each gene with approximately 25 shRNAs as well as containing shRNAs as negative controls. The reporter cell line was transduced with the shRNAs, and shRNA-expressing cells were subsequently stressed through treatment with thapsigargin in either the presence or the absence of ISRIB. Knockdown of two genes in particular, two subunits of eIF2B, eIF2B4 and eIF2B5, significantly reduced the sensitivity of cells to ISRIB. Hence, eIF2B was hypothesized as the target of ISRIB, and more specifically that ISRIB acts as an activator of eIF2B. In addition, SAR studies of the twofold symmetric molecule ISRIB suggested that both ends of ISRIB may bind to its target in a symmetrical manner and that both halves of the compound are involved in similar binding interactions with its target, suggesting that ISRIB binds to eIF2B in a twofold symmetric interface which stabilizes the dimer. It was subsequently shown that ISRIB enhances thermal stability of eIF2B4 upon binding as measured by CETSA (cell-based thermal shift assay), suggesting target engagement, and also promotes dimerization of eIF2B in cells, leading to enhanced functional activity. This is an elegant example combining small-molecule phenotypic screening with genetic screens, chemistry, and cell biology to elucidate the molecular target of a valuable tool compound with a novel mode of action and emphasizes the importance of a range of mode of action methods to more fully understand the MMoA of compounds discovered from phenotypic screening. Simultaneously, a second disclosure reported the independent identification of the molecular target eIF2B using a different MMoA elucidation approach.[47] With this understanding in place more focused screens can now be performed to find alternative chemical starting points to activate eIF2B for use to study neurological disease.

10. APPLYING THE PRINCIPLES OF FRAGMENT-BASED SCREENING

Despite advances in the synthesis of small-molecule libraries only a small fraction of the human proteome has chemical ligands, so there is a gap in the biology that can be modulated by screening this modality. To address this, groups including Parker et al. have moved to fragment-based ligand discovery combined with quantitative chemical proteomics to identify small molecule–protein interactions in human cells.[48] The benefit of fragment-based screening is that it uses a smaller number of low-molecular weight compounds that, due to their limited size, may represent a larger fraction of accessible chemical space, thus aiding in identification of structurally simple hit compounds that can then be optimized for increased potency.[49] In the example demonstrated by Parker et al., they used a small library of 14 fragment probes functionalized with alkyne and diazirine moieties, with an average molecular weight of 176 Da that represented structural motifs founds in many biologically active natural products and clinically approved drugs. These probes were used to label their binding partners by photoactivation of the diaziridine ring followed by coupling to a tetramethylrhodamine-azide via a click cycloaddition reaction and allow quantitative chemical proteomics to reveal small-molecule protein interactions directly in human cells. By this method over 2000 protein targets were identified including both soluble and membrane proteins, where only a small fraction of which (17%) had previously known ligands. An expanded library of slightly larger chemical probes with an average weight of 267 Da was then used to assay for effects on adipogenesis and hits were identified that were not direct agonists of PPARγ, the positive control. One hit, compound 25, was selected due to its effects on lipid accumulation and induction of specific adipogenic markers, and this compound was also specific for effects in adipose cells. In addition, a structurally similar molecule 26 was identified, which showed significantly less activity on adipogenesis, a very useful comparator in subsequent experiments. Using quantitative proteomics approaches, the authors were able to identify the poorly characterized transmembrane protein progesterone receptor membrane component 2 (PGRMC2) as a key binding partner. Further experiments utilizing shRNAs suggested that 25 acts on PGRMC2 in a gain-of-function manner to promote adipogenesis (Fig. 5). This approach combines the identification of active ligands in human cells with the identification of their molecular target. The functionalized fragments were promiscuous

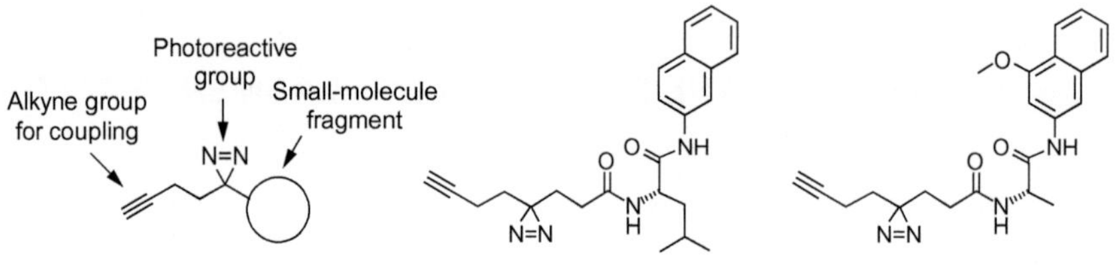

Functionalized fragment probe **25** **26**

Hit compound 25 ⟹
- Differentiation of 3T3-L1 preadipocytes: lipid accumulation and induction of key markers
- Quantitative proteomics revealed PGRMC2 as a key binding partner
- shRNA experiments showed the 25 causes PGRMC2 gain of function

Fig. 5 Schematic illustration of the functionalized fragments utilized by Parker et al. Structure of the hit molecule 25 and the significantly less active control 26 and key steps to understand the mode of action of 25.

in their binding to proteins, which makes lower abundance proteins more difficult to detect. However, this technique is potentially transformative, which could streamline phenotypic screening and target deconvolution efforts.

11. UTILIZATION OF OTHER MODALITIES

There are now an increasing number of alternative modalities beyond small molecules which can be used to interrogate the biology of interest. RNA interference (RNAi), or blocking the translation of RNA into protein via targeting the RNA with shRNA, short interfering (si) or micro (mi) RNA, allows for identification of genes that are relevant to biological function by examining cellular phenotype after gene modulation. This approach has been used by Amgen to identify modulators of autophagy.[50] They performed a phenotypic screen in a cell line that expressed fluorescent SQSTM1, a ubiquitin-binding autophagy receptor involved in cargo recognition, and examined turnover in normal nutritive conditions. They queried more than 12,000 genes with 4–8 individual siRNAs per gene to avoid off-target effects common to RNAi screening. The readouts were reduction in SQSTM1 fluorescence in combination with a change in localization with lysosomes to identify autophagic flux, which allowed for removal of hits that affected fluorescence due to effects on protein synthesis and allowed for enrichment with genes more likely to be involved in autophagy. The subsequent screening cascade removed hits that caused a reduction in cell viability and confirmed the knockdown with gene expression as well as examining autophagic flux in a kinetic assay, and in the end 10 hits remained. Of these hits, three increased autophagic flux without affecting mTOR signaling, two of which were novel discoveries, and one, CSNK1A1, had previously been identified as a modulator of autophagy.

Noncoding RNAs such as miRNAs have emerged as important regulators of healthy and disease states through the regulation of translation of messenger RNAs by the ribosomal machinery. Hence, either inhibiting miRNAs using anti-miRNAs or mimicking miRNAs to enhance their function can affect the regulation of many proteins and hence impact cellular phenotypes significantly. An example by Eulalio et al. where they performed an HCS-based on fluorescence microscopy using a miRNA library consisting of 875 miRNA mimics highlights this approach.[51] The screen identified 204 miRNAs, which enhanced proliferation in the cellular assay. The hits were triaged using a number of cellular and high-content assays to leave two, miR-590-3p and miR-199a-3p, which passed the assay criteria. Synthetic miR-590-3p and miR-199a-3p were subsequently tested in vivo and resulted in an increase in cardiomyocyte proliferation and when expressed with AAV vectors after a myocardial infarction led to beneficial effects on cardiac function.

Gene editing approaches have become increasingly popular as a means to assess the role of genes in biology of interest, including CRISPR/Cas9 technology. This approach has the advantage of being easily reprogrammable by changing the 20 base pairs of homology between the single-guide RNA (sgRNA) and the genomic target. These screens were initially pooled screens where the entire library is delivered to a single vessel of cells at a low multiplicity of infection to increase the probability that each cell will be infected with only one sgRNA. However, arrayed screens are now feasible where individual sgRNAs are in separate wells, which is more amenable to high-content imaging and understanding complex phenotypes (Fig. 6). Sidik et al. employed a genome-wide CRISPR screen to disrupt the genes in *Toxoplasma gondii* parasites during infection of human fibroblasts using a library of sgRNAs containing 10 guides against each of the over 8000 predicted protein-coding genes.[52] The changes in sgRNA representation after exposure of the parasite to a toxin allowed measure of a gene's contribution to fitness. After validation of the findings by comparing the hit genes with those that had been previously reported to be dispensable or essential for growth, they focused their efforts on the 200 genes identified that lacked

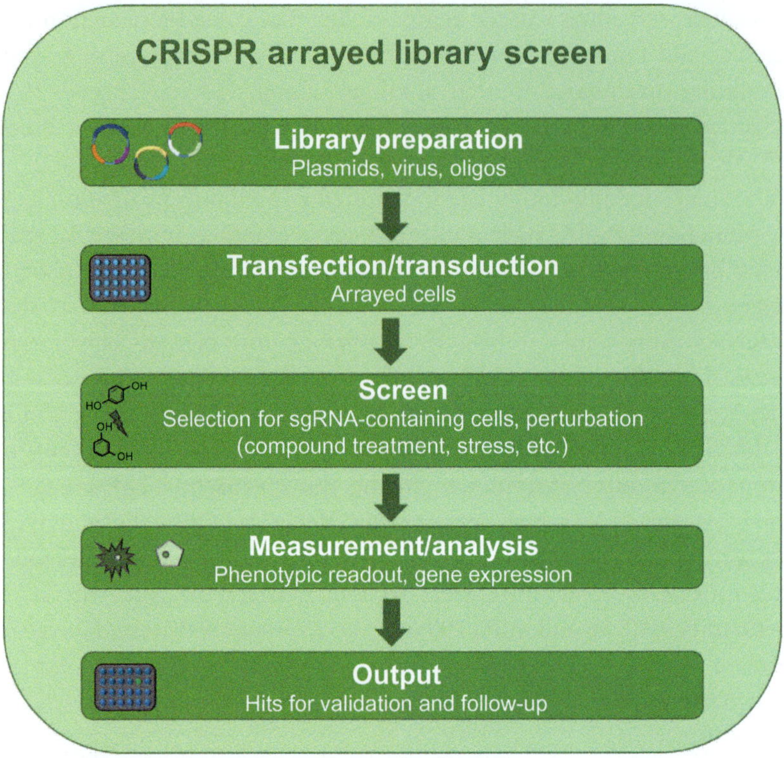

Fig. 6 Screening workflow for an arrayed CRISPR screen.

functional annotation. The functional validation included screening for invasion, plaque formation, and conservation across the coccidian genomes, including two other parasites. The genes were further characterized in terms of cellular localization, where a preponderance was found to localize to the mitochondria. One gene, *Clamp*, was shown to be required for invasion of host cells but did not affect motility or invasion and sheds light into new potential interventions that can control these parasitic infections.

A different strategy to identify potential therapeutic entry points into diseases can be achieved by screening for gene networks that influence disease or progression. This genetic screening approach has the benefit of providing high confidence that the genes play a role in disease, though understanding the underlying mechanism is not always straightforward. An example of this is the work by the Zoghbi group, who ran parallel cell and *Drosophila*-based genetic screens focused on spinocerebellar ataxia type 1 (SCA1).[53] Their approach revealed that downregulation of several components in the

RAS–MAPK–MSK1 pathway decreased levels of ATXN1, a key driver of neurodegeneration. Using this genetic methodology has revealed new potential therapeutic entry points for SCA1.

12. INFERENCE METHODS FOR TARGET IDENTIFICATION

After phenotypic screens have been run, additional challenges remain and one of the most critical steps is to identify the target(s) of the compounds of interest. There are many methods to achieve this, all with their own sets of benefits and limitations. Target deconvolution is covered in more detail in chapter "Chemical biology in drug discovery" by Castaldi et al., but initial target profiling will be discussed here due to its link to the selection of the screening set. A variety of computational methods are now available to analyze HTS and profiling data to generate hypotheses around the MMoA of compounds which are discovered in phenotypic screens. A computational approach will unlikely be sufficient, but the generated target hypotheses can then be tested experimentally using other techniques. These methods can be further utilized to uncover new targets of existing drugs to enable potential repositioning or finding off-targets which are responsible for toxicity or enhancing efficacy of the compound itself. The methods include clustering based on biology, ligand-based methods to search for similarity across chemical structures, and structure-based methods to predict protein–ligand interactions. In cases where the initial screen can only be prosecuted on a limited number of compounds, the initial hits are often chemical singletons or members of small chemical clusters (<5 compounds). This is in contrast to HTS campaigns where clusters of greater than 100 members can be found. The singleton or small cluster must be transformed into a larger chemical cluster, ideally showing a dynamic SAR. This can be achieved by testing chemically similar molecules, or near neighbors,[54] selected by multiple fingerprint methods combined with text-based molecular similarity searching.[55]

One of the first methods to generate hypotheses around the MMoA of compounds is by examining chemical similarity. To improve on chemical similarity database searches and their limitations on throughput, Lo et al. developed CSNAP (Chemical Similarity Network Analysis Pulldown) that utilizes chemical similarity networks for chemotype recognition via consensus chemical patterns.[56] This method has an improved accuracy of target prediction over previous approaches and could integrate biological knowledge databases for target validation. To test the system, they identified the major mitotic targets of hit compounds from a cell-based screen on cancer cells. From the active compounds with unknown cellular targets, a bioactivity database search was performed to identify structurally similar reference compounds with known target annotations. Structurally diverse ligands could then be clustered into chemical similarity subnetworks, and then network-based target scoring was used to guide and quantify the protein–ligand interactions for target prediction. The predicted targets were then validated experimentally by comparing RNAi with compound-induced cellular phenotypes and testing protein–ligand interactions in vitro. Using this technique, the authors were able to identify novel compounds that target microtubules.

Another technique was disclosed by Reker et al. for inferring the potential targets of natural products.[57] This method dissects natural products into natural product-derived fragments and compares the pharmacophoric features of the fragments to reference molecules with known targets. This statistical analysis leads to hypotheses of the biological targets of the fragments and therefore of the overall natural product. The authors applied this technique in a prospective validation to the antitumor macrolide natural product archazolid A (ArcA) and showed that fragments of the molecule contain important information regarding its polypharmacology. ArcA is known as a potent inhibitor of the ubiquitously expressed ATP-driven ion pump vacuolar-type H^+-ATPase (V-ATPase) but has long been suspected to have additional activity. ArcA was dissected into four fragments, and analysis of the fragments with known molecules indicated potential activity in the processing and recognition of arachidonic acid (AA). The predictions were confirmed by subsequent biochemical and biophysical evaluation, including inhibition of 5-lipoxygenase (5-LO) and potent FXR agonism (Fig. 7). These results using natural product-derived fragments for target prediction highlight the potential of this computational approach to infer targets for natural products and could therefore allow for the systematic identification of their targets.

Other computational inference methods can now be used to generate hypotheses on the targets of molecules by creating bioactivity profiles for tested compounds by utilizing the data generated in historical screening assays and subsequently clustering screening hits based on their biological profile.[58] One such example was disclosed by Bornot et al. where they used known bioactivity data of screened compounds, obtained from both in-house AstraZeneca data and external sources such as CHEMBL (https://www.ebi.ac.uk/chembl/), GoStar, and BioPrint databases, and performed an enrichment analysis.[59] In this case a statistical enrichment of

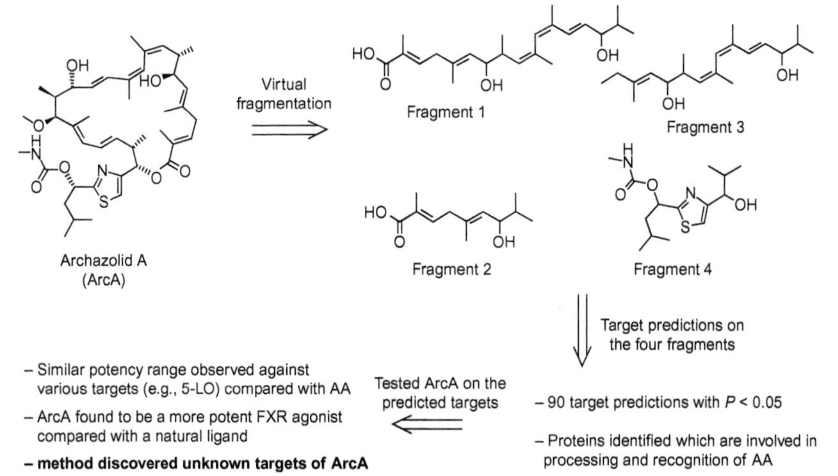

Fig. 7 Workflow to infer the molecular targets of natural products based on virtual fragmentation, computational prediction, and biochemical and biophysical confirmation.

targets can generate early hypotheses of targets which may be driving the phenotypic effect. This approach therefore has the advantage in that it combines the known bioactivity data across the whole screening set which provides additional evidence for or against a certain biological pathway being involved in the phenotypic response. On the other hand, the method relies on the availability, and accuracy, of the data used in the evaluation. In addition, targets for which there are more hits and therefore perhaps more well known are also more likely to be enriched whereas a novel target which does not have much bioactivity data will not be identified as being enriched. This method could infer putative targets, pathways, and biological processes that are consistent with the observed phenotypic response, leading to the selection of additional chemically diverse structures which modulate the enriched targets for further screening. The relevance of the enriched targets can subsequently also be investigated using gene knockdown methods such as siRNA or CRISPR–Cas9 to replicate the phenotype.

These inference methods using computational approaches often rely on the use of public data. The range of public data available to aid these analyses is advantageous as more data are provided than any company or group can have on their own. However, the data needs to be utilized with care as it often comes from different experimental settings and assays and is therefore not always comparable. These data also rely on the accurate annotation of the activity of the molecules as well as a requirement of straightforward access to the underlying data for an effective follow-up process. As these publicly available databases are structured further and the underlying data are enhanced with scientific rigor and depth, including more accurate descriptions of a compound's biological activity, then even more effective use of the data to enhance our efforts to generate more meaningful hypotheses will be facilitated.

Further methods are available to infer the target of compounds originating from a phenotypic screen. These include, for example, the use of gene expression profiling to compare an active compound, with a structurally related but much less active, or inactive, analogue to reveal differentially expressed genes which could be the target, or regulated via the molecular target of the active compound. Other methods also rely on the comparison of gene expression profiles after treatment of cells with compound such as using the NCI60 cancer cell line (CCL) collection and the Connectivity Map. A recent example profiled 823 CCLs with 481 small molecules.[60] The correlation in the sensitivity to the small molecules with the basal transcript level in the CCLs led to many examples where the MMoA of compounds could be inferred. For example, sensitivity of CCLs to ML239, a compound identified in a phenotypic screen using breast cancer stem cells, was highly correlated with high expression of the gene encoding delta(6) fatty acid desaturase, FADS2. Further studies revealed that ML239 activates FADS2, an unexpected finding.

Unraveling the precise MMoA of a compound discovered in a phenotypic screen is a challenging and often time-consuming and resource-intensive task. No one approach is likely to be sufficient and complementary approaches, including both experimental and computational, will be beneficial to identify the precise MMoA. This is especially true as cellular phenotypes and model systems become more complex and these methods must continue to evolve to provide novel and much-needed biological targets of relevance to human disease.

13. DRIVING SAR USING THE PRIMARY ASSAY IN THE ABSENCE OF A KNOWN TARGET

In situations where target deconvolution is difficult or has not produced results, optimization of the hit compounds in the absence of a known biological target can be performed. One pathway that has been investigated heavily from a phenotypic screening standpoint, initially in the search for new anticancer targets, is the Wnt pathway. A pioneering example was the discovery of the tankyrase 1 and 2 inhibitor XAV939.[61] A second example was the discovery of GNF-1331, which was shown to inhibit Wnt signaling through the inhibition of Porcupine. Subsequent medicinal chemistry led to LGK974, which inhibited Wnt signaling both in vitro and in vivo.[62] More recently, a cell-based reporter assay of Wnt pathway activity was used to profile greater than 60,000 compounds.[63] One of the most attractive hit molecules was a 3,4,5-trisubstituted pyridine (Fig. 8). As this molecule was the only member of this structural class in the screening set and as its biological target was unknown, it was elaborated using a hypothesis-driven ligand-based medicinal chemistry approach with the cell-based assay as the SAR-driving assay. Three hypotheses were formulated: (1) the 3,5-dichloro phenyl disubstitution pattern led to a twisted conformation about the piperidine–pyridine ring system; (2) the H-bond donor and acceptor properties of the primary amide would be essential for activity; and (3) the pyridine nitrogen atom would also be essential for activity. Each of these hypotheses was tested through the chemical synthesis of key analogues and testing in the cellular reporter gene assay. In addition, design criteria were applied to maintain the proposed cell permeability and improve the metabolic stability of the compounds. This elegant medicinal chemistry strategy led to CCT251545. CCT251545 was shown to be a potent inhibitor that displayed tumor growth inhibition in vivo after oral administration. The molecular targets of CCT251545 were subsequently disclosed in a separate publication where a chemical proteomics approach was applied using Cellular Target Profiling (Evotec, https://www.evotec.com/).[64] This approach for target deconvolution is exemplified in the chemical biology chapter. The targets with the strongest affinity to an active analogue of CCT251545 were the kinases CDK8 and CDK19. This work is an excellent

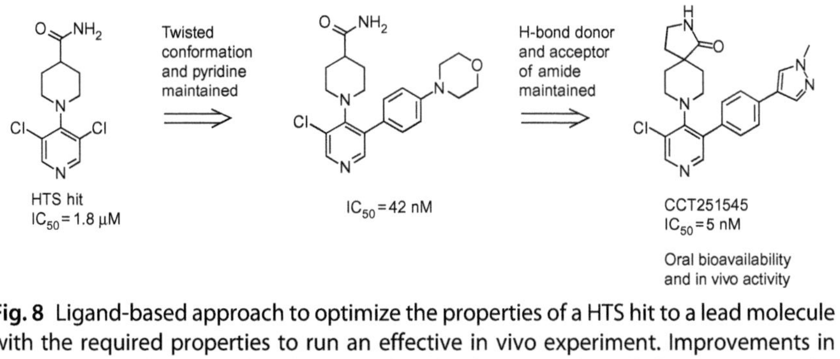

Fig. 8 Ligand-based approach to optimize the properties of a HTS hit to a lead molecule with the required properties to run an effective in vivo experiment. Improvements in potency were driven in the absence of molecular target knowledge but using the cell-based assay.

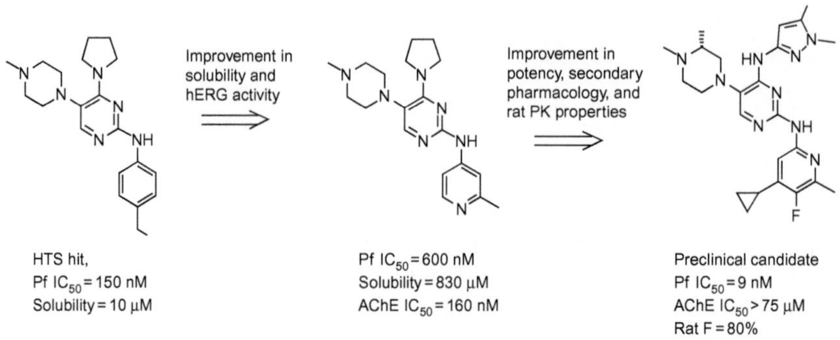

Fig. 9 Evolution of a hit from a HTS on blood stages of *Plasmodium falciparum* to optimize the chemical series to provide a preclinical candidate with excellent potency, improved secondary pharmacology, and appropriate pharmacokinetic properties for in vivo dosing.

example of how to deliver novel chemical probes, new targets, and a broader biological understanding through the combination of phenotypic screening, chemical biology, medicinal chemistry, and molecular and cellular biology. Chemical probes discovered through phenotypic screening can then be used to explore different biological processes where a pathway may be implicated. The Wnt pathway is a classic case where, for example, the tankyrase inhibitor XAV939 is now used extensively in stem cell differentiation protocols, including the production of iPS-derived cardiomyocytes from iPS cells.

A second example was the discovery of a series of triaminopyrimidines through performing a phenotypic screen on blood stages of *P. falciparum* using 500,000 compounds from the AstraZeneca corporate collection.[65] The hit compound formed the start point for a lead optimization program where the chemistry was driven using the primary cell-based assay and physicochemical, pharmacokinetic, and secondary pharmacology parameters such as activity against the hERG channel. The medicinal chemistry optimization led to a preclinical candidate, which showed good in vivo activity in a mouse malaria model and an acceptable safety margin in a 3-day rat toxicology study to support the progression of the compound into more advanced preclinical toxicological evaluation (Fig. 9).

14. CONCLUSIONS

Phenotypic screening and discovery have been shown to be a powerful approach to discover new molecular targets and pathways, leading to the discovery of drugs which impact positively on the lives of patients. The reemergence of phenotypic screening in both academic and industrial institutions has gathered speed with the development and accessibility to new cell sources, such as primary or iPS-derived cells, and new technologies to perform and measure the most relevant screening endpoints. These technologies include high-content biology, more sensitive image analysis techniques, and whole-genome expression profiling. With the advances being made in the field of microphysiological systems and with the heightened hurdles placed on differentiation of new therapies over standard of care, it is likely that screening on more complex and human disease-relevant systems will increase. Phenotypic discovery will be especially important to probe emerging biology where few, or no, relevant targets are known. While small molecules will remain important molecules to test in phenotypic assays, the availability of different modalities for screening such as genome-wide CRISPR–Cas9 libraries will create opportunities for the discovery of highly novel biological targets which can be tackled using the most appropriate of the diverse arsenal of available methods to modulate their function.

As the sensitivity of measurement techniques increases, an increasing amount of information will be generated from primary screening campaigns. This could include measurements of the key phenotype along with gene expression, pathway modulation, and potentially endpoints related to adverse events. The challenge is choosing relevant endpoints, where the benefit of this rich data source is the improved selection of the most biologically relevant compounds and targets and avoidance of failure at a later stage of the screening cascade. Although difficult, the key will be ensuring translation to the human setting and focusing compound hit and target selection on those offering the best chance of success against all of the parameters which will ultimately be required in a medicine.

Analyzing multiple parameters at the screening stage will place a large burden on our ability to effectively visualize and mine the data set and link the generated data to all known data, both from internal and external sources, to extract the full extent of knowledge on the compound hits and targets. This will be especially important for the inference of potential targets which are driving the phenotypic endpoint. These target hypotheses will then need to be tested effectively through the use of orthogonal methods such as gene knockdown or activation or by pharmacological methods. Target identification is still a major challenge, and advances will need to be made to ensure that phenotypic discovery can be as effective as possible. The advances in this area include the use of CETSA-MS and this and others will be discussed in the chapter on chemical biology. In addition, the recent example by Parker et al. using fragments functionalized with reactive groups also provides a potentially exciting glimpse to the future of identifying a target directly from the screen itself.

In conclusion, the application of cross-disciplinary phenotypic discovery to probe emerging biology is a powerful and exciting approach to discover new biological targets and molecules. Despite differences in the definition of phenotypic screening, it has been shown to make a significant impact on drug discovery and will continue to do so. The advances in screening technology and increased biological relevance along with effective practices will increase the clinical success of new candidate drugs and ultimately deliver transformative medicines to impact the lives of patients.

REFERENCES

1. Arrowsmith, J.; Miller, P. Trial Watch: Phase II and Phase III Attrition Rates 2011–2012. *Nat. Rev. Drug Discov.* **2013**, *12*, 569.
2. van der Worp, H. B.; Howells, D. W.; Sena, E. S.; Porritt, M. J.; Rewell, S.; O'Collins, V.; MacLeod, M. R. Can Animal Models of Disease Reliably Inform Human Studies? *PLoS Med.* **2010**, *7*, e1000245.
3. Ellis, L. M.; Fidler, I. J. Finding the Tumor Copycat. Therapy Fails, Patients Don't. *Nat. Med.* **2010**, *16*, 974–975.
4. Swinney, D. C.; Anthony, J. How Were New Medicines Discovered? *Nat. Rev. Drug Discov.* **2011**, *10*, 507–519.
5. Eder, J.; Sedrani, R.; Wiesmann, C. The Discovery of First-in-Class Drugs: Origins and Evolution. *Nat. Rev. Drug Discov.* **2014**, *13*, 577–587.
6. Hopkins, A. L. Network Pharmacology: The Next Paradigm in Drug Discovery. *Nat. Chem. Biol.* **2008**, *4*, 682–690.
7. Drawnel, F. M.; Boccardo, S.; Prummer, M.; Delobel, F.; Graff, A.; Weber, M.; Gerard, R.; Badi, L.; Kam-Thong, T.; Bu, L.; Jiang, X.; Hoflack, J. C.; Kiialainen, A.; Jeworutzki, E.; Aoyama, N.; Carlson, C.; Burcin, M.; Gromo, G.; Boehringer, M.; Stahlberg, H.; Hall, B. J.; Magnone, M. C.; Kolaja, K.; Chien, K. R.; Bailly, J.; Iacone, R. Disease Modeling and Phenotypic Drug Screening for Diabetic Cardiomyopathy Using Human Induced Pluripotent Stem Cells. *Cell Rep.* **2014**, *9*, 810–820.
8. Drowley, L.; Koonce, C.; Peel, S.; Jonebring, A.; Plowright, A. T.; Kattman, S. J.; Andersson, H.; Anson, B.; Swanson, B. J.; Wang, Q.-D.; Brolen, G. Human Induced Pluripotent Stem Cell-Derived Cardiac Progenitor Cells in Phenotypic Screening: A Transforming Growth Factor-β Type 1 Receptor Kinase Inhibitor Induces Efficient Cardiac Differentiation. *Stem Cells Transl. Med.* **2016**, *5*, 164–174.
9. Pagliuca, F. W.; Millman, J. R.; Guertler, M.; Segel, M.; Van Dervort, A.; Ryu, J. H.; Peterson, Q. P.; Greiner, D.; Melton, D. A. Generation of Functional Human Pancreatic β Cells In Vitro. *Cell* **2014**, *159*, 428–439.
10. Khurana, V.; Tardiff, D. F.; Chung, C.-Y.; Lindquist, S. Toward Stem Cell-Based Phenotypic Screens for Neurodegenerative Diseases. *Nat. Rev. Neurol.* **2015**, *11*, 339–350.
11. Sances, S.; Bruijn, L. I.; Chandran, S.; Eggan, K.; Ho, R.; Klim, J. R.; Livesy, M. R.; Lowry, E.; Macklis, J. D.; Rushton, D.; Sadegh, C.; Sareen, D.; Wichterle, H.; Zhang, S. C.; Svendsen, C.; Modeling, N. ALS With Motor Neurons Derived From Human Induced Pluripotent Stem Cells. *Nat. Neurosci.* **2016**, *19*, 542–553.

12. Eglen, R.; Reisine, T. Primary Cells and Stem Cells in Drug Discovery: Emerging Tools for High-Throughput Screening. *Assay Drug Dev. Technol.* **2011**, *9*, 108–124.
13. Gobel, J.; Gartland, M.; Gurley, S. H.; Kadwell, S.; Gillie, D.; Moore, C.; Goetz, A. A Phenotypic High-Throughput Screen With RSV-Infected Primary Human Small Airway Epithelial Cells (SAECs). *J. Biomol. Screen.* **2015**, *20*, 729–738.
14. Wang, S. S.; Ehrlich, D. J. Image-Based Phenotypic Screening With Human Primary T Cells Using One-Dimensional Imaging Cytometry With Self-Tuning Statistical-Gating Algorithms. *SLAS Discov.* **2017**, (Online early Access). https://doi.org/10.1177/2472555217705953.
15. Paunovic, A. I.; Drowley, L.; Nordqvist, A.; Ericson, E.; Mouchet, E.; Jonebring, A.; Gronberg, G.; Kvist, A. J.; Engkvist, O.; Brown, M. R.; Goumans, M.-J.; Wang, Q.-D.; Plowright, A. T. Phenotypic Screen for Cardiac Regeneration Identifies Molecules With Differential Activity in Primary Epicardium-Derived Cells Compared to Cardiac Fibroblasts. *ACS Chem. Biol.* **2017**, *12*, 132–141.
16. Kenny, H. A.; Lal-Nag, M.; White, E. A.; Shen, M.; Chiang, C.-Y.; Mitra, A. K.; Zhang, Y.; Curtis, M.; Schryver, E. M.; Bettis, S.; Jadhav, A.; Boxer, M. B.; Li, Z.; Ferrer, M.; Lengyel, E. Quantitative High Throughput Screening Using a Primary Human Three-Dimensional Organotypic Culture Predicts in vivo Efficacy. *Nat. Commun.* **2015**, *6*, 6220.
17. Horvath, P.; Aulner, N.; Bickle, M.; Davies, A. M.; Del Nery, E.; Ebner, D.; Montoya, M. C.; Östling, P.; Pietiäinen, V.; Price, L. S.; Shorte, S. L.; Turcatti, G.; von Schantz, C.; Carragher, N. O. Screening Out Irrelevant Cell-Based Models of Disease. *Nat. Rev. Drug Discov.* **2016**, *15*, 751–769.
18. Gustafsdottir, S. M.; Ljosa, V.; Sokolnicki, K. L.; Wilson, J. A.; Walpita, D.; Kemp, M. M.; Seiler, K. P.; Carrel, H. A.; Golub, T. R.; Schreiber, S. L.; Clemons, P. A.; Carpenter, A. E.; Shamji, A. F. Multiplex Cytological Profiling Assay to Measure Diverse Cellular States. *PLoS One* **2013**, *8*, e80999.
19. Reisen, F.; Sauty de Chalon, A.; Pfeifer, M.; Zhang, X.; Gabriel, D.; Selzer, P. Linking Phenotypes and Modes of Action Through High-Content Screen Fingerprints. *Assay Drug Dev. Technol.* **2015**, *13*, 415–427.
20. Hakkim, A.; Fuchs, T. A.; Martinez, N. E.; Hess, S.; Prinz, H.; Zychlinsky, A.; Waldmann, H. Activation of the Raf-MEK-ERK Pathway Is Required for Neutrophil Extracellular Trap Formation. *Nat. Chem. Biol.* **2011**, *7*, 75–77.
21. Vincent, F.; Loria, P.; Pregel, M.; Stanton, R.; Kitching, L.; Nocka, K.; Doyonnas, R.; Steppan, C.; Gilbert, A.; Schroeter, T.; Peakman, M. C. Developing Predictive Assays: The Phenotypic Screening "Rule of 3". *Sci. Transl. Med.* **2015**, *7*, 1–5.
22. Engle, S. J.; Vincent, F. Small Molecule Screening in Human Induced Pluripotent Stem Cell-Derived Terminal Cell Types. *J. Biol. Chem.* **2014**, *289*, 4562–4570.
23. Piccini, I.; Rao, J.; Seebohm, G.; Greber, B. Human Pluripotent Stem Cell-Derived Cardiomyocytes: Genome-Wide Expression Profiling of Long-Term in vitro Maturation in Comparison to Human Heart Tissue. *Nature* **2013**, *498*, 325–331.
24. Chimenti, I.; Pagano, F.; Angelini, F.; Siciliano, C.; Mangino, G.; Picchio, V.; De Falco, E.; Peruzzi, M.; Carnevale, R.; Ibrahim, M.; Biondi-Zoccai, G.; Messina, E.; Frati, G. Human Lung Spheroids as In Vitro Niches of Lung Progenitor Cells With Distinctive Paracrine and Plasticity Properties. *Stem Cells Transl. Med.* **2017**, *6*, 767–777.
25. Pointon, A.; Pilling, J.; Dorval, T.; Wang, Y.; Archer, C.; Pollard, C. High-Throughput Imaging of Cardiac Microtissues for the Assessment of Cardiac Contraction During Drug Discovery. *Toxicol. Sci.* **2017**, *155*, 444–457.
26. Santoro, M.; Lamhamedi-Cherradi, S. E.; Menegaz, B. A.; Ludwig, J. A.; Mikos, A. G. Flow Perfusion Effects on Three-Dimensional Culture and Drug Sensitivity of Ewing Sarcoma. *PNAS* **2015**, *112*, 10304–10309.
27. Kim, J. Y.; Fluri, D. A.; Marchan, R.; Boonen, K.; Mohanty, S.; Singh, P.; Hammad, S.; Landuyt, B.; Hengstler, J. G.; Kelm, J. M.; Hierlemann, A.; Frey, O. 3D Spherical

Microtissues and Microfluidic Technology for Multi-Tissue Experiments and Analysis. *J. Biotechnol.* **2015**, *205*, 24–35.

28. Esch, E. W.; Bahinski, A.; Huh, D. Organs-on-Chips at the Frontiers of Drug Discovery. *Nat. Rev. Drug Discov.* **2015**, *14*, 248–260.
29. Liu, Y.; Asnani, A.; Zou, L.; Bentley, V. L.; Yu, M.; Wang, Y.; Dellaire, G.; Sarkar, K. S.; Dai, M.; Chen, H. H.; Sosnovik, D. E.; Shin, J. T.; Haber, D. A.; Berman, J. N.; Chao, W.; Peterson, R. T. Visnagin Protects Against Doxorubicin-Induced Cardiomyopathy Through Modulation of Mitochondrial Malate Dehydrogenase. *Sci. Transl. Med.* **2014**, *266,* 266ra170.
30. Markstein, M.; Dettorre, S.; Cho, J.; Neumeuller, R. A.; Craig-Mueller, S.; Perrimon, N. Systematic Screen of Chemotherapeutics in Drosophila Stem Cell Tumors. *Proc. Natl. Acad. Sci. U. S. A.* **2014**, *111*, 4530–4535.
31. Yu, P. B.; Hong, C. C.; Sachidanandan, C.; Babitt, J. L.; Deng, D. Y.; Hoyng, S. A.; Lin, H. Y.; Bloch, K. D.; Peterson, R. T. Dorsomorphin Inhibits BMP Signals Required for Embryogenesis and Iron Metabolism. *Nat. Chem. Biol.* **2008**, *4*, 33–41.
32. Rihel, J.; Prober, D. A.; Arvanites, A.; Lam, K.; Zimmerman, S.; Jang, S.; Haggarty, S. J.; Kokel, D.; Rubin, L. L.; Peterson, R. T.; Schier, A. F. Zebrafish Behavioral Profiling Links Drugs to Biological Targets and Rest/Wake Regulation. *Science* **2010**, *327*, 348–351.
33. Shalem, O.; Sanjana, N. E.; Hartenian, E.; Shi, X.; Scott, D. A.; Mikkelsen, T. S.; Heckl, D.; Ebert, B. L.; Root, D. E.; Doench, J. G.; Zhang, F. Genome-Scale CRISPR-Cas9 Knockout Screening in Human Cells. *Science* **2014**, *6166*, 84–87.
34. Kakeya, H. Natural Products-Prompted Chemical Biology: Phenotypic Screening and a New Platform for Target Identification. *Nat. Prod. Rep.* **2016**, *33*, 648–654.
35. Van Hattum, H.; Waldmann, H. Biology-Oriented Synthesis: Harnessing the Power of Evolution. *J. Am. Chem. Soc.* **2014**, *136*, 11853–11859.
36. Lewis, R. J.; Garcia, M. L. Therapeutic Potential of Venom Peptides. *Nat. Rev. Drug Discov.* **2003**, *2*, 790–802.
37. Sandercock, A. M.; Rust, S.; Guillard, S.; Sachsenmeier, K. F.; Holoweckyj, N.; Hay, C.; Flynn, M.; Huang, Q.; Yan, K.; Herpers, B.; Price, L. S.; Soden, J.; Freeth, J.; Jermutus, L.; Hollingsworth, R.; Minter, R. Identification of Anti-Tumour Biologics Using Primary Tumour Models, 3-D Phenotypic Screening and Image-Based Multi-Parameter Profiling. *Mol. Cancer* **2015**, *14*, 147.
38. Petersen, D. N.; Hawkins, J.; Ruangsiriluk, W.; Stevens, K. A.; Maguire, B. A.; O'Connell, T. N.; Rocke, B. N.; Boehm, M.; Ruggeri, R. B.; Rolph, T.; Hepworth, D.; Loria, P. M.; Carpino, P. A. A Small-Molecule Anti-Secretagogue of PCSK9 Targets the 80S Ribosome to Inhibit PCSK9 Protein Translation. *Cell Chem. Biol.* **2016**, *23*, 1362–1371.
39. Howe, J. A.; Wang, H.; Fischmann, T. O.; Balibar, C. J.; Xiao, L.; Galgoci, A. M.; Malinverni, J. C.; Mayhood, T.; Villafania, A.; Nahvi, A.; Murgolo, N.; Barbieri, C. M.; Mann, P. A.; Carr, D.; Xia, E.; Zuck, P.; Riley, D.; Painter, R. E.; Walker, S. S.; Sherbourne, B.; de Jesus, R.; Pan, W.; Plotkin, M. A.; Wu, J.; Ringden, D.; Cummings, J.; Garlisi, C. G.; Zhang, R.; Sheth, P. R.; Gill, C. J.; Tang, H.; Roemer, T. Selective Small-Molecule Inhibition of an RNA Structural Element. *Nature* **2015**, *526*, 672–677.
40. Rottmann, M.; McNamara, C.; Yeung, B. K. S.; Lee, M. C. S.; Zou, B.; Russell, B.; Seitz, P.; Plouffe, D. M.; Dharia, N. V.; Tan, J.; Cohen, S. B.; Spencer, K. R.; González-Páez, G. E.; Lakshminarayana, S. B.; Goh, A.; Suwanarusk, R.; Jegla, T.; Schmitt, E. K.; Beck, H.-P.; Brun, R.; Nosten, F.; Renia, L.; Dartois, V.; Keller, T. H.; Fidock, D. A.; Winzeler, E. A.; Diagana, T. T. Spiroindolones, a Potent Compound Class for the Treatment of Malaria. *Science* **2010**, *329*, 1175–1180.
41. Bohacek, R. S.; McMartin, C.; Guida, W. C. The Art and Practice of Structure-Based Drug Design: A Molecular Modeling Perspective. *Med. Res. Rev.* **1996**, *16*, 3–50.

42. Kauvar, L. M.; Higgins, D. L.; Villar, H. O.; Sportsman, J. R.; Engqvist-Goldstein, A.; Bukar, R.; Bauer, K. E.; Dilley, H.; Rocke, D. M. Predicting Ligand Binding to Proteins by Affinity Fingerprinting. *Chem. Biol.* **1995**, *2*, 107–118.
43. Petrone, P. M.; Simms, B.; Nigsch, F.; Lounkine, E.; Kutchukian, P.; Cornette, A.; Deng, Z.; Davies, J. W.; Jenkins, J. L.; Glick, M. Rethinking Molecular Similarity: Comparing Compounds on the Basis of Biological Activity. *ACS Chem. Biol.* **2012**, *7*, 1399–1409.
44. Wawer, M. J.; Li, K.; Gustafsdottier, S. M.; Ljosa, V.; Bodycombe, N. E.; Marton, M. A.; Sokolnicki, K. L.; Bray, M. A.; Kemp, M. M.; Winchester, E.; Taylor, B.; Grant, G. B.; Hon, C. S.-Y.; Duvall, J. R.; Wilson, J. A.; Bittker, J. A.; Dancik, V.; Narayan, R.; Subranmanian, A.; Winckler, W.; Golub, T. R.; Carpenter, A. E.; Shamji, A. F.; Schreiber, S. L.; Clemons, P. A. Toward Performance-Diverse Small-Molecule Libraries for Cell-Based Phenotypic Screening Using Multiplexed High-Dimensional Profiling. *Proc. Natl. Acad. Sci. U. S. A.* **2014**, *111*, 10911–10916.
45. Sidrauski, C.; Acosta-Alvear, D.; Khoutorsky, A.; Vedantham, P.; Hearn, B. R.; Li, H.; Gamache, K.; Gallagher, C. M.; Ang, K. K.-H.; Wilson, C.; Okreglak, V.; Ashkenazi, A.; Hann, B.; Nader, K.; Arkin, M. R.; Renslo, A. R.; Sonenberg, N.; Walter, P. Pharmacological brake-release of mRNA translation enhances cognitive memory. *eLife* **2013**, *2*, e00498.
46. Sidrauski, C.; Tsai, J. C.; Kampmann, M.; Hearn, B. R.; Vedantham, P.; Jaishankar, P.; Sokabe, M.; Mendez, A. S.; Newton, B. W.; Tang, E. L.; Verschueren, E.; Johnson, J. R.; Krogan, N. J.; Fraser, C. S.; Weissman, J. S.; Renslo, A. R.; Walter, P. Pharmacological Dimerization and Activation of the Exchange Factor eIF2B Antagonizes the Integrated Stress Response. *eLife* **2015**, *4*, e07314.
47. Sekine, Y.; Zyryanova, A.; Crespillo-Casado, A.; Fischer, P. M.; Harding, H. P.; Ron, D. Mutations in a Translation Initiation Factor Identify the Target of a Memory-Enhancing Compound. *Science* **2015**, *348*, 1027–1030.
48. Parker, C. G.; Galmozzi, A.; Wang, Y.; Correia, B. E.; Sasaki, K.; Joslyn, C. M.; Kim, A. S.; Cavallaro, C. L.; Lawrence, R. M.; Johnson, S. R.; Narvaiza, I.; Saez, E.; Cravatt, B. F. Ligand and Target Discovery by Fragment-Based Screening in Human Cells. *Cell* **2017**, *168*, 527–541.
49. Bembenek, S. D.; Tounge, B. A.; Reynolds, C. H. Ligand Efficiency and Fragment-Based Drug Discovery. *Drug Discov. Today* **2009**, *14*, 278–283.
50. Hale, C. M.; Cheng, Q.; Ortuno, D.; Huang, M.; Nojima, D.; Kassner, P. D.; Wang, S.; Ollmann, M. M.; Carlisle, H. J. Identification of Modulators of Autophagic Flux in an Image-Based High Content siRNA Screen. *Autophagy* **2016**, *12*, 713–726.
51. Eulalio, A.; Mano, M.; Dal Ferro, M.; Zentilin, L.; Sinagra, G.; Zacchigna, S.; Giacca, M. Functional Screening Identifies miRNAs Inducing Cardiac Regeneration. *Nature* **2012**, *492*, 376–381.
52. Sidik, S. M.; Huet, D.; Ganesan, S. M.; Huynh, M. H.; Wang, T.; Nasamu, A. S.; Thiru, P.; Saeji, J. P.; Carruthers, V. B.; Niles, J. C.; Lourido, S. A Genome-Wide CRISPR Screen in Toxoplasma Identified Essential Apicomplexan Genes. *Cell* **2016**, *167*, 1423–1435.
53. Park, J.; Al-Ramahi, I.; Tan, Q.; Mollema, N.; Diaz-Garcia, J. R.; Gallego-Flores, T.; Lu, H. C.; Lagalwar, S.; Duvick, L.; Kang, H.; Lee, Y.; Jafar-Nejad, P.; Sayegh, L. S.; Richman, R.; Liu, X.; Gao, Y.; Shaw, C. A.; Arthur, J. S.; Orr, H. T.; Westbrook, T. F.; Botas, J.; Zoghbi, H. Y. Genetic Screens Reveal RAS/MAPK/MSK1 Modulate Ataxin 1 Protein Levels and Toxicity in SCA1. *Nature* **2013**, *7454*, 325–331.
54. Shanmugasundaram, V.; Maggiora, G. M.; Lajiness, M. S. Hit-Directed Nearest-Neighbor Searching. *J. Med. Chem.* **2005**, *48*, 240–248.
55. Kogej, T.; Engkvist, O.; Blomberg, N.; Muresan, S. Multifingerprint Based Similarity Searches for Targeted Class Compound Selection. *J. Chem. Inf. Model.* **2006**, *46*, 1201–1213.

56. Lo, Y. C.; Senese, S.; Li, C.-M.; Hu, Q.; Huang, Y.; Damoiseaux, R.; Torres, J. Z. Large-Scale Chemical Similarity Networks for Target Profiling of Compounds Identified in Cell-Based Chemical Screens. *PLoS Comput. Biol.* **2015**, *11,* e1004153.
57. Reker, D.; Perna, A. M.; Rodrigues, T.; Schneider, P.; Reutlinger, M.; Mönch, B.; Koeberle, A.; Lamers, C.; Gabler, M.; Steinmetz, H.; Müller, R.; Schubert-Zsilavecz, M.; Werz, O.; Schneider, G. Revealing the Macromolecular Targets of Complex Natural Products. *Nat. Chem.* **2014**, *6*, 1072–1078.
58. Helal, K. Y.; Maciejewski, M.; Gregori-Puigjane, E.; Glick, M.; Wassermann, A. M. Public Domain HTS Fingerprints: Design and Evaluation of Compound Bioactivity Profiles From PubChem's Bioassay Repository. *J. Chem. Inf. Model.* **2016**, *56*, 390–398.
59. Bornot, A.; Blackett, C.; Engkvist, O.; Murray, C.; Bendtsen, C. The Role of Historical Bioactivity Data in the Deconvolution of Phenotypic Screens. *J. Biomol. Screen.* **2014**, *9*, 696–706.
60. Rees, M. G.; Seashore-Ludlow, B.; Cheah, J. H.; Adams, D. J.; Price, E. V.; Shubhroz, G.; Javaid, S.; Coletti, M. E.; Jones, V. L.; Bodycombe, N. E.; Soule, C. K.; Alexander, B.; Li, A.; Montgomery, P.; Kotz, J. D.; Hon, C. S.-Y.; Munoz, B.; Liefeld, T.; Dancik, V.; Haber, D. A.; Clish, C. B.; Bittker, J. A.; Palmer, M.; Wagner, B. K.; Clemons, P. A.; Shamji, A. F.; Schreiber, S. L. Correlating Chemical Sensitivity and Basal Gene Expression Reveals Mechanism of Action. *Nat. Chem. Biol.* **2016**, *2*, 109–116.
61. Huang, S.-M. A.; Mishina, Y. M.; Liu, S.; Cheung, A.; Stegmeier, F.; Michaud, G. A.; Charlet, O.; Wiellette, E.; Zhang, Y.; Wiessner, S.; Hild, M.; Shi, X.; Wilson, C. J.; Mickanin, C.; Myer, V.; Fazal, A.; Tomlinson, R.; Serluca, F.; Shao, W.; Cheng, H.; Shultz, M.; Rau, C.; Schirle, M.; Schegl, J.; Ghidelli, S.; Fawell, S.; Lu, C.; Curtis, D.; Kirschner, M. W.; Lengauer, C.; Finan, P. M.; Tallarico, J. A.; Bouwmeester, T.; Porter, J. A.; Bauer, A.; Cong, F. Tankyrase Inhibition Stabilizes Axin and Antagonizes Wnt Signaling. *Nature* **2009**, *461*, 614–620.
62. Liu, J.; Pan, S.; Hsieh, M. H.; Ng, N.; Sun, F.; Wang, T.; Kasibhatla, S.; Schuller, A. G.; Li, A. G.; Cheng, D.; Li, J.; Tompkins, C.; Pferdekamper, A. M.; Steffy, A.; Cheng, J.; Kowal, C.; Phung, V.; Guo, G.; Wang, Y.; Graham, M. P.; Flynn, S.; Brenner, J. C.; Li, C.; Villaroel, M. C.; Schultz, P. G.; Wu, X.; McNamara, P.; Sellers, W. R.; Petruzzelli, L.; Boral, A. L.; Seidel, H. M.; McLaughlin, M. E.; Che, J.; Carey, T. E.; Vanasse, G.; Harris, J. L. Targeting Wnt-Driven Cancer Through the Inhibition of Porcupine by LGK974. *Proc. Natl. Acad. Sci. U. S. A.* **2013**, *110*, 20224–20229.
63. Mallinger, A.; Crumpler, S.; Pichowicz, M.; Waalboer, D.; Stubbs, M.; Adeniji-Popoola, O.; Wood, B.; Smith, E.; Thai, C.; Henley, A. T.; Georgi, K.; Court, W.; Hobbs, S.; Box, G.; Ortiz-Ruiz, M.-J.; Valenti, M.; De Haven Brandon, A.; TePoele, R.; Leuthner, B.; Workman, P.; Aherne, W.; Poeschke, O.; Dale, T.; Wienke, D.; Esdar, C.; Rohdich, F.; Raynaud, F.; Clarke, P. A.; Eccles, S. A.; Stieber, F.; Schiemann, K.; Blagg, J. Discovery of Potent, Orally Bioavailable, Small-Molecule Inhibitors of WNT Signaling From a Cell-Based Pathway Screen. *J. Med. Chem.* **2015**, *58*, 1717–1735.
64. Dale, T.; Clarke, P. A.; Esdar, C.; Waaldoer, D.; Adeniji-Popoola, O.; Ortiz-Ruiz, M.-J.; Mallinger, A.; Samant, R. S.; Czodrowski, P.; Musil, D.; Schwarz, D.; Schneider, K.; Stubbs, M.; Ewan, K.; Fraser, E.; TePoele, R.; Court, W.; Box, G.; Valenti, M.; De Haven Brandon, A.; Gowan, S.; Rohdich, F.; Raynaud, F.; Schneider, R.; Poeschke, O.; Blaukat, A.; Workman, P.; Schiemann, K.; Eccles, S. A.; Wienke, D.; Blagg, J. A Selective Chemical Probe for Exploring the Role of CDK8 and CDK19 in Human Disease. *Nat. Chem. Biol.* **2015**, *11*, 973–980.

65. Hameed, P. S.; Solapure, S.; Patil, V.; Henrich, P. P.; Magistrado, P. A.; Bharath, S.; Murugan, K.; Viswanath, P.; Puttur, J.; Srivastava, A.; Bellale, E.; Panduga, V.; Shanbag, G.; Awasthy, D.; Landge, S.; Morayya, S.; Koushik, K.; Saralaya, R.; Raichurkar, A.; Rautela, N.; Choudhury, N. R.; Ambady, A.; Nandishaiah, R.; Reddy, J.; Prabhakar, K. R.; Menasinakai, S.; Rudrapatna, S.; Chatterji, M.; Jiménez-Díaz, M. B.; Martínez, M. S.; Sanz, L. M.; Coburn-Flynn, O.; Fidock, D. A.; Lukens, A. K.; Wirth, D. F.; Bandodkar, B.; Mukherjee, K.; McLaughlin, R. E.; Waterson, D.; Rosenbrier-Ribeiro, L.; Hickling, K.; Balasubramanian, V.; Warner, P.; Hosagrahara, V.; Dudley, A.; Iyer, P. S.; Narayanan, S.; Kavanagh, S.; Sambandamurthy, V. K. Triaminopyrimidine is a fast-killing and long-acting antimalarial clinical candidate. *Nat. Commun.* **2015**, *6*, 6715.

CHAPTER NINE

Targeted Protein Degradation

Nello Mainolfi*, Tim Rasmusson†,[1]
*Kymera Therapeutics LLC, Cambridge, MA, United States
†AstraZeneca Pharmaceuticals LP, Waltham, MA, United States
[1]Corresponding author: e-mail address: timothy.rasmusson@astrazeneca.com

Contents

1. INTRODUCTION TO THE UBIQUITIN PROTEASOME SYSTEM (UPS)

1.1 Protein Homeostasis and UPS

Protein homeostasis, also known as proteostasis, refers to a delicate intracellular balance between generation of newly synthesized proteins and timely disposal of damaged and misfolded proteins that are beyond repair or refolding.[1] Protein homeostasis is essential for a living cell and more broadly, an organism to maintain its normal cellular function as dysregulation of cellular protein homeostasis can often lead to various disease states.[2] Therefore, living organisms, especially eukaryotic cells, have developed several

Annual Reports in Medicinal Chemistry, Volume 50
ISSN 0065-7743
https://doi.org/10.1016/bs.armc.2017.08.005

sophisticated mechanisms to degrade targeted proteins, a process that is also part of overall protein quality control. One such mechanism is the UPS, in which a protein substrate targeted for degradation is covalently modified by polyubiquitin chains and directed to the proteasome for proteolysis.[3] It should be noted that in addition to disposing damaged or misfolded intracellular proteins, UPS also plays a critical role in various cellular pathways such as cell cycle progression[4] and signaling pathways,[5] where selective degradation of key protein factors in a temporally and spatially controlled fashion is essential to ensuring normal cellular functions.

The UPS comprises a series of finely orchestrated enzymatic events that ultimately lead to protein polyubiquitination and degradation by the 26S proteasome in eukaryotic cells. Protein ubiquitination is mediated by the E1–E2–E3 enzymatic cascade (see Fig. 1). In the first step, ubiquitin, which is a highly conserved 76 amino acid protein, is activated by the E1 enzyme in the presence of ATP to form an E1-ubiquitin covalent intermediate.[6] In this intermediate, a conserved cysteine residue in E1 forms a thioester with the carboxylate group of ubiquitin's C-terminal glycine residue. In the second step, E1-ubiquitin thioester passes the ubiquitin onto a cysteine residue in E2, also called ubiquitin conjugating enzyme, through transthiolation reaction.[7] In the last step of the cascade, an E3 ubiquitin ligase catalyzes the ubiquitin transfer from the E2-ubiquitin thioester to the specific lysine

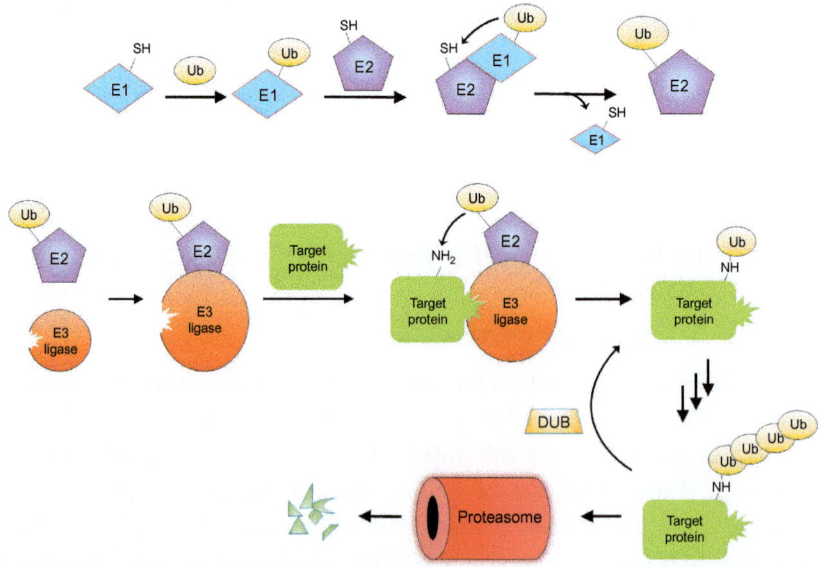

Fig. 1 The ubiquitin proteasome system (UPS).

residue on the protein substrate. In humans, there are two classes of ubiquitin activating E1 enzymes, 30–40 E2s and ~600 E3s. Polyubiquitin chains of different linkages (attaching to different internal lysine residues or the N-terminal NH_2 group within ubiquitin) have been shown to play distinct roles. It is generally believed that Ub chains conjugated to lysine-48 (or K-48) are involved in proteasomal degradation[8] while lysine-63 (or K-63) conjugated chains mediate protein–protein interactions (PPIs).[9] Physiological functions of other linkages such as linear, K11, K29, K33, or mixed linkages are also emerging.[10] Understanding substrate specificity and regulation of the ubiquitin transfer process remains an active area of research for the past 20 years.

1.2 Structures and Mechanisms of E3 Ligases

Ubiquitin E3 ligases represent a collection of more than 600 proteins in the human proteome.[11] Functionally, all E3 ligases interact with both E2-ubiquitin thioester and the substrate protein and catalyze efficient ubiquitin transfer to the lysine residue of a target protein (polyubiquitin chain initiation) or ubiquitin in a growing chain. Based on distinct structural and mechanistic features, these E3s can be categorized into three families: the homologous to the E6AP carboxyl terminus domain (HECT) family, the really interesting new gene (RING) and RING-related family, and the RBR family with a characteristic RING1-(in-between-Ring)-RING2 tripartite motif. Structural and mechanistic studies on these three subfamilies E3s have provided tremendous insight into the roles of E3 ligases in regulating diverse biological pathways.

1.2.1 The HECT E3 Family

The founding HECT E3 family member in human, E6-associated protein (E6AP), was discovered by Huibregtse et al.[12] Since then, a total of 29 HECT E3s have been identified in the human proteome. These HECT E3s have been shown to play diverse roles in cellular pathways including signal transduction, protein localization, and growth and proliferation. The HECT E3s contain a conserved, ~350 amino acid HECT domain in the C-terminal part of the enzyme and more diverse N-terminal domains of various sizes that are responsible for substrate protein recognition. A common feature of HECT E3-catalyzed ubiquitination involves transfer of ubiquitin from E2-ubiquitin thioester to an internal conserved cysteine residue in the HECT domain to form an E3-ubiquitin thioester intermediate. The other domains of HECT E3s help to recruit substrate proteins and mediate subsequent handover of ubiquitin from the E3 thioester to the lysine residue on the substrate protein.

The HECT domain can further be divided into two substructures: an N-terminal lobe (N-lobe) that provides a platform to bind ubiquitin-loaded E2 thioester and a C-terminal lobe (C-lobe) that contains the absolutely conserved cysteine to accept ubiquitin from E2. The crystal structure of E6AP in complex with its E2 (UBCH7) suggests that the C-lobe exists in a more open conformation relative to E2 and the two catalytic cysteine residues in E2 and E3's HECT domain are more than 30 Å apart.[13] This observation suggests that a large conformational change must take place during the transthiolation step to bring the two cysteine residues to close proximity. The proposed "closed" conformation was confirmed in an elegant study by Kamadurai et al.[14] In that study, the NEDD4L HECT domain in complex with ubiquitin-modified UBCH5B (E2) was chemically trapped in the "closed" conformation and the crystal structure shows the C-lobe rotates almost 180 degrees compared to that in the open conformation and makes a close contact to UbcH5B. As a result, the distance between the two catalytic cysteines in E2 and E3 is only 8 Å. By comparing the conformation of the open and closed states in these structures, we can conclude that the flexible linker connecting the N- and C-lobes plays an important role in facilitating ubiquitin transfer from E2 to the cysteine residue in the HECT domain. Finally, HECT E3s are shown to be capable of synthesizing both K-48 and K-63 ubiquitin chains although it is not well understood how the linkage specificity is determined. One study on the yeast HECT E3 Rsp5 suggests that the chain specificity is entirely determined by Rsp5 HECT domain and that the identity of E2s or substrates plays a minimum role.[15] Structural studies on several WW domain-containing HECT E3s including Rsp5, Nedd4, and Smurf2, and demonstrate that N-lobe contains a ubiquitin binding site and may play a role in facilitating polyubiquitin chain formation by juxtaposing the distal ubiquitin in the growing chain to the active site cysteine in the C-lobe loaded with ubiquitin thioester.[16] It is not clear, however, whether other HECT E3s use the same strategy to control the ubiquitin chain specificity.

1.2.2 The RING E3 Family

The RING E3 family represents the largest group of the E3 ligases with ~600 members. All RING E3s contain a structurally conserved cross-braced motif that coordinates two Zn^{2+} ions. A consensus RING motif usually contains a linear arrangement of eight Cys and His residues in a pattern of C3H2C3 or C3HC4 with variations in some RING E3s. In RING-like U-box E3s, a tertiary structure motif exists that is very similar to RING but has no coordination of zinc cation. Instead, the RING-like motif is stabilized by noncovalent interactions among conserved residues. The RING and U-box motif provide a common platform for E3s to interact with E2 ubiquitin conjugating enzymes. While many RING E3s function as monomers, many well-studied RING E3s exist as homo- and hetero-dimers and multisubunit protein complexes such as cullin-RING ligases. Dimerization of RING E3s occur either directly through the RING motif itself or by adjacent structural elements and in heterodimer cases one RING E3 partner is often inactive such as in the case of BRCA1–BARD1.[17] Dimerization of RING E3 often provides a mechanism of regulation of its activity. For example, in cIAP1, the dimerization interface is normally secluded in an inactive conformation until an antagonist protein such as SMAC binds and triggers the conformational change that allows RING dimerization to occur, leading to binding to E2 and subsequent ubiquitin transfer.[18] For cullin-RING E3 ligases, the RING motif that recruits E2 exists in a small protein (RBX1) that binds to a scaffold protein (cullins). Cullins also binds to a variety of adaptor proteins. These adaptor proteins and their binding partners play a crucial role in recognizing the substrate proteins to be ubiquitinated. Other multisubunit RING E3s are anaphase promoting complex/cyclosome (APC/C)[19] and Fanconi anemia (FANC)[20] which are important players in cell cycle control and DNA repair, respectively.

Unlike HECT E3s, RING E3s lack a conserved catalytic cysteine and it's been thought that they act as a scaffold to bring substrates and E2s in close proximity. This notion seems to be validated by available E2–E3 complex structures, where the RING motif-binding area in E2 is far away from the active site Cys that forms ubiquitin thioester.[21] Therefore, for a long time it has not been clear if RING E3 exerted any influence on ubiquitin transfer from E2 to substrate proteins. More recently, in several structural studies involving RING E3s (RNF4, E4B, and BIRC7) with ubiquitin-loaded E2 (UbcH5s–Ub), a more closed conformation is revealed that includes crucial interactions between residues from both E3 and E2–Ub required for efficient ubiquitin transfer.[22] These and other studies suggest that RING E3s function not only as a molecular scaffold, but also promoting ubiquitin transfer by stabilizing relevant E2–ubiquitin conformations allosterically and/or actively participating in catalysis by providing crucial side chain interactions during ubiquitin transfer.

1.2.3 The RING-in-Between-RING-RING (RBR) E3 Family

In humans, there are about 13 E3 ligases that have two RING motifs with a Cys-rich region often referred to as "in-between-RING." This type of E3s has been shown to employ a hybrid RING–HECT mechanism where the first RING motif recruits ubiquitin-loaded E2 and transfers ubiquitin to a conserved cysteine in the second RING domain.[23] Notable members of this family include HOIP and HOIL-1L, that are part of linear ubiquitin chain assembly complex (LUBAC), Parkin, that is associated with Parkinson's disease, and HHARI. Most recently, Schulman's group showed that ARIH1, a RBR E3, facilitates cullin-RING ligase-mediated polyubiquitin chain assembly by catalyzing monoubiquitination (also called "priming") of protein substrates.[24]

1.3 E3s and Human Diseases

Because E3 ligases are involved in diverse biological pathways, it is not surprising that they are linked to many human diseases. One prominent example is Parkin, a RBR E3, which is mutated in more than 50% of patients with autosomal recessive juvenile Parkinsonism and has been shown to directly cause the progression of this neurodegenerative disorder.[25] E3s such as MDM2, BRCA1, FBXO11, FBXW7, and VHL are also involved in a variety of tumor malignancies.[26] Interestingly, viruses have evolved to hijack host cell's E3 machinery to their own advantage. The oncogenic human papillomavirus (HPV) encodes E6 protein that recruits the host E3 ligase E6AP (hence the name) to target p53 and trigger its degradation.[27] In addition, the HIV virus's genome contains an adaptor protein, Vpu, that when expressed in host cells, direct cullin E3 βTrCP to bind and degrade cell surface viral receptor CD4 in order to prevent further infection and to promote assembly of virion particles.[28] HIV also encodes another viral adaptor protein, Vpr, that binds VprBP(DCAF1)–DDB1–Cullin4 E3 complex and redirects this E3 machinery to degrade a panel of host proteins such as UNG2 and SMUG1 to evade host defensive mechanisms.[29] The strategy used by viruses to modulate E3's substrate specificity bears striking similarity to the concept of targeted protein degradation via small molecules, which is the main topic of this chapter.

2. SMALL MOLECULE-DRIVEN PROTEOSTASIS

Synthesis and degradation of proteins maintain cellular metabolic integrity and proliferation. The proteasome, a multimeric protease complex, plays a key role in cellular protein regulation by degrading proteins, thus regulating many pathways.[30,31] The ubiquitin–proteasome system was initially discovered by studying the degradation of denatured globin in reticulocyte lysates.[32] Two seminal papers showed that proteins were targeted for degradation in an ATP-dependent manner by conjugation of multiple molecules later identified as ubiquitins.[33,34] Ubiquitination is reversed by ~100 deubiquitinating enzymes (DUBs) that are proteases consisting of five subfamilies: ubiquitin-specific proteases (USPs), ubiquitin carboxyl-terminal hydrolases (UCH), ovarian tumor-like (OTUs), Machado–Joseph disease (MJD) protein domain proteases, and the JAMM (qJAB1/MPN/MOV34) metalloprotease family.[35]

The critical role of the ubiquitin–proteasome system in cellular physiology and its dysregulation in many diseases including cancer, neurodegenerative diseases, infections, metabolic disorders, and inflammation, has resulted in diverse efforts in targeting different components of this pathway. Malignant cells are more dependent upon the proteasome to remove misfolded or damaged proteins due to their genetic instability and rapid proliferation. Accumulation of ubiquitinated proteins and/or misfolded proteins can lead to apoptosis. Proteins involved in cell division, proliferation, and apoptosis are substrates for the ubiquitin–proteasome pathway as are misfolded and mutated proteins. Proteasomes are abundant and are located in the cytoplasm and nuclei and of cells. The proteasome refers to the 26S proteasome and includes a 20S core catalytic complex with 19S regulatory subunits on each end. Proteasomes enter the nucleus through pores or postmitosis during development of the nuclear envelope. Proteasomal degradation is enabled by conjugation of multiple ubiquitin units to a protein substrate. Ubiquitin is a 76 amino acid protein that is covalently linked to proteins by an enzymatic reaction requiring Ub activating (E1), Ub conjugating (E2), and Ub ligating (E3) enzymes acting sequentially. At least four ubiquitin units must be linked to the target protein for proteasome degradation.[36] The S19 regulatory particle guides ubiquitinated proteins into the 20S particle. The 20S core complex houses the proteolytic chamber. The 20S complex includes four ring structures and three proteolytic activities in the β rings, specifically caspase-like (C-L), trypsin-like (T-L), and chymotrypsin-like (CT-L) catalytic proteases. This process results in 3–22 amino acid peptides and free ubiquitin protein units after proteolysis.

2.1 Proteasome Inhibitors

Given the key role of the ubiquitin–proteasome system in cellular protein homeostasis regulation, and the fact that malignant cells are more dependent upon the proteasome to remove misfolded or damaged proteins due to their genetic instability and rapid proliferation, many groups investigated synthetic and natural products as modulators of proteasome function.[37,38] Since the early 1990s, after the seminal work from Etlinger and Goldberg,[39] proteasome inhibitors that have been investigated include: lactacystin, a streptomyces metabolite, which is metabolized to lactacystin ß-lactone, the active proteasome inhibitor[38,40]; peptide aldehydes, such as carbobenzoxyl–leucinyl–leucinyl–leucinal-H (MG-132); and boronic acid peptides.[38,41,42] Orally active dipeptide boronic acid derivatives, potent proteasome inhibitors, have also been identified.[9,43] It was empirically found that while boronic acid peptides were generally serine proteases inhibitors,[44,45] dipeptide boronates had a high degree of selectivity for the proteasome and were not potent inhibitors of many common proteases. Bortezomib (Fig. 2), a selective inhibitor of the chymotryptic activity of the 20S proteasome showed a potent cytotoxic effect first in MCF-7 human breast carcinoma line and then in a series of liquid and solid tumor lines when tested in the NCI-60 panel. Given the broad toxicity mechanism of proteasome inhibition, the goal of finding the right

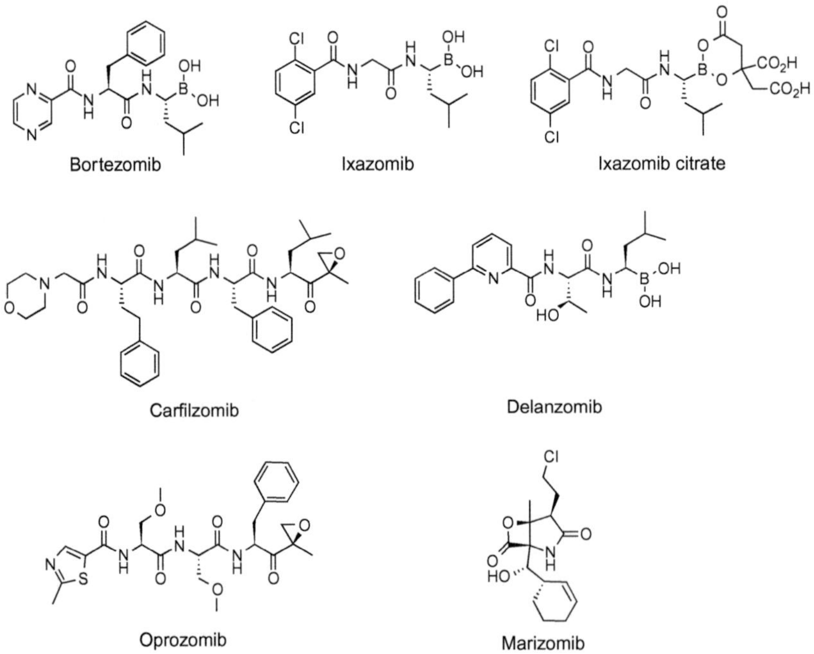

Fig. 2 Small molecule proteasome inhibitors.

context to exploit the therapeutic opportunity became paramount. NF-κB was recognized as a protumor cell survival protein in multiple myeloma, and its activity was shown to be increased in multiple myeloma cells vs normal hematopoietic cells. It was shown later that bortezomib blocks nuclear translocation of NF-κB thus providing the mechanistic rationale for clinical exploration.[46] In 2003, bortezomib was approved by the US FDA for the treatment of refractory multiple myeloma. Several other agents were developed over the years to explore the therapeutic potential of proteasome inhibition (Fig. 2). Carfilzomib, more selective for the chymotrypsin-like activity of the proteasome, is an irreversible inhibitor.[47] Marizomib, a naturally occurring γ-lactam-β-lactone is another covalent inhibitor and inhibits all three proteasome activities.[48] Ixazomib is a more recent orally bioactive inhibitor.[49] Oprozomib, a novel proteasome irreversible inhibitor, is a tripeptide epoxyketone and can be described as an orally bioavailable analog of carfilzomib.[50] Delanzomib is a further example of orally bioavailable inhibitor of the chymotrypsin-like activity that decreases NF-κB activity.

2.2 Molecular Glues

For decades, thalidomide has been highlighted as a typical example of risk associated with the old drug discovery paradigm of phenotypic driven drug discovery. In fact, it was withdrawn from the market after reports of birth deformities in infants born to women that took the antimorning sickness compound while pregnant. In the meanwhile, this—in many ways—"wonder drug" has been found to be effective in leprosy and multiple myeloma due to its antiangiogenic, immunomodulatory, and antiinflammatory properties.[31,51–54]

Only in 2010, was thalidomide's molecular target identified by seminal work in Handa's group showing that the drug binds to CRBN and inhibits its ubiquitination.[55] Growing interest in the immunomodulatory effect of thalidomide lead to the discovery of other more potent and selective analogs (lenalidomide, pomalidomide, Fig. 3) and to the broad definition of this class as immunomodulatory imide drugs (IMiD's).

The field had to wait until 2014, when studies from the Ebert, Kaelin, and Chopra groups showed that IMiD's binding to CRBN has an effect on the E3 ligase substrate preference which results in recruitment, ubiquinination, and degradation of the CRBN neosubstrate transcription factors Ikaros (IKZF1) and Aiolos (IKZF3) (Fig. 4).[56–58]

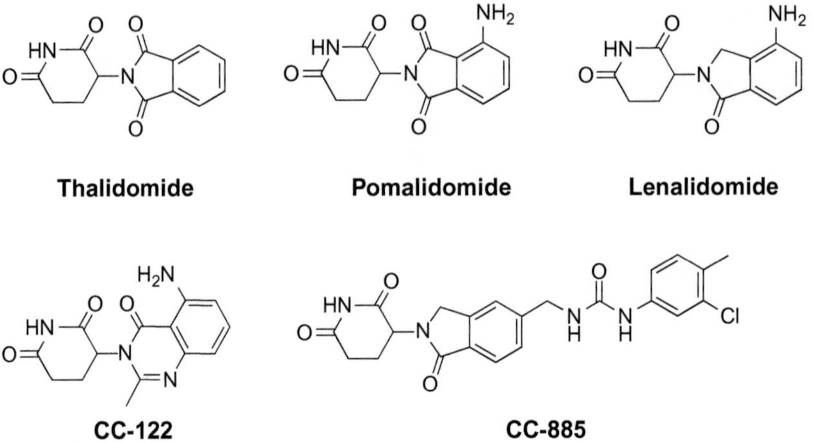

Fig. 3 Selected IMiDs.

It was also shown that proteasomal degradation of Ikaros and Aiolos resulted in downregulation of c-MYC and decreases in interferon regulatory factor 4 expression. In addition, degradation of Aiolos and Ikaros in T-cells[27] is thought to be responsible for the activation of the immune system seen in patients receiving IMiD compounds. More recently, a close thalidomide analog, lenalidomide, was found to recruit CK1α to CRBN and effect its proteasomal degradation. In fact, treatment with lenalidomide resulted in close to complete degradation of CK1α, in cells from patients with myelodysplastic syndrome. Recent reports of crystal structures of DDB1–CRBN complex bound to thalidomide, lenalidomide, and pomalidomide have elucidated differences in CRBN E3 ligase substrate specificity.[59,60] The glutarimide ring of thalidomide and its analogs bind into a hydrophobic tri-tryptophan pocket, termed the thalidomide-binding domain (TBD), which is very evolutionarily conserved. The phthalimide ring is exposed on the surface of the CRBN protein and alters the surface of the E3 ligase substrate receptor to enable interaction with new substrates. In a 2.45 Å crystal structure of DDB1–CRBN bound to lenalidomide and CK1α,[61] CRBN, and lenalidomide jointly provide the binding interface for a CK1α β-hairpin loop located in the kinase N-lobe. The presence of Gly40 in the β-hairpin loop is important for mediating the degradation of target substrates, implying that this residue may form part of the protein degradation motif (degron) recognized by the IMiD–CRBN complex.

After the identification of neosubstrates, two endogenous substrates were identified: MEIS2 and glutamine synthetase (GS).[62] Interestingly, while MEIS

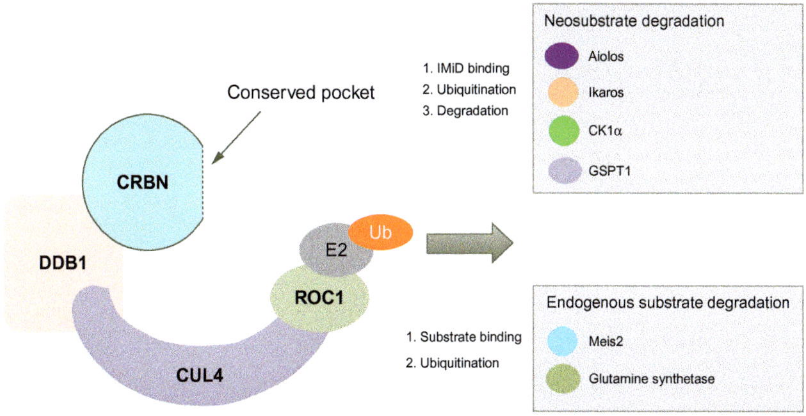

Fig. 4 Molecular glues for Cereblon (CRBN).

ubiquitination is inhibited by IMiDs, resulting in increased protein level, GS degradation is increased by IMiD–CRBN complex. Recently, two more thalidomide analogs were identified and characterized: CC-122[63] and CC-885[64] (Fig. 4). CC-122 is a pleiotropic pathway modifier that also binds CRBN and promotes degradation of Aiolos and Ikaros in diffuse large B-cell lymphoma (DLBCL) T-cells in in vitro and in vivo models and in patients, resulting in both cell autonomous and immune-stimulatory effects. The compound is entering phase II/III studies. In addition to inducing CRBN-mediated degradation of Ikaros as seen with other thalidomide analogs, CC-885 promotes the degradation of the translation termination factor GSPT1, resulting in cytotoxicity. Lenalidomide and pomalidomide do not degrade GSPT1, possibly due to their lack of the extended urea moiety of CC-885 that enables additional interactions with CRBN and GSPT1.

Very recently, two reports[65,66] uncovered, that a known selective anticancer agent, indisulam (Fig. 5) and its analogs, act as a "molecular glues" to recruit a neosubstrate RBM39 to a different DCAF protein: CUL4-DCAF15 (an E3 ligase in the same class as CRBN). Interestingly, two different groups reached the same conclusion using orthogonal target identification approaches: proteomics vs mutational analysis in resistant lines.

The opportunity to use small molecules to recruit neosubstrates to E3 ligases resulting in substrate degradation is clearly an exciting and expanding field. The ability to explore diverse target classes without restriction of inhibition/occupancy-driven pharmacology could have a transformational effect on modern drug discovery. Clearly the field is still driven by a characterization of pharmacologically interesting phenotypes to deconvolute

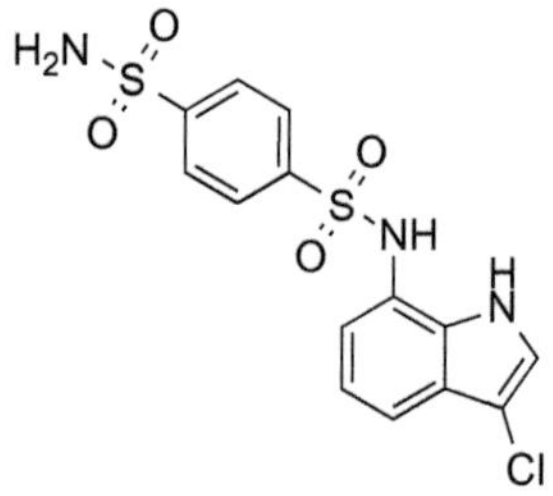

Fig. 5 Indisulam.

mechanism of action. Rational design of small molecules degrons that have the ability to recruit substrates to E3 ligases is the obvious next step and the goal of many academic labs and small and large drug discovery companies.

2.3 DUBs Inhibitors

There are ~100 DUBs known in the human genome playing a key role in the ubiquitin–proteasome system as well as other biological processes.[67] DUBs are responsible for: (a) liberation of ubiquitin from protein substrates (e.g., to remove degradation signal), (b) editing of polyubiquitin signal on protein substrates to change the fate of the protein,[68] (c) disassembling polyubiquitin chains to free up ubiquitin monomers, and (d) cleaving ubiquitin precursors or adducts to regenerate active ubiquitin.[69] Of the DUBs classes mentioned in Section 1, UCHs and USPs are the best characterized; in fact USPs represent more than half of the known human DUBs.[70] In the past 10 years, increased biological understanding has led to numerous DUBs being implicated in various diseases spanning oncology, neurodegeneration, hematology, and infectious diseases.[71,72] Because of the key role of this class of enzymes in maintaining protein homeostasis, DUB antagonists are being "hunted" in both academic institutions and biotech/pharmaceutical companies alike. This is illustrated through chemically diverse small molecules that have been reported to inhibit one or more of the UCH and USP family members. Some representative examples will be described here while a more comprehensive review can be found elsewhere.[68]

USP7 is one of the DUBs that have received most attention as it has been found to be an indirect regulator of the tumor suppressor p53.[73] Several inhibitors of USP7 have been reported, of which HBX 41108 (Fig. 6) shows an IC_{50} of 0.42 μM with an uncompetitive mechanism and high selectivity against other DUBs.[74]

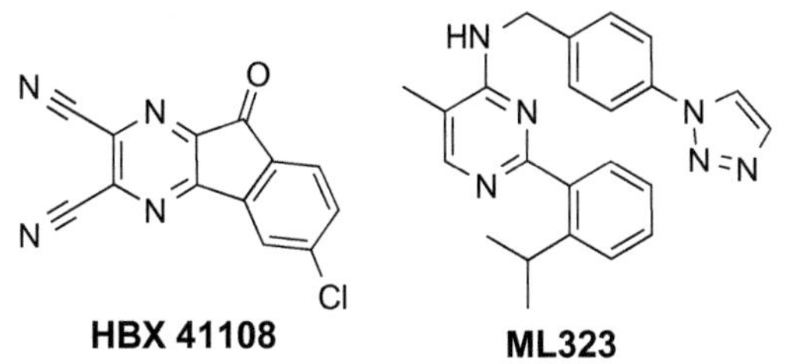

Fig. 6 USP1 inhibitors.

USP1 is one of the deubiquitinating enzyme implicated in the DNA damage response (DDR) mechanism, and for this reason it is gaining more attention from the biotech and the pharma industry. ML323, reported in 2014,[75] was found to have excellent potency (IC_{50}=76 nM) against USP1/UAF1 complex.

Because of the conserved nature of the catalytic triad (cysteine, histidine, and aspartic acid), questions about ability to develop specific inhibitors is still relevant, although recent advances in the understanding of UPS biology as well as structural biology of this class is providing a promising path forward for drug discovery efforts in this area.

2.4 Receptor Degraders

Nuclear receptors, and especially the estrogen receptor (ER), have provided a therapeutically viable avenue to explore the concept of small molecule (proteasome-dependent) degradation. ERα signaling in breast cancer has been well characterized since the early days of selective ER modulators (SERM) such as tamoxifen.[76,77] In the 1990s, a high affinity competitive antagonist of ER that also targets the receptor for proteasome-dependent degradation was identified: fulvestrant (Fig. 7). Because of its unique pharmacological profile, fulvestrant is a first-in-class selective ER downregulator (SERD). Unfortunately, fulvestrant has significant pharmaceutical liabilities (requiring intramuscular injection) that have negatively impacted its widespread use.[78] This has led to development of a second generation or SERDs of which GDC-0810 is probably the most advanced in clinical investigation.[79]

Early models describing the pharmacology of ER were quite simple. In the absence of hormone, the receptor was maintained in an inactive state, agonist binding would induce a conformational change in the receptor resulting in its dimerization, nuclear translocation, and subsequent

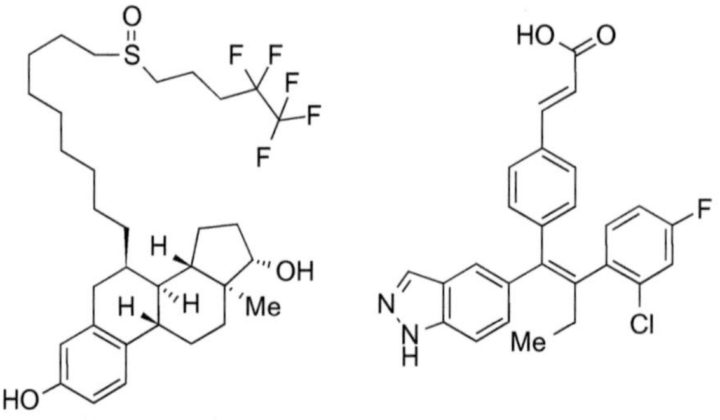

Fig. 7 Selective estrogen receptor degraders (SERDs).

interaction with specific DNA sequences within the regulatory regions of target genes. However, in studies first performed in rodents and subsequently in humans, it became apparent that depending on the target organ, tamoxifen could function either as an antagonist or as an agonist.[80] This observation led to the reclassification of tamoxifen, and most other early "ER antagonists," as SERMs, reflecting their tissue selective agonist/antagonist properties. Subsequent insight into the mechanism of ER biology lead to the generation of three basic principles: (1) the overall shape of ER is influenced by the nature of the ligand to which it is bound; (2) receptor conformation influences the presentation of PPI surfaces that allow it to interact with either positive or negative acting "coregulators"; and (3) the functional activity of coregulators differs between cells. With deeper structural and mechanistic understanding of the SERD class of ER antagonist, such as crystallographic analysis of the ERα-GW7604 (similar in structure to GDC-0810) complex, it was quite clear that the carboxylic acid group in this compound was the key "warhead" and that its direct interaction with the peptide backbone of the receptor induced a conformational change that exposed a hydrophobic surface on the receptor that targeted it for degradation.[81]

More importantly, in the broader drug discovery field, this novel concept provided an opportunity to evaluate its applicability to other member within the nuclear receptor class or even other classes or targets. Reports of small molecule androgen receptor degraders (SARDs) have started to appear in the literature fueling the excitement in the field.

2.5 Hydrophobic Tagging-Mediated Degradation

Building on the SERDs and their ability to destabilize regions of the protein to trigger the unfolded protein response (UPR), a new degradation approach has evolved that uses bulky hydrophobic regions added to a small molecule ligand that either causes unfolding intentionally, or simply mimics unfolded protein with its hydrophobic nature. This has become known as hydrophobic tagging (HyT), but can be traced back to fulvestrant[82] with a long fluoroalkyl–sulfinyl chain attached to the steroid oestradiol.

Targeted examples of hydrophobic tagging started with bulky groups such as adamantyl and biphenyl groups tethered to a chloroalkane group that is covalently attached to the halotag fusion protein and triggers degradation of the fused target protein.[83] These early tool compounds were able to degrade target proteins at submicromolar levels over 24 h and are selective for only the halotag fused protein expressed in the cellular system. This system showed temporally controlled degradation of the target protein with the addition of a HyT molecule. An interesting observation from the HyT–halotag system is that longer chains spacing the hydrophobic groups reduce the maximum degradation achieved, while addition of branching groups at the hydrophobic region increases degradation.[84] This idea was extended to a selective covalent tyrosine kinase inhibitor for the kinase Her3: HyT TX2-121-1 (Fig. 8).[85] This compound contained a cysteine targeting acrylamide on one end of the molecule, and a tethered adamantane on the other end; loss of either of these groups reduces its biological activity and ablated degradation of Her3. These TKI-HyTs were able to induce antiproliferative effects in Her3-dependent cells at around 1 μM levels, despite Her3 having little kinase activity and functioning more as a scaffolding pseudokinase.

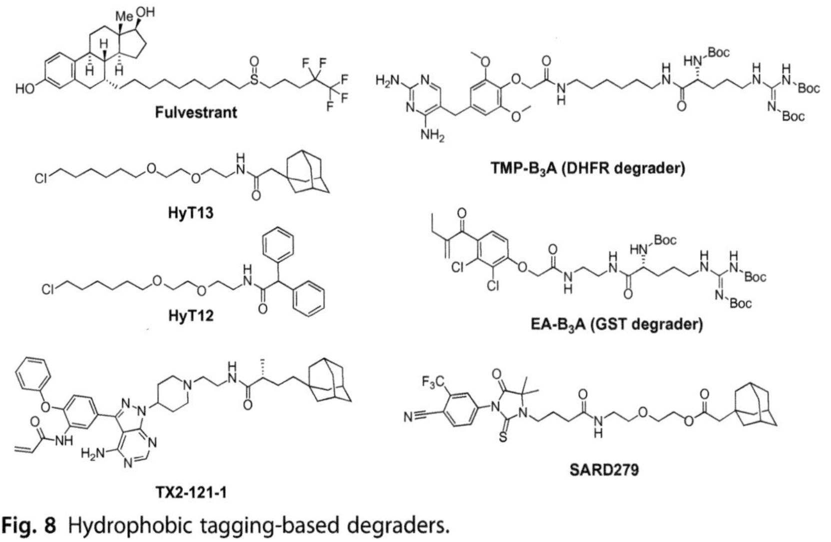

Fig. 8 Hydrophobic tagging-based degraders.

This shows the promise of using hydrophobic tagging over typical enzyme functional inhibitors for future therapies aimed at "undruggable" targets.

In addition to adamantane groups, tri-Boc arginine (Boc_3-Arg) has been shown to induce significant degradative effects.[86] Boc_3-Arg was tagged to covalent ligands for glutathione *S*-transferase (GST) and dihydrofolate reductase (DHFR), however, unlike for adamantyl tagged HyTs, Boc_3-Arg did not cause ubiquitination of the target protein through HSP70 or HSP90, nor does it destabilized the protein.[87] Rather, it appears to direct the target protein to the 20S proteasome directly.

An elegant proof of concept of hydrophobic tagging was represented by the design of a SARD built with a noncovalent androgen receptor (AR) agonist attached to an adamantyl group.[88] These SARDs showed degradation at micromolar levels but no longer needed the covalent functionality to enable the ubiquitination effect. In AR-dependent prostate cancer cells, these HyTs showed significant antiproliferative effects, even in the presence of high levels of natural substrate which is known to out compete the effects of the selective androgen receptor modulator (SARM) enzalutamide. This showed that the transiently formed ternary complex is capable of inducing ubiquitination and degradation whereas the antagonist SARM requires receptor occupancy to drive its biological effect. These early examples show reasons to believe that hydrophobic tagging can become a useful tool for targeted degradation, but the physicochemical properties associated with a bulky hydrophobic molecule as well as their limited degradation kinetics may limit their potential applications.

3. HETEROBIFUNCTIONAL DEGRADERS: AN EXCITING NEW THERAPEUTIC MODALITY

The ability to potentially direct the UPS to target and degrade any protein of interest with a small molecule has become possible by combining approaches previously described here. A modality that allows the transformational ability to target the remaining 80% of the proteome (noncatalytic in nature) that is not presently accessible via traditional, occupancy-based modulators, and unlock novel and impactful biology is now within grasp. These heterobifunctional molecules that have been called PROTACs (PROteolysis TArgeting Chimeras), SNIPERs (specific and nongenetic IAP-dependent protein erasers), or Degronimids have recently grown into an exciting and potentially new therapeutic modality in both academia and industry.[89]

3.1 Mechanism of Action

Heterobifunctional small molecule directed protein degradation is a chemical knockdown strategy where a heterobifunctional molecule binds to a target protein and an E3 ligase (concurrently) generating a key ternary complex. Spatial proximity facilitated by the ternary complex trigger ubiquitin transfer from the E3 ligase complex to a surface lysine (or likely a population of surface lysines) of the target protein. Subsequent polyubiquitination leads to proteasomal recognition of the tagged protein and to its degradation (Fig. 9). Unlike typical small molecule–protein inhibitory processes that are controlled purely by occupancy for both kinetics of protein inhibition and duration of PD/efficacy, a heterobifunctional degrader can allow irreversible and catalytic protein degradation. In fact, upon protein degradation the heterobifunctional degrader remains and is able to degrade more of the target protein. Concentrations needed for degradation of a target protein are in fact often much lower than the binding affinity toward the protein target.

3.2 History and Early Heterobifunctional Degraders

The first heterobifunctional degraders appeared in 2001, with ovalicin, a small molecule ligand of methionine aminopeptidase 2 (MetAP2), tethered through an alkyl chain to phosphopeptide that bound to the E3 ligase complex $SCF^{\beta\text{-}TrCP}$ (Fig. 10).[90] This early compound, coined PROTAC, was able to visibly degrade MetAP2 in *Xenopus* egg extracts, albeit at high doses (50 μM), showing promise as a new mechanism of action. Follow-up molecules utilized the same approach to recruit SCF to degrade both AR and

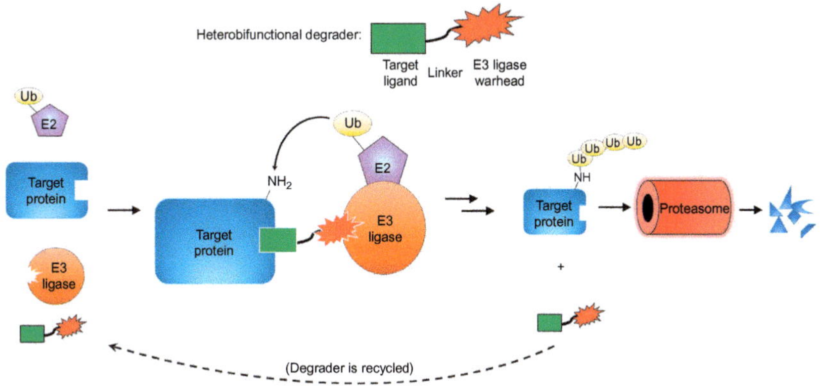

Fig. 9 Heterobifunctional molecule-driven degradation.

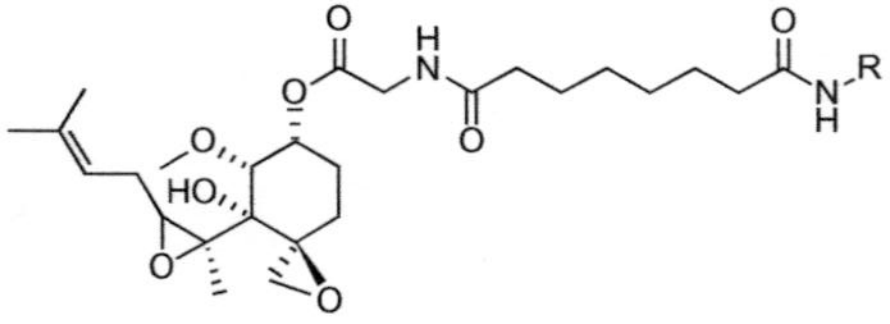

Ovalicin-SCF (R = GGGGGGDRHDS*GLDS*M)

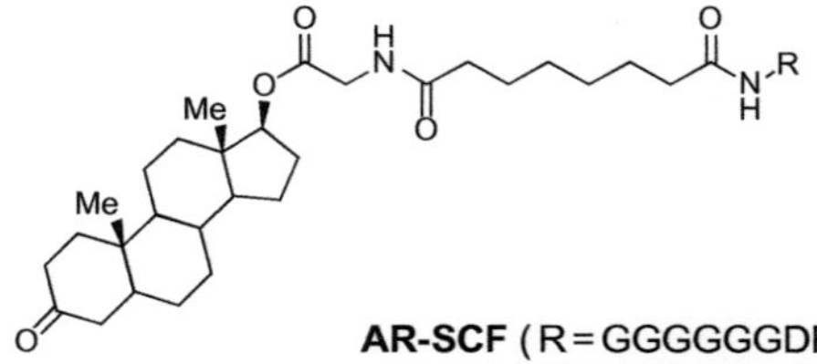

AR-SCF (R = GGGGGGDRHDS*GLDS*M)
AR-VHL (R = ALAHypYIP-(D-R)$_8$)

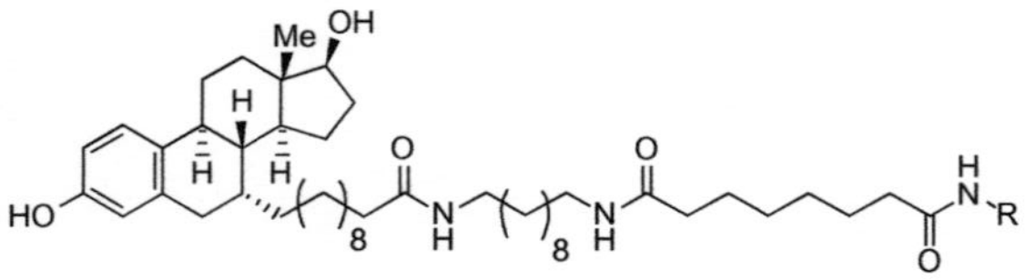

ER-SCF (R = GGGGGGDRHDS*GLDS*M)

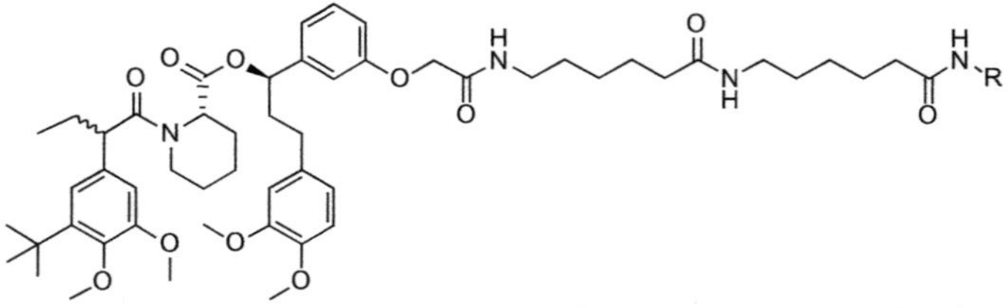

FKBP12-VHL (R = ALAHypYIP-(D-R)$_8$)

Fig. 10 Peptide-based PROTAC examples.

ER with appreciable success at 10 μM.[91] Because of the peptidic nature of early E3 ligase ligands, early heterobifunctional degraders suffered from poor cellular permeability, therefore, permeabilizing groups such as poly-arginine were added. Such strategy was exemplified utilizing peptide-based ligands for the E3 ligase Von Hippel–Lindau (VHL) to degrade target proteins such as FKBP12 and AR in cells at ~25 μM.[92]

While successfully validating the concept of the technology, peptidic degraders represented a big hurdle toward advancing this promising technology toward a novel therapeutic modality. Despite the abundance of E3 ligases known (~600), very few small molecule ligands have been identified over the years. In fact the first small molecule (nonpeptidic) degrader was synthesized in 2008 and combined an AR ligand tethered to a nutlin (Fig. 11), a ligand for the E3 ligase MDM2 that inhibits its interaction with its substrate p53.[93] This nonpeptidic degrader was able to markedly degrade AR in HeLa cells at a concentration of 10 μM. More recently, all small molecule-based degraders were reported using the ligand bestatin to recruit the E3 ligase cellular inhibitor of apoptosis protein (cIAP1) and led to a series of degraders known as SNIPERs. Various SNIPERs have been able to effectively degrade a range of targets such as cellular retinoic acid-binding proteins (CRABP),[94] hormone receptors: retinoic acid receptor (RAR), AR, and ER,[95] ERRα,[96] and transforming acidic coiled-coil-3 (TACC3) (Fig. 11).[97] These SNIPERs began to demonstrate significant degradation at single digit micromolar concentrations and even demonstrated in vivo target degradation[98] and cell death in MCF-7 cells through generation of reactive oxygen species, showing the promise of a heterobifunctional degrader as an anticancer therapeutic. While exciting degradation levels were reached, the promiscuity of bestatin and its limitation to micromolar efficacy required new E3 ligands to bring the field to the forefront of therapeutic options for degradation.

3.3 Recent Advances

In 2015, the directed protein degradation field took a significant leap forward with the concurrent advancement of chemistry toward two new E3 ligases: Von Hippel–Lindau (VHL), with the development of a peptidomimetic small molecule ligand,[99,100] and cereblon (CRBN), thanks to the identification of thalidomide and its analogs (IMiDs) as binders.[55] The first examples of a degrader with nanomolar potency utilized this VHL ligand tethered to either BET family bromodomain inhibitor(+)-JQ1 (MZ1),[101] an ERRα

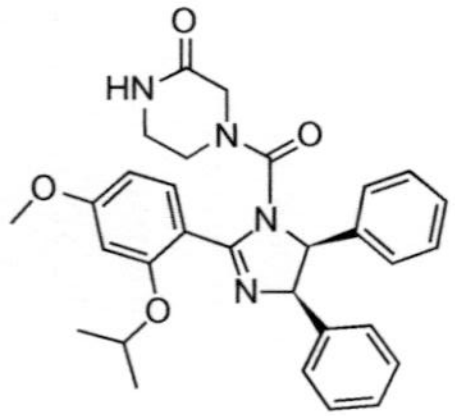

Nutlin-3, targets MDM2

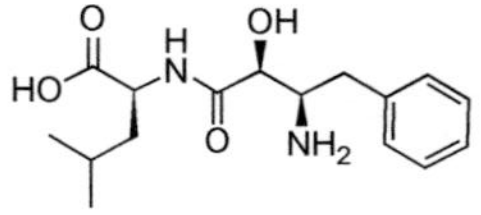

Bestatin, targets cIAP1

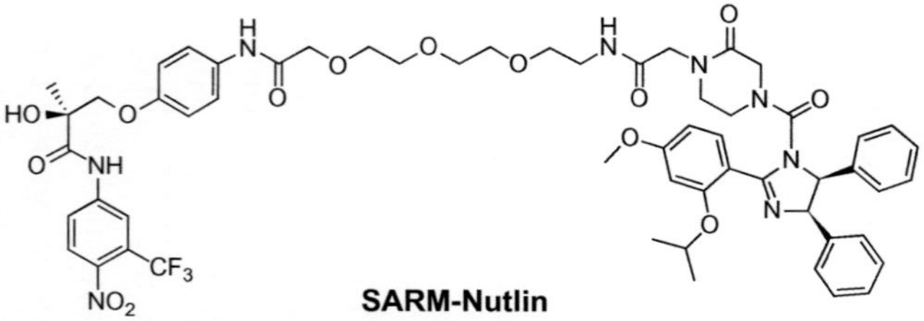

SARM-Nutlin

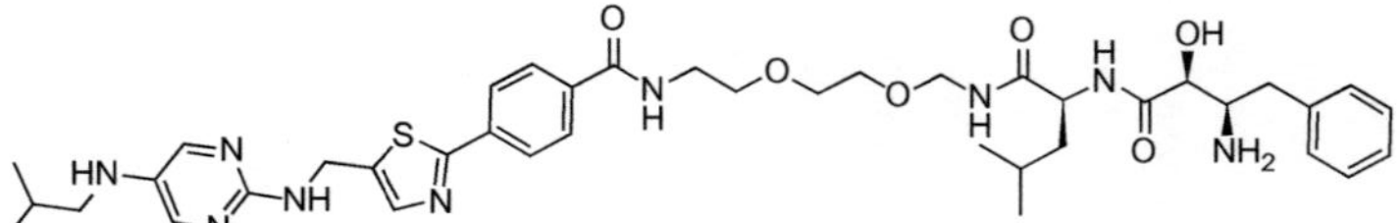

TACC3-SNIPER

Fig. 11 Early small molecule-based heterobifunctional degraders.

ligand, or a kinase inhibitor for RIPK2 (see Fig. 13).[102] Each PROTAC was able to degrade its targeted protein up to 90% within 4–8 h, with a DC_{50} (concentration for 50% degradation effect) of 1.4 nM for the RIPK2 compound. Degradation of target ERRα was also demonstrated in vivo by treating mice with a PROTAC (100 mg/kg, 3 × daily IP) and measuring protein levels in harvested tissue samples. Elegant means to control for the experiment to function through a ternary complex-driven degradation can be: (1) utilize an inactive PROTAC containing the negative epimer of the VHL ligand, preventing recruitment of the E3 ligase complex and (2) the active degrader is added to cells pretreated with a proteasome inhibitor and/or with an excess of E3 ligase ligand, that effectively either shuts down degradation for the former, or saturates the E3 ligase site practically preventing ternary complex formation for the latter (see Fig. 12). Building on this hypothesis, Bondeson et al. were able to demonstrate also that the degraders were catalytic by measuring the stoichiometry of ubiquitinated RIPK2 relative to the degrader concentration in vitro, indicating a ubiquitination ratio of ∼3.4 in this case. Another interesting feature is the observed "Hook effect" when using high levels of potent heterobifunctional molecule; in fact self-competition occurs and a bell shape curve appears in the degradation assay due to saturation of each component of the ternary complex. Finally, Bondeson et al. also demonstrated the selective nature of degradation via proteomics due to the specific arrangement of the ternary complex in transferring Ub. Meanwhile, Zengerle et al. showed that

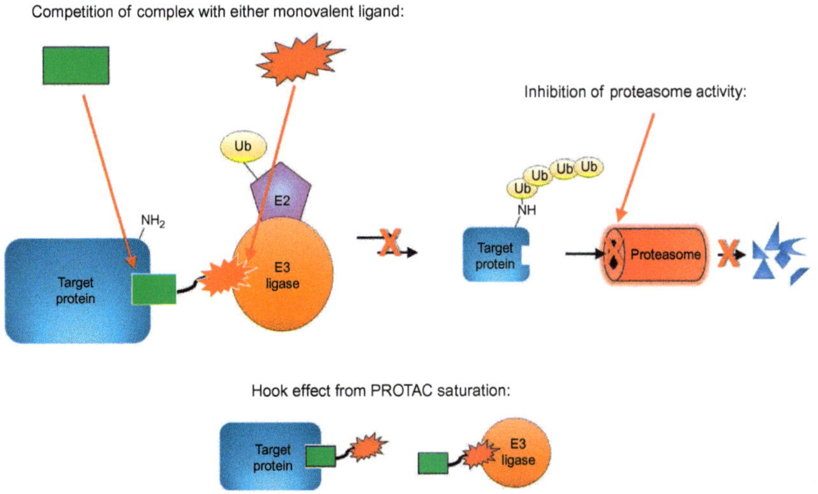

Fig. 12 Mechanistic features for blocking target degradation.

different linkers give different degradation patterns among the BET bromodomains pointing to an impact of linker chemistry to the selectivity of ubiquitin transfer.[101] In fact, ternary complex-driven selectivity, i.e., driven by PPI and or lysine targeting (different from small molecule occupancy-based selectivity) is a clearly an area of high interest in the field with some early promise.

Immediately following this work, two reports were published near simultaneously using thalidomide as an E3 warhead to recruit CRBN to degrade BET family bromodomains with the small molecules (+)-JQ1 and OTX-015.[103,104] Both degraders dBET1 and ARV-825 showed significant degradation of BRD4, with ARV-825 showing a DC_{50} at sub-nanomolar concentrations. ARV-825 shows that potent cellular effects that can be achieved by catalytically degrading the protein rather than simple binding-induced inhibition; with degradation occurring at 2 log units lower concentration than the binding affinity of JQ1 alone (IC_{50} = 33–66 nM in alpha screen assay for both bromodomains of BRD4).[105] dBET1 showed in vivo efficacy in mice, attenuating tumor volume over 14 days of treatment without observed adverse effects (50 mg/kg daily, IP) and showing higher levels of induced apoptosis compared to parent JQ1. The selectivity of dBET1 was also confirmed by proteomics, showing markedly decreased levels of only the BET family bromodomains along with MYC, whose transcription is regulated by BRD4 and is also downregulated by JQ1 treatment alone.

The selectivity of PROTAC molecules was further investigated in a report on various linkers attached to different kinase inhibitors, with both VHL and thalidomide E3 ligands.[106] Using tyrosine kinase inhibitors imatinib, bosutinib, and dasatinib in combination with linkers of varying length and hydrophobicity, different degradation patterns were observed between kinases BCL-ABL and c-ABL. This demonstrated again the importance of all three components: the E3 ligase ligand, the target protein ligand, and the linker in the formation of a ternary complex specific to each protein target. Recently, an X-ray crystal structure was solved of the ternary complex of the BRD4-PROTAC MZ1 bound simultaneously to both the E3 ligase (VHL) and target protein.[107] This structure showed that the linker can play a role driving cooperativity through PPI, which was confirmed with site mutagenesis and proximity assays. Furthermore, the structure informed design of a more selective linker site on the VHL ligand, giving rise to degrader AT1 which shows unique selectivity of BRD4 degradation vs other BET family bromodomains by proteomics. Further optimization of

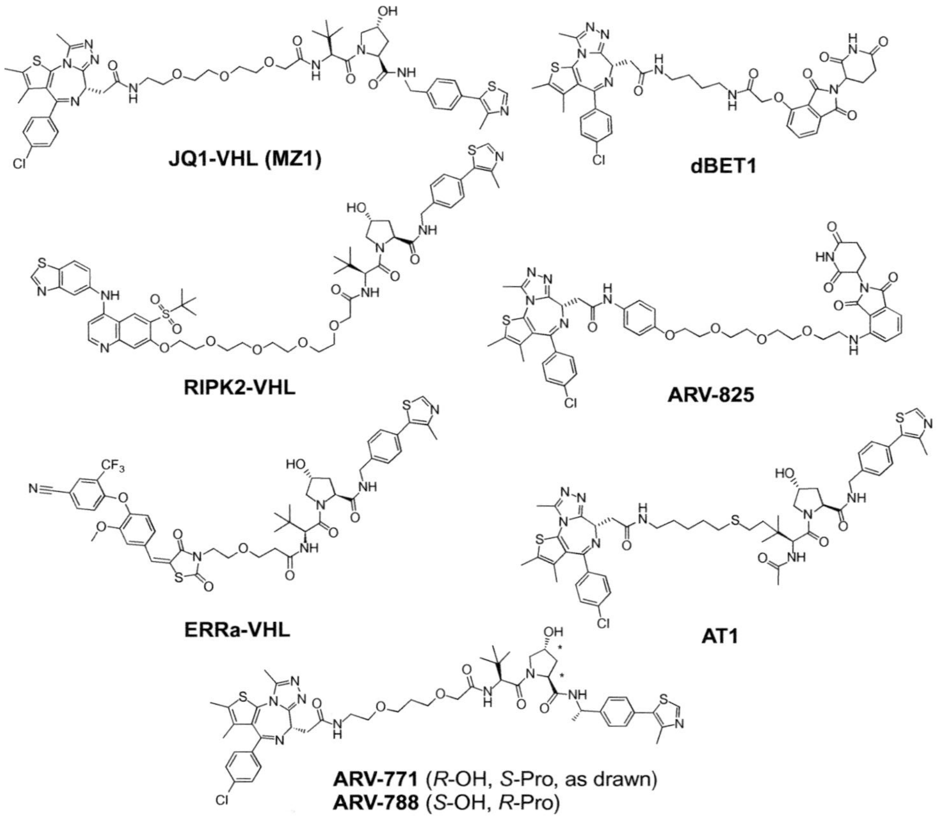

Fig. 13 Heterobifunctional degraders with nanomolar potency.

the JQ1-VHL degrader MZ1, led to the identification of ARV-771, potent BRD4 degrader (DC_{50} of <1 nM) with reasonable in vivo properties.[108] In fact this degrader had significant antiproliferative effects in various prostate cancer cell lines over parent BRD4 inhibitors and induced tumor regression in 22Rv1 tumor bearing mice when dosed at only 30 mg/kg daily SC. This effect was not seen with BRD4 inhibitor treatment alone, nor by treatment with the inactive enantiomer control ARV-766, showing again the substantial benefit that can be achieved with a protein degradation vs occupancy-based inhibition pharmacology (Fig. 13).

The search for novel E3 ligase binders to use in this technology has been hampered by the lack of small molecule ligands reported, and by the general low tractability of E3 ligases. However, while the industry as a whole has historically searched for E3 modulators (with high degree of difficulty in developing good assays), the heterobifunctional degradation field is instead searching for E3 ligase binders. Advent of new technologies, such as DNA-encoded libraries, fragment screens, etc., provides good affinity-based screening platforms that could potentially advance hit finding campaigns. Recently, a new use of IAP ligands has been reported by GlaxoSmithKline, showing these can also functions as degraders at submicromolar levels when targeting Bruton's tyrosine kinase (BTK), ER, and RIPK2 (Fig. 14).[109]

The application of PROTACs has also extended to new targets recently, with potent PROTACs reported for BRD9,[110] TANK binding kinase (TBK1),[111] and Sirtuin 2.[112] dBRD9 is an interesting example; it is able to highlight a previously unreported phenotype of IMiD's-based degraders. While the compound is extremely selective for BRD9 degradation, the inactive control with a BRD9-dead ligand attached still showed antiproliferative effects in multiple myeloma cells. This is believed to be due to the inherent activity of the pomalidomide warhead degrading Icarus family zinc finger (IKZF) proteins over the course of the experiment (24 h). A BRD4 PROTAC with a different BRD4 ligand (RX37) was also reported with a remarkably short alkyl linker to lenalidomide with picomolar degradation of BRD4.[113]

4. EMPIRICAL VS RATIONAL DESIGN

The potential impact of this novel therapeutic modality in drug discovery is without a doubt apparent to any scientific observer. A high level comparison with occupancy-based pharmacology of small molecules delivers very powerful points (see Fig. 15).

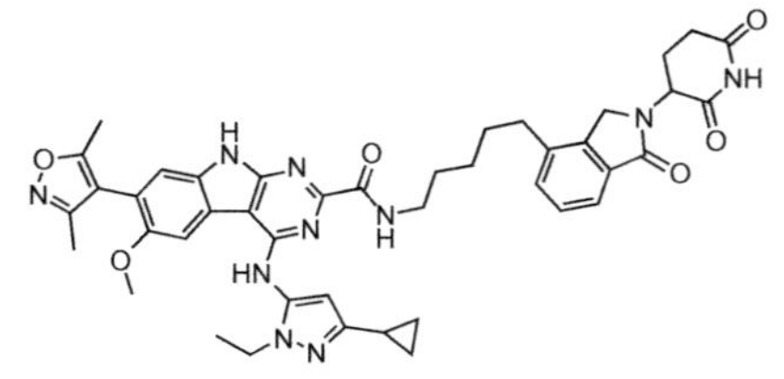

RX37-Lenalidomide

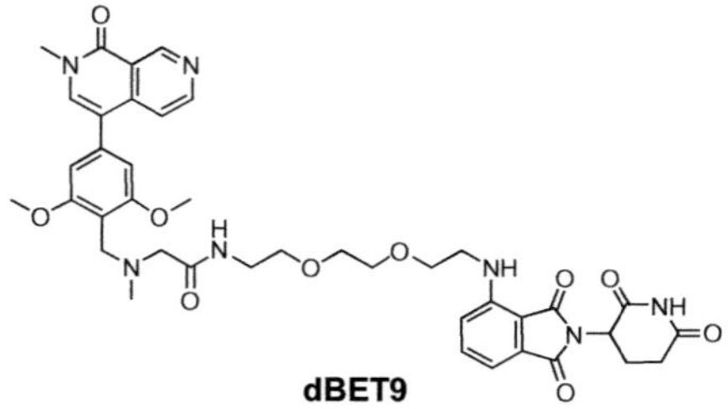

dBET9

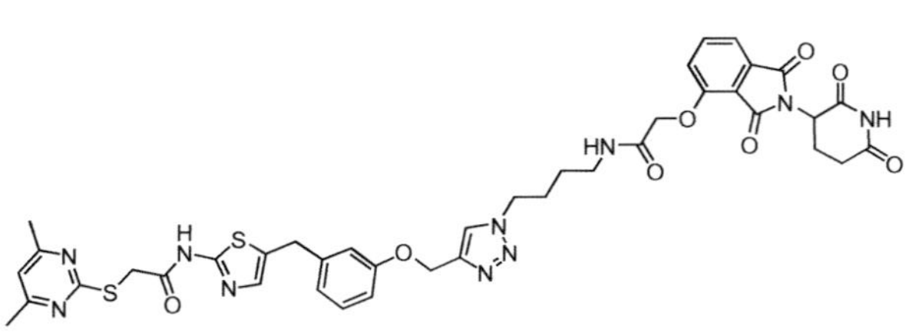

SIRT2-Thalidomide

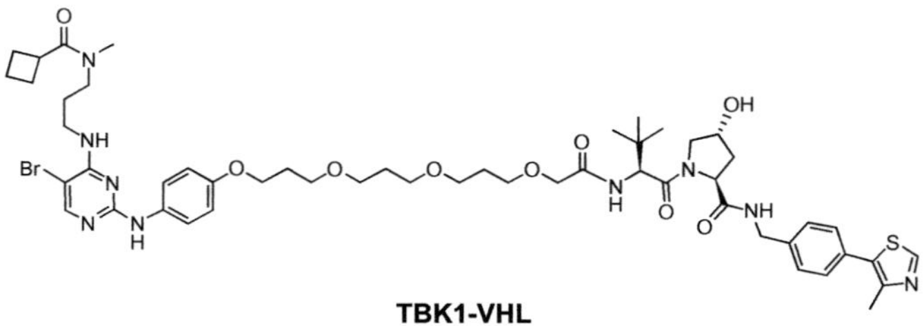

TBK1-VHL

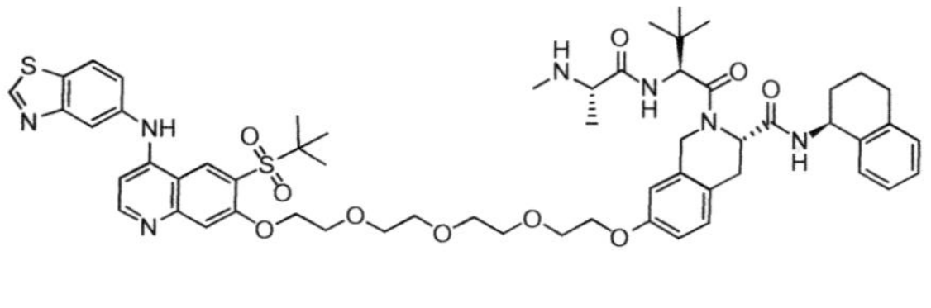

RIPK2-IAP

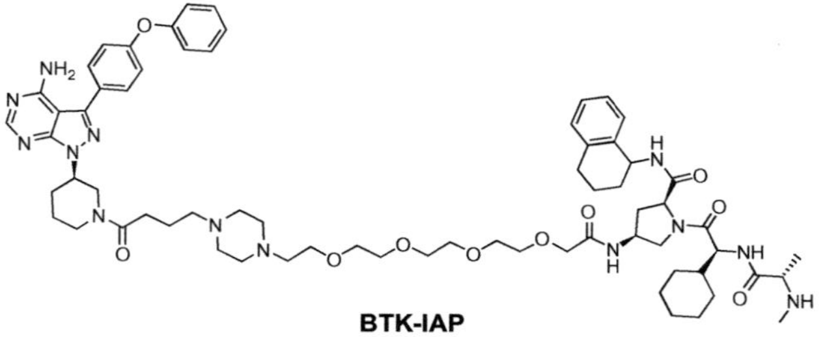

BTK-IAP

Fig. 14 Recently disclosed heterobifunctional degraders.

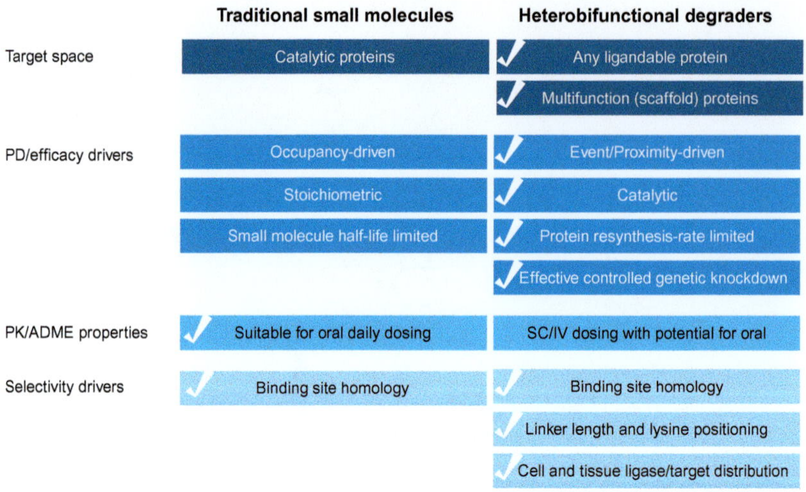

Fig. 15 Comparison of traditional small molecules to heterobifunctional degraders.

In principle, the ability to drug ligandable proteins gives access to a plethora of novel critical biological targets that have been historically untractable: i.e., transcription factors, protein complexes, pseudokinases, etc. The irreversible and catalytic event of degradation allows one to decouple PK and PD in a favorable way: in principle an effective degrader (i.e., >80% degradation in <4 h) would have to be exposed to the target of interest for a limited amount of time (<4) and with lower exposures (catalytic) while PD and efficacy would be driven by the basal protein resynthesis rate of the target protein. The challenges that might arise from these typically larger small molecules (typically in the range of 700–1100 Da) in terms of ADME/DMPK properties can be potentially offset by the PK–PD decoupling. The ability to use E3 ligase expression profiles as a means to achieve tissue or cell population selectivity, in addition to the ternary complex-driven selectivity discussed before, is another exciting feature. Small molecule degraders also hold advantages over other new modalities in nontraditional therapeutic space such as antisense oligonucelotides (ASOs) and RNA interfering molecules (RNAi), which also have a catalytic mechanism of effect and the ability to target "undruggable" targets. These modalities, however, have limiting PK properties and are not currently amenable to oral delivery. They are also currently limited to specific tissues and mainly target the liver or require an additional tissue targeting moiety.[114]

The excitement in this field triggered from seminal papers described above delivering a preclinically validated, potentially transformational, therapeutic modality has sparked interest in academia and industry. Several small biotech companies have been created to advance both the platform as well as exemplifying the technology with new clinical agents: spinoffs from academic institutions such as Arvinas Therapeutics (Yale), C4 Therapeutics (DFCI), or VC founded/backed, such as Kymera Therapeutics. The field has been so far driven by mostly empirical findings while a rational drug discovery approach to understand the key principles that drive this new technology are still evasive.

Designing compounds with good cellular permeability and good ADME properties toward oral-driven efficacy appears to be not guided by traditional "small molecule" principles but rather higher-order factors that the industry is not used to working with (Fig. 16).

While all the published papers have showed efficacy in animal models achieved with IP or SC dosing, Arvinas Therapeutics has recently reported an ERRα degrader that not only shows tumor growth inhibition in mouse xenografts (30 mg/kg, QD, SC), but also has showed in vivo knockdown of

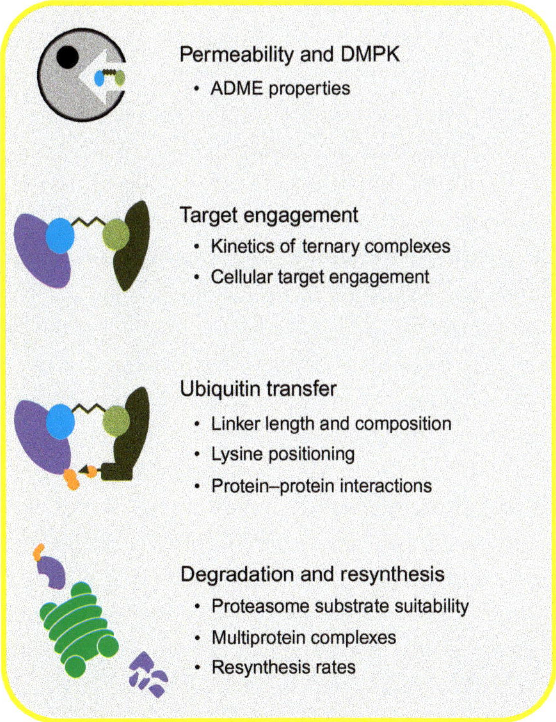

Fig. 16 Caveats of heterobifunctional degradation approach.

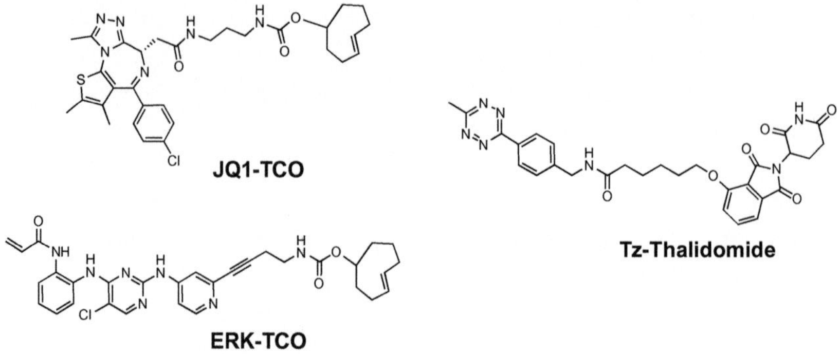

Fig. 17 CLIPTACs for self-assembly within cells.

ERR when orally administered (30 mg/kg TID, PO).[115] Arvinas has also presented in vivo antitumor activity with an orally dosed AR PROTAC.[116]

In an intriguing attempt to circumnavigate the permeability/ADME issues posed by the size of typical degraders, Lebraud et al. were able to assemble a functional degrader within the cell by supplying each ligand end with a matched pair of bioorthogonal click chemical moieties attached.[117] These so-called click-formed proteolysis targeting chimeras (CLIPTACs, Fig. 17) were able to keep the molecular weight of the parent molecules to a reasonable range (~600), that was able to form the full degrader molecule in cells and successfully degrade target proteins BRD4 and ERK1/2 with nanomolar concentrations of either component. Preassembled CLIPTACs were found to be inactive in cells, confirming the assembly occurs within cells. Considering the ability of any unclicked reagent to function as a degrader competitor, these results are still remarkable.

Ternary complex formation is a key rate limiting step toward successful ubiquitination/degradation. Understanding of what drives efficient ternary complex formation and its impact on rate and selectivity of ubiquitination in cellular system would represent a much needed advancement in the field. Early data from Ciulli et al. on key components such as cooperativity and PPI is starting to show promise.[107]

Probably, the most important aspect of this technology that remains elusive in scientific reports is the principle behind what makes a protein target more or less suited for targeted degradation. Advancement in this aspect could lead to more productive, capital efficient, and successful drug discovery campaigns in the new world of heterobifunctional molecules-driven targeted protein degradation.

REFERENCES

1. Powers, E. T.; Morimoto, R. I.; Dillin, A.; Kelly, J. W.; Balch, W. E. *Annu. Rev. Biochem.* **2009**, *78*, 959–991.
2. (a) Balch, W. E.; Morimoto, R. I.; Dillin, A.; Kelly, J. W. *Science* **2008**, *319*, 916–919; (b) Hetz, C.; Glimcher, L. H. *Curr. Opin. Cell Biol.* **2011**, *23*, 123–125.
3. Coux, O.; Tanaka, K.; Goldberg, A. L. *Annu. Rev. Biochem.* **1996**, *65*, 801–847.
4. Bassermann, F.; Eichner, R.; Pagano, M. *Biochim. Biophys. Acta* **2014**, *1843*, 150–162.
5. Chen, Z. J. *Immunol. Rev.* **2012**, *246*, 95–106.
6. Haas, A. L.; Rose, I. A. *J. Biol. Chem.* **1982**, *257*, 10329–10337.
7. Chen, Z.; Pickart, C. M. *J. Biol. Chem.* **1990**, *265*, 21835–21842.
8. Chau, V.; Tobias, J. W.; Bachmair, A.; Marriott, D.; Ecker, D. J.; Gonda, D. K.; Varshavsky, A. *Science* **1989**, *243*, 1576–1583.
9. Mukhopadhyay, D.; Riezman, H. *Science* **2007**, *315*, 201–205.
10. Komander, D. *Biochem. Soc. Trans.* **2009**, *37*, 937–953.
11. Nguyen, V. N.; Huang, K. Y.; Weng, J. T.; Lai, K. R.; Lee, T. Y. *Database (Oxford)* **2016**, baw054. https://doi.org/10.1093/database/baw054 27114492.
12. Huibregtse, J. M.; Scheffner, M.; Beaudenon, S.; Howley, P. M. *Proc. Natl. Acad. Sci. U.S.A.* **1995**, *92*, 2563–2567.
13. Huang, L.; Kinnucan, E.; Wang, G.; Beaudenon, S.; Howley, P. M.; Huibregtse, J. M.; Pavletich, N. P. *Science* **1999**, *286*, 1321–1326.
14. Kamadurai, H. B.; Souphron, J.; Scott, D. C.; Duda, D. M.; Miller, D. J.; Stringer, D.; Piper, R. C.; Schulman, B. A. *Mol. Cell* **2009**, *36*, 1095–1102.
15. Kim, H. C.; Huibregtse, J. M. *Mol. Cell. Biol.* **2009**, *29*, 3307–3318.
16. (a) French, M. E.; Kretzmann, B. R.; Hicke, L. *J. Biol. Chem.* **2009**, *284*, 12071; (b) Maspero, E.; Valentini, E.; Mari, S.; Cecatiello, V.; Soffientini, P.; Pasqualato, S.; Polo, S. *Nat. Struct. Mol. Biol.* **2013**, *20*, 696–701; (c) Ogunjimi, A. A.; Briant, D. J.; Pece-Barbara, N.; Le Roy, C.; Di Guglielmo, G. M.; Kavsak, P.; Rasmussen, R. K.; Seet, B. T.; Sicheri, F.; Wrana, J. L. *Mol. Cell* **2005**, *19*, 297–308.
17. Brzovic, P. S.; Rajagopal, P.; Hoyt, D. W.; King, M. C.; Klevit, R. E. *Nat. Struct. Biol.* **2001**, *8*, 833–837.
18. Dueber, E. C.; Schoeffler, A. J.; Lingel, A.; Elliott, J. M.; Fedorova, A. V.; Giannetti, A. M.; Zobel, K.; Maurer, B.; Varfolomeev, E.; Wu, P.; Wallweber, H.; Hymowitz, S. G.; Deshayes, K.; Vucic, D.; Fairbrother, W. J. *Science* **2011**, *334*, 376–380.
19. Schreiber, A.; Stengel, F.; Zhang, Z.; Enchev, R. I.; Kong, E. H.; Morris, E. P.; Robinson, C. V.; da Fonseca, P. C.; Barford, D. *Nature* **2011**, *470*, 227–232.
20. Kee, Y.; Kim, J. M.; D'Andrea, A. D. *Genes Dev.* **2009**, *23*, 555–560.
21. (a) Yin, Q.; Lin, S. C.; Lamothe, B.; Lu, M.; Lo, Y. C.; Hura, G.; Zheng, L.; Rich, R. L.; Campos, A. D.; Myszka, D. G.; Lenardo, M. J.; Darnay, B. G.; Wu, H. *Nat. Struct. Mol. Biol.* **2009**, *16*, 658–666; (b) Eddins, M. J.; Carlile, C. M.; Gomez, K. M.; Pickart, C. M.; Wolberger, C. *Nat. Struct. Mol. Biol.* **2006**, *13*, 915–920.
22. Page, R. C.; Pruneda, J. N.; Amick, J.; Klevit, R. E.; Misra, S. *Biochemistry* **2012**, *51*, 4175–4187.
23. Wenzel, D. M.; Lissounov, A.; Brzovic, P. S.; Klevit, R. E. *Nature* **2011**, *474*, 105–108.
24. Scott, D. C.; Rhee, D. Y.; Duda, D. M.; Kelsall, I. R.; Olszewski, J. L.; Paulo, J. A.; de Jong, A.; Ovaa, H.; Alpi, A. F.; Harper, J. W.; Schulman, B. A. *Cell* **2016**, *166*, 1198–1214.
25. Kitada, T.; Asakawa, S.; Hattori, N.; Matsumine, H.; Yamamura, Y.; Minoshima, S.; Yokochi, M.; Mizuno, Y.; Shimizu, N. *Nature* **1998**, *392*, 605–608.
26. Nakayama, K. I.; Nakayama, K. *Nat. Rev. Cancer* **2006**, *6*, 369–381.

27. (a) Kumar, S.; Talis, A. L.; Howley, P. M. *J. Biol. Chem.* **1999**, *274*, 18785–18792; (b) Martinez-Zapien, D.; Ruiz, F. X.; Poirson, J.; Mitschler, A.; Ramirez, J.; Forster, A.; Cousido-Siah, A.; Masson, M.; Vande Pol, S.; Podjarny, A.; Travé, G.; Zanier, K. *Nature* **2016**, *529*, 541–545.
28. Margottin, F.; Bour, S. P.; Durand, H.; Selig, L.; Benichou, S.; Richard, V.; Thomas, D.; Strebel, K.; Benarous, R. *Mol. Cell* **1988**, *1*, 565–574.
29. Romani, B.; Cohen, E. A. *Curr. Opin. Virol.* **2012**, *2*, 755–763.
30. Palombella, V. J.; Rando, O. J.; Goldberg, A. L.; Maniatis, T. *Cell* **1994**, *78*, 773–785.
31. Collins, I.; Wang, H.; Caldwell, J. J.; Chopra, R. *Biochem. J.* **2017**, *474*, 1127–1147.
32. (a) Ciehanover, A.; Hod, Y.; Hershko, A. *Biochem. Biophys. Res. Commun.* **1978**, *81*, 1100–1105; (b) Hershko, A.; Ciechanover, A.; Rose, I. A. *Proc. Natl. Acad. Sci. U.S.A.* **1979**, *76*, 3107–3110.
33. Bett, J. S. *Essays Biochem.* **2016**, *60*, 143–151.
34. Komander, D.; Rape, M. *Annu. Rev. Biochem.* **2012**, *81*, 203–229.
35. McClurg, U. L.; Robson, C. N. *Oncotarget* **2015**, *6*(12), 9657–9668.
36. Demarchi, F.; Brancolini, C. *Drug Res. Updates* **2005**, *8*, 359–368.
37. Adams, J. Nature Rev. *Cancer* **2004**, *4*, 349–360.
38. Orlowski, R. Z.; Kuhn, D. J. *Clin. Cancer Res.* **2008**, *14*, 1649–1657.
39. Etlinger, J. D.; Goldberg, A. L. *Proc. Natl. Acad. Sci. U.S.A.* **1977**, *74*(1), 54–58.
40. Imajoh-Ohmi, S.; Kawaguchi, T.; Sugiyama, S.; Tanaka, K.; Omura, S.; Kikuchi, H. *Biochem. Biophys. Res. Commun.* **1995**, *217*, 1070–1077.
41. Adams, J.; Palombella, V. J.; Sausville, E. A.; Johnson, J.; Destree, A.; Lazarus, D. D.; Maas, J.; Pien, C. S.; Prakash, S.; Elliott, P. J. *Cancer Res.* **1999**, *59*, 2615–2622.
42. Orlowski, R. Z.; Eswara, J. R.; Lafond-Walker, A.; Grever, M. R.; Orlowski, M.; Dang, C. V. *Cancer Res.* **1998**, *58*, 4342–4348.
43. Adams, J.; Ma, Y.; Stein, R.; Baevsky, M.; Grenier, L.; Plamondon, L. Boronic Ester and Acid Compounds, Synthesis and Uses. Int. Patent appl. WO1996013266 A1, May 9, 1996.
44. Shenvi, A. *Biochemistry* **1986**, *25*, 1286–1291.
45. Kinder, D.; Elstad, C.; Meadows, G.; Ames, M. *Invasion Metastasis* **1992**, *12*, 309–319.
46. Sunwoo, J. B.; Chen, Z.; Dong, G.; Yeh, N.; Bancroft, C. C.; Sausville, E.; Adams, J.; Elliott, P.; Van Waes, C. *Clin. Cancer Res.* **2001**, 7, 1419–1428.
47. Demo, S. D.; Kirk, C. J.; Aujay, M. A.; Buchholz, T. J.; Dajee, M.; Ho, M. N.; Jiang, J.; Laidig, G. J.; Lewis, E. R.; Parlati, F.; Shenk, K. D.; Smyth, M. S.; Sun, C. M.; Vallone, M. K.; Woo, T. M.; Molineaux, C. J.; Bennett, M. K. *Cancer Res.* **2007**, *67*, 6383–6391.
48. Chauhan, D.; Hideshima, T.; Anderson, K. C.; Brit, J. *Cancer* **2006**, *95*, 961–965.
49. Chauhan, D.; Tian, Z.; Zhou, B.; Kuhn, D.; Orlowski, R.; Raje, N.; Richardson, P.; Anderson, K. C. *Clin. Cancer Res.* **2011**, *17*, 5311–5321.
50. Chauhan, D.; Singh, A. J.; Aujay, M.; Kirk, C. J.; Bandi, M.; Ciccarelli, B.; Raje, N.; Richardson, P.; Anderson, K. C. A Novel Orally Active Proteasome Inhibitor ONX0912 Triggers In Vitro and In Vivo Cytotoxicity in Multiple Myeloma. *Blood* **2010**, *116*, 4906–4915.
51. Marriott, J. B.; Muller, G.; Dalgleish, A. G. *Immunol. Today* **1999**, *20*, 538–540.
52. Raje, N.; Anderson, K. N. *J. Med. Med.* **1999**, *341*, 1606–1609.
53. Stephens, T. D.; Bunde, C. J. W.; Fillmore, B. J. *Biochem. Pharmacol.* **2000**, *59*, 1489–1499.
54. Bartlett, J. B.; Dredge, K.; Dalgleish, A. G. *Nat. Rev. Cancer* **2004**, *4*, 314–322.
55. Ito, T.; Ando, H.; Suzuki, T.; Ogura, T.; Hotta, K.; Imamura, Y.; Yamaguchi, Y.; Handa, H. *Science* **2010**, *327*, 1345–1350.
56. Gandhi, A. K.; Kang, J.; Havens, C. G.; Conklin, T.; Ning, Y.; Wu, L.; Ito, T.; Ando, H.; Waldman, M. F.; Thakurta, A.; Klippel, A.; Handa, H.; Daniel, T. O.; Schafer, P. H.; Chopra, R. *Br. J. Haematol.* **2014**, *164*, 811–821.

57. Krönke, J.; Udeshi, N. D.; Narla, A.; Grauman, P.; Hurst, S. N.; McConkey, M.; Svinkina, T.; Heckl, D.; Comer, E.; Li, X.; Ciarlo, C.; Hartman, E.; Munshi, N.; Schenone, M.; Schreiber, S. L.; Carr, S. A.; Ebert, B. L. *Science* **2014**, *343*, 301–305.
58. Lu, G.; Middleton, R. E.; Sun, H.; Naniong, M.; Ott, C. J.; Mitsiades, C. S.; Wong, K. K.; Bradner, J. E.; Kaelin, W. G., Jr. *Science* **2014**, *343*, 305–309.
59. Chamberlain, P. P.; Lopez-Girona, A.; Miller, K.; Carmel, G.; Pagarigan, B.; Chie-Leon, B.; Rychak, E.; Corral, L. G.; Ren, Y. J.; Wang, M.; Riley, M.; Delker, S. L.; Ito, T.; Ando, H.; Mori, T.; Hirano, Y.; Handa, H.; Hakoshima, T.; Daniel, T. O.; Cathers, B. E. *Nat. Struct. Mol. Biol.* **2014**, *21*, 803–809.
60. Fischer, E. S.; Böhm, K.; Cavadini, S.; Lingaraju, G. M.; Thoma, N. H.; Lydeard, J. R.; Harper, J. W.; Yang, H.; Nagel, J.; Serluca, F.; Tichkule, R. B.; Schebesta, M.; Forrester, W. C.; Schirle, M.; Hild, M.; Beckwith, R. E. J.; Jenkins, J. L.; Stadler, M. B.; Acker, V.; Hassiepen, U.; Ottl, J. *Nature* **2014**, *512*, 49–53.
61. Petzold, G.; Fischer, E. S.; Thomä, N. H. *Nature* **2016**, *532*, 127–130.
62. Nguyen, T. V.; Lee, J. E.; Sweredoski, M. J.; Yang, S. J.; Jeon, S. J.; Harrison, J. S.; Yim, J. H.; Lee, S. G.; Handa, H.; Kuhlman, B.; Jeong, J. S.; Reitsma, J. M.; Park, C. S.; Hess, S.; Deshaies, R. *J. Mol. Cell* **2016**, *61*, 809–820.
63. Hagner, P. R.; Man, H. W.; Fontanillo, C.; Wang, M.; Couto, S.; Breider, M.; Bjorklund, C.; Havens, C. G.; Lu, G.; Rychak, E.; Raymon, H.; Narla, R. K.; Barnes, L.; Khambatta, G.; Chiu, H.; Kosek, J.; Kang, J.; Amantangelo, M. D.; Waldman, M.; Lopez-Girona, A.; Cai, T.; Pourdehnad, M.; Trotter, M.; Daniel, T. O.; Schafer, P. H.; Klippel, A.; Thakurta, A.; Chopra, R.; Gandhi, A. K. *Blood* **2015**, *126*, 779–789.
64. Matyskiela, M. E.; Lu, G.; Ito, T.; Pagarigan, B.; Lu, C. C.; Miller, K.; Fang, W.; Wang, N. Y.; Nguyen, D.; Houston, J.; Carmel, G.; Tran, T.; Riley, M.; Nosaka, L.; Lander, G. C.; Gaidarova, S.; Xu, S.; Ruchelman, A. L.; Handa, H.; Carmichael, J.; Daniel, T. O.; Cathers, B. E.; Lopez-Girona, A.; Chamberlain, P. P. *Nature* **2016**, *535*, 252–257.
65. Han, T.; Goralski, M.; Gaskill, N.; Capota, E.; Kim, J.; Ting, T.C.; Xie, Y.; Williams, N. S.; Nijhawan, D. Science, 2017, *356*, [online early access]. DOI: https://doi.org/10.1126/science.aal3755. Published online: March 16, 2017. http://science.sciencemag.org/content/early/2017/03/15/science.aal3755 (accessed May 31, 2017).
66. Uehara, T.; Minoshima, Y.; Sagane, K.; Sugi, H. H.; Mitsuhashi, K. O.; Yamamoto, N.; Kamiyama, H.; Takahashi, K.; Kotake, Y.; Uesugi, M.; Yokoi, A.; Inoue, A.; Yoshida, T.; Mabuchi, M.; Tanaka, A.; Owa, T. *Nat. Chem. Biol.* **2017**, *13*, 675–680.
67. Ndubaku, C.; Tsui, V. *J. Med. Chem.* **2015**, *58*(4), 1581–1595.
68. Wertz, I. E.; O'Rourke, K. M.; Zhou, H.; Eby, M.; Aravind, L. *Nature* **2004**, *430*, 694–699.
69. Komander, D.; Clague, M. J.; Urbé, S. *Nat. Rev. Mol. Cell Biol.* **2009**, *10*, 550–563.
70. Sowa, M. E.; Bennett, E. J.; Gygi, S. P.; Harper, J. W. *Cell* **2009**, *138*, 389–403.
71. Zhang, W.; Sidhu, S. S. *FEBS Lett.* **2014**, *588*, 356–367.
72. Clague, M. J.; Barsukov, I.; Coulson, J. M.; Liu, H.; Rigden, D. J.; Urbe, S. *Physiol. Rev.* **2013**, *93*, 1289–1315.
73. Meulmeester, E.; Jochemsen, A. *Curr. Cancer Drug Targets* **2008**, *8*, 87–97.
74. Colland, F.; Formstecher, E.; Jacq, X.; Reverdy, C.; Planquette, C.; Conrath, S.; Trouplin, V.; Bianchi, J.; Aushev, V. N.; Camonis, J.; Calabrese, A.; Borg-Capra, C.; Sippl, W.; Collura, V.; Boissy, G.; Rain, J. C.; Guedet, P.; Delasorne, R.; Daviet, L. *Mol. Cancer Ther.* **2009**, *8*, 2286–2295.
75. Liang, Q.; Dexheimer, T. S.; Zhang, P.; Rosenthal, A. S.; Villamil, M. A.; You, C.; Zhang, Q.; Chen, J.; Ott, C. A.; Sun, H.; Luci, D. K.; Yuan, B.; Simeonov, A.;

Jadhav, A.; Xiao, H.; Wang, Y.; Maloney, D. J.; Zhuang, Z. *Nat. Chem. Biol.* **2014**, *10*, 298–304.
76. Johnston, S. R. *Clin. Cancer Res.* **2001**, 7, 4376–4387.
77. McDonnell, D. P.; Wardell, S. E.; Norris, J. D. *J. Med. Chem.* **2015**, *58*, 4883–4887.
78. Robertson, J. F. R. *Oncologist* **2007**, *12*, 774–784.
79. Lai, A.; Kahraman, M.; Govek, S.; Nagasawa, J.; Bonnefous, C.; Julien, J.; Douglas, K.; Sensintaffar, J.; Lu, N.; Lee, K.-j.; Aparicio, A.; Kaufman, J.; Qian, J.; Shao, G.; Prudente, R.; Moon, M. J.; Joseph, J. D.; Darimont, B.; Brigham, D.; Grillot, K.; Heyman, R.; Rix, P. J.; Hager, J. H.; Smith, N. D. *J. Med. Chem.* **2015**, *58*, 4888–4904.
80. Gottardis, M. M.; Robinson, S. P.; Satyaswaroop, P. G.; Jordan, V. C. *Cancer Res.* **1988**, *48*, 812–815.
81. Wu, Y. V. L.; Yang, X.; Ren, Z.; McDonnell, D. P.; Norris, J. D.; Willson, T. M. *Mol. Cell* **2005**, *18*, 413–424.
82. Dauvois, S.; Danielian, P. S.; White, R.; Parker, M. G. *Proc. Natl. Acad. Sci. U.S.A.* **1992**, *89*, 4037–4041.
83. Neklesa, T. K.; Tae, H. S.; Schneekloth, A. R.; Stulberg, M. J.; Corson, T. W.; Sundberg, T. B.; Raina, K.; Holley, S. A.; Crews, C. M. *Nat. Chem. Biol.* **2011**, 7, 538–543.
84. Tae, H. S.; Sundberg, T. B.; Neklesa, T. K.; Noblin, D. J.; Gustafson, J. L.; Roth, A. G.; Raina, K.; Crews, C. M. *ChemBioChem* **2012**, *13*, 538–541.
85. Xie, T.; Lim, S. M.; Westover, K. D.; Dodge, M. E.; Ercan, D.; Ficarro, S. B.; Udayakumar, D.; Gurbani, D.; Tae, H. S.; Riddle, S. M.; Sim, T.; Marto, J. A.; Jänne, P. A.; Crews, C. M.; Gray, N. S. *Nat. Chem. Biol.* **2014**, *10*, 1006–1012.
86. Long, M. J. C.; Gollapalli, D. R.; Hedstrom, L. *Chem. Biol.* **2012**, *19*, 629–637.
87. Shi, Y.; Long, M. J. C.; Rosenberg, M. M.; Li, S.; Kobjack, A.; Lessans, P.; Coffey, R. T.; Hedstrom, L. *ACS Chem. Biol.* **2016**, *11*, 3328–3337.
88. Gustafson, J. L.; Neklesa, T. K.; Cox, C. S.; Roth, A. G.; Buckley, D. L.; Tae, H. S.; Sundberg, T. B.; Stagg, D. B.; Hines, J.; McDonnell, D. P.; Norris, J. D.; Crews, C. M. *Angew. Chem. Int. Ed.* **2015**, *54*, 9659–9662.
89. Tinworth, C. P.; Lithgow, H.; Churcher, I. *Med. Chem. Commun.* **2016**, 7, 2206–2216.
90. Sakamoto, K.; Kim, K. B.; Kumagai, A.; Mercurio, F.; Crews, C. M.; Deshaies, R. J. *Proc. Natl. Acad. Sci. U.S.A.* **2001**, *98*, 8554–8559.
91. Sakamoto, K.; Kim, K. B.; Verma, R.; Ransick, A.; Stein, B.; Crews, C. M.; Deshaies, R. J. *Mol. Cell. Proteomics* **2003**, *2*, 1350–1358.
92. Schneekloth, J. S., Jr.; Fonseca, F. N.; Koldobskiy, M.; Mandal, A. K.; Deshaies, R. J.; Sakamoto, K.; Crews, C. M. *J. Am. Chem. Soc.* **2004**, *126*, 3748–3754.
93. Schneekloth, A. R.; Pucheault, M.; Tae, H. S.; Crews, C. M. *Bioorg. Med. Chem. Lett.* **2008**, *18*, 5904–5908.
94. Itoh, Y.; Ishikawa, M.; Naito, M.; Hashimoto, Y. *J. Am. Chem. Soc.* **2010**, *132*, 5820–5826.
95. Itoh, Y.; Kitaguchi, R.; Ishikawa, M.; Naito, M.; Hashimoto, Y. *Bioorg. Med. Chem.* **2011**, *19*, 6768–6778.
96. Okuhira, K.; Demizu, Y.; Hattori, T.; Ohoka, N.; Shibata, N.; Nishimaki-Mogami, T.; Okuda, H.; Kurihara, M.; Naito, M. *Cancer Sci.* **2013**, *104*, 1492–1498.
97. Ohoka, N.; Nagai, K.; Hattori, T.; Okuhira, K.; Shibata, N.; Cho, N.; Naito, M. *Cell Death Dis.* **2014**, *5, e1513.*
98. Ohoka, N.; Okuhira, K.; Ito, M.; Nagai, K.; Shibata, N.; Hattori, T.; Ujikawa, O.; Shimokawa, K.; Sano, O.; Koyama, R.; Fujita, H.; Teratani, M.; Matsumoto, H.; Imaeda, Y.; Nara, H.; Cho, N.; Naito, M. *J. Biol. Chem.* **2017**, *292*(11), 4556–4570.

99. Buckley, D. L.; Gustafson, J. L.; Van Molle, I.; Roth, A. G.; Tae, H. S.; Gareiss, P. C.; Jorgensen, W. L.; Ciulli, A.; Crews, C. M. *Angew. Chem. Int. Ed.* **2012**, *51*, 11463–11467.
100. Galdeano, C.; Gadd, M. S.; Soares, P.; Scaffidi, S.; Van Molle, I.; Birced, I.; Hewitt, S.; Dias, D. M.; Ciulli, A. *J. Med. Chem.* **2014**, *57*, 8657–8663.
101. Zengerle, M.; Chan, K. H.; Ciulli, A. *ACS Chem. Biol.* **2015**, *10*, 1770–1777.
102. Bondeson, D. P.; Mares, A.; Smith, I. E. D.; Ko, E.; Campos, S.; Miah, A. H.; Mulholland, K.; Routly, N.; Buckley, D. L.; Gustafson, J. L.; Zinn, N.; Grandi, P.; Shimamura, S.; Bergamini, G.; Faelth-Savitski, M.; Bantscheff, M.; Cox, C.; Gordon, D. A.; Willard, R. R.; Flanagan, J. J.; Casillas, L. N.; Votta, B. J.; den Besten, W.; Famm, K.; Kruidenier, L.; Carter, P. S.; Harling, J. D.; Churcher, I.; Crews, C. M. *Nat. Chem. Biol.* **2015**, *11*, 611–617.
103. Lu, J.; Qian, Y.; Altieri, M.; Dong, H.; Wang, J.; Raina, K.; Hines, J.; Winkler, J. D.; Crew, A. P.; Coleman, K.; Crews, C. M. *Chem. Biol.* **2015**, *22*, 755–763.
104. Winter, G. E.; Buckley, D. L.; Paulk, J.; Roberts, J. M.; Souza, A.; Dhe-Paganon, S.; Bradner, J. E. *Science* **2015**, *348*, 1376–1381.
105. Filippakopoulos, P.; Qi, J.; Picaud, S.; Shen, Y.; Smith, W. B.; Fedorov, O.; Morse, E. M.; Keates, T.; Hickman, T. T.; Felletar, I.; Philpott, M.; Munro, S.; McKeown, M. R.; Wang, Y.; Christie, A. L.; West, N.; Cameron, M. J.; Schwartz, B.; Heightman, T. D.; La Thangue, N.; French, C. A.; Wiest, O.; Kung, A. L.; Knapp, S.; Bradner, J. E. *Nature* **2010**, *468*, 1067–1073.
106. Lai, A. C.; Toure, M.; Hellerschmied, D.; Salami, J.; Jaime-Figueroa, S.; Ko, E.; Hines, J.; Crews, C. M. *Angew. Chem. Int. Ed.* **2016**, *55*, 807–810.
107. Gadd, M. S.; Testa, A.; Lucas, X.; Chan, K. H.; Chen, W.; Lamont, D.; Zengerle, M.; Ciulli, A. *Nat. Chem. Biol.* **2017**, *13*, 514–521.
108. Raina, K.; Lu, J.; Qian, Y.; Altieri, M.; Gordon, D.; Rossi, A. M. K.; Wang, J.; Chen, X.; Dong, H.; Siu, K.; Winkler, J. D.; Crew, A. P.; Crews, C. M.; Coleman, K. G. *Proc. Natl. Acad. Sci. U.S.A.* **2016**, *113*, 7124–7129.
109. Harling, J. D.; Smith, I. E. D. IAP E3 Ligase Directed Proteolysis Targeting Chimeric Molecules, Int. Patent Appl. WO 2016169989 A1, Oct. 27, 2016.
110. Remillard, D.; Buckley, D. L.; Paulk, J.; Brien, G. L.; Sonnett, M.; Seo, H. S.; Dastjerdi, S.; Wühr, M.; Dhe-Paganon, S.; Armstrong, S. A.; Bradner, J. E. *Angew. Chem. Int. Ed.* **2017**, *56*, 5738–5743.
111. Crew, A. P.; Wang, J.; Dong, H.; Qian, Y.; Siu, K.; Ferraro, C.Crews, C.M Tank-Binding Kinase-1 PROTACs and Associated Methods of use, Int. Patent Appl. WO 2016197114 A1, Dec. 8, 2016.
112. Schiedel, M.; Herp, D.; Hammelmann, S.; Swyter, S.; Lehotzky, A.; Robaa, D.; Oláh, J.; Ovádi, J.; Sippl, W.; Jung, M. J. Med. Chem. 2017, [Online early access]. DOI: https://doi.org/10.1021/acs.jmedchem.6b01872.Published online: April 5, 2017. http://pubs.acs.org/doi/abs/10.1021/acs.jmedchem.6b01872 (accessed May 25, 2017).
113. Zhou, B.; Hu, J.; Xu, F.; Chen, Z.; Bai, L.; Fernandez-Salas, E.; Lin, M.; Liu, L.; Yang, C. Y.; Zhao, Y.; McEachern, D.; Przybranowski, S.; Wen, B.; Sun, D.; Wang, S. J. Med. Chem. 2017, [Online early access]. DOI: https://doi.org/10.1021/acs.jmedchem.6b01816. Published online: March 24, 2017. http://pubs.acs.org/doi/abs/10.1021/acs.jmedchem.6b01816 (accessed May 25, 2017).
114. Neklesa, T. K.; Winkler, J. D.; Crews, C. M. *Pharmacol. Ther.* **2017**, *174*, 138–144.
115. Flanagan, J. J.; Qian, Y.; Rossi, A. M. K.; Andreoli, M.; Willard, R.; Gordon, D.; Harling, J.; Churcher, I.; Smith, I.; Zinn, N.; Bantscheff, M.; Crews, C. M.; Taylor, I.; Crew, A.; Coleman, K.; Winkler, J. In: *Targeted and selective degradation of ERα by PROTACs*, Presented at: Proceedings of the 39th San Antonio Breast Cancer Symposium, December 6–10, 2016; San Antonio, TX, 2016.

116. Neklesa, T. K.; Snyder, L. B.; Altieri, M.; Bookbinder, M.; Chen, X.; Crew, A. P.; Crews, C. M.; Dong, H.; Gordon, D. A.; Macaluso, J.; Raina, K.; Rossi, A. M. K. Taylor, I.; Vitale, N.; Wang, G.; Wang, J.; Willard, R. R.; Zimmermann, K. An oral androgen receptor PROTAC degrader for prostate cancer, Presented at: Proceedings of the American Association for Cancer Research, April 1–5, 2017, Washington DC.
117. Lebraud, H.; Wright, D. J.; Johnson, C. N.; Heightman, T. D. *ACS Cent. Sci.* **2016**, *2*(12), 927–934.

CHAPTER TEN

Chemical Biology in Drug Discovery

M. Paola Castaldi*[,1], Andrea Zuhl*, Piero Ricchiuto†, J. Adam Hendricks*

*Discovery Sciences, IMED Biotech Unit, AstraZeneca, Waltham, MA, United States
†Discovery Sciences, IMED Biotech Unit, AstraZeneca, Cambridge, United Kingdom
[1]Corresponding author: e-mail address: paola.castaldi@astrazeneca.com

Contents

Chemical biology represents the discipline at the intersection of synthetic chemistry, cellular and molecular biology, genomics, proteomics, biochemistry, biophysics, and computational science[1] (Fig. 1). Advances in chemical biology techniques over the past 10 years, together with the power of high-resolution mass spectrometry (MS) and bioinformatics, have demonstrated the value that these approaches can bring to clinical research.[2,3]

Annual Reports in Medicinal Chemistry, Volume 50
ISSN 0065-7743
https://doi.org/10.1016/bs.armc.2017.08.009

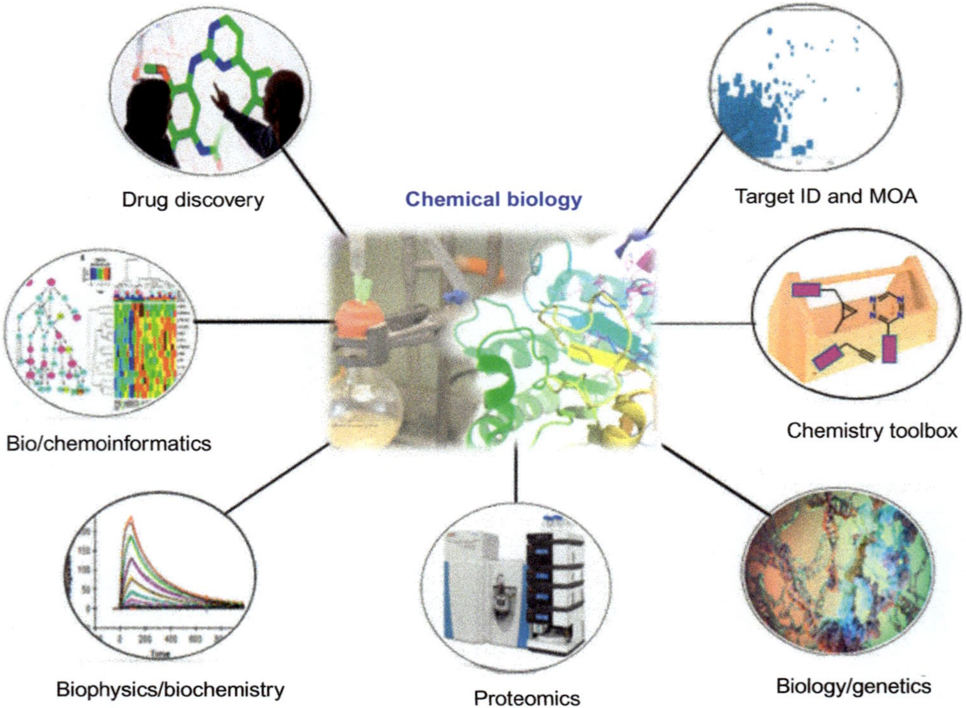

Fig. 1 Chemical biology disciplines.

The synergy between modern chemical biology and drug discovery relies on the identification and extensive characterization of novel small molecules that potently and selectively modulate the functions of molecular targets ultimately influencing human diseases.[4]

The evolution of chemical biology tools has enabled the scientific community to move away from artificial model systems that only marginally replicate relevant biology of an organism toward the analysis of complex physiological biological systems. Cellular potent and selective chemical probes will play a critical role in preclinical target validation to ensure that the desired biological effect in disease-relevant phenotypic assays is supported by evidences of direct drug–target engagement in cells[5] and modulation of specific biomarkers.[6] The application of these approaches to a larger number of drug discovery efforts in future will contribute to the discovery of novel drug candidates that are more likely to deliver the required efficacy and safety demands in the clinic ultimately reducing attrition rates in clinical trials, a major challenge faced in pharmaceutical research and development.

In this chapter, we focus on describing methods for target identification, target engagement, and target validation with demonstrative examples from the last few years.

1. CHEMICAL PROTEOMICS FOR TARGET DECONVOLUTION AND MECHANISM OF ACTION STUDIES

Preclinical drug discovery requires a multitude of biochemical, enzymatic, cellular, and in vivo assays in order to characterize the effects of drug candidates on cellular systems and model organisms.[7] Even though precise knowledge of the biological target(s) may not be broadly required for advancing molecules into clinical trials, shedding light into this mechanistic black box holds tremendous value when conducting phenotypic drug discovery. The process of identifying the molecular targets of active hits, also called "target deconvolution," is an essential step for understanding compound mechanism of action and for using the identified hits as tools for further dissection of a given biological process.

Dramatic technological improvements in mass spectrometry-based proteomic and chemoproteomic strategies, coupled with in silico approaches and the reduced cost of whole genome sequencing, have greatly improved the workflow of target deconvolution.[8] For both phenotype-based workflows and/or target-based approaches, proteomics enables a multitude of investigations addressing different questions in the process. These applications can be roughly grouped into (1) those characterizing direct or indirect drug–target and off-target interactions for target deconvolution and selectivity profiling, (2) those aimed at elucidating the mechanism of action (MoA) by which a drug exerts its pharmacological effect.

1.1 Affinity-Based Proteomics

The earliest chemical proteomic approach for identifying drug targets was affinity-based proteomics with reversible inhibitors.[9] In a typical experiment, a small molecule is first covalently linked to a solid support, such as a sepharose bead. An appropriate linker site may be apparent based on known SAR information, or cycles of probe synthesis and biological testing in relevant assays may be required to identify a suitable site. Incubation of the immobilized drug with a cell or tissue lysate results in noncovalent drug–protein interactions. After gentle washing of the beads to remove the unbound proteins, the resulting enriched proteins are identified and quantified by MS. Because proteins can also be enriched nonspecifically by binding to the linker or bead, control experiments are necessary to define true interactors. A classical control includes repeating the experiment with an inactive analogue, and removing hits enriched by the inactive probe matrix. Alternatively, competition experiments can be used in which free compound is added to the lysate before immobilized probe incubation, and specifically bound proteins are identified as those with reduced binding in the presence of free compound compared to a vehicle control (Fig. 2). Specific proteins interactions can include both direct binders and components of enriched protein complexes. The identified targets are then validated by orthogonal means such as a genetic modification (e.g., knockdown/knockout, overexpression, site-directed mutagenesis), biophysical and biochemical assays, or through the use of tool compounds.

This approach has been used in combination with phenotypic screening to discover novel mechanisms for disrupting the Wnt pathway, which is

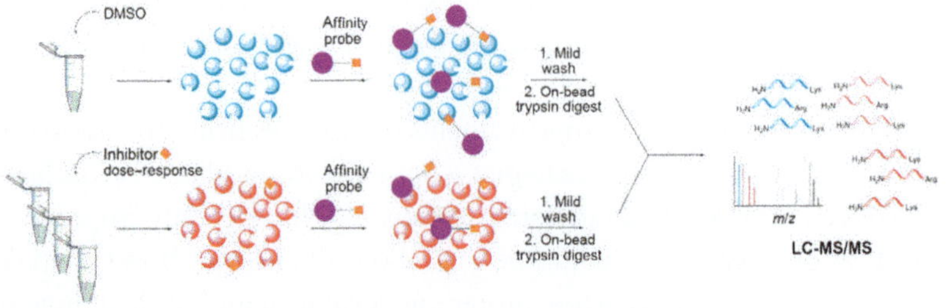

Fig. 2 Schematic workflow of affinity-based chemical proteomics.

upregulated in many cancers.[10] Wnt activation disrupts a multiprotein β-catenin destruction complex, leading to accumulation of β-catenin in the nucleus and transcription of genes that promote cell growth and survival. XAV939 was identified as an inhibitor of Wnt-signaling using a luciferase reporter assay in HEK293 cells. To determine the mechanism of action of XAV939, a bioactive analogue was immobilized on sepharose beads, and competition experiments with free XAV939 or an inactive control compound were performed. Specifically competed targets included members of poly (ADP-ribose) polymerases family including PARP1, PARP2, tankyrase1 (TNKS1), and tankyrase 2 (TNKS2). Biochemical and biophysical studies designated both tankyrases as the higher affinity targets. SiRNA-mediated codepletion of both TNKS1 and TNKS2, but not PARP1 and PARP2, or either tankyrase by itself, copied the phenotype of XAV939 treatment. Further experiments revealed that tankyrase 1/2 inhibition by XAV939 decreased degradation of axin, the concentration-limiting constituent of the β-catenin destruction complex. This provided not only a new avenue for treating Wnt-dependent cancers, but also a novel insight into axin regulation, highlighting the ability of chemoproteomics to generate new mechanisms for disease understanding and treatment.

A similar approach was used to annotate CCT251545 as a CDK8/19 inhibitor that modulates Wnt signaling.[11] (For further discussion on the discovery of CCT251545, see chapter "Phenotypic Screening" by Plowright and Drowley.) In this example, CCT251545, an additional active compound, and an inactive control were incubated with a CCT2514545-based probe matrix at a range of concentrations in a cellular lysate. The concentration–response information was used to calculate apparent K_d values of all three compounds against the probe-enriched proteins. The most potently competed proteins by the active compounds were subunits of the Mediator complex including, cyclin C, and cyclin C-dependent kinases 8 and 19 (CDK8 and CDK19). The prominence of kinases among the entire list of significantly competed targets (18 of 53 identified proteins), suggested that CCT251545 binds to CDK8/19, and the Mediator complex proteins were copurified. Direct binding to CDK8 and CDK19 was confirmed with several experiments including binding assays with purified proteins, a cellular thermal shift assay (CETSA, see Section 3), and altering the levels of CDK8/19 gene expression in cells. Crystallographic studies of CCT251545 bound to CDK8–cyclin C revealed a unique type I binding conformation that potentially explains the remarkable kinase selectivity of this compound, as well as its differentiated cellular activity compared to type II CDK8 inhibitors. The Mediator complex is a transcriptional coactivator. CCT251545 was found to modulate not only Wnt-responsive genes, but also several other pathways, although it was not a general transcription inhibitor. The potency and selectivity of CCT251545 should enable further characterization of the biological functions of CDK8 and CDK19.

Affinity-based chemoproteomics has been used not only to deconvolute phenotypic screens, but also to identify drivers of drug toxicity as reported for thalidomide.[12] Thalidomide was prescribed widely in the mid-20th century as a sedative and for morning sickness, but its use was subsequently discontinued after it was found to cause severe birth defects in children whose mothers used the drug during pregnancy. While this discovery had immediate impacts on drug regulation, an understanding of the direct binding targets of thalidomide or a molecular mechanism of action for the birth defects remained elusive for more than 50 years. To address this question, thalidomide-modified beads were incubated with various cell lysates, bound proteins were eluted with excess free thalidomide, and the eluate was separated by SDS-PAGE. Two protein bands were competed by pretreatment with thalidomide. Excision of those bands, trypsin digestion, and MS analysis identified the proteins as cereblon (CRBN) and DNA-binding protein 1 (DDB1). Purified CRBN bound to the thalidomide beads, while DDB1 did not, suggesting CRBN was a direct binding target and DDB1 was copurified. This was confirmed through coimmunoprecipitation experiments using a tagged CRBN bait. To further validate CRBN as the primary binding target, an orthologous gene in zebrafish was knocked down and found to duplicate many effects of thalidomide treatment. While CRBN was not well characterized at the time of these studies, DDB1 was known to be a component of E3 ubiquitin ligase complexes that assist in ubiquitination of various substrate proteins, consequently facilitating proteosomal degradation of those substrates. Expression of a CRBN point mutant capable of forming E3 ubiquitin ligase complexes, but incapable of binding thalidomide, abrogated the effects of thalidomide treatment in cells, zebrafish, and chicks, further validating CRBN as a necessary driver of thalidomide teratogenicity. Additional experiments revealed that thalidomide inhibits the E3-ligase function of CRBN-containing complexes, leading to downstream changes in critical developmental signaling pathways.

1.2 Family Affinity Matrices

A major bottleneck for affinity-based proteomics is the generation of an appropriately active probe matrix. If detailed SAR information for the molecule of interest is not available, time-consuming synthesis and activity testing of various linker sites can be required. Even when an appropriate linker site is available, any modifications to the molecule can create false positives and negatives that obscure difficult to capture targets. One solution is to use shared substrates, cofactors, or promiscuous inhibitors in the affinity matrix that enrich particular protein classes. In this approach, beads are first optimized to achieve the broadest possible coverage of the protein family. After identification of an appropriate matrix, unmodified compounds can then be tested against the matrix to determine selectivity against the captured portion of the target class. The major advantage of this approach compared to conventional selectivity profiling is that proteins are in a physiologically relevant context, and many proteins can be screened in the same assay.

The most extensive protein family affinity matrix work has been done with kinases.[9] More than 500 human kinases share a relatively conserved ATP-binding site, which is often exploited for inhibitor development. Due to this structural homology and shared mechanism of inhibition, selectively targeting individual protein kinases can be extremely difficult. To probe kinase selectivity in a native context, a "kinobead®" matrix was generated that quantitatively measures competition against seven immobilized ATP-competitive kinase inhibitors in a relevant cell or tissue lysate.[13] The promiscuity of the inhibitors attached to the matrix allowed for more than 200 protein kinases to be assayed simultaneously. Profiling of three ABL inhibitors gave IC_{50} values that were in line with reported cellular activities. In addition, several novel kinase and nonkinase off-targets were identified, which provided information that could be important for understanding drug safety and polypharmacology. There have been several generations of kinobeads®,[9] with a recent version, KBγ (Fig. 3) that assays more than 250 kinases by probing a mixture of four cell lysates with a five compound matrix.[14] Using more compounds or lysates reduced the number of identified kinases by biasing the matrix toward abundant and high affinity targets. The outlined optimization steps including using in silico prediction, careful choice of cell line(s), and sample preparation modification should be a useful guideline for future work in other protein families.

Affinity probes can enrich not only their direct binding targets, but also protein complexes associated with those targets. This phenomenon was exploited to define novel complex members associated with histone deacetylases (HDACs).[15] HDACs are members of megadalton multiprotein complexes that repress gene transcription and have therapeutic potential in a number of diseases including cancer, neurodegeneration,

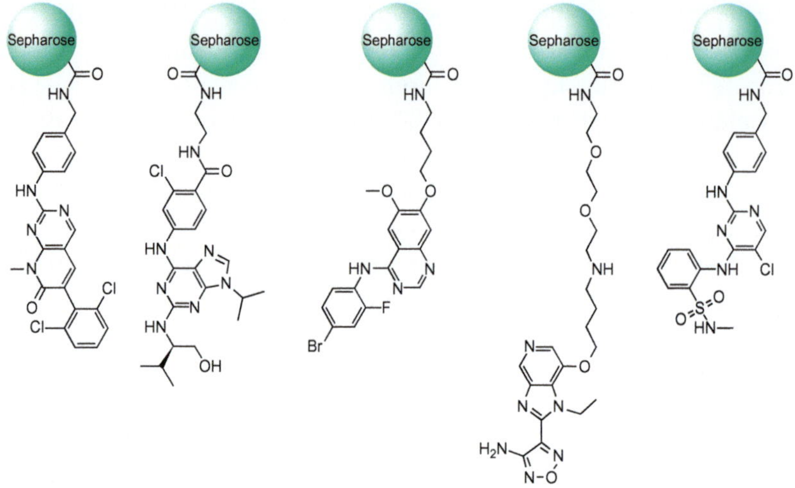

Fig. 3 Structures of the probes constituting the KBγ version of kinobeads®.[14]

and autoimmunity. Two nonselective hydroxamate inhibitors were used to capture class I and class IIb HDACs, along with a majority of their previously reported subunits, from cell and tissue lysates. Individual HDACs and their corresponding subunits generated matching dose–response profiles in competition experiments, which allowed novel HDAC complexes to be identified. These protein associations were then confirmed with coimmunoprecipitation studies. Several inhibitors were found to be more selective than in previously reported enzyme assay studies, and some inhibitors had different affinities for their direct binding targets depending on the composition of the larger protein complex. Collectively, these results emphasize the importance of profiling inhibitors in a physiologically relevant context.

1.3 Activity-Based Protein Profiling (ABPP)

Probes can be targeted toward particular enzyme families not just through shared binding elements, but also through shared reactivity. These probes are called activity-based probes (ABPs), because they react with or bind specifically to the catalytic residues of active enzymes.[16,17] ABPs are comprised of three elements: a reactive warhead that binds the enzyme active site, an affinity group that directs the warhead, and a linked reporter tag. The reporter tag can be a fluorophore for visualization, biotin for enrichment and identification by MS, or a click-chemistry compatible (clickable) handle that allows for bioorthogonal attachment of various reporter groups.[18,19] ABPs react only with the active form of enzymes, and therefore can be used to determine the functional state of bound protein targets. This is a distinct advantage compared to affinity-based probes, which require follow-up studies to determine if binding to target proteins has a functional consequence. In addition, because ABPs modify their protein targets covalently, they can be applied directly in live cells and animals. There have been successful examples of affinity-based chemoproteomics in which live cells or animals are first treated with inhibitors, and binding is assessed after a gentle lysis.[13] However, target engagement may be altered during lysis. Some integral membrane proteins, proteins requiring coregulators, or proteins belonging multisubunit protein complexes can also lose their appropriate binding conformation during lysate preparation. ABPP (Activity-Based Protein Profiling) has been developed for several enzyme classes including serine hydrolases, cysteine proteases, kinases, phosphatases, and glycosidases among others.[16,17]

ABPP was used to identify cysteine and serine proteases that are involved in host cell rupture after infection by the malaria parasite *Plasmodium falciparum*.[20] Host cell rupture is essential to pathogen propagation. However, at the time of this study there was relatively little understanding of this process compared to host cell invasion. As cysteine and serine protease have been implicated in host cell rupture, a library of more than 1200 covalent, irreversible inhibitors of both protease families was screened for disruption of the rupture of red blood cells after *P. falciparum* infection. A modestly potent biotinylated chloroisocoumarin inhibitor ($EC_{50}=22\ \mu M$), JCP104, was identified. Enrichment of JCP104-treated cells and MS analysis identified the serine protease PfSUB1 as a target of this probe compound. JCP104 inhibited recombinant PfSUB1 in a substrate assay with roughly equivalent potency to the rupture assay, suggesting that PfSUB1 mediates the activity of JCP104. Two cysteine protease inhibitors also disrupted host cell rupture. Target deconvolution of these inhibitors was more difficult because both were found to inhibit several *P. falciparum* proteases related to human cathepsin G, the target for which these compounds were initially synthesized. Therefore, a small library of compounds was synthesized that had differential selectivity for the various potential targets. Testing of this library identified an optimized inhibitor, SAK1, which acted through selective inhibition of the cysteine protease DPAP3. SAK1 inhibited DPAP3 with roughly equivalent potency to the nonselective hits from the rupture assay, suggesting that DPAP3 drives the activity of these compounds. Initial results suggest that DPAP3 regulates the maturity of PfSUB1, which in turn activates SERA5, a previously identified critical protein for host cell rupture. Overall, this study was significantly aided by the decision to use covalent

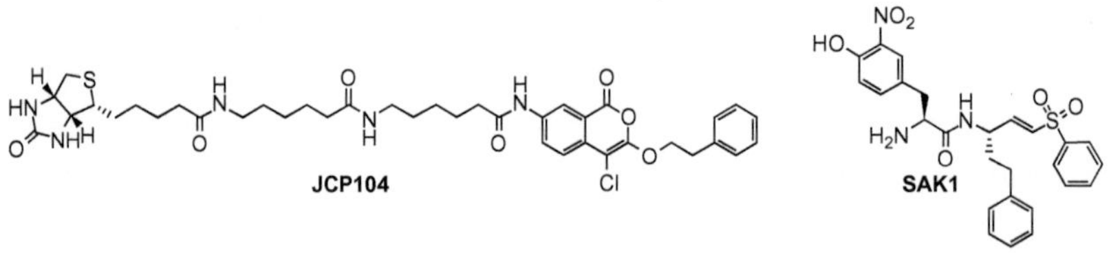

Fig. 4 Structures of the serine protease inhibitor JCP104 and cysteine protease inhibitor SAK1.

inhibitors designed for individual protein families, both by allowing protein enrichment with modestly active hits, and also by narrowing target hypotheses after screening (Fig. 4).

ABPP has primarily been used to study irreversible inhibitors. This is because covalent probe labeling can increasingly displace reversible inhibitors over time, resulting in a loss of information about engaged protein targets. However, this technology can be extended to reversible inhibitors.[16,17] The first example of ABPP of reversible inhibitors used a fluorophosphonate probe to monitor the potency and selectivity of inhibitors of the serine hydrolase fatty acid amide hydrolase (FAAH) in lysates.[21] Careful optimization of the probe labeling conditions was required to monitor selectivity among a significant number of serine hydrolases, without displacing the noncovalent FAAH inhibitors. In a more recent example, a kinetically tuned probe was used to determine target engagement of reversible serine hydrolase inhibitors in vivo.[22] This study required the development of a bioavailable probe with moderate reactivity. The resulting clickable triazole urea probe was used to confirm the in vivo selectivity of inhibitors of lysophospholipases 1 and 2. The ability to monitor target interactions of both reversible and irreversible inhibitors directly in vivo is a significant advantage of ABPP compared to other chemical proteomic methods.

1.4 Cellular Photoaffinity Labelling

While the covalent binding of ABPP provides significant advantages, it cannot be extended to proteins that lack conserved reactive groups. More recently, photoaffinity probes have been developed to extend covalent probe labeling to the broader proteome.[23,24] These probes integrate a photocrosslinking group that is activated by UV light to generate a transient, high-energy intermediate that reacts through space to form a covalent bond. The three most commonly used photocrosslinking moieties are diazirines, benzophenones, and aryl azides. Photoaffinity probes also contain a reporter tag for downstream analysis similar to those used with ABPs.

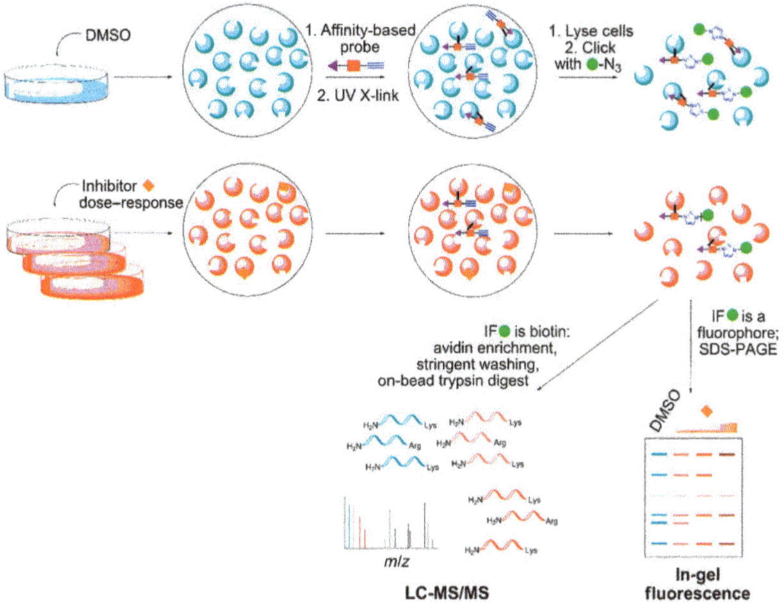

Fig. 5 Schematic workflow of an in-cell chemical proteomics experiment with a clickable photoaffinity-probe. *Reproduced from Zuhl, A.M.; Nolan, C.E.; Brodney, M.A.; Niessen, S.; Atchison, K.; Houle, C.; Karanian, D.A.; Ambroise, C.; Brulet, J.W.; Beck, E.M.; Doran, S.D.; O'Neill, B.T.; Am Ende, C.W.; Chang, C.; Geoghegan, K.F.; West, G.M.; Judkins, J.C.; Hou, X.; Riddell, D.R.; Johnson, D.S. Chemoproteomic profiling reveals that cathepsin D off-target activity drives ocular toxicity of β-secretase inhibitors, Nat. Commun.* ***7****, 2016, 13042.*

(Fig. 5). Click-chemistry reporters are the most common, since many photoaffinity probes are developed specifically for use in live cells, and these reporters are the most likely to allow for passive cell permeability of the probe. Because of this cellular labeling, photoaffinity probes can access targets, such as integral membrane proteins, which are typically not captured during traditional affinity-based chemoproteomic experiments in lysates. However, unlike ABPs, photoaffinity probes cannot be used in vivo due to the requirement for UV light. Additionally, while some photoaffinity probes bind selectively to active enzymes,[17] labeling is not necessarily activity-dependent, and further studies on the functional consequences of probe binding should be performed.

A clickable photoaffinity probe was recently applied to deconvolute the mechanism of ocular toxicity associated with inhibitors of β-secretase BACE1.[25] BACE1 is a major target for the treatment of Alzheimer's disease. BACE1 inhibitor development has been hampered by safety liabilities, including an ocular toxicity observed in the retinal pigment epithelium (RPE) layer of animals treated with clinical candidate LY2811376, as well as several advanced preclinical candidates, such as PF-9283. To determine the cause of this toxicity, probe PF-7802 was generated, where PF-9283 was modified by incorporation of a benzophenone group for photocrosslinking and an alkyne moiety for click chemistry. Photoaffinity labeling in live RPE cells and conjugation to reporter tags identified the primary target of PF-7802 as cathepsin D (CatD). CatD was already a known off-target of some BACE1 inhibitors. However, the BACE1 inhibitors used in this study such as PF-9283 and LY2811376 were not believed to have significant cross reactivity with CatD. Therefore, PF-7802 was further leveraged to measure cellular target engagement of CatD. BACE1 inhibitors were found to have substantially enhanced potency for CatD in live cells compared to in vitro assays with purified proteins. This was hypothesized to have been due to accumulation of the weakly basic compounds in the acidic lysosome, where CatD is highly expressed. Exposure–response analysis of in vivo studies with several BACE1 inhibitors confirmed that this level of target engagement was sufficient to drive ocular toxicity. These results emphasize the value of profiling activity in live cells.

In a recent application of photoaffinity labeling, a series of lipid-based probes were used to discover novel cellular lipid-binding proteins.[26] Several fatty acids were derivitized with diazirines for photocrosslinking, and alkynes for click chemistry. Cells were incubated with each probe, UV irradiated, lysed, and probe-labeled proteins were conjugated to biotin for enrichment and identification by MS. Greater than 1000 targets were identified including many known lipid-binding proteins such as enzymes involved in fatty acid uptake, transport, biosynthesis, and catabolism. When the data from multiple probes were viewed globally, a roughly equal distribution of membrane and soluble proteins were identified. These targets originated from varied compartments within the cell, and belonged to diverse protein families. The data set was highly enriched in proteins also found in the DrugBank database, suggesting that some of the poorly characterized proteins enriched by the lipid probes may also be druggable. This was established by the development of a selective ligand for the poorly characterized protein nucleobinding 1 (NUCB1); this ligand was further used to establish that NUCB1 plays a role in fatty acid amide metabolism. The probes were also leveraged to determine target engagement and selectivity of drugs to their known binding targets, many of which were integral membrane proteins that are difficult to assay by other means. Most drugs were found to have additional targets beyond those which are annotated, which may have important implications for drug safety and drug repurposing. The diversity of targets identified and assayed in this study highlights the ability of photoaffinity probes to broadly survey the proteome.

2. TARGET IDENTIFICATION BY COMPUTATIONAL APPROACHES

The increasing amount of data of the past decade enabled the application of fit-to-purpose machine learning (ML) algorithms and Naïve Bayes (NB) classifiers for target prediction in drug discovery. The impact of these new predictive tools together with the advances in high-throughput technologies has shifted the predominant paradigm of "one target, one drug" that has been difficult to achieve to "one drug, multiple targets."[27]

In general, cheminformatics methods applied for target identification integrate different levels of information to enhance the reliability of data prediction outcomes, while network-based methods integrate different levels of information in drug–protein and protein–disease network[28] to predict new targets or to identify a different disease where the same drug might be active on, also known as drug repurposing. Recent alternative methods use gene expression profiles induced by drugs, such as Connectivity Map (CMap) (http://www.broadinstitute.org/ccle/home),[29] NCI-60 cell lines (http://dtp.nci.nih.gov/), LINCS (http://lincs.hms.harvard.edu/db/), and CCLE (http://www.broadinstitute.org/ccle/home), to predict drug targets based on the transcriptome data. The cheminformatics methods, routinely and successfully implemented[30] are usually classified in four groups based on the nature of the source information (i) chemical structure similarity searches,[31] (ii) data mining/machine learning,[32] (iii) panel docking,[33] and (iv) bioactivity spectra-based algorithms.[34] Table 1 provides a comprehensive list of the cheminformatics software/algorithms followed by their main concepts.

2.1 Chemical Similarity Searches

The generally valid[55] assumption of the chemical similarity searches is that "small molecules with similar structures may have similar properties." Similarity searches use stored information under in-house or public databases to infer the potential gene/protein targets starting from a "query" molecule. The similarity measure (coefficient) such as distance, probabilistic, correlation, and association is a quantitative measure of similarity used to rank the molecules in the database that are similar to the "query" molecule. The most popular similarity measure for comparing chemical structures represented by means of fingerprints is the Tanimoto (or Jaccard) coefficient "*T*." The Tanimoto index spans from 0 to 1 equating to no-similarity or maximum similarity (identical structures) between molecules, respectively. Computer

Table 1 Summary of Developed Tools for In Silico Target Identification

	Algorithms/Software	Pros	Cons	References
(1) Chemical similarity searches	*Topological 2D:* TOPOSIM,[35] LaSSI,[36] Daylight[web1], Certara[web2], MDL Keys,[37] Similogs,[38] *E*-states[39] *3D structural similarities:* MATRAS,[40] CSNAP3D[41], Relibase,[42] TV,[43] FLOG[44]	No prior knowledge on compound's activity is needed	Not always biological relevant	35–44 web1: http://www.daylight.com/products/ web2: https://www.certara.com/
(2) Data mining/ML	PASS,[45] NIWA,[46] Bayesian Models on chemogenomics DB targets,[32] connectivity fingerprints[web3]	High accuracy, ranked list of targets	Predictions are limited to the used training set, inconsistencies in targets names will affect the outcomes	32,45,46 web3: http://www.scitegic.com/
(3) Panel docking	TarFisDock,[33] INVDOCK,[47] SIFts[48]	Ligand information are not necessary	Time-consuming, the accuracy and versatility of the scoring functions are the main issues	33,48,49
(4) Bioactivity spectra-based algorithms	SPiDER,[50] SuperPred,[51] SwissTargetPrediction,[52] DINIES,[53] iDrug[54]	Connect chemical and biological information and predict multiple targets	Extensive analysis of the results due to the nature of bioactivity databases	50–54

algorithms that search for similarity will scan drug databases for topological, steric, electronic, and/or physical properties and put them in relation with the query molecule(s). For instance, topological searches only look at compounds that share at least one functional group, i.e., substructure searches,[56] or the same frame/scaffold with various functional groups, i.e., Markush search.[57] These searches are also classified as 2D-methods based on the general principle that two molecules are projected into corresponding graphical representations and the algorithms search for the maximum correspondence between the atoms of the two graphs. The 3D conformations of two molecules may also be used to define their similarity, where the steric static and pharmacophore[58] properties are considered. That is, algorithms search for the best superposition of the molecules as 3D objects, which might or might not take flexibility into account. The advantages of chemical similarity searches are that little information is necessary to formulate a reasonable query, no assumptions are needed about which part(s) or conformation of the query molecule confers activity and they are relatively inexpensive for searching routinely large databases. Thus, these methods are mostly useful at the beginning of a drug discovery project when the information about the target is limited. For brevity of this paragraph, the reader is referred to more comprehensive reviews.[59–63]

2.2 Data Mining/Machine Learning

Databases of compounds with annotated targets are mined for target prediction purposes. Data mining and Machine Learning (ML) algorithms are used to circumvent the extensive analysis of the large amount of data available in these databases and extract automatically only the relevant information that connect the targets and chemical substructures. The outcome is usually multiple-target models that reflect the thousands of possible molecule-to-target interactions. In comparison to the previously described similarity searching in databases, that uses all information around chemical fingerprint to find similar ligands, models built from ML retain only the relevant information and ignore those common to both active and inactive molecule.

2.3 Panel Docking

Traditional docking algorithms are applied to find ligands and binding conformations at a receptor site close to what they could be in vivo. Hence in traditional docking, also named inverse-docking, one protein is tested for multiple molecules.[47] In this approach, the 3D information about the target protein is used to perform ligand–target docking to a wide panel of proteins and to identify, based on the scoring function, the most probable interaction partner in vivo. Thus, these algorithms can be applied only when the 3D structure of the target protein is available and no information is needed on the ligand. The binding affinity data are then compared across different proteins which requires a normalization step that, is not trivial and is key for obtaining the "absolute" binding partner. Different software will be able to account for protein or ligand flexibility within docking simulation as described by Sliwoski[49] and coworkers. Finally, inherent limitations of this methodology are related to the availability of the input protein structures and to the time and resource intensive computations.

2.4 Bioactivity Spectra-Based Algorithms

Bioactivity spectra are an extensive class of methods that transforms 2D or 3D structures into one or more "spectra" or histograms and then compute the overlap of the histograms.[64,65] For instance, in case of a new hit whose target is not known, these methods that rely on the assumption that similar ligands show similar activity, use the bioactivity databases which store chemical data in combination with biological data to predict the possible targets or to identify promiscuous molecules (false-positive results based on reactivity) that often occur as hits in screening campaigns.[66] Although, it is a very promising step forward through the integration of different data sources, the massive amount of data require automatic data curation which could lead to inaccurate information which are then used for target prediction. In addition, data coming from different sources may have been obtained under different assay conditions and hence not directly comparable. Thus, any bioactivity data must be handled with care and the underlying publications need to be read for data confirmation. Ensuring more consistent and reliable curation of data will enable more reliable analyses and provide greater impact in future. For a description of all the bioactivity database, the reader is referred to a more comprehensive review.[67]

All the above cheminformatics methods calculate an interaction probability of a compound against a given target but, none of them use inactive bioactivity data.[68] In a recent publication,[69] the authors have demonstrated that inactive data held in chemogenomics repositories can indeed be used for target prediction. Yet, in this publication the authors compiled the *inactive data set* by mining PubChem Compound and PubChem BioAssay databases and were able to extract 194 million protein–ligand pairs, including over 640,000 distinct compounds. ChEMBL[70] was used for determining the *active data set* encompassing over 295,000 bioactivities covering 1080 protein classes. A comparison between the inactivity-inclusive models and the activity-only based approach showed that the former has a statistically significant benefit in including negative bioactivity data when building the target prediction models.

These novel approaches which leverage negative bioactivity information for target prediction require that the active or nonactive labels are attributed correctly. Unfortunately, this is a challenging task when dealing with the known issue of poor data reproducibility. However, some solutions have already emerged including the putative inactive selection method based on the sphere exclusion algorithm,[69] or the diversity selection algorithms proposed by Hudson et al.[71] and Gobbi et al.[72]

3. TARGET ENGAGEMENT

Confirming cellular target engagement of small molecules is a critical component of the drug discovery process as it helps correlate small molecule–protein interactions in living cells. The first section of this chapter discussed techniques used to generate target hypotheses for compounds of interest with unknown mechanisms of action or identifying off-targets with potential safety liabilities via chemical proteomics. In this section, we will focus on emerging technologies that can be used to either validate the engagement between compounds and the putative targets identified from chemoproteomic studies or to study target engagement in a more relevant biological setting with an annotated compound for determining mechanism of action or study more complex biological processes.

3.1 Cellular Thermal Shift Assay

Thermal shift assays have been used for many years in drug discovery and are a useful biophysical means to probe the binding between a recombinant protein and ligand.[73,74] Thermal shift assays probe this interaction by monitoring the induced thermal stabilization of the target protein as a result of ligand binding. Recently, the work of Nordlund and colleagues expanded this concept beyond biochemical profiling into more relevant biologically relevant samples, cell lysate, and live cells, using Western blot to monitor the interaction between a number of targets and validated inhibitors.[5] This CETSA has two main experiments: first is the generation of a melt curve in which the induced stabilization of the target protein is monitored at a single compound concentration (≥10 μM) across a range of temperatures (37–75°C) and compared with untreated samples (Fig. 6). The second experiment is an isothermal dose–response fingerprint (ITDRF) in which the induced

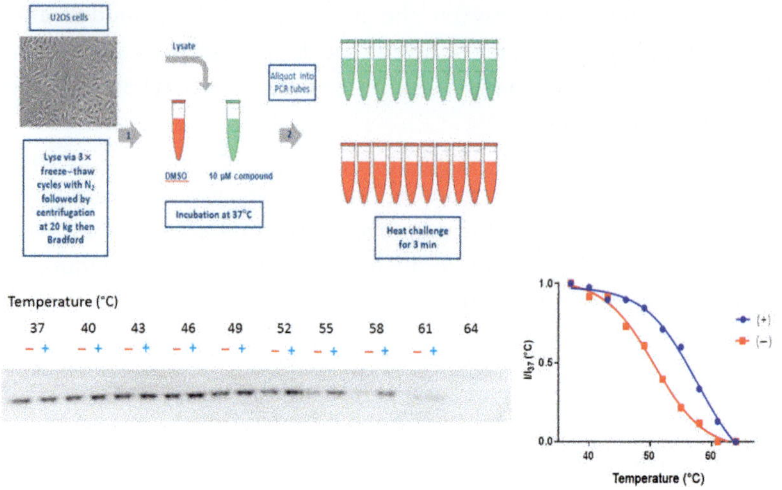

Fig. 6 Description of the CETSA workflow.

stabilization is looked at varying compound concentrations for a single temperature, determined from the melt curve, to estimate relative differences in drug concentration needed to achieve a similar degree of target engagement.

A key example from this seminal work was the study of poly [ADP-ribose] polymerase-1 (PARP-1) with iniparib which was withdrawn from the clinic due to lack of efficacy and subsequently shown to have no activity against PARP-1 in living cells[75] and olaparib, an FDA approved PARP-1 inhibitor. While olaparib induced a large thermal shift of PARP-1, iniparib failed to induce any shift when compared to vehicle, a finding corroborated in the ITDRF experiment. This highlights how CETSA can be employed to verify interaction between a small molecule and its putative target in a biologically relevant context rather than via cell-based reporter assays and demonstrates its potential value early in screening cascades. Another interesting application of CETSA was its ability to monitor the more complex process of cellular uptake of the antifolate cancer drug, raltitrexed, by monitoring the thermal modification of thymidylate synthase (TS). After the observation that raltitrexed induced a large thermal shift, they compared the ITDRF in-lysate to the live cell and observed a three orders of magnitude shift to a higher value in-lysate. This result signified active transport of the drug across the membrane, resulting in a high concentration of raltitrexed inside the cell. This is an important consideration when thinking of observed differences between in-lysate and live cell experiments. Further probing the active uptake of TS inhibitors, Nordlund and colleagues conducted an elegant study of nitrobenzylmercaptopurine riboside (NBMPR), an inhibitor of fluorouracil (5-FU) uptake. The ITDRF for TS in cells treated with 5-FU demonstrates that its import is abrogated by the addition of NBMPR as illustrated by a lack of observed stabilization.

The main challenge that prevented CETSA from being employed in a primary screening cascade for a reasonably sized compound library is the throughput of Western blots. This problem was addressed by adapting CETSA to a microplate-based format that uses homogeneous detection technologies such as AlphaScreen™.[76,77] Nordlund and Lundbäck validated their high-throughput CETSA (HT-CETSA) method by screening a diversity library of almost 11,000 compounds against thymidylate synthase in K562 cells,[78] which was known to be amenable to CETSA as described earlier. Importantly, the screen did not provide any false negatives as all known TS interactors were identified in addition to 65 novel compounds. Of particular note from this screen was the identification of a compound, CBK115334, a compound differentiated from known TS inhibitors which was not previously identified as a hit across a variety of different assays. However, additional CETSA experiments confirmed its binding to TS and a subsequent crystal structure revealed binding in the folate pocket via a novel binding mode for antifolates. This example further argues for the inclusion of HT-CETSA into screening cascades by illustrating its potential to identify new tractable chemical equity that other screens might miss. Furthermore, due to the increase in throughput which HT-CETSA provided they were able to build on their uptake studies by looking at time-trace ITDRFs for floxuridine, another known TS inhibitor, and 5-FU and correlate this with reactive metabolite formation. Both compounds act via a common reactive metabolite that is formed by different enzymatic pathways.[79] CETSA data for floxuridine showed stabilization of TS at low concentrations (nM) after 10 min and are consistent with the measurable appearance of the metabolite. 5-FU on the other hand had a CETSA response that increased slowly over time, despite the concentration of 5-FU remaining constant. Taken together, these results suggest that the enzymatic conversion, and thus pathway, of floxuridine is significantly faster than 5-FU in K562 cells under these conditions and reflected by the CETSA results.

A drawback of both the classic and HT versions of CETSA is that they require validated and selective antibodies for monitoring target response and the proteins are predefined. Combining CETSA with quantitative mass spectrometry (CETSA-MS) would eliminate the need of an antibody and allows for the target agnostic study the affect a small molecule has on the thermal profile of the cellular proteome. This approach essentially allows one to screen a desired compound against ~6000 proteins at once. Furthermore, for in-cell experiments, this approach may provide additional information such as downstream events that are modulated upon compound treatment such as phosphorylation and/or target downregulation. This can allow one to look at potential small molecule target interactions for which direct monitoring isn't possible due to lack of reagent or potentially the nature of the interaction. Further, this could provide a nice correlation between target engagement and cellular response. The initial CETSA-MS study was performed by GlaxoSmithKline (GSK), in collaboration with others, looking at the thermal profile of several broad-specificity kinase inhibitors.[80] They identified shifts in the melting temperatures of many kinase targets and other proteins, highlighting ferrochelatase as an off-target of several kinase inhibitors, most notably vemurafenib, implicating it for the observed photosensitivity.[81] Arguably even more revealing was that for dasatinib, an ABL inhibitor, they observed thermal shifts in K562 cells for several proteins downstream of BCR–ABL, including CRKL, which was confirmed as nondirect binders via an in-lysate experiment in which no thermal shift was observed. K562 cells possess the fusion protein BCR–ABL, for which a thermal shift was not detected and suggests a lack of binding. However, looking in Jurkat cells, which are BCR–ABL deficient, revealed that the fusion protein was less thermostable than ABL1 and possibly not amenable to CETSA. Thus, the ability to look at the downstream effectors provides a means to identify proteins that might otherwise be missed, albeit indirectly.

In a subsequent paper, looking at HDACs, the team at GSK devised what they termed two dimensional thermal proteome profiling (2D-TPP) that measures proteome-wide protein stability at 12 temperatures and 5 different compound concentrations, essentially combining the melt curve and ITDR into a more streamline experiment that provides greater depth.[82] Profiling Novartis' pan-HDAC inhibitor panobinostat, they identified four proteins which were dose-dependently stabilized, both in-lysate and live cell, indicating that these proteins were direct interactors. One of these proteins, phenylalanine hydroxylase (PAH), showed increased thermal stability at low panobinostat concentrations (pEC_{50} 7.2). When using an orthogonal bead-based affinity chemoproteomics approach they noticed that PAH binding was abrogated by detergents but under detergent-free conditions they saw good agreement with the 2D-TPP experiments. However, this was only true when they used panobinostat loaded on beads for affinity enrichment while the original HDAC affinity matrix did not identify PAH as a target of panobinostat.[15]

The lack of detergent in the original CETSA-MS gave it an inherent bias toward cytosolic (soluble) proteins making the study of membrane proteins a blind spot for the technique. To expand the scope of CETSA-MS to include membrane proteins two groups published papers around the same time demonstrating that the inclusion of a small percentage of detergent to the buffer could enable this.[83,84] The same GSK group performed a thorough evaluation of different detergents used to lyse cells using ATP as the ligand and identified 0.4% NP-40 as sufficient to solubilize many membrane proteins, without causing resolubilization of precipitated proteins and importantly, given the above example, did so without interfering with protein–compound affinities established in previous affinity-based proteomics studies.[13] In the presence of 0.4% NP-40, they noticed proteins aggregating at lower temperatures, 2.9°C lower on average, indicating increased thermal susceptibility without substantially affecting the shifts in melting temperatures induced by the addition of physiological concentrations of ligand to the cell extract, indicating that the ligand–protein interaction was not affected by the detergent. The use of 0.4% NP-40 resulted in the identification of 371 membrane proteins with melt curves from extracts a 495% increase over detergent-free conditions (75 proteins) and 748 proteins in the intact cells (971% increase) making the study of transmembrane proteins more feasible.

CETSA represents a significant advance in target engagement studies by allowing one to measure drug–target(s) interactions on full-length protein in a physiological and disease relevant environment with any potential binding partners present and without the need for chemistry to modify the compound of interest. Furthermore, the application of MS techniques allows CETSA to be used in a target discovery mode, particularly off-targets, or in cases where a good antibody doesn't exist. While still early in CETSA's adoption, in particular CETSA-MS and the study of membrane targets, the initial few years have already seen significant advances.

3.2 Fluorescence Polarization, FRET, and nanoBRET™

A variety of techniques exist to study ligand–protein interactions by proximity assays between a ligand and its target protein where one must attach a fluorescent reporter on the small molecule which can be exploited for fluorescence or bioluminescence resonance energy transfer measurements.

Fluorescence polarization assays, much like the thermal shift assay, are used routinely in a biochemical setting to monitor small molecule binding and K_D determination via the displacement of a small molecule tracer by the compound of interest.[85,86] However, until recently this was only performed using a recombinant protein. Weissleder and colleagues have demonstrated that this binding assay can be used in live-cells[87] and even more remarkably, in vivo,[88] to map drug–target interaction in real time at subcellular resolution with high-resolution spatial and temporal mapping of bound and unbound drug distribution via a multiphoton fluorescence anisotropy microscopy imaging technique. Using ibrutinib,[88] a covalent irreversible Bruton's tyrosine kinase (BTK) inhibitor and olaparib, a PARP1 inhibitor, they synthesized corresponding fluorescent probes of each drug labeled with the fluorescent dye BODIPY-FL. While these probes enabled the authors to profile the parent compounds against their corresponding target but, these probes could be used to profile other drugs as well. What is particularly exciting about the potential of this new application will be the ability to assess the microenvironment and heterogeneity for the in vivo experiments afforded by the high spatial resolution allowing single cell analysis.

Fluorescence resonance energy transfer (FRET) experiments designed to study target engagement require a ligand linked to a fluorophore at a site that does not negatively affect binding and also the target protein expressed with a suitable fluorescent protein. These two components can then generate a FRET pair. Measuring alterations in the fluorescence lifetime of the donor using fluorescence lifetime imaging microscopy (FLIM) allows monitoring of target engagement and can map sites of interaction within a cell.[89] FRET–FLIM was used to confirm the direct binding of deltarasin to PDEδ in live PANC-1 cells supporting the hypothesis that inhibition of the PDEd–KRAS interaction by binding of deltarasin to PDEd leads to a loss of KRAS spatial organization.[90] To note, a similar study wherein both the proteins are labeled would be sufficient to support the linkage between protein–protein disruption and phenotype. However, that would not provide information about where the small molecule is binding.

Bioluminescence resonance energy transfer (BRET) is the process of energy transfer between a luminescent donor and a fluorescent acceptor wherein the luciferase enzyme provides the requisite resonance energy by oxidizing a substrate resulting in light at 480 nm. As in FRET, when a suitable acceptor is in close proximity a longer wavelength of light is emitted. Promega Corporation have applied their expertise in luciferase assay systems to develop their nanoBRET™ system which combines a small (19 kDa) and bright luciferase they termed Nanoluc™ to tag intracellular proteins that can probe target engagement between the tagged protein and a long-wavelength fluorophore tagged small molecule.[91] A powerful advantage of this approach is that a single tracer can be used to study multiple isoforms of an enzyme class and in addition binding kinetics can be determined in live cells. Promega demonstrated the effectiveness of this approach by studying HDACs,[92] using Vorinostat®, a potent class I HDAC inhibitor from Merck, as its tracer to study all class I HDAC isoforms that had been individually fused with NanoLuc™ against seven structurally diverse HDAC inhibitors. The aforementioned ability to study binding kinetics allowed them to confirm a long intracellular residence time for one of the inhibitors, FK228, on HDAC1, explaining its extended intracellular efficacy.[93] The ability to also perform these experiments in cell lysates confirmed that the observed slow off-rate was not a result of intracellular trapping.

DiscoverX has developed two target engagement assays, InCell Hunter™ and InCell Pulse™ based on their enzyme fragment complementation technology. These two technologies require the protein of interest to be fused with a β-galactosidase enzyme donor tag that when combined with a β-galactosidase enzyme acceptor fragment create an active β-galactosidase enzyme, which then converts its substrate into a chemiluminescent signal.

InCell Hunter™ measures the protein levels and how they are altered upon compound treatment. This technology was used to demonstrate the engagement of a small molecule, SGC707, to protein arginine methyltransferase 3 (PRMT3) in cells. SGC707 stabilized the PRMT3 levels in two different cell lines in a dose-dependent manner.[94] An interesting note of this study is that SGC707 is an allosteric inhibitor highlighting that InCELL Hunter™ works with nonactive site binders. InCell Hunter has also been used to study another epigenetic target, BRD4.[95] InCell Pulse™, like CETSA, is also based on the alteration of a protein thermal stability upon ligand binding but provides an internal luminescence readout through the fusion-protein. InCell Pulse™ uses pulse denaturation of proteins via repeated intervals of heating in comparison to a single heat shock step. No published examples were found given the recent development of InCell Pulse™.

This section has aimed to provide a highlight of a number of emerging ways to study target engagement studies for small molecules. In planning how to probe target engagement one must think of their experimental requirements and the information they hope to obtain in determining which approach(es) to use. The above examples highlight the range of needs and information that are required and can be learned, respectively, from each technology.

4. TARGET VALIDATION

Target validation is one of the most important steps of the drug discovery workflow. Ensuring that molecular target modulation is responsible for the desired phenotype to ultimately achieve therapeutic efficacy can be achieved using a combination of complementary approaches described here.[96] As described in chapter "Phenotypic Screening" by Plowright and Drowley, when starting a novel screen it is recommended to choose a disease relevant cell and phenotype. Following the identification of high-quality chemical tools, pharmacoproteomics can be used to assess the effects of drug targeting on the proteome, highlighting potential off-target effects, and identifying potential biomarkers of therapeutic efficacy.[7] Chemical proteomics experiments can then inform on binding event and selectivity profiling in physiological environments; CETSA will provide evidence for cellular target engagement[5]; PROteolysis TArgeting Chimeras (PROTACs) will allow monitoring of the cellular phenotype upon small molecule-driven protein degradation under spatial and temporal control[97]; genetic techniques involving RNA interference (siRNA, shRNA),[98] clustered regularly interspaced short palindromic repeats (CRISPR)–Cas systems,[99,100] transcription activator-like effector nucleases,[101] and zinc-finger nucleases[102] will enable the efficient introduction of engineered alterations into the genome to monitor gene functions; and generation of cell lines expressing mutated target proteins will inform on small-molecule sensitivity.[6,103]

A typical workflow including the overall strategy for target identification in the context of phenotypic screens followed by several orthogonal validation methods is depicted in Figs. 7 and 8.

In this session, we will describe exemplary examples of target validation using the PROTACs technology and genetic methods.

4.1 Target Validation by Protein Degradation Approaches

PROTACs are heterobifunctional molecules capable of recruiting the cellular proteasome system to degrade proteins.[97] Tim Rasmusson, Nello Mainolfi (For further details, see chapter "Targeted Protein Degradation" by Mainolfi and Rasmusson.) They consist of three components: a target protein-binding moiety, a degradation machinery recruiting unit, and a linker region that couples these two functionalities. Typically, the utilized degradation machinery is the ubiquitin–proteasome system which functions through the recruitment of an E3 ubiquitin ligase resulting in ubiquitination

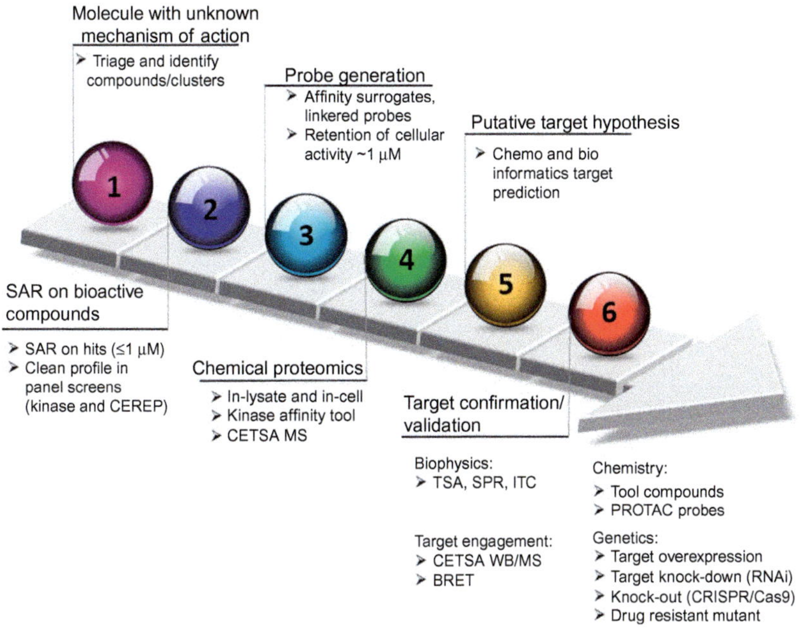

Fig. 7 Target validation in the context of target deconvolution.

of the target protein and its subsequent degradation by the proteasome (Fig. 9). The technology can be used as a target validation method to study the biological effects of knocking down targets that cannot be readily modulated by siRNA/CRISPR or other genetic approaches. Small molecule chemical probes or inhibitors acting at the posttranslational level hold several advantages for target validation over genetic techniques such as dominant-negative mutants or knockouts and RNAi knockdowns, including affording spatial and temporal control in a reversible fashion. A general limitation associated with conventional occupancy-driven target inhibition is that it often demands full target engagement, requiring sustained high concentration of a potent small molecule inhibitor over a prolonged time. This in turn enhances off-target effects and can lead to unwanted side effects or toxicity confounding cellular readouts for target validation. To provide an alternative small molecule approach that could address these issues, Ciulli and collaborators designed a PROTAC molecule (MZ1) that can remove BET proteins entirely from the cell as opposed to just inhibit them, yielding new tools for studying BET bromodomain proteins and validating them as drug targets.[104] BET (Bromodomain and Extra-Terminal) inhibitors show no selectivity for individual BET family members, thereby limiting

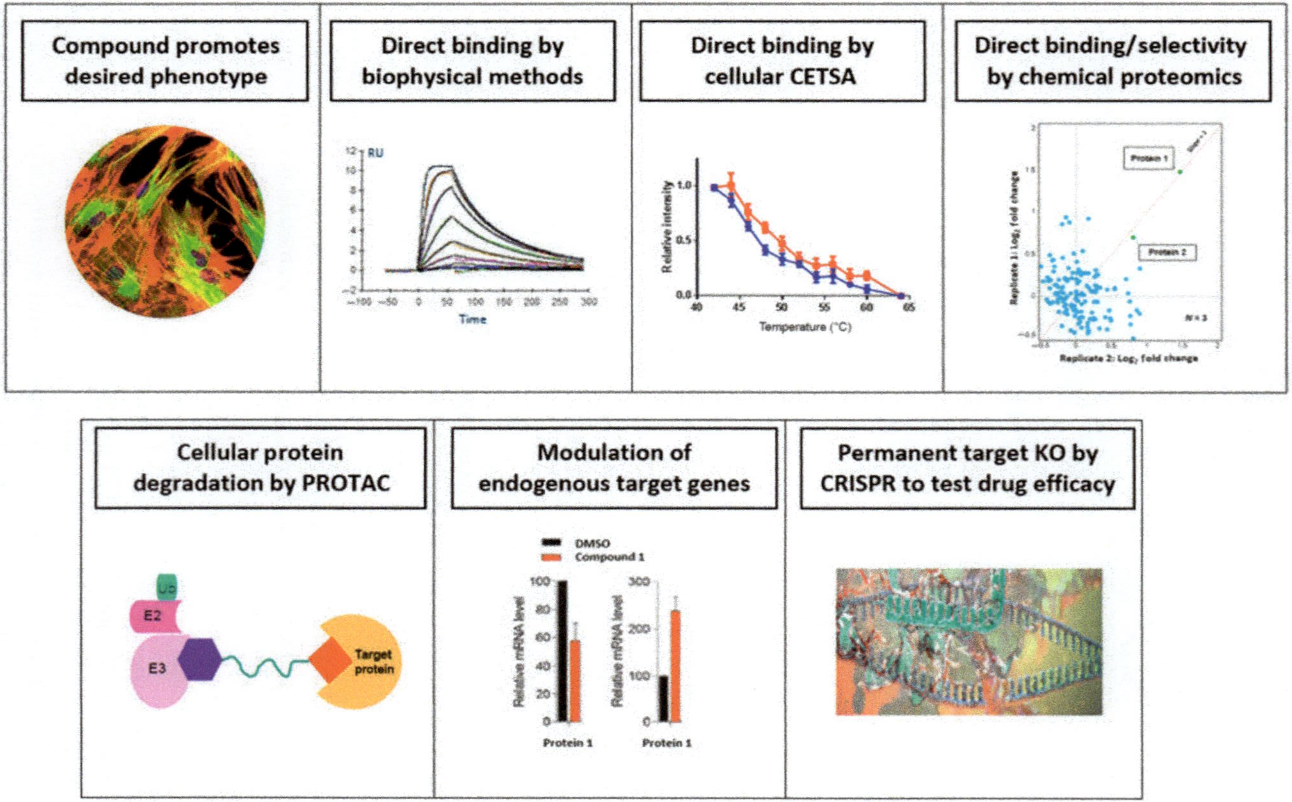

Fig. 8 Complementary approaches for target validation.

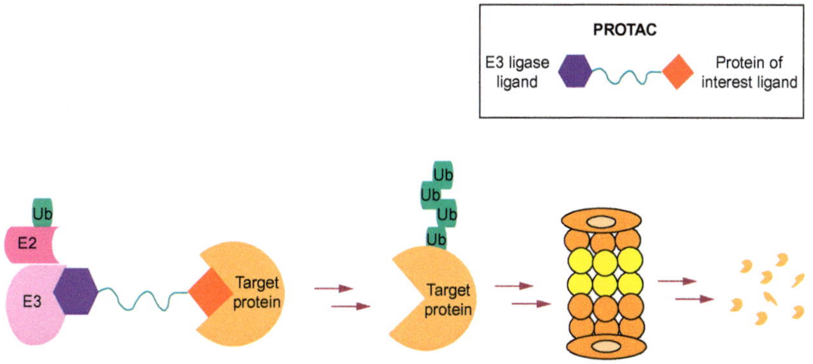

Fig. 9 Schematic representation of PROTACs.

their scope as chemical probes for validating the roles of individual BET targets in physiology and disease. Therapeutically, the effects of BET inhibitors on different transcriptional pathways have raised concerns about the safety and tolerability of BET inhibitors in humans. Crucially, none of the inhibitors described to date is selective for binding BRD4 bromodomains over BRD2 and BRD3. Compound MZ1 potently and rapidly induced reversible, long-lasting, and unexpectedly selective removal of BRD4 over BRD2 and BRD3. Gene expression profiles of selected cancer-related genes responsive to JQ1 revealed distinct and more limited transcriptional responses induced by MZ1, consistent with selective suppression of BRD4.

4.2 Target Validation by Genetic Methods

CRISPR is a method of gene editing that utilizes the Cas9 protein and specific guide RNAs to either disrupt genes or insert sequences of interest. Initially used in bacteria as an adaptive immunity response, CRISPR has since been utilized in the biological field as a new alternative to genome engineering.[99,100]

Cas9 (CRISPR-associated protein 9) is a naturally occurring enzyme found in some bacteria that is used for immunity. Cas9 works by using guide RNA with short sequences complimentary to potential foreign DNA to break the DNA strand and combating infection. This mechanism has similarities to RNA interference found in many eukaryotes. Due to its capabilities, Cas9 has recently been used in experiments to serve as a genome-editing tool.

In the context of CRISPR, Cas9 with designed RNAs are introduced into the cells as donor templates. Cas9 searches through the cells for a

DNA sequence that corresponds to the sequence in the designed guide RNA. Once the sequence is found, Cas9 is able to unwind the DNA helix creating a double-stranded break in the cell's DNA and repair enzymes then repair the double-stranded break while integrating new experimental DNA.

In addition to its value for target identification, CRISPR can have a major impact on target validation by enabling the efficient introduction of engineered alterations into the genome.

A recent illustrative and impactful example of CRISPR target validation was applied to the MTH1 enzyme.[105] MTH1 is one of the "housekeeping" enzymes that are responsible for hydrolyzing damaged nucleotides in cells and thus prevent them from being incorporated into DNA. Recent papers claimed a role for MTH1 in the general survival of cancer cells and described experiments to support a "cancer lethal" phenotype.[106] These findings were later contradicted by several other scientific reports indicating that selective MTH1 inhibitors can potently inhibits MTH1 activity in cancer cells but do not actually kill the cells. These observations were ultimately validated by genetic studies which showed that cancer cells remained viable after silencing the MTH1 gene using siRNA or CRISPR.[105,107]

5. CONCLUSION AND FUTURE DIRECTIONS IN CHEMICAL BIOLOGY

Chemical biology has contributed many important tools which have increased our ability to measure and perturb biological functions with a degree of specificity that was simply inaccessible less than a decade ago. From merely an academic exercise, we have witnessed the application of chemical biology approaches in large pharmaceutical organizations and more recently in biotech/start-up companies.

This chapter is intended to provide guidance to the audience for the application of the available technologies to address specific questions during the drug development process in order to reduce attrition rates in the clinic. In particular it is considered highly valuable to thoroughly validate biological targets at the beginning of projects using a combination of genetic means and high quality chemical probes bearing the desired selectivity for the intended cellular activity. Chemical proteomics and CETSA techniques can be utilized for phenotypic screen hits target deconvolution alongside computational approaches. In addition, mass spectrometry-based approaches can be utilized for advanced lead and candidate drugs to inform on drug selectivity and off-target liabilities.

Over the next decade, we anticipate that chemical biology will become even further embedded within the pharmaceutical industry as a core discipline for all drug-discovery projects and translational research.

Application of chemoproteomics, CETSA and PROTACs will be advancing from preclinical research to clinical settings to understand drug–target interactions and to identify and validate new therapeutically relevant targets in the physiologically relevant environments of cells, tissue, or whole organisms.

REFERENCES

1. Bucci, M.; Goodman, C.; Sheppard, T. L. A Decade of Chemical Biology. *Nat. Chem. Biol.* **2010**, *6*(12), 847–854.
2. Cohen, M. S.; Hadjivassiliou, H.; Taunton, J. A Clickable Inhibitor Reveals Context-Dependent Autoactivation of p90 RSK. *Nat. Chem. Biol.* **2007**, *3*(3), 156–160.
3. Bunnage, M. E.; Chekler, E. L.; Jones, L. H. Target Validation Using Chemical Probes. *Nat. Chem. Biol.* **2013**, *9*(4), 195–199.
4. Bunnage, M. E. Getting Pharmaceutical R&D Back on Target. *Nat. Chem. Biol.* **2011**, 7, 335–339.
5. Martinez Molina, D.; Jafari, R.; Ignatushchenko, M.; Seki, T.; Larsson, E. A.; Dan, C.; Sreekumar, L.; Cao, Y.; Nordlund, P. Monitoring Drug Target Engagement in Cells and Tissues Using the Cellular Thermal Shift Assay. *Science* **2013**, *341*(6141), 84–87.
6. Bunnage, M. E.; Gilbert, A. M.; Jones, L. H.; Hett, E. C. Know Your Target, Know Your Molecule. *Nat. Chem. Biol.* **2015**, *11*(6), 368–372.
7. Schirle, M.; Bantscheff, M.; Kuster, B. Mass Spectrometry-Based Proteomics in Preclinical Drug Discovery. *Chem. Biol.* **2012**, *19*(1), 72–84.
8. Lee, J.; Bogyo, M. Target Deconvolution Techniques in Modern Phenotypic Profiling. *Curr. Opin. Chem. Biol.* **2013**, *17*(1), 118–126.
9. Zinn, N.; Hopf, C.; Drewes, G.; Bantscheff, M. Mass Spectrometry Approaches to Monitor Protein-Drug Interactions. *Methods* **2012**, *57*(4), 430–440.
10. Huang, S. M.; Mishina, Y. M.; Liu, S.; Cheung, A.; Stegmeier, F.; Michaud, G. A.; Charlat, O.; Wiellette, E.; Zhang, Y.; Wiessner, S.; Hild, M.; Shi, X.; Wilson, C. J.; Mickanin, C.; Myer, V.; Fazal, A.; Tomlinson, R.; Serluca, F.; Shao, W.; Cheng, H.; Shultz, M.; Rau, C.; Schirle, M.; Schlegl, J.; Ghidelli, S.; Fawell, S.; Lu, C.; Curtis, D.; Kirschner, M. W.; Lengauer, C.; Finan, P. M.; Tallarico, J. A.; Bouwmeester, T.; Porter, J. A.; Bauer, A.; Cong, F. Tankyrase Inhibition Stabilizes Axin and Antagonizes Wnt Signalling. *Nature* **2009**, *461*(7264), 614–620.
11. Dale, T.; Clarke, P. A.; Esdar, C.; Waalboer, D.; Adeniji-Popoola, O.; Ortiz-Ruiz,-M. J.; Mallinger, A.; Samant, R. S.; Czodrowski, P.; Musil, D.; Schwarz, D.; Schneider, K.; Stubbs, M.; Ewan, K.; Fraser, E.; TePoele, R.; Court, W.; Box, G.; Valenti, M.; de Haven Brandon, A.; Gowan, S.; Rohdich, F.; Raynaud, F.; Schneider, R.; Poeschke, O.; Blaukat, A.; Workman, P.; Schiemann, K.; Eccles, S. A.; Wienke, D.; Blagg, J. A Selective Chemical Probe for Exploring the Role of CDK8 and CDK19 in Human Disease. *Nat. Chem. Biol.* **2015**, *11*(12), 973–980.
12. Ito, T.; Ando, H.; Suzuki, T.; Ogura, T.; Hotta, K.; Imamura, Y.; Yamaguchi, Y.; Handa, H. Identification of a Primary Target of Thalidomide Teratogenicity. *Science* **2010**, *327*(5971), 1345–1350.
13. Bantscheff, M.; Eberhard, D.; Abraham, Y.; Bastuck, S.; Boesche, M.; Hobson, S.; Mathieson, T.; Perrin, J.; Raida, M.; Rau, C.; Reader, V.; Sweetman, G.;

Bauer, A.; Bouwmeester, T.; Hopf, C.; Kruse, U.; Neubauer, G.; Ramsden, N.; Rick, J.; Kuster, B.; Drewes, G. Quantitative Chemical Proteomics Reveals Mechanisms of Action of Clinical ABL Kinase Inhibitors. *Nat. Biotechnol.* **2007**, *25*(9), 1035–1044.

14. Médard, G.; Pachl, F.; Ruprecht, B.; Klaeger, S.; Heinzlmeir, S.; Helm, D.; Qiao, H.; Ku, X.; Wilhelm, M.; Kuehne, T.; Wu, Z.; Dittmann, A.; Hopf, C.; Kramer, K.; Kuster, B. Optimized Chemical Proteomics Assay for Kinase Inhibitor Profiling. *J. Proteome Res.* **2015**, *14*(3), 1574–1586.
15. Bantscheff, M.; Hopf, C.; Savitski, M. M.; Dittmann, A.; Grandi, P.; Michon, A. M.; Schlegl, J.; Abraham, Y.; Becher, I.; Bergamini, G.; Boesche, M.; Delling, M.; Dümpelfeld, B.; Eberhard, D.; Huthmacher, C.; Mathieson, T.; Poeckel, D.; Reader, V.; Strunk, K.; Sweetman, G.; Kruse, U.; Neubauer, G.; Ramsden, N. G.; Drewes, G. Chemoproteomics Profiling of HDAC Inhibitors Reveals Selective Targeting of HDAC Complexes. *Nat. Biotechnol.* **2011**, *29*(3), 255–265.
16. Yang, P.; Liu, K. Activity-Based Protein Profiling: Recent Advances in Probe Development and Applications. *Chembiochem* **2015**, *16*(5), 712–724.
17. Sanman, L. E.; Bogyo, M. Activity-Based Profiling of Proteases. *Annu. Rev. Biochem.* **2014**, *83*, 249–273.
18. Martell, J.; Weerapana, E. Applications of Copper-Catalyzed Click Chemistry in Activity-Based Protein Profiling. *Molecules* **2014**, *19*(2), 1378–1393.
19. Rostovtsev, V. V.; Green, L. G.; Fokin, V. V.; Sharpless, K. B. A Stepwise Huisgen Cycloaddition Process: Copper(I)-Catalyzed Regioselective "Ligation" of Azides and Terminal Alkynes. *Angew. Chem. Int. Ed. Engl.* **2002**, *41*(14), 2596–2599.
20. Arastu-Kapur, S.; Ponder, E. L.; Fonović, U. P.; Yeoh, S.; Yuan, F.; Fonović, M.; Grainger, M.; Phillips, C. I.; Powers, J. C.; Bogyo, M. Identification of Proteases that Regulate Erythrocyte Rupture by the Malaria Parasite Plasmodium falciparum. *Nat. Chem. Biol.* **2008**, *4*(3), 203–213.
21. Leung, D.; Hardouin, C.; Boger, D. L.; Cravatt, B. F. Discovering Potent and Selective Reversible Inhibitors of Enzymes in Complex Proteomes. *Nat. Biotechnol.* **2003**, *21*(6), 687–691.
22. Adibekian, A.; Martin, B. R.; Chang, J. W.; Hsu, K. L.; Tsuboi, K.; Bachovchin, D. A.; Speers, A. E.; Brown, S. J.; Spicer, T.; Fernandez-Vega, V.; Ferguson, J.; Hodder, P. S.; Rosen, H.; Cravatt, B. F. Confirming Target Engagement for Reversible Inhibitors in vivo by Kinetically Tuned Activity-Based Probes. *J. Am. Chem. Soc.* **2012**, *134*(25), 10345–10348.
23. Lapinsky, D. J.; Johnson, D. S. Recent Developments and Applications of Clickable Photoprobes in Medicinal Chemistry and Chemical Biology. *Future Med. Chem.* **2015**, 7(16), 2143–2171.
24. Smith, E.; Collins, I. Photoaffinity Labeling in Target- and Binding-Site Identification. *Future Med. Chem.* **2015**, 7(2), 159–183.
25. Zuhl, A. M.; Nolan, C. E.; Brodney, M. A.; Niessen, S.; Atchison, K.; Houle, C.; Karanian, D. A.; Ambroise, C.; Brulet, J. W.; Beck, E. M.; Doran, S. D.; O'Neill, B. T.; Am Ende, C. W.; Chang, C.; Geoghegan, K. F.; West, G. M.; Judkins, J. C.; Hou, X.; Riddell, D. R.; Johnson, D. S. Chemoproteomic Profiling Reveals that Cathepsin D Off-Target Activity Drives Ocular Toxicity of β-Secretase Inhibitors. *Nat. Commun.* **2016**, 7, 13042.
26. Niphakis, M. J.; Lum, K. M.; Cognetta, A. B.; Correia, B. E.; Ichu, T. A.; Olucha, J.; Brown, S. J.; Kundu, S.; Piscitelli, F.; Rosen, H.; Cravatt, B. F. A Global Map of Lipid-Binding Proteins and Their Ligandability in Cells. *Cell* **2015**, *161*(7), 1668–1680.
27. Koutsoukas, A.; Simms, B.; Kirchmair, J.; Bond, P. J.; Whitmore, A. V.; Zimmer, S.; Young, M. P.; Jenkins, J. L.; Glick, M.; Glen, R. C.; Bender, A. From In Silico Target Prediction to Multi-Target Drug Design: Current Databases, Methods and Applications. *J. Proteomics* **2011**, *74*(12), 2554–2574.

28. Guney, E.; Menche, J.; Vidal, M.; Barábasi, A. L. Network-Based In Silico Drug Efficacy Screening. *Nat. Commun.* **2016**, *7 . 10331.*
29. Lamb, J. The Connectivity Map: A New Tool for Biomedical Research. *Nat. Rev. Cancer* **2007**, 7(1), 54–60.
30. Rognan, D. Structure-Based Approaches to Target Fishing and Ligand Profiling. *Mol. Inf.* **2010**, *29*(3), 176–187.
31. Keiser, M. J.; Setola, V.; Irwin, J. J.; Laggner, C.; Abbas, A. I.; Hufeisen, S. J.; Jensen, N. H.; Kuijer, M. B.; Matos, R. C.; Tran, T. B.; Whaley, R.; Glennon, R. A.; Hert, J.; Thomas, K. L.; Edwards, D. D.; Shoichet, B. K.; Roth, B. L. Predicting New Molecular Targets for Known Drugs. *Nature* **2009**, *462*(7270), 175–181.
32. Nidhi; Glick, M.; Davies, J. W.; Jenkins, J. L. Prediction of Biological Targets for Compounds Using Multiple-Category Bayesian Models Trained on Chemogenomics Databases. *J. Chem. Inf. Model.* **2006**, *46*(3), 1124–1133.
33. Li, H.; Gao, Z.; Kang, L.; Zhang, H.; Yang, K.; Yu, K.; Luo, X.; Zhu, W.; Chen, K.; Shen, J.; Wang, X.; Jiang, H. TarFisDock: A Web Server for Identifying Drug Targets With Docking Approach. *Nucleic Acids Res.* **2006**, *34*(Web Server issue), W219–24.
34. Cheng, T.; Li, Q.; Wang, Y.; Bryant, S. H. Identifying Compound-Target Associations by Combining Bioactivity Profile Similarity Search and Public Databases Mining. *J. Chem. Inf. Model.* **2011**, *51*(9), 2440–2448.
35. Simon, K. K.; Susan, S.; Eugene, M. F.; Joseph, D. A.; Ralph, T. M.; Robert, P. S. Chemical Similarity Using Physiochemical Property Descriptors. *J. Chem. Inf. Comput. Sci.* **1996**, *36*(1), 118–127.
36. Hull, R. D.; Singh, S. B.; Nachbar, R. B.; Sheridan, R. P.; Kearsley, S. K.; Fluder, E. M. Latent Semantic Structure Indexing (LaSSI) for Defining Chemical Similarity. *J. Med. Chem.* **2001**, *44*(8), 1177–1184.
37. Durant, J. L.; Leland, B. A.; Henry, D. R.; Nourse, J. G. Reoptimization of MDL Keys for Use in Drug Discovery. *J. Chem. Inf. Comput. Sci.* **2002**, *42*(6), 1273–1280.
38. Schuffenhauer, A.; Floersheim, P.; Acklin, P.; Jacoby, E. Similarity Metrics for Ligands Reflecting the Similarity of the Target Proteins. *J. Chem. Inf. Comput. Sci.* **2003**, *43*(2), 391–405.
39. Roy, K.; Mitra, I. Electrotopological State Atom (E-State) Index in Drug Design, QSAR, Property Prediction and Toxicity Assessment. *Curr. Comput. Aided Drug Des.* **2012**, *8*(2), 135–158.
40. Kawabata, T. MATRAS: A Program for Protein 3D Structure Comparison. *Nucleic Acids Res.* **2003**, *31*(13), 3367–3369.
41. Lo, Y. C.; Senese, S.; Damoiseaux, R.; Torres, J. Z. 3D Chemical Similarity Networks for Structure-Based Target Prediction and Scaffold Hopping. *ACS Chem. Biol.* **2016**, *11*(8), 2244–2253.
42. Hendlich, M.; Bergner, A.; Günther, J.; Klebe, G. Relibase: Design and Development of a Database for Comprehensive Analysis of Protein-Ligand Interactions. *J. Mol. Biol.* **2003**, *326*(2), 607–620.
43. Sheridan, R. P.; Nachbar, R. B.; Bush, B. L. Extending the Trend Vector: The Trend Matrix and Sample-Based Partial Least Squares. *J. Comput. Aided Mol. Des.* **1994**, *8*(3), 323–340.
44. Miller, M. D.; Kearsley, S. K.; Underwood, D. J.; Sheridan, R. P. FLOG: A System to Select 'quasi-flexible' Ligands Complementary to a Receptor of Known Three-Dimensional Structure. *J. Comput. Aided Mol. Des.* **1994**, *8*(2), 153–174.
45. Poroikov, V. V.; Filimonov, D. A. How to Acquire New Biological Activities in Old Compounds by Computer Prediction. *J. Comput. Aided Mol. Des.* **2002**, *16*(11), 819–824.
46. Geronikaki, A. A.; Dearden, J. C.; Filimonov, D.; Galaeva, I.; Garibova, T. L.; Gloriozova, T.; Krajneva, V.; Lagunin, A.; Macaev, F. Z.; Molodavkin, G.;

Poroikov, V. V.; Pogrebnoi, S. I.; Shepeli, F.; Voronina, T. A.; Tsitlakidou, M.; Vlad, L. Design of New Cognition Enhancers: From Computer Prediction to Synthesis and Biological Evaluation. *J. Med. Chem.* **2004**, *47*(11), 2870–2876.

47. Chen, Y. Z.; Zhi, D. G. Ligand-Protein Inverse Docking and Its Potential Use in the Computer Search of Protein Targets of a Small Molecule. *Proteins* **2001**, *43*(2), 217–226.
48. Deng, Z.; Chuaqui, C.; Singh, J. Structural Interaction Fingerprint (SIFt): A Novel Method for Analyzing Three-Dimensional Protein-Ligand Binding Interactions. *J. Med. Chem.* **2004**, *47*(2), 337–344.
49. Sliwoski, G.; Kothiwale, S.; Meiler, J.; Lowe, E. W., Jr. Computational Methods in Drug Discovery. *Pharmacol. Rev.* **2014**, *66*(1), 334–395.
50. Reker, D.; Rodrigues, T.; Schneider, P.; Schneider, G. Identifying the Macromolecular Targets of De Novo-Designed Chemical Entities Through Self-Organizing Map Consensus. *Proc. Natl. Acad. Sci. U. S. A.* **2014**, *111*(11), 4067–4072.
51. Dunkel, M.; Günther, S.; Ahmed, J.; Wittig, B.; Preissner, R. SuperPred: Drug Classification and Target Prediction. *Nucleic Acids Res.* **2008**, *36*(Web Server issue), W55–9.
52. Gfeller, D.; Michielin, O.; Zoete, V. Shaping the Interaction Landscape of Bioactive Molecules. *Bioinformatics* **2013**, *29*(23), 3073–3079.
53. Yamanishi, Y.; Kotera, M.; Moriya, Y.; Sawada, R.; Kanehisa, M.; Goto, S. DINIES: Drug-Target Interaction Network Inference Engine Based on Supervised Analysis. *Nucleic Acids Res.* **2014**, *42*(Web Server issue), W39–W45.
54. Xiao, X.; Min, J. L.; Lin, W. Z.; Liu, Z.; Cheng, X.; Chou, K. C. iDrug-Target: Predicting the Interactions Between Drug Compounds and Target Proteins in Cellular Networking Via Benchmark Dataset Optimization Approach. *J. Biomol. Struct. Dyn.* **2015**, *33*(10), 2221–2233.
55. Patterson, D. E.; Cramer, R. D.; Ferguson, A. M.; Clark, R. D.; Weinberger, L. E. Neighborhood Behavior: A Useful Concept for Validation of "Molecular Diversity" Descriptors. *J. Med. Chem.* **1996**, *39*(16), 3049–3059.
56. Engel, T. Basic Overview of Chemoinformatics. *J. Chem. Inf. Model.* **2006**, *46*(6), 2267–2277.
57. Yan, X.; Liao, C.; Liu, Z.; Hagler, A. T.; Gu, Q.; Xu, J. Chemical Structure Similarity Search for Ligand-Based Virtual Screening: Methods and Computational Resources. *Curr. Drug Targets* **2016**, *17*(14), 1580–1585.
58. Cross, S.; Baroni, M.; Goracci, L.; Cruciani, G. GRID-Based Three-Dimensional Pharmacophores I: FLAPpharm, a Novel Approach for Pharmacophore Elucidation. *J. Chem. Inf. Model.* **2012**, *52*(10), 2587–2598.
59. Bajorath, J. Molecular Similarity Concepts for Informatics Applications. *Methods Mol. Biol.* **2017**, *1526*, 231–245.
60. Downs, G. M.; Willett, P. *Similarity Searching in Databases of Chemical Structures*; John Wiley & Sons, Inc.: Hoboken, NJ, USA, 1995
61. Peter, W.; Barnard, J. M.; Geoffrey, M. D. Chemical Similarity Searching. *J. Chem. Inf. Comput. Sci.* **1998**, *38*, 983–996.
62. Bajorath, J. Selected Concepts and Investigations in Compound Classification, Molecular Descriptor Analysis, and Virtual Screening. *J. Chem. Inf. Comput. Sci.* **2001**, *41*(2), 233–245.
63. Cheng, A.; Diller, D. J.; Dixon, S. L.; Egan, W. J.; Lauri, G.; Merz, K. M. Computation of the Physio-Chemical Properties and Data Mining of Large Molecular Collections. *J. Comput. Chem.* **2002**, *23*(1), 172–183.
64. Schuur, J. H.; Selzer, P.; Gasteiger, J. The Coding of the Three-Dimensional Structure of Molecules by Molecular Transforms and Its Application to Structure-Spectra Correlations and Studies of Biological Activity. *J. Chem. Inf. Comput. Sci.* **1996**, *36*(2), 334–344.

65. Ginn, C. M. R.; Willett, P.; Bradshaw, J. *Combination of Molecular Similarity Using Data Fusion Perspect*; Kluwer Academic Publishers: Dordrecht, 2000.
66. Huang, Q.; Jin, H.; Liu, Q.; Wu, Q.; Kang, H.; Cao, Z.; Zhu, R. Proteochemometric Modeling of the Bioactivity Spectra of HIV-1 Protease Inhibitors by Introducing Protein-Ligand Interaction Fingerprint. *PLoS One* **2012**, 7(7) . e41698.
67. Humbeck, L.; Koch, O. What Can We Learn From Bioactivity Data? Chemoinformatics Tools and Applications in Chemical Biology Research. *ACS Chem. Biol.* **2017**, *12*(1), 23–35.
68. Koutsoukas, A.; Lowe, R.; Kalantarmotamedi, Y.; Mussa, H. Y.; Klaffke, W.; Mitchell, J. B.; Glen, R. C.; Bender, A. In Silico Target Predictions: Defining a Benchmarking Data Set and Comparison of Performance of the Multiclass Naïve Bayes and Parzen-Rosenblatt Window. *J. Chem. Inf. Model.* **2013**, *53*(8), 1957–1966.
69. Mervin, L. H.; Afzal, A. M.; Drakakis, G.; Lewis, R.; Engkvist, O.; Bender, A. Target Prediction Utilising Negative Bioactivity Data Covering Large Chemical Space. *J. Chem.* **2015**, 7, 51.
70. Gaulton, A.; Bellis, L. J.; Bento, A. P.; Chambers, J.; Davies, M.; Hersey, A.; Light, Y.; McGlinchey, S.; Michalovich, D.; Al-Lazikani, B.; Overington, J. P. ChEMBL: A Large-Scale Bioactivity Database for Drug Discovery. *Nucleic Acids Res.* **2012**, *40*(Database issue), D1100–7.
71. Hudson, B. D.; Hyde, R. M.; Elizabeth, R. Parameter Based Methods for Compound Selection From Chemical Databases. *Quant. Struct. Act. Relat.* **1996**, *15*, 285–289.
72. Gobbi, A.; Lee, M. L. DISE: Directed Sphere Exclusion. *J. Chem. Inf. Comput. Sci.* **2003**, *43*(1), 317–323.
73. Pantoliano, M. W.; Petrella, E. C.; Kwasnoski, J. D.; Lobanov, V. S.; Myslik, J.; Graf, E.; Carver, T.; Asel, E.; Springer, B. A.; Lane, P.; Salemme, F. R. High-Density Miniaturized Thermal Shift Assays as a General Strategy for Drug Discovery. *J. Biomol. Screen.* **2001**, *6*(6), 429–440.
74. Lo, M. C.; Aulabaugh, A.; Jin, G.; Cowling, R.; Bard, J.; Malamas, M.; Ellestad, G. Evaluation of Fluorescence-Based Thermal Shift Assays for Hit Identification in Drug Discovery. *Anal. Biochem.* **2004**, *332*(1), 153–159.
75. Liu, X.; Shi, Y.; Maag, D. X.; Palma, J. P.; Patterson, M. J.; Ellis, P. A.; Surber, B. W.; Ready, D. B.; Soni, N. B.; Ladror, U. S.; Xu, A. J.; Iyer, R.; Harlan, J. E.; Solomon, L. R.; Donawho, C. K.; Penning, T. D.; Johnson, E. F.; Shoemaker, A. R. Iniparib Nonselectively Modifies Cysteine-Containing Proteins in Tumor Cells and Is Not a Bona Fide PARP Inhibitor. *Clin. Cancer Res.* **2012**, *18*(2), 510–523.
76. Peppard, J.; Glickman, F.; He, Y.; Hu, S. I.; Doughty, J.; Goldberg, R. Development of a High-Throughput Screening Assay for Inhibitors of Aggrecan Cleavage Using Luminescent Oxygen Channeling (AlphaScreen). *J. Biomol. Screen.* **2003**, *8*(2), 149–156.
77. Eglen, R. M.; Reisine, T.; Roby, P.; Rouleau, N.; Illy, C.; Bossé, R.; Bielefeld, M. The Use of AlphaScreen Technology in HTS: Current Status. *Curr. Chem. Genomics* **2008**, *1*, 2–10.
78. Almqvist, H.; Axelsson, H.; Jafari, R.; Dan, C.; Mateus, A.; Haraldsson, M.; Larsson, A.; Martinez Molina, D.; Artursson, P.; Lundbäck, T.; Nordlund, P. CETSA Screening Identifies Known and Novel Thymidylate Synthase Inhibitors and Slow Intracellular Activation of 5-Fluorouracil. *Nat. Commun.* **2016**, 7, 11040.
79. Longley, D. B.; Harkin, D. P.; Johnston, P. G. 5-Fluorouracil: Mechanisms of Action and Clinical Strategies. *Nat. Rev. Cancer* **2003**, *3*(5), 330–338.
80. Savitski, M. M.; Reinhard, F. B.; Franken, H.; Werner, T.; Savitski, M. F.; Eberhard, D.; Martinez Molina, D.; Jafari, R.; Dovega, R. B.; Klaeger, S.; Kuster, B.; Nordlund, P.; Bantscheff, M.; Drewes, G. Tracking Cancer Drugs in Living Cells by Thermal Profiling of the Proteome. *Science* **2014**, *346*(6205), 1255784.

81. Gelot, P.; Dutartre, H.; Khammari, A.; Boisrobert, A.; Schmitt, C.; Deybach, J. C.; Nguyen, J. M.; Seité, S.; Dréno, B. Vemurafenib: An Unusual UVA-Induced Photosensitivity. *Exp. Dermatol.* **2013**, *22*(4), 297–298.
82. Becher, I.; Werner, T.; Doce, C.; Zaal, E. A.; Tögel, I.; Khan, C. A.; Rueger, A.; Muelbaier, M.; Salzer, E.; Berkers, C. R.; Fitzpatrick, P. F.; Bantscheff, M.; Savitski, M. M. Thermal Profiling Reveals Phenylalanine Hydroxylase as an Off-Target of Panobinostat. *Nat. Chem. Biol.* **2016**, *12*(11), 908–910.
83. Reinhard, F. B.; Eberhard, D.; Werner, T.; Franken, H.; Childs, D.; Doce, C.; Savitski, M. F.; Huber, W.; Bantscheff, M.; Savitski, M. M.; Drewes, G. Thermal Proteome Profiling Monitors Ligand Interactions with Cellular Membrane Proteins. *Nat. Methods* **2015**, *12*(12), 1129–1131.
84. Huber, K. V.; Olek, K. M.; Müller, A. C.; Tan, C. S.; Bennett, K. L.; Colinge, J.; Superti-Furga, G. Proteome-Wide Drug and Metabolite Interaction Mapping by Thermal-Stability Profiling. *Nat. Methods* **2015**, *12*(11), 1055–1057.
85. Lea, W. A.; Simeonov, A. Fluorescence Polarization Assays in Small Molecule Screening. *Expert Opin. Drug Discovery* **2011**, *6*(1), 17–32.
86. Parker, G. J.; Law, T. L.; Lenoch, F. J.; Bolger, R. E. Development of High Throughput Screening Assays Using Fluorescence Polarization: Nuclear Receptor-Ligand-Binding and Kinase/Phosphatase Assays. *J. Biomol. Screen.* **2000**, *5*(2), 77–88.
87. Dubach, J. M.; Vinegoni, C.; Mazitschek, R.; Fumene Feruglio, P.; Cameron, L. A.; Weissleder, R. In Vivo Imaging of Specific Drug-Target Binding at Subcellular Resolution. *Nat. Commun.* **2014**, *5*, 3946.
88. Dubach, J. M.; Kim, E.; Yang, K.; Cuccarese, M.; Giedt, R. J.; Meimetis, L. G.; Vinegoni, C.; Weissleder, R. Quantitating Drug-Target Engagement in Single Cells in vitro and In Vivo. *Nat. Chem. Biol.* **2017**, *13*(2), 168–173.
89. Kumar, S.; Alibhai, D.; Margineanu, A.; Laine, R.; Kennedy, G.; McGinty, J.; Warren, S.; Kelly, D.; Alexandrov, Y.; Munro, I.; Talbot, C.; Stuckey, D. W.; Kimberly, C.; Viellerobe, B.; Lacombe, F.; Lam, E. W.; Taylor, H.; Dallman, M. J.; Stamp, G.; Murray, E. J.; Stuhmeier, F.; Sardini, A.; Katan, M.; Elson, D. S.; Neil, M. A.; Dunsby, C.; French, P. M. FLIM FRET Technology for Drug Discovery: Automated Multiwell-Plate High-Content Analysis, Multiplexed Readouts and Application In Situ. *Chemphyschem* **2011**, *12*(3), 609–626.
90. Zimmermann, G.; Papke, B.; Ismail, S.; Vartak, N.; Chandra, A.; Hoffmann, M.; Hahn, S. A.; Triola, G.; Wittinghofer, A.; Bastiaens, P. I.; Waldmann, H. Small Molecule Inhibition of the KRAS-PDEδ Interaction Impairs Oncogenic KRAS Signalling. *Nature* **2013**, *497*(7451), 638–642.
91. Machleidt, T.; Woodroofe, C. C.; Schwinn, M. K.; Méndez, J.; Robers, M. B.; Zimmerman, K.; Otto, P.; Daniels, D. L.; Kirkland, T. A.; Wood, K. V. NanoBRET—A Novel BRET Platform for the Analysis of Protein-Protein Interactions. *ACS Chem. Biol.* **2015**, *10*(8), 1797–1804.
92. Robers, M. B.; Dart, M. L.; Woodroofe, C. C.; Zimprich, C. A.; Kirkland, T. A.; Machleidt, T.; Kupcho, K. R.; Levin, S.; Hartnett, J. R.; Zimmerman, K.; Niles, A. L.; Ohana, R. F.; Daniels, D. L.; Slater, M.; Wood, M. G.; Cong, M.; Cheng, Y. Q.; Wood, K. V. Target Engagement and Drug Residence Time Can Be Observed in Living Cells With BRET. *Nat. Commun.* **2015**, *6 . 10091.*
93. Ito, T.; Umehara, T.; Sasaki, K.; Nakamura, Y.; Nishino, N.; Terada, T.; Shirouzu, M.; Padmanabhan, B.; Yokoyama, S.; Ito, A.; Yoshida, M. Real-Time Imaging of Histone H4K12-Specific Acetylation Determines the Modes of Action of Histone Deacetylase and Bromodomain Inhibitors. *Chem. Biol.* **2011**, *18*(4), 495–507.
94. Kaniskan, H.; Szewczyk, M. M.; Yu, Z.; Eram, M. S.; Yang, X.; Schmidt, K.; Luo, X.; Dai, M.; He, F.; Zang, I.; Lin, Y.; Kennedy, S.; Li, F.; Dobrovetsky, E.; Dong, A.;

Smil, D.; Min, S. J.; Landon, M.; Lin-Jones, J.; Huang, X. P.; Roth, B. L.; Schapira, M.; Atadja, P.; Barsyte-Lovejoy, D.; Arrowsmith, C. H.; Brown, P. J.; Zhao, K.; Jin, J.; Vedadi, M. A Potent, Selective and Cell-Active Allosteric Inhibitor of Protein Arginine Methyltransferase 3 (PRMT3). *Angew. Chem. Int. Ed. Engl.* **2015**, *54*(17), 5166–5170.

95. Schulze, J.; Moosmayer, D.; Weiske, J.; Fernández-Montalván, A.; Herbst, C.; Jung, M.; Haendler, B.; Bader, B. Cell-Based Protein Stabilization Assays for the Detection of Interactions Between Small-Molecule Inhibitors and BRD4. *J. Biomol. Screen.* **2015**, *20*(2), 180–189.
96. Butler, G. S.; Overall, C. M. Proteomic Identification of Multitasking Proteins in Unexpected Locations Complicates Drug Targeting. *Nat. Rev. Drug Discov.* **2009**, *8*(12), 935–948.
97. Lai, A. C.; Crews, C. M. Induced Protein Degradation: An Emerging Drug Discovery Paradigm. *Nat. Rev. Drug Discov.* **2017**, *16*(2), 101–114.
98. Boutros, M.; Ahringer, J. The Art and Design of Genetic Screens: RNA Interference. *Nat. Rev. Genet.* **2008**, *9*(7), 554–566.
99. Cong, L.; Ran, F. A.; Cox, D.; Lin, S.; Barretto, R.; Habib, N.; Hsu, P. D.; Wu, X.; Jiang, W.; Marraffini, L. A.; Zhang, F. Multiplex Genome Engineering Using CRISPR/Cas Systems. *Science* **2013**, *339*(6121), 819–823.
100. Haurwitz, R. E.; Jinek, M.; Wiedenheft, B.; Zhou, K.; Doudna, J. A. Sequence- and Structure-Specific RNA Processing by a CRISPR Endonuclease. *Science* **2010**, *329*(5997), 1355–1358.
101. Wood, A. J.; Lo, T. W.; Zeitler, B.; Pickle, C. S.; Ralston, E. J.; Lee, A. H.; Amora, R.; Miller, J. C.; Leung, E.; Meng, X.; Zhang, L.; Rebar, E. J.; Gregory, P. D.; Urnov, F. D.; Meyer, B. J. Targeted Genome Editing Across Species Using ZFNs and TALENs. *Science* **2011**, *333*(6040), 307.
102. Sander, J. D.; Dahlborg, E. J.; Goodwin, M. J.; Cade, L.; Zhang, F.; Cifuentes, D.; Curtin, S. J.; Blackburn, J. S.; Thibodeau-Beganny, S.; Qi, Y.; Pierick, C. J.; Hoffman, E.; Maeder, M. L.; Khayter, C.; Reyon, D.; Dobbs, D.; Langenau, D. M.; Stupar, R. M.; Giraldez, A. J.; Voytas, D. F.; Peterson, R. T.; Yeh, J. R.; Joung, J. K. Selection-Free Zinc-Finger-Nuclease Engineering by Context-Dependent Assembly (CoDA). *Nat. Methods* **2011**, *8*(1), 67–69.
103. Schenone, M.; Dančík, V.; Wagner, B. K.; Clemons, P. A. Target Identification and Mechanism of Action in Chemical Biology and Drug Discovery. *Nat. Chem. Biol.* **2013**, *9*(4), 232–240.
104. Zengerle, M.; Chan, K. H.; Ciulli, A. Selective Small Molecule Induced Degradation of the BET Bromodomain Protein BRD4. *ACS Chem. Biol.* **2015**, *10*(8), 1770–1777.
105. Kettle, J. G.; Alwan, H.; Bista, M.; Breed, J.; Davies, N. L.; Eckersley, K.; Fillery, S.; Foote, K. M.; Goodwin, L.; Jones, D. R.; Käck, H.; Lau, A.; Nissink, J. W.; Read, J.; Scott, J. S.; Taylor, B.; Walker, G.; Wissler, L.; Wylot, M. Potent and Selective Inhibitors of MTH1 Probe Its Role in Cancer Cell Survival. *J. Med. Chem.* **2016**, *59*(6), 2346–2361.
106. Gad, H.; Koolmeister, T.; Jemth, A. S.; Eshtad, S.; Jacques, S. A.; Ström, C. E.; Svensson, L. M.; Schultz, N.; Lundbäck, T.; Einarsdottir, B. O.; Saleh, A.; Göktürk, C.; Baranczewski, P.; Svensson, R.; Berntsson, R. P.; Gustafsson, R.; Strömberg, K.; Sanjiv, K.; Jacques-Cordonnier, M. C.; Desroses, M.; Gustavsson, A. L.; Olofsson, R.; Johansson, F.; Homan, E. J.; Loseva, O.; Bräutigam, L.; Johansson, L.; Höglund, A.; Hagenkort, A.; Pham, T.; Altun, M.; Gaugaz, F. Z.; Vikingsson, S.; Evers, B.; Henriksson, M.; Vallin, K. S.; Wallner, O. A.; Hammarström, L. G.; Wiita, E.; Almlöf, I.; Kalderén, C.; Axelsson, H.; Djureinovic, T.; Puigvert, J. C.; Häggblad, M.; Jeppsson, F.; Martens, U.; Lundin, C.; Lundgren, B.; Granelli, I.; Jensen, A. J.; Artursson, P.;

Nilsson, J. A.; Stenmark, P.; Scobie, M.; Berglund, U. W.; Helleday, T. MTH1 Inhibition Eradicates Cancer by Preventing Sanitation of the dNTP Pool. *Nature* **2014**, *508*(7495), 215–221.

107. Petrocchi, A.; Leo, E.; Reyna, N. J.; Hamilton, M. M.; Shi, X.; Parker, C. A.; Mseeh, F.; Bardenhagen, J. P.; Leonard, P.; Cross, J. B.; Huang, S.; Jiang, Y.; Cardozo, M.; Draetta, G.; Marszalek, J. R.; Toniatti, C.; Jones, P.; Lewis, R. T. Identification of Potent and Selective MTH1 Inhibitors. *Bioorg. Med. Chem. Lett.* **2016**, *26*(6), 1503–1507.

CHAPTER ELEVEN

Fragment-Based Lead Discovery

Ben J. Davis[1], **Stephen D. Roughley**
Vernalis (R&D) Ltd, Granta Park, Cambridge, United Kingdom
[1]Corresponding author: e-mail address: b.davis@vernalis.com

Contents

1. INTRODUCTION

In the 20 years since the seminal "SAR-by-NMR" paper,[1] fragment-based lead discovery (FBLD) has evolved into a mature, well-validated approach for chemical biology and drug discovery.[2–8] Two marketed drugs have been developed using FBLD,[9,10] and at least 30 others are in various

Annual Reports in Medicinal Chemistry, Volume 50
ISSN 0065-7743
https://doi.org/10.1016/bs.armc.2017.07.002

stages of clinical development.[11] In this chapter, we present and summarize the salient points for the nonspecialist, in order to allow an understanding of the technique. Additionally, many excellent reviews and books have been published over the last decade discussing the approaches and philosophy underlying FBLD,[2,12–16] and we urge the reader to refer to these for additional details.

Various names and acronyms are used, more or less widely, to refer to the approach we will discuss as FBLD. Originally termed SAR-by-NMR ("structure–activity relationships by NMR"),[1] the more generic term FBDD ("fragment-based drug discovery") came into widespread use in 2004.[17,18] The acronym FBLD (fragment-based lead discovery, or sometimes fragment-based ligand discovery, fragment-based lead development, and so on) evolved rapidly out of this term to reflect the distinction between leads (or ligands) and drugs.[6] Other terms in less common use are FBHG (fragment-based hit generation),[19] FADD (fragment-assisted drug discovery),[20] and various other combinations of similar acronyms. We will use FBLD in a general sense to refer to all of these approaches except where specific examples are required.

FBLD usually refers to the identification, and subsequent evolution, of low molecular weight molecules which bind to the target of interest. In essence, *FBLD is based on the reductionist concept of attributing a set of interactions (and thus an interaction energy) to a particular moiety or group present in a fragment, and of maintaining these interactions into larger molecules.* The small size of these initial compounds ("fragments") is the key feature of FBLD and is underpinned by two key concepts—molecular complexity[21,22] and the size of drug-like chemical space.[23,24] The development of FBLD is further supported by an earlier body of work around the additivity and dissection of molecular interactions stretching back to the 1980s.[25–28]

Owing to their small size (typically less than 18 nonhydrogen ("heavy") atoms), these initial fragments often have a relatively low interaction energy (ΔG_{bind}) with the target and hence relatively low affinity. This low affinity puts many constraints on the experimental approaches used to identify and characterize these interactions. The parameter "binding energy per heavy atom" is used extensively in FBLD and is typically referred to as "ligand efficiency" (LE).[29] Many alternative formats of this parameter are used (and are the source of much discussion within the FBLD community), but the simple LE value is most widely used.

2. HISTORY AND OVERVIEW OF FBLD

The first published example of the approach which has evolved into FBLD was the work on FKBP by the Fesik group, then at Abbott.[1] This paper introduced the phrase "SAR-by-NMR" to describe the approach of identifying ligands which bound to a target with low affinity, with the intention of subsequently linking two of these small ligands together to yield a larger, high-affinity ligand. Subsequent papers by the same group then built on this approach, applying it to a range of protein targets and using a range of alternative NMR techniques.[30–33]

In the original SAR-by-NMR approach, a library of approximately 10,000 small compounds (average MW 213 Da) were screened at mM concentration using protein-observed NMR (2D ^{15}N–^{1}H HSQC (heteronuclear single-quantum coherence)). In order to reduce the requirement for large amounts of isotopically labeled protein, mixtures of 10 compounds were screened and subsequently deconvoluted to identify the specific hit. Compounds that interacted with the target protein were identified using perturbations in the HSQC spectrum and further characterized by NMR and X-ray crystallography. Structural information was then used to guide the linking of two fragments in such a way as to generate a potent molecule. The rationale for linking two molecules was to dramatically increase the effective number of molecules screened from a relatively small library, and to combine the ΔG_{bind} of both low-affinity fragments in an additive manner. In the FKBP example mentioned previously, this additivity was used to generate an impressively potent 19 nM ligand from two initial hits with affinities of 2 and 800 μM, respectively.

A key feature of this SAR-by-NMR approach was the use of biophysical methods to identify small molecules which interacted with the protein target with low affinity. The binding assay, 2D ^{15}N–^{1}H HSQC NMR, was chosen both to be sensitive to these low-affinity interactions and to have minimal interference from artifacts associated with the high concentrations of compounds present. Many features of this initial work have gone on to form the core of the FBLD approach, notably the use of biophysical methods and of medicinal chemistry driven, structure-guided evolution from the initial low-affinity fragment to more "classical" potent hit or lead molecule.

Initially, few other groups were able to apply the SAR-by-NMR technique successfully. There were various reasons for this slow uptake, ranging from the prosaic (relatively few targets have two distinct ligand-binding sites in close proximity) to the more mundane (SAR-by-NMR, and FBLD, requires close integration and collaboration between biophysicists and medicinal chemists, something which was rather unusual at the time). However, the approach attracted a great deal of attention, being both intellectually attractive and conceptually straightforward, and requiring a relatively low investment compared to more traditional lead identification techniques such as high-throughput screening (HTS). This ease of application led to much of the evolution of the SAR-by-NMR approach into what is now recognized as FBLD being carried out by small biotech companies (such as Astex, Vernalis, Plexxikon, and SGC) and by small groups within larger pharmaceutical companies (such as the "Needles" approach at Roche[34]).

As the initial SAR-by-NMR approach was extended and evolved into FBLD, other biophysical techniques were demonstrated as suitable for fragment-based screening (FBS), the initial step of robustly identifying low-affinity ligands. The Abbott group first applied both 1D "ligand-observed" NMR techniques[35] and crystallographic[36,37] methods, two approaches which have become mainstays of FBS. Surface plasmon resonance (SPR), a technique which evolved and advanced rapidly over this period, was found to be particularly adept at FBS.[38–40] Other techniques, such as thermal shift assay (TSA)[41–43] and more recently microscale thermophoresis (MST),[44] have also been used successfully. Robust biochemical assays have also been successfully used for FBS,[45,46] although the risk of compound interference and artifacts associated with the typically high concentrations used for FBS has resulted in the continued dominance of biophysical methods for FBS.[47]

During the early 2000s, FBLD was often applied to targets which had failed more conventional HTS approaches, particularly in larger companies.[48] However, a key validation of the approach came with the acceptance in 2011 by the FDA of Vemurafenib, an inhibitor of mutant B-Raf V600E developed by Plexxikon using FBLD methods.[9,49] This demonstrated that the perceived hurdles in going from mM to nM affinity were not insurmountable, as had been widely assumed, and that FBLD could generate clinically proven molecules with high efficacy in a timely manner. The approval of the Abbott Bcl-2 inhibitor Venetoclax in 2015,[50] as well as clinical trials with FBLD derived molecules against "undruggable" targets such as BACE,[51] has served to further support the widespread applicability of the FBLD technique.

A theoretical basis for this approach was developed more thoroughly in work by Hann and coworkers at GSK,[21,22] initially in response to the

HTS and combinatorial chemistry paradigms of the 1990s. A simple but insightful model of molecular complexity was developed where interactions between a ligand and a receptor were described as "favorable" or "unfavorable"; analysis of this model for ligands of increasing length demonstrated that, although the strength of the interaction would increase with ligand size, the probability of a ligand matching the receptor site fell dramatically with size (Fig. 1). This decreased probability results from a reduction in the number of ways the ligand features can those of the receptor as the

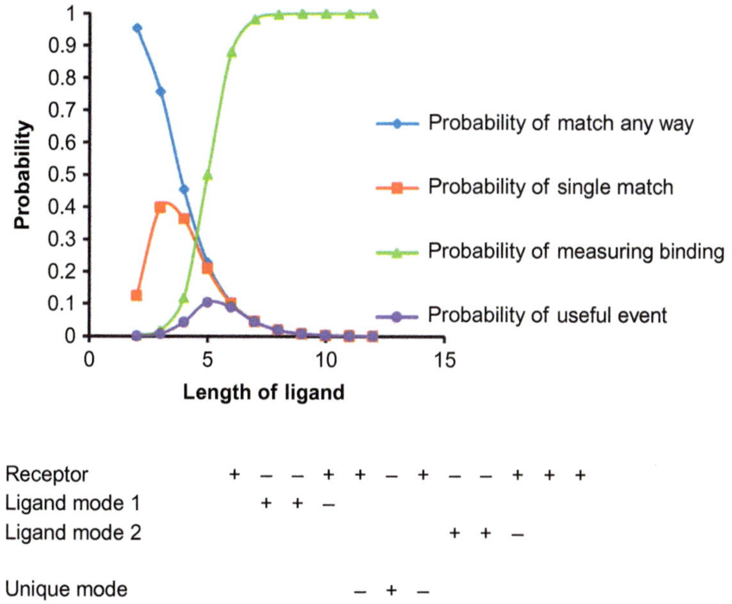

Current Opinion in Chemical Biology

Fig. 1 The complexity model. Ligands and their receptors are represented as linear strings of +/–. A match corresponds to an exact correspondence between the complementary features of ligand and receptor (*top*, which also illustrates that some ligands can match more than one way and other ligands only show a unique mode). The *graph* shows data for receptors of length 12. It is found that the probability of a ligand binding to a randomly chosen receptor in at least one mode decreases rapidly with ligand length (*blue line*). The probability of matching just one way goes through a maximum (*red curve*; in this case, for ligand length 3). The probability of physically detecting a binding event is represented by the *sigmoidal curve* (*green*), giving in *purple* the overall probability of a "useful event" (where the ligand binds in a single-binding mode and can be detected). *Figure reprinted from Leach, A.R.; Hann, M.M. Molecular Complexity and Fragment-Based Drug Discovery: Ten Years on. Curr. Opin. Chem. Biol.* ***2011,*** *15, 489–496, Copyright (2011), with permission from Elsevier.*

size (or complexity) of the ligand increases. However, the chance of identifying a suitable ligand passes through a maximum, dependent on the sensitivity of the binding assay used; as weaker and weaker interactions are detected, the probability of identifying a ligand increases. With a sensitive binding assay, therefore, the model implied that more ligands could be found with a lower size and affinity than was the case with larger, more elaborate ligands.

A second important theoretical tool, which has developed over the same period as FBLD, is the concept of LE.[29,52,53] This parameter is an attempt to normalize the affinity by the size of the molecule. This allows a direct comparison of ligands of widely different affinities and molecular sizes; this in turn can be used to assess whether a fragment is a viable start point for the desired final molecule, to track improvements in affinity during fragment evolution, and to guide the choice and design of an assay used in the FBS stage of an FBLD campaign (Fig. 2). While many variations of the parameter exist, LE in its simplest form describes the binding energy (ΔG_{bind}) per heavy atom.

The way in which fragments are utilized in the drug-discovery program has also evolved substantially since the initial SAR-by-NMR approach. While the initial emphasis was on developing potent hits by linking low-affinity fragments, other evolution strategies (such as fragment growth, morphing, or merging) have become more widespread and more generic; these

K_D

LE (kcal/mol/HA)	10 mM	1 mM	100 μM	10 μM	1 μM	100 nM	10 nM	1 nM
0.15	18	27	36	45	55	64	73	82
0.20	14	20	27	34	41	48	55	61
0.25	11	16	22	27	33	38	44	49
0.30	9	14	18	23	27	32	36	41
0.35	8	12	16	19	23	27	31	35
0.40	7	10	14	17	20	24	27	31
0.45	6	9	12	15	18	21	24	27
0.50	5	8	11	14	16	19	22	25
0.55	5	7	10	12	15	17	20	22
0.60	5	7	9	11	14	16	18	20

HAC

Fig. 2 The calculated size that a molecule with given ligand efficiency must be in order to bind with a range of affinities. Typical fragment sizes are shown in *black*. This calculation can be used effectively to inform the choice of an assay for a fragment-based screening campaign; for example, a library of 14–18 HA screened against a target with an intrinsic LE of approximately 0.3 kcal/HA/mol would be expected to yield hits with affinities in the range of 1–100 mM as shown.

approaches will be discussed in detail later. The incorporation of fragment chemotypes with ligands from other sources (such as HTS, substrate analogues, or existing literature inhibitors) has also proved to be extremely successful.[2,8,54]

3. THE FBLD PROCESS

FBLD has proven to be a robust and generic method for identifying novel chemical hits, which can then be evolved into hits, leads, and ideally approved drugs. As such, FBLD can be considered to be a mainstay of early-stage hit identification. However, there are a number of factors which can affect the suitability of a target for an FBLD campaign. In particular, FBLD approaches often involve biophysical identification and characterization of ligands; although not crucial, it is a significant advantage if the target is amenable to biophysical study. Crystallography is often regarded as a "gate" for an FBLD campaign, but this has been shown to not be the case in all (or even most) circumstances.[55] FBLD campaigns often require significant (mg) quantities of pure, stable protein, and this can also prove to be a hurdle where protein expression is a limiting factor. The output from an FBLD campaign can be applied for both hit identification and elaboration, but also in later stages of the drug-discovery process (such as lead optimization) in order to avoid or bypass issues such as toxicity, synthetic, or intellectual property hurdles.[8,48,56]

In general, an FBLD campaign consists of three major components—the library of compounds to be screened (the "fragment library"), the assay used to identify low-affinity compounds (the "screening method"), and a strategy for evolving the fragment hits into more potent ligands which can be developed using conventional medicinal chemistry techniques. Each of these components is vital to the success of the strategy and will be discussed in detail in the following sections.

The majority of FBLD campaigns are progressed using an approach similar to that shown in Fig. 3.[57] The fragment library, typically a relatively small library of diverse, carefully selected and well-characterized compounds, is screened for binding to the pure target molecule, typically a protein. This screen is often biophysical in nature,[47] since biophysical assays have consistently proven to be reliable and robust at identifying low-affinity ligands; conventional biochemical assays, in contrast, are prone to artifacts and errors when screening high concentrations of ligands, and this can seriously impact

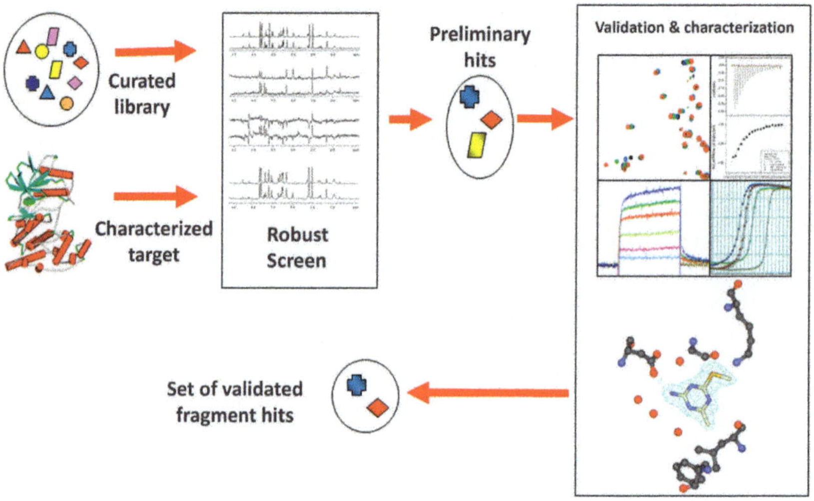

Fig. 3 A typical workflow for an FBS campaign. A curated library of defined, carefully selected fragments are experimentally screened for binding against a well-characterized discrete molecular target. The screening method chosen must be robust and sensitive over the expected range of affinities (see Fig. 2). A preliminary set of hits of obtained, which undergo further validation and characterization to give a final set of validated fragment hits. These form the core for subsequent evolution strategies into high-affinity ligands.

the chances of a successful FBLD campaign. Hits from this screen are then typically validated using an orthogonal technique, usually one which is more stringent but which has lower throughput or is more resource intensive. Validated hits are characterized further in terms of potency and binding mode, and where feasible a high-resolution three-dimensional structure of the fragment: protein complex is obtained. Chemical space around the initial fragment hit is then explored ("near-neighbor analysis"), and an optimized fragment or series of fragments is identified.

These optimized fragments can then be integrated into drug-discovery programs in a number of strategies, from straightforward fragment growth or linking to merging with ligands identified using other hit ID techniques such as HTS or DNA-encoded library technology. Some of the most successful applications of FBLD have been in conjunction with other hit ID techniques, and the synergy observed when applying multiple hit ID techniques to the same therapeutic target is a key feature which has evolved over the past decade.[8]

4. FRAGMENT LIBRARIES

Numerous excellent papers have been published on the design of fragment libraries,[48,57–60] many of which adopt a common approach where a diverse set of small molecules with physical chemical properties (size, polarity, chemical moieties, and so on) are identified from commercial and/or in-house collections. Molecules identified by cheminformatics approaches are then typically subjected to manual inspection by medicinal chemists, and a subset of preferred compounds tested experimentally for purity, stability, and other issues which may give rise to artifacts during the screening process.[58] This latter step is key to the quality of the fragment library and the hits obtained from it, since (as discussed later) low affinity-binding assays are fraught with potential pitfalls and these will reduce the utility of the hits identified.

Alternative approaches to the selection of fragments on simple cheminformatics grounds, such as diversity-oriented synthesis fragments,[61,62] chemically "poised" fragments,[63,64] and others, have also been proposed, although these are currently less widely used than the more generic approach detailed later and will not be discussed further in this review.

4.1 Fragment Library Design

The criteria applied by a range of practitioners of FBLD to select compounds for incorporation in their fragment libraries have been published and discussed in a number of excellent reviews.[48,57–60] In general, compounds of an appropriate size are selected, and filters applied to remove chemical functionalities which are deemed undesirable. Additional filters are also applied, for example, to limit the number of rotatable bonds, to restrict the number of ring systems to a certain range, or to reduce the number of highly planar compounds. Finally, a diversity analysis is performed to ensure the fragments which meet the desired criteria cover as much of chemical space as possible.[65,66] This final diverse set of fragments is then typically validated experimentally before being used in FBS campaigns.

The size of compounds to be included in the fragment library is possibly the most important parameter to be discussed and is what distinguishes FBLD from other screening approaches such as HTS; by "size," we refer to the number of nonhydrogen ("heavy") atoms present in the molecule. A series of exhaustive computational studies have suggested that the number

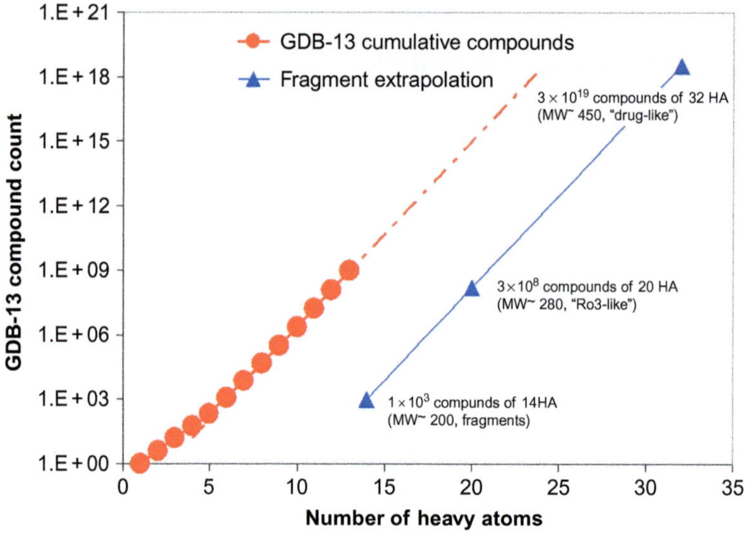

Fig. 4 Plot showing how the number of compounds enumerated in the GDB-13 database varies with the heavy atom count. The number of compounds increases approximately eightfold per heavy atom added. By extrapolation, a library of 1×10^3 compounds of 14 heavy atoms (~200 Da, typical fragment-sized molecules) correspond to approximately 3×10^8 compounds of 20 heavy atoms (~280 Da, "rule-of-three"- or "scaffold-like" molecules) and to 3×10^{19} compounds of 32 heavy atoms (~450 Da, "drug-like" or "HTS-like" molecules). *Based on the data in Blum, L. C.; Reymond, J. L. 970 Million Druglike Small Molecules for Virtual Screening in the Chemical Universe Database GDB-13.* J. Am. Chem. Soc. ***2009,*** 131 *(25), 8732–8733; Ruddigkeit, L.; van Deursen, R.; Blum, L. C.; Reymond, J. L. Enumeration of 166 Billion Organic Small Molecules in the Chemical Universe Database GDB-17.* J. Chem. Inf. Model. ***2012,*** 52 *(11), 2864–2875.*

of lead-like, synthetically achievable molecules increases approximately eightfold for every heavy atom added.[23,24] This dramatic increase has a profound effect on the degree of coverage of available chemical space by libraries of different molecular sizes. Although there are some approximations, a library of 1000 fragments averaging 14 heavy atoms are equivalent (in terms of chemical space coverage) to a library of 10^8 compounds of 21 heavy atoms, and to a library of 10^{18} compounds of 33 heavy atoms (Fig. 4). This constitutes a strong argument for restricting the size of fragments included in a library; indeed, the "maximum recommended size" has shifted from 300 Da (originally part of the so-called Rule of 3[67,68]) to 16–18 heavy atoms (~240–260 Da) or even smaller, and "molecular obesity" is a well-recognized issue with compound libraries for all applications.[69]

The lower limit on fragment size is dominated by the sensitivity of the assay technique used to identify ligands. Obviously, as the number of heavy atoms decreases, so the affinity for a site of given efficiency will drop. Few assay techniques will reliably detect binding weaker than 10 mM; this implies that, except for exceptionally efficient binding sites, fragments smaller than eight or so heavy atoms will not readily be detected (Fig. 2). This gives an effective range for the size of fragments in most fragment libraries of approximately 8–18 heavy atoms.

In addition to the size of the fragment, the nature of the chemical functionalities present is of key importance. Highly reactive moieties such as acyl chlorides, Michael acceptors, and anhydrides are typically excluded from the library[58] in order to avoid stability and reactivity issues. Other functional groups such as amides, alcohols, and amines are frequently found to interact with proteins and as such are often well represented in fragment libraries.[70]

An important class of compounds containing undesirable functionalities are the "pan-assay interference compounds" or PAINS.[71] These compounds are identified as frequently giving rise to artifacts in assays which are readily mistaken for the productive binding or inhibition expected from a useful ligand; unless rapidly identified, these PAINS can prove to be misleading and resource draining. Filters to exclude these compounds are typically applied at the library design stage,[72] but these filters are not infallible and it is productive to remain aware of potential artifacts during the fragment screening and characterization stages of a project.[73]

A further criterion to be considered in the design of a fragment library is the solubility of the compounds in aqueous solution. Since the fragments will typically be of high μM or low mM affinity, they must be soluble to high concentration—typically greater than 1 mM—in aqueous solution. While the prediction of aqueous solubility is notoriously difficult for small molecules, the prediction for fragments has generally been found to be more reliable.[58] However, experimental confirmation of fragment solubility remains one of the key steps involved in the characterization of potential fragments prior to inclusion in a library.

A final step, often applied at this stage, is visual inspection of the proposed fragment library by a set of medicinal chemists. Although time-consuming, this is an important phase; there is little point having compounds present in the fragment library which a medicinal chemist would not be happy to work on. Indeed, this was frequently found to be a limitation with some early fragment libraries, and close communication between chemists, modelers, and experimentalists has proven to be a key driver for the improvement and use of fragment hits. A widely adopted approach to this visual inspection step is to poll the opinion of a small group of medicinal chemists (typically 5–7) as to whether the compound is a viable start point for a campaign; a simple majority decision is used to include or exclude each fragment.

4.1.1 Commercially Available Fragment Libraries

Numerous chemical suppliers now offer individual fragments or entire fragment libraries for sale. These libraries are viable alternatives to designing and purchasing a bespoke "in-house" library and obviously simplify the cheminformatics and medicinal chemistry input required at the early stages of setting up an FBS approach. None of these libraries are necessarily "better" than any other, and if this route is to be taken, then careful inspection of the chemical matter is recommended. Furthermore, any library should still be experimental characterized as discussed later; although this takes time and resource, it is small compared to the effort which will be applied to evolving the fragments into leads or even drugs, and careful examination of the library at this point can avoid many artifact-related issues later in the campaign.

4.2 Experimental Characterization of Fragment Libraries

While the selection of compounds for inclusion in a fragment library is key, considerable effort must also be put into the experimental validation and characterization of these compounds. Since the compounds will typically be used in a range of experimental methods, and since (as will be discussed later) the conditions required to detect weak interactions often push assay technologies to their limits, it is vital that any compounds included in the library be as "well behaved" as possible. This means that in addition to having the correct molecular structure, compounds must be pure, soluble, stable, and free of functionalities identified to frequently generate artifacts in experimental assays.

Validation of molecular structure and purity is typically carried out using standard analytical techniques, namely LC–MS and/or NMR. Although this step may seem superfluous, it is key; it has been estimated that 10%–20% of compounds purchased for incorporation into a fragment library proved to be either impure or the incorrect structure.[74] Verifying the chemical structure of the compound before inclusion in the library is therefore vital.

The requirement for compound purity is particularly stringent for fragments. When screening at low μM concentrations (as is typically the case in HTS and biochemical assays), a 1% contaminant will be present at approximately 10 nM; few compounds will interact in an artifactual manner with proteins at this concentration. However, when screening fragments at mM concentrations, a 1% contaminant will be present at approximately 10 μM, a concentration where particulate formation and other experimental artifacts are a major issue.[75,76]

The solubility of the compound, in both organic and aqueous solvents, is also critical. Given that many fragments will bind with affinities in the high μM or low mM range, aqueous solubility of low mM or higher is typically required in order for detection in a binding assay. Additionally, even low levels of aggregate or precipitate can interfere with assays, and this is particularly relevant when screening fragments.[76] The determination of solubility of compounds in relevant aqueous buffer, either by NMR, light scattering, or other techniques, is therefore a key step in the characterization of compounds before inclusion in the fragment library.

In order to screen compounds at high concentration (often high μM or mM) at a cosolvent level which is tolerated by the protein, the concentration of the stock solution in organic solvent (usually DMSO) must also be high, often 100–200 mM. This is particularly the case where mixtures of compounds are screened (for example, using NMR or crystallographic screening).

At these high concentrations, long-term stability and solubility in DMSO can be a major issue, and regular assessment of the stock solutions is required in order to confirm the quality of the library. DMSO is a mild oxidant, and compound degradation in DMSO solution can occur readily over days, months, or years. This is particularly apparent where DMSO stocks undergo freeze/thaw cycles. DMSO is also hygroscopic, and the buildup of water in DMSO samples can result in compound degradation and/or precipitation.[77,78] It is therefore vital that regular analysis and quality control of library stock solutions, typically by NMR and/or LC–MS, are undertaken.

In addition to compound stability in DMSO stock solutions, aqueous stability should also be verified. Many of the assays used to detect fragment binding take a significant length of time to run, particularly biophysical assays and crystallography. Under these conditions, compounds can be held in aqueous solution for hours or even days. Water is a reactive solvent, and a significant proportion of fragments are not stable over this time period. If NMR or LC–MS is used to verify the fragment solubility, the same samples can be readily rerun to confirm aqueous stability.

The exclusion of "PAINS" compounds at the library design stage has been discussed previously. However, issues with reactivity or interactions with aqueous solvent or buffer components can also become apparent during experimental assessment of the library, and it is good practice to maintain an awareness of these potential problems during the characterization and screening steps. Again, careful characterization and quality control of the library should be performed in order to identify these issues before they impact on any screening campaign.

4.3 Fragment-Based Screening

The first challenge facing an FBLD campaign is simply that of finding fragments which can be confidently identified as binding to the target. Having confidence in the validity of a fragment hit is key, particularly since the risks of being misled by experimental artifacts are so much greater for fragments than when identifying high-affinity ligands (as might be the case in an HTS campaign). Since fragments generally have low affinities for their targets—sometimes weaker than mM—it is essential to have sensitive and robust methods for detecting weak interactions. As demonstrated in several polls conducted on the Practical Fragments blog,[47] biophysical methods dominate the techniques used by the FBLD community in order to confidently identify low-affinity ligands (Fig. 5). More classical biochemical assays are also used, where appropriate, along with computational methods.

An important consideration, when deciding which technique to use to conduct an FBS campaign, is an estimate of the range of affinities expected for a fragment-sized molecule binding to the target protein. The ligandability of a binding site is reflected in the range of LEs of compounds which bind to this site[52]; for example, a "low ligandability" site as might be found on a protein/protein interaction (PPIs) surface may produce hits with a maximal LE of around 0.3 kcal/mol/HA, while a "high ligandability" site (such a well-defined ATP site on a kinase) might produce hits with ligand efficiencies of up to 0.6 kcal/mol/HA or even higher. Since the size range of compounds in the fragment-screening library is well defined (for example, all compounds might be in the range of 14–18 HA), it is straightforward to calculate the expected affinity for these compounds against the target of interest (Fig. 2).

Having determined an approximate expected affinity for the fragments against the target, a suitable binding assay can then be chosen. As mentioned previously, biophysical methods dominate FBS; this is in large part due to the robustness of these experiments in the presence of high concentrations of ligands and the sensitivity of the methods to detecting low-affinity interactions. Biochemical methods are sometimes used for fragment screening,

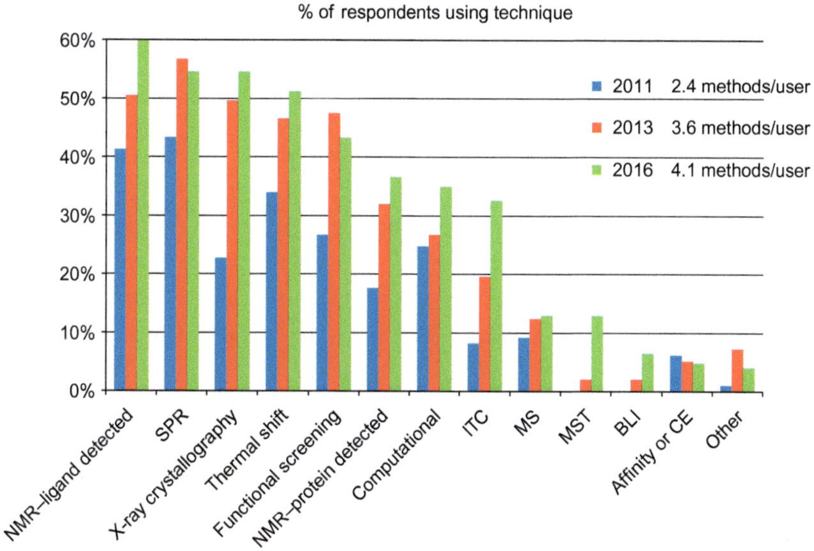

Fig. 5 Techniques used for fragment-based screening (FBS) campaigns, according to respondents to a poll on the Practical Fragments blog[REF]. The percentage of respondents who use each technique are shown for the three years (2011, 2013, and 2016) that the poll has been conducted. Biophysical techniques can be seen to dominate the methods used, with ligand-detected NMR, SPR, and X-ray crystallographic methods being the most commonly used. An additional point of interest is the number of methods used per respondent, which has increased dramatically since 2011 and reflects the widespread use of orthogonal methods for fragment validation and characterization. *Figure reprinted from Erlanson, D. Poll Results: What Structural Information Is Needed to Optimize Fragments Practical Fragments, 2016, http://practicalfragments.blogspot.co.uk/, Copyright (2016), with permission from D. Erlanson.*

but typically for high efficiency sites where affinities are expected to be in the mid- to low-μM range and where these assays are most reliable.

The following is a brief summary of the main characteristics and considerations for each of the most frequently used fragment-screening approaches.

4.3.1 Summary of FBS Methods

4.3.1.1 Nuclear Magnetic Resonance

NMR was the first experimental method used to detect the binding of low-affinity ligands to proteins and remains one of the most frequently used techniques. There are many reasons for this: NMR is a truly label-free technique, requiring no chemical labeling or modification of the system; it is exquisitely sensitive to low populations of bound ligand, making it ideal for detecting weak binding; it is a solution phase, directly observed method, thereby reducing the risk of being misled by experimental artifacts. However, it is a highly specialized technique where many approaches to both experimental design and interpretation are possible, and is demanding of both resource and material. Since NMR remains the most widely used experimental technique for FBS, and since many different methods can be applied, we will discuss them in some detail later.

Many spectral parameters are affected when a rapidly tumbling, low molecular weight ligand interacts with a slowly tumbling protein or other macromolecule. The chemical environment (and hence chemical shift) of both ligand and protein is changed, the tumbling rate of the ligand is dramatically different in the bound state compared to the free state, and other parameters such as diffusion, relaxation, and the transfer of magnetization between resonances are modulated by these changes. In general, experiments use the modulation of these parameters to distinguish the presence of ligands, either through observation of the protein (or other macromolecular species) or through observation of the ligands themselves.

4.3.1.2 Protein-Observed NMR

In the original SAR-by-NMR studies published by Fesik and coworkers in the mid-1990s,[1] 2D ^{15}N–^{1}H HSQC NMR spectra were used to directly observe the protein. Mixtures of compounds were added to the protein, and changes in the HSQC spectra used to identify mixtures (and, by deconvolution, individual ligands) which interacted with the protein. In the ^{15}N–^{1}H HSQC experiment, magnetization is transferred between ^{1}H and ^{15}N nuclei, giving rise to one resonance for each backbone amide (along with signal from the side-chain amides of glutamine and asparagine residues, although these are typically overlapped and infrequently used to detect binding). The chemical shift of both amide nitrogen and proton is highly sensitive to subtle changes in the local chemical environment (for example, the presence of a bound ligand or changes in backbone conformation) and serve as probes for interaction between protein and ligand.

This technique remains one of the most reliable, widely used and robust methods to detect low-affinity ligands. Although there is a size limitation for directly observing protein resonances (of approximately 25–35 kDa), variations of the HSQC experiment (such as TROSY or ^{13}C–^{1}H HMQC experiments) are available which can extend the applicability to systems of 80 kDa or higher. Unlike ligand-observed methods, protein-observed methods are not limited by a requirement for weak binding (fast on/off rates), although this is rarely an issue when carrying out a primary fragment screen. Rather, protein-observed methods are typically limited by compound solubility;

with sufficiently soluble molecules, ligands can be identified readily with mM affinity. A further key feature of the protein-observed NMR-binding experiments is that they give an indication of the binding site of the ligand. This can range from the simple identification of a subset of perturbed peaks (in the case of an unassigned protein) through to the identification of a specific binding site at an atomic level of resolution (where the protein is fully assigned). Even in the absence of a full assignment, fragments may be clustered by "sets" of perturbed residues, allowing the identification of distinct binding sites without requiring crystallographic support at this point.

It is important to mention that, as discussed earlier, protein-observed methods detect changes in the chemical environment of the protein, and at high concentrations of ligand, it is vital to confirm that these changes arise from ligand binding rather than from, say, the presence of high concentrations of DMSO or small changes in the pH of the sample. This is a commonly observed artifact, but is readily identified with careful control experiments.

There are also a number of more serious issues with protein-observed methods which have led to the more widespread use of ligand-observed methods for fragment screening. In particular, the requirement for large amounts of labeled protein can be demanding. Protein-observed NMR experiments are typically acquired with 20–100 μM protein; with even a modestly sized fragment library, and using mixtures of compounds in each sample, this can equate to hundreds of mg of isotopically labeled protein. While this is feasible for some proteins, for many others it is not, and for these targets alternative approaches must be used.

4.3.1.3 Ligand-Observed NMR

The most widespread use of NMR in FBS campaigns is one or more of the ligand-detected NMR-binding experiments. As discussed previously, many NMR spectroscopic parameters are highly sensitive to the motion of the molecule in solution; typically this is related to the rotational correlation time, although the translational correlation time can also be used in some experiments. The ligand-observed NMR-binding experiments rely on this differential behavior of slowly and rapidly tumbling molecules to distinguish ligands from nonligands; when a ligand is bound to a slowly tumbling macromolecule, its behavior is more like that of a macromolecule than a small molecule, and thus it is differentiated from small molecules which are moving freely in solution.

With some NMR parameters, such as relaxation or transfer of magnetization, this differential behavior persists after the ligand has dissociated from the receptor. Thus, the population detected is not the "bound population" per se, but the population which has bound and has then dissociated during the time course of the experiment. This has several significant implications for the detection of ligands. First, since multiple ligands may bind and dissociate over the course of the experiment, the population of ligands which have bound is significantly greater than the bound population at any specific time. This gives rise to signal amplification, where even a small bound population can generate a significant signal. Second, since the detected species is the population which has dissociated from the receptor, the ligand must be in fast exchange between the bound and free populations on the NMR timescale. This imposes a restriction on the affinity range which can be detected; molecules that bind too tightly, and therefore are not in fast exchange, will give little or no signal in the ligand-observed NMR-binding experiments. As discussed previously, this is in contrast to the situation where the protein is the observed species. This limit is rarely observed in an initial fragment screen, but is an important restriction that must be considered when progressing hits from a ligand-observed fragment screen using ligand-observed NMR methods.

Although many ligand-observed NMR experiments have been published, a relatively small number have become widely used. These will be discussed briefly below.

Saturation Transfer Difference

Perhaps the most commonly used ligand-observed NMR-binding experiment, the saturation transfer difference (STD) experiment[79] relies on direct magnetization transfer between the protein and any bound ligand. A series of selective inversion pulses, typically centered around −0.3 to 0.3 ppm, are used to saturate the hydrophobic core of the protein. This saturation then propagates across the protein, and to any bound ligands, via dipole–dipole interactions. This is referred to as the "on-resonance" spectrum. A second, "off-resonance" spectrum is then acquired with identical parameters, but with the selective saturation pulses set to a frequency where no protein or ligand resonances are present (for example, 40 ppm).

By subtracting the "off-resonance" spectrum from the "on-resonance" spectrum, a difference spectrum is obtained corresponding to those components in the sample which are saturated by the on-resonance selective excitation pulses (i.e., protein, ligands, but not nonbinding compounds). Samples are usually prepared with a large molar excess (often 50–100 times) of ligand over protein, in order to maximize the population amplification effect discussed earlier.

Water-Ligand Observed Through Gradient Spectroscopy

Like the STD experiment, the water-ligand observed through gradient spectroscopy (water-LOGSY) experiment relies on a magnetization transfer step to differentiate ligands from nonbinding compounds.[80] However, the initial excitation step is selective for the bulk water resonance, and magnetization is then transferred from water molecules to all components present in solution. Since the sign and magnitude of the magnetization transfer step depend on the tumbling rate of the molecule, differential signal is obtained for ligands bound to proteins (which give rise to a large positive signal via interactions both directly with bound waters in the binding site and indirectly via excitation of the protein itself) and for nonbinding molecules free in solution (which give rise to a small negative signal directly via interactions with the hydration shell).

Since the free and bound populations give rise to signal (of opposite signs), some care has to be taken both with sample conditions and interpretation of the experiment. As with the STD experiment, excess ligand over protein will give rise to a signal enhancement effect. However, too high a molar ratio will give rise to a negative spectrum as the signal from the free population begins to dominate. Thus, a negative observed spectrum does not necessarily mean that the compound is not binding, merely that the signal from the free population is larger than that of the bound; this may be due to a low population or due to other factors such as inefficient magnetization transfer resulting from few bound water molecules in the binding site of the ligand. As a result, the most reliable method for interpreting water-LOGSY spectra is to reacquire the experiment after perturbing the populations (for example, by adding a competitor, as discussed later) and to identify spectra which shift toward the more negative (i.e., higher free population) after displacement.

Relaxation-Filtered Experiments (^1H and ^{19}F)

Experiments based around the differential relaxation rates and mechanisms of rapidly and slowly tumbling molecules are widely used to distinguish ligands from nonbinding molecules.[35,81] Both $T1\rho$ and T2 relaxation experiments are common, using spin-lock or CPMG sequences in order to specifically attenuate signal from slowly tumbling molecules. This "relaxation filter" then leaves a spectrum containing signal only from nonbinding molecules. Relaxation-filtered experiments are optimal with a relatively low molar excess of ligand over protein, in order to observe the effect of a large bound fraction of the total ligand concentration; in particular, ^{19}F relaxation-filtered experiments are typically configured in this manner, with a ligand: protein ratio of perhaps as low as 5:1.[82,83] ^{1}H-observed relaxation-filtered experiments, where the ligand signal might be obscured by protein resonances, are more typically run at a higher molar ratio.

Combining Ligand-Observed NMR Experiments

Since the ligand-observed NMR experiments discussed earlier are relatively rapid, and use different physical mechanisms in order to distinguish ligands from nonbinding compounds, a commonly used and robust approach is to simply acquire a combination of experiments (for example, STD, T2 filtered, and water LOGSY) on the same sample, and to compare the results for each experiment. Typically, compounds identified as binding in several NMR-binding experiments are more likely to be identified as validated ligands by other biophysical techniques than those which are hits in only a subset of the experiments acquired.[57,76,84,85]

Competition

One major drawback with all ligand-observed NMR-binding experiments is that, unlike protein-observed NMR-binding experiments, they do not yield any indication of the binding site of the ligand. Rather, they merely indicate that the ligand has interacted with a slowly tumbling macromolecule, and cannot resolve specific binding to a site of interest from binding to an allosteric site or, worse, nonspecific interactions. Spectroscopic methods have been proposed to distinguish specific from nonspecific binders,[86] but a straightforward and robust experimental approach remains a simple competition experiment. This is only possible if a potent ligand is known to bind to the site of interest, but where this is the case the identification of ligands which bind, and are then displaced by a known competitor, is far more reliable than the simple observation of binding.

However, any potent competitor molecule must be characterized in terms of both affinity and stability; ideally, the binding site should also be determined by crystallography, and binding stoichiometry confirmed by ITC. In our experience, not all potent tool compounds are as potent, or as stable, as may be claimed in the literature, and degradation of competitor molecule during the course of fragment screen is often the cause of a high rate of false negatives.

4.3.1.4 X-Ray Crystallography

X-ray crystallography was one of the earliest methods used to detect low-affinity fragments and remains a key approach. Crystallographic approaches have two key advantages over other techniques—first, the three-dimensional structure of the protein:ligand complex is, by definition, available from the outset, and second, that soaking of crystals can be performed using very high concentrations of compound (to the extent that some crystal-soaking experiments have been described as "slurries"). This allows the observation of very weakly binding compounds; indeed, some of the earliest crystallographic "fragment"-soaking experiments were actually performed using cosolvents such as DMSO or acetamide rather than more conventional compounds.[87,88] Further, advantages in technology over the past decade have reduced the experimental burden significantly, such that screening a fragment library by crystallography is no longer a significant hurdle.[89–92] This has been the case in crystal handling and preparation, data acquisition and processing, and analysis. Indeed, the redundancy present in many empty datasets has allowed the development of novel data processing methods such as PANDDA.[93]

However, there remain significant issues with fragment screening by X-ray crystallography. The first and most obvious is the requirement that the target protein crystallizes in a form suitable for screening. This is rather more of an issue than simply crystallizing the protein, since the crystal form must be amenable to soaking, rather than requiring cocrystallization; the crystal lattice must be sufficiently open to allow the access to the binding site of interest, and the binding site itself must not be occluded by lattice contacts. The crystallization conditions can also limit suitability, for example, where high salt or nonnative pH is required in order to crystallize the protein. In these instances, it has generally proven worthwhile to invest time and resource developing an alternative crystal form, more suitable for fragment screening, rather than attempting to screen under suboptimal conditions.

4.3.1.5 Surface Plasmon Resonance

Another technique which is widely used to identify fragments is surface plasmon resonance, SPR. Since the advent of FBLD in the late 1990s and the early 2000s, SPR technology has advanced dramatically, and the sensitivity and throughput of modern SPR instruments are far in advance of the earliest machines. The fundamental technology, however, remains unchanged: the target of interest is immobilized onto a sensor chip, and buffer containing the test compound flows across the surface of the chip. On the reverse side of the chip, a laser is reflected from a gold surface; the refractive index of this gold surface is affected by interactions between the immobilized protein on the chip and compounds present in the buffer. A key feature of this technology is that, in addition to identifying binding, the association and dissociation kinetics of the interaction are determined. This ability to resolve kinetic, as well as equilibrium, parameters has proven to be one of the most important aspects of SPR.

Fragment screening using SPR is typically carried out using direct binding of the fragments to immobilized protein. A "clean screen," where artifactual interactions between the fragments and the chip surface are identified, is key to improving the reliability of the method. In addition, one or more known ligands are typically run periodically in order to confirm the viability of the immobilized protein; optimization of regeneration or immobilization conditions are sometimes required in order to allow a robust screen to be performed, since this can often take several days to run. An initial binding level screen is run in order to identify interacting compounds; this is followed by a dose–response screen of hits to confirm the validity of the single-point assay and to determine a K_D value for the ligand. A useful recent development is the "ABA" type of injection, where one compound can be "spiked" during the injection of another compound; in a manner analogous to the NMR competition step discussed earlier, this allows the identification of ligands binding specifically to the site of interest.

As with all techniques, SPR suffers from a number of limitations. In particular, the heterogeneous nature of the immobilized protein on the chip surface can generate significant artifacts; these can be identified and controlled for, but care must be taken. The method of immobilization can also have a profound effect on the nature of the protein: chip interface and development of a robust immobilization strategy can prove to be an obstacle to running a fragment screen using SPR.

4.3.1.6 Thermal Shift Assay

A widely used and accessible method for identifying low-affinity compounds is the thermal shift assay (TSA), also known as differential scanning fluorimetry or "thermal melts." In this assay, a solution of protein (typically at low μM) is heated in the presence of a fluorescent dye. This dye binds to hydrophobic surfaces with an increase in fluorescence. At a certain temperature (the "melting point") the protein undergoes thermal denaturation, and the fluorescence of the sample increases as the dye interacts with the exposed hydrophobic residues which were buried in the folded state. If a ligand binds to, and stabilizes, the protein, a shift in the melting point of the protein will be observed.

TSA has proven useful for screening fragment libraries[41–43] and also for identifying ligands which stabilize a protein for crystallization.[94] It has also found widespread applicability in the identification of buffer conditions which stabilize the protein.[95]

This technique has several favorable features, not least that it is fast, relatively generic and can be performed on a simple instrument such as a qPCR machine. It also uses small amounts of protein, particularly compared to techniques such as NMR and crystallography. However, although the method is relatively robust for identifying more potent ligands, it has a high false-negative rate when used for identifying fragments;[84,96] a weakly binding fragment does not necessarily stabilize the protein significantly. There are also issues associated with screening compounds at high concentration; at mM concentrations, the compounds themselves can act as buffer components and stabilize or destabilize the protein irrespective of binding. Additionally, since the fragments are themselves often hydrophobic, competition occurs between the dye and fragment molecules for the exposed hydrophobic surfaces after protein denaturation. However, given these caveats, TSA is a useful and inexpensive method for screening fragment libraries.

4.3.1.7 Biochemical Assays

For proteins with relatively druggable binding sites, the expected affinities for fragment-sized molecules can fall in a range where a simple biochemical assay is robust (Fig. 2). Although many assays are compromised by high concentrations of compounds, several (such as some functional enzyme assays, displacement of a fluorescently labeled probe, and so on) have been found to be sufficiently resilient to be used as primary fragment screens and to show good correlation with biophysical techniques applied to the same target.[84,96] Where possible, however, fragments identified using biochemical assays should always be validated using an orthogonal biophysical technique.[8]

4.3.1.8 Mass Spectrometry

The technology and methodology for mass spectrometry, and in particular native-state mass spectrometry (MS), have advanced dramatically over the past decade. In native-state MS, relatively gentle ionization conditions are used which preserve noncovalent complexes from solution into gas state. As long as the protein:ligand complex survives excitation, this allows the direct identification of noncovalent complexes present in solution with no labeling or indirect measurement required.[97–99]

Although this technique is powerful, it requires a great deal of optimization of experimental parameters (such as sample preparation, voltage, vacuum pressure, and so on), and as such typically requires both an expert user and a dedicated instrument. It is also demanding of both ligand and protein behavior and lacks the sensitivity to low-affinity interactions which some other techniques can provide. As such, it remains an infrequently used screening method, although further advances may help to make native-state MS more widely used.

A more widely used application of mass spectrometry has been the detection of covalently bound fragments.[100] These have ranged from the "tethered" fragment approach of Sunesis and others[18] to the more general approach pioneered by NovAliX.[101]

4.3.1.9 Microscale Thermophoresis

Although a relatively new technique, MST, has already become widely used[44] and is finding applicability to FBS. The approach monitors the mobility of fluorescently labeled molecules in the presence of thermal gradients ("thermophoretic mobility"). Thermophoretic mobility has been shown to be sensitive to changes to both protein conformation and particularly the hydration shell, and thus to molecular interactions.

One particular advantage of this technique is that, although a fluorophore is required to be attached to the target, this attachment can be remote from the binding site and therefore have little or no impact on ligand binding. Direct observation of the fluorescently labeled species also allows ready identification of artifacts such as protein precipitation or interaction with the capillary wall; this direct observation reduces the risk of false positives associated with these modes of action.

4.3.1.10 Other Approaches

A plethora of alternative approaches and technologies have been suggested, and used, for identifying low-affinity ligands binding to proteins. These include second harmonic generation,[102,103] surface acoustic waves,[104] graphene-based field effect Biosensors,[105] and capillary electrophoresis.[106] Additionally, some proprietary methods (such as the TINS approach[107]) exist which require specialized instrumentation and have not found widespread applicability.

One relatively recent method which may prove to be of general interest is so-called a weak affinity chromatography (WAC) method.[84] In this method, the target protein is immobilized onto a silica gel and loaded onto an LC column. Fragments are then flowed over the immobilized protein before detection of the compound using mass spectrometry. Compounds that interact with the protein are retarded when compared to a "blank" column, which typically consists of an immobilized unrelated protein. Further controls using potent competitor molecules, where available, can further increase the reliability of the screen. As with any heterogeneous phase system, care must be taken to avoid restricting or impairing the active site of the target protein during the immobilization process. It is also vital to rule out nonspecific interactions with the column matrix. However, with careful implementation, WAC has been shown to be a robust screening method for FBS; in a direct comparison with other widely used biophysical techniques, WAC had a similar hit rate and identified the same set of compounds as ligand-observed NMR, SPR, and X-ray crystallography.

In general, no particular technique has proven to be the "best-in-class" approach, but it is useful to be aware of, and consider, all available options.

4.4 Fragment Validation

Regardless of the primary screening method used, a key step in any FBS campaign is validation of any hits identified. As discussed previously, assays used to detect these low-affinity ligands are typically being pushed to their limits of reliability, and as such experimental errors will arise. False-negative hits are unavoidable, and best practice is typically to minimize the false-negative rate by setting the threshold for identifying hits to be low. This inevitably means that the pool of primary fragment hits will include false positives; depending on the assay and target class, this false-positive rate can be high.

Before taking hits from the primary screen forward into the resource intensive areas of fragment characterization, exploration, and evolution (see below), it is therefore vital to validate these "preliminary hits" as thoroughly as possible in order to avoid wasting resource on experimental artifacts.

There has been debate over reported differences in hits obtained using different biophysical methods, both in terms of the observed hit rate and in the nature of the fragment hits identified. In principle, all assay techniques should identify the same set of hits from a library for a given target, provided that the assay conditions are equivalent. However, different methods have different sensitivities and are prone to different artifacts; different experimental conditions are also required for different assays (for example, high protein concentration for protein-observed NMR, a crystal lattice for crystallography and protein immobilized onto a chip surface for SPR). The combination of differential sensitivity, assay conditions, and experimental artifacts may well account for differences which are sometimes observed in the output from different biophysical assays. As these effects become more pronounced, less and less overlap between different techniques is expected; this appears to be the case for some targets, particularly where assay sensitivity is low.

In our experience, for a well-behaved protein with well-characterized assays and a well-behaved fragment library, a high degree of correlation (greater than 70%–80%) is observed in the output from a range of biophysical methods.[84] Where the correlation between techniques was observed to be lower, this was attributable to factors specific to the relevant assay (such as the stringent sample solubility requirements resulting in a high false-negative rate in ITC, and the lower signal/noise in TSA resulting in a high false-positive rate).

One key point from this assay was that, in order to identify an optimal set of validated fragment hits, as many orthogonal methods should be applied as is reasonably practicable; this can be seen to be the case in real-world applications (Fig. 4). However, the output from these assays should not be regarded as a simple "yes/no" filter, since this simply reduces all assays to that with the lowest hit rate; rather, the differences between assay output should be considered and evaluated where possible. This evaluation of orthogonal methods, although complex, is valuable and can identify (and importantly rectify) issues with the fragment-screening campaign; for an extensive discussion of this point, see discussion elsewhere in Ref. 108. This in turn can save a great deal of effort further down the drug-discovery process.

4.5 Fragment Characterization

When a final set of validated fragment hits have been determined, the next step is to characterize the fragment hits as fully as possible. In the first instance, this typically consists of acquiring any data missing from the validation steps—i.e., each fragment hit should, ideally, have an affinity (determined by two or more biophysical methods) and a knowledge of the ligand-binding mode. This binding mode is typically a high-resolution crystal structure, but where this is not possible, NMR structures or docking models incorporating NMR data have proven to be viable and valuable alternatives.[109–113]

A second, often overlooked, element of fragment characterization is the examination of chemical "near neighbors" around the original fragment hits. As discussed previously, the sampling of chemical space by a fragment library is sparse, with each fragment representing a large area of chemical space. As such, the likelihood of any initial fragment hit being the "optimal" fragment is small; rather, near neighbors of the initial hit may exist which have an improved affinity or other valuable features, such as vector accessibility, chemical tractability, or toxicity. Exploration of these near neighbors, often via "SAR by catalog," can identify closely related compounds with significantly improved potency or other physicochemical properties.[96,114] Characterization of the initial fragment hit to identify an optimized fragment start point can dramatically improve the progression of a fragment-based medicinal chemistry campaign.

5. FRAGMENT EVOLUTION

Having obtained validated fragment hits, evolution of those hits into leads and ultimately clinical candidates is required. The exact approach depends on the target, the hits, and the available additional information (e.g., 3D-bound structures, second site-binding fragments, literature tool compounds/endogenous ligands/HTS hits/published compounds) and resources. We start this section with a brief summary of the approaches typically used, and their benefits and difficulties before showing some published examples to demonstrate the real-world application. In all cases, bound structures of the fragments bound to the target (ideally X-ray crystal structures, but NMR structures and those from molecular modeling may also be used) are either essential or highly beneficial to obtain the full advantages of the rapid progress possible in the early stages of fragment evolution.

The individual approaches are summarized below:

- *Fragment optimization*—Perhaps often overlooked, this is an important initial phase of fragment evolution, in which the interactions from fragment hits are optimized while keeping within fragment-like chemical space—usually with reference to structural information. Information from multiple overlapping fragments may be incorporated, either to modify slightly an existing fragment or to design a totally new fragment combining features from others (see also Section 5.4).
- *Fragment linking*—The original SAR-by-NMR paper and approach was predicated on the concept of linking fragments bound in adjacent, non-overlapping pockets within the binding site.[1] This approach has the

possibility of rapid gains in potency, but is often difficult to achieve in practice due to the requirement to find an optimal linker. A combination of fragment growing and subsequent merging with a known second site fragment is often more successful in practice.

- *Fragment growing*—Of all the approaches to fragment evolution, which is perhaps most similar to conventional medicinal chemistry. The approach consists of exploring SAR around multiple vectors, but the need to track *LE* (see below) is perhaps more critical with smaller fragments if the result is not to be large hydrophobic structures with relatively weak binding. The low potency of fragment hits (generally 100 μM to 10 mM) is often an initial block when presented to practitioners not used to working with fragments, which can be helped by comparing the LE of fragment hits and hits in more conventional lead-like or drug-like space for the same target. Structural information is helpful but not essential in the early stages.
- *Fragment merging*—Fragments that show some overlap in their binding mode with other compounds (fragments or otherwise—see below) can be incorporated, either in their entirety or in part, in the other structure. When both compounds are fragments and they overlap significantly in the binding site, then this approach is akin to fragment optimization.

We discuss each of these in more detail later showing examples (and counterexamples), before introducing some relevant chemical technologies and considerations. We conclude with some examples combining multiple approaches.

5.1 Fragment Optimization

Before expanding the initial fragment hits to larger molecules with improved binding activity, it is worth spending some time optimizing the fragment hits themselves. At this stage, the aim is to increase the binding efficiency (potency and LE) by optimizing the interactions of the fragment with the target while keeping the molecules within the fragment space. This may involve inspecting one or multiple overlaid bound structures to see where additional interactions from one fragment (or indeed from a more elaborated compound structure) may be added to another fragment, increasing its binding efficiency and potentially also modifying the available vectors for subsequent elaboration. A further benefit is that novelty can be introduced at this early stage. This stage may be undertaken by the purchase of commercially available near neighbors ("SAR by catalog") or the synthesis of small numbers of carefully designed bespoke molecules (which may well be suitable for incorporation into the fragment library for future screening campaigns).

In a recent example, Igoe et al.[63] disclosed the very simple optimization of an initial fragment bromodomain binder **1** by the iterative addition of an $-NH_2$ 6-position substituent intended to provide a synthetic vector and a "magic methyl"[115,116] 3-position substituent (Fig. 6). The loss of LE on introduction of the $-NH_2$ group suggests that this may not be optimal; however, it provides the possibility of derivatization to improve selectivity and potency. Introduction of the 3-position $-CH_3$ group more than recovered the LE lost by the initial derivatization.

Similarly, Lerner et al.[117] looked for near neighbors of fragment **4** binding in the *S*-adenosyl methionine pocket of catechol *O*-methyl transferase (COMT) and identified a thiazole analogue **5** with improved LE. Further optimization showed that the methyl group in the R2 position was optimal, and the fragment could be improved further by truncating the ethyl chain (R1 = Me) to a methyl chain (R1 = H; **6**). From here the group embarked upon a fragment-growing strategy to generate a series of potent COMT inhibitors.

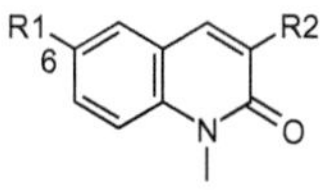

1 R1 = R2 = H IC_{50} 4.8 μM LE 0.60
2 R1 = NH_2 R2 = H IC_{50} 2.1 μM LE 0.28
3 R1 = NH_2 R2 = Me IC_{50} 190 nM LE 0.65

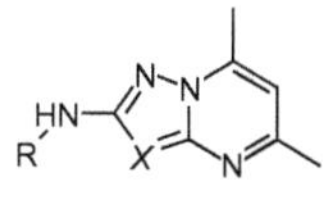

9 R = HX = CH IC_{50} 915 μM LE 0.36
10 R = HX = N IC_{50} > 1600 μM LE < 0.32
11 R = AcX = CH IC_{50} 19 μM LE 0.46

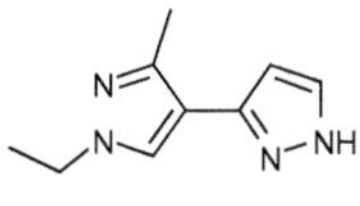

4 IC_{50} 69 μM LE 0.44

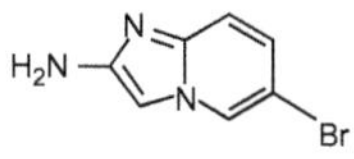

12 IC_{50} 156 μM LE 0.47

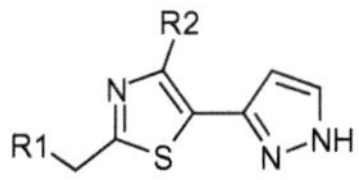

5 R1 = Me, R2 = Me IC_{50} 21 μM LE 0.49
6 R1 = H, R2 = Me IC_{50} 16 μM LE 0.54
7 R1 = H, R2 = H IC_{50} > 625 μM
8 R1 = H, R2 = ^{c}Pr IC_{50} > 625 μM

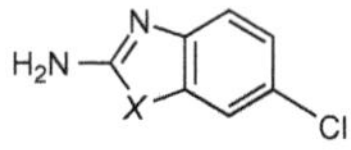

13 X = –O– IC_{50} 380 μM LE 0.42
14 X = –N = N– IC_{50} 53 μM LE 0.49

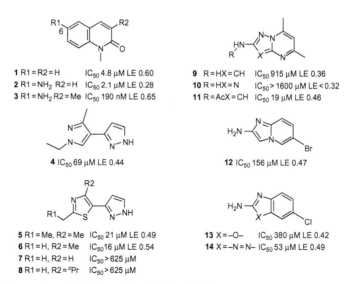

Fig. 6 Fragment optimizations. See text for details.

Hughes et al.[118] demonstrated the utility of the approach during their search for isoform-selective PI3 kinase inhibitors. The original hit **9** had an LE of 0.36 and formed a donor–acceptor interaction with the kinase hinge as indicated by the arrows shown. A focussed screen of near-neighbor fragments led to the screening of around 20 fragments, which showed some steep SAR around the bicyclic ring. Replacing a single CH with an N (**10**) rendered the fragment essentially inactive, while other changes to the 6,5-fused motif showed increases in LE (compounds **12** and **13**). Changing to the 6,6-fused system gave even further increases in LE (compound **14**, LE 0.49); however, this was not pursued as they were unable to obtain a crystal structure of this fragment in complex with the protein. A simple elaboration (still within the fragment space) of the original fragment gave the acetyl derivative **11**, which also showed a significant increase in LE, and this fragment was pursued for further elaboration (see Section 5.4).

In all the cases shown earlier, the changes made have been relatively simple changes to aromatic heterocycles or addition of simple substituents. However, more profound changes are also possible, as highlighted in the case of a series of thieno[2,3-*d*]pyrimidine inhibitors of Hsp90 reported by Brough et al.,[54,119] as described in Section 5.8.

Pros

- Improvement of binding motif and interactions
- Optimal starting point for elaboration
- Enhanced intellectual property (IP) position
- Improved vectors

Cons

- Investment without significant apparent gains in absolute potency
- Binding mode may change—structures of every active compound may be required for meaningful interpretation at this stage. Changes in the binding mode may provide useful new chemical information; however, radical changes in the binding mode can occur, and if not identified can provide misleading SAR.

5.2 Fragment Linking

The original SAR-by-NMR fragment approach from Abbott[1] was based on the theory that linking two proximally bound fragments would give a molecule for which the binding was greater than the sum of the parts. In this conceptually simple approach, weak-binding fragments in adjacent sites can be rapidly converted to high-potency elaborated compounds. In theory, the linked fragments "A–B" have dissociation constants relative to the individual fragments "A" and "B" obeying the following relationship:

$$K_d(A-B) = K_d(A) \times K_d(B) \times E,$$

where E is a linking coefficient, which will be less than 1.0 if linking has improved potency relative to the individual fragments.[27,120,121] This effect has been rationalized in part on the basis of the reduction in the entropic penalty of restricting the translational and rotational degrees of freedom from two particles to one. One complication which may result in less improvement than expected is the possibility that two proximal, cobound fragments might bind cooperatively even without linking, resulting in the apparent increase in potency on linking being diminished relative to that expected. In terms of the above linking coefficient equation, the cobound state "A · B" already has a value of $E<1$, and so the apparent E value of "A–B" is reduced by a factor of this amount. As we will show, when linking works, E values considerably less than 1 may be obtained, and if suitably oriented fragments are found, then it is certainly well worth attempting.

In order to understand the linking process, a number of retrospective "unlinking" studies have been published. The Fesik group at Abbott[31] showed (by fragmenting **15**; Fig. 7A) in an analysis of the effect of fragment linking in the matrix metalloproteinase Stromelysin that the coefficient E in the equation above was in the region of 0.018 (i.e., ~50-fold improvement over the cobound components) when considering the binding of **17** in the presence of saturating AHA (**16**), but the true value of E was in the region of 0.0016 (i.e., ~615-fold improvement over the individual components).

The observed cooperativity was attributed to enthalpic interactions between the ligands; the linking effect was a combination of similar enthalpic and entropic effects, with a relatively small entropic effect compared to that expected on going from ternary to binary complexes. The effect of linker length was shown to be negligible on entropy, suggesting that a strategy of initial linking with a relatively flexible $-(CH_2)_n-$ linker followed by rigidification or addition of additional functionality on the basis of the resulting 3D structure.

More recently, Toone et al.[121] reanalyzed compound **15**, attempting to remove the solvation and protonation effects of changing functional group. To this end, they deconstructed the fragment by breaking one bond to give

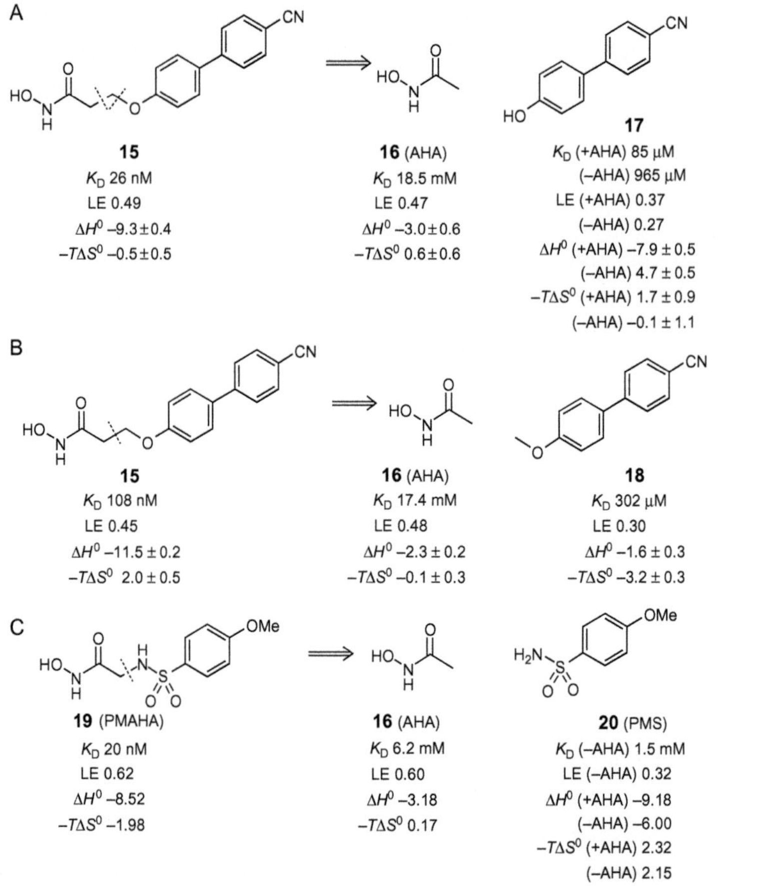

Fig. 7 Deconstruction of linked fragments. (A) Abbott deconstruction of Stromelysin inhibitor **15**.[31] (B) Toone deconstruction of **15**.[121] (C) Luchinat deconstruction of MMP-12 inhibitor PMAHA (**19**).[122] *Broken bond(s)* shown with a *dashed* line enthalpies (ΔH^0) and entropies ($-T\Delta S^0$) are in units of kcal/mol.

16 and **18** (in which the ether functionality is preserved) with an *E* value of 0.021 (Fig. 7B). The results obtained were consistent with those of Fesik et al.,[31] except that the enthalpic contribution to the linking was even higher than the entropic effect.

The MMP-12 inhibitor PMAHA **19** was similarly investigated by Luchinat et al.[122] In this case, the value of *E* was 0.00215 or a ~465-fold increase in potency relative to the individual fragments **16** and **20** (Fig. 7C). The workers found that in this case the linking effect was primarily due to entropic effects, which is in line with the expectation. It is an interesting aside that in all three examples a small, highly ligand efficient fragment (**16**) is linked with a larger, less ligand efficient fragment to give a link compounds in which the LE is close to that of the small, highly efficient fragment.

Another retrospective study by workers at Sanofi-Aventis looked at the effect of linker rigidity, by breaking selected bonds in inhibitors to generate pairs of fragments in a similar manner to that described earlier, concluding that more rigid linkers may not result in more potent molecules as non-favorable interactions with the linker, and suboptimal orientation of the binding elements may result.[123] Workers at Evotec[120] reviewed the literature of fragment linking in 2011 and concluded that in order to maximize the chance of achieving successful fragment linking, a fragment pair should consist of "one fragment that binds by strong H-bonds (or non-classical equivalents) and a second fragment that is more tolerant of changes in binding mode (hydrophobic or vdW binders)," and that

> *The keys to achieving super-additivity upon linking are to maintain the binding modes of the parent fragments, not introduce both entropy and solvation penalties while designing the linker, and also make any interactions with the intervening protein surface that need to be made.*
>
> ***Ichihara et al.***[120]

Caution should be exercised around these conclusions, however, as there is only a relatively small dataset.

Ward et al.[124] provide an excellent example of the fragment-linking approach applied to lactate dehydrogenase A (LDHA) inhibitors—a target for which they had failed to obtain tractable hits using an HTS approach (Fig. 8). An initial ligand-observed NMR screen resulted in hits binding to the adenine pocket of LDHA, including **21**. A search for analogues to optimize the fragment identified **22** as the minimal binding motif retaining LE, along with compounds which retained LE while extending toward the phosphate pocket. A second screen was performed by a library targeting the substrate or NADH pocket using SPR, which also gave a range of hits including **23**. Overlaying the bound crystal structures of **21** and **23** showed the distance between the nearest atoms to be 6.5 Å (Fig. 8C), which the authors decided to be too far to link directly. Instead, they chose to grow **21** toward **23** first, making a variety of related analogues including **24**, which they were able to crystallize with **23** also bound (Fig. 8D), and which retained LE. Now the distance to be linked was 3.0 Å, and the authors decided to attempt to link the fragments in various manners including **25**

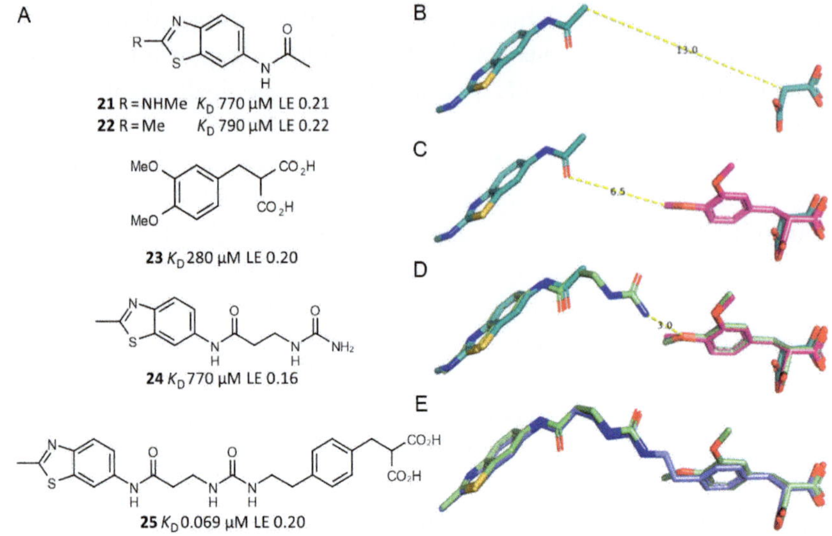

Fig. 8 Linking lactate dehydrogenase A (LDHA) binders. (A) Structures, SPR K_D values and LEs (NB LE quoted here is pK_d/HAC). (B) **21** and malonate cobound to LDHA (PDB 4AJ1). (C) Overlay of **21** and **23** (*magenta*; PDB 4AJI) bound to LDHA. (D) Overlay of **21** and **23** cobound to LDHA (*green*; PDB 4AJJ) with **21** (*cyan*) and **23** (*magenta*). (E) Overlay of linked compound **25** (*purple*; PDB 4AJN) and cobound **21** and **23** (*green*).

in which LE was still retained. The E value comparing **21** and **23** with this compound was 0.320, based on SPR K_D values. As the second screen was performed with a focussed library around the site of the malonate binding, this could be considered an example of linking **21** to malonate, over a distance of 13.0 Å, although the authors do not make this claim. As can be seen from Fig. 8B–E, the fragments and elaborated, and ultimately, linked compounds closely maintained their binding modes throughout.

Finally, interligand NOEs (ILOEs) between fragments in the presence of protein are a commonly used method of detecting adjacent site binders,[125,126] particularly in the absence of X-ray crystallography. However, Abell et al.[127] showed that care in such instances needs to be taken to ensure that the ILOEs observed are due to specific protein binding, and not due to aggregation effects or the intermediacy of the protein itself. The same group have also disclosed a direct comparison of fragment-growing and -linking strategies, in this case at least showing that similar evolved molecules resulted from either approach,[128] although those from linking had lower LE values, and the aromatic groups bounds in different pockets. An alternative NMR-based approach is to modify one component with a spin label and then look for fragments which bind closely (see Section 5.5.2 for an example).

Pros

- Linking of two optimized fragments can allow rapid generation of highly potent elaborated compounds

Cons

- Requires bound fragments to be found in suitable proximal sites
- The linker needs to be optimal in order to achieve gains, which in turn require suitably oriented functional handles on the bound fragments or potentially extensive synthetic chemistry resource
- Structural data are almost essential (it may be possible to link fragments binding sufficiently closely to show ILOEs without full structural information about the binding modes)
- Potential gains in potency are often not realized in practice

5.3 Fragment Growing

The fragment-growing approach is conceptually most closely related to "traditional" medicinal chemistry approached; i.e., groups are added and the compounds tested, and further analogues designed based on the results; as such, we only present a single example in this section. However, while early leaps in the absolute potency are often relatively easily obtained, it is important to also track the binding efficiency, as the addition of groups which are far from optimal can still potentially give large leaps in potency but without maintaining efficiency. While efficiency may be regained by further careful adjustment of the added group, e.g., changing to a different aromatic ring, or adding alternative substituents to a ring), this is not always the case, and it is important for project teams not to be seduced by inefficient leaps in binding. Additionally, in the early stages of elaboration, the same challenges of assaying compounds may be experienced as for the initial fragment screen.

Vernalis have disclosed a series of resorcinol-based Hsp90 inhibitors which have their origins in both fragment and virtual screens.[119,129,130] In this section, we will concentrate only on the fragment-based aspect of the series, although various approaches were initiated in parallel. A fragment screen of around 790 compounds (the original Vernalis fragment library) yielded a number of resorcinol-containing hits (**26–30**; Fig. 9). It is

26 R = CO_2Me IC_{50} 400 μM LE 0.39
27 R = OMe IC_{50} 1200 μM LE 0.40
28 R = CH_2CO_2Me IC_{50} 490 μM LE 0.35
29 X = –CH_2– IC_{50} 570 μM LE 0.37
30 X = –O– IC_{50} ~1 mM LE 0.34

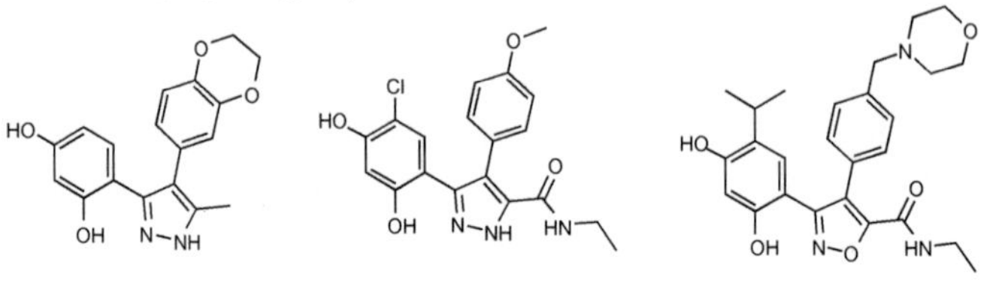

31
IC_{50} 0.28 μM LE 0.37
GI_{50} 6 μM

32 (VER-49009)
IC_{50} 0.025 μM 0.38
GI_{50} 0.26 μM

33 (VER-52296/NVP-AUY-922)
IC_{50} 0.021 μM LE 0.27
GI_{50} 0.016 μM

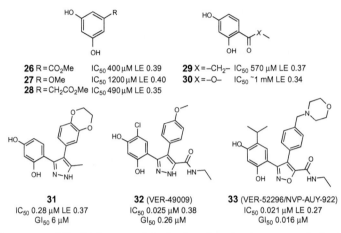

Fig. 9 Structures of Hsp90 fragment hits (*top*) and evolved compounds (*bottom*).

interesting to consider the IC_{50} and LE values for compounds **26**–**28**; compounds **26** and **27** show a significant change in IC_{50}, but the LE reveals them to be very similar, whereas compounds **26** and **28** have comparable IC_{50}s, but **28** is considerably less efficient. Initial chemistry around these hits resulted in compounds with affinities in the range of 1–10 μM, which led to an improved understanding of the SAR and binding of the resorcinol motif, although again none of the resulting series were progressed. An "SAR-by-catalog" approach identified in excess of 1000 compounds containing the resorcinol motif in an in-house virtual library,[58] including the pyrazole-containing **31**, which showed some cellular potency in addition to submicromolar binding to Hsp90. Further optimizations[130,131] led to VER-49009 (**32**) which showed cellular effect including reduction in levels of Raf-1 and CDK4 and upregulation of Hsp70 and subsequently to a clinical candidate (VER-52296/NVP-AUY922; **33**) was identified, in which the pyrazole had been replaced by an isoxazole (which had the modest effect on potency, but showed a significant increase if the dissociation rate, k_d), the resorcinol chloro substituent replaced by an ^{i}Pr group, and a solubilizing morpholine group introduced.

Pros

- Of all the FBLD approaches, this is the most amenable to progressing fragment hits in the absence of structural data (see below)

Cons

- May require considerable resource in the absence of structural data, particularly in the earlier stages

5.4 Fragment Merging

Fragment merging is the process of combining the features of two fragments or a fragment and a larger molecule, in which at least one atom overlaps in their respective binding modes. Because of this overlap, cobinding is not possible, and so X-ray structural data are almost essential. In addition to "pure" FBLD approaches where overlapping fragments are merged, merging can also be used to "fix" liabilities in other existing series by replacing the "defective" part, or to "scaffold hop," replacing a patented core with an alternative to escape IP issues. We show an example of each here. Further merging examples are shown in Section 5.8.

As the ligand structures overlap, merging is conceptually simpler than linking, as the need to design a suitable linking unit is dispensed with. However, as Abell et al.[132] demonstrate (Fig. 10), merging can still present

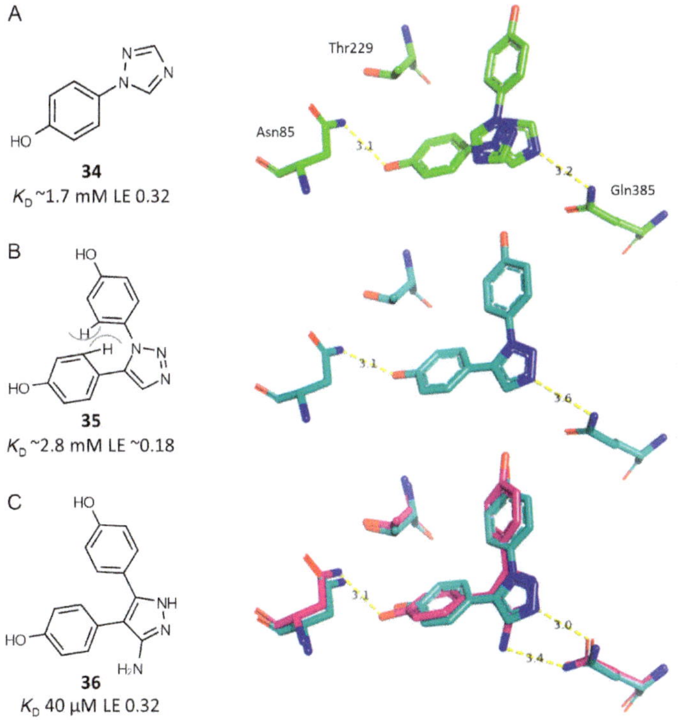

Fig. 10 Merging of Cyp121 fragments. (A) Fragment **34** with two overlapping binding modes (PDB 4G47). (B) "Broken" merged compound **35** with *ortho*-H clashes indicated with *red arcs* (PDB 4G2G). (C) "Fixed" merged compound **36** (*magenta*) overlaid with **35**, showing additional H-bond to Gln385 and greater twist between phenol rings; H-bonds are shown as *yellow dashes*.

significant challenges. In this case, fragment **34** bound in two separate, overlapping orientations in the binding site of the *Mycobacterium tuberculosis* target CYP121. Attempts to perform a seemingly simple "merge" of the two fragments led only to weakly binding compounds such as **35**. Crystallography and quantum mechanics calculations showed this to be due to the strain imparted on the binding mode by the clashing *ortho*-hydrogens (Fig. 10B), which, it was predicted, could be relaxed by strengthening the hydrogen bond from the 5-membered ring to Gln385. Thus, addition of an $-NH_2$ group with an additional H-bond donor gave **36**, in which the strain has now been released as hypothesized, and the LE restored.

There are numerous other examples of trouble-free merging in the literature, and we present only a few examples here (Fig. 11). Hughes et al.[118]

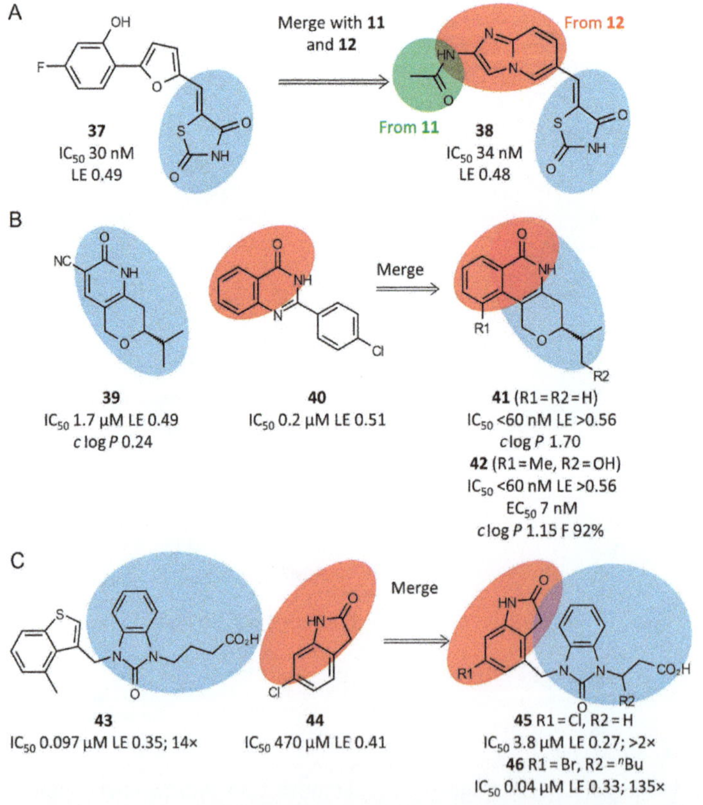

Fig. 11 Fragment merging. (A) Merging two fragments with the literature PI3γ kinase inhibitor (see Fig. 6 for structures of fragments).[118] (B) Merging 2 TNKS2 fragment hits to new core.[133] (C) Merging fragment with advanced Chymase lead to resolve metabolic issues (selectivities vs Cathepsin G shown as, e.g., 135 ×).[134]

noted that their optimized imidazopyridine fragment **12** (see Section 5.1) overlapped in the binding site with the known literature PI3ɣ inhibitor **37**.[135] Synthesis of the merged compound **38**, in which the thiazolidinedione head group from **37** has been incorporated but the phenyl-furanyl group replaced with the imidazopyridine group, with the *N*-acetyl substituent from the original fragment hit **11** also incorporated showed an increase in potency and retained an LE of ~0.48. X-ray crystallography (not shown) confirmed that the binding mode was retained.

de Vicente et al.[133] showed an example where two hits from a fragment screen against tankyrase (TNKS) were merged to generate a novel core **41** with even higher LE than the starting fragments **39** and **40**. Incorporation of a "magic methyl" group and optimization of the physicochemical properties led to compound **42**, which was "a promising candidate for *in vivo* proof-of-concept studies." In this case, the binding mode of **39** was obtained by crystallography and that of **40** by modeling based on literature structures.

In a final example (see Section 5.8), Taylor et al.[134] merged information from a fragment hit (**44**) with a "literature-to-lead" compound **43**, which was a potent inhibitor of Chymase, with moderate selectivity compared to Cathepsin G, but which suffered metabolic liabilities, particularly around the benzothiophene motif. A fragment screen found a number of less lipophilic compounds which overlaid with the benzothiophene motif in the S1 pocket. Merging and some further elaboration led to a series of compounds with good potency and improved selectivity, with "the liabilities frequently associated with large lipophilic molecules diminished." The authors also note that the use of the fragments allowed the incorporation of more polar, soluble motifs into an established scaffold and "demonstrated that dogma about what is tolerated [in the 'lipophilic' S1 pocket] can be challenged."

We have seen a small set of examples where fragment merging has been applied to replace parts of a competitor compound, generate a novel core from multiple in-house fragment hits, and "fix" compound liabilities from an advanced in-house series, demonstrating the utility and versatility of this approach.

Pros

- Simpler than linking
- Allows combining the results of a fragment screen with existing knowledge to escape from IP or compound issues

Cons

- Optimizing the merged structure can still present difficulties
- Requires good quality structural data

5.5 Target-Templated Chemistry

While not the primary goal of this review, mention should be made of approaches to fragment evolution in which chemistry is performed in the presence of a protein target to act as a template.[136,137] While there is nothing intrinsically fragment based about these approaches, the small starting points offered in FBLD and the need to elaborate them rapidly to larger potent binders means that the methods described have potential synergies with the fragment-based methodology. However, approaches such as Click chemistry and dynamic combinatorial chemistry (DCC) have shown mixed results in the FBLD field to date. The application of these approaches to fragments and associated technical challenges has been recently reviewed,[138] and we provide here a brief summary of some attempts to use these methods to elaborate fragments to illustrate their applicability and some potential pitfalls to consider during their application.

5.5.1 Dynamic Combinatorial Chemistry

In DCC,[139] a number of reagents which can react *reversibly* to form products are present in a single reaction vessel in the presence of a biomolecule of interest, resulting in theory in an enrichment of those products of higher affinity for the target. Many of the early DCC approaches used the reversible formation of (acyl) hydrazones or of disulfides as the equilibration step in the presence of proteins. While these are not the medicinal chemists favored groups, it is conceivable that they could be replaced in linked molecules. We show a small number of representative examples here.

One of the earliest examples of DCC in the fragment arena was published by Lehn [136] (Fig. 12A), targeting carbonic anhydrase II (CAII). A mixture of three aldehydes (**47a–c**) and four primary amines (**48a–d**) were allowed to react in the presence or absence of CAII (14 days in the presence of enzyme, 24 h without), and "quenched" with $NaBH_3CN$ to "lock" the equilibration. One amine product **50** was found to be enhanced in the presence of the enzyme, and its binding rationalized with reference to known CAII SAR and crystallography. In this approach, it is important to note that the templating preference is for the *intermediate imine* **49** rather than the reduced product.

Thiol–disulfide equilibration has also been the subject of various DCC efforts. Abell et al.[140] targeted the adenosine-binding site of pantothenate synthetase from *M. tuberculosis* using a modified thiol-bearing adenosine **52** and a mixture of eight thiols (**51a–h**; designed to target either the

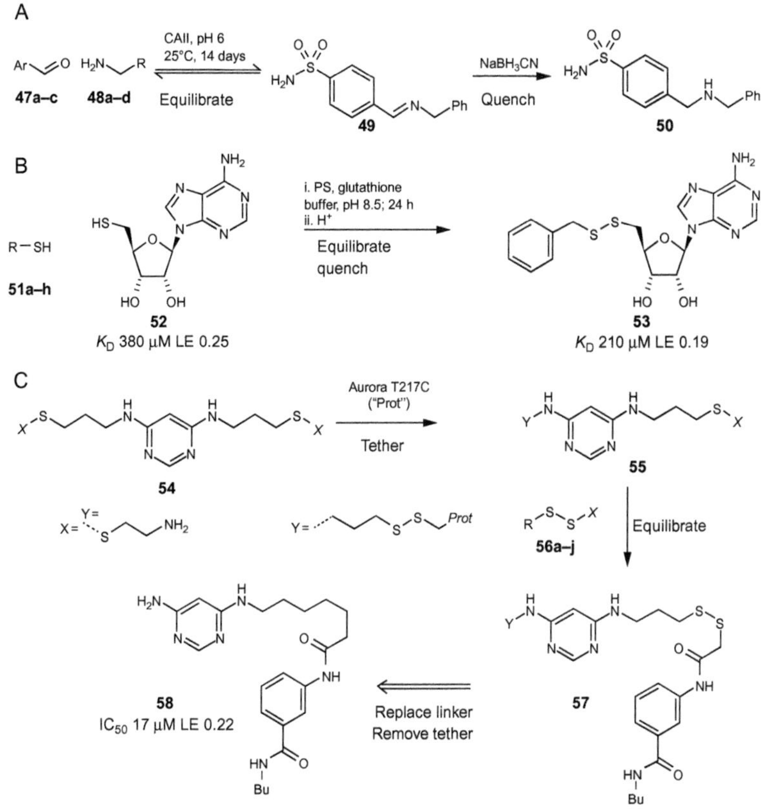

Fig. 12 Dynamic combinatorial chemistry (DCC) applied to fragments. (A) Imine equilibration and quench in the presence of carbonic anhydrase II (CAII). (B) Thiol–disulfide equilibration and quench in the presence of pantothenate synthetase (PS). (C) Sunesis "tethering" approach, disulfide equilibration, and ligand design for mutated Aurora A kinase T217C.

phosphate or glycerol/pantoate pockets) in the presence of the protein and glutathione redox buffer at pH 8.5 for 24 h followed by acidification to "lock" the equilibration. The mixture was enriched in a single disulfide **53** (Fig. 12B). A more complex variant on this theme has been reported by Erlanson et al.[141] in which the "head" group **54** was first tethered to a modified thiol-bearing protein (Aurora A kinase in this case) and then equilibrated with a set of 10 disulfide fragments (**56a–j**; Fig. 12C; in an approach the authors refer to as "site-directed DCC"). Again only a single doubly tethered example **57** was observed. The protein tether was removed and the disulfide linked replaced with a simple alkyl chain to give the elaborated compound **58**.

5.5.2 Click Chemistry

In Click chemistry,[137] a pool of reagents can react, normally irreversibly, to form products in the presence of a biomolecule, but with a sufficiently low background rate as to require binding to colocalized the reacting partners, and thus template the product from the reactants in the active site. Early efforts in this field focussed on the Huisgen Azide–Alkyne cycloaddition reaction,[142] and that is also reflected in the fragment literature in this area.

Ernst et al.[143] took the known myelin-associated glycoprotein (MAG) ligand **59**[144] and labeled it with a spin label (a group containing an unpaired electron; compound **61**; Fig. 13A), which causes a change in the relaxation rate within its close surroundings by the effect known as "paramagnetic relaxation enhancement" (PRE).[148,149] This effect has a large distance dependency, such that different protons in a fragment may show different degrees of effect if bound near the label. In this case, 5-nitroindole **63** was seen to bind with the 5-membered ring nearest the spin label, and so a set of azide derivatives **64a–c** with various length spacers were incubated with a set of derivatives of **60a–d** with the spin label replaced by alkynes, also with varying length spacers, all in the presence of the protein for 3 days at 37° C. LC–MS analysis of the resulting mixture showed a single major product, which was confirmed as **62** by resynthesis (the reaction can produce two regioisomeric products—the other product, not shown, is around 10-fold less active). It is noteworthy that the product retains fully the LE of the starting sialic acid derivative. However, when the same group applied this approach to E-selectin, they saw no products formed.[150] To investigate why, they synthesized a library of 20 triazole-linked products and found 5 compounds binding with nanomolar potency, a result which was rationalized by the suggestion that the flat binding site of E-selectin does not provide sufficient preorganization of the reactants to enable to reaction to occur.

Another example by Suzuki et al.[145] attempted to react azides with histone deacetylase-8 (HDAC8)-binding hydroxamic acids such as **65** (Fig. 13B), which gave the *anti* product **67** as the only observed product. The researchers were alerted by the unusually high yield of this product, the fact that it showed no improvement in binding compared with the original hydroxamic acid fragment **65**, and that the unobserved *syn* isomer **68** was around 10-fold more potent. In this case, it appears that copper contaminants in the protein—most likely from the synthesis of **65**—appear to be responsible for catalyzing the formation of **67**. (While copper or ruthenium can

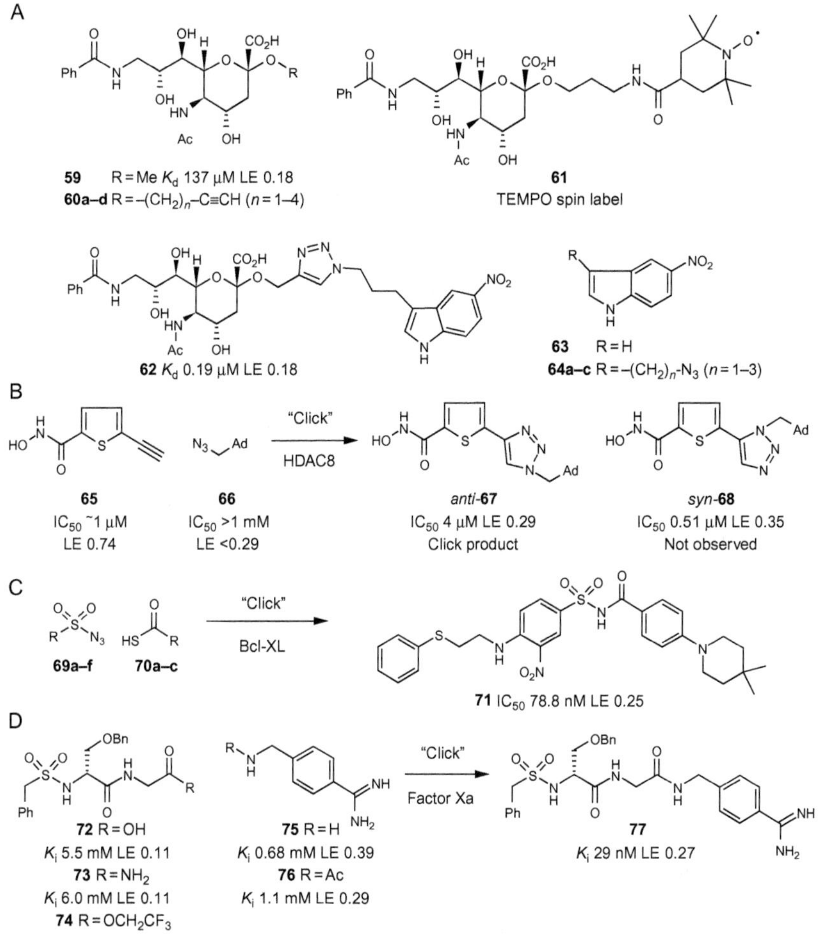

Fig. 13 Click chemistry. (A) Structures for Click precursors, spin label, and selection product from myelin-associated glycoprotein (MAG).[143] (B) Click reaction in the presence of HDAC8 and nonobserved product (*Ad*=1-adamantyl).[145] (C) Click amide formation in the presence of Bcl-XL.[146] (D) Click amide formation from 2,2,2-trifluoroethyl ester in the presence of Factor Xa showing superadditivity.[147]

catalyze the room temperature Huisgen reaction,[151] in this case neither copper alone nor protein with all copper removed catalyzed the formation of **67**.)

Other approaches have attempted to use more commonly found drug-like groups in the linker group. Manetsch et al.[146] described the generation of a known Bcl-XL binder from a pool of 6 sulfonylazides and 3 thioacids (18 binary mixtures), in an unusual amidation reaction. Only a single compound

71 showed enrichment in the presence of the protein compared with the control reactions without protein present (Fig. 13C). Synthesis of the other possible products showed that they all showed IC_{50} values >5 μM. This successful approach is most limited by the relatively low numbers of commercially available reacting partners.

In a more recent approach to click amide formation, Rademann et al.[147] investigated a variety of esters and other activated carboxylic acid derivatives for their potential to form amides in the presence of protein (in this case, Factor Xa was used as a model). They found the 2,2,2-trifluoroethyl ester to have the desired balance of lack of reactivity under the incubation conditions in the absence of protein, with reactivity in the presence of the protein. The product itself shows superadditivity of binding compared to the acid **72** and amine **75** ($E=0.00775$), which was confirmed not to be a feature of the linking amide group interacting with the protein by comparison with the capped fragments **73** and **76** ($E=0.00439$).

Pros

- Can directly steer the product formed toward the best binders without synthesis/testing of large number of library members

Cons

- Limited chemistries available
- Requires deconvolution of enriched product
- Can give misleading results

5.6 Evolving Fragments in the Absence of Structural Information

Fragment evolution is easiest when good quality structural information is available, the "gold standard" being X-ray crystallography of ligand-bound complexes. However, it is still possible to evolve fragments into high-potency leads in the absence of crystallography. NMR can provide varying degrees of structural information ("full structure," HSQC chemical shift perturbation patterns, NOE constraints (e.g., to selectively ^{13}C-labeled protein), ILOE, PRE) which can also be used to guide initial evolution, and when this is not possible, homology models, docking, and other modeling techniques may also provide insight. Furthermore, as ligands become more elaborated and more potent, it may become possible to obtain crystallographic structures where this was not possible at the early fragment evolution phase of a project.

In the absence of a crystal structure, it may be possible to gain some initial SAR by screening carefully selected near-neighbor compounds to get a sense of what may be tolerated around the fragment-binding site (bearing in mind, however, the caveat about binding mode mentioned in Section 5.7.2) where those analogues are available. It becomes inevitable, however, that new analogues will need to be synthesized, while at this stage in a fragment program there is likely still to be considerable uncertainty in the absence of structural information. A method described recently by workers at Vernalis[4,11] demonstrates that SPR dissociation rate constants (k_{off} or k_d) can be obtained when screening crude reaction products, as the dissociation event is a unimolecular process which is independent of ligand concentration (and thus of reaction yield). As has been noted, k_{off} relates to residence time, and also potency ($K_D = k_{off}/k_{on}$), and there are potential additional benefits to optimizing k_{off}.[152–157] In the first disclosure of this approach,[11] the workers demonstrated that the known SAR of a series of Hsp90 inhibitors **80a–n** (mean variation between crude and pure 5.8%, max 12%; Fig. 14A), and a more challenging series of Pin1 inhibitors **83a–g** (mean variation 7.8%, max 18%; Fig. 14B), could be reproduced in this manner. In the case of the Pin1 inhibitors, the SAR could be reproduced using racemic *rac*-**82** (1 step from the commercial material, not shown) in lieu of the enantiomerically pure reagent (*R*)-**82** which required a five-step synthesis. The approach, referred to as off-rate screening (ORS), was tolerant of the conditions used in the most commonly used reactions favored by medicinal chemists[158,159] and could be performed in 96-well plates. The methodology allows the rapid SAR screening of a wide variety of substituents without the bottleneck of purification and is thus particularly applicable to early fragment evolution when structural data are not routinely available, although it can also be usefully used in structurally enable targets.

In a more recent example, a prospective application was reported for the PDHK family of ATPases (Fig. 14C).[4] This family is closely related to Hsp90 (both are members of the GHKL superfamily[160]), and selectivity between these two targets was required. There was a lot of overlap in the fragment hits for PDHK with those seen previously for Hsp90 (see Fig. 9), and a near-neighbor screen based around the resorcinol fragment **84** identified the resorcinol amide **85** as a promising starting point. At this point, the ORS approach was used to explore PDHK SAR and Hsp90 selectivity in parallel, generating both novel chemical space (the series identified was closely related to other series of published Hsp90 inhibitors, e.g., **88**[161,162] and **89**[163,164]) and more impressively selectivity between the two targets (compounds were disclosed which were both selective for PDHK, e.g., **86** or selective for Hsp90, e.g., **87**). A fringe benefit of the approach was that

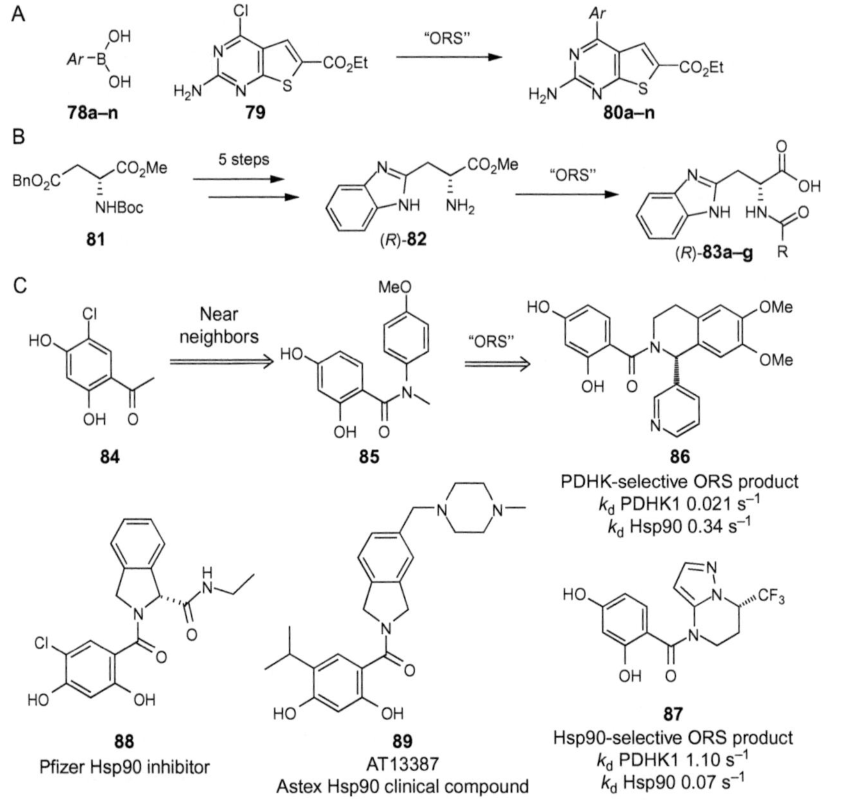

Fig. 14 Off-rate screening (ORS) approach. (A) Hsp90 Suzuki coupling library—ORS conditions: $NaHCO_3$, $PdCl_2(PPh_3)_2$, DMF–H_2O, microwave 100°C, 10 min.[11]. (B) Pin1 library (two steps, enantiopure and racemic versions)—ORS conditions: 1. RCO_2H, COMU, Et_3N, DMF; 2. LiOH, MeOH, H_2O.[11] (C) PDHK example structures (ORS conditions: i. 2,4-Diacetoxybenzoic acid, COMU R_2NH, Et_3N; ii. NH_3–MeOH).[4]

the chemistry needed some modification in order to run in a parallel manner in a 96-well plate (previous syntheses used benzyl protection of the phenolic groups, which was changed to acetate protection, enabling simultaneous quench of amide formation and deprotection with methanolic ammonia), which also improved the downstream synthetic tractability of the series.

5.7 Factors to Consider During Fragment Evolution

5.7.1 Conserved ("Structural") Waters

It is not uncommon for structures of bound fragments to also contain water molecules in the binding site, as might be expected for a small ligand in a larger site. There is growing understanding that while some water molecules should be displaced during the fragment evolution process, for others ("structural" or "conserved" waters), the energy required to displace them is too great and cannot be compensated for by the interaction energy gained. In these cases, the individual water molecules are best considered to be part of the binding site—in effect, an extension of the protein structure. There are a number of computational tools which attempt to identify those waters which should be retained using various approaches.[165] Ultimately, however, careful examination of the crystal structures involved, and attempting to displace or interact with (both firmly held waters, and to stabilize the positioning of intermediate waters) various water molecules with suitably modified compounds, particularly during a fragment optimization stage, is the only way to be certain of the outcome.

Workers at Vernalis[166] targeting Chk1 kinase described the optimization and evolution of the benzimidazolyl-pyridone fragment **90** to **93** (Fig. 15), in which an amide has been added in order to form a hydrogen bond to the conserved water network shown—noting that the water molecule forming the interaction with the ligand was not visible in the structure of the initial fragment hit, but "was placed by modelling based on the positions seen in other Chk1 structures"—the water molecule is then visible in the structure of the elaborated fragment **91** and grown compound **92**, suggesting that its placement has been *stabilized* by the additional interaction with the ligand. Reversing the amide (**93**) maintained the hydrogen bond and improved potency further.

Hsp90 also has a well-known network of four conserved water molecules in the adenine-binding site around Asp93. In this case, many publications *retain* the water network in its entirety, as shown for the example thienopyrimidine structure **94** (Fig. 16A),[54] which interacts with the protein via multiple hydrogen bonds with the water network. However, workers at Vernalis also demonstrated in a related series of compounds that one of the conserved water molecules could be *displaced*.[50] The purine fragment hit **95** also retains the water network, with a slightly modified hydrogen bonding pattern in which the C–H of the imidazole ring acts as a donor in place of the second donor from the $-NH_2$ group of **94** (Fig. 16B). Applying the known arene N("=N–") to C–C≡N replacement[167] (Fig. 16D) from the original fragment **95** and merging in the 1,3-disubsituted aromatic ring from **94** gave the novel cyanopyrrolopyrimidine **96** in which one of the water molecules has been displaced by the –C≡N nitrogen atom (Fig. 16C).

To summarize, all water molecules within a binding site must be considered for one of three fates—displacement or retention with or without

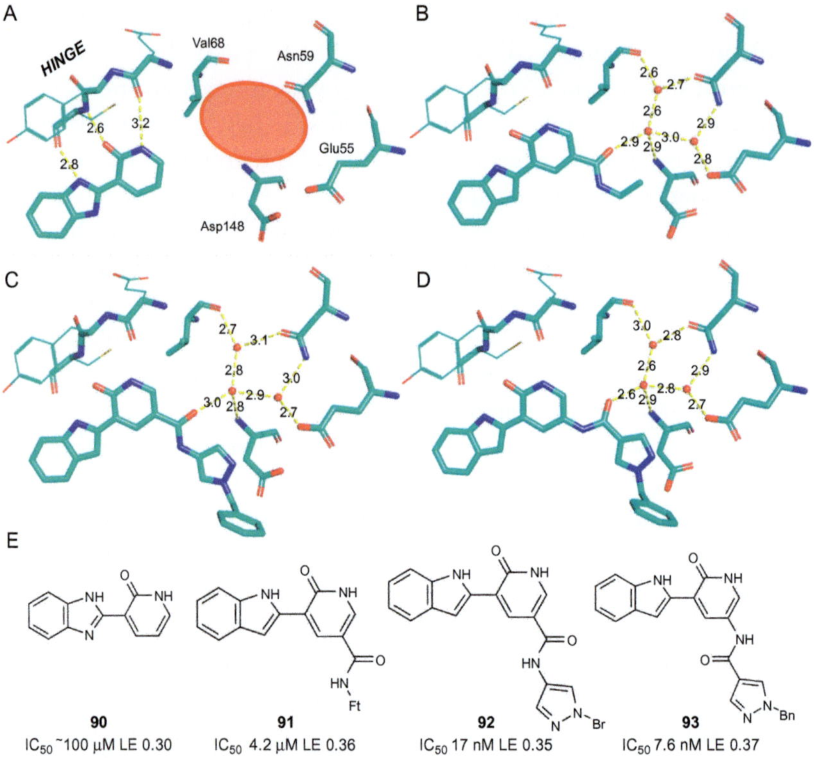

Fig. 15 Stabilizing water in Chk1. (A) Structure of fragment hit **90** showing hinge interactions. Water pocket with disordered waters highlighted in *red*, and residue names shown. (B) Elaborated fragment **91** with the amide hydrogen bond acceptor stabilizing water network. (C) Elaborated amide **92**. (D) Reversed elaborated amide **93** retaining interactions with water network. (E) Ligand structures and potencies. Hinge residue side chains shown as *lines* for clarity.

stabilization by the ligand. For many water molecules, displacement is the obvious option; however, for others, it may be less clear, and comparison of multiple structures and careful modeling may be required to assist in the decision.

5.7.2 Binding Mode

As fragments are small ligands, with potentially many interaction sites, it should not be surprising that multiple binding modes are possible—indeed workers at Vernalis showed that the same fragment (**97**) forms different interactions at the kinase hinge in the kinases PDPK1, Chk1, and CDK2

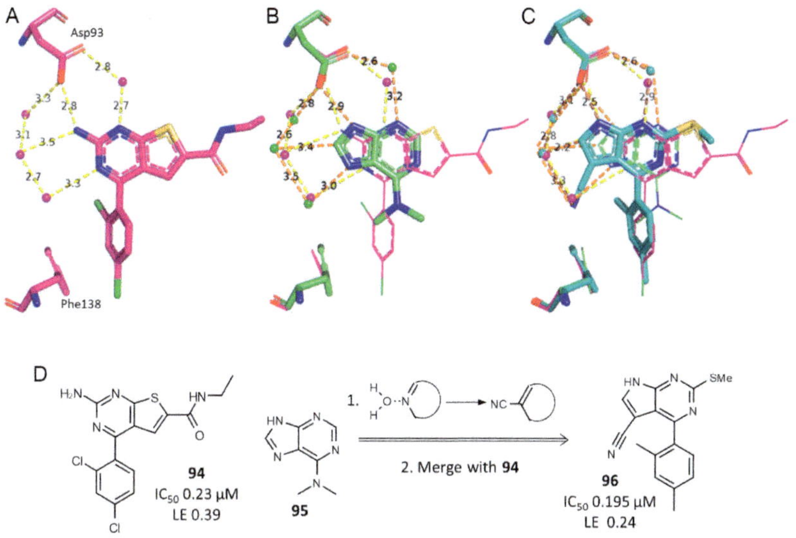

Fig. 16 Displacement of water from Hsp90. Waters are shown as *small spheres* in the *same color* as parent carbons. (A) Four conserved water molecules' network shown with complex to thienopyrimidine **94** (PDB 2WI6). (A) Overlay of **94** (*magenta lines*) with **95** (PDB 4FCP). (C) Overlay of **94** (*magenta lines*) and **95** (*green lines*) with the cyanopyrrolopyrimidine **96** (PDB 4FCQ). (D) Structures and genesis of **96** from fragment hit **95**.

(Fig. 17A–D)[57]—in the case of Chk1 and PDPK1, the D–A–D triad to the hinge backbone being completed with a C–H donor from the pyrimidine C–H. The pyrazole C–H is directed toward the backbone carbonyl in CDK2, but at 4.2 Å the C···O distance is too great to form an H-bond. Addition of a *para*-substituted phenyl ring demonstrates further the care needed; in CDK2 and Chk1, the binding mode of the fragment remains essentially unchanged. However, in PDPK1, the pyrazolopyrimidine motif flips (Fig. 17D–E), using the same three protons to form the D–A–D triad as in Chk1, but in the *reverse* orientation. A similar example has also been reported by workers at Merck Serono.[168]

Perhaps slightly more surprising is the observation that the binding mode of the fragment **102** in Hsp90 is dependent on the conditions under which the crystals were obtained.[54] Ligand-soaking experiments of **102** with Hsp90 gave an unusual binding mode, which was not consistent with other known aminopyrimidine and aminotriazine inhibitors, whereas cocrystallization of **102** with Hsp90 gave the expected binding mode, which was then maintained throughout the elaboration to the clinical candidate NVP-BEP800/VER-82576 **118** (see Section 5.8).

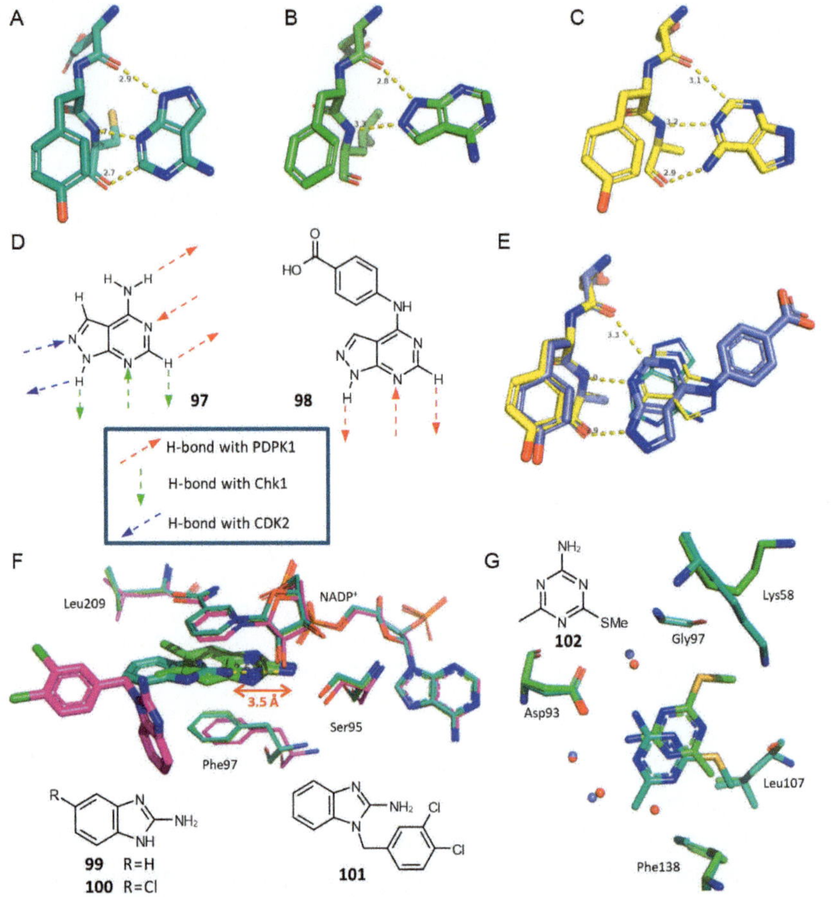

Fig. 17 Fragment-binding modes. (A)–(E) **97** and **98** with Chk1, CDK2, and PDPK1 kinases. All structures aligned on the backbone of the three hinges residues shown. (A) **97** complexed with Chk1. (B) **97** complexed with CDK2. (C) **97** complexed with PDPK1. (D) Ligand structures and schematic view of hinge H-bonds. (E) Elaborated compound **98** complexed with PDPK1, overlaid with **97** in PDPK1 (*yellow*) and Chk1 (*cyan*). (F) Compounds **99** (*cyan*), **100** (*green*), and **101** (*magenta*) overlaid in pteridine reductase (*red arrow* shows **99**/**100** shift). (G) **102** in complex with Hsp90 by soaking (*cyan/slate waters*) or cocrystallization (*green/red waters*).

Brenk's group also saw shifts in the binding mode of three closely related inhibitors of pteridine reductase 1 (Fig. 17F).[169] The aminobenzimidazole **99** fragment shifts ~3.5 Å toward Ser95 on the addition of a 5-chloro substituent (**100**). In both cases, the aromatic system forms a stack between the NADP+ cofactor and the side chain of Phe97. However, elaboration by the addition of a substituent to one of the ring N-atoms changes the binding mode so that the ring is no longer stacked in this manner (**101**).

A detailed analysis of the binding mode stability of fragments has recently been published, highlighting further examples and attempting to provide rationalization.[170] A case study using Hsp90[171] showed that, regardless of the possibility for changing binding modes, at least in this case, fragments do sufficiently maintain similar binding modes as to be able to account for the known published inhibitors at the time. A more general study by Drwal et al.[172] draws similar conclusion for fragments and small crystallographic additives, and drug-like ligands across four diverse targets (cyclin-dependent kinase-2 (CDK2), β-secretase-1, CAII, and trypsin).

The key points to take from this are that regular checking of binding mode is essential throughout the fragment evolution process, particularly if SAR is inexplicable from the current state of structural knowledge. Closely related proteins (e.g., kinases) may show surprisingly different binding modes for the same or similar fragments, while subtle changes to the fragment structure may also result in unexpected binding mode changes in the same protein-binding site.

5.7.3 Potency, Efficiency, and Metrics Thereof

It is to be expected that the small size of fragments may lead to weak binding, and this can make it difficult to decide which fragment(s) to pursue, and indeed whether a small change in potency is the result of a successful modification or not. The oft-cited, but rarely observed "magic methyl" effect,[115,116,173] for example, in which the addition of a single heavy atom can have a profound effect on potency; Schönherr and Cernak[116] showed examples with >2000-fold improvement. The problem is even greater when comparing for fragment hits against HTS hits and literature compounds, where the heavy atom count (HAC) may be double or more. To counter this, a number of metrics have been proposed, which all "normalize'"in some way the potency by accounting for the molecular size (HAC or molecular weight (MW)). The most widely used measure is *LE*, defined as the Gibb's free energy of binding per heavy atom[29,174]

$$\mathrm{LE} = \Delta G/\mathrm{HAC}$$

or in terms of K_D,

$$\mathrm{LE} = -RT\ \ln K_\mathrm{d}/\mathrm{HAC}.$$

LE is almost invariably quoted without units, in which case, the units should be assumed to be kcal/mol/non-H atom and was originally defined with a temperature of 300K, although 298K (i.e., 25°C) is also widely used in practice. An alternative metric also encountered is *Binding Efficiency Index* (BEI) and the closely related *Percentage Efficiency Index* (PEI) (for single-point inhibitions).[(175)]

$$\mathrm{BEI} = \mathrm{p}K_\mathrm{i}\ (\mathrm{or}\ \mathrm{p}K_\mathrm{d}\ \mathrm{or}\ \mathrm{pIC50})/\mathrm{MW},$$
$$\mathrm{PEI} = \%\ \mathrm{inhibition\ at\ a\ given\ compound}/\mathrm{MW},$$

where MW is in kDa and pXX is defined as $-\log(\mathrm{XX})$. In both cases, it is possible to back-calculate from a target potency for a molecule of, e.g., 10 nM and MWt ~ 500 (HAC ~35), the BEI is 16.0 and the LE is 0.31. Therefore, in order to achieve this, a fragment with HAC 15/MWt 200 needs to have a potency of ~630 μM (using BEI 16.0) or ~390 μM (using LE 0.31) (see Fig. 2).

Not all targets are equal, and the anticipated LE for a buried binding site in which much of the ligand is surrounded by the target protein, e.g., serine–threonine and tyrosine kinases, is likely to be higher than, for example, a PPI with a large, relatively open binding site, in which one face of the molecule simply cannot interact with the protein.

Additionally, not all heavy atoms are equal; Hughes et al.[118] observe that "*it should be noted that the LE of bromine-containing compounds does tend to be higher than methyl analogues, as bromine is more lipophilic and capable of achieving a greater binding affinity than a carbon atom.*" To compensate for this effect, a variety of metrics related to LE and BEI have been proposed. The simplest are *Surface-binding Efficiency Index* (SEI),[175] which is related to BEI and PEI, relating instead to the topological polar surface area (TPSA), and *Lipophilic Ligand Efficiency* (LLE,[176] aka *Lipophilic Efficiency*, LipE[177,178]) and less commonly *Ligand Efficiency-Dependent Lipophilicity* (LELP)[179] which relate to LE. These are defined as follows:

$$\mathrm{SEI} = \mathrm{pIC50}/(\mathrm{TPSA}/100),$$
$$\mathrm{LLE} = \mathrm{LipE} = \mathrm{pIC50} - \log P,$$
$$\mathrm{LELP} = \log P/\mathrm{LE}.$$

While these are the simplest and most useful, they have some limitations which have been discussed elsewhere, and a variety of alternative metrics,[177] of varying degrees of utility and esotericity, including *Group Efficiency*[180,181] (which is reminiscent of the earlier Andrews Energy[28]), *Fit-quality scaled ligand efficiency,*[182–184] *Size-Independent Ligand Efficiency* (SILE),[185] and *Ligand Lipophilicity Index* (LLE_{AT})[186] have been proposed. The relative merits and applicability of these alternatives is beyond the scope of this review—we suggest that the use of LE (or BEI/PEI) and the corresponding LLE (or SEI) covers most cases adequately.[53] Numerous reviews discuss the wide variety of metrics in this family and their application.[52,177,187,188]

5.8 Combined Approaches and Integration With Other Hit ID Methods

While there are many examples in the literature of the "pure" application of fragment-based approaches, in reality FBLD does not exist in a vacuum and is one of the many tools in the drug-discovery arsenal. Fragment-based approaches can be used in combination with HTS hits, literature compounds, published crystal structures, and any other available sources of information. We have already seen some examples in the sections on fragment optimization and fragment merging of combining fragments with information from other sources. We will conclude here with a few examples from Vernalis showing combined approaches.

In the first example, hits from various sources were combined in a program to target the kinase PDK1.[57] The known CDK2 inhibitor **106**[189] was seen to be a reasonably potent PDK1 inhibitor, and an X-ray structure was obtained, showing the cyclohexyl ring to lie on a partially hydrophobic, solvent-exposed area of the protein surface. A search of our rCat[58] database of commercially available compounds for compounds containing both a hydrophobic ring and an imidazole moiety (from the fragment hit **107**) identified the symmetrically substituted imidazole **108** in which one of the phenyl amides occupies the region of the hydrophobic ring of **106** and the other retains that from **107** in which it forms part of an unusual D–A–D triad with the hinge. **108** retains most of the LE of its parents. In this case, we have combined a pharmacophoric feature from a literature compound targeting a homologous protein, with a structural feature from a fragment hit, and obtained the grown fragment **108** from commercial compound space. Limited optimization of **108** using structural information from X-ray crystallography led to **109**, in which a solubilizing piperidine group has been introduced, again retaining LE.

In parallel, a crystal structure of the known Chk1 kinase inhibitor **103**[190] bound to PDK1 was obtained, and a fragment **104** designed (NB this was a designed fragment, not a hit from our fragment screen against PDK1) retaining the hinge-binding D–A–D motif. With multiple X-ray structures of these compounds and other fragment hits including **105**, it became possible to design new ligands combining features from these molecules, such as **110** by merging fragment and other hits. It is interesting to note as an aside that the intramolecular hydrogen bond forming a *pseudo*-ring[191] in **110** has reversed its direction from **107–109**, resulting in a small shift in the position of the piperidine ring in the binding site. Compound **110** was subsequently optimized to potent compounds tolerated in vivo with PD marker changes in xenograft studies consistent with PDK1 inhibition (Fig. 18).

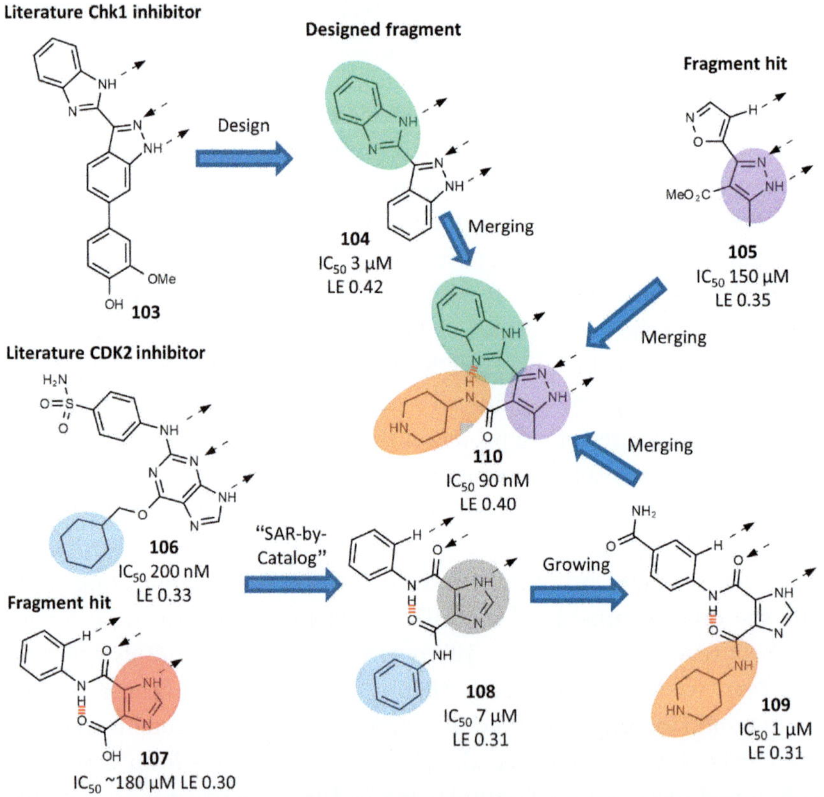

Fig. 18 Combined approach to potent PDK1 inhibitor **110** showing evolution of features from literature compounds and fragment hits via designed fragments, "SAR by catalog" and merging. All structures are orientated in the same direction as determined by X-ray crystallography, and the D–A–D triad with the hinge indicated by *dashed arrows*. Intramolecular H-bonds are indicated with a *dashed red line*.

In the second example[54,119] a follow-up series of Hsp90 inhibitors was required following the resorcinol series of compounds described in Section 5.3, preferably with good oral bioavailability. As such, there was considerable in-house data from a fragment screen, medium-throughput screen, virtual screen, and medicinal chemistry program, along with a growing body of literature in the public domain. Information from crystal structures of a number of fragment hits was combined with that from crystal structures of virtual screening hits to design a new fragment **116**, following some early fragment elaboration work in which the original fragment hit **111** was substituted with an amide-bearing thioether **112**, giving a significant increase in potency (Fig. 19). In this case the researchers noted the relatively poor LE for this optimized designer fragment but "were confident from both structural information and the SAR available" that they would be able to progress this and improve upon the LE during optimization. A crystal structure of **116** confirmed the design hypothesis. This is a more complex example of fragment optimization to those shown in Section 5.1, in which two fragments have been merged with features from a larger hit compound to generate an optimized fragment.

Following this, an aryl substituent was introduced to replace the 4-amino substituent, based on the observations of an aryl substituent in this position in the virtual screening hits **114** and **115**, and in the ~18-fold increase in potency seen for the elaborated fragment **113**. A simple phenyl substituent in this position (**117**) resulted in a >200-fold leap in potency; incorporation of small, lipophilic substituents in the 2- and 4-positions as seen in the virtual screening hit **115** gave a further 25-fold leap (compound **94**). In particular, the 2-position substituent was seen in the virtual screening hit **115** to occupy a small lipophilic pocket from which it displaced a water molecule (Fig. 19A and B; green circle), and to improve the torsional orientation of the phenyl substituent relative to the thienopyrimidine core. In this second phase, a viable lead series was obtained by merging features from virtual screening hits with our optimized fragment.

Finally, the compounds' physicochemical properties, particularly solubility, were tackled. Comparison with the clinical candidate lead compound **33** (VER-52296/NVP-AUY922)-bound structure suggested that a solubilizing group might be introduced from the 5-position of the phenyl substituent, with the pyrrolidin-1-yl-ethyl ether **118** being optimal (Fig. 19C). Compound **118** was shown to be efficacious in BT474 human breast tumor xenograft studies with oral dosing in preclinical in vivo studies and was

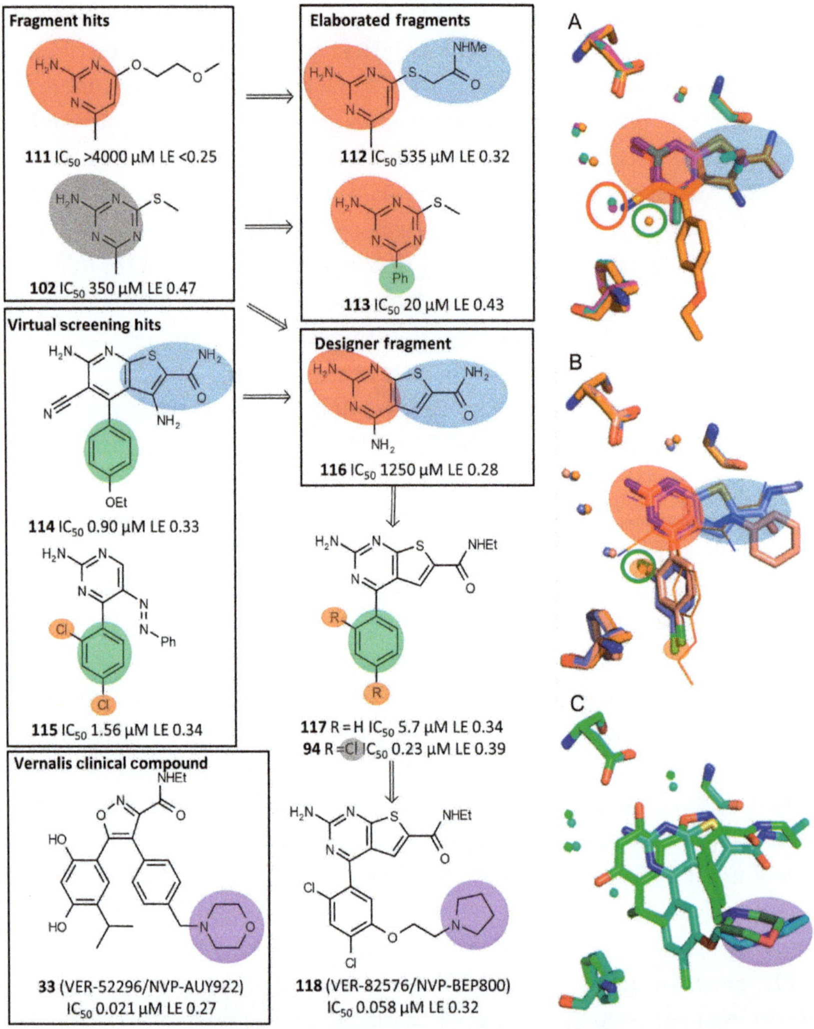

Fig. 19 Evolution of VER-82576 (**118**) from fragment and virtual screening hits and clinical candidate—structures on the *left* are "inputs," those on the right evolved compounds. (A) Overlay of **102** (*magenta*), **111** (*cyan*), and **114** (*orange*) highlighting conserved water displaced by cyano group of **114** (*red circle*) and conserved water displaced later (*green circle*). (B) Overlay of **94** (*slate*), **114** (*orange*; *thin lines* for clarity), and **115** (*salmon*), highlighting water displaced by *ortho*-substituent (*green circle*). (C) Overlay of **33** (*green*) and **118** (*cyan*). Waters are shown as *small spheres* in the *same color* as parent carbons.

selected for further development as VER-82576/NVP-BEP800. In this final stage, a physicochemical properties issue was addressed by merging information from another clinical compound to suggest a vector for fragment growth. The Vernalis Hsp90 program has been described in an informative review.[119]

6. FUTURE CHALLENGES

While fragment-based techniques have developed rapidly, with a wide array of biophysical screening techniques now available, there are significant challenges still remaining. The weak binding and often high ligand or target concentrations required to detect that binding increased the likelihood of artifacts resulting from assay interference from fluorescence, aggregation, or other mechanisms.[76] Developments in biophysical techniques are enabling more accurate detection of weak binding, and more rapid or higher quality structural data to be obtained, but obtaining suitable soakable crystal forms remains challenging, and NMR-based structural methods require significant investment of time and resources.

Much of the work on developing fragment-based methods to date has focussed on structurally enabled targets, particularly proteins which remain folded in solution, for example, protein kinases, proteases, and other enzymes. However, a significant proportion of marketed drugs target membrane-bound GPCRs, and a significant proportion of GPCRs remain as orphan receptors, or undrugged.[192] Advances in the application of fragment-based and structure-based methods to GPCRs have been made in recent years (GPCRdb currently lists 190 crystal structures of 43 GPCR receptors across 24 receptor families in 3 GPCR classes[193–195]), but many challenges still remain in applying fragment-based approaches more widely to membrane-bound proteins.[196–200] Another area of growing interest is in PPIs;[201–203] often a large shallow-binding site is present, possible with small pockets induced by binding of the second protein partner to accommodate specific amino acid side chains. The shallow-binding sites mean that lower LEs are likely, and often crystallography is not possible until elaborated ligands have been developed inducing conformational changes or rigidifying a bound form.

Chemistry can be a frustratingly slow process, with seemingly simple changes to a fragment requiring relatively long synthetic routes, and despite many advances in purification equipment, purification of final compounds remains a bottleneck. We have highlighted above an approach to tackle this in the earlier stages of fragment elaboration, but this can create new informatics challenges to handle the data and track the library members. Other approaches are required to quicken the turnaround of ideas into screening samples, such as the direct functionalization chemistries of Macmillan[204,205] and Baran[206–216]—Schönherr et al. highlighted the need for reliable methods for the direct introduction of a methyl group in their analysis of the "magic methyl" effect,[116] and Roughley and Jordan called for "new methodologies or improvements to existing transformations, making them more generally applicable and amenable to parallel chemistry".[158] The application of late-stage functionalizations to drug-like molecules has recently been reviewed.[217]

With data coming from myriad chemical, analytical, and biophysical techniques, informatics tools are required to both streamline the data processing, and possibly even sample submission processes, and join data from multiple sources. Data pipelining tools such as KNIME[218–220] allow powerful manipulation and reporting of data from multiple sources, including access to multiple scripting languages (e.g., Python, Java, R, ImageJ) and chemical toolkits (e.g., RDKit, CDK, Indigo, etc.) to manipulate those data, but many instruments still produce their data in proprietary formats which require interactive processing by the user before exporting to more accessible formats (XML, JSON, delimited ASCII text files, spreadsheets, etc).

7. CONCLUSIONS

Over the past two decades, FBLD has proven to be a valuable and robust approach for identifying and evolving novel chemical matter into "lead" compounds suitable for the preclinical study. Indeed, FBLD is now typically one of the core technologies applied to a new drug-discovery program, particularly where the target is novel. The FBLD approach has continued to both mature and converge, with a common framework now widely accepted for the identification and characterization of small (and hence typically low affinity) ligands, and the evolution of these initial fragment ligands into more potent hits or lead molecules. These evolved molecules are suitable for progression into preclinical and clinical candidates using conventional medicinal chemistry techniques, although the wealth of data provided by the fragment approach can continue to inform and guide drug discovery even in the later stages of lead optimization. Using these approaches, two fragment-derived molecules are now approved for clinical use, and a plethora of additional compounds are currently in various stages of clinical trials; the next decade will hopefully see many additional drugs approved which have been discovered and developed using the FBLD-based approach.

REFERENCES

1. Shuker, S. B.; Hajduk, P. J.; Meadows, R. P.; Fesik, S. W. Discovering High-Affinity Ligands for Proteins: SAR by NMR. *Science* **1996**, *274*(5292), 1531–1534.
2. Erlanson, D. A.; Fesik, S. W.; Hubbard, R. E.; Jahnke, W.; Jhoti, H. Twenty Years on: The Impact of Fragments on Drug Discovery. *Nat. Rev. Drug Discov.* **2016**, *15*(9), 605–619.

3. Erlanson, D. A. Introduction to Fragment-Based Drug Discovery. *Top. Curr. Chem.* **2012**, *317*, 1–32.
4. Fischer, M.; Hubbard, R. E. Fragment-Based Ligand Discovery. *Mol. Interv.* **2009**, *9*(1), 22–30.
5. Schulz, M. N.; Hubbard, R. E. Recent Progress in Fragment-Based Lead Discovery. *Curr. Opin. Pharmacol.* **2009**, *9*(5), 615–621.
6. Rees, D. C.; Congreve, M.; Murray, C. W.; Carr, R. Fragment-Based Lead Discovery. *Nat. Rev. Drug Discov.* **2004**, *3*(8), 660–672.
7. Congreve, M.; Chessari, G.; Tisi, D.; Woodhead, A. J. Recent Developments in Fragment-Based Drug Discovery. *J. Med. Chem.* **2008**, *51*(13), 3661–3680.
8. Kutchukian, P. S.; Wassermann, A. M.; Lindvall, M. K.; et al. Large Scale Meta-Analysis of Fragment-Based Screening Campaigns: Privileged Fragments and Complementary Technologies. *J. Biomol. Screen.* **2015**, *20*(5), 588–596.
9. Bollag, G.; Tsai, J.; Zhang, J.; et al. Vemurafenib: The First Drug Approved for BRAF-Mutant Cancer. *Nat. Rev. Drug Discov.* **2012**, *11*(11), 873–886.
10. Souers, A. J.; Leverson, J. D.; Boghaert, E. R.; et al. ABT-199, a Potent and Selective BCL-2 Inhibitor, Achieves Antitumor Activity While Sparing Platelets. *Nat. Med.* **2013**, *19*(2), 202–208.
11. Erlanson, D. *Fragments in the Clinic: 2016 Edition*, 2016, Accessed 10 August 2017. http://practicalfragments.blogspot.co.uk/2016/07/fragments-in-clinic-2016-edition.html.
12. Chessari, G.; Woodhead, A. J. From Fragment to Clinical Candidate—A Historical Perspective. *Drug Discov. Today* **2009**, *14*(13–14), 668–675.
13. Howard, S.; Abell, C., Eds.; *Fragment-Based Drug Discovery*: The Royal Society of Chemistry: London, 2015.
14. Erlanson, D. A.; Jahnke, W., Eds.; *Fragment-Based Drug Discovery: Lessons and Outlook*; Wiley: New York, NY, 2015.
15. Czechtizky, W.; Hamley, P., Eds.; *Small Molecule Medicinal Chemistry: Strategies and Technologies*, 1st ed.; Wiley: New York, NY, 2015.
16. Magee, T. V. Progress in Discovery of Small-Molecule Modulators of Protein-Protein Interactions via Fragment Screening. *Bioorg. Med. Chem. Lett.* **2015**, *25*(12), 2461–2468.
17. Erlanson, D. A.; McDowell, R. S.; O'Brien, T. Fragment-Based Drug Discovery. *J. Med. Chem.* **2004**, *47*(14), 3463–3482.
18. Erlanson, D. A.; Wells, J. A.; Braisted, A. C. Tethering: Fragment-Based Drug Discovery. *Annu. Rev. Biophys. Biomol. Struct.* **2004**, *33*, 199–223.
19. Zartler, E. R. Fragonomics: The -Omics With Real Impact. *ACS Med. Chem. Lett.* **2014**, *5*(9), 952–953.
20. Whittaker, M. Picking Up the Pieces With FBDD or FADD: Invest Early for Future Success. *Drug Discov. Today* **2009**, *14*(13–14), 623–624.
21. Hann, M. M.; Leach, A. R.; Harper, G. Molecular Complexity and Its Impact on the Probability of Finding Leads for Drug Discovery. *J. Chem. Inf. Comput. Sci.* **2001**, *41*(3), 856–864.
22. Leach, A. R.; Hann, M. M. Molecular Complexity and Fragment-Based Drug Discovery: Ten Years on. *Curr. Opin. Chem. Biol.* **2011**, *15*(4), 489–496.
23. Blum, L. C.; Reymond, J. L. 970 Million Druglike Small Molecules for Virtual Screening in the Chemical Universe Database GDB-13. *J. Am. Chem. Soc.* **2009**, *131*(25), 8732–8733.
24. Ruddigkeit, L.; van Deursen, R.; Blum, L. C.; Reymond, J. L. Enumeration of 166 Billion Organic Small Molecules in the Chemical Universe Database GDB-17. *J. Chem. Inf. Model.* **2012**, *52*(11), 2864–2875.
25. Goodford, P. J. A Computational Procedure for Determining Energetically Favorable Binding Sites on Biologically Important Macromolecules. *J. Med. Chem.* **1985**, *28*(7), 849–857.

26. Wade, R. C.; Clark, K. J.; Goodford, P. J. Further Development of Hydrogen Bond Functions for Use in Determining Energetically Favorable Binding Sites on Molecules of Known Structure. 1. Ligand Probe Groups With the Ability to Form Two Hydrogen Bonds. *J. Med. Chem.* **1993**, *36*(1), 140–147.
27. Jencks, W. P. On the Attribution and Additivity of Binding Energies. *Proc. Natl. Acad. Sci. U. S. A.* **1981**, *78*(7), 4046–4050.
28. Andrews, P. R.; Craik, D. J.; Martin, J. L. Functional Group Contributions to Drug-Receptor Interactions. *J. Med. Chem.* **1984**, *27*(12), 1648–1657.
29. Hopkins, A. L.; Groom, C. R.; Alex, A. Ligand Efficiency: A Useful Metric for Lead Selection. *Drug Discov. Today* **2004**, *9*(10), 430–431.
30. Hajduk, P. J.; Dinges, J.; Miknis, G. F.; et al. NMR-Based Discovery of Lead Inhibitors That Block DNA Binding of the Human Papillomavirus E2 Protein. *J. Med. Chem.* **1997**, *40*(20), 3144–3150.
31. Olejniczak, E. T.; Hajduk, P. J.; Marcotte, P. A.; et al. Stromelysin Inhibitors Designed From Weakly Bound Fragments: Effects of Linking and Cooperativity. *J. Am. Chem. Soc.* **1997**, *119*(25), 5828–5832.
32. Hajduk, P. J.; Zhou, M. M.; Fesik, S. W. NMR-Based Discovery of Phosphotyrosine Mimetics That Bind to the Lck SH2 Domain. *Bioorg. Med. Chem. Lett.* **1999**, *9*(16), 2403–2406.
33. Hajduk, P. J.; Dinges, J.; Schkeryantz, J. M.; et al. Novel Inhibitors of Erm Methyltransferases From NMR and Parallel Synthesis. *J. Med. Chem.* **1999**, *42*(19), 3852–3859.
34. Boehm, H. J.; Boehringer, M.; Bur, D.; et al. Novel Inhibitors of DNA Gyrase: 3D Structure Based Biased Needle Screening, Hit Validation by Biophysical Methods, and 3D Guided Optimization. A Promising Alternative to Random Screening. *J. Med. Chem.* **2000**, *43*(14), 2664–2674.
35. Hajduk, P. J.; Meadows, R. P.; Fesik, S. W. Discovering High-Affinity Ligands for Proteins. *Science* **1997**, *278*(5337), 497–499.
36. Nienaber, V. L.; Richardson, P. L.; Klighofer, V.; Bouska, J. J.; Giranda, V. L.; Greer, J. Discovering Novel Ligands for Macromolecules Using X-Ray Crystallographic Screening. *Nat. Biotechnol.* **2000**, *18*(10), 1105–1108.
37. Blundell, T. L.; Jhoti, H.; Abell, C. High-Throughput Crystallography for Lead Discovery in Drug Design. *Nat. Rev. Drug Discov.* **2002**, *1*(1), 45–54.
38. Perspicace, S.; Banner, D.; Benz, J.; Müller, F.; Schlatter, D.; Huber, W. Fragment-Based Screening Using Surface Plasmon Resonance Technology. *J. Biomol. Screen.* **2009**, *14*(4), 337–349.
39. Danielson, U. H. Fragment Library Screening and Lead Characterization Using SPR Biosensors. *Curr. Top. Med. Chem.* **2009**, *9*(18), 1725–1735.
40. Giannetti, A. M. From Experimental Design to Validated Hits a Comprehensive Walk-Through of Fragment Lead Identification Using Surface Plasmon Resonance. *Methods Enzymol.* **2011**, *493*, 169–218.
41. Silvestre, H. L.; Blundell, T. L.; Abell, C.; Ciulli, A. Integrated Biophysical Approach to Fragment Screening and Validation for Fragment-Based Lead Discovery. *Proc. Natl. Acad. Sci. U. S. A.* **2013**, *110*(32), 12984–12989.
42. Ciulli, A. Biophysical Screening for the Discovery of Small-Molecule Ligands. *Methods Mol. Biol.* **2013**, *1008*, 357–388.
43. Kranz, J. K.; Schalk-Hihi, C. Protein Thermal Shifts to Identify Low Molecular Weight Fragments. *Methods Enzymol.* **2011**, *493*, 277–298.
44. Linke, P.; Amaning, K.; Maschberger, M.; et al. An Automated Microscale Thermophoresis Screening Approach for Fragment-Based Lead Discovery. *J. Biomol. Screen.* **2016**, *21*(4), 414–421.
45. Barker, J.; Courtney, S.; Hesterkamp, T.; Ullmann, D.; Whittaker, M. Fragment Screening by Biochemical Assay. *Expert Opin. Drug Discov.* **2006**, *1*(3), 225–236.

46. Boettcher, A.; Ruedisser, S.; Erbel, P.; et al. Fragment-Based Screening by Biochemical Assays: Systematic Feasibility Studies With Trypsin and MMP12. *J. Biomol. Screen.* **2010**, *15*(9), 1029–1041.
47. Erlanson, D. *Poll Results: Affiliation, Metrics, and Fragment-Finding Methods*, 2016, Accessed 10 August 2017. http://practicalfragments.blogspot.co.uk/2016/10/poll-results-affiliation-metrics-and.html.
48. Albert, J. S.; Blomberg, N.; Breeze, A. L.; et al. An Integrated Approach to Fragment-Based Lead Generation: Philosophy, Strategy and Case Studies From AstraZeneca's Drug Discovery Programmes. *Curr. Top. Med. Chem.* **2007**, 7(16), 1600–1629.
49. Kim, G.; McKee, A. E.; Ning, Y. M.; et al. FDA Approval Summary: Vemurafenib for Treatment of Unresectable or Metastatic Melanoma With the BRAFV600E Mutation. *Clin. Cancer Res.* **2014**, *20*(19), 4994–5000.
50. FDA. FDA Approves New Drug for Chronic Lymphocytic Leukemia in Patients With a Specific Chromosomal Abnormality, 2016, Accessed 10 August 2017. https://www.fda.gov/newsevents/newsroom/pressannouncements/ucm495253.htm.
51. Stamford, A.; Strickland, C. Inhibitors of BACE for Treating Alzheimer's Disease: A Fragment-Based Drug Discovery Story. *Curr. Opin. Chem. Biol.* **2013**, *17*(3), 320–328.
52. Hopkins, A. L.; Keseru, G. M.; Leeson, P. D.; Rees, D. C.; Reynolds, C. H. The Role of Ligand Efficiency Metrics in Drug Discovery. *Nat. Rev. Drug Discov.* **2014**, *13*(2), 105–121.
53. Murray, C. W.; Erlanson, D. A.; Hopkins, A. L.; et al. Validity of Ligand Efficiency Metrics. *ACS Med. Chem. Lett.* **2014**, *5*(6), 616–618.
54. Brough, P. A.; Barril, X.; Borgognoni, J.; et al. Combining Hit Identification Strategies: Fragment-Based and In Silico Approaches to Orally Active 2-Aminothieno[2,3-d] Pyrimidine Inhibitors of the Hsp90 Molecular Chaperone. *J. Med. Chem.* **2009**, *52*(15), 4794–4809.
55. Erlanson, D. *Poll Results: What Structural Information Is Needed to Optimize Fragments*, 2017, Accessed 10 August 2017. http://practicalfragments.blogspot.co.uk/2017/06/poll-results-what-structural.html.
56. Fuller, N.; Spadola, L.; Cowen, S.; et al. An Improved Model for Fragment-Based Lead Generation at AstraZeneca. *Drug Discov. Today* **2016**, *21*(8), 1272–1283.
57. Hubbard, R. E.; Davis, B.; Chen, I.; Drysdale, M. J. The SeeDs Approach: Integrating Fragments Into Drug Discovery. *Curr. Top. Med. Chem.* **2007**, 7(16), 1568–1581.
58. Baurin, N.; Baker, R.; Richardson, C.; et al. Drug-Like Annotation and Duplicate Analysis of a 23-Supplier Chemical Database Totalling 2.7 Million Compounds. *J. Chem. Inf. Comput. Sci.* **2004**, *44*(2), 643–651.
59. Keserű, G. M.; Erlanson, D. A.; Ferenczy, G. G.; Hann, M. M.; Murray, C. W.; Pickett, S. D. Design Principles for Fragment Libraries: Maximizing the Value of Learnings From Pharma Fragment-Based Drug Discovery (FBDD) Programs for Use in Academia. *J. Med. Chem.* **2016**, *59*(18), 8189–8206.
60. Lau, W. F.; Withka, J. M.; Hepworth, D.; et al. Design of a Multi-Purpose Fragment Screening Library Using Molecular Complexity and Orthogonal Diversity Metrics. *J. Comput. Aided Mol. Des.* **2011**, *25*(7), 621–636.
61. Schreiber, S. L. Target-Oriented and Diversity-Oriented Organic Synthesis in Drug Discovery. *Science* **2000**, *287*(5460), 1964–1969.
62. Hung, A. W.; Ramek, A.; Wang, Y.; et al. Route to Three-Dimensional Fragments Using Diversity-Oriented Synthesis. *Proc. Natl. Acad. Sci. U. S. A.* **2011**, *108*(17), 6799–6804.
63. Igoe, N.; Bayle, E. D.; Fedorov, O.; et al. Design of a Biased Potent Small Molecule Inhibitor of the Bromodomain and PHD Finger-Containing (BRPF) Proteins Suitable for Cellular and In Vivo Studies. *J. Med. Chem.* **2017**, *60*(2), 668–680.
64. Cox, O. B.; Krojer, T.; Collins, P.; et al. A Poised Fragment Library Enables Rapid Synthetic Expansion Yielding the First Reported Inhibitors of PHIP(2), an Atypical Bromodomain. *Chem. Sci.* **2016**, 7(3), 2322–2330.

65. Chen, I. J.; Hubbard, R. E. Lessons for Fragment Library Design: Analysis of Output From Multiple Screening Campaigns. *J. Comput. Aided Mol. Des.* **2009**, *23*(8), 603–620.
66. Schulz, M. N.; Landström, J.; Bright, K.; Hubbard, R. E. Design of a Fragment Library That Maximally Represents Available Chemical Space. *J. Comput. Aided Mol. Des.* **2011**, *25*(7), 611–620.
67. Congreve, M.; Carr, R.; Murray, C.; Jhoti, H. A 'Rule of Three' for Fragment-Based Lead Discovery? *Drug Discov. Today* **2003**, *8*(19), 876–877.
68. Jhoti, H.; Williams, G.; Rees, D. C.; Murray, C. W. The 'Rule of Three' for Fragment-Based Drug Discovery: Where Are We Now? *Nat. Rev. Drug Discov.* **2013**, *12*(8), 644–645.
69. Hann, M. M. Molecular Obesity, Potency and Other Addictions in Drug Discovery. *Med. Chem. Commun.* **2011**, *2*(5), 349–355.
70. Bissantz, C.; Kuhn, B.; Stahl, M. A Medicinal Chemist's Guide to Molecular Interactions. *J. Med. Chem.* **2010**, *53*(14), 5061–5084.
71. Baell, J. B.; Holloway, G. A. New Substructure Filters for Removal of Pan Assay Interference Compounds (PAINS) From Screening Libraries and for Their Exclusion in Bioassays. *J. Med. Chem.* **2010**, *53*(7), 2719–2740.
72. Saubern, S.; Guha, R.; Baell, J. B. KNIME Workflow to Assess PAINS Filters in SMARTS Format. Comparison of RDKit and Indigo Cheminformatics Libraries. *Mol. Inf.* **2011**, *30*(10), 847–850.
73. Capuzzi, S. J.; Muratov, E. N.; Tropsha, A. Phantom PAINS: Problems With the Utility of Alerts for Pan-Assay INterference CompoundS. *J. Chem. Inf. Model.* **2017**, *57*(3), 417–427.
74. Begley, D. *Caveat Emptor*, 2014, Accessed 10 August 2017. http://practicalfragments.blogspot.co.uk/2014/10/caveat-emptor.html.
75. McGovern, S. L.; Caselli, E.; Grigorieff, N.; Shoichet, B. K. A Common Mechanism Underlying Promiscuous Inhibitors From Virtual and High-Throughput Screening. *J. Med. Chem.* **2002**, *45*(8), 1712–1722.
76. Davis, B. J.; Erlanson, D. A. Learning From Our Mistakes: The 'Unknown Knowns' in Fragment Screening. *Bioorg. Med. Chem. Lett.* **2013**, *23*(10), 2844–2852.
77. Kozikowski, B. A.; Burt, T. M.; Tirey, D. A.; et al. The Effect of Freeze/Thaw Cycles on the Stability of Compounds in DMSO. *J. Biomol. Screen.* **2003**, *8*(2), 210–215.
78. Kozikowski, B. A.; Burt, T. M.; Tirey, D. A.; et al. The Effect of Room-Temperature Storage on the Stability of Compounds in DMSO. *J. Biomol. Screen.* **2003**, *8*(2), 205–209.
79. Mayer, M.; Meyer, B. Characterization of Ligand Binding by Saturation Transfer Difference NMR Spectroscopy. *Angew. Chem. Int. Ed.* **1999**, *38*(12), 1784–1788.
80. Dalvit, C.; Fogliatto, G.; Stewart, A.; Veronesi, M.; Stockman, B. WaterLOGSY as a Method for Primary NMR Screening: Practical Aspects and Range of Applicability. *J. Biomol. NMR* **2001**, *21*(4), 349–359.
81. Hajduk, P. J.; Meadows, R. P.; Fesik, S. W. NMR-Based Screening in Drug Discovery. *Q. Rev. Biophys.* **1999**, *32*(3), 211–240.
82. Jordan, J. B.; Poppe, L.; Xia, X.; et al. Fragment Based Drug Discovery: Practical Implementation Based on ^{19}F NMR Spectroscopy. *J. Med. Chem.* **2012**, *55*(2), 678–687.
83. Jordan, J. B.; Whittington, D. A.; Bartberger, M. D.; et al. Fragment-Linking Approach Using (19)F NMR Spectroscopy to Obtain Highly Potent and Selective Inhibitors of β-Secretase. *J. Med. Chem.* **2016**, *59*(8), 3732–3749.
84. Meiby, E.; Simmonite, H.; le Strat, L.; et al. Fragment Screening by Weak Affinity Chromatography: Comparison With Established Techniques for Screening Against HSP90. *Anal. Chem.* **2013**, *85*(14), 6756–6766.

85. Davis, B. Screening Protein-Small Molecule Interactions by NMR. *Methods Mol. Biol.* **2013**, *1008*, 389–413.
86. Cala, O.; Krimm, I. Ligand-Orientation Based Fragment Selection in STD NMR Screening. *J. Med. Chem.* **2015**, *58*(21), 8739–8742.
87. English, A. C.; Done, S. H.; Caves, L. S.; Groom, C. R.; Hubbard, R. E. Locating Interaction Sites on Proteins: The Crystal Structure of Thermolysin Soaked in 2% to 100% Isopropanol. *Proteins* **1999**, *37*(4), 628–640.
88. English, A. C.; Groom, C. R.; Hubbard, R. E. Experimental and Computational Mapping of the Binding Surface of a Crystalline Protein. *Protein Eng.* **2001**, *14*(1), 47–59.
89. Patel, D.; Bauman, J. D.; Arnold, E. Advantages of Crystallographic Fragment Screening: Functional and Mechanistic Insights From a Powerful Platform for Efficient Drug Discovery. *Prog. Biophys. Mol. Biol.* **2014**, *116*(2–3), 92–100.
90. Schiebel, J.; Krimmer, S. G.; Röwer, K.; et al. High-Throughput Crystallography: Reliable and Efficient Identification of Fragment Hits. *Structure* **2016**, *24*(8), 1398–1409.
91. Schiebel, J.; Radeva, N.; Krimmer, S. G.; et al. Six Biophysical Screening Methods Miss a Large Proportion of Crystallographically Discovered Fragment Hits: A Case Study. *ACS Chem. Biol.* **2016**, *11*(6), 1693–1701.
92. Krojer, T.; Talon, R.; Pearce, N.; et al. The XChemExplorer Graphical Workflow Tool for Routine or Large-Scale Protein-Ligand Structure Determination. *Acta Crystallogr. Sect. D* **2017**, *73*(3), 267–278.
93. Pearce, N. M.; Bradley, A. R.; Krojer, T.; Marsden, B. D.; Deane, C. M.; von Delft, F. Partial-Occupancy Binders Identified by the Pan-Dataset Density Analysis Method Offer New Chemical Opportunities and Reveal Cryptic Binding Sites. *Struct. Dyn.* **2017**, *4*(3), 032104.
94. Niesen, F. H.; Berglund, H.; Vedadi, M. The Use of Differential Scanning Fluorimetry to Detect Ligand Interactions That Promote Protein Stability. *Nat. Protoc.* **2007**, *2*(9), 2212–2221.
95. Vedadi, M.; Niesen, F. H.; Allali-Hassani, A.; et al. Chemical Screening Methods to Identify Ligands That Promote Protein Stability, Protein Crystallization, and Structure Determination. *Proc. Natl. Acad. Sci. U. S. A.* **2006**, *103*(43), 15835–15840.
96. Hubbard, R. E.; Murray, J. B. Experiences in Fragment-Based Lead Discovery. *Methods Enzymol.* **2011**, *493*, 509–531.
97. Riccardi Sirtori, F.; Caronni, D.; Colombo, M.; et al. Establish an Automated Flow Injection ESI-MS Method for the Screening of Fragment Based Libraries: Application to Hsp90. *Eur. J. Pharm. Sci.* **2015**, *76*, 83–94.
98. El-Hawiet, A.; Kitova, E. N.; Arutyunov, D.; Simpson, D. J.; Szymanski, C. M.; Klassen, J. S. Quantifying Ligand Binding to Large Protein Complexes Using Electrospray Ionization Mass Spectrometry. *Anal. Chem.* **2012**, *84*(9), 3867–3870.
99. Kitova, E. N.; El-Hawiet, A.; Schnier, P. D.; Klassen, J. S. Reliable Determinations of Protein-Ligand Interactions by Direct ESI-MS Measurements. Are We There yet? *J. Am. Soc. Mass Spectrom.* **2012**, *23*(3), 431–441.
100. Ameriks, M. K.; Bembenek, S. D.; Burdett, M. T.; et al. Diazinones as P2 Replacements for Pyrazole-Based Cathepsin S Inhibitors. *Bioorg. Med. Chem. Lett.* **2010**, *20*(14), 4060–4064.
101. Vivat Hannah, V.; Atmanene, C.; Zeyer, D.; Van Dorsselaer, A.; Sanglier-Cianférani, S. Native MS: An 'ESI' Way to Support Structure- and Fragment-Based Drug Discovery. *Future Med. Chem.* **2010**, *2*(1), 35–50.
102. Moree, B.; Connell, K.; Mortensen, R. B.; Liu, C. T.; Benkovic, S. J.; Salafsky, J. Protein Conformational Changes Are Detected and Resolved Site Specifically by Second-Harmonic Generation. *Biophys. J.* **2015**, *109*(4), 806–815.

103. Moree, B.; Yin, G.; Lázaro, D. F.; et al. Small Molecules Detected by Second-Harmonic Generation Modulate the Conformation of Monomeric α-Synuclein and Reduce Its Aggregation in Cells. *J. Biol. Chem.* **2015**, *290*(46), 27582–27593.
104. Klumpers, F.; Götz, U.; Kurtz, T.; Herrmann, C.; Gronewold, T. M. A. Conformational Changes at Protein–Protein Interaction Followed Online With an SAW Biosensor. *Sens. Actuators B* **2014**, *203*, 904–908.
105. Sarkar, D.; Liu, W.; Xie, X.; Anselmo, A. C.; Mitragotri, S.; Banerjee, K. MoS2 Field-Effect Transistor for Next-Generation Label-Free Biosensors. *ACS Nano* **2014**, *8*(4), 3992–4003.
106. Austin, C.; Pettit, S. N.; Magnolo, S. K.; et al. Fragment Screening Using Capillary Electrophoresis (CEfrag) for Hit Identification of Heat Shock Protein 90 ATPase Inhibitors. *J. Biomol. Screen.* **2012**, *17*(7), 868–876.
107. Vanwetswinkel, S.; Heetebrij, R. J.; van Duynhoven, J.; et al. TINS, Target Immobilized NMR Screening: An Efficient and Sensitive Method for Ligand Discovery. *Chem. Biol.* **2005**, *12*(2), 207–216.
108. Davis, B. J.; Giannetti, A. M. The Synthesis of Biophysical Methods in Support of Robust Fragment-Based Lead Discovery. In: *Fragment-Based Drug Discovery Lessons and Outlook*; Erlanson, D., Jahnke, W., Eds.; Wiley: New York, NY, 2016; pp 119–138.
109. Williamson, M. P. Using Chemical Shift Perturbation to Characterise Ligand Binding. *Prog. Nucl. Magn. Reson. Spectrosc.* **2013**, *73*, 1–16.
110. Barile, E.; Pellecchia, M. NMR-Based Approaches for the Identification and Optimization of Inhibitors of Protein-Protein Interactions. *Chem. Rev.* **2014**, *114*(9), 4749–4763.
111. Constantine, K. L.; Davis, M. E.; Metzler, W. J.; Mueller, L.; Claus, B. L. Protein-Ligand NOE Matching: A High-Throughput Method for Binding Pose Evaluation That Does Not Require Protein NMR Resonance Assignments. *J. Am. Chem. Soc.* **2006**, *128*(22), 7252–7263.
112. Constantine, K. L.; Mueller, L.; Metzler, W. J.; et al. Multiple and Single Binding Modes of Fragment-Like Kinase Inhibitors Revealed by Molecular Modeling, Residue Type-Selective Protonation, and Nuclear Overhauser Effects. *J. Med. Chem.* **2008**, *51*(19), 6225–6229.
113. Wälti, M. A.; Riek, R.; Orts, J. Fast NMR-Based Determination of the 3D Structure of the Binding Site of Protein-Ligand Complexes With Weak Affinity Binders. *Angew. Chem. Int. Ed. Engl.* **2017**, *56*(19), 5208–5211.
114. Hubbard, R. Fragment Approaches in Structure-Based Drug Discovery. *J. Synchrotron Radiat.* **2008**, *15*(3), 227–230.
115. Leung, C. S.; Leung, S. S.; Tirado-Rives, J.; Jorgensen, W. L. Methyl Effects on Protein-Ligand Binding. *J. Med. Chem.* **2012**, *55*(9), 4489–4500.
116. Schönherr, H.; Cernak, T. Profound Methyl Effects in Drug Discovery and a Call for New C-H Methylation Reactions. *Angew. Chem. Int. Ed. Engl.* **2013**, *52*(47), 12256–12267.
117. Lerner, C.; Jakob-Roetne, R.; Buettelmann, B.; Ehler, A.; Rudolph, M.; Rodríguez Sarmiento, R. M. Design of Potent and Druglike Nonphenolic Inhibitors for Catechol O-Methyltransferase Derived From a Fragment Screening Approach Targeting the S-Adenosyl-L-Methionine Pocket. *J. Med. Chem.* **2016**, *59*(22), 10163–10175.
118. Hughes, S. J.; Millan, D. S.; Kilty, I. C.; et al. Fragment Based Discovery of a Novel and Selective PI3 Kinase Inhibitor. *Bioorg. Med. Chem. Lett.* **2011**, *21*(21), 6586–6590.
119. Roughley, S.; Wright, L.; Brough, P.; Massey, A.; Hubbard, R. E. Hsp90 Inhibitors and Drugs From Fragment and Virtual Screening. *Top. Curr. Chem.* **2012**, *317*, 61–82.
120. Ichihara, O.; Barker, J.; Law, R. J.; Whittaker, M. Compound Design by Fragment-Linking. *Mol. Inf.* **2011**, *30*(4), 298–306.

121. Wilfong, E. M.; Du, Y.; Toone, E. J. An Enthalpic Basis of Additivity in Biphenyl Hydroxamic Acid Ligands for Stromelysin-1. *Bioorg. Med. Chem. Lett.* **2012**, *22*(20), 6521–6524.
122. Borsi, V.; Calderone, V.; Fragai, M.; Luchinat, C.; Sarti, N. Entropic Contribution to the Linking Coefficient in Fragment Based Drug Design: A Case Study. *J. Med. Chem.* **2010**, *53*(10), 4285–4289.
123. Nazaré, M.; Matter, H.; Will, D. W.; et al. Fragment Deconstruction of Small, Potent Factor Xa Inhibitors: Exploring the Superadditivity Energetics of Fragment Linking in Protein–Ligand Complexes. *Angew. Chem. Int. Ed.* **2012**, *51*(4), 905–911.
124. Ward, R. A.; Brassington, C.; Breeze, A. L.; et al. Design and Synthesis of Novel Lactate Dehydrogenase a Inhibitors by Fragment-Based Lead Generation. *J. Med. Chem.* **2012**, *55*(7), 3285–3306.
125. Becattini, B.; Culmsee, C.; Leone, M.; et al. Structure-Activity Relationships by Interligand NOE-Based Design and Synthesis of Antiapoptotic Compounds Targeting Bid. *Proc. Natl. Acad. Sci. U. S. A.* **2006**, *103*(33), 12602–12606.
126. Chen, J.; Zhang, Z.; Stebbins, J. L.; et al. A Fragment-Based Approach for the Discovery of Isoform-Specific p38alpha Inhibitors. *ACS Chem. Biol.* **2007**, *2*(5), 329–336.
127. Sledz, P.; Silvestre, H. L.; Hung, A. W.; Ciulli, A.; Blundell, T. L.; Abell, C. Optimization of the Interligand Overhauser Effect for Fragment Linking: Application to Inhibitor Discovery against Mycobacterium Tuberculosis Pantothenate Synthetase. *J. Am. Chem. Soc.* **2010**, *132*(13), 4544–4545.
128. Hung, A. W.; Silvestre, H. L.; Wen, S.; Ciulli, A.; Blundell, T. L.; Abell, C. Application of Fragment Growing and Fragment Linking to the Discovery of Inhibitors of Mycobacterium tuberculosis Pantothenate Synthetase. *Angew. Chem. Int. Ed. Engl.* **2009**, *48*(45), 8452–8456.
129. Brough, P. A.; Aherne, W.; Barril, X.; et al. 4,5-Diarylisoxazole Hsp90 Chaperone Inhibitors: Potential Therapeutic Agents for the Treatment of Cancer. *J. Med. Chem.* **2008**, *51*(2), 196–218.
130. Dymock, B. W.; Barril, X.; Brough, P. A.; et al. Novel, Potent Small-Molecule Inhibitors of the Molecular Chaperone Hsp90 Discovered Through Structure-Based Design. *J. Med. Chem.* **2005**, *48*(13), 4212–4215.
131. Brough, P. A.; Barril, X.; Beswick, M.; et al. 3-(5-Chloro-2,4-Dihydroxyphenyl)-Pyrazole-4-Carboxamides as Inhibitors of the Hsp90 Molecular Chaperone. *Bioorg. Med. Chem. Lett.* **2005**, *15*(23), 5197–5201.
132. Hudson, S. A.; Surade, S.; Coyne, A. G.; et al. Overcoming the Limitations of Fragment Merging: Rescuing a Strained Merged Fragment Series Targeting Mycobacterium Tuberculosis CYP121. *ChemMedChem* **2013**, *8*(9), 1451–1456.
133. de Vicente, J.; Tivitmahaisoon, P.; Berry, P.; et al. Fragment-Based Drug Design of Novel Pyranopyridones as Cell Active and Orally Bioavailable Tankyrase Inhibitors. *ACS Med. Chem. Lett.* **2015**, *6*(9), 1019–1024.
134. Taylor, S. J.; Padyana, A. K.; Abeywardane, A.; et al. Discovery of Potent, Selective Chymase Inhibitors via Fragment Linking Strategies. *J. Med. Chem.* **2013**, *56*(11), 4465–4481.
135. Pomel, V.; Klicic, J.; Covini, D.; et al. Furan-2-Ylmethylene Thiazolidinediones as Novel, Potent, and Selective Inhibitors of Phosphoinositide 3-Kinase Gamma. *J. Med. Chem.* **2006**, *49*(13), 3857–3871.
136. Huc, I.; Lehn, J. M. Virtual Combinatorial Libraries: Dynamic Generation of Molecular and Supramolecular Diversity by Self-Assembly. *Proc. Natl. Acad. Sci. U. S. A.* **1997**, *94*(6), 2106–2110.
137. Kolb, H. C.; Finn, M. G.; Sharpless, K. B. Click Chemistry: Diverse Chemical Function From a Few Good Reactions. *Angew. Chem. Int. Ed. Engl.* **2001**, *40*(11), 2004–2021.

138. Jaegle, M.; Wong, E. L.; Tauber, C.; Nawrotzky, E.; Arkona, C.; Rademann, J. Protein-Templated Fragment Ligations—From Molecular Recognition to Drug Discovery. *Angew. Chem. Int. Ed. Engl.* **2017**, *56*, 7358–7378.
139. Corbett, P. T.; Leclaire, J.; Vial, L.; et al. Dynamic Combinatorial Chemistry. *Chem. Rev.* **2006**, *106*(9), 3652–3711.
140. Scott, D. E.; Dawes, G. J.; Ando, M.; Abell, C.; Ciulli, A. A Fragment-Based Approach to Probing Adenosine Recognition Sites by Using Dynamic Combinatorial Chemistry. *Chembiochem* **2009**, *10*(17), 2772–2779.
141. Cancilla, M. T.; He, M. M.; Viswanathan, N.; et al. Discovery of an Aurora Kinase Inhibitor Through Site-Specific Dynamic Combinatorial Chemistry. *Bioorg. Med. Chem. Lett.* **2008**, *18*(14), 3978–3981.
142. Huisgen, R. Centenary Lecture—1,3-Dipolar Cycloadditions. *Proc. Chem. Soc.* **1961**, 357–396.
143. Shelke, S. V.; Cutting, B.; Jiang, X.; et al. A Fragment-Based In Situ Combinatorial Approach to Identify High-Affinity Ligands for Unknown Binding Sites. *Angew. Chem. Int. Ed. Engl.* **2010**, *49*(33), 5721–5725.
144. Mesch, S.; Moser, D.; Strasser, D. S.; et al. Low Molecular Weight Antagonists of the Myelin-Associated Glycoprotein: Synthesis, Docking, and Biological Evaluation. *J. Med. Chem.* **2010**, *53*(4), 1597–1615.
145. Suzuki, T.; Ota, Y.; Kasuya, Y.; et al. An Unexpected Example of Protein-Templated Click Chemistry. *Angew. Chem. Int. Ed. Engl.* **2010**, *49*(38), 6817–6820.
146. Hu, X.; Sun, J.; Wang, H. G.; Manetsch, R. Bcl-XL-Templated Assembly of Its Own Protein-Protein Interaction Modulator From Fragments Decorated With Thio Acids and Sulfonyl Azides. *J. Am. Chem. Soc.* **2008**, *130*(42), 13820–13821.
147. Jaegle, M.; Steinmetzer, T.; Rademann, J. Protein-Templated Formation of an Inhibitor of the Blood Coagulation Factor Xa Through a Background-Free Amidation Reaction. *Angew. Chem. Int. Ed. Engl.* **2017**, *56*(13), 3718–3722.
148. Solomon, I. Relaxation Processes in a System of Two Spins. *Phys. Rev.* **1955**, *99*(2), 559–565.
149. Jahnke, W. Spin Labels as a Tool to Identify and Characterize Protein–Ligand Interactions by NMR Spectroscopy. *Chembiochem* **2002**, *3*(2–3), 167–173.
150. Egger, J.; Weckerle, C.; Cutting, B.; et al. Nanomolar E-Selectin Antagonists With Prolonged Half-Lives by a Fragment-Based Approach. *J. Am. Chem. Soc.* **2013**, *135*(26), 9820–9828.
151. Hein, J. E.; Fokin, V. V. Copper-Catalyzed Azide–Alkyne Cycloaddition (CuAAC) and Beyond: New Reactivity of Copper(I) Acetylides. *Chem. Soc. Rev.* **2010**, *39*(4), 1302–1315.
152. Copeland, R. A.; Pompliano, D. L.; Meek, T. D. Drug-Target Residence Time and Its Implications for Lead Optimization. *Nat. Rev. Drug Discov.* **2006**, *5*(9), 730–739.
153. Copeland, R. A. The Dynamics of Drug-Target Interactions: Drug-Target Residence Time and Its Impact on Efficacy and Safety. *Expert Opin. Drug Discov.* **2010**, *5*(4), 305–310.
154. Copeland, R. A. Drug-Target Interaction Kinetics: Underutilized in Drug Optimization?*Future Med. Chem.* **2016**, *8*(18), 2173–2175.
155. Copeland, R. A. The Drug-Target Residence Time Model: A 10-Year Retrospective. *Nat. Rev. Drug Discov.* **2016**, *15*(2), 87–95.
156. Lu, H.; Tonge, P. J. Drug-Target Residence Time: Critical Information for Lead Optimization. *Curr. Opin. Chem. Biol.* **2010**, *14*(4), 467–474.
157. Holdgate, G. A.; Gill, A. L. Kinetic Efficiency: The Missing Metric for Enhancing Compound Quality?*Drug Discov. Today* **2011**, *16*(21–22), 910–913.
158. Roughley, S. D.; Jordan, A. M. The Medicinal Chemist's Toolbox: An Analysis of Reactions Used in the Pursuit of Drug Candidates. *J. Med. Chem.* **2011**, *54*(10), 3451–3479.

159. Schneider, N.; Lowe, D. M.; Sayle, R. A.; Tarselli, M. A.; Landrum, G. A. Big Data From Pharmaceutical Patents: A Computational Analysis of Medicinal Chemists' Bread and Butter. *J. Med. Chem.* **2016**, *59*(9), 4385–4402.
160. Dutta, R.; Inouye, M. GHKL, an Emergent ATPase/Kinase Superfamily. *Trends Biochem. Sci.* **2000**, *25*(1), 24–28.
161. Kung, P. P.; Funk, L.; Meng, J.; et al. Dihydroxylphenyl Amides as Inhibitors of the Hsp90 Molecular Chaperone. *Bioorg. Med. Chem. Lett.* **2008**, *18*(23), 6273–6278.
162. Kung, P. P.; Huang, B.; Zhang, G.; et al. Dihydroxyphenylisoindoline Amides as Orally Bioavailable Inhibitors of the Heat Shock Protein 90 (hsp90) Molecular Chaperone. *J. Med. Chem.* **2010**, *53*(1), 499–503.
163. Murray, C. W.; Carr, M. G.; Callaghan, O.; et al. Fragment-Based Drug Discovery Applied to Hsp90. Discovery of Two Lead Series With High Ligand Efficiency. *J. Med. Chem.* **2010**, *53*(16), 5942–5955.
164. Woodhead, A. J.; Angove, H.; Carr, M. G.; et al. Discovery of (2,4-Dihydroxy-5-Isopropylphenyl)-[5-(4-Methylpiperazin-1-Ylmethyl)-1,3-Dihydroisoindol-2-Yl] Methanone (AT13387), a Novel Inhibitor of the Molecular Chaperone Hsp90 by Fragment Based Drug Design. *J. Med. Chem.* **2010**, *53*(16), 5956–5969.
165. Graves, A. P.; Wall, I. D.; Edge, C. M.; et al. A Perspective on Water Site Prediction Methods for Structure Based Drug Design, *Curr. Top. Med. Chem.* **2017**, 1873-4294, *17*. https://doi.org/10.2174/1568026617666170427095035. Electronic (ePub ahead of print).
166. Massey, A. J.; Stokes, S.; Browne, H.; et al. Identification of Novel, In Vivo Active Chk1 Inhibitors Utilizing Structure Guided Drug Design. *Oncotarget* **2015**, *6*(34), 35797–35812.
167. Southall, N. T.; Ajay. Kinase Patent Space Visualization Using Chemical Replacements. *J. Med. Chem.* **2006**, *49*(6), 2103–2109.
168. Czodrowski, P.; Hölzemann, G.; Barnickel, G.; Greiner, H.; Musil, D. Selection of Fragments for Kinase Inhibitor Design: Decoration Is Key. *J. Med. Chem.* **2015**, *58*(1), 457–465.
169. Mpamhanga, C. P.; Spinks, D.; Tulloch, L. B.; et al. One Scaffold, Three Binding Modes: Novel and Selective Pteridine Reductase 1 Inhibitors Derived From Fragment Hits Discovered by Virtual Screening. *J. Med. Chem.* **2009**, *52*(14), 4454–4465.
170. Malhotra, S.; Karanicolas, J. When Does Chemical Elaboration Induce a Ligand to Change Its Binding Mode?*J. Med. Chem.* **2017**, *60*(1), 128–145.
171. Roughley, S. D.; Hubbard, R. E. How Well Can Fragments Explore Accessed Chemical Space? A Case Study From Heat Shock Protein 90. *J. Med. Chem.* **2011**, *54*(12), 3989–4005.
172. Drwal, M. N.; Jacquemard, C.; Perez, C.; Desaphy, J.; Kellenberger, E. Do Fragments and Crystallization Additives Bind Similarly to Drug-Like Ligands?*J. Chem. Inf. Model.* **2017**, *57*, 1197–1209.
173. Lowe, D. *In the Pipeline—More Magic Methyls, Please.* http://blogs.sciencemag.org/pipeline/archives/2013/10/30/more_magic_methyls_please, 2013. Accessed 22 May 2017.
174. Kuntz, I. D.; Chen, K.; Sharp, K. A.; Kollman, P. A. The Maximal Affinity of Ligands. *Proc. Natl. Acad. Sci. U. S. A.* **1999**, *96*(18), 9997–10002.
175. Abad-Zapatero, C.; Metz, J. T. Ligand Efficiency Indices as Guideposts for Drug Discovery. *Drug Discov. Today* **2005**, *10*(7), 464–469.
176. Leeson, P. D.; Springthorpe, B. The Influence of Drug-Like Concepts on Decision-Making in Medicinal Chemistry. *Nat. Rev. Drug Discov.* **2007**, *6*(11), 881–890.
177. Shultz, M. D. Setting Expectations in Molecular Optimizations: Strengths and Limitations of Commonly Used Composite Parameters. *Bioorg. Med. Chem. Lett.* **2013**, *23*(21), 5980–5991.

178. Shultz, M. D. The Thermodynamic Basis for the Use of Lipophilic Efficiency (LipE) in Enthalpic Optimizations. *Bioorg. Med. Chem. Lett.* **2013**, *23*(21), 5992–6000.
179. Keserü, G. M.; Makara, G. M. The Influence of Lead Discovery Strategies on the Properties of Drug Candidates. *Nat. Rev. Drug Discov.* **2009**, *8*(3), 203–212.
180. Verdonk, M. L.; Rees, D. C. Group Efficiency: A Guideline for Hits-to-Leads Chemistry. *ChemMedChem* **2008**, *3*(8), 1179–1180.
181. Hung, A. W.; Silvestre, H. L.; Wen, S.; et al. Optimization of Inhibitors of Mycobacterium tuberculosis Pantothenate Synthetase Based on Group Efficiency Analysis. *ChemMedChem* **2016**, *11*(1), 38–42.
182. Reynolds, C. H.; Bembenek, S. D.; Tounge, B. A. The Role of Molecular Size in Ligand Efficiency. *Bioorg. Med. Chem. Lett.* **2007**, *17*(15), 4258–4261.
183. Reynolds, C. H.; Tounge, B. A.; Bembenek, S. D. Ligand Binding Efficiency: Trends, Physical Basis, and Implications. *J. Med. Chem.* **2008**, *51*(8), 2432–2438.
184. Orita, M.; Ohno, K.; Niimi, T. Two 'Golden Ratio' Indices in Fragment-Based Drug Discovery. *Drug Discov. Today* **2009**, *14*(5–6), 321–328.
185. Nissink, J. W. Simple Size-Independent Measure of Ligand Efficiency. *J. Chem. Inf. Model.* **2009**, *49*(6), 1617–1622.
186. Mortenson, P. N.; Murray, C. W. Assessing the Lipophilicity of Fragments and Early Hits. *J. Comput. Aided Mol. Des.* **2011**, *25*(7), 663–667.
187. Reynolds, C. H. Ligand Efficiency Metrics: Why All the Fuss?*Future Med. Chem.* **2015**, 7(11), 1363–1365.
188. Schultes, S.; de Graaf, C.; Haaksma, E. E. J.; de Esch, I. J. P.; Leurs, R.; Kramer, O. Ligand Efficiency as a Guide in Fragment Hit Selection and Optimization. *Drug Discov. Today Technol.* **2010**, 7(3), e147–202.
189. Davies, T. G.; Bentley, J.; Arris, C. E.; et al. Structure-Based Design of a Potent Purine-Based Cyclin-Dependent Kinase Inhibitor. *Nat. Struct. Biol.* **2002**, *9*(10), 745–749.
190. Kania, R. S.; Bender, S. L.; Borchardt, A. J.; et al. *Indazole Compounds and Pharmaceutical Compositions for Inhibiting Protein Kinases, and Methods for Their Use*; Google Patents, 2001.
191. Kuhn, B.; Mohr, P.; Stahl, M. Intramolecular Hydrogen Bonding in Medicinal Chemistry. *J. Med. Chem.* **2010**, *53*(6), 2601–2611.
192. Filmore, D. It's a GPCR World. *Mod. Drug Discov.* **2004**, 7(11), 24–28.
193. GPCRdb2017. Accessed 08 June 2017.
194. Isberg, V.; Mordalski, S.; Munk, C.; et al. GPCRdb: An Information System for G Protein-Coupled Receptors. *Nucleic Acids Res.* **2016**, *44*(D1), D356–364.
195. Munk, C.; Isberg, V.; Mordalski, S.; et al. GPCRdb: The G Protein-Coupled Receptor Database—An Introduction. *Br. J. Pharmacol.* **2016**, *173*(14), 2195–2207.
196. Congreve, M.; Marshall, F. The Impact of GPCR Structures on Pharmacology and Structure-Based Drug Design. *Br. J. Pharmacol.* **2010**, *159*(5), 986–996.
197. Congreve, M.; Dias, J. M.; Marshall, F. H. Chapter One—Structure-Based Drug Design for G Protein-Coupled Receptors. In *Progress in Medicinal Chemistry*; Lawton, G., Witty, D. R., Eds.; Vol. 53; Elsevier: New York, NY, 2014; pp 1–63.
198. Shepherd, C. A.; Hopkins, A. L.; Navratilova, I. Fragment Screening by SPR and Advanced Application to GPCRs. *Prog. Biophys. Mol. Biol.* **2014**, *116*(2), 113–123.
199. Lawson, A. D. G Protein-Coupled Receptors—Targets for Fragment-Based Drug Discovery. *Curr. Top. Med. Chem.* **2015**, *15*, 2523–2527.
200. Andrews, S. P.; Brown, G. A.; Christopher, J. A. Structure-Based and Fragment-Based GPCR Drug Discovery. *ChemMedChem* **2014**, *9*(2), 256–275.
201. Wells, J. A.; McClendon, C. L. Reaching for High-Hanging Fruit in Drug Discovery at Protein-Protein Interfaces. *Nature* **2007**, *450*(7172), 1001–1009.
202. Valkov, E.; Sharpe, T.; Marsh, M.; Greive, S.; Hyvönen, M. Targeting Protein–Protein Interactions and Fragment-Based Drug Discovery. In: *Fragment-Based Drug*

Discovery and X-Ray Crystallography; Davies, T. G., Hyvönen, M., Eds.; Springer Berlin Heidelberg: Berlin, Heidelberg, 2012; pp 145–179.
203. Scott, D. E.; Ehebauer, M. T.; Pukala, T.; et al. Using a Fragment-Based Approach to Target Protein–Protein Interactions. *Chembiochem* **2013**, *14*(3), 332–342.
204. Jin, J.; MacMillan, D. W. Alcohols as Alkylating Agents in Heteroarene C-H Functionalization. *Nature* **2015**, *525*(7567), 87–90.
205. Nagib, D. A.; MacMillan, D. W. Trifluoromethylation of Arenes and Heteroarenes by Means of Photoredox Catalysis. *Nature* **2011**, *480*(7376), 224–228.
206. Fujiwara, Y.; Dixon, J. A.; O'Hara, F.; et al. Practical and Innate Carbon-Hydrogen Functionalization of Heterocycles. *Nature* **2012**, *492*(7427), 95–99.
207. Gianatassio, R.; Kawamura, S.; Eprile, C. L.; et al. Simple Sulfinate Synthesis Enables C-H Trifluoromethylcyclopropanation. *Angew. Chem. Int. Ed. Engl.* **2014**, *53*(37), 9851–9855.
208. Gui, J.; Zhou, Q.; Pan, C. M.; et al. C-H Methylation of Heteroarenes Inspired by Radical SAM Methyl Transferase. *J. Am. Chem. Soc.* **2014**, *136*(13), 4853–4856.
209. Horn, E. J.; Rosen, B. R.; Baran, P. S. Synthetic Organic Electrochemistry: An Enabling and Innately Sustainable Method. *ACS Cent. Sci.* **2016**, *2*(5), 302–308.
210. Ji, Y.; Brueckl, T.; Baxter, R. D.; et al. Innate C-H Trifluoromethylation of Heterocycles. *Proc. Natl. Acad. Sci. U. S. A.* **2011**, *108*(35), 14411–14415.
211. Kawamata, Y.; Yan, M.; Liu, Z.; et al. Scalable, Electrochemical Oxidation of Unactivated C-H Bonds. *J. Am. Chem. Soc.* **2017**, *139*(22), 7448–7451.
212. O'Brien, A. G.; Maruyama, A.; Inokuma, Y.; Fujita, M.; Baran, P. S.; Blackmond, D. G. Radical C-H Functionalization of Heteroarenes Under Electrochemical Control. *Angew. Chem. Int. Ed. Engl.* **2014**, *53*(44), 11868–11871.
213. O'Hara, F.; Baxter, R. D.; O'Brien, A. G.; et al. Preparation and Purification of Zinc Sulfinate Reagents for Drug Discovery. *Nat. Protoc.* **2013**, *8*(6), 1042–1047.
214. O'Hara, F.; Blackmond, D. G.; Baran, P. S. Radical-Based Regioselective C-H Functionalization of Electron-Deficient Heteroarenes: Scope, Tunability, and Predictability. *J. Am. Chem. Soc.* **2013**, *135*(32), 12122–12134.
215. Rodriguez, R. A.; Pan, C. M.; Yabe, Y.; Kawamata, Y.; Eastgate, M. D.; Baran, P. S. Palau'chlor: A Practical and Reactive Chlorinating Reagent. *J. Am. Chem. Soc.* **2014**, *136*(19), 6908–6911.
216. Zhou, Q.; Ruffoni, A.; Gianatassio, R.; et al. Direct Synthesis of Fluorinated Heteroarylether Bioisosteres. *Angew. Chem. Int. Ed. Engl.* **2013**, *52*(14), 3949–3952.
217. Cernak, T.; Dykstra, K. D.; Tyagarajan, S.; Vachal, P.; Krska, S. W. The Medicinal Chemist's Toolbox for Late Stage Functionalization of Drug-Like Molecules. *Chem. Soc. Rev.* **2016**, *45*(3), 546–576.
218. Berthold, M. R.; Cebron, N.; Dill, F.; et al. In *KNIME: The Konstanz Information Miner*; Preisach, C., Burkhardt, H., Schmidt-Thieme, L., Decker, R., Eds.; *Data Analysis, Machine Learning and Applications: Proceedings of the 31st Annual Conference of the Gesellschaft für Klassifikation e.V., Albert-Ludwigs-Universität Freiburg, March 7–9, 2007*; Springer Berlin Heidelberg: Berlin, Heidelberg, 2008; pp 319–326.
219. Berthold, M. R.; Cebron, N.; Dill, F.; et al. KNIME—The Konstanz Information Miner: Version 2.0 and beyond. *SIGKDD Explor. Newsl.* **2009**, *11*(1), 26–31.
220. *KNIME: Open for Innovation*. https://www.knime.org/. Accessed 14 June 2017.

CHAPTER TWELVE

Antibody-Drug Conjugates

Amit Kumar, Jason White, R. James Christie, Nazzareno Dimasi, Changshou Gao[1]
Antibody Discovery and Protein Engineering, MedImmune, Gaithersburg, MD, United States
[1]Corresponding author: e-mail address: GaoC@MedImmune.com

Contents

1. INTRODUCTION

Chemotherapeutic strategies targeting rapidly dividing cells are traditionally employed for treatment of a broad range of cancers. Unfortunately,

Annual Reports in Medicinal Chemistry, Volume 50
ISSN 0065-7743
https://doi.org/10.1016/bs.armc.2017.08.002

rapid proliferation is shown by many nonmalignant cell types as well, leading to systemic side effects. An approach that imparts target specificity to a chemotherapeutic agent has potential to reduce systemic side effects and also increase the therapeutic window. Antibody–drug conjugates (ADCs) are such an approach, where the antibody provides specificity for a tumor target antigen and the drug confers the cytotoxicity. In recent years, extensive effort has been devoted to development of ADCs that deliver highly potent cytotoxic drugs directly to the tumor site.

An ADC construct is typically comprised of three components, namely, the antibody, linker, and cytotoxic drug (Fig. 1). The antibody component of an ADC targets an antigen that is highly upregulated within the tumor. Cytotoxic drug is attached to the antibody through cleavable or noncleavable linkers designed to achieve desired stability and drug release properties.

Historical perspective: The notion of combining a cytotoxic drug with a targeting agent is not new and was first proposed by the German physician and scientist Paul Ehrlich, more than 100 years ago.[1] Ehrlich coined the term "magic bullet" to describe his vision; however, it took nearly 50 years to put Ehrlich's concept into practice. In 1958 methotrexate (MTX), an immunosuppressive drug was attached to an antibody targeting leukemia cells in mice.[2] This provided the first example of an antibody being used to selectively deliver drugs to the target cells. Early research utilized polyclonal antibodies and synthesized ADCs by covalently or noncovalently linking cytotoxic drugs.[3,4] Invention of monoclonal antibodies (mAbs) in 1975

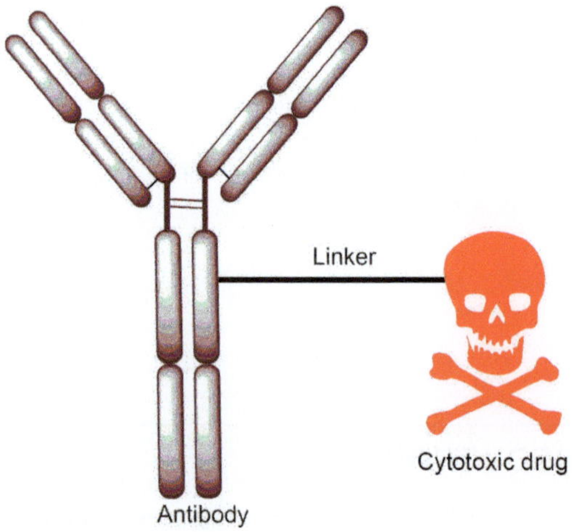

Fig. 1 Components of an ADC.

by Kohler and Milstein that made mAbs freely available to scientists thereby greatly advancing the field of ADCs.[5] It took another decade for the first human clinical trial using the antimitotic vinca alkaloid, vindesine, to be conducted.[6] The rapid advancement in antibody technology in past two decades has yielded some of the most commercially successful anticancer antibody therapeutics including rituximab, trastuzumab, cetuximab, and bevacizumab.[7]

First-generation ADCs were mAbs conjugated to clinically approved drugs with well-established mechanisms of action (MOAs), such as antimetabolites (MTX and 5-fluorouracil), DNA cross-linkers (mitomycin), and antimicrotubule agents (vinblastine).[8] These ADCs faced a host of issues including immunogenicity due to use of murine mAbs, low drug potency, and instability of the linker attaching the cytotoxic drug to the mAb. For example, BR96-DOX, a first-generation ADC with an anti-Lewis^{Y} mAb conjugated with doxorubicin via a hydrazone linker failed in Phase II trials due to the low potency of the doxorubicin payload and linker instability. Lesson learned from such failures led to improved ADC designs with better target selection and higher drug potency. Such improvements led to the approval of gemtuzumab ozogamicin (Mylotarg®) by the US Food and Drug Administration (FDA) in 2000. However, Mylotarg® was voluntarily withdrawn from the market in 2010 owing to a lack of improvement in overall survival.[9] Efforts toward improvement in ADC design by including drugs with higher potency, better target selection, improved linker technology have continued and have led to the FDA approval of Adcetris® (brentuximab vedotin) and Kadcyla® (ado-trastuzumab emtansine, T-DM1).

2. MABS SELECTION FOR ADC

2.1 General Overview of mAbs

2.1.1 Full-Length Antibodies

Antibodies, also known as immunoglobulins, are secreted from the B-cell receptors. They are Y-shaped molecules (Fig. 2) and have a molecular weight of approximately 150 kDa. Antibodies circulate in soluble form or can be bound to the B-cell membrane as part of the B-cell receptor. Antibodies are composed of two unique polypeptide chains commonly distinguished as heavy chain (approximately 50 kDa) and light chain (approximately 25 kDa). Antibodies are symmetric molecules comprised of two heavy–light chain pairs. Individual heavy–light chain pairs are covalently linked through

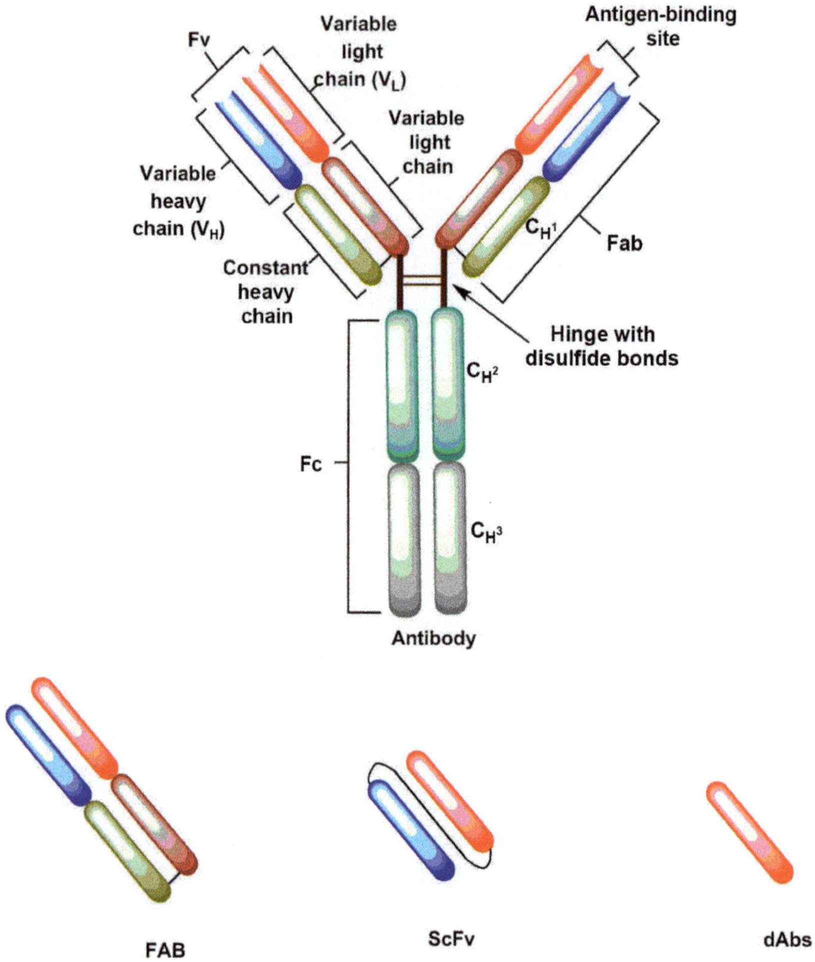

Fig. 2 The structure of a full-length IgG and antibody fragments.

a disulfide bond, and the full molecule is connected through a pair of heavy chain disulfides in the hinge region. Antibody light chains can be subdivided into categories, namely, lambda (λ) and kappa (κ). An IgG can be comprised of either kappa (κ) or lambda (λ) light chains, but never one of each. Functionally, IgGs bearing either λ or κ light chains have been found to be identical. Based on its heavy chains, antibodies can be subdivided into five main classes or isotypes, namely, immunoglobulin M (IgM), immunoglobulin D (IgD), immunoglobulin G (IgG), immunoglobulin A (IgA), and immunoglobulin E (IgE). The heavy chains of each isotype are denoted by the corresponding lower-case Greek letter, namely, μ, δ, γ, α, and ε, respectively. Among all the antibody isotypes, IgG is by far the most abundant immunoglobulin with several subclasses, for example, four subclasses of IgG can be found in humans, distinguished as IgG1, 2, 3, and 4. IgG accounts for approximately 80% of the antibodies found in humans, and with a long circulation half-life between 7 and 21 days. IgG is also the most common class of antibodies used as the structural basis for the production/generation of therapeutic mAbs and ADCs.[10]

Heavy chains of an antibody are made up of polypeptide chain of about 440 amino acids and can be subdivided into four domains each comprising of about 110 amino acid units. Light chains of an antibody are composed of polypeptide chain of about 220 amino acids, which can be subdivided into two domains each comprising of about 110 amino acid units. The first 110 amino acids (the amino-terminal domain) in the first domain of the heavy and light chains can vary significantly, whereas the rest remain constant. The variable domains of the heavy and light chains are designated as V_H and V_L (Fig. 2), respectively, and make up the variable region of the antibody. Taken together, the V_H and V_L bestow an antibody with the ability to bind a specific antigen. The constant domains of the heavy and light chains are designated as C_H and C_L, respectively, and make up the constant region. Heavy chain constant domains are further divided into $C_H{}^1$, $C_H{}^2$, $C_H{}^3$ and are numbered from the amino-terminus to the carboxy-terminus.

Structurally, the IgG can be divided into three sections: two "fragment antigen binding" or F(ab) regions and a "fragment crystalline" or Fc region (Fig. 2). The Fab domain is comprised of the entire light chain, and the V_H and $C_H{}^1$ domain of the heavy chain. The Fc region is comprised of the $C_H{}^2$ and $C_H{}^3$ domains of the heavy chains. While the F(ab) regions contain the variable domains that are responsible for antigen specificity, the Fc fragment provides a binding site for endogenous Fc receptors on the surface of lymphocytes. Within the Fab region, the variable domains contain three hypervariable amino acid sequences responsible for the antibody specificity (the "complementary determining regions" or CDRs) embedded into largely constant "framework regions."

2.1.2 Antibody Fragments

Fabs are the oldest known class of antibody fragment and are generated by cleavage of an intact antibody through enzymatic treatment with papin.[11,12] The resulting cleavage product yields two monovalent Fab fragments, each consisting of complete light chains paired with the V_H and $C_H{}^1$ domains of the heavy chains via a disulfide linkage (Fig. 2). Each Fab fragment has a molecular weight of approximately 50 kDa and displays a single antigen-binding site.

Single-chain Fv fragments (scFvs) have a molecular weight of approximately 25 kDa and consist of V_H and a V_L domains separated by a short, flexible amino acid spacer, typically 12–20 amino acids in length. The flexible linker enables the V_H and V_L domains to form a single antigen-binding site. Structurally, scFvs and Fabs retain the complete antigen-binding site of their parental antibody molecule but lack the Fc region, and thus are unable to initiate effector functions.[12]

Single-domain antibody fragments (sdAbs) are the smallest functional antibody fragments that preserve the full antigen-binding specificity and have a molecular weight of approximately 12–15 kDa. SdAbs consist of either a V_H or V_L domain with only three of the six CDRs from the parent antibody and are remarkably stable under harsh conditions of temperature, pressure, and denaturing chemicals.[13,14]

2.2 Properties of Antibodies Used in ADCs

An ADC confers its effectiveness by specifically binding to the surface of antigen-positive cells creating an antigen–antibody interaction, which subsequently leads to internalization of the ADC.[15,16] Once inside the cell, the ADC is transported into the appropriate intracellular compartment, typically the lysosome, for subsequent degradation and release of the cytotoxic drug. The choice of the antibody and its (1) binding affinity, (2) internalization rate, and (3) intracellular localization postinternalization are other key considerations in designing an effective ADC. While it is difficult to predict the efficacy of an ADC by correlating it with binding affinities and/or internalization rates, it is commonly believed that a high binding affinity can improve internalization efficiency, thereby increasing the overall effectiveness of the ADC.[17] However, an antibody with high-affinity binding to the target is less desirable because this may inhibit the tumor penetration of the ADC due to slow rates of dissociation leading to decrease in the local concentration of diffusible, free antibody.[18,19] Ideally, the antibody selected for the basis of an ADC should exhibit rapid internalization upon antigen binding. However, there are no direct evidences to support a correlation between internalization rate and efficacy of ADCs, and a slow internalizing or a noninternalizing ADC may still be as effective, as seen with Kadcyla. Kadcyla is very potent at inhibiting tumor growth, and its component antibody, trastuzumab (Herceptin), has a high binding affinity but has a relatively slow internalization rate as compared to other antibodies used for ADCs.[20,21] As mentioned earlier, postinternalization, ADCs are transported to the correct intracellular compartment for subsequent processing and release of the active drug. Conditions found inside a cell, either in the cytoplasm or in the lysosome, can be harnessed to release the active drug. Thus, cleavable vs noncleavable linker designs also impact ADC potency in addition to binding affinity and internalization rate.

It should be noted that rigorous screening of an antibody for high-affinity binding and rapid internalization may not yield an effective ADC. Conjugation of a thoroughly screened antibody with a payload may disrupt or adversely affect binding and internalization, thereby diminishing its overall effectiveness. Potential changes to an antibody caused by payload conjugation therefore require extensive evaluation in both in vitro and in vivo models to evaluate targeting, efficacy, potency, and safety of the ADC.

3. DRUG ATTACHMENT AND RELEASE

Conjugation describes the process of covalently attaching drug–linker to an antibody and is a necessary step for production of ADCs. Choice of conjugation chemistries can impact the overall process of ADC formulation (buffer, pH, temperature, etc.), the type of drug–linkers available, and stability of the resulting construct. Many types of chemical reactions have been applied for conjugation of drugs and/or other entities to proteins as outlined in several comprehensive reviews found elsewhere.[22–27] However, relatively few conjugation reactions are routinely employed for production of ADCs. Traditional conjugation strategies utilize naturally occurring amino acids such as lysine and cysteine to conjugate drug–linker via nucleophilic displacement or Michael addition reactions, respectively. These approaches are utilized for production of the only two FDA-approved ADCs brentuximab vedotin (Adcetris) and ado-trastuzumab emtansine (Kadcyla). Other approaches employ enzyme-based reactions, or target drug–linker attachment to sugars at N-glycosylation sites or utilize nonnatural amino acids. A brief overview of conjugation chemistry for ADC production is shown in Fig. 3.

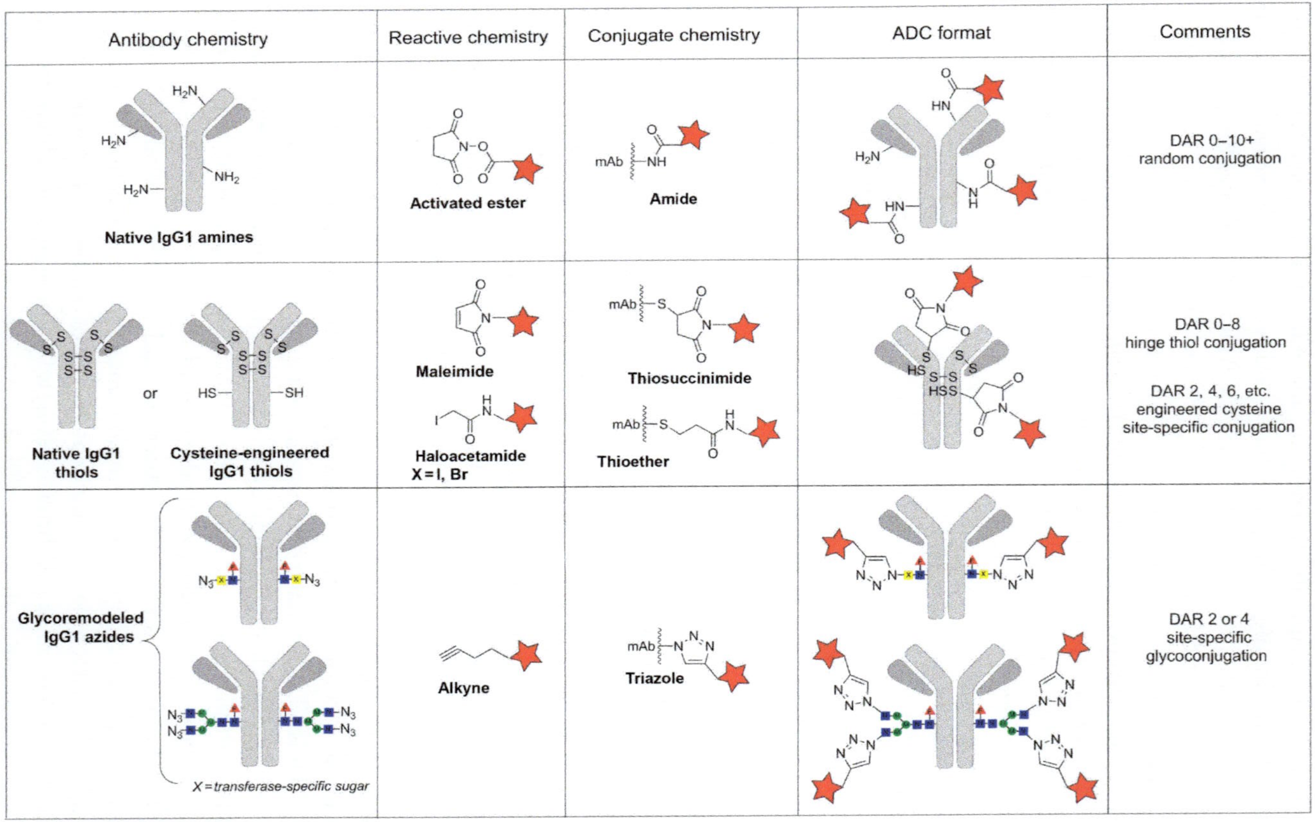

Fig. 3 Chemistries used to generate antibody–drug conjugates.

3.1 Conjugation to Lysine

Coupling payloads to lysine occurs through the epsilon amine in the amino acid, which can react with electrophiles such as esters activated with the electron-withdrawing group *N*-hydroxysuccinimide (NHS). Lysine coupling to NHS esters occurs under mildly basic conditions (typically pH ~8), which balances lysine reactivity (pK_a ~9) with NHS ester hydrolysis half-life ($T_{1/2}$ ~1 h at pH 8.0, 25°C).[28] Lysine is reactive as a nucleophile only when the amine is deprotonated (i.e., pH ~9), while NHS esters are susceptible to nucleophilic attack by the hydroxide ion at high pH. Thus, choosing appropriate reaction conditions is essential to maximize conjugation efficiency with this approach. There are approximately 40 surface-exposed lysine amines available for antibody conjugation, of which only a fraction are targeted for conjugation (typically 2–8).[27] Reaction products are heterogeneous with multiple species obtained that contain different degrees of lysine modification. Overall drug:antibody ratio (DAR) is typically reported as an average value, with the desired DAR achieved by controlling reaction stoichiometry between free amines and active esters.

In the case of Kadcyla, lysine amines are first modified with the NHS ester/maleimide heterobifunctional cross-linker *N-succinimidyl-4-(N-maleimidomethyl)cyclohexane-1-carboxylate* (SMCC) to introduce maleimides into the antibody structure. Thiol-containing mertansine (DM1) is then reacted with maleimides of the cross-linker-modified mAb to form the ADC. This process results in an average of DAR 3.5 in the resulting ADC, with drug randomly attached to lysines.[20]

3.2 Conjugation to Cysteine

Cysteine-based conjugations target sulfur atoms (thiols) in the side chain for nucleophilic reactions with Michael acceptors such as maleimides or electrophiles such as haloacetamides.[29] In terms of simplicity, thiol–maleimide chemistry is attractive because the reaction occurs in both slightly acidic and basic conditions (pH 6–8), the kinetics are sufficient to allow complete conjugation in minutes, the maleimide functional group can be incorporated into a wide range drug–linkers, and there are relatively few naturally occurring cysteines available for conjugation.

Cysteine–haloacetamide conjugation reactions are conducted above pH 8 (thiolate anion pK_a ~8), and drug–linker solubility can be significantly impacted by the bromine or iodine halogen atom. Haloacetamide reactions require more specific reaction conditions than maleimide reactions and exhibit slower kinetics; however, they do have the advantage of high stability of the thioether conjugation product.[30] Maleimide reactions generate a thiosuccinimide group that is susceptible to the retro-Michael reaction, which can lead to loss of drug–linker from the antibody. The retro-Michael reaction can be chemically prevented by hydrolyzing the thiosuccinimide to form a ring-opened thioether which stabilizes the thiol-antibody linkage.[31–34]

Conjugation to cysteine thiols affords more control of DAR than conjugation to lysine amines, as there are fewer cysteines available for conjugation in native antibodies (e.g., there are eight cysteines available in IgG1), and cysteines can be introduced by engineering approaches to allow site-specific conjugation. Conjugation to native IgG1 involves reduction of the four interchain disulfides in the antibody hinge region to generate reactive sulfhydryls for subsequent reaction with drug–linker. Site-specific conjugation is achieved by mutating a natural amino acid to cysteine, or, inserting a cysteine amino acid into the polypeptide chain.[35,36] In this approach introducing a single cysteine into the antibody heavy or light chain sequence results in two additional cysteines in the final antibody, introducing two cysteines into the antibody heavy or light chain sequence results in four additional cysteines in the final antibody and so on, since antibodies are symmetrical molecules built from two identical heavy–light chain pairs. Recent reports have demonstrated the benefits of site-specific conjugation, as high DAR species produced by lysine, or native hinge cysteine conjugation can have lower serum stability leading to faster clearance, lower exposure, and reduced efficacy.[37] Additionally, ADC performance is linked to the site of drug attachment, which can be precisely controlled by site-specific attachment to engineered cysteines.[38]

3.3 Conjugation to Sugars

Carbohydrates provide an additional opportunity to attach drug–linker to an antibody through either natural or modified sugars. There is one conserved N-glycosylation site in a native IgG1 antibody at position N297. The sugar structure is extended from N297 in the general sequence: *N*-acetylglucosamine-*N*-acetylglucosamine-mannose (branch point)-mannose-*N*-acetylglucosamine-galactose-sialic acid; however, the exact carbohydrate structure obtained in manufactured antibodies is variable and cell line dependent. ADCs can be generated in DAR 2 or DAR 4 formats, depending on if the drug is attached before or after the sugar branch point.

Early attempts to conjugate to sugars employed an oxidizing agent, such as sodium cyanoborohydride to generate aldehydes in sugar moieties, which could then react with hydrazide- or aminooxy-containing payload to generate an ADC.[39,40] However, chemical liabilities such as conjugate instability and nonspecific oxidation are possibilities with this approach. More recent approaches aim to incorporate chemically modified sugars into an antibody, which can be achieved by several approaches referred to as metabolic engineering, glycoengineering, glycan remodeling, and chemo-enzymatic conjugation.[41] In the metabolic engineering approach, a sugar containing the reactive handle (such as azide or thiol) is incorporated during expression of the antibody. This approach requires a cell line capable of producing homogeneous mAb glycoforms if homogenous ADCs are desired. For example, incorporation of thiofucose is ~70% efficient, resulting in production of ADCs with a DAR of ~1.4 following conjugation of maleimide-MMAE.[42] An example of glycan remodeling involves use of enzymes to artificially modify mAb glycans and incorporate reactive sugar moieties. Homogeneity is achieved by either a specific cell line to produce uniform N-glycan (e.g., G0F containing IgG1 produced by CHO-LEC8 cells) or pretreating mAbs with a glycosidase that trims sugar chains back to a common starting point (e.g., galactosidase to trim back to GlcNAc) and then attaching a chemically modified sugar with the appropriate transferase.[43–45] Using this method, sugars containing reactive ketone groups (2-acetonyl-2-deoxy-galactose) and reactive azide groups (*N*-acetyl-azido-galactosamine) have been incorporated into antibodies. Ketone-functionalized antibodies can further react with hydrazide and aminooxy drug–linkers, whereas azide-functionalized antibodies can react with alkynes through copper-catalyzed (CuAAC) or strain-promoted (SPAAC) azide–alkyne cycloaddition "click" reactions (Fig. 1). ADCs prepared by incorporation of azidosugars and subsequent attachment of alkyne-drug–linker by a SPAAC conjugation achieved DARs >1.8 with in vitro activities similar to Kadcyla.[46] Further discussion of the click reaction is found in the following section.

3.4 Enzyme-Based Conjugation to Amino Acids

Covalent attachment of drug–linker to antibodies can also be facilitated by enzymatic processes, either by direct action on the drug–linker or by modification of the antibody to generate a unique reactive group that can subsequently be reacted with drug–linker.

Enzymes used to directly couple the drug–linker to antibodies include transglutaminase and sortase A. Transglutaminase catalyzes amide bond formation between a side-chain amide group of the amino acid glutamine (Q) side chain and an alkyl primary amine contained on drug–linker.[47] The transglutaminase enzyme recognizes the sequence LL**Q**G on the antibody (termed Q-tag) as well as Q295 in a glycosylated IgG1 and also the N297S and/or N297Q mutations in IgG1.[48] Transglutaminase conjugations are reported to be quite efficient, with complete conversion achieved after 16 h reaction using 10-fold excess payload in the presence of 2% w/v bacterial transglutaminase enzyme.[49] Also, the enzyme can be removed using a single purification step ceramic hydroxyapatite chromatography. ADCs prepared with transglutaminase-catalyzed conjugation of MMAE (DAR 2) were stable in serum and potent in murine tumor models, with MEDs similar to ADCs prepared with cysteine–maleimide conjugation (~1.5–3 mg/kg).[49] Transglutaminase has also been used to introduce conjugation handles, such as azides, which can be subsequently reacted with payloads with appropriate orthogonal chemistry.[50]

Drug–linker conjugation to antibodies can also be achieved using sortase A.[51–54] However, unlike transglutaminase, which utilizes glutamine amino acid side chains for direct coupling to amines, sortase A ligation inserts drug–linker into a polypeptide sequence and removes one glycine amino acid in the process (i.e., transpeptidation). Sortase A ligates the peptide sequence LPXTG to an oligo glycine (G)*n*, with $n=3$–5 repeating glycine units. The LPXTG sortase A tag is incorporated into the antibody sequence at either the N- or C-terminus, in proteins, whereas the oligo glycine peptide is incorporated into the drug–linker intended for conjugation. ADC constructs generally incorporate sortase A tags onto the C-terminus of IgG1 heavy or light chains (for DAR 2 ADCs) or the C-terminus of both heavy and light chains for DAR 4 ADCs. Drug–linker conjugation to an antibody can occur in ~4 h at room temperature in the presence of 3 μM sortase A and calcium-containing buffer.[54] Sortase A-based ligation has been used to generate ADCs with MMAE and doxorubicin-based drug–linkers, with similar or slightly better conjugation efficiency and similar potency as ADCs prepared with thiol–maleimide conjugation.[51,52,54]

Enzymes can also be used to introduce a unique chemical handle for attachment of drug–linker, rather than catalyzing the conjugation reaction itself. Such is the case with the formylglycine-generating enzyme, which converts the cysteine thiol in the peptide sequence **C**XPXR into an aldehyde group.[55] This aldehyde then serves as a reactive group toward aminooxy

Table 1 Summary of Peptide-Based Enzyme Conjugation Technologies

Enzyme	Recognition Sequence	Drug–Linker Reactive Group	Comment
Transglutaminase	LL**Q**G	Free amine	Drug–linker attached to amino acid side chain
Sortase	LPXTG	Oligo glycine	Drug–linker inserted into polypeptide chain
Formylglycine-generating enzyme	**C**XPXR	Aminooxy or hydrazine	Drug–linker attached via nonnatural aldehyde amino acid side chain

Note: amino acids modified by enzyme activity are shown in *bold*.

and hydrazine-isoPictet–Spengler chemistry.[56] The formylglycine amino acid tag has been placed at the heavy chain C-terminus and inserted into the heavy chain or light chain framework. In general, the tag is well tolerated; however, insertion between residues D283/E285 (C_H2) or G361/E366 (C_H3) leads to significant aggregation.[57] Conversion of cysteine to formylglycine is achieved by expressing antibodies in CHO-S cells that also overexpress the formylglycine-generating enzyme. Thus, aldehyde-tagged antibody is obtained directly from the cell culture with typical cysteine conversion to formal glycine efficiency of 86%–92%. Conjugation of drug to aldehyde-tagged antibody is achieved with Pictet–Spengler chemistry, which uses indole-substituted hydrazide (or aminooxy) groups to generate a hydrolytically stable conjugate.[58] High conjugate stability in combination with site-specific conjugation (DAR 2) of a maytansine drug–linker resulted in ADCs with similar or slightly higher tumor growth inhibition activity compared to Kadcyla[57] (Table 1).

3.5 Click Conjugation

Click conjugation refers to application of the classic Huisgen azide–alkyne 1,3-dipolar cycloaddition reaction to produce bioconjugates.[59] This reaction has gained popularity for bioconjugation applications due to its specificity, or "biorthogonal" property. Reaction of azides with alkynes can be performed in complex biological milieu without interference from naturally occurring substrates.[60] Early applications utilized Cu(I) to catalyze the reaction between azides and alkynes (CuAAC), whereas more recently strained cyclic alkynes have been developed that react with azides without need for a copper catalyst (SPAAC).[61] CuAAC click reactions exhibit second-order rate constants of 10–200 M^{-1}/s, while SPAAC reactions are slower at 0.002–1.0 M^{-1}/s, depending on the structure of the cyclic alkyne.[60,61]

Click functional groups are incorporated into antibodies via glycoengineering (as discussed earlier), linkers, or nonnatural amino acids.[62–64] Incorporation of click functional groups via nonnatural amino acids offers the advantage of minimal antibody manipulation to produce ADCs. Description of nonnatural amino acid technology is beyond the scope of this chapter, and reviews are available elsewhere.[65,66] Antibodies produced bearing *para*-azidomethyl-L-phenylalanine (pAMF) incorporated at position S136 were used to generate DAR 2 ADCs by SPAAC conjugation to monomethyl auristatin F (MMAF) bearing a dibenzyl cyclooctyne (DBCO) functional group. SPAAC conjugation efficiency was ~95% after 16 h at RT in the presence of 10 equivalents DBCO-MMAF. Resulting ADCs were evaluated in vitro and showed sub-nM EC_{50} values, which is in the expected potency range for DAR 2 ADCs produced with MMAF.[62] Another study incorporated *N*6-((2-azidoethoxy)carbonyl)-L-lysine into a single heavy chain (H274, H359) or light chain (L70, L81) position (Kabat numbering) to generate DAR 2 ADCs via CuAAC bioconjugation to auristatin F (AF) or pyrrolobenzodiazepine (PBD) drug–linkers bearing alkyne functional groups. CuAAC conjugation was found be efficient (>90%) at each conjugation site and resulting ADCs demonstrated EC_{50} values in the nM range in receptor-positive cells. Interestingly, the PBD CuAAC conjugate showed significantly reduced in vitro potency. However, strong antitumor activity for the CuAAC PBD ADC was demonstrated in vivo, with complete tumor regression (100 days) and similar potency to an analogous Herceptin-maleimide-PBD construct achieved at a single dose of 1 mg/kg ADC. Altogether, click-based ADCs generated with nonnatural amino acids offer great potential for production of potent ADCs with simplified conjugation procedures.

Other recent advances: Conjugation chemistries continue to evolve and adapt as opportunities for improvement are identified. For example, observation of thiol–maleimide conjugate instability due to the retro-Michael addition reaction resulted in development of several different methods to hydrolyze the thiosuccinimide ring and generate a stable thioether. These methods ranged from physical approaches (temperature, pH, time) to chemical methods (substituted maleimides, phenyl maleimides), offering several options to generate stable thiol–maleimide conjugates.[31–34]

Other approaches aim to develop new chemistries (or newly apply old chemistries) that offer site-specific and stable conjugation modalities to natural or nonnatural substrates. For example, 2-arylproprionitriles can be applied for stable thiol conjugation,[67] norbornene–tetrazine reactions are emerging as a nonnatural amino acid conjugation platform,[68,69] and other strategies utilizing an antibodies natural structure are exploiting hinge thiols to produce ADCs with controlled DARs via chemical bridging.[70–72]

3.6 Chemically Labile Linkers

A linker, which covalently attaches the antibody to the cytotoxic drug, plays an important role in determining the toxicity, PK properties, and the therapeutic index of an ADC. Ideally a linker is designed to be stable in the bloodstream preventing premature drug release from the ADC. Upon internalized within the tumor, linkers are designed to efficiently release the active free drug. Based on the drug release mechanism, ADC linkers can be subdivided into two categories, namely, cleavable and noncleavable linkers. Cleavable linker strategy employs two major types of release mechanism which are discussed later.

3.6.1 Acid-Labile Linkers (Hydrazones)

After internalization into a target cell, an ADC is trafficked to the lysosomal compartment and subjected to low pH. Acid-sensitive linkers like hydrazones utilize low pH to release drug. Hydrozones have a relatively longer plasma half-life at pH 7 as compared to pH 5, suggesting that they are labile under conditions existing in the lysosome. The downside to this strategy is the nonspecific drug release due to the acidic condition found in various places in the body and the inherent instability of hydrozones when subjected to prolonged circulation. The first-generation ADC gemtuzumab ozogamicin (Mylotarg) containing a hydrozone linker had a similar problem and was withdrawn from the US market due to toxicities partially attributed to plasma instability of hydrazone (Fig. 4). The ADC produced an increase in fatalities in patients treated with Mylotarg in combination with chemotherapy as opposed to patients treated with chemotherapy alone.[73] Inotuzumab ozogamicin (anti-CD22 calicheamicin conjugate)[65,66] is another example of an ADC with a hydrazone-based linker that was recently withdrawn from a Phase III clinical trial owing to a lack of improvement in overall survival.[40,74]

3.6.2 Disulfide Linkers

Glutathione concentration inside of cells is much higher than within the bloodstream, and glutathione-sensitive likers have been designed to take

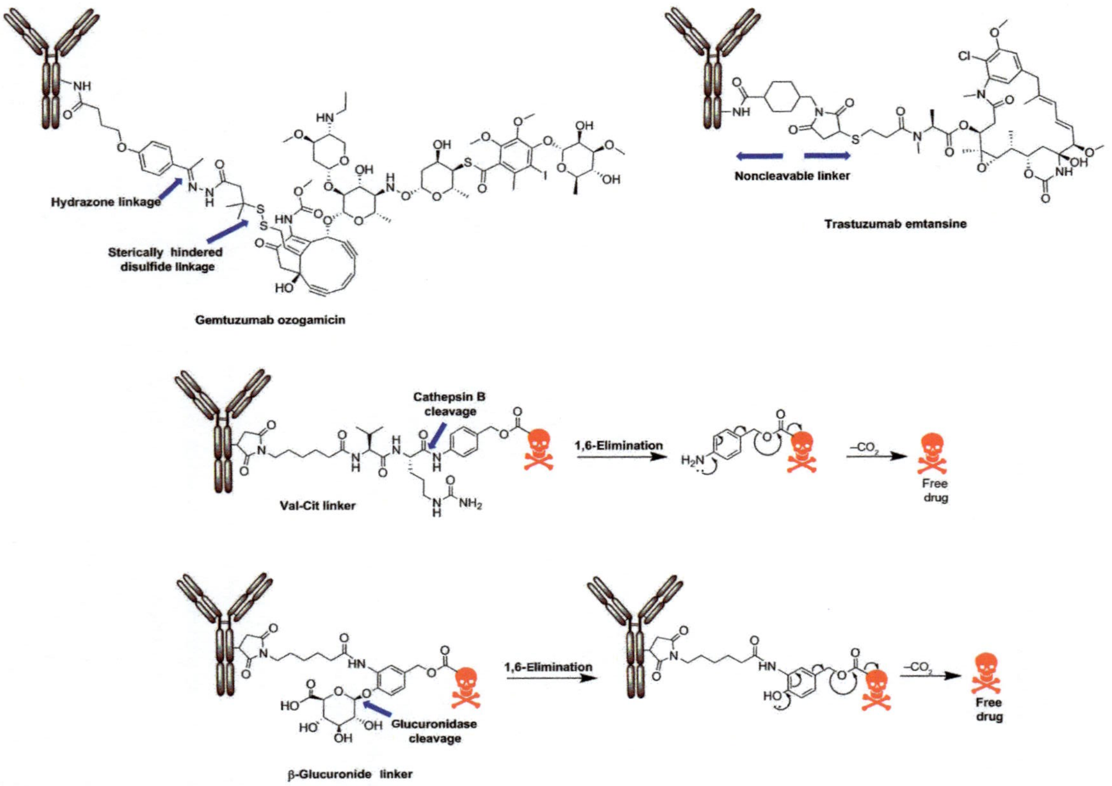

Fig. 4 Examples of cleavable and noncleavable linkers.

advantage of this gradient. Glutathione-based linkers utilize the redox reaction between thiols and disulfides and thus target this gradient of free thiol (glutathione) between the intracellular and extracellular space. Disulfides are relatively stable in circulation, but upon internalization, free drug is released by the reducing environment existing in the cell. The stability of disulfide bonds in circulation can be further enhanced by introducing methyl groups (Fig. 4) that flank the disulfide thereby increasing steric hindrance and slowing cleavage rate in the presence of low free thiol concentrations.[75] Sterically hindered disulfide linkers have been used in several clinical candidates such as AVE9633 (anti-CD33 maytansine conjugate), SAR3419 (anti-CD19 maytansine conjugate), and IMGN901 (anti-CD56 maytansine conjugate).[74] Erickson et al. have investigated the intracellular cleavage mechanism of the disulfide linker and have found an abundance of lysine-bound, disulfide-linked drug among the metabolites of ADC degradation suggesting a proteolytic degradation of the antibody followed by drug release via glutathione reduction.[76]

3.7 Enzymatically Cleavable Linkers

3.7.1 Peptide Linkers

Enzymatically cleavable linkers are gaining significant attention in ADC development due to superior plasma stability and specificity of the release mechanism. Among all the enzymatic cleavable linkers reported, cathepsin B-responsive linkers are the most popular. Cathepsin B is a lysosomal protease that is expressed almost exclusively in all mammalian lysosomes and is found to be overexpressed in many cancer cells. Thus, cathepsin B can be exploited to degrade programmed peptidic cleavage points in ADC drug–linkers to release free drug.[77] Considerable work has been done toward identifying cathepsin B cleavable peptide sequences. Dubowchik and coworkers have screened a library of dipeptide linkers carrying doxorubicin drug for cathepsin B-mediated drug release.[78,79] Their results showed that the dipeptide, phe-lys, was the most rapidly cleaving dipeptide with a half-life of 8 min, followed by val-lys with a half-life of 9 min. The dipeptide, val-cit, was more stable and showed a half-life of 240 min. Another study using auristatin derivative MMAE linked via dipeptide linkers val-cit and phe-lys categorized the former dipeptide to be substantially more stable than the latter in human plasma. The mentioned work has contributed immensely in popularizing the val-cit linker.[80]

The most popular cleavable linker currently used in ADC utilizes the dipeptide val-cit in combination with a self-immolative *para*-aminobenzyl alcohol (PABA) spacer. Once the ADC is internalized through endocytosis and trafficked to lysosomes, cathepsin B selectively cleaves this linker and the cytotoxic drug is released from the ADC. The PABA functions as a spacer between the val-cit moiety and the payload, allowing cathepsin B to cleave the linker connected to a bulky warhead. Cleavage of the amide-linked PABA triggers a 1,6-elimination of carbon dioxide and simultaneous release of the warhead in parent amine form (Fig. 4).[81]

3.7.2 *β*-Glucuronide Linkers

A glucuronide linker incorporates a hydrophilic saccharide group that can be enzymatically hydrolyzed by the lysosomal enzyme, β-glucuronidase. The glucuronide-based linkers contain the β-glucuronic acid linked to a self-immolative linker via a glycosidic bond (Fig. 4). The β-glucuronidase triggers the cleavage of β-glucuronic acid from the phenolic backbone leading to the self-immolation of a *para*-aminobenzyl group to releasing the free drug (Fig. 4). Burke and coworkers compared glucuronide PAB- and dipeptide PAB-linked ADCs side by side to determine the effect of linker on ADC aggregation and efficacy. They found that glucuronide, owing mainly to its hydrophilic character, showed nominal aggregation (less than 5%) as compared to dipeptide-linked conjugates, which demonstrated higher aggregation (up to 80%). The in vitro efficacy of the glucuronide linker-based drug was found to be better than the dipeptide-based ADC in vitro; however, it was not well tolerated in vivo.[81]

3.8 Noncleavable Linkers

Noncleavable linkers consist of stable chemical bonds and are designed to be resistant to proteolytic degradation. These linkers offer increased plasma stability and increased specificity of drug release as compared to the cleavable linkers. ADCs prepared using noncleavable linkers rely on full degradation of the antibody by cytosolic and lysosomal proteases, which eventually liberates the drug molecule linked to an amino acid residue derived from the degraded antibody (Fig. 4). An ADC equipped with a noncleavable linker exhibits limited "bystander" effects. Reduced "bystander" effects can be explained by the decreased permeability of the drugs (resulting from the attached amino acid residue) thereby limiting their ability to kill nearby cells. A popular example of an ADC comprising a noncleavable linker is Kadcyla®.

4. PAYLOADS FOR ADC

Drugs commonly used in the development of ADCs can be divided into several distinct categories according to their MOA. In the following sections, warheads with cytotoxic properties based on tubulin inhibition, DNA-damaging mechanisms, and RNA inhibition will be introduced and discussed.

4.1 Tubulin Inhibitors

4.1.1 Auristatins

The largest class of cytotoxic warheads found in ADCs undergoing clinical development are those based on auristatins. Auristatins are synthetic analogs of the natural antimitotic agent dolastatin 10, which was first isolated from the sea hare *Dolabella auricularia* by Pettit and colleagues in 1987.[82] The potent cytotoxicity of dolastatin 10 is derived from its ability to inhibit microtubule assembly and tubulin-dependent GTP hydrolysis, which eventually leads to cell cycle arrest and apoptosis.[83] Dolastatin 10 was initially investigated in clinical trials; however, despite its cytotoxic properties, significant toxicities were observed at dose levels which were not sufficient to achieve clinical efficacy. Modifications to dolastatin 10 were soon developed, resulting in monomethyl auristatin E (MMAE) and MMAF, each of which included a secondary amine at their N-terminus.[80,84] The pioneering work of Senter et al. exploited the presence of this free amine within MMAE to enable linker attachment and subsequent conjugation to mAbs with mc-MMAE (Fig. 5).[80,85] This important work led to the discovery of brentuximab vedotin (Adcetris) an FDA-approved ADC used to treat Hodgkin lymphoma and anaplastic large cell lymphoma.[85]

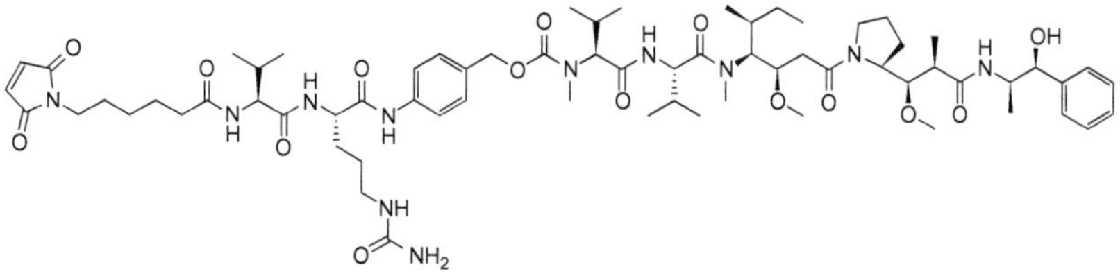

Fig. 5 Chemical structure of the mc-MMAE payload.

4.1.2 Maytansinoids

Maytansinoids represent a second class of microtubulin polymerization inhibitors derived from the naturally occurring maytansine, a benzoansamacrolide which is isolated from the bark of the African shrub *Maytenus ovatus*.[86,87] Maytansine binds tubulin at the vinca-binding site, similar to vinca alkaloids, thereby depolymerizing tubulin and inducing mitotic arrest.[88,89] Similar to auristatin, maytansine in its original form yielded a narrow therapeutic window due to associated neurological and gastrointestinal toxicities.[90,91] Consequently, synthetic maytansine derivatives were synthesized that possessed 100–1000-fold increases in potency over drug maytansinoids (DMs) while also including modifications to enable conjugation to an antibody.[92] The first and only maytansinoid-based ADC, trastuzumab emtansine (Kadcyla), was approved in 2013 for HER2-positive metastatic breast cancer.[93]

4.1.3 Tubulysins

Tubulysins are cytostatic compounds that were first isolated from the broth of the myxobacteria strain *Archangium gephyra* by Sasse and colleagues in 2000.[94] These antimitotic peptides are close relatives to dolastatin 10 but inhibit tubulin polymerization more effectively and exhibit potent cytotoxicity on a variety of cancer cell lines including, but not limited to, colon, breast, lung, and ovarian. Similar to the aurastatins and maytansinoids, tubulysins have not been successfully translated to the clinic as stand-alone drugs due to their inherent toxicities. However, much attention has been paid to the use of tubulysins as cytotoxic agents in targeted therapeutics such as ADCs. For example, MedImmune/AstraZeneca has recently developed a derivative of tubulysin, AZ13599185, which when conjugated to engineered cysteines within a biparatopic antibody, is currently being evaluated for safety and efficacy in Phase I clinical trials (Fig. 6). This ADC, MEDI4276, targets two separate epitopes on the extracellular domain of the HER2 receptor and is being investigated in patients with local or metastatic breast and gastric cancers which are refractory, or those who are ineligible for current HER2-based therapies.[95]

4.1.4 Cryptophycins

Cryptophycins are a class of dioxadiazacyclohexadecenetetrone cytotoxins with a potent ability to induce tubulin depolymerization. The cryptophycins were first isolated from the cultures of the *Nostoc* cyanobacteria in the early 1990s.[96,97] Similar to the maytansinoids, cryptophycins bind microtubules at

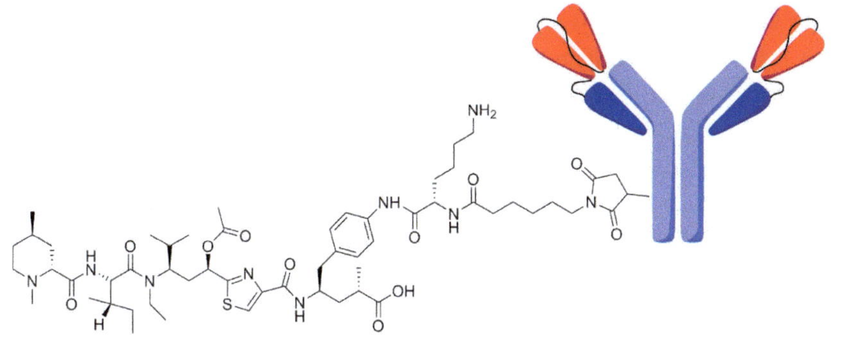

Fig. 6 Schematic representation of MEDI4276, a biparatopic ADC containing a tubulysin warhead.

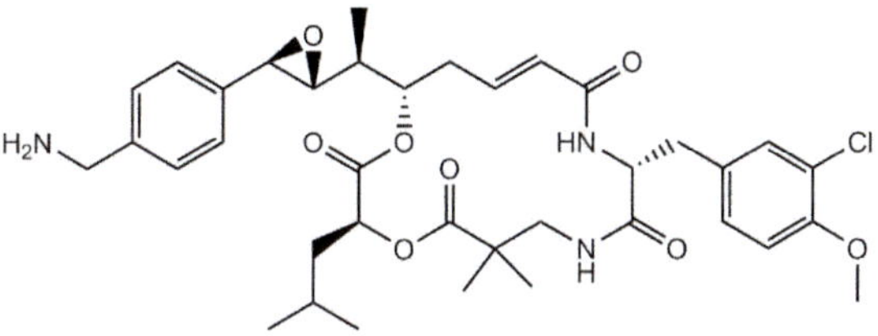

Fig. 7 Structure of cryptophycin analog 1.

the vinca-binding site eventually leading to mitotic arrest. Initial preclinical data with synthetic versions of cryptophycins such as LY355703 revealed promising antitumor effects in mammary and prostate xenograft models, which facilitated the transition of LY355703 into human clinical trials. However, as was the demise of several other tubulin inhibitors in clinical trials, the doses required for LY355703 to achieve therapeutic efficacy elicited significant toxicities, thus precluding its use as a stand-alone therapeutic agent.[98] Before cryptophycin could be converted into a warhead for an ADC, it first had to be modified with a suitable handle for linker attachment. Bouchard et al. synthesized a variant of cryptophycin, cryptophycin analog 1, which included an amine handle while retaining potency similar to that of the parental cryptophycin (Fig. 7). Later, cleavable (valine–citrulline) and noncleavable (bromoacetamide) linkers were installed onto cryptophycin analog 1 by Verma and colleagues and the payloads were conjugated to engineered cysteine residues within anti-HER2 and anti-CD22 antibodies.[99] Initial in vitro work confirmed the potency of these cryptophycin-based ADCs, which were superior in cytotoxicity when compared to their MMAE counterparts. Continued investigations are ongoing to evaluate this unique class of ADCs in vivo.

4.2 DNA-Damaging Agents

4.2.1 Pyrrolobenzodiazepines

Despite their discovery as potent antitumor agents over 50 years ago, PBDs such as anthramycin and sibiromycin have only recently emerged as cytotoxic warheads for the development of ADCs.[100] Naturally occurring PBD monomers are tricyclic systems consisting of an aromatic A-ring, a 1,4-diazepin-5-one B-ring, and a pyrrolidine C-ring, which differ from one another by the position and type of substituents in the A- and C-rings, and the degree and position of points of unsaturation in the C-ring. In their monomeric form, PBDs derive their biological activity by selectively binding the minor groove of DNA through the formation of covalent bonds with the exocyclic amino group of the guanine base.[101] Synthetic PBD monomers have been synthesized in which the A- and C-rings have been partially modified in efforts to improve cytotoxicity; however, only two C2-aryl-substituted monomeric PBDs have been tested in vivo (SG2738 and SG2042) with modest antitumor activity observed in colon, renal, and breast cancer models in mice.[102,103]

This early work led to the design and synthesis of PBD dimers, the rationale being that a dimer could form two covalent bonds with guanine bases and thus enable the PBD dimer to span greater lengths of DNA via cross-linking.[104] Early generations of dimerized PBD warheads such as SG2000 have evolved over time to incorporate more desirable properties such as enhanced solubility and improved DNA cross-linking ability while retaining potency.[105] For example, Gregson et al. demonstrated that SG2057, which includes a longer 5'carbon linker between the two PBD monomers, produced >3400-fold increase in cytotoxicity and >10-fold DNA cross-linking ability compared to its 3'carbon-linked PBD counterpart.[106] These characteristics were incorporated into the synthesis of the SG3199 warhead by Tiberghien and colleagues, which included replacement of the C2-aryls with methyl groups for improved water solubility (Fig. 8). The SG3199 warhead is linked to a maleimidocaproyl-PEG_8 spacer via self-immolative val-ala dipeptide at the N10 position to generate the SG3249 payload.[107] Currently, SG3249 is being evaluated in ADC clinical trials for the treatment

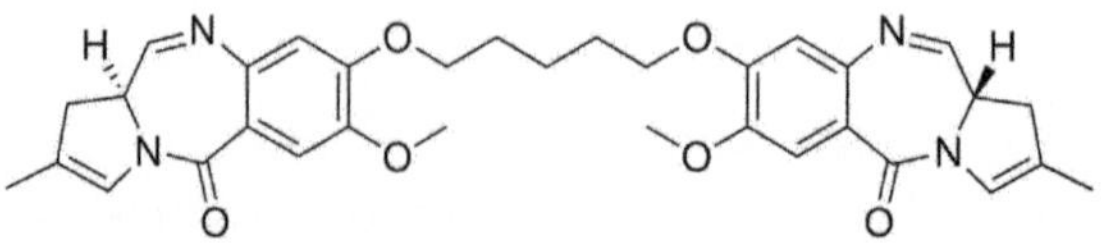

Fig. 8 Structure of the SG3199 warhead.

of small cell lung cancer and refractory Hodgkin and non-Hodgkin lymphomas.[108,109]

4.2.2 Doxorubicin

Doxorubicin is a DNA interchelator that inhibits topoisomerase II thereby inhibiting cancer cell growth. It is a potent anthracycline antibiotic first discovered from the actinobacteria *Streptomyces peucetius* in the 1960s. It is routinely used in the clinic as a chemotherapeutic agent for the treatment for various cancers including both solid and hematological malignancies. To reduce the systemic toxicity caused by doxorubicin, Bristol Myers Squibb and Seattle Genetics collaborated to develop the first doxorubicin-based ADC, BMS-182248, using an anti-Lewis^Y cBR96 antibody (Fig. 9).[110] Limited clinical efficacy combined with a narrow safety profile ultimately prevented this ADC from gaining FDA approval. Milatuzumab doxorubicin is an ADC comprised of an anti-CD74 antibody linked to a doxorubicin derivative consisting of a pH-sensitive cleavage mechanism. The rationale behind this unique approach relies on the acidic pH of the tumor microenvironment to enable rapid release of the doxorubicin warhead upon cellular internalization of the ADC. This ADC is currently under investigation in Phase II clinical trials for treatment of chronic lymphocytic leukemia and non-Hodgkin lymphoma.[111]

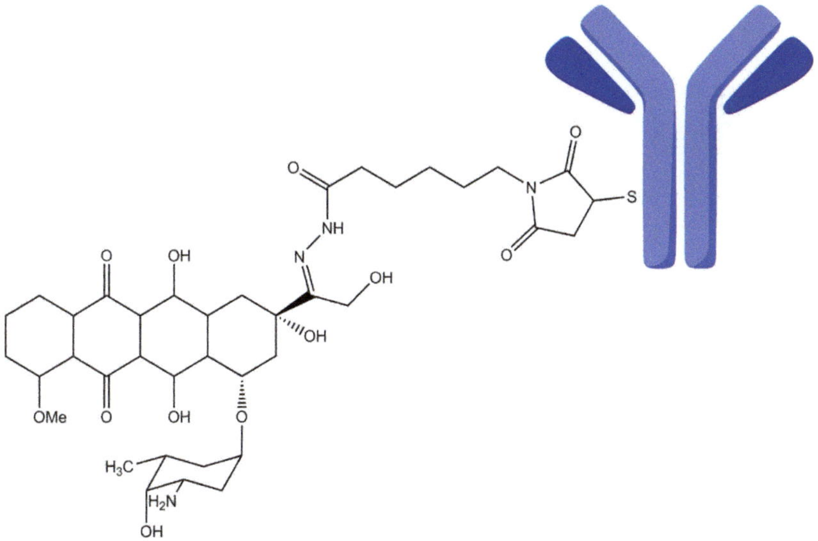

Fig. 9 Schematic representation of BMS-182248 consisting of the anti-Lewis^Y antibody conjugated with doxorubicin.

4.2.3 Duocarmycins

Duocarmycins are another class of DNA-alkylating agents that were first isolated in the 1970s from the culture broth of *Streptomyces* bacteria.[112] Duocarmycins bind the minor groove of DNA and alkylate the adenine residues at the N3 position.[113] The first member of the duocarmycin family to be evaluated in vivo was CC-1065, and despite showing moderate antitumor activity, hepatic toxicity limited its effectiveness.[114,115] In efforts to improve the therapeutic index of duocarmycin-based therapeutics, several ADCs have been developed including BMS-936561 (anti-CD70) and SYD985 (anti-HER2). BMS-936561 was first analyzed in patients with advanced clear cell carcinoma and B-cell non-Hodgkin lymphoma; however, the clinical trial was stopped during Phase I despite being tolerated at doses up to 8 mpk.[116] More recently, Synthon generated SYD985 (trastuzumab duocarmycin), which utilizes a duocarmycin prodrug known as seco-DUBA conjugated with a cleavable linker to trastuzumab (Fig. 10), as an alternative to ado-trastuzumab emtansine.[117] Phase I trials are currently ongoing with STD985 in patients with breast and gastric cancers.

4.2.4 Calicheamicins

Calicheamicins are a group of potent antitumor antibiotics that cleave DNA in a site-specific, double-stranded manner. This antibiotic was originally isolated from the actinomycete *Micromonospora echinospora* found in a soil sample from Texas.[118] Calicheamicin is approximately 4000 times more active than doxorubicin, often destroying the DNA of normal cells, and consequently cannot be used as a stand-alone treatment in patients. A team of researchers at Pfizer has developed several ADCs based on the calicheamicin, most

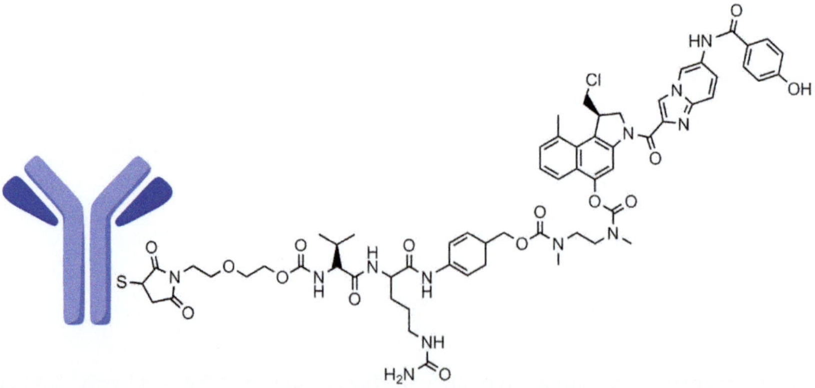

Fig. 10 Schematic representation of the trastuzumab duocarmycin (SYD985) ADC.

notably gemtuzumab ozogamicin and inotuzumab ozogamicin. The former consists of an anti-CD33 IgG4 antibody linked to a synthetic calicheamicin derivative using a 4-(4-acetylphenoxy)butanoic acid linker and was introduced into clinical trials in the United States for patients presenting with acute myeloid leukemia.[119] In 2000, the FDA granted marketing approval for gemtuzumab ozogamicin, which was later expanded to Europe and Japan. Unfortunately, the postapproval study failed to improve remission rates and fatal incidences of toxicity were significantly higher than chemotherapy alone, thus prompting Pfizer to voluntarily withdraw gemtuzumab ozogamicin from the US and European markets; however, commercial use is still available in Japan.[120,121] Inotuzumab ozogamicin consists of an anti-CD22 IgG4 antibody linked to a derivative of calicheamicin and recently completed Phase I clinical trials for patients with B-cell non-Hodgkin lymphoma with response rates ranging from 39% to 53% (Fig. 11). A combination Phase I–II trial consisting of inotuzumab ozogamicin and rituximab therapies is currently ongoing.[119]

4.3 RNA Inhibitors

α-Amanitin is a highly toxic cyclic octopeptide found in genus of mushrooms known as *Amanita*, including *Amanita phalloides*, *Amanita verna*, and *Amanita virosa* (Fig. 12). The cytotoxicity found in amanitin is the result of inhibition of RNA polymerases, in particular RNA polymerase II, which precludes mRNA synthesis.[122] Although no clinical studies have been performed with

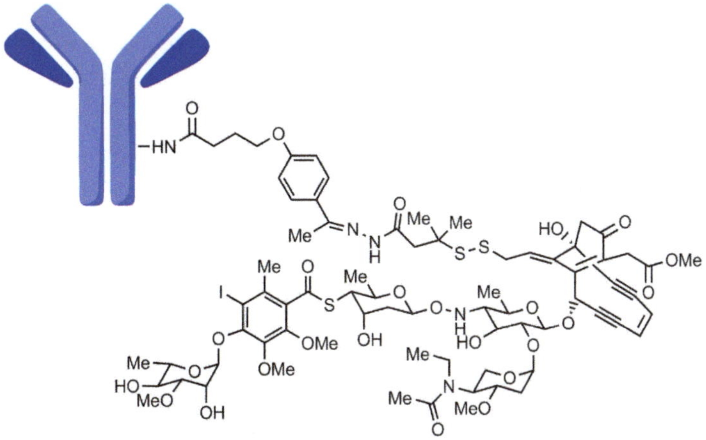

Fig. 11 Schematic representation of inotuzumab ozogamicin consisting of an anti-CD22 IgG4 and a calicheamicin warhead.

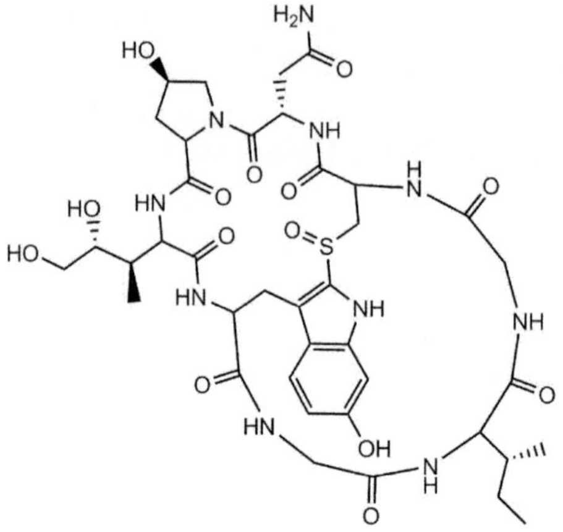

Fig. 12 Structure of α-amanitin.

α-amanitin, several preclinical investigations have proven the effectiveness of this potential ADC warhead as an antitumor agent. A chimeric antiepithelial cell adhesion molecule (EPCAM) antibody (chHEA125) was conjugated with α-amanitin from *A. phalloides* using glutaric anhydride and dicyclohexylcarbodiimide cross-linking chemistry to exposed lysine residues by Moldenhauer and colleagues.[123] Preliminary results indicated antiproliferative activity toward pancreatic (BxPC-3), breast (MCF-7), colorectal (Colo205), and bile duct (OZ) cancer cell lines in vitro, with in vivo results showing tumor regression in 9/10 mice at doses of 100 μg/kg with respect to α-amanitin in BxPC-3 tumor-bearing mice. Heidelberg Pharma is currently utilizing α-amanitin as a warhead to develop ADCs toward the anti-B-cell maturation antigen (BCMA), among other targets. In this case, the cytotoxic warhead is conjugated to lysine or cysteine residues present within anti-BCMA antibodies via stable or protease cleavable linkers. Initial preclinical data with HDP-101 revealed dose-dependent tumor regression in mice bearing multiple myeloma tumor xenografts with good tolerability observed in cynomolgus monkeys at doses as high as 4 mg/kg.

5. ANALYTICAL CHARACTERIZATION OF ADCs

ADCs are highly complex drugs comprising of an antibody, linker, and cytotoxic warhead (Fig. 13). Because of this complexity multiple orthogonal analytical methods are employed to enable full characterization.[124,125] Separation-based techniques are commonly used to characterize ADCs, where differences in molecular properties such as hydrophobicity

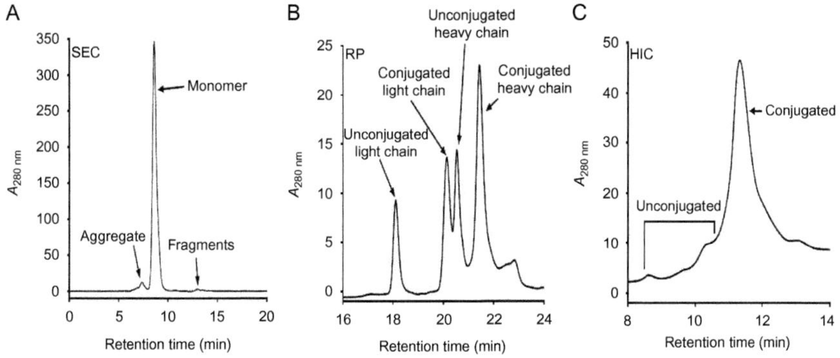

Fig. 13 Examples of data from the analytical characterization of a PBD-based ADC. (A) Size-exclusion chromatography. Corresponding aggregate, monomer, and fragments are shown. (B) Reduced reverse-phase chromatography. Unconjugated and conjugated light and heavy chains are identified. (C) Hydrophobic interaction chromatography. Unconjugated antibody and ADCs are labeled.

translate into separation of different species (i.e., conjugated vs nonconjugated). Size-exclusion (SEC), reversed-phase (RP), and hydrophobic interaction (HIC) are chromatographic methods frequently used to characterize monomeric content, free drug, and drug loading of ADCs, respectively.

RP coupled with mass spectrometry (LCMS) is used to confirm conjugation of payload and determine DAR. RP-HPLC LCMS provides molecular information not obtained using other separation-based methods, which rely on UV–Vis for detection. Generally, analytical characterizations are carried out with nonreduced or reduced ADCs that have conserved N-oligosaccharide in the Fc domain. However, in order to reduce heterogeneity, LCMS is often carried out after the removal of the conserved N-oligosaccharide in the Fc domain using enzymatic methods.

SEC analysis provides information about monomer content, aggregate, and fragments present in ADC preparations (Fig. 13A). SEC is carried out in nondenaturing aqueous conditions, except IPA (10%, v/v) is commonly used in the elution buffer to decrease nonspecific interactions between the stationary phase and the hydrophobic payload, which results in peak tailing and poor resolution. SEC analysis is also the primary methods employed to detect ADC fragmentation during stability studies.

Separation-based methods that rely on molecule hydrophobicity include RP and HIC. RP is used to determine drug distribution, load, and heterogeneity (Fig. 13B). RP is a versatile chromatographic technique that can be used to quantify free drug in the ADC reaction mixture as well as to study the stability of the payload during the storage of the ADC. RP is often used to characterize reduced antibodies and confirm if drug–linker is attached to the antibody heavy or light chains. Similar to RP, HIC is an analytical technique that utilizes the hydrophobic properties of the ADCs (Fig. 13C). HIC is commonly used to quantify drug load distribution because ADCs with higher DARs exhibit higher hydrophobicities when compared to ADCs with lower DARs, therefore making HIC a useful analytical method to characterize drug loading.

6. THERAPEUTIC INDEX

6.1 Mechanism of ADC Therapeutic Action

ADCs consist of a highly potent anticancer drug coupled via a suitable linker to a high-affinity antibody, which selectively binds to tumor specific/associate antigens at the surface of the cancer cells. The complex of ADC/receptor is then internalized and trafficked to the lysosome, where the active drug is released to kill cancer cells, depending on its MOA. Targeted delivery of highly potent drugs to tumors has great potential to maximize the antitumor effect of ADCs while minimizing normal tissue exposure, therefore broadening the therapeutic index. Therapeutic index represents the value difference between the maximum tolerated dose (MTD) and the minimum effective dose (MED). The optimal goal for developing a successful ADC is to increase the MTD while maintaining similar and/or decrease MED for improved safety profiles.

6.2 Mechanisms of ADC Toxicities

An ideal ADC for therapeutic applications should: (1) retain the selectivity, favorable pharmacokinetic, and functional properties of the parent antibody; (2) be stable and nontoxic in circulation; (3) be preferentially activated in the tumor microenvironment and efficiently kill the target tumor cells at a dose less than MTD. However, the majority of ADCs in clinical trial have very narrow or no therapeutic index such as the suggested dose for the FDA-approved ado-trastuzumab emtansine and brentuximab vedotin is 3.6 and 1.8 mg/kg, respectively, which is same as their MTD in patients.[93,126–128] The observed ADC toxicities in clinic can be mediated by both target-dependent (on-target but off-tissue) and target-independent (off-target).

The specific (target-dependent) and nonspecific (target-independent) cell uptake of ADCs can result in intracellular trafficking of ADCs to lysosome to release the active drug for cell killing regardless tumor or normal cells. ADC-induced toxicities can be contributed by all ADC components: antibodies, linkers, and cytotoxic warheads.

6.2.1 On-Target Toxicity

The ideal ADC target should be specifically expressed on tumor cells with no expression on normal cells. However, most target proteins have some level of normal tissue expression, which can be a major obstacle to ADC development.[129] The low-level expression of a target antigen on normal cells may mediate antigen-specific toxicities. A well-known example of on-target toxicity is with the use of bivatuzumab mertansine, an antihuman CD44v6 antibody conjugated to DM1, where development of a skin rash was one of the observed dose-limiting toxicities. One patient in the dose escalation developed grade 4 epidermal necrolysis at 140 mg/m^2 and subsequently died, which is thought to be associated with CD44v6 expression in epithelial cells of the skin as evidenced by immunohistochemistry studies.[130,131] This highlighted the importance of thoroughly evaluating normal tissue expression of a target antigen prior to clinical trials.

6.2.2 Off-Target Toxicity

The most frequently observed toxicity for ADCs is thought to be derived from linkers and warheads and are independent of target expression.[132] Off-target toxicity is a major cause of dose-limiting toxicity and probably the most significant obstacle hampering the progression of ADCs in clinical development. Target-independent uptake (e.g., pinocytosis, FcγRs and FcRn binding, etc.) into normal cells and catabolism of ADCs, premature free drug release due to deconjugation (e.g., drug exchange and liker instability), and bystander activity (free drug release following ADC catabolism) can all contribute to off-target toxicities,[133] which could lead to systemic toxicity and a lower therapeutic index.

6.3 Therapeutic Index of ADCs in Clinical Development

ADCs have been considered as the magic bullets for cancer therapeutics of the 21st century. However, in reality, the magic bullet theory has experienced only incremental success as only two FDA-approved ADC products have entered the market so far: brentuximab vedotin and ado-trastuzumab emtansine. To fully realize the magic bullet theory, ADCs must be administered at a therapeutically effective dose to cancer patients without evoking significant toxicity. Actually, only a very small portion (<1%) of an injected ADC accumulates within targeted tumors, while the remainder is widely distributed to normal organs and tissues, which can induce broad toxicities.[134] Most ADCs must be used near their MTD to achieve a clinically meaningful therapeutic effect. MTDs for the majority of ADCs based on tubulin inhibitors range from 2 to 3 mg/kg for MMAE and MMAF and 3 to 6.5 mg/kg for maytansinoids DM1 and DM4 regardless of target expression.[135] For instance, the MTD and suggested dose of ado-trastuzumab emtansine are 3.6 mg/kg dosed once every 3 weeks in humans.[93,126] The observed tumor regression in mouse models of breast cancer for ado-trastuzumab emtansine is 3 mg/kg or above. However, more potent in vivo efficacy was achieved at 15 mg/kg, indicating that ado-trastuzumab emtansine may achieve better clinical efficacy at the dose higher than MTD (3.6 mg/kg).[20,136] Similarly, the MTD and suggested clinic dose for brentuximab vedotin are 1.8 mg/kg administrated once every 3 weeks in patient.[127,128] The lowest dose in vivo mouse model to induce partial tumor regression is usually above 1 mg/kg, suggesting very narrow therapeutic index for brentuximab vedotin.[137] Another example is rovalpituzumab tesirine, an anti-DLL3 antibody conjugated to a PBD payload. The observed MTD is 0.4 mg/kg dosed every 3 weeks, with the dose of 0.3 mg/kg every 6 weeks moving forward into Phase II study.[138,139] In preclinical models of small cell lung cancer and large cell neuroendocrine carcinoma, rovalpituzumab tesirine eradicated tumor or induced tumor regulation at the dose of 1 mg/kg,[140] again demonstrating the narrow therapeutic index. It would be interesting to know the tumor inhibition in a preclinical animal model using the clinic dose 0.3 mg/kg in a preclinical animal model to demonstrate the therapeutic index for rovalpituzumab tesirine. Additional examples can be found in the review article by de Goeij and Lambert.[135]

In summary, the therapeutic index for ADCs currently in clinic development is fairly narrow, thus there is a great need to improve therapeutic index for clinic success. Only a small portion of injected ADCs (<1%) is able to accumulate in tumor, indicating the vast majority of ADCs (>99% of injected dose) is distributed into whole body to induce systemic toxicity.[130] Future focus on designing antibodies for selective binding to antigens within the tumor environment and/or payload design for preferred killing to tumor cells over normal cells might enhance ADC safety, therefore improving therapeutic index.

7. CHALLENGES AND PERSPECTIVE

Although great progress has been made in the design of ADCs since their inception by Paul Ehrlich more than 100 years ago, a better understanding of various parameters such as target biology, target identification, internalization efficiency, binding affinity, pharmacokinetics, stable and controlled drug conjugation, payload release mechanism, and drug potency is needed to design next-generation ADC molecules with improved performance. ADCs have long been plagued by linker instability and product heterogeneity leading to suboptimal efficacy and undesired therapeutic windows. Progress in site-specific conjugation technology and optimization of linkers with balanced stability have enabled construction of homogeneous ADCs with predictable properties. Technological developments in site-directed conjugation chemistry, along with antibody engineering and drug design, will continue to emerge for the development of homogenous ADCs. In our pursuit toward designing a better ADC we continue to investigate what's the optimal antibodies, linkers, and cytotoxic warheads for ADCs and development of better conjugation methods that will lead to further improvement of ADCs for cancer therapeutics and potentially other diseases.

REFERENCES

1. Himmelweit, F. The Collected Papers of Paul Ehrlich. *Proc. R. Soc. Med.* **1957**, *50*(3), 210.
2. Mathé, G.; Loc, T. B.; Bernard, J. C. C. Effet sur la leucemie L1210 de la souris d'une combinaison par diazotation d'A-methopterine et de gamma-globulines de hamsters porteur de cette leucemie par heterogreffe. *C. R. Acad. Sci.* **1958**, *246*, 145–146.
3. Ghose, T.; Cerini, M.; Carter, M.; Nairn, R. C. Immunoradioactive Agent Against Cancer. *Br. Med. J.* **1967**, *1*(5532), 90–93.
4. Rowland, G. F.; O'Neill, G. J.; Davies, D. A. L. Suppression of Tumour Growth in Mice by a Drug-Antibody Conjugate Using a Novel Approach to Linkage. *Nature* **1975**, *255*(5508), 487–488.
5. Kohler, G.; Milstein, C. Continuous Cultures of Fused Cells Secreting Antibody of Predefined Specificity. *Nature* **1975**, *256*(5517), 495–497.
6. Ford, C. H.; Newman, C. E.; Johnson, J. R.; Woodhouse, C. S.; Reeder, T. A.; Rowland, G. F.; Simmonds, R. G. Localisation and Toxicity Study of a Vindesine-Anti-CEA Conjugate in Patients With Advanced Cancer. *Br. J. Cancer* **1983**, *47*(1), 35–42.
7. Scott, A. M.; Allison, J. P.; Wolchok, J. D. Monoclonal Antibodies in Cancer Therapy. *Cancer Immun.* **2012**, *12*, 14.
8. Pietersz, G.; Krauer, K. Antibody-Targeted Drugs for the Therapy of Cancer. *J. Drug Target.* **1994**, *2*(3), 183–215.
9. Linenberger, M. L.; Hong, T.; Flowers, D.; Sievers, E. L.; Gooley, T. A.; Bennett, J. M.; Berger, M. S.; Leopold, L. H.; Appelbaum, F. R.; Bernstein, I. D. Multidrug-Resistance Phenotype and Clinical Responses to Gemtuzumab Ozogamicin. *Blood* **2001**, *98*(4), 988–994.

10. Keizer, R. J.; Huitema, A. D. R.; Schellens, J. H. M.; Beijnen, J. H. Clinical Pharmacokinetics of Therapeutic Monoclonal Antibodies. *Clin. Pharmacokinet.* **2010**, *49*(8), 493–507.
11. Porter, R. R. The Hydrolysis of Rabbit γ-Globulin and Antibodies With Crystalline Papain. *Biochem. J.* **1959**, *73*(1), 119–127.
12. Devaux, C.; Moreau, E.; Goyffon, M.; Rochat, H.; Billiald, P. Construction and Functional Evaluation of a Single-Chain Antibody Fragment That Neutralizes Toxin AahI From the Venom of the Scorpion Androctonus australis Hector. *Eur. J. Biochem.* **2001**, *268*(3), 694–702.
13. Ward, E. S.; Gussow, D.; Griffiths, A. D.; Jones, P. T.; Winter, G. Binding Activities of a Repertoire of Single Immunoglobulin Variable Domains Secreted From Escherichia coli. *Nature* **1989**, *341*(6242), 544–546.
14. Dumoulin, M.; Conrath, K.; Van Meirhaeghe, A.; Meersman, F.; Heremans, K.; Frenken, L. G.; Muyldermans, S.; Wyns, L.; Matagne, A. Single-Domain Antibody Fragments With High Conformational Stability. *Protein Sci.* **2002**, *11*(3), 500–515.
15. Alley, S. C.; Okeley, N. M.; Senter, P. D. Antibody-Drug Conjugates: Targeted Drug Delivery for Cancer. *Curr. Opin. Chem. Biol.* **2010**, *14*(4), 529–537.
16. Polson, A. G.; Calemine-Fenaux, J.; Chan, P.; Chang, W.; Christensen, E.; Clark, S.; de Sauvage, F. J.; Eaton, D.; Elkins, K.; Elliott, J. M.; Frantz, G.; Fuji, R. N.; Gray, A.; Harden, K.; Ingle, G. S.; Kljavin, N. M.; Koeppen, H.; Nelson, C.; Prabhu, S.; Raab, H.; Ross, S.; Slaga, D. S.; Stephan, J. P.; Scales, S. J.; Spencer, S. D.; Vandlen, R.; Wranik, B.; Yu, S. F.; Zheng, B.; Ebens, A. Antibody-Drug Conjugates for the Treatment of Non-Hodgkin's Lymphoma: Target and Linker-Drug Selection. *Cancer Res.* **2009**, *69*(6), 2358–2364.
17. Thomas, A.; Teicher, B. A.; Hassan, R. Antibody-Drug Conjugates for Cancer Therapy. *Lancet Oncol.* **2016**, *17*(6), e254–262.
18. Adams, G. P.; Schier, R.; McCall, A. M.; Simmons, H. H.; Horak, E. M.; Alpaugh, R. K.; Marks, J. D.; Weiner, L. M. High Affinity Restricts the Localization and Tumor Penetration of Single-Chain fv Antibody Molecules. *Cancer Res.* **2001**, *61*(12), 4750–4755.
19. Fujimori, K.; Covell, D. G.; Fletcher, J. E.; Weinstein, J. N. A Modeling Analysis of Monoclonal Antibody Percolation Through Tumors: A Binding-Site Barrier. *J. Nucl. Med.* **1990**, *31*(7), 1191–1198.
20. Lewis Phillips, G. D.; Li, G.; Dugger, D. L.; Crocker, L. M.; Parsons, K. L.; Mai, E.; Blattler, W. A.; Lambert, J. M.; Chari, R. V.; Lutz, R. J.; Wong, W. L.; Jacobson, F. S.; Koeppen, H.; Schwall, R. H.; Kenkare-Mitra, S. R.; Spencer, S. D.; Sliwkowski, M. X. Targeting HER2-Positive Breast Cancer With Trastuzumab-DM1, an Antibody-Cytotoxic Drug Conjugate. *Cancer Res.* **2008**, *68*(22), 9280–9290.
21. Hommelgaard, A. M.; Lerdrup, M.; van Deurs, B. Association With Membrane Protrusions Makes ErbB2 an Internalization-Resistant Receptor. *Mol. Biol. Cell* **2004**, *15*(4), 1557–1567.
22. Boutureira, O.; Bernardes, G. J. Advances in Chemical Protein Modification. *Chem. Rev.* **2015**, *115*(5), 2174–2195.
23. Hu, Q.-Y.; Berti, F.; Adamo, R. Towards the Next Generation of Biomedicines by Site-Selective Conjugation. *Chem. Soc. Rev.* **2016**, *45*(6), 1691–1719.
24. Kline, T.; Steiner, A. R.; Penta, K.; Sato, A. K.; Hallam, T. J.; Yin, G. Methods to Make Homogenous Antibody Drug Conjugates. *Pharm. Res.* **2015**, *32*(11), 3480–3493.
25. McCombs, J. R.; Owen, S. C. Antibody Drug Conjugates: Design and Selection of Linker, Payload and Conjugation Chemistry. *AAPS J.* **2015**, *22*, 339–351.
26. Jain, N.; Smith, S. W.; Ghone, S.; Tomczuk, B. Current ADC Linker Chemistry. *Pharm. Res.* **2015**, *32*(11), 3526–3540.

27. Agarwal, P.; Bertozzi, C. R. Site-Specific Antibody-Drug Conjugates: The Nexus of Bioorthogonal Chemistry, Protein Engineering, and Drug Development. *Bioconjug. Chem.* **2015**, *26*(2), 176–192.
28. Hermanson, G. T. *Bioconjugate Techniques*, 2nd ed.; Elsevier Academic Press Inc.: San Diego, CA USA, 2008
29. Cal, P. M.; Bernardes, G. J.; Gois, P. M. Cysteine-Selective Reactions for Antibody Conjugation. *Angew. Chem. Int. Ed.* **2014**, *53*(40), 10585–10587.
30. Alley, S. C.; Benjamin, D. R.; Jeffrey, S. C.; Okeley, N. M.; Meyer, D. L.; Sanderson, R. J.; Senter, P. D. Contribution of Linker Stability to the Activities of Anticancer Immunoconjugates. *Bioconjug. Chem.* **2008**, *19*(3), 759–765.
31. Lyon, R. P.; Setter, J. R.; Bovee, T. D.; Doronina, S. O.; Hunter, J. H.; Anderson, M. E.; Balasubramanian, C. L.; Duniho, S. M.; Leiske, C. I.; Li, F.; Senter, P. D. Self-Hydrolyzing Maleimides Improve the Stability and Pharmacological Properties of Antibody-Drug Conjugates. *Nat. Biotechnol.* **2014**, *32*(10), 1059–1062.
32. Tumey, L. N.; Charati, M.; He, T.; Sousa, E.; Ma, D.; Han, X.; Clark, T.; Casavant, J.; Loganzo, F.; Barletta, F.; Lucas, J.; Graziani, E. I. Mild Method for Succinimide Hydrolysis on ADCs: Impact on ADC Potency, Stability, Exposure, and Efficacy. *Bioconjug. Chem.* **2014**, *25*(10), 1871–1880.
33. Fontaine, S. D.; Reid, R.; Robinson, L.; Ashley, G. W.; Santi, D. V. Long-Term Stabilization of Maleimide-Thiol Conjugates. *Bioconjug. Chem.* **2015**, *26*(1), 145–152.
34. Christie, R. J.; Fleming, R.; Bezabeh, B.; Woods, R.; Mao, S.; Harper, J.; Joseph, A.; Wang, Q.; Xu, Z. Q.; Wu, H.; Gao, C.; Dimasi, N. Stabilization of Cysteine-Linked Antibody Drug Conjugates With N-aryl Maleimides. *J. Control. Release* **2015**, *220*, 660–670.
35. Bhakta, S.; Raab, H.; Junutula, J. R. Engineering THIOMABs for Site-Specific Conjugation of Thiol-Reactive Linkers. *Methods Mol. Biol.* **2013**, *1045*, 189–203.
36. Dimasi, N.; Fleming, R.; Zhong, H.; Bezabeh, B.; Kinneer, K.; Christie, R. J.; Fazenbaker, C.; Wu, H.; Gao, C. Efficient Preparation of Site-Specific Antibody-Drug Conjugates Using Cysteine Insertion. *Mol. Pharm.* **2017**, *14*(5), 1501–1516.
37. Lyon, R. P.; Bovee, T. D.; Doronina, S. O.; Burke, P. J.; Hunter, J. H.; Neff-LaFord,-H. D.; Jonas, M.; Anderson, M. E.; Setter, J. R.; Senter, P. D. Reducing Hydrophobicity of Homogeneous Antibody-Drug Conjugates Improves Pharmacokinetics and Therapeutic Index. *Nat. Biotechnol.* **2015**, *33*(7), 733–735.
38. Shen, B. Q.; Xu, K.; Liu, L.; Raab, H.; Bhakta, S.; Kenrick, M.; Parsons-Reponte,-K. L.; Tien, J.; Yu, S. F.; Mai, E.; Li, D.; Tibbitts, J.; Baudys, J.; Saad, O. M.; Scales, S. J.; McDonald, P. J.; Hass, P. E.; Eigenbrot, C.; Nguyen, T.; Solis, W. A.; Fuji, R. N.; Flagella, K. M.; Patel, D.; Spencer, S. D.; Khawli, L. A.; Ebens, A.; Wong, W. L.; Vandlen, R.; Kaur, S.; Sliwkowski, M. X.; Scheller, R. H.; Polakis, P.; Junutula, J. R. Conjugation Site Modulates the in vivo Stability and Therapeutic Activity of Antibody-Drug Conjugates. *Nat. Biotechnol.* **2012**, *30*(2), 184–189.
39. Hamann, P. R.; Hinman, L. M.; Beyer, C. F.; Lindh, D.; Upeslacis, J.; Flowers, D. A.; Bernstein, I. An Anti-CD33 Antibody-Calicheamicin Conjugate for Treatment of Acute Myeloid Leukemia. Choice of Linker. *Bioconjug. Chem.* **2002**, *13*(1), 40–46.
40. Hamann, P. R.; Hinman, L. M.; Hollander, I.; Beyer, C. F.; Lindh, D.; Holcomb, R.; Hallett, W.; Tsou, H. R.; Upeslacis, J.; Shochat, D.; Mountain, A.; Flowers, D. A.; Bernstein, I. Gemtuzumab Ozogamicin, a Potent and Selective Anti-CD33 Antibody-Calicheamicin Conjugate for Treatment of Acute Myeloid Leukemia. *Bioconjug. Chem.* **2002**, *13*(1), 47–58.
41. Qasba, P. K. Glycans of Antibodies as a Specific Site for Drug Conjugation Using Glycosyltransferases. *Bioconjug. Chem.* **2015**, *26*(11), 2170–2175.

42. Okeley, N. M.; Toki, B. E.; Zhang, X.; Jeffrey, S. C.; Burke, P. J.; Alley, S. C.; Senter, P. D. Metabolic Engineering of Monoclonal Antibody Carbohydrates for Antibody-Drug Conjugation. *Bioconjug. Chem.* **2013**, *24*(10), 1650–1655.
43. Li, X.; Fang, T.; Boons, G. J. Preparation of Well-Defined Antibody-Drug Conjugates Through Glycan Remodeling and Strain-Promoted Azide-Alkyne Cycloadditions. *Angew. Chem. Int. Ed.* **2014**, *53*(28), 7179–7182.
44. Boeggeman, E.; Ramakrishnan, B.; Pasek, M.; Manzoni, M.; Puri, A.; Loomis, K. H.; Waybright, T. J.; Qasba, P. K. Site Specific Conjugation of Fluoroprobes to the Remodeled Fc N-Glycans of Monoclonal Antibodies Using Mutant Glycosyltransferases: Application for Cell Surface Antigen Detection. *Bioconjug. Chem.* **2009**, *20*(6), 1228–1236.
45. Thompson, P.; Ezeadi, E.; Hutchinson, I.; Fleming, R.; Bezabeh, B.; Lin, J.; Mao, S.; Chen, C.; Masterson, L.; Zhong, H.; Toader, D.; Howard, P.; Wu, H.; Gao, C.; Dimasi, N. Straightforward Glycoengineering Approach to Site-Specific Antibody-Pyrrolobenzodiazepine Conjugates. *ACS Med. Chem. Lett.* **2016**, 7(11), 1005–1008.
46. van Geel, R.; Wijdeven, M. A.; Heesbeen, R.; Verkade, J. M.; Wasiel, A. A.; van Berkel, S. S.; van Delft, F. L. Chemoenzymatic Conjugation of Toxic Payloads to the Globally Conserved N-Glycan of Native mAbs Provides Homogeneous and Highly Efficacious Antibody-Drug Conjugates. *Bioconjug. Chem.* **2015**, *26*(11), 2233–2242.
47. Jeger, S.; Zimmermann, K.; Blanc, A.; Grunberg, J.; Honer, M.; Hunziker, P.; Struthers, H.; Schibli, R. Site-Specific and Stoichiometric Modification of Antibodies by Bacterial Transglutaminase. *Angew. Chem. Int. Ed.* **2010**, *49*(51), 9995–9997.
48. Farias, S. E.; Strop, P.; Delaria, K.; Galindo Casas, M.; Dorywalska, M.; Shelton, D. L.; Pons, J.; Rajpal, A. Mass Spectrometric Characterization of Transglutaminase Based Site-Specific Antibody-Drug Conjugates. *Bioconjug. Chem.* **2014**, *25*(2), 240–250.
49. Strop, P.; Tran, T. T.; Dorywalska, M.; Delaria, K.; Dushin, R.; Wong, O. K.; Ho, W. H.; Zhou, D.; Wu, A.; Kraynov, E.; Aschenbrenner, L.; Han, B.; O'Donnell, C. J.; Pons, J.; Rajpal, A.; Shelton, D. L.; Liu, S. H. RN927C, a Site-Specific Trop-2 Antibody-Drug Conjugate (ADC) With Enhanced Stability, Is Highly Efficacious in Preclinical Solid Tumor Models. *Mol. Cancer Ther.* **2016**, *15*(11), 2698–2708.
50. Dennler, P.; Chiotellis, A.; Fischer, E.; Bregeon, D.; Belmant, C.; Gauthier, L.; Lhospice, F.; Romagne, F.; Schibli, R. Transglutaminase-Based Chemo-Enzymatic Conjugation Approach Yields Homogeneous Antibody-Drug Conjugates. *Bioconjug. Chem.* **2014**, *25*(3), 569–578.
51. Beerli, R. R.; Hell, T.; Merkel, A. S.; Grawunder, U. Sortase Enzyme-Mediated Generation of Site-Specifically Conjugated Antibody Drug Conjugates With High in vitro and in vivo Potency. *PLoS One* **2015**, *10*(7), e0131177.
52. Pan, L.; Zhao, W.; Lai, J.; Ding, D.; Zhang, Q.; Yang, X.; Huang, M.; Jin, S.; Xu, Y.; Zeng, S.; Chou, J. J.; Chen, S. Sortase A-Generated Highly Potent Anti-CD20-MMAE Conjugates for Efficient Elimination of B-Lineage Lymphomas. *Small* **2017**, *13*(6), 1–12, 1602267.
53. Chen, L.; Cohen, J.; Song, X.; Zhao, A.; Ye, Z.; Feulner, C. J.; Doonan, P.; Somers, W.; Lin, L.; Chen, P. R. Improved Variants of SrtA for Site-Specific Conjugation on Antibodies and Proteins With High Efficiency. *Sci. Rep.* **2016**, *6, 31899.*
54. Stefan, N.; Gebleux, R.; Waldmeier, L.; Hell, T.; Escher, M.; Wolter, F. I.; Grawunder, U.; Beerli, R. R. Highly Potent, Anthracycline-Based Antibody-Drug Conjugates Generated by Enzymatic, Site-Specific Conjugation. *Mol. Cancer Ther.* **2017**, *16*(5), 879–892.
55. Frese, M. A.; Dierks, T. Formylglycine Aldehyde Tag—Protein Engineering Through a Novel Post-Translational Modification. *Chembiochem* **2009**, *10*(3), 425–427.

56. Appel, M. J.; Bertozzi, C. R. Formylglycine, a Post-Translationally Generated Residue With Unique Catalytic Capabilities and Biotechnology Applications. *ACS Chem. Biol.* **2015**, *10*(1), 72–84.
57. Drake, P. M.; Albers, A. E.; Baker, J.; Banas, S.; Barfield, R. M.; Bhat, A. S.; de Hart, G. W.; Garofalo, A. W.; Holder, P.; Jones, L. C.; Kudirka, R.; McFarland, J.; Zmolek, W.; Rabuka, D. Aldehyde Tag Coupled With HIPS Chemistry Enables the Production of ADCs Conjugated Site-Specifically to Different Antibody Regions With Distinct in vivo Efficacy and PK Outcomes. *Bioconjug. Chem.* **2014**, *25*(7), 1331–1341.
58. Agarwal, P.; van der Weijden, J.; Sletten, E. M.; Rabuka, D.; Bertozzi, C. R. A Pictet-Spengler Ligation for Protein Chemical Modification. *Proc. Natl. Acad. Sci. U.S.A.* **2013**, *110*(1), 46–51.
59. Amblard, F.; Cho, J. H.; Schinazi, R. F. Cu(I)-Catalyzed Huisgen Azide-Alkyne 1,3-Dipolar Cycloaddition Reaction in Nucleoside, Nucleotide, and Oligonucleotide Chemistry. *Chem. Rev.* **2009**, *109*(9), 4207–4220.
60. Sletten, E. M.; Bertozzi, C. R. Bioorthogonal Chemistry: Fishing for Selectivity in a Sea of Functionality. *Angew. Chem. Int. Ed.* **2009**, *48*(38), 6974–6998.
61. McKay, C. S.; Finn, M. G. Click Chemistry in Complex Mixtures: Bioorthogonal Bioconjugation. *Chem. Biol.* **2014**, *21*(9), 1075–1101.
62. Zimmerman, E. S.; Heibeck, T. H.; Gill, A.; Li, X.; Murray, C. J.; Madlansacay, M. R.; Tran, C.; Uter, N. T.; Yin, G.; Rivers, P. J.; Yam, A. Y.; Wang, W. D.; Steiner, A. R.; Bajad, S. U.; Penta, K.; Yang, W.; Hallam, T. J.; Thanos, C. D.; Sato, A. K. Production of Site-Specific Antibody-Drug Conjugates Using Optimized Non-Natural Amino Acids in a Cell-Free Expression System. *Bioconjug. Chem.* **2014**, *25*(2), 351–361.
63. VanBrunt, M. P.; Shanebeck, K.; Caldwell, Z.; Johnson, J.; Thompson, P.; Martin, T.; Dong, H.; Li, G.; Xu, H.; D'Hooge, F.; Masterson, L.; Bariola, P.; Tiberghien, A.; Ezeadi, E.; Williams, D. G.; Hartley, J. A.; Howard, P. W.; Grabstein, K. H.; Bowen, M. A.; Marelli, M. Genetically Encoded Azide Containing Amino Acid in Mammalian Cells Enables Site-Specific Antibody-Drug Conjugates Using Click Cycloaddition Chemistry. *Bioconjug. Chem.* **2015**, *26*(11), 2249–2260.
64. Zhou, Q.; Gui, J.; Pan, C. M.; Albone, E.; Cheng, X.; Suh, E. M.; Grasso, L.; Ishihara, Y.; Baran, P. S. Bioconjugation by Native Chemical Tagging of C-H Bonds. *J. Am. Chem. Soc.* **2013**, *135*(35), 12994–12997.
65. Link, A. J.; Mock, M. L.; Tirrell, D. A. Non-canonical Amino Acids in Protein Engineering. *Curr. Opin. Biotechnol.* **2003**, *14*(6), 603–609.
66. Chin, J. W.; Cropp, T. A.; Anderson, J. C.; Mukherji, M.; Zhang, Z.; Schultz, P. G. An Expanded Eukaryotic Genetic Code. *Science* **2003**, *301*(5635), 964–967.
67. Koniev, O.; Leriche, G.; Nothisen, M.; Remy, J. S.; Strub, J. M.; Schaeffer-Reiss, C.; Van Dorsselaer, A.; Baati, R.; Wagner, A. Selective Irreversible Chemical Tagging of Cysteine With 3-Arylpropiolonitriles. *Bioconjug. Chem.* **2014**, *25*(2), 202–206.
68. Knall, A. C.; Slugovc, C., Inverse Electron Demand Diels-Alder (iEDDA)-Initiated Conjugation: A (High) Potential Click Chemistry Scheme. Chem. Soc. Rev. 42 (12), 5131–5142.
69. Lang, K.; Davis, L.; Torres-Kolbus, J.; Chou, C.; Deiters, A.; Chin, J. W. Genetically Encoded Norbornene Directs Site-Specific Cellular Protein Labelling via a Rapid Bioorthogonal Reaction. *Nat. Chem.* **2012**, *4*(4), 298–304.
70. Balan, S.; Choi, J. W.; Godwin, A.; Teo, I.; Laborde, C. M.; Heidelberger, S.; Zloh, M.; Shaunak, S.; Brocchini, S. Site-Specific PEGylation of Protein Disulfide Bonds Using a Three-Carbon Bridge. *Bioconjug. Chem.* **2007**, *18*(1), 61–76.
71. Badescu, G.; Bryant, P.; Bird, M.; Henseleit, K.; Swierkosz, J.; Parekh, V.; Tommasi, R.; Pawlisz, E.; Jurlewicz, K.; Farys, M.; Camper, N.; Sheng, X.; Fisher, M.; Grygorash, R.; Kyle, A.; Abhilash, A.; Frigerio, M.; Edwards, J.; Godwin, A. Bridging Disulfides for Stable and Defined Antibody Drug Conjugates. *Bioconjug. Chem.* **2014**, *25*(6), 1124–1136.

72. Maruani, A.; Smith, M. E.; Miranda, E.; Chester, K. A.; Chudasama, V.; Caddick, S. A Plug-and-Play Approach to Antibody-Based Therapeutics via a Chemoselective Dual Click Strategy. *Nat. Commun.* **2015**, *6*, 6645.
73. Ducry, L.; Stump, B. Antibody-Drug Conjugates: Linking Cytotoxic Payloads to Monoclonal Antibodies. *Bioconjug. Chem.* **2010**, *21*(1), 5–13.
74. Sapra, P.; Hooper, A. T.; O'Donnell, C. J.; Gerber, H. P. Investigational Antibody Drug Conjugates for Solid Tumors. *Expert Opin. Investig. Drugs* **2011**, *20*(8), 1131–1149.
75. Saito, G.; Swanson, J. A.; Lee, K.-D. Drug Delivery Strategy Utilizing Conjugation via Reversible Disulfide Linkages: Role and Site of Cellular Reducing Activities. *Adv. Drug Deliv. Rev.* **2003**, *55*(2), 199–215.
76. Erickson, H. K.; Park, P. U.; Widdison, W. C.; Kovtun, Y. V.; Garrett, L. M.; Hoffman, K.; Lutz, R. J.; Goldmacher, V. S.; Blattler, W. A. Antibody-Maytansinoid Conjugates Are Activated in Targeted Cancer Cells by Lysosomal Degradation and Linker-Dependent Intracellular Processing. *Cancer Res.* **2006**, *66*(8), 4426–4433.
77. Gondi, C. S.; Rao, J. S. Cathepsin B as a Cancer Target. *Expert Opin. Ther. Targets* **2013**, *17*(3), 281–291.
78. Dubowchik, G. M.; Firestone, R. A.; Padilla, L.; Willner, D.; Hofstead, S. J.; Mosure, K.; Knipe, J. O.; Lasch, S. J.; Trail, P. A. Cathepsin B-Labile Dipeptide Linkers for Lysosomal Release of Doxorubicin From Internalizing Immunoconjugates: Model Studies of Enzymatic Drug Release and Antigen-Specific in vitro Anticancer Activity. *Bioconjug. Chem.* **2002**, *13*(4), 855–869.
79. Dubowchik, G. M.; Radia, S.; Mastalerz, H.; Walker, M. A.; Firestone, R. A.; Dalton King, H.; Hofstead, S. J.; Willner, D.; Lasch, S. J.; Trail, P. A. Doxorubicin Immunoconjugates Containing Bivalent, Lysosomally-Cleavable Dipeptide Linkages. *Bioorg. Med. Chem. Lett.* **2002**, *12*(11), 1529–1532.
80. Doronina, S. O.; Toki, B. E.; Torgov, M. Y.; Mendelsohn, B. A.; Cerveny, C. G.; Chace, D. F.; DeBlanc, R. L.; Gearing, R. P.; Bovee, T. D.; Siegall, C. B.; Francisco, J. A.; Wahl, A. F.; Meyer, D. L.; Senter, P. D. Development of Potent Monoclonal Antibody Auristatin Conjugates for Cancer Therapy. *Nat. Biotechnol.* **2003**, *21*(7), 778–784.
81. Burke, P. J.; Senter, P. D.; Meyer, D. W.; Miyamoto, J. B.; Anderson, M.; Toki, B. E.; Manikumar, G.; Wani, M. C.; Kroll, D. J.; Jeffrey, S. C. Design, Synthesis, and Biological Evaluation of Antibody-Drug Conjugates Comprised of Potent Camptothecin Analogues. *Bioconjug. Chem.* **2009**, *20*(6), 1242–1250.
82. Pettit, G. R.; Singh, S. B.; Hogan, F.; Lloyd-Williams, P.; Herald, D. L.; Burkett, D. D.; Clewlow, P. J. Antineoplastic Agents. Part 189. The Absolute Configuration and Synthesis of Natural (-)-Dolastatin 10. *J. Am. Chem. Soc.* **1989**, *111*(14), 5463–5465.
83. Bai, R.; Pettit, G. R.; Hamel, E. Structure-Activity Studies With Chiral Isomers and With Segments of the Antimitotic Marine Peptide Dolastatin 10. *Biochem. Pharmacol.* **1990**, *40*(8), 1859–1864.
84. Miyazaki, K.; Kobayashi, M.; Natsume, T.; Gondo, M.; Mikami, T.; Sakakibara, K.; Tsukagoshi, S. Synthesis and Antitumor Activity of Novel Dolastatin 10 Analogs. *Chem. Pharm. Bull.* **1995**, *43*(10), 1706–1718.
85. Senter, P. D.; Sievers, E. L. The Discovery and Development of Brentuximab Vedotin for Use in Relapsed Hodgkin Lymphoma and Systemic Anaplastic Large Cell Lymphoma. *Nat. Biotechnol.* **2012**, *30*(7), 631–637.
86. Kupchan, S. M.; Komoda, Y.; Court, W. A.; Thomas, G. J.; Smith, R. M.; Karim, A.; Gilmore, C. J.; Haltiwanger, R. C.; Bryan, R. F. Tumor Inhibitors. LXXIII. Maytansine, a Novel Antileukemic Ansa Macrolide From Maytenus ovatus. *J. Am. Chem. Soc.* **1972**, *94*(4), 1354–1356.

87. Kupchan, S. M.; Komoda, Y.; Branfman, A. R.; Sneden, A. T.; Court, W. A.; Thomas, G. J.; Hintz, H. P. J.; Smith, R. M.; Karim, A. Tumor Inhibitors. 122. The Maytansinoids. Isolation, Structural Elucidation, and Chemical Interrelation of Novel Ansa Macrolides. *J. Org. Chem.* **1977**, *42*(14), 2349–2357.
88. Bhattacharyya, B.; Wolff, J. Maytansine Binding to the Vinblastine Sites of Tubulin. *FEBS Lett.* **1977**, *75*(1–2), 159–162.
89. Mandelbaum-Shavit, F.; Wolpert-DeFilippes, M. K.; Johns, D. G. Binding of Maytansine to Rat Brain Tubulin. *Biochem. Biophys. Res. Commun.* **1976**, *72*(1), 47–54.
90. Issell, B. F.; Crooke, S. T. Maytansine. *Cancer Treat. Rev.* **1978**, *5*(4), 199–207.
91. Cabanillas, F.; Bodey, G. P.; Burgess, M. A.; Freireich, E. J. Results of a Phase II Study of Maytansine in Patients With Breast Carcinoma and Melanoma. *Cancer Treat Rep.* **1979**, *63*(3), 507–509.
92. Chari, R. V. J.; Martell, B. A.; Gross, J. L.; Cook, S. B.; Shah, S. A.; Blättler, W. A.; McKenzie, S. J.; Goldmacher, V. S. Immunoconjugates Containing Novel Maytansinoids: Promising Anticancer Drugs. *Cancer Res.* **1992**, *52*, 127–131.
93. Lambert, J. M.; Chari, R. V. J. Ado-Trastuzumab Emtansine (T-DM1): An Antibody–Drug Conjugate (ADC) for HER2-Positive Breast Cancer. *J. Med. Chem.* **2014**, *57*(16), 6949–6964.
94. Sasse, F.; Steinmetz, H.; Heil, J.; Hofle, G.; Reichenbach, H. Tubulysins, New Cytostatic Peptides From Myxobacteria Acting on Microtubuli. Production, Isolation, Physico-Chemical and Biological Properties. *J. Antibiot.* **2000**, *53*(9), 879–885.
95. Li, J. Y.; Perry, S. R.; Muniz-Medina, V.; Wang, X.; Wetzel, L. K.; Rebelatto, M. C.; Hinrichs, M. J. M.; Bezabeh, B. Z.; Fleming, R. L.; Dimasi, N.; Feng, H.; Toader, D.; Yuan, A. Q.; Xu, L.; Lin, J.; Gao, C.; Wu, H.; Dixit, R.; Osbourn, J. K.; Coats, S. R. A Biparatopic HER2-Targeting Antibody-Drug Conjugate Induces Tumor Regression in Primary Models Refractory to or Ineligible for HER2-Targeted Therapy. *Cancer Cell* **2016**, *29*(1), 117–129.
96. Schwartz, R. E.; Hirsch, C. F.; Sesin, D. F.; Flor, J. E.; Chartrain, M.; Fromtling, R. E.; Harris, G. H.; Salvatore, M. J.; Liesch, J. M.; Yudin, K. Pharmaceuticals From Cultured Algae. *J. Ind. Microbiol. Biotechnol.* **1990**, *5*(2), 113–123.
97. Smith, C. D.; Zhang, X.; Mooberry, S. L.; Patterson, G. M. L.; Moore, R. E. Cryptophycin: A New Antimicrotubule Agent Active Against Drug-Resistant Cells. *Cancer Res.* **1994**, *54*(14), 3779.
98. Sessa, C.; Weigang-Köhler, K.; Pagani, O.; Greim, G.; Mora, O.; De Pas, T.; Burgess, M.; Weimer, I.; Johnson, R. Phase I and Pharmacological Studies of the Cryptophycin Analogue LY355703 Administered on a Single Intermittent or Weekly Schedule. *Eur. J. Cancer* **2002**, *38*(18), 2388–2396.
99. Verma, V. A.; Pillow, T. H.; DePalatis, L.; Li, G.; Phillips, G. L.; Polson, A. G.; Raab, H. E.; Spencer, S.; Zheng, B. The Cryptophycins as Potent Payloads for Antibody Drug Conjugates. *Bioorg. Med. Chem. Lett.* **2015**, *25*(4), 864–868.
100. Hurley, L. H. Pyrrolo(1,4)Benzodiazepine Antitumor Antibiotics. Comparative Aspects of Anthramycin, Tomaymycin and Sibiromycin. *J. Antibiot.* **1977**, *30*(5), 349–370.
101. Hertzberg, R. P.; Hecht, S. M.; Reynolds, V. L.; Molineux, I. J.; Hurley, L. H. DNA Sequence Specificity of the Pyrrolo[1,4]Benzodiazepine Antitumor Antibiotics. Methidiumpropyl-EDTA-Iron(II) Footprinting Analysis of DNA Binding Sites for Anthramycin and Related Drugs. *Biochemistry* **1986**, *25*(6), 1249–1258.
102. Antonow, D.; Kaliszczak, M.; Kang, G.-D.; Coffils, M.; Tiberghien, A. C.; Cooper, N.; Barata, T.; Heidelberger, S.; James, C. H.; Zloh, M.; Jenkins, T. C.; Reszka, A. P.; Neidle, S.; Guichard, S. M.; Jodrell, D. I.; Hartley, J. A.; Howard, P. W.; Thurston, D. E. Structure–Activity Relationships of Monomeric C2-Aryl Pyrrolo[2,1-c][1,4]Benzodiazepine (PBD) Antitumor Agents. *J. Med. Chem.* **2010**, *53*(7), 2927–2941.

103. Burger, A. M.; Loadman, P. M.; Thurston, D. E.; Schultz, R.; Fiebig, H. H.; Bibby, M. C. Preclinical Pharmacology of the Pyrrolobenzodiazepine (PBD) Monomer DRH-417 (NSC 709119). *J. Chemother.* **2007**, *19*(1), 66–78.
104. Smellie, M.; Bose, D. S.; Thompson, A. S.; Jenkins, T. C.; Hartley, J. A.; Thurston, D. E. Sequence-Selective Recognition of Duplex DNA Through Covalent Interstrand Cross-Linking: Kinetic and Molecular Modeling Studies With Pyrrolobenzodiazepine Dimers. *Biochemistry* **2003**, *42*(27), 8232–8239.
105. Gregson, S. J.; Howard, P. W.; Thurston, D. E.; Jenkins, T. C.; Kelland, L. R. Synthesis of a Novel C2/C2[Prime or Minute]-Exo Unsaturated Pyrrolobenzodiazepine Cross-Linking Agent With Remarkable DNA Binding Affinity and Cytotoxicity. *Chem. Commun.* **1999**, 797–798.
106. Gregson, S. J.; Howard, P. W.; Gullick, D. R.; Hamaguchi, A.; Corcoran, K. E.; Brooks, N. A.; Hartley, J. A.; Jenkins, T. C.; Patel, S.; Guille, M. J.; Thurston, D. E. Linker Length Modulates DNA Cross-Linking Reactivity and Cytotoxic Potency of C8/C8′ Ether-Linked C2-Exo-Unsaturated Pyrrolo[2,1-c][1,4] Benzodiazepine (PBD) Dimers. *J. Med. Chem.* **2004**, *47*(5), 1161–1174.
107. Tiberghien, A. C.; Levy, J.-N.; Masterson, L. A.; Patel, N. V.; Adams, L. R.; Corbett, S.; Williams, D. G.; Hartley, J. A.; Howard, P. W. Design and Synthesis of Tesirine, a Clinical Antibody–Drug Conjugate Pyrrolobenzodiazepine Dimer Payload. *ACS Med. Chem. Lett.* **2016**, 7(11), 983–987.
108. *Study of ADCT-301 in Patients With Relapsed or Refractory Hodgkin and Non-Hodgkin Lymphoma.* https://www.clinicaltrials.gov/ct2/show/NCT02432235Study.
109. *SC16LD6.5 in Recurrent Small Cell Lung Cancer.* https://clinicaltrials.gov/ct2/show/NCT01901653.
110. Bouchard, H.; Viskov, C.; Garcia-Echeverria, C. Antibody–Drug Conjugates—A New Wave of Cancer Drugs. *Bioorg. Med. Chem. Lett.* **2014**, *24*(23), 5357–5363.
111. Govindan, S. V.; Cardillo, T. M.; Sharkey, R. M.; Tat, F.; Gold, D. V.; Goldenberg, D. M. Milatuzumab-SN-38 Conjugates for the Treatment of CD74+ Cancers. *Mol. Cancer Ther.* **2013**, *12*(6), 968.
112. Hanka, L. J.; Dietz, A.; Gerpheide, S. A.; Kuentzel, S. L.; Martin, D. G. CC-1065 (NSC-298223), a New Antitumor Antibiotic. Production, in vitro Biological Activity, Microbiological Assays and Taxonomy of the Producing Microorganism. *J. Antibiot.* **1978**, *31*(12), 1211–1217.
113. Boger, D. L.; Johnson, D. S. CC-1065 and the Duocarmycins: Understanding Their Biological Function Through Mechanistic Studies. *Angew. Chem. Int. Ed.* **1996**, *35*(13–14), 1438–1474.
114. Martin, D. G.; Biles, C.; Gerpheide, S. A.; Hanka, L. J.; Krueger, W. C.; McGovren, J. P.; Mizsak, S. A.; Neil, G. L.; Stewart, J. C.; Visser, J. CC-1065 (NSC 298223), a Potent New Antitumor Agent Improved Production and Isolation, Characterization and Antitumor Activity. *J. Antibiot.* **1981**, *34*(9), 1119–1125.
115. McGovren, J. P.; Clarke, G. L.; Pratt, E. A.; DeKoning, T. F. Preliminary Toxicity Studies With the DNA-Binding Antibiotic, CC-1065. *J. Antibiot.* **1984**, *37*(1), 63–70.
116. Owonikoko, T. K.; Hussain, A.; Stadler, W. M.; Smith, D. C.; Kluger, H.; Molina, A. M.; Gulati, P.; Shah, A.; Ahlers, C. M.; Cardarelli, P. M.; Cohen, L. J. First-in-Human Multicenter Phase I Study of BMS-936561 (MDX-1203), an Antibody-Drug Conjugate Targeting CD70. *Cancer Chemother. Pharmacol.* **2016**, 77(1), 155–162.
117. Elgersma, R. C.; Coumans, R. G. E.; Huijbregts, T.; Menge, W. M. P. B.; Joosten, J. A. F.; Spijker, H. J.; de Groot, F. M. H.; van der Lee, M. M. C.; Ubink, R.; van den Dobbelsteen, D. J.; Egging, D. F.; Dokter, W. H. A.; Verheijden, G. F. M.; Lemmens, J. M.; Timmers, C. M.; Beusker, P. H. Design, Synthesis, and Evaluation of Linker-Duocarmycin Payloads: Toward Selection of HER2-Targeting Antibody–Drug Conjugate SYD985. *Mol. Pharm.* **2015**, *12*(6), 1813–1835.

118. Zein, N.; Sinha, A. M.; McGahren, W. J.; Ellestad, G. A. Calicheamicin Gamma 1I: An Antitumor Antibiotic That Cleaves Double-Stranded DNA Site Specifically. *Science* **1988**, *240*(4856), 1198.
119. Ricart, A. D. Antibody-Drug Conjugates of Calicheamicin Derivative: Gemtuzumab Ozogamicin and Inotuzumab Ozogamicin. *Clin. Cancer Res.* **2011**, *17*(20), 6417.
120. Petersdorf, S.; Kopecky, K.; Stuart, R. K.; Larson, R. A.; Nevill, T. J.; Stenke, L.; Slovak, M. L.; Tallman, M. S.; Willman, C. L.; Erba, H.; Appelbaum, F. R. Preliminary Results of Southwest Oncology Group Study S0106: An International Intergroup Phase 3 Randomized Trial Comparing the Addition of Gemtuzumab Ozogamicin to Standard Induction Therapy Versus Standard Induction Therapy Followed by a Second Randomization to Post-Consolidation Gemtuzumab Ozogamicin Versus No Additional Therapy for Previously Untreated Acute Myeloid Leukemia. *Blood* **2015**, *114*(22), 790.
121. Burnett, A. K.; Hills, R. K.; Milligan, D.; Kjeldsen, L.; Kell, J.; Russell, N. H.; Yin, J. A. L.; Hunter, A.; Goldstone, A. H.; Wheatley, K. Identification of Patients With Acute Myeloblastic Leukemia Who Benefit From the Addition of Gemtuzumab Ozogamicin: Results of the MRC AML15 Trial. *J. Clin. Oncol.* **2011**, *29*(4), 369–377.
122. Lindell, T. J.; Weinberg, F.; Morris, P. W.; Roeder, R. G.; Rutter, W. J. Specific Inhibition of Nuclear RNA Polymerase II by α-Amanitin. *Science* **1970**, *170*(3956), 447.
123. Moldenhauer, G.; Salnikov, A. V.; Lüttgau, S.; Herr, I.; Anderl, J.; Faulstich, H. Therapeutic Potential of Amanitin-Conjugated Anti-Epithelial Cell Adhesion Molecule Monoclonal Antibody Against Pancreatic Carcinoma. *J. Natl. Cancer Inst.* **2012**, *104*(8), 622–634.
124. Alley, S. C.; Anderson, K. E. Analytical and Bioanalytical Technologies for Characterizing Antibody–Drug Conjugates. *Curr. Opin. Chem. Biol.* **2013**, *17*(3), 406–411.
125. Wakankar, A.; Chen, Y.; Gokarn, Y.; Jacobson, F. S. Analytical Methods for Physicochemical Characterization of Antibody Drug Conjugates. *MAbs* **2011**, *3*(2), 161–172.
126. Peddi, P. F.; Hurvitz, S. A. Ado-Trastuzumab Emtansine (T-DM1) in Human Epidermal Growth Factor Receptor 2 (HER2)-Positive Metastatic Breast Cancer: Latest Evidence and Clinical Potential. *Ther. Adv. Med. Oncol.* **2014**, *6*(5), 202–209.
127. Deng, C.; Pan, B.; O'Connor, O. A. Brentuximab Vedotin. *Clin. Cancer Res.* **2013**, *19*(1), 22–27.
128. Ansell, S. M. Brentuximab Vedotin. *Blood* **2014**, *124*(22), 3197–3200.
129. Cheever, M. A.; Allison, J. P.; Ferris, A. S.; Finn, O. J.; Hastings, B. M.; Hecht, T. T.; Mellman, I.; Prindiville, S. A.; Viner, J. L.; Weiner, L. M.; Matrisian, L. M. The Prioritization of Cancer Antigens: A National Cancer Institute Pilot Project for the Acceleration of Translational Research. *Clin. Cancer Res.* **2009**, *15*(17), 5323–5337.
130. Tijink, B. M.; Buter, J.; de Bree, R.; Giaccone, G.; Lang, M. S.; Staab, A.; Leemans, C. R.; van Dongen, G. A. A Phase I Dose Escalation Study With Anti-CD44v6 Bivatuzumab Mertansine in Patients With Incurable Squamous Cell Carcinoma of the Head and Neck or Esophagus. *Clin. Cancer Res.* **2006**, *12*(20 Pt. 1), 6064–6072.
131. Mackay, C. R.; Terpe, H. J.; Stauder, R.; Marston, W. L.; Stark, H.; Gunthert, U. Expression and Modulation of CD44 Variant Isoforms in Humans. *J. Cell Biol.* **1994**, *124*(1–2), 71–82.
132. Hinrichs, M. J.; Dixit, R. Antibody Drug Conjugates: Nonclinical Safety Considerations. *AAPS J.* **2015**, *17*(5), 1055–1064.
133. Polakis, P. Antibody Drug Conjugates for Cancer Therapy. *Pharmacol. Rev.* **2016**, *68*(1), 3–19.
134. Donaghy, H. Effects of Antibody, Drug and Linker on the Preclinical and Clinical Toxicities of Antibody-Drug Conjugates. *MAbs* **2016**, *8*(4), 659–671.
135. de Goeij, B. E.; Lambert, J. M. New Developments for Antibody-Drug Conjugate-Based Therapeutic Approaches. *Curr. Opin. Immunol.* **2016**, *40*, 14–23.

136. Jumbe, N. L.; Xin, Y.; Leipold, D. D.; Crocker, L.; Dugger, D.; Mai, E.; Sliwkowski, M. X.; Fielder, P. J.; Tibbitts, J. Modeling the Efficacy of Trastuzumab-DM1, an Antibody Drug Conjugate, in Mice. *J. Pharmacokinet. Pharmacodyn.* **2010**, *37*(3), 221–242.
137. Francisco, J. A.; Cerveny, C. G.; Meyer, D. L.; Mixan, B. J.; Klussman, K.; Chace, D. F.; Rejniak, S. X.; Gordon, K. A.; DeBlanc, R.; Toki, B. E.; Law, C. L.; Doronina, S. O.; Siegall, C. B.; Senter, P. D.; Wahl, A. F. cAC10-vcMMAE, an Anti-CD30-Monomethyl Auristatin E Conjugate With Potent and Selective Antitumor Activity. *Blood* **2003**, *102*(4), 1458–1465.
138. Rossi, A. Rovalpituzumab Tesirine and DLL3: A New Challenge for Small-Cell Lung Cancer. *Lancet Oncol.* **2017**, *18*(1), 3–5.
139. Rudin, C. M.; Pietanza, M. C.; Bauer, T. M.; Ready, N.; Morgensztern, D.; Glisson, B. S.; Byers, L. A.; Johnson, M. L.; Burris, H. A., 3rd; Robert, F.; Han, T. H.; Bheddah, S.; Theiss, N.; Watson, S.; Mathur, D.; Vennapusa, B.; Zayed, H.; Lally, S.; Strickland, D. K.; Govindan, R.; Dylla, S. J.; Peng, S. L.; Spigel, D. R. Rovalpituzumab Tesirine, a DLL3-Targeted Antibody-Drug Conjugate, in Recurrent Small-Cell Lung Cancer: A First-in-Human, First-in-Class, Open-Label, Phase 1 Study. *Lancet Oncol.* **2017**, *18*(1), 42–51.
140. Saunders, L. R.; Bankovich, A. J.; Anderson, W. C.; Aujay, M. A.; Bheddah, S.; Black, K.; Desai, R.; Escarpe, P. A.; Hampl, J.; Laysang, A.; Liu, D.; Lopez-Molina, J.; Milton, M.; Park, A.; Pysz, M. A.; Shao, H.; Slingerland, B.; Torgov, M.; Williams, S. A.; Foord, O.; Howard, P.; Jassem, J.; Badzio, A.; Czapiewski, P.; Harpole, D. H.; Dowlati, A.; Massion, P. P.; Travis, W. D.; Pietanza, M. C.; Poirier, J. T.; Rudin, C. M.; Stull, R. A.; Dylla, S. J. A DLL3-Targeted Antibody-Drug Conjugate Eradicates High-Grade Pulmonary Neuroendocrine Tumor-Initiating Cells In Vivo. *Sci. Transl. Med.* **2015**, 7(302), 302–136.

CHAPTER THIRTEEN

Antibody-Recruiting Small Molecules: Synthetic Constructs as Immunotherapeutics

Patrick J. McEnaney*[,2], Christopher G. Parker†[,2], Andrew X. Zhang‡[,1,2]

*The Scripps Research Institute, Jupiter, FL, United States
†The Skaggs Institute for Chemical Biology, The Scripps Research Institute, La Jolla, CA, United States
‡Discovery Biology, Discovery Sciences, IMED Biotech Unit, AstraZeneca, Waltham, MA, United States
[1]Corresponding author: e-mail address: andrew.zhang@astrazeneca.com

Contents

[2] All authors contributed equally to this work.

Annual Reports in Medicinal Chemistry, Volume 50
ISSN 0065-7743
https://doi.org/10.1016/bs.armc.2017.08.008

1. INTRODUCTION TO THE IMMUNE SYSTEM, IMMUNOTHERAPY, AND ANTIBODY-RECRUITING SMALL MOLECULES

1.1 Harnessing the Human Immune Response to Treating Diseases

The human immune system is a sophisticated network of biomolecules and cells responsible for responding to and defending our body against internal and external pathogenic agents, including bacterial and viral infections, as well as cancer. There are two types of immune response: innate immunity and adaptive immunity.[1] Innate immunity is generated through the recognition of pathogen-associated molecular patterns (PAMPs), such as glycolipids or nucleic acids, by pattern recognition receptors on the surface of macrophages, neutrophils, and dendritic cells.[1,2] The onset of innate immunity is rapid, but the response is not pathogen specific and subsequent exposures to the same PAMPs result in the same level of immune response. The adaptive immune response is generated against a specific antigen or small-molecule hapten based on recognition of chemical epitopes and functionalities.[1,3] Antibodies make up a critical component of the adaptive immune response through their ability to recognize surface exposed chemical epitopes on foreign antigens with high specificity and affinity. The humoral arm of the adaptive immune system generates antibody repertoires through the maturation and hypermutation of B cell receptors after exposure to foreign pathogenic epitopes along with the appropriate costimulatory signals generated during the innate response.[4] The structure of an antibody can be divided into two parts (Fig. 1), the bivalent fragment antigen binding (Fab) and the fragment crystallizable (Fc). Specificity is conferred through the Fab portion of the antibody that has undergone hypermutation for affinity maturation against the immunogenic epitopes. Upon binding to the antigen or small-molecule hapten, circulating antibodies direct effector

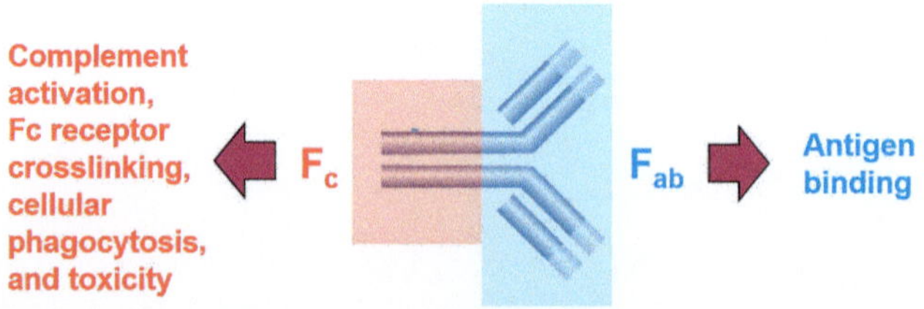

Fig. 1 Anatomy of an antibody and functions of the F_{ab} and F_c in pathogen clearance.

functions, including opsonization, pathogen neutralization, complement-dependent cytotoxicity (CDC),[5] antibody-dependent cellular cytotoxicity (ADCC), and phagocytosis (ADCP), through the Fc portion,[1,6] which is constant across a particular antibody isotype. The initial antibody-mediated immune response can then engage the cellular arm of adaptive immunity in a variety of mechanisms, including activating T cells against additional antigens,[7] which can lead to direct cytotoxicity on the pathogenic cell and/or activation of other immune cells.[8] The specificity and memory of an antibody-mediated immune response are maintained after clearance of the pathogen by circulating memory B-cells, such that responses to subsequent exposures of the pathogen presenting the same antigen occur more rapidly and elicit a larger response.[1,9,10]

Immunotherapy is the method of treating, or preventing, diseases by harnessing the capabilities of our immune system. The first lauded example modern immunotherapy was Edward Jenner's use of the cowpox virus to impart immunity against the much more dangerous smallpox virus, a discovery that led to the eradication of one of the most deadly diseases in history and laid the groundwork for vaccination as a powerful therapeutic strategy.[11] This success demonstrated that our immune system is capable of eliciting a response against foreign antigens and retaining memory against that antigen, such that subsequent challenge will generate a strong enough immune response to suppress, or even eradicate, the pathogen. Subsequent studies have shown that antibodies in the serum represent the key component responsible for the adaptive immune response.[12–14] In the 1970s, Milstein and Kohler reported the ability to manipulate the process of antibody production by B cells to engineer antibodies to recognize a specific antigen of interest.[15] This allowed the generation of monoclonal antibodies (mAbs), a breakthrough in immunotherapy representing a new modality for treating serious diseases. These mAbs provided a new hope for modern medicine for their potential to target a specific pathogen associated antigen, while sparing nondiseased cells and thereby reducing unwanted side effects normally associated with less specific chemotherapeutics of that era.[16,17] mAb-based therapeutics have become a mainstay in therapeutics in the past 30 years, starting with Muromonab-CD3, the first mAb drug, which was approved by the FDA for treating transplant rejection by targeting the CD3 receptor on T cells in 1986.[18] By the end of 2014, 47 mAbs have been approved with an average of four new approvals per year.[19] As of 2016, 52 mAbs are undergoing Phase 3 clinical trials and collectively, the therapeutic antibody market is projected to exceed \$125 billion by 2020.[20] Owing to the success of mAbs, variants, and derivatives, such as bispecific antibodies, which are antibodies engineered to recognize two epitopes,[21,22] or more simplified single domain antibodies fragments[23–26] and antibody drug conjugates (ADCs), have been extensively researched and utilized as promising therapeutics (reviewed in Chapter "Antibody-drug conjugates with this issue of the annual reviews in medicinal chemistry" by Gao et al.).[27,28]

While mAbs have demonstrated tremendous success in clinical applications, they are associated with inherent challenges associated with high molecular weight, peptidic biomolecules. These challenges have, at times, compromised their efficacy and increased the incidence of serious side effects due to the inherent immunogenicity of the antibody.[17] While advancements, including the humanization of antibodies, have drastically improved safety profiles, adverse events related to the physical infusion, as well as on-target toxicity still persist.[29] Notably, the success of angiogenesis inhibitor Avastin® has been significantly stymied by side effects such as prolonged bleeding and wound healing.[30] Unlike their biologic counterparts, small molecules are often orally available and nonimmunogenic.[31] Thus small molecules that possess the capacity to redirect, or recapitulate, the functions of antibodies have the potential to overcome many of these obstacles and see significant clinical impact.[32,33] In this chapter, we present the concept, development, and application of antibody-recruiting small molecules (ARMs), which are synthetic, bifunctional small molecules capable of directing an endogenous, antibody-mediated immune response to a pathogenic target of interest.[34] While protein engineering and synthetic biology-based methods[35] for immune modulation and pathogen targeting have been highly impactful,[36] we will focus our chapter primarily on the use of synthetic constructs, such as ARMs, to achieve the same goals.

1.2 Introduction and Evolution of ARMs

ARMs are binfunctional molecules composed of two main components (Fig. 2), (1) a target-binding terminus (TBT) that recognizes a pathogen-associated cell surface protein and (2) an antibody-binding terminus

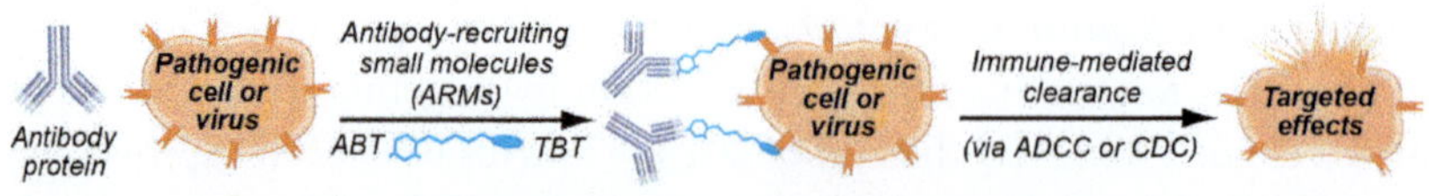

Fig. 2 Structure and function of ARMs. *Figure adapted with permission from McEnaney, P. J.; Parker, C.G.; Zhang, A.X.; Spiegel, D.A.* ACS Chem. Biol. ***2012,*** 7 *(7), 1139–1151. https://doi.org/10.1021/cb300119g. Copyright 2012 American Chemical Society.*

(ABT) that can bind antihapten antibodies. These two moieties enable ARMs to form a molecular bridge between a target protein and antihapten antibodies, and this ternary complex will, in turn, recruit downstream effector responses such as CDC, ADCC, and ADCP. By design, ARMs maintain the specificity for a target protein while bypassing many of the previously mentioned side effects of mAbs. The concept and application of ARMs have been thoroughly discussed in recent reviews and perspectives.[34] In this chapter, we will summarize landmark innovations in this field, discuss recent advances, including many novel synthetic constructs beyond, but inspired by ARMs, and lastly, provide commentary about future goals and challenges faced by ARM-based strategies.

The manipulation of antibody-mediated immune responses with rationally designed molecules is not a new concept. In the 1970s, researchers demonstrated that decoration of the surface of an otherwise nonpathogenic cell with dinitrophenyl (DNP) or trinitrophenyl (TNP) moiety elicited CDC[37] and ADCC[38,39] in the presence of anti-DNP or anti-TNP antiserum, respectively. While the labeling was not specific to a protein on the cell surface, this work demonstrated that simple chemical functionalities, such as nitroarenes, can redirect antibodies to the surface of cells that otherwise would not have been recognized. In the late 1980s, a series of publications disclosed bifunctional protein chimers consisting of the Fc portion of IgG or IgM fused with CD4, the target of the HIV protein gp120.[40–42] This fusion protein was able to redirect complement proteins and immune effector cells to act on HIV-infected cells, and represented the first examples of using an engineered bifunctional molecule to induce an immune response specifically to pathogenic cells. Subsequently, researchers in Peter Schultz's group demonstrated that the CD4 labeled with DNP could form a ternary complex with gp120 and anti-DNP antibodies, and as a result, recruit C1q, the first component of the complement pathway.[43]

1.3 Features of ABTs and TBTs

Antibody motifs used in ARMs have fallen into two general classes: endogenous and nonendogenous (Fig. 3). The former class comprises antibodies already present in the bloodstream with recognition properties that are highly conserved among human beings, including nitroarenes and carbohydrates, while the latter must be elicited by exposure of an individual to appropriate immunizing hapten, such as fluorescein. Nitroarene antibodies (e.g., anti-DNP antibodies) have been found to be present in sera across all

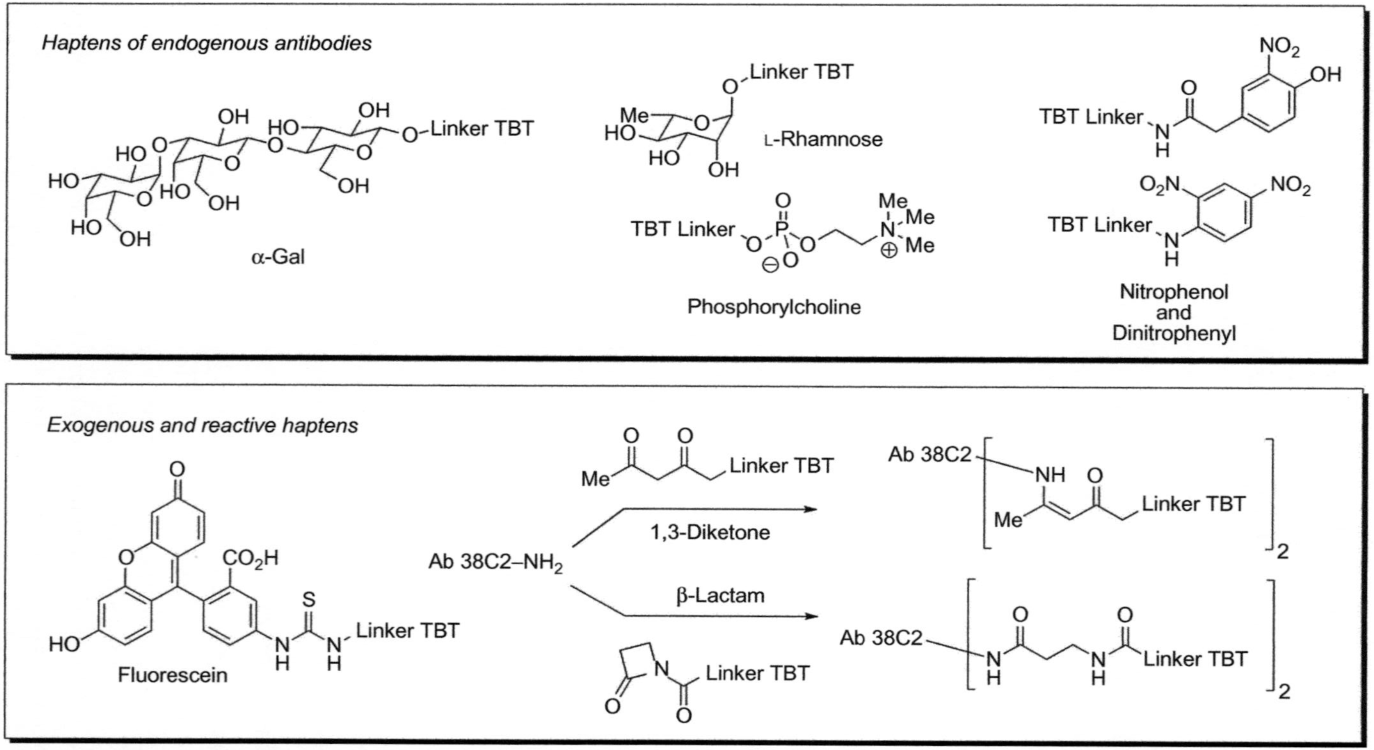

Fig. 3 Structures of haptens used for ABTs.

genders, ethnicities, and ages among Americans.[44,45] While the reason for their presence is not fully understood, possible explanations include environmental exposure to nitroarenes, as well as exposures to nitroarenes generated during the cooking process.[46] Additionally, the release of NO radicals during immune responses[47] open up the possibility of nitrating an aromatic amino acid side chain, thereby generating a "nonself" motif on endogenous proteins. Another well-studied class of naturally occurring haptens is carbohydrates, the most well known being the ABO antigens found on red blood cells. Another is galactosyl-(1–3)-galactose (α-Gal) moiety is found on membrane proteins and lipids of nonprimate mammals.[48,49] Over the course of evolution, old world primates and subsequently more advanced organisms lost this epitope due to the loss of the α-1,3-galactotransferase activity,[50] thus leading to these moieties being recognized as "nonself" and development of endogenous antibodies against α-Gal.[51] More recently, rhamnose has become another popular carbohydrate hapten used in ARMs due to its relatively simple structure compared to α-Gal.[52–54]

An alternative ABT strategy utilizes antibodies that are produced upon exposure to an exogenous hapten. Fluorescein is a commonly used hapten against which antibodies have been readily generated.[55–59] Despite not being endogenous, antifluorescein antibodies generated through immunization are fully capable of mediating antibody-dependent immune effector responses. In addition to functioning as ABTs to redirect immune effector responses, fluorescein-based ARMs can be simultaneously used for fluorescence imaging of tumors. In an alternative strategy that has proven efficacious, Barbas and coworkers exploit the unique properties of a class of catalytic aldolase antibodies, often referred to in this context as "chemically programmed" (cpAb) antibodies. Specific or "programmed" cpAbs can be generated by reacting the monoclonal catalytic aldolase antibody 38C2 with electrophilic groups such as the 1,3-diketone or β-lactams to form a stable, covalent bond. Conjugation to small molecules in this fashion maintains the ability of 38C2 to redirect immune effector functions such as ADCC.[60–64] These antibodies can also be generated in vivo through reactive immunization with a 1,3-diketone-KLH conjugate to function as an aldolase enzyme.[65] The rest of the chapter will review the applications of these different ABTs in greater depth.

In addition to the employment and exploration of various structures and classes of haptens as ABTs, it has been shown that multivalent display can have profound effects ARM activity. Multivalency plays an important role in the activation of immune functions, particularly with respect to antigen presentation, where a multivalent interaction can increase the affinity from otherwise weak interactions[66] and promote receptor clustering (immunological synapse formation), a critical step in the signaling responsible for effector responses.[67–69] The Kiessling group was one the first to explore the effects of multivalent display of haptens in ARM-based strategies.[70,71] Here, an ARM was generated with a TBT targeting the αvβ3 integrin, which belongs to a family of cell surface receptors that are overexpressed on cancer cells, and correlates with tumor invasion and metastasis,[72] and the trisaccharide α-Gal ABT. Although the affinity between α-Gal and endogenous α-Gal antibodies is relatively weak ($K_d \sim 10^{-5}$ M), the multivalent display of α-Gal on the cell surface allowed for significant anti-α-Gal antibody recruiting and subsequent CDC. However, due to this relatively weak affinity it was found that there was a threshold of αvβ3 integrin expression required for antibody recruiting and subsequent CDC, allowing clearance of cells highly overexpressing the integrin. Interestingly, a similar bifunctional construct targeting αvβ3 but conjugated to doxorubicin killed cells indiscriminately, regardless of the levels of integrin expression. This result demonstrated a distinct feature of ARMs, that through multivalent display effects, can be more selective than other targeted cytotoxic drug conjugates in targeting cells with an overexpressed receptor, while sparing cells that express a normal amount of the same receptor. In subsequent work exploring the role of multivalency, O'Reilly and coworkers constructed a bifunctional molecule containing a glycan sequence known to bind to CD22, a cell surface regulator of B cell signaling, and a target for B cell lymphoma therapy,[73] and incorporated an *o*-nitro phenol group as the ABT. This nitroarene motif binds endogenous antinitrophenol (anti-NP) antibodies with a 1 μM K_d.[74] In the presence of decavalent anti-NP IgM, a multivalent display of the CD22 binding glycan was achieved, resulting in a two-order magnitude K_d enhancement, vs. in the absence of anti-NP IgM. In follow up studies in an CD22-targeted ARM, Cui and coworkers demonstrated that by increasing the valency at the TBT, by displaying it on a polymeric scaffold, their bifunctional molecule recruited 100-fold higher levels of anti-NP IgM to targeted cells.[75] Although the employment of multivalent hapten displays have shown to improve the activity of ARMs in various cases, it still remains unclear as to whether confounding issues, such as multivalent-induced antibody aggregation could lead to nonspecific immune response initiation in the absence of cellular targets. Certainly these studies warrant further investigation into the use of ARMs with multivalent displays in more complex in vivo settings.

2. SELECTED APPLICATION OF ARMs TO CANCER AND VIRAL INFECTIONS

Despite being a relatively new strategy, there are several notable successes of both in vitro and in vivo applications of ARMs, raising hopes for its potential clinical applicability. We invite our readers to read the comprehensive review of earlier examples of ARMs in disease relevant settings.[34] Here, we intend to highlight some a few landmark advances of ARMs, particularly those demonstrating in vivo efficacy in disease-relevant settings and as well as novel applications.

2.1 In Vivo Application of ARMs in Oncology

The Barbas lab at the Scripps Research Institute has been pioneers in the development of small molecules to modulate immune response, particularly through their development of the "chemically programmed" antibodies (cpAb).[76] cpAbs have two potential pharmacological mechanisms: immune modulation and/or prolonging the serum half life of the small-molecule component, both of which will be highlighted. In preliminary studies, Rader and coworkers generated cp38C2 by conjugating mAb 38C2 to SCS873 (Fig. 4, Compound **1**), a bifunctional molecule containing an αvβ3 integrin targeting functionality (mimicking the RGD motif known to target integrins[77]) and a 1,3-diketone.[62] It was found that cp38C2 inhibited the proliferation and migration of αvβ3 and αvβ5 human melanoma cells, and also dramatically increased the serum half life of SCS-873 from 15 min to 3 days (with unconjugated 38C2 possessing a half life of 4 days). Furthermore, cp38C2 demonstrated remarkable enhancement of the in vivo efficacy of **1** through inhibition of tumor growth in human melanoma, Karposi's sarcoma, and colon cancer xenograft models; over 50% tumor growth inhibition was observed with cp38C2 treatment compared to treatment with **1** alone. Treatment with cp38C2 in M21 melanoma models (lung metastases) showed a prolonged median survival by over 100 days in a 200-day study compared with unconjugated **1**, even though both **1** and cp38C2 have comparable inhibitory potencies of αvβ3- and αvβ5-expressing human KS cells at 1 μM. No weight loss or changes in behavior was observed upon treatment with cp38C2, suggesting lowered adverse reaction side effects compared to conventional chemotherapeutics. Subsequent mechanism of action studies on M21 cells showed that cp38C2 induces CDC (up to 60% cell killing) and ADCC (up to 30% cell killing).[78]

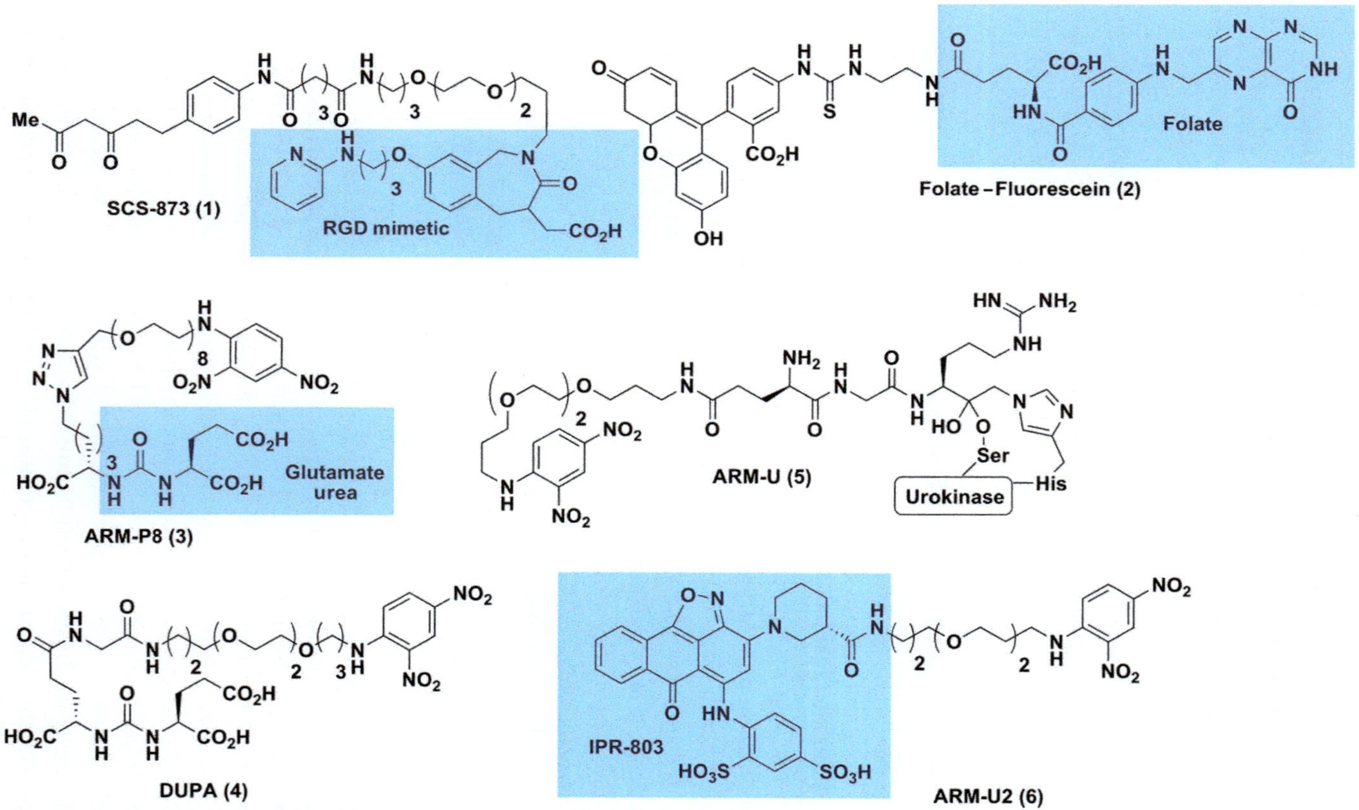

Fig. 4 Structures of anticancer ARM-based strategies that have showed in vivo efficacy.

In an extension of this work Popkov et al.,[65] demonstrated that endogenous catalytic antibodies could be generated through reactive immunization. Here, the immunogen KLH was decorated with the 1,3-diketone functionality, and the resulting conjugate, JW-KLH, was used to immunize either BALB/C (CT26 colon cancer syngeneic), C57BL6 (B16 melanoma syngeneic model), or FcγRIII knockout mice. As a result, the mice developed endogenous catalytic antibodies that can recognize, *and covalently react with*, the 1,3-diketone hapten. In contrast to previous studies, the full cp38C2 was no longer required to inhibit tumor growth in immunized mice; as **1** alone resulted in tumor inhibition in both BALB/C and C57BL6 mice by at least 70%, indicating the formation of the cp38C2 in vivo. Importantly, tumor inhibition was not observed in C57BL6—FcγRIII knockout mice, suggesting that ADCC through FcγRIII-expressing natural killer (NK) cells may be a significant contributor in the efficacy of **1**. Taken together, these studies represent an early demonstration of the power of combining small molecules with immunobiologics. Furthermore, the development of cp38C2 laid important conceptual framework for the expansion of single molecule, bifunctional synthetic constructs that directly engage immune effector cells.

Moving away from reactive functionalities, the Low group at Purdue University generated fully synthetic, nonbiologic, ARMs with low molecular weight haptens as the ABT, and was among the first to demonstrate their in vivo efficacy.[56,79] In landmark studies, Low and colleagues demonstrated that ARMs are not only able to mediate ADCC and ADCP, but that ARMs are able to elicit T-cell-mediated immunity and immunological memory. Lu et al. demonstrated this concept by synthesizing a folate receptor-targeting ARM consisting of a folic acid (FA) linked to fluorescein (Fig. 4, Compound **2**) and testing it in M109 (syngeneic lung cancer model) tumor bearing mice immunized using fluorescein KLH.[56] Treatment with **2** alone yielded a modest enhancement in median survival of less than 25% compared to vehicle control. Importantly, concurrent dosing of **2** with IL-2 enhanced survival by 250%, while treatment with IL-2 and IFN-α enhanced survival by at least 300% and resulted in the mice being cured (complete eradication of the tumor), indicating that increased efficacy can be achieved by providing the appropriate activating immune costimulatory signals. Notably, **2** enhanced median survival with virtually no off-target toxicity. Subsequent mechanism of action studies confirmed the role of ADCP and ADCC on folate receptor-positive cancer cells.[57] Additionally, the efficacy of **2** was heavily dependent on both CD4+ and CD8+ T cells, as mice depleted of CD4+ or CD8+

T cells showed no response to therapy. Notably, in the case of cured mice, upon rechallenging of folate receptor-positive tumors, the mice rejected the tumors even in the absence of **2**. It was further demonstrated that CD8+ T cells isolated from M109-cured mice induced over twice the amount cytotoxicity on M109 cells compared to CD8+ T cells isolated from mice unexposed to M109 tumors, suggesting the role of CD8+ T cells in the memory response. Importantly, the Low group is the first to show the role of T cells in tumor eradication upon ARM treatment and in subsequent long-lasting immunity to the same tumor.

The Spiegel group at Yale University has generalized ARM-based strategies by expanding the approach to multiple disease-related targets, exploring endogenous antibodies repertoires, and also incorporating alternative modalities in place of the ABT to direct different components of the immune response. The Spiegel group has utilized ARMs to target multiple types of cancers[80–83] as well as HIV.[84,85] The first cancer targeting ARM developed by the Spiegel group, ARM-P8 (ARM-targeting prostate cancer, Fig. 4, Compound **3**), contains the DNP hapten at the ABT and a glutamate urea moiety at the TBT for targeting the prostate-specific membrane antigen (PSMA), an overexpressed membrane biomarker of prostate cancer.[86,87] ARM-P8 induced ternary complex formation between PSMA expressing prostate cancer cells and anti-DNP antibodies in the presence of PBMCs resulting in PSMA-specific ADCC. During optimization efforts of ARM–Ps, it was discovered that six oxyethylene units represent the minimum linker length required to simultaneously engage both PSMA and anti-DNP antibodies, and eight oxyethylene (ARM–P8) units make the optimal length. Furthermore, ARM-P8 exhibited an autoinhibitory effect in ADCC: where high concentrations of ARM-P8 decreased the amount of immune effector functions. This phenomenon is a distinct feature of ternary complex formation, as excess bifunctional molecules can compete with either the TBT or ABT ends, ultimately shifting the equilibrium back to binary complexes, and inspired an extensive follow-up effort in characterizing ternary complex equilibria using mathematical models.[88]

Schultz and coworkers followed up on these initial results with in vivo studies in a NOD/SCID engrafted with human lymphocytes (hu-PBL-NOD/SCID) and PSMA+ prostate cancer xenograft.[81] Here, the mice were immunized with DNP-KLH to generate human anti-DNP antibodies. An analog of ARM-P8, DUPA (Fig. 4, Compound **4**), prolonged the mean survival of mice implanted with PSMA+ xenograft only in the presence of DNP immunization. DUPA (formulation 75% PEG300, 25% of 5%

dextrose in distilled water) showed a 90 min serum half life and distributed highly into the kidney, but there was no evidence of cytotoxic lymphocyte infiltration- or immune-mediated toxicity. Immunostaining for immune cell surface markers, however, confirmed the presence of NK cells (CD56, CD16) and T cells (CD3, CD8). Furthermore, tumor tissue stained positively for granzyme B, an enzyme secreted by NK cells during ADCC, as well as for caspase 3, indicative of onset of apoptotic pathways. Xenograft models lacking T cells and NK showed no tumor inhibition in the presence of DNP immunization and DUPA, indicating that likely NK and T-cell-mediated mechanisms are directly responsible for tumor inhibition. Specifically, NK-cell-mediated ADCC provided the onset of cytotoxicity, and this process then increased the immunogenicity of the tumor environment and brought about T cell responses.

The Spiegel group targeted the urokinase plasminogen activator receptor (uPaR)–urokinase (uPa) system, an important modulator of cell signaling and interactions with its extracellular matrix. Aberrant expression and signaling in this pathway have been implicated in numerous forms of cancer, correlating with tumor invasion and metastasis, and previous research has shown uPaR to be an important therapeutic target for metastatic cancer.[89–91] In targeting the uPa–uPaR system, researchers generated ARM-U (Fig. 4, Compound **5**), a protein–small molecule conjugate consisting of uPa labeled with DNP.[92] ARM-U was obtained by through a covalent bond formation between uPa with a bifunctional molecule containing a chloromethylketone targeted to the uPa catalytic domain linked to a DNP moiety. The resulting conjugate binds uPaR and recruits anti-DNP antibodies, and in doing so, mediates ADCC and ADCP on uPaR expressing cells. Following up on these initial results, researchers eliminated the requirement for the formation of the uPa conjugate by utilizing a high affinity small molecule against uPaR. Computationally guided design facilitated the generation of ARM-U2 (Fig. 4, Compound **6**): IPR-803, a known small-molecule inhibitor of the uPa/uPaR protein–protein interaction, was derivatized and linked to a DNP group via a 4-oxyethylene linker.[93] ARM-U2 maintained high potency against uPaR and mediated ADCP on A172 glioblastoma cells. In B16 mouse melanoma allograft models immunized to boost anti-DNP antibody titers demonstrated that ARM-U2 at 20 mpk (mg/kg) inhibited tumor growth comparable with doxorubicin at 1 mpk and increased median survival by 100%. Importantly, in contrast with 1 mpk doxorubicin treatment, ARM-U2 at 100 mpk did not cause weight loss. Taken together, these studies demonstrate the possibility of designing ARMs based on activity in addition to affinity, and more specifically, the potential for the uPa–uPaR system as a relevant cancer target for the ARM strategy.

Thus far, ARMs targeting cancer has generated a significant amount of data that speak to its potential as a therapeutic. ARMs redirect antibodies as hypothesized, and as a result redirect immune effector functions, including ADCC and CDC, and subsequently increase the immunogenicity of the tumor environment in the same manner as mAbs would. These results are promising and critical in taking ARMs from the initial research stages of conception into clinical application.

2.2 Application of ARMs to Viral Infections

The current repertoire of antivirals primarily consists of small-molecule disruptors of viral replication; however, such strategies are impeded by issues related to resistance and efficacy.[64,94–96] Despite the historical and modern therapeutic success of vaccine strategies against viral disease, there still remains a significant number of viruses that have proven obstinate to such approaches (e.g., HIV, herpes simplex viruses, etc.). mAb therapy targeting viruses have had modest therapeutic successes, with only one clinical approved antiviral mAb (Palivizumab targeting RSV).[97] Likely this shortcoming can be attributed to unique challenges found in viral lifecycles, including shielding of epitopes by host-derived carbohydrates, the increased rate of spontaneous mutations,[98] as well as challenges associated with translating the success in animal models to efficacy in humans.[99] ARMs have the potential to overcome some of these immune-evasion mechanisms by converting dynamic structures that are critical to the viral lifecycle, but are generally poorly immunogenic, to synthetically controllable epitopes that are recognized by the human immune system. Despite the numerous successes of ARM-based strategies against cancers, there are relatively few examples of ARMs targeting viruses or virus-infected cells.

The first example of an ARM-based approach against a virus was conducted by Wang and colleagues.[100] Using chemoenzymatic synthesis, these researchers prepared a bifunctional molecule (Fig. 5, Compound **7**) designed to redirect endogenous anti-α-Gal antibodies to HIV. Compound **7** consists of the trisaccharide α-Gal epitope linked to the 36-amino acid gp41 fusion inhibitory peptide, T-20. Here the authors demonstrated that functionalization of T-20 had minimal effects on its fusion inhibitory properties, and that the immobilized bifunctional glycopeptide could bind anti-α-Gal

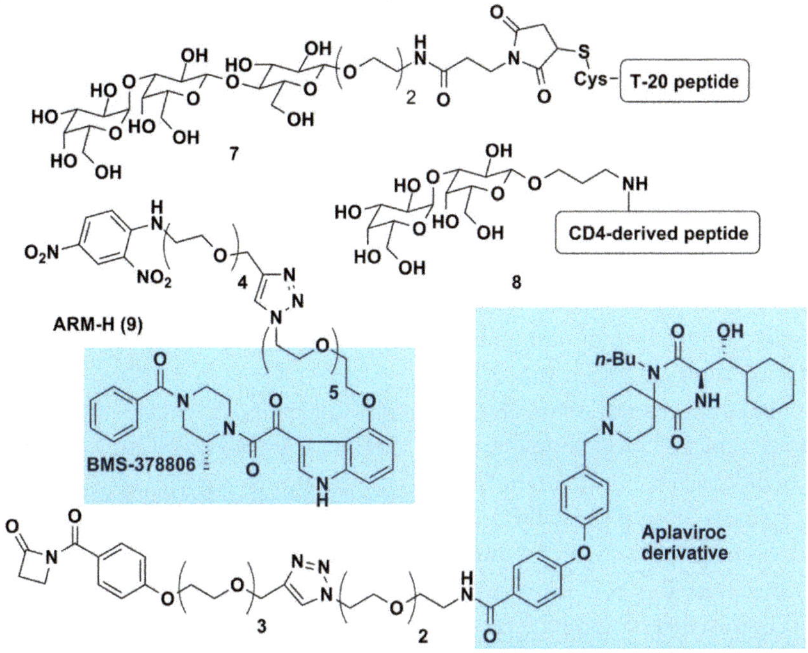

Fig. 5 Structures of antiviral ARMs.

IgG and IgM antibodies from human serum. More recently, Vahlne et al.[101] developed a series bifunctional glycopeptides (Fig. 5, Compound **8**) that are capable of mediating an immune response to HIV-infected cells. These molecules were constructed by chemically linking the α-Gal disaccharide to a series of 15-mer oligopeptides derived from the gp120-binding region of CD4. Using ELISA and immunofluorescence microscopy, the authors demonstrated that these bifunctional glycopeptides could redirect endogenous anti-α-gal antibodies from human serum to both immobilized and cell surface-expressed gp120. Infectivity, syncytia inhibition, and neutralization assays demonstrated that the presence of human antibodies derived from heat-inactivated serum with **8** enhanced the fusion inhibitory activity and were shown to mediate immune responses against chronically HIVIIIB/LAV-infected ACH2 cells in the presence of human serum and isolated NK cells through ADCC, although some analogs proved cytotoxic in the absence of NK cells.

The Spiegel lab group developed the first nonpeptidic ARM designed to target viruses and virus-infected cells, called ARM-H (antibody recruiting molecule targeting HIV, Fig. 5, Compound **9**).[84] This bifunctional small molecule incorporates a derivative of the known fusion inhibitor BMS-378806 at the TBT,[102] along with the DNP motif at the ABT. ARM-H-mediated formation of a ternary complex with anti-DNP antibodies and Env-expressing cells was shown to induce complement-dependent destruction of these cells. Furthermore, ARM-H binds gp120 competitively with CD4, and can also inhibit the entry of HIV virus into human T cells. In follow-up work from the Spiegel lab,[85] medicinal chemistry efforts resulted in a newer generation of ARM-H's with ~1000-fold increased potency, and related studies from the Barbas lab demonstrated that employment of their chemically programmed antibody strategy resulted in HIV-targeting ARMs with enhanced potency.[103]

In an alternate application, the Barbas lab explored the possibility of using the chemically programmable 38C2 aldolase antibody to improve the pharmacological properties of the HIV entry inhibitor Aplaviroc, which has been previously shown to be hepatotoxic.[104] In this work, a bifunctional derivative of Aplaviroc, which binds to the coreceptor CCR5, was designed and synthesized (Fig. 5, Compound **10**). Compound **10** was shown to be capable of chemically engaging the reactive site of 38C2 via a β-lactam moiety while maintaining an antiviral activity similar to that of Aplaviroc. It was further demonstrated that a rhodamine-38C2 analog possessed robust serum stability, suggesting that this antibody conjugation approach has the potential to improve pharmacological properties of small molecules.

Although the ARM-based approach for targeting viral pathogens is still in its infancy, these studies reveal their potential as promising alternatives to available antiviral immunotherapies.

3. STRUCTURE–ACTIVITY RELATIONSHIP STUDIES FOR OPTIMIZING TBTs

The development of ARMs has not only provided a new small-molecule immunomodulatory strategy to treat disease, but also often have been tools themselves in uncovering novel insights into the target protein structure as well as corresponding ligand design. In the process of optimizing PSMA-targeting ARM-Ps, an intriguing trend was observed between linker length and inhibitory potency against PSMA: having 2–4 oxyethylene units between the PSMA targeting glutamate urea and the DNP group afforded up to two orders of magnitude increase in potency compared to the parent glutamate urea's. A combination of biochemical, computational, and structural studies revealed an arene-binding site on PSMA that could accommodate electron poor aromatic rings through mainly a π-stacking interaction with a tryptophan residue.[105] Here, 2–4 oxyethylene units presented the optimal length for accessing this secondary binding interaction without paying entropic penalties. While simple, this interaction increased the potency of the parent compound by 1–2 orders of magnitude. With this insight the development of trifunctional analogs of ARM-P8 have been explored, installing a radioactive ^{125}I atom on the triazole adjacent to the linker to function as a dual therapeutic with immunotherapy and radiotherapy.[106] This approach faced two distinct challenges: the ARM must recruit anti-DNP antibodies and immune effector functions, but then be uptaken into the tumor. Again, the linker length was shown to play a very important role, in that ^{125}I-labeled ARM-P8 showed poor tumor uptake, and tumor uptake could only be improved by removing the oxyethylene linker. However, removing the oxyethylene linker would render the ARM unable to recruit antibodies, in agreement with previous linker SAR demonstrating the importance of linker length in recruiting effector function. This concept, while compelling, faces the challenge of having to balance out two fundamentally opposing forces: the requirement for cell surface presentation of the DNP group, which requires the oxyethylene linker, and the requirement for tumor uptake for the radiotherapeutic effects.

In recent studies focused on the optimization of ARMs targeting the HIV glycoprotein gp120 (ARM-H), computationally guided SAR revealed a previously undescribed hydrophobic cleft near the CD4 binding site on gp120.[85] The first ARM targeting HIV was based on the attachment inhibitor BMS-378806,[84] which was found to be a potent (~32 nM) inhibitor of infection through the blockade of the gp120–CD4 interaction. However, upon attachment of a PEG linker and ABT at the C4 position of the azaindole, a 300-fold drop in activity was observed. This prompted subsequent modeling studies and SAR efforts, which revealed that the parent attachment inhibitor likely interacts with gp120 through two distinct binding modes, and that attachment of the linker and ABT locks the TBT into one of the two modes, and in one of the modes, the ligand can access a secondary hydrophobic binding pocket beneath the well-described CD4 Phe43 binding site. These efforts led to an ARM with a ~1000-fold increase in potency from the first generation and were consistent with subsequent studies proposing a novel mechanism of action for the homologous attachment inhibitor BMS-626529 currently in clinical trials for the treatment of HIV.[107,108]

4. NEW STRUCTURAL MODALITIES

The traditional structure of an ARM has been that of a bifunctional molecule with a well-defined TBT against a cell surface protein, and an ABT that binds antibodies. Newer modalities and strategies have emerged to expand upon the structures, targeting moieties, and effector functions of ARMs. These modalities include different chemical entities used for targeting antihapten antibodies to the surface of cancer cells, which, in many cases, delivered ARMs against new pathogens by targeting new cell surface proteins and pathways. Furthermore, researchers have explored changing the ABT to more general effector-directing terminus, capable of redirecting immune effector cells without the need for antibodies.

4.1 Activity-Based and Metabolic Labeling as a Strategy to Target Bacterial Infections

Small-molecule, natural product based, antibiotics have traditionally been the therapeutic of choice against bacterial infections. While most infections can be fully eradicated with antibiotics, this approach is not without its challenges. Specifically, the emergence of antibiotic resistance, particularly to multiple classes of antibiotics, greatly limit the treatment options, and efficacy for infections; strains of *Staphylococcus aureus* resistant to vancomycin,[109] often used only as a final line of treatment, appeared in the early 2000s.[110] Antibody-based antimicrobial agents have been explored as an alternate method to antibiotics,[111,112] and just last year, Bezlotoxumab[113] was approved to treat *Clostridium difficile* as the first mAb against bacterial infections. Bezlotoxumab functions by neutralizing the *C. difficile* exotoxin TcdB, and inhibiting its interactions with host cells.[114] However, no evidence has yet been put forward to indicate that Bezlotoxumab is able to redirect immune effector responses. Given that bacteria have mechanisms to naturally evade the host immune system,[115] approaches to overcome such immune evasion can enhance the treatment of bacterial infections. ARMs have been developed against bacterial infections since the early 1990s, with the first efforts targeting the mannose receptor of *Escherichia coli* with antiavidin[116,117] or anti-α-Gal antibodies.[118] Subsequent efforts on Gram-positive bacteria utilized polymers displaying vancomycin and fluorescein to recruit antifluorescein antibodies and demonstrated that these constructs can induce opsonization and phagocytosis.[58,59]

New strategies have recently emerged to take advantage of the bacteria's own metabolic pathways to insert ABTs onto the bacterial surface by incorporating simple, synthetic haptens as building blocks within biosynthetic metabolic pathways. Furthermore, by utilizing a critical pathway for survival, these approaches serve as a deterrent to resistance in that failure to incorporate the hapten-labeled building blocks likely would mean bacterial cell death. The Pires group incorporated DNP-labeled D-amino acids into the peptidoglycan of Gram-positive bacteria using the bacteria's transpeptidases, coining the term DART (D-amino acid antibody recruitment therapy).[119–122] Scanning different DNP-labeled D-amino acids found that incorporating D-Lys(DNP)-OH (Fig. 6, Compound **11**) was the most efficient at recruiting anti-DNP antibodies, and labeling of *B. subtilis* doubled the amount of phagocytosis by macrophages in the presence of anti-DNP antibodies compared to background phagocytosis (in the absence of D-Lys (DNP) labeling or in the absence of anti-DNP antibodies).[119] Furthermore, replacement of D-Lys(DNP)-OH with the carboxamide analog, D-Lys (DNP)-NH_2 (Fig. 6, Compound **12**) increased the efficiency of anti-DNP antibody-dependent opsonization across multiple species of Gram-positive bacteria.[120] In subsequent follow-up studies,[121,122] the Pires group utilized a two-step labeling process to incorporate these haptens into the

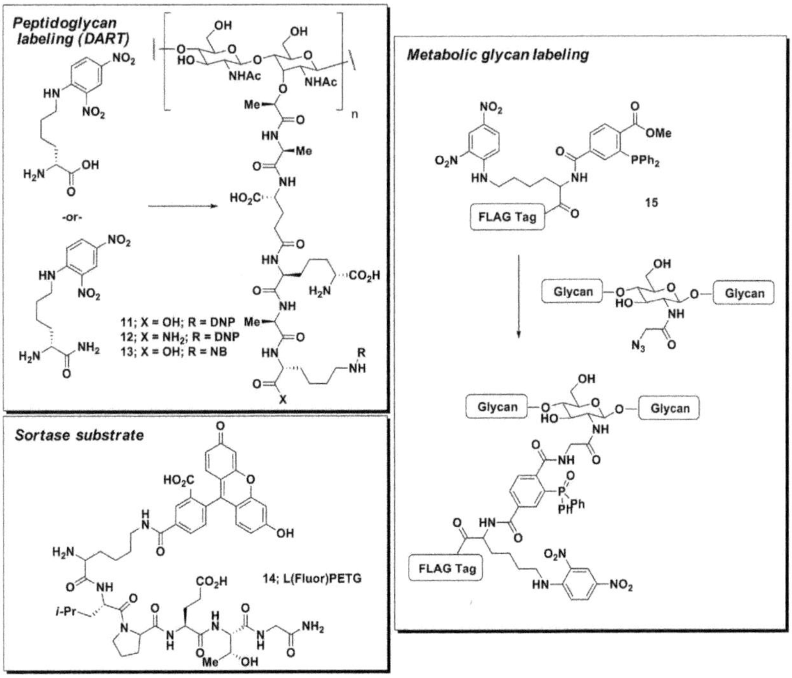

Fig. 6 Methods for incorporating small-molecule haptens into bacterial cell surface.

bacterial cell wall. In the first step, instead of directly labeling the peptidoglycan with DNP, an artificial D-amino acid containing a reactive handle (e.g., norbornene-NB, Fig. 6, Compound **13**) was incorporated. The second step involves using a bifunctional molecule containing the hapten linked to a bioorthogonally reactive moiety (e.g., tetrazine) that will react with the handle installed in the first step (e.g., maleimide for thiol, tetrazine for alkene). Presence of a linker resulted in increased antibody opsonization, as the linker could extend the hapten further out of the peptidoglycan, thereby increasing its accessibility. Importantly, strategies like DART that target critical, evolutionarily conserved processes in bacteria using targeting moieties highly similar to the natural substrates should increase the barrier to developing resistance.

In a related vein of cell wall biosynthesis, sortase is an enzyme used by Gram-positive bacteria to anchor proteins to the cell wall using a conserved LPXTG recognition motif.[123] This reaction has been widely utilized for engineering biologics,[124–126] including joining antibody fragments together to produce bispecific antibodies[127] and enhancing the immunogenicity of antigens.[128] Additionally, small-molecule functionalities such as fluorescein have been incorporated into the *S. aureus* cell wall using sortase through L(Fluor)PXTG (Fig. 6, Compound **14**) and antifluorescein antibodies were subsequently recruited to the surface of the cell wall.[129] However, to date, this strategy has not yet been shown to induce antibody opsonization and ADCP or ADCC. Spiegel and coworkers attempted to use sortase as a means of incorporating haptens on the surfaces of Gram-positive bacteria, and in the process, discovered that wall teichoic acids (WTA) on the cell wall acts as an "immunological cloak" by preventing antibody recruitment.[130] After labeling with fluorescein or Alexafluor488, antifluorescein or anti-Alexa488 antibodies were redirected at a significantly higher level in the absence of WTA (either through tunicamycin treatment or through mutants lacking the ability to synthesize WTA). These results agreed with previously highlighted work done in the Pires lab where similarly, tunicamycin treatment resulted in increased anti-DNP antibody recruiting.[119] Taken together, these results bring significant insights toward applying the ARM strategy toward Gram-positive bacteria in terms of the types of targets available and methods to increase the efficiency of antibody recruiting.

Bacterial cell surface glycans are pathogenic biomarkers that can be harnessed for diagnosing and treating bacterial infections. For example, metabolic oligosaccharide engineering (MOE) is a method for incorporating bioorthogonal handles onto cell surface glycoproteins.[131] The Dube lab utilized MOE to incorporate azidosugars into the cell surface glycan of *Helicobacter pylori*, a species of Gram-negative bacteria.[132] DNP was incorporated onto the cell surface with a phosphine-DNP construct (Fig. 6, Compound **15**) via the Staudinger ligation. Macrophage-directed cell killing occurred only upon metabolic labeling with azidosugars, treatment with phosphine-DNP construct, and presence of anti-DNP antibodies, demonstrating the requirement of each component. Importantly, using ARMs to target bacterial cell surface glycans is a suitable strategy for Gram-negative bacteria, given that their cell wall is not surface exposed and thus targeting cell wall biosynthesis would not be appropriate.

4.2 New ARM Modalities

The structure of ARMs has evolved beyond the traditional bifunctional molecule as researchers have begun to investigate the incorporation of alternative modalities. Aptamers are nucleic acids sequences generated through directed evolution-based approaches to have low nanomolar affinities to protein targets.[133,134] Accordingly, aptamers for cell surface targets are possible candidates as ARM TBTs. While highly potent, aptamers often have poor pharmacokinetic properties, and thus need to be modified or decorated with larger molecules (e.g., antibodies,[135] liposomes,[136] and lipids[137]) for delivery. To this end, the Barbas group have functionalized cell surface-targeting aptamers with the β-lactam or 1,3-diketone groups, which in turn allows for conjugation to mAb 38C2. In the case of ARC245, a VEGF-targeting aptamer conjugated to 38C2 through a β-lactam, an improved serum half life was observed, from mere minutes to 21 h.[138] Recently, Altermune Technologies (acquired by Centauri Therapeutics) has developed a type of ARM construct termed an "alphamer," which consists of an aptamer as TBT linked to α-Gal,[139] mostly to target Gram-positive and Gram-negative bacteria. Alphamers targeting group A *Streptococcus* surface anchored M protein was conjugated to α-Gal and was shown to redirect anti-α-Gal antibodies. This construct mediated enhanced phagocytosis as well as increased bacterial clearance from human blood.

Researchers from the Spiegel and McNaughton groups developed an ARM strategy to target the HER2 receptor,[140,141] an important target for antibody and small molecule-based cancer therapeutics.[142] Here, a HER2 binding nanobody (low molecular weight, peptidic binding domains generated by directed evolution) was engineered at the N terminus to contain a substrate for a mutant lipoic acid lipase.[143] The lipoic acid lipase transformed the N terminus to incorporate an aldehyde, which was subsequently functionalized with DNP-hydrazine. Doing so allowed the nanobody to redirect anti-DNP antibodies and mediated ADCC on HER2+ SK-BR-3 cells.

Nanoparticles have been explored as a means of targeting payloads specifically to a pathogenic cell based on its cell surface receptors through relative ease of installing different functionalities.[144] Recently, Li and coworkers generated a liposome with two independent molecules displayed on the surface, FA, and rhamnose (Rha), to redirect endogenous anti-rhamnose antibodies to FA receptor expressing breast cancer cells.[145] The targeted liposomes were taken up by cells expressing FA receptors, and upon fusion of the membrane, rhamnose would effectively be surface expressed. The goal was to then take advantage of multivalent effects of rhamnose imparted by the liposomes in redirecting antirhamnose antibodies. A liposome with a FA:Rha ratio of at least 1:2500 was found to mediate maximum complement-mediated cytotoxicity. Targeted liposomes demonstrated in vivo efficacy in delaying tumor growth compared to nontargeted (FA or Rha functionalized) liposomes in Balb/c mice challenged with 4T1 breast cancer cells and immunized for antirhamnose antibodies.

4.3 Beyond ARMs: Synthetic Bifunctional Constructs for Redirecting Immune Effector Cells

The development of bifunctional small molecules to effectively reprogram endogenous antibodies to target and induce the immune-mediated clearance of diseased cells has established a foundation to extend this concept more generally. One can perhaps envision the development of alternative classes of heterobifunctional molecules that can directly engage innate immune receptors, such as Toll-like receptors, or even directly engage Fc receptors, bypassing the requirement of engaging circulating antibodies altogether. Doing so would, in essence, yield a small molecule "mimic" of an antibody, maintaining the functions but significant decreasing its size and complexity. This particular idea has evolved in the protein engineering and antibody–small-molecule conjugation fields, where a portion of an antibody is effectively replaced with a synthetic construct, which is then conjugated directly to remainder of a normal antibody. Similar to the work of the Barbas lab, Francis and coworkers utilized bioconjugation chemistry to the N-terminus of the Fc protein in order to site specifically attach targeting aptamers.[146] In this research two model aptamers were utilized, targeting protein tyrosine kinase 7 and TD05.1, and targeting membrane bound IgM. The constructs were found to selectively bind only to cells overexpressing the protein of interest on their surface. This research paves the way for utilization of SELEX to affinity mature aptamers against proteins of interest and targets the effector region of an antibody directly to pathogenic cells.

In a similar vein, the recent discovery of "meditopes," molecules capable of binding to pockets on the Fab region of an antibody, has enabled the direct attachment of various linkers for drug conjugation and/or directing effects, in tandem with the original designed function of the mAb.[147] These binding pockets have either been engineered or have been found to be naturally present, on the surfaces of current therapeutic antibodies. This allows for potential of coadministration of therapeutics for combined therapy, imaging, and targeted delivery. Recently, a cyclic peptide was found to bind to a unique cavity formed by the light and heavy chains of the Fab domain of Cetuximab (an anti-EGFR monoclonal for colorectal and head-and-neck cancers).[148] Furthermore, grafting this site to other mAbs (Trastuzumab) allowed for use of the same cyclic peptide. By creating a bivalent construct (Fc fragment with two cyclic peptides displayed) enhanced the binding to cell-associated cetuximab, taking advantage of avidity and multivalency effects.[149] Furthering the utility of this approach, the initial low affinity cyclic peptide was affinity matured using mutational scanning to yield a binding partner with a 15 nM affinity.[150]

Schultz and coworkers have designed semisynthetic bifunctional constructs for redirecting cytotoxic T cells to the surface of cancer cells through the T cell marker CD3. In this work, a CD3 targeting Fab was engineered to include a *p*-acetylphenylalanine residue using unnatural amino acid mutagenesis for site-specific conjugation with TBTs via oxime ligation chemistry. This approach was applied to two systems, FA receptor[151] and PSMA.[152] Kularatne et al. demonstrated that the anti-CD3 Fab–folate conjugate-mediated T-cell-dependent killing of FA receptor-positive (FR^+) KB cells and OV90 cells in the presence of PBMCs, but not on FR^- Caki cells.[151] In vivo, the anti-CD3 Fab–folate conjugate had comparable serum half life to the unconjugated anti-CD3 Fab, suggesting that the conjugation did not alter the stability of the parent Fab. The anti-CD3 Fab–folate conjugate demonstrated in vivo efficacy in NOD-SCID mice implanted with KB xenograft and human PBMCs, effectively eradicating tumors to an undetectable level in the presence of 10:1 activated PBMC: tumor cell. In the presence of inactivated PBMC (100:1 effector:target), the tumor was not completely eradicated, but nonetheless significant tumor reduction and growth inhibition was observed. In a parallel study, Kim et al. conjugated the anti-CD3 Fab to the PSMA-binding glutamate urea for targeting prostate cancer.[152] After studying the dependence of in vitro efficacy on linker length, composition, and conjugation site, the optimized construct, P-SMAC, was taken into a NOD-SCID model implanted with PSMA+ C4-2 xenograft and human PBMCs. Similar to the anti-CD3 Fab–folate conjugate, P-SMAC inhibited the growth of C4-2 xenografts at levels of 1–2 mg/kg without showing weight loss or toxicity. As a whole, T-cell therapy has generated significant promise as a means of fighting cancer, as evidenced by recent work on the chimeric antigen receptor–T cells (CAR-T).[153] Here, antibody-binding domains for cancer-specific biomarkers are engineered into a T-cell-specific cell surface protein (e.g., the transmembrane domains of CD3), and in turn, will redirect T cells to the surface of cancer cells, much like the anti-CD3 conjugates. Recently, Endocyte Inc. reported small molecule–drug conjugates (SMDC), which, in this instance, are effectively ARMs to redirect hapten-specific CAR-T based on the TBT, potentially broadening the scope and increasing the modularity of CAR-T therapy.[154]

Expanding upon their ARM studies, the Spiegel group recently developed fully synthetic constructs to by-pass the need to recruit antibodies in order to elicit a response. These synthetic antibody mimetics (SyAMs, Fig. 7)[155–157] were developed using the PSMA-targeting glutamate urea derivative, discussed above, with a low nM affinity to PSMA coupled to a previously described cyclic peptide capable of engaging FcγRIa. These constructs have a molecular weight of roughly 7 kDa, only 5% of the molecular weight of a full-length antibody (150 kDa). Initial generations of the molecules in a bivalent display yielded promising results against ideal bead

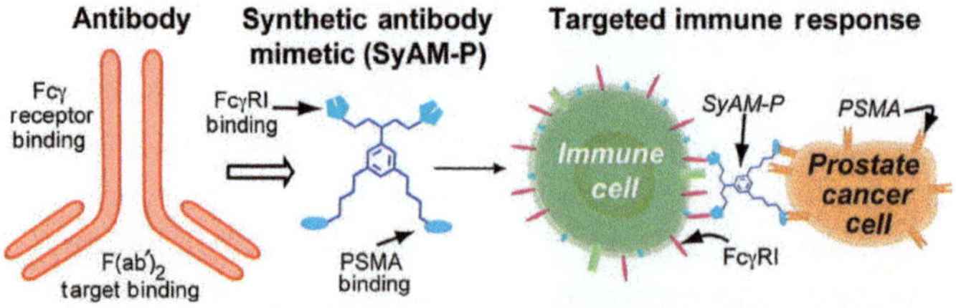

Fig. 7 Overview of the structure and intended function of synthetic antibody mimics (SyAMs). *Reprinted with permission from McEnaney, P.J.; Fitzgerald, K.J.; Zhang, A.X.; Douglass, E.F.; Shan, W.; Balog, A.; Kolesnikova, M.D.; Spiegel, D.A.* J. Am. Chem. Soc. ***2014,*** 136 *(52), 18034–18043. http://pubs.acs.org/doi/full/10.1021/ja509513c. Copyright 2014 American Chemical Society.*

targets displaying PSMA, however, the molecule lacked efficacy in a cancer cell model. Subsequent generations utilized, and validated, the three-body binding equilibria model,[88] and by expanding the valency of both of the binding partners, generating a nearly symmetric tetravalent molecule (containing a bivalent TBT and a bivalent effector directing terminus) efficacy in an in vitro cancer cell model was achieved. The modular nature of construct allows for rapid redevelopment for targeting new surface antigens or eliciting alternative immune responses. This is also a significant improvement over the fully biological bispecific antibodies, as it overcomes protein heterogeneity, purification, and storage limitations inherent to biologics. SyAMs provided compelling evidence for the simplification of mAbs into synthetic constructs capable of directly engage immune effector responses. Achieving this goal requires additional efforts around finding small-molecule ligands for the Fc receptors that can induce receptor crosslinking and replace cyclic peptide constructs.

5. ARMs IN THE CLINIC

To date, ARMs have demonstrated tremendous potential in various in vitro and in vivo studies. These initial successes have attracted the interest of the pharmaceutical industry in advancing ARMs into the clinic. The cpAb approach was one of the earliest examples of ARMs advancing to clinical trials. Given its modularity, the cpAb approach has been broadly applicable across numerous disease areas, including cancer, infectious diseases, pain, and diabetes.[76] Recently, through the company CovX (since acquired by Pfizer), several cpAbs have progressed into clinical trials for cancer (CVX-045, an antagonist of angiogenesis,[158] CVX-060-targeting angiopoietin-2,[159] and CVX-241,[160] a bispecific cpAb-targeting angiopoietin-2 and VEGF), mostly to take advantage of the enhancement in serum stability. These trials, however, have since been discontinued and the results are pending publication.

Owing to the success of in vitro and in vivo studies around the folate-fluorescein ARM, clinical trials were conducted by Endocyte, Inc. In contrast to the Cov-X work, where the main goal was to enhanced serum half life of the Cov-X bodies, these clinical trials were centered around the immunomodulatory potential of ARMs, representing the first such examples of ARMs in the clinic. In recent years two clinical trials, a Phase I clinical trial[161] and a Phase I/Ib clinical trial,[162] were conducted around the safety and efficacy of the folate–fluorescein ARM. In these trials, patients with metastatic renal cell carcinoma were first immunized for antifluorescein antibodies using E90, a KLH–fluorescein conjugate much like what was used in the in vivo studies, then injected with E17, the folate–fluorescein ARM. Similar to the in vivo studies, patients were also treated with IL2 and IFN-α. While this treatment regimen was generally well tolerated with no severe side effects, partial response was observed in less than 5% of the patients in both trials, with another 30% and 54% showing stable disease in the Phase I/Ib trial and Phase I trial, respectively. Although these results from the clinic show some initial promises for ARMs as therapeutics, limits in efficacy underscore that significant challenges remain.

6. CONCLUSIONS AND OUTLOOK: EXPANDING THE REACH OF ARMs TO IDENTIFY NEW TBTs AND ABTs

ARM-based strategies have seen applications across multiple disease areas, including cancer and infectious diseases. In this section, we comment on potential scenarios where utilizing ARMs would be appropriate, as well as some of the challenges still facing this field.

6.1 Applicability of ARMs

ARMs require three key conditions to function: (1) a cell surface target that is selectively displayed on diseased cells over healthy cells, (2) a specific small molecule that binds to the disease-specific cell surface target with high specificity and potency, and (3) flexible SAR that guides attachment of a linker between the TBT and ABT that does not dramatically alter the binding of the small molecule to the cell surface target. Once these conditions are fulfilled, there are specific scenarios wherein an ARM-based strategy might prove advantageous over traditional interventions. First, we have discussed several studies demonstrating that ARMs have the potential to better discriminate various target expression levels compared to cytotoxic drug conjugates, and thus might serve as an alternative. Second, ARMs have the adaptability to incorporate previously discovered and developed small molecules against cell surface targets which were unable to produce the desired pharmacological phenotypes, allowing for compound repurposing. This case was most apparent in ARM-P, as the PSMA-targeting glutamate urea, despite its high potency, did not result in any phenotypic effects on prostate cancer cells unless derivatized as an ARM. Lastly, one can tailor which ABT to employ based on the levels and isotype of endogenous antibodies in a specific patient; alternatively, the response could be adjusted, or even significantly enhanced, through immunization.

6.2 Identification of Novel Cell Surface Targets and TBT Ligands

The advancement and general application of ARM constructs necessitate the identification of selective, high affinity ligands for both cell surface target and endogenous antibodies. To this end, the utilization of various high-throughput screening technologies and the libraries therein has provided an unbiased approach for ligand discovery. While this subject is far too broad, and beyond the scope, to be effectively discussed in this chapter, it is important to briefly mention the utility of compound libraries and HTS campaigns in the discovery of new motifs.

Screening for ligands capable of recognizing surface expressed proteins is a nontrivial matter, as the majority of surface expressed proteins require interactions with the cell membrane to adopt proper folded conformations. Advances in the field of OBOC screening against fully intact cells have yielded promising preliminary results. Kodadek and coworkers utilized a peptoid OBOC library and a magnetic screening approach against the surface expressed Orexin receptor 1 (OxR1), a GPCR with drug candidate potential in the treatment of insomnia, diabetes, and drug addiction.[163] CHO cells, either expressing OxR1 or not were labeled with either red or green quantum dots, respectively, these cells were then incubated with a million compound peptoid OBOC library, followed by incubation with a rabbit anti-OxR1 antibody and an iron-oxide antirabbit antibody the beads bound to cells were removed through magnetic pull down. Following the pull down, the beads were analyzed under a fluorescent microscope to eliminate any false-positive beads that had been carried through the initial isolation. After screening several hits were identified, with the best being (Fig. 8, Compound **16**) which was capable of antagonizing OxR1 with a modest IC50 (50 μM). In another example, Kodadek and coworkers screened against cells expressing vascular endothelial growth factor receptor 2 (VEGFR2) using a two color microscope method against a peptoid library and isolated several peptoids capable of binding in the low micromolar range (~2 μM). It was found that by dimerization of one of these hits and optimization of the linker distance (Fig. 8, Compound **17**) significantly improved the affinity to mid-nanomolar range (~30 nM) and was capable of reducing tumor growth in a mouse model, again demonstrating the utility of multivalent designs. In another example, Lewis and coworkers have utilized the complex object parametrix analyzer and sorter (COPAS) large particle sorter to optimize the screening conditions required for isolation of ligands against αvβ3 integrin.[164] However, it was found that during sorting the interaction between cells and RGD displaying was rapidly reversible, causing a loss of

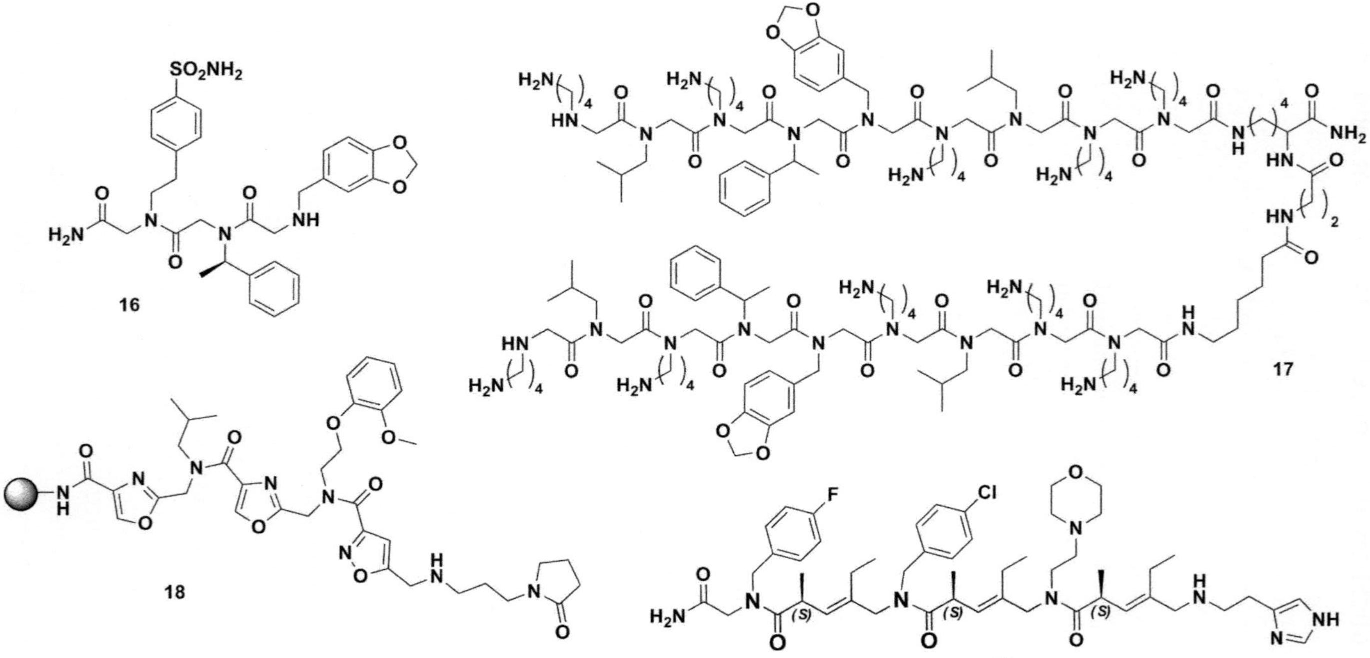

16 17 18 19

Fig. 8 Structures of potential TBT ligands for novel ARM targets.

isolation, to overcome this the cells and beads were reversibly crosslinked with paraformaldehyde, allowing for significant enrichment of the RGD displaying beads during isolation. While the targeting portion of antibody-recruiting molecules would benefit from noncompetitive (non-agonistic or antagonistic) binding events, this research shows molecular discovery in a rapid and unbiased screening campaign.

Using one-bead-one compound (OBOC) libraries the Kodadek lab has shown the utility of biomarker discovery through screening of disease serum, a potentially new source for personalized ABT development. The screening allows for the identification of ligands capable of functioning as "antigen surrogates," to pull down disease-specific antibodies. These antibodies are markers capable of differentiating diseases and their various activation states for the rapid and reliable diagnosis.[165] While these ligands are not the native epitopes to the antibodies, they can also be used as probe molecules to elucidate the native antigens. Doran and coworkers identified a peptoid antigen surrogate capable of identifying murine type-1 diabetes. The subsequent experiments identified that the native epitope of the disease-specific antibody was peripherin. Interestingly, the phosphorylation and dimerization of the native protein were required for antibody recognition. Utilizing a DNA-encoded OBOC platform, Mendes and coworkers screened serum isolated from either active or latent tuberculosis with a two-color fluorescent-activated cell sorting (FACS) method. This led to the discovery of compound **18** (Fig. 8) which was capable of differentiating between the activation state of TB. Using the discovered ligand, the disease-specific serum antibodies were pulled down and utilized in Western blotting to identify the native antigen, Ag85. It is important to note that the specific source of the antigen protein was incredibly important, as *E. coli* expressed protein showed no binding to the antibodies, while TB excreted Ag85 was recognized. In an interesting bridge between cell surface target and antibody biomarker screening, The Kodadek lab performed screening to isolate compounds capable of binding to the antigen pocket of a FAB fragment of a common, highly over abundant, antibody found in chronic lymphocytic leukemia (CLL). These antibodies are expressed on the surface of the over proliferated B-cells and the compound is therefore able to bind to the surface of disease cells. Using a combined magnetic and microscope screening platform, compound **19** (Fig. 8) was discovered and found to be an ~500 nM affinity binder in solution to CLL 169 IgG.[166] However, it was found that this compound-labeled cells expressing a surface CLL 169 antibody poorly, to enhance the labeling an multivalent display approach with dextran polymer[68] was utilized, the multivalency and avidity effects enhanced the compounds soluble affinity by approximately 10-fold showed significant increases in surface bound display.

OBOC screening is a vast improvement from previous antibody biomarker screening endeavors utilizing peptide and protein arrays, avoiding the complications of secondary and tertiary antigenic epitopes, and those responses dependent on posttranslational modifications. While the selected examples above primarily focus on utilization of these molecules as antagonists of surface receptors or as diagnostics, the principles regarding design and unbiased selection of synthetic molecules capable of recruiting antibodies to a cell surface targets holds. This has implications for the unbiased identification of personalized antigens for therapeutic development of antibody-recruiting molecules. The current research is focusing on the development and deployment of techniques to increase throughput and further diversify the structural elements. The field has progressed from the previously utilized isolation process of either magnetic, microscope, or COPAS isolation of beads toward utilization of flow cytometry and DNA-encoded libraries[167] to accelerate the pace of discovery for unbiased ligands.

6.3 Identification of Suitable ABTs to Maximize Endogenous Antibody Recruiting

The success of ARM-based strategies is not only dependent on the ability of the bifunctional molecule to bind disease-specific surface proteins, but also the ability to bind and direct endogenous, functional antibodies. Ideally, the ABT would be composed of a small-molecule hapten that is recognized by antibodies present in a large fraction of the human population, and not require additional preimmunization or addition of exogenous antibodies. In addition, the ABT hapten would be recognized, depending on the specific application, and desired effector mechanism(s), by specific functional isotypes (e.g., IgG and/or IgM). Furthermore, the ABT hapten would be readily scalable, modifiable, and modulatable through standard synthetic methods, enabling conjugation to various TBTs with ease. As discussed, the hapten dinitrophenyl (DNP) was one of the first small molecules to be used in generating specific and directed endogenous antibody-mediated responses. DNP is chemically very simple and given its historical context, perhaps unsurprisingly has been widely used by various ARM-based approaches.[43,56,57,80,82,84,85] Other successful ARM-based approaches have made use of well-known glycan-based haptens, such as α-Gal, however, their structural and synthetic complexity presents a significant challenge to their broad integration into antibody-recruiting strategies. While different small-molecule haptens have been incorporated into ARMs, there had not yet been a study to compare the titers of different antihapten antibodies (as well as the titers of each isotype) in the same setting. Thus, at this time, it is not possible to definitively identify the best ABT to incorporate to elicit a specific immune effector function and maximize the efficiency of the ARM strategy.

In order to begin more broadly assessing the utility of various small-molecule haptens in antibody-directing strategies, the Spiegel lab recently developed a cell-based screening platform to assess the functional capacity of various antigen–antibody pairs. Although ELISA- and microarray-based screening approaches have been broadly utilized to compare relative levels of antibodies in serum,[53,168] strategies to evaluate relative immunomodulatory function of endogenous antibodies remain limited. In this work, the small-molecule antigens DNP, phosphorylcholine (PC), and β-1-(*S*)-rhamnose were tagged with *N*-hydroxysuccinimide (NHS) esters, which can form covalent adducts with N-termini and lysines found on endogenous proteins. NHS–haptens were then used to decorate proteins found on the surface of living A172 glioblastoma cells and subsequently relative levels of corresponding IgG1, IgG2, IgG3 IgG4, and IgGM antibodies recruited from pooled human serum were measured by flow cytometry. It was found that PC was found to recruit substantially more IgM antibody compared to the other haptens while the rhamnose antigen was found to bind significantly more IgG, particularly from the IgG1 and IgG2 subclasses, followed by PC and then DNP. ELISA further confirmed that these results across 122 individual serum samples, suggesting antirhamnose IgG antibodies are widely prevalent across a broader population. The authors subsequently showed that the relative recruitment of IgG, but not IgM, antibodies were mirrored in the relative levels of CDC of haptenized cells. It was suggested since PC is also known to bind C-reactive protein (CRP), which is known to suppress later stages of complement recruitment, that perhaps PC-mediated CDC was blocked. The authors also demonstrated that by using chemically bivalent haptens (e.g., bis-rhamnose) not only were more antibodies directed to cell surfaces, but also significantly higher effector (CDC) responses were observed, consistent with previously published studies regarding multivalent antigen display. Similar observations were made in subsequent studies from the Kiessling lab,[54] in which a comparison of serum-derived antibody binding to α-Gal, rhamnose, and DNP haptens by SPR revealed a strong preference for rhamnose over DNP and α-Gal.

Furthermore, they demonstrated that when rhamnose was conjugated to a lipid tail, it would insert into the membranes of M21 melanoma cells, resulting in directed CDC in the presence of human serum.

Methods to compare not only the relative abundance and distribution of endogenous antibodies that recognize small-molecule haptens, but also their ability to induce a directed, robust immune-response are essential to the advancement and expansion of ARM-based strategies. The studies described above that explore the utility of various known small-molecule haptens, lay-out an initial framework toward this end. However, the vast repertoire of available antibodies in the human population, which are well suited for ARM-based strategies, remains unrealized. Thus, methods similar to those above that can not only compare haptens, but also identify new ones, would further expand the reach and utility of ARMs.

REFERENCES

1. Murphy, K. P.; Travers, P.; Walport, M.; Janeway, C. *Janeway's Immunobiology*; Garland Science: New York, 2008.
2. Iwasaki, A.; Medzhitov, R. *Nat. Immunol.* **2015**, *16*(4), 343–353.
3. Warrington, R.; Watson, W.; Kim, H. L.; Antonetti, F. R. *Allergy Asthma Clin. Immunol.* **2011**, 7(Suppl. 1), S1.
4. Goodnow, C. C.; Vinuesa, C. G.; Randall, K. L.; Mackay, F.; Brink, R. *Nat. Immunol.* **2010**, *11*(8), 681–688.
5. Dunkelberger, J. R.; Song, W. C. *Cell Res.* **2010**, *20*(1), 34–50.
6. Nimmerjahn, F.; Ravetch, J. V. *Nat. Rev. Immunol.* **2008**, *8*(1), 34–47.
7. Dhodapkar, K. M.; Krasovsky, J.; Williamson, B.; Dhodapkar, M. V. *J. Exp. Med.* **2002**, *195*(1), 125–133.
8. Cioca, D. P.; Deak, E.; Cioca, F.; Paunescu, V. *J. Immunother.* **2006**, *29*(1), 41–52.
9. Crotty, S.; Ahmed, R. *Semin. Immunol.* **2004**, *16*(3), 197–203.
10. Rosenblum, M. D.; Way, S. S.; Abbas, A. K. *Nat. Rev. Immunol.* **2016**, *16*(2), 90–101.
11. Riedel, S. *Proc. (Baylor. Univ. Med. Cent.)* **2005**, *18*(1), 21–25.
12. Ipsen, J. *J. Immunol.* **1946**, *54*(4), 325–347.
13. Looney, J. M.; Edsall, G.; Ipsen, J., Jr.; Chasen, W. H. *N. Engl. J. Med.* **1956**, *254*(1), 6–12.
14. Pulendran, B.; Ahmed, R. *Nat. Immunol.* **2011**, *12*(6), 509–517.
15. Kohler, G.; Milstein, C. *Nature* **1975**, *256*(5517), 495–497.
16. Weiner, L. M.; Surana, R.; Wang, S. *Nat. Rev. Immunol.* **2010**, *10*(5), 317–327.
17. Hansel, T. T.; Kropshofer, H.; Singer, T.; Mitchell, J. A.; George, A. J. *Nat. Rev. Drug Discov.* **2010**, *9*(4), 325–338.
18. Hooks, M. A.; Wade, C. S.; Millikan, W. J., Jr. *Pharmacotherapy* **1991**, *11*(1), 26–37.
19. Ecker, D. M.; Jones, S. D.; Levine, H. L. *mAbs* **2015**, 7(1), 9–14.
20. Reichert, J. M. *mAbs* **2017**, *9*(2), 167–181.
21. Chames, P.; Baty, D. *Curr. Opin. Drug Discov. Devel.* **2009**, *12*(2), 276–283.
22. Spiess, C.; Zhai, Q.; Carter, P. J. *Mol. Immunol.* **2015**, *67*(2 Pt A), 95–106.
23. Holliger, P.; Hudson, P. J. *Nat. Biotechnol.* **2005**, *23*(9), 1126–1136.
24. Muyldermans, S. *Annu. Rev. Biochem.* **2013**, *82*, 775–797.
25. Nelson, A. L. *mAbs* **2010**, *2*(1), 77–83.
26. Nelson, A. L.; Reichert, J. M. *Nat. Biotechnol.* **2009**, *27*(4), 331–337.

27. Beck, A.; Goetsch, L.; Dumontet, C.; Corvaia, N. *Nat. Rev. Drug Discov.* **2017**, *16*, 315–337.
28. Diamantis, N.; Banerji, U. *Br. J. Cancer* **2016**, *114*(4), 362–367.
29. Liu, L.; Li, Y. *Drugs Today (Barc.)* **2014**, *50*(1), 33–50.
30. Sathish, J. G.; Sethu, S.; Bielsky, M. C.; de Haan, L.; French, N. S.; Govindappa, K.; Green, J.; Griffiths, C. E.; Holgate, S.; Jones, D.; Kimber, I.; Moggs, J.; Naisbitt, D. J.; Pirmohamed, M.; Reichmann, G.; Sims, J.; Subramanyam, M.; Todd, M. D.; Van Der Laan, J. W.; Weaver, R. J.; Park, B. K. *Nat. Rev. Drug Discov.* **2013**, *12*(4), 306–324.
31. Imai, K.; Takaoka, A. *Nat. Rev. Cancer* **2006**, *6*(9), 714–727.
32. Spiegel, D. A. *Nat. Chem. Biol.* **2010**, *6*(12), 871–872.
33. Spiegel, D. A. *Expert. Rev. Clin. Pharmacol.* **2013**, *6*(3), 223–225.
34. McEnaney, P. J.; Parker, C. G.; Zhang, A. X.; Spiegel, D. A. *ACS Chem. Biol.* **2012**, *7*(7), 1139–1151.
35. Roybal, K. T.; Lim, W. A. *Annu. Rev. Immunol.* **2017**, *35*, 229–253.
36. Geering, B.; Fussenegger, M. *Trends Biotechnol.* **2015**, *33*(2), 65–79.
37. Six, H. R.; Uemura, K. I.; Kinsky, S. C. *Biochemistry* **1973**, *12*(20), 4003–4011.
38. Shearer, G. M. *Eur. J. Immunol.* **1974**, *4*(8), 527–533.
39. Nelson, D. L.; Poplack, D. G.; Holiman, B. J.; Henkart, P. A. *Clin. Exp. Immunol.* **1979**, *35*(3), 447–453.
40. Capon, D. J.; Chamow, S. M.; Mordenti, J.; Marsters, S. A.; Gregory, T.; Mitsuya, H.; Byrn, R. A.; Lucas, C.; Wurm, F. M.; Groopman, J. E.; et al. *Nature* **1989**, *337*(6207), 525–531.
41. Traunecker, A.; Schneider, J.; Kiefer, H.; Karjalainen, K. *Nature* **1989**, *339*(6219), 68–70.
42. Byrn, R. A.; Mordenti, J.; Lucas, C.; Smith, D.; Marsters, S. A.; Johnson, J. S.; Cossum, P.; Chamow, S. M.; Wurm, F. M.; Gregory, T.; Groopman, J. E.; Capon, D. J. *Nature* **1990**, *344*(6267), 667–670.
43. Shokat, K. M.; Schultz, P. G. *J. Am. Chem. Soc.* **1991**, *113*, 1861–1862.
44. Farah, F. S. *Immunology* **1973**, *25*(2), 217–226.
45. Ortega, E.; Kostovetzky, M.; Larralde, C. *Mol. Immunol.* **1984**, *21*(10), 883–888.
46. Lauer, K. *Mol. Immunol.* **1990**, *27*(7), 697–698.
47. Bogdan, C. *Nat. Immunol.* **2001**, *2*(10), 907–916.
48. Galili, U.; Macher, B. A.; Buehler, J.; Shohet, S. B. *J. Exp. Med.* **1985**, *162*(2), 573–582.
49. Galili, U. *Immunology* **2013**, *140*(1), 1–11.
50. Galili, U.; Swanson, K. *Proc. Natl. Acad. Sci. U.S.A.* **1991**, *88*(16), 7401–7404.
51. Galili, U.; Clark, M. R.; Shohet, S. B.; Buehler, J.; Macher, B. A. *Proc. Natl. Acad. Sci. U.S.A.* **1987**, *84*(5), 1369–1373.
52. Jakobsche, C. E.; Parker, C. G.; Tao, R. N.; Kolesnikova, M. D.; Douglass, E. F., Jr.; Spiegel, D. A. *ACS Chem. Biol.* **2013**, *8*(11), 2404–2411.
53. Chen, W.; Gu, L.; Zhang, W.; Motari, E.; Cai, L.; Styslinger, T. J.; Wang, P. G. *ACS Chem. Biol.* **2011**, *6*(2), 185–191.
54. Sheridan, R. T.; Hudon, J.; Hank, J. A.; Sondel, P. M.; Kiessling, L. L. *ChemBioChem* **2014**, *15*(10), 1393–1398.
55. Lussow, A. R.; Buelow, R.; Fanget, L.; Peretto, S.; Gao, L.; Pouletty, P. *J. Immunother. Emphasis Tumor Immunol.* **1996**, *19*(4), 257–265.
56. Lu, Y.; Low, P. S. *Cancer Immunol. Immunother.* **2002**, *51*(3), 153–162.
57. Lu, Y.; Sega, E.; Low, P. S. *Int. J. Cancer* **2005**, *116*(5), 710–719.
58. Metallo, S. J.; Kane, R. S.; Holmlin, R. E.; Whitesides, G. M. *J. Am. Chem. Soc.* **2003**, *125*(15), 4534–4540.
59. Krishnamurthy, V. M.; Quinton, L. J.; Estroff, L. A.; Metallo, S. J.; Isaacs, J. M.; Mizgerd, J. P.; Whitesides, G. M. *Biomaterials* **2006**, *27*(19), 3663–3674.

60. Wagner, J.; Lerner, R. A.; Barbas, C. F., III. *Science (Washington, DC)* **1995**, *270*(5243), 1797–1800.
61. Barbas, C. F.; Heine, A.; Zhong, G.; Hoffmann, T.; Gramatikova, S.; Bjrnestedt, R.; List, B.; Anderson, J.; Stura, E. A.; Wilson, I. A.; Lerner, R. A. *Science (Washington, DC)* **1997**, *278*(5346), 2085–2092.
62. Rader, C.; Sinha, S. C.; Popkov, M.; Lerner, R. A.; Barbas, C. F., III. *Proc. Natl. Acad. Sci. U.S.A.* **2003**, *100*(9), 5396–5400.
63. Guo, F.; Das, S.; Mueller, B.; Barbas, C.; Lerner, R.; Sinha, S. *Proc. Natl. Acad. Sci. U.S.A.* **2006**, *103*(29), 11009–11014.
64. Gavrilyuk, J.; Wuellner, U.; Barbas, C. *Bioorg. Med. Chem. Lett.* **2009**, *19*(5), 1421–1424.
65. Popkov, M.; Gonzalez, B.; Sinha, S. C.; Barbas, C. F., 3rd. *Proc. Natl. Acad. Sci. U.S.A.* **2009**, *106*(11), 4378–4383.
66. Krishnamurthy, V. M.; Estroff, L. A.; Whitesides, G. M. Multivalency in ligand design. In *Fragment-Based Approaches in Drug Discovery*; Jahnke, W., Erlanson, D. A., Eds.; WILEY-VCH Verlag GmbH & Co. KGaA: Weinheim, Germany, 2006; pp 11–53.
67. Jayaraman, N. *Chem. Soc. Rev.* **2009**, *38*(12), 3463–3483.
68. Morimoto, J.; Sarkar, M.; Kenrick, S.; Kodadek, T. *Bioconjug. Chem.* **2014**, *25*(8), 1479–1491.
69. Bennett, N. R.; Zwick, D. B.; Courtney, A. H.; Kiessling, L. L. *ACS Chem. Biol.* **2015**, *10*(8), 1817–1824.
70. Carlson, C. B.; Mowery, P.; Owen, R. M.; Dykhuizen, E. C.; Kiessling, L. L. *ACS Chem. Biol.* **2007**, *2*(2), 119–127.
71. Owen, R. M.; Carlson, C. B.; Xu, J.; Mowery, P.; Fasella, E.; Kiessling, L. L. *ChemBioChem* **2007**, *8*(1), 68–82.
72. Wong, N. C.; Mueller, B. M.; Barbas, C. F.; Ruminski, P.; Quaranta, V.; Lin, E. C.; Smith, J. W. *Clin. Exp. Metastasis* **1998**, *16*(1), 50–61.
73. Crocker, P. R.; Paulson, J. C.; Varki, A. *Nat. Rev. Immunol.* **2007**, 7(4), 255–266.
74. O'Reilly, M. K.; Collins, B. E.; Han, S.; Liao, L.; Rillahan, C.; Kitov, P. I.; Bundle, D. R.; Paulson, J. C. *J. Am. Chem. Soc.* **2008**, *130*(24), 7736–7745.
75. Cui, L.; Kitov, P. I.; Completo, G. C.; Paulson, J. C.; Bundle, D. R. *Bioconjug. Chem.* **2011**, *22*(4), 546–550.
76. Rader, C. *Trends Biotechnol.* **2014**, *32*(4), 186–197.
77. Ruoslahti, E. *Annu. Rev. Cell Dev. Biol.* **1996**, *12*, 697–715.
78. Popkov, M.; Rader, C.; Gonzalez, B.; Sinha, S. C.; Barbas, C. F., 3rd. *Int. J. Cancer* **2006**, *119*(5), 1194–1207.
79. Lu, Y.; You, F.; Vlahov, I.; Westrick, E.; Fan, M.; Low, P. S.; Leamon, C. P. *Mol. Pharm.* **2007**, *4*(5), 695–706.
80. Murelli, R. P.; Zhang, A. X.; Michel, J.; Jorgensen, W. L.; Spiegel, D. A. *J. Am. Chem. Soc.* **2009**, *131*(47), 17090–17092.
81. Dubrovska, A.; Kim, C.; Elliott, J.; Shen, W.; Kuo, T. H.; Koo, D. I.; Li, C.; Tuntland, T.; Chang, J.; Groessl, T.; Wu, X.; Gorney, V.; Ramirez-Montagut, T.; Spiegel, D. A.; Cho, C. Y.; Schultz, P. G. *ACS Chem. Biol.* **2011**, *6*(11), 1223–1231.
82. Jakobsche, C. E.; McEnaney, P. J.; Zhang, A. X.; Spiegel, D. A. *ACS Chem. Biol.* **2012**, 7(2), 316–321.
83. Rullo, A. F.; Fitzgerald, K. J.; Muthusamy, V.; Liu, M.; Yuan, C.; Huang, M.; Kim, M.; Cho, A. E.; Spiegel, D. A. *Angew. Chem. Int. Ed. Engl.* **2016**, *55*(11), 3642–3646.
84. Parker, C. G.; Domaoal, R. A.; Anderson, K. S.; Spiegel, D. A. *J. Am. Chem. Soc.* **2009**, *131*(45), 16392–16394.
85. Parker, C. G.; Dahlgren, M. K.; Tao, R. N.; Li, D. T.; Douglass, E. F., Jr.; Shoda, T.; Jawanda, N.; Spasov, K. A.; Lee, S.; Zhou, N.; Domaoal, R. A.; Sutton, R. E.; Anderson, K. S.; Jorgensen, W. L.; Krystal, M.; Spiegel, D. A. *Chem. Sci.* **2014**, *5*(6), 2311–2317.

86. Holmes, E. H.; Greene, T. G.; Tino, W. T.; Boynton, A. L.; Aldape, H. C.; Misrock, S. L.; Murphy, G. P. *Prostate Suppl.* **1996**, 7, 25–29.
87. Mohammed, A. A.; Shergill, I. S.; Vandal, M. T.; Gujral, S. S. *Expert Rev. Mol. Diagn.* **2007**, 7(4), 345–349.
88. Douglass, E. F., Jr.; Miller, C. J.; Sparer, G.; Shapiro, H.; Spiegel, D. A. *J. Am. Chem. Soc.* **2013**, *135*(16), 6092–6099.
89. Mazar, A. P. *Clin. Cancer Res.* **2008**, *14*(18), 5649–5655.
90. Smith, H. W.; Marshall, C. J. *Nat. Rev. Mol. Cell Biol.* **2010**, *11*(1), 23–36.
91. Kenny, H. A.; Leonhardt, P.; Ladanyi, A.; Yamada, S. D.; Montag, A.; Im, H. K.; Jagadeeswaran, S.; Shaw, D. E.; Mazar, A. P.; Lengyel, E. *Clin. Cancer Res.* **2011**, *17*(3), 459–471.
92. Jakobsche, C. E.; McEnaney, P. J.; Zhang, A. X.; Spiegel, D. A. *ACS Chem. Biol.* **2011**, 7(2), 316–321.
93. Rullo, A. F.; Fitzgerald, K. J.; Muthusamy, V.; Liu, M.; Yuan, C.; Huang, M.; Kim, M.; Cho, A. E.; Spiegel, D. A. *Angew. Chem.* **2016**, *128*(11), 3706–3710.
94. Carreiras, F.; Denoux, Y.; Staedel, C.; Lehmann, M.; Sichel, F.; Gauduchon, P. *Gynecol. Oncol.* **1996**, *62*(2), 260–267.
95. Wuellner, U.; Gavrilyuk, J.; Barbas, C. *Angew. Chem. Int. Ed.* **2010**, *49*(34), 5934–5937.
96. Gasparini, G.; Brooks, P. C.; Biganzoli, E.; Vermeulen, P. B.; Bonoldi, E.; Dirix, L. Y.; Ranieri, G.; Miceli, R.; Cheresh, D. A. *Clin. Cancer Res.* **1998**, *4*(11), 2625–2634.
97. De Clercq, E.; Li, G. D. *Clin. Microbiol. Rev.* **2016**, *29*(3), 695–747.
98. Dessain, S. K.; Adekar, S. P.; Berry, J. D. *Curr. Top. Microbiol. Immunol.* **2008**, *317*, 155–183.
99. Pelegrin, M.; Naranjo-Gomez, M.; Piechaczyk, M. *Trends Microbiol.* **2015**, *23*(10), 653–665.
100. Naicker, K. P.; Li, H.; Heredia, A.; Song, H.; Wang, L.-X. *Org. Biomol. Chem.* **2004**, *2*(5), 660–664.
101. Perdomo, M. F.; Levi, M.; Saellberg, M.; Vahlne, A. *Proc. Natl. Acad. Sci. U.S.A.* **2008**, *105*(34), 12515–12520.
102. Wang, T.; Zhang, Z.; Wallace, O. B.; Deshpande, M.; Fang, H.; Yang, Z.; Zadjura, L. M.; Tweedie, D. L.; Huang, S.; Zhao, F.; Ranadive, S.; Robinson, B. S.; Gong, Y. F.; Ricarrdi, K.; Spicer, T. P.; Deminie, C.; Rose, R.; Wang, H. G.; Blair, W. S.; Shi, P. Y.; Lin, P. F.; Colonno, R. J.; Meanwell, N. A. *J. Med. Chem.* **2003**, *46*(20), 4236–4239.
103. Sato, S.; Inokuma, T.; Otsubo, N.; Burton, D. R.; Barbas, C. F. *ACS Med. Chem. Lett.* **2013**, *4*(5), 460–465.
104. Gavrilyuk, J.; Uehara, H.; Otsubo, N.; Hessell, A.; Burton, D. R.; Barbas, C. F., 3rd. *ChemBioChem* **2010**, *11*(15), 2113–2118.
105. Zhang, A. X.; Murelli, R. P.; Barinka, C.; Michel, J.; Cocleaza, A.; Jorgensen, W. L.; Lubkowski, J.; Spiegel, D. A. *J. Am. Chem. Soc.* **2010**, *132*(36), 12711–12716.
106. Genady, A. R.; Janzen, N.; Banevicius, L.; El-Gamal, M.; El-Zaria, M. E.; Valliant, J. F. *J. Med. Chem.* **2016**, *59*(6), 2660–2673.
107. Langley, D. R.; Kimura, S. R.; Sivaprakasam, P.; Zhou, N. N.; Dicker, I.; McAuliffe, B.; Wang, T.; Kadow, J. F.; Meanwell, N. A.; Krystal, M. *Proteins* **2015**, *83*(2), 331–350.
108. Moraca, F.; Acharya, K.; Melillo, B.; Smith, A. B.; Chaiken, I.; Abrams, C. F. *J. Chem. Inf. Model.* **2016**, *56*(10), 2069–2079.
109. Gardete, S.; Tomasz, A. *J. Clin. Invest.* **2014**, *124*(7), 2836–2840.
110. Sievert, D. M.; Rudrik, J. T.; Patel, J. B.; McDonald, L. C.; Wilkins, M. J.; Hageman, J. C. *Clin. Infect. Dis.* **2008**, *46*(5), 668–674.
111. Bebbington, C.; Yarranton, G. *Curr. Opin. Biotechnol.* **2008**, *19*(6), 613–619.

112. DiGiandomenico, A.; Sellman, B. R. *Curr. Opin. Microbiol.* **2015**, *27*, 78–85.
113. Morrison, C. *Nat. Rev. Drug Discov.* **2015**, *14*(11), 737–738.
114. Orth, P.; Xiao, L.; Hernandez, L. D.; Reichert, P.; Sheth, P. R.; Beaumont, M.; Yang, X.; Murgolo, N.; Ermakov, G.; DiNunzio, E.; Racine, F.; Karczewski, J.; Secore, S.; Ingram, R. N.; Mayhood, T.; Strickland, C.; Therien, A. G. *J. Biol. Chem.* **2014**, *289*(26), 18008–18021.
115. Janeway, C.; Travers, P.; Walport, M.; Schlomchik, M. *Immunobiology: The Immune System in Health and Disease*, 5th ed.; Garland Science: New York, 2001.
116. Bertozzi, C. R.; Bednardski, M. D. *J. Am. Chem. Soc.* **1992**, *114*, 2242–2245.
117. Bertozzi, C. R.; Bednardski, M. D. *J. Am. Chem. Soc.* **1992**, *114*, 5543–5546.
118. Li, J.; Zacharek, S.; Chen, X.; Wang, J.; Zhang, W.; Janczuk, A.; Wang, P. G. *Bioorg. Med. Chem.* **1999**, 7(8), 1549–1558.
119. Fura, J. M.; Sabulski, M. J.; Pires, M. M. *ACS Chem. Biol.* **2014**, *9*(7), 1480–1489.
120. Fura, J. M.; Pires, M. M. *Biopolymers* **2015**, *104*(4), 351–359.
121. Pidgeon, S. E.; Pires, M. M. *Chem. Commun. (Camb.)* **2015**, *51*(51), 10330–10333.
122. Fura, J. M.; Pidgeon, S. E.; Birabaharan, M.; Pires, M. M. *ACS Infect. Dis.* **2016**, *2*(4), 302–309.
123. Spirig, T.; Weiner, E. M.; Clubb, R. T. *Mol. Microbiol.* **2011**, *82*(5), 1044–1059.
124. Mao, H.; Hart, S. A.; Schink, A.; Pollok, B. A. *J. Am. Chem. Soc.* **2004**, *126*(9), 2670–2671.
125. Theile, C. S.; Witte, M. D.; Blom, A. E.; Kundrat, L.; Ploegh, H. L.; Guimaraes, C. P. *Nat. Protoc.* **2013**, *8*(9), 1800–1807.
126. Guimaraes, C. P.; Witte, M. D.; Theile, C. S.; Bozkurt, G.; Kundrat, L.; Blom, A. E.; Ploegh, H. L. *Nat. Protoc.* **2013**, *8*(9), 1787–1799.
127. Wagner, K.; Kwakkenbos, M. J.; Claassen, Y. B.; Maijoor, K.; Bohne, M.; van der Sluijs, K. F.; Witte, M. D.; van Zoelen, D. J.; Cornelissen, L. A.; Beaumont, T.; Bakker, A. Q.; Ploegh, H. L.; Spits, H. *Proc. Natl. Acad. Sci. U.S.A.* **2014**, *111*(47), 16820–16825.
128. Swee, L. K.; Guimaraes, C. P.; Sehrawat, S.; Spooner, E.; Barrasa, M. I.; Ploegh, H. L. *Proc. Natl. Acad. Sci. U.S.A.* **2013**, *110*(4), 1428–1433.
129. Nelson, J. W.; Chamessian, A. G.; McEnaney, P. J.; Murelli, R. P.; Kazmierczak, B. I.; Spiegel, D. A. *ACS Chem. Biol.* **2010**, *5*(12), 1147–1155.
130. Gautam, S.; Kim, T.; Lester, E.; Deep, D.; Spiegel, D. A. *ACS Chem. Biol.* **2016**, *11*(1), 25–30.
131. Tra, V. N.; Dube, D. H. *Chem. Commun. (Camb.)* **2014**, *50*(36), 4659–4673.
132. Kaewsapsak, P.; Esonu, O.; Dube, D. H. *ChemBioChem* **2013**, *14*(6), 721–726.
133. Zhou, J.; Rossi, J. *Nat. Rev. Drug Discov.* **2017**, *16*(3), 181–202.
134. Darmostuk, M.; Rimpelova, S.; Gbelcova, H.; Ruml, T. *Biotechnol. Adv.* **2015**, *33*(6 Pt. 2), 1141–1161.
135. Heo, K.; Min, S. W.; Sung, H. J.; Kim, H. G.; Kim, H. J.; Kim, Y. H.; Choi, B. K.; Han, S.; Chung, S.; Lee, E. S.; Chung, J.; Kim, I. H. *J. Control. Release* **2016**, *229*, 1–9.
136. Willis, M. C.; Collins, B. D.; Zhang, T.; Green, L. S.; Sebesta, D. P.; Bell, C.; Kellogg, E.; Gill, S. C.; Magallanez, A.; Knauer, S.; Bendele, R. A.; Gill, P. S.; Janjic, N. *Bioconjug. Chem.* **1998**, *9*(5), 573–582.
137. Lee, C. H.; Lee, S. H.; Kim, J. H.; Noh, Y. H.; Noh, G. J.; Lee, S. W. *Mol. Ther. Nucleic Acids* **2015**, *4,* e254.
138. Wuellner, U.; Gavrilyuk, J. I.; Barbas, C. F., 3rd. *Angew. Chem. Int. Ed. Engl.* **2010**, *49*(34), 5934–5937.
139. Kristian, S. A.; Hwang, J. H.; Hall, B.; Leire, E.; Iacomini, J.; Old, R.; Galili, U.; Roberts, C.; Mullis, K. B.; Westby, M.; Nizet, V. *J. Mol. Med. (Berl.)* **2015**, *93*(6), 619–631.

140. Slamon, D. J.; Clark, G. M.; Wong, S. G.; Levin, W. J.; Ullrich, A.; McGuire, W. L. *Science* **1987**, *235*(4785), 177–182.
141. Menard, S.; Casalini, P.; Campiglio, M.; Pupa, S.; Agresti, R.; Tagliabue, E. *Ann. Oncol.* **2001**, *12*(Suppl. 1), S15–9.
142. Arteaga, C. L.; Sliwkowski, M. X.; Osborne, C. K.; Perez, E. A.; Puglisi, F.; Gianni, L. *Nat. Rev. Clin. Oncol.* **2011**, *9*(1), 16–32.
143. Gray, M. A.; Tao, R. N.; DePorter, S. M.; Spiegel, D. A.; McNaughton, B. R. *ChemBioChem* **2016**, *17*(2), 155–158.
144. Davis, M. E.; Chen, Z. G.; Shin, D. M. *Nat. Rev. Drug Discov.* **2008**, 7(9), 771–782.
145. Li, X.; Rao, X.; Cai, L.; Liu, X.; Wang, H.; Wu, W.; Zhu, C.; Chen, M.; Wang, P. G.; Yi, W. *ACS Chem. Biol.* **2016**, *11*(5), 1205–1209.
146. Netirojjanakul, C.; Witus, L. S.; Behrens, C. R.; Weng, C.-H.; Iavarone, A. T.; Francis, M. B. *Chem. Sci.* **2013**, *4*(1), 266–272.
147. Avery, K. N.; Zer, C.; Bzymek, K. P.; Williams, J. C. *Sci. Rep.* **2015**, *5*, 7817.
148. Donaldson, J. M.; Zer, C.; Avery, K. N.; Bzymek, K. P.; Horne, D. A.; Williams, J. C. *Proc. Natl. Acad. Sci.* **2013**, *110*(43), 17456–17461.
149. Bzymek, K. P.; Ma, Y.; Avery, K. A.; Horne, D. A.; Williams, J. C. *Acta Crystallogr. Sec. F: Struct. Biol. Commun.* **2016**, *72*(6), 434–442.
150. van Rosmalen, M.; Janssen, B. M.; Hendrikse, N. M.; van der Linden, A. J.; Pieters, P. A.; Wanders, D.; de Greef, T. F.; Merkx, M. *J. Biol. Chem.* **2017**, *292*, 1477–1489.
151. Kularatne, S. A.; Deshmukh, V.; Gymnopoulos, M.; Biroc, S. L.; Xia, J.; Srinagesh, S.; Sun, Y.; Zou, N.; Shimazu, M.; Pinkstaff, J.; Ensari, S.; Knudsen, N.; Manibusan, A.; Axup, J. Y.; Kim, C. H.; Smider, V. V.; Javahishvili, T.; Schultz, P. G. *Angew. Chem. Int. Ed. Engl.* **2013**, *52*(46), 12101–12104.
152. Kim, C. H.; Axup, J. Y.; Lawson, B. R.; Yun, H.; Tardif, V.; Choi, S. H.; Zhou, Q.; Dubrovska, A.; Biroc, S. L.; Marsden, R.; Pinstaff, J.; Smider, V. V.; Schultz, P. G. *Proc. Natl. Acad. Sci. U.S.A.* **2013**, *110*(44), 17796–17801.
153. Fesnak, A. D.; June, C. H.; Levine, B. L. *Nat. Rev. Cancer* **2016**, *16*(9), 566–581.
154. Lee, Y. G.; Chu, H.; Tenneti, S.; Kanduluru, A. K.; Low, P. S. *Cancer Res.* **2016**, *76*(14 Suppl). Abstract no. LB-254.
155. Adams, J. J.; Sidhu, S. S. *Curr. Opin. Struct. Biol.* **2014**, *24*, 1–9.
156. McEnaney, P. J.; Fitzgerald, K. J.; Zhang, A. X.; Douglass, E. F.; Shan, W.; Balog, A.; Kolesnikova, M. D.; Spiegel, D. A. *J. Am. Chem. Soc.* **2014**, *136*(52), 18034–18043.
157. Rader, C. *Nature* **2015**, *518*(7537), 38–39.
158. Li, L.; Leedom, T. A.; Do, J.; Huang, H.; Lai, J.; Johnson, K.; Osothprarop, T. F.; Rizzo, J. D.; Doppalapudi, V. R.; Bradshaw, C. W.; Lappe, R. W.; Woodnutt, G.; Levin, N. J.; Pirie-Shepherd, S. R. *Transl. Oncol.* **2011**, *4*(4), 249–257.
159. Huang, H.; Lai, J. Y.; Do, J.; Liu, D.; Li, L.; Del Rosario, J.; Doppalapudi, V. R.; Pirie-Shepherd, S.; Levin, N.; Bradshaw, C.; Woodnutt, G.; Lappe, R.; Bhat, A. *Clin. Cancer Res.* **2011**, *17*(5), 1001–1011.
160. Doppalapudi, V. R.; Huang, J.; Liu, D.; Jin, P.; Liu, B.; Li, L.; Desharnais, J.; Hagen, C.; Levin, N. J.; Shields, M. J.; Parish, M.; Murphy, R. E.; Del Rosario, J.; Oates, B. D.; Lai, J. Y.; Matin, M. J.; Ainekulu, Z.; Bhat, A.; Bradshaw, C. W.; Woodnutt, G.; Lerner, R. A.; Lappe, R. W. *Proc. Natl. Acad. Sci. U.S.A.* **2010**, *107*(52), 22611–22616.
161. Amato, R. J.; Shetty, A.; Lu, Y.; Ellis, R.; Low, P. S. *J. Immunother.* **2013**, *36*(4), 268–275.
162. Amato, R. J.; Shetty, A.; Lu, Y.; Ellis, P. R.; Mohlere, V.; Carnahan, N.; Low, P. S. *J. Immunother.* **2014**, *37*(4), 237–244.
163. Qi, X.; Astle, J.; Kodadek, T. *Mol. Biosyst.* **2010**, *6*(1), 102–107.

164. Cho, C.-F.; Behnam Azad, B.; Luyt, L. G.; Lewis, J. D. *ACS Comb. Sci.* **2013**, *15*(8), 393–400.
165. Doran, T. M.; Sarkar, M.; Kodadek, T. *J. Am. Chem. Soc.* **2016**, *138*(19), 6076–6094.
166. Sarkar, M.; Liu, Y.; Morimoto, J.; Peng, H.; Aquino, C.; Rader, C.; Chiorazzi, N.; Kodadek, T. *Chem. Biol.* **2014**, *21*(12), 1670–1679.
167. MacConnell, A. B.; McEnaney, P. J.; Cavett, V. J.; Paegel, B. M. *ACS Comb. Sci.* **2015**, *17*(9), 518–534.
168. Oyelaran, O.; McShane, L. M.; Dodd, L.; Gildersleeve, J. C. *J. Proteome Res.* **2009**, *8*(9), 4301–4310.

CHAPTER FOURTEEN

A Decade of Deuteration in Medicinal Chemistry

Julie F. Liu[1], Scott L. Harbeson, Christopher L. Brummel, Roger Tung, Robert Silverman, Dario Doller
Concert Pharmaceuticals, Inc., Lexington, MA, United States
[1]Corresponding author: e-mail address: jliu@concertpharma.com

Contents

1. INTRODUCTION

Medicinal chemists are constantly seeking new tools to design optimized development candidates, with the goal of providing the highest likelihood of clinical success, in order to deliver improved therapeutic options to patients. One tool which is rising in prominence takes advantage of the differences in properties between protium (^{1}H) and deuterium (^{2}H). Deuterium is a stable (nonradioactive) isotope of hydrogen and is among the few isotopes in the periodic table to have its own symbol, D. The natural abundance of deuterium in ocean water is 0.0156%, with protium accounting for 99.98% of the hydrogen present.[1] This percentage may have played a role in the development of life from its very beginning. A comparison of the chemical

Annual Reports in Medicinal Chemistry, Volume 50
ISSN 0065-7743
https://doi.org/10.1016/bs.armc.2017.08.010

constituents of the complex mixture produced during a Miller–Urey spark discharge experiment (an experiment intended to simulate the creation of the primordial soup which produced the first "biopolymers") conducted using deuterated substrates (e.g., D_2O, D_2, ND_3, CD_4) instead of their hydrogen-containing counterparts, provided interesting results. The deuterated experiment formed mostly the same amino acids, but also resulted in 43 new, unique compounds not seen with the hydrogen-based control, thus highlighting the similarities and differences that deuterium can afford.[2] D_2O itself is considered relatively safe to ingest, and a number of clinical applications of heavy water are listed in the clinicaltrials.gov database. For example, D_2O has been explored as a safe "tracer" for biomarker studies of diseases of the brain and spinal cord. The protocol called for ingesting heavy water orally for up to 6 weeks.[3] The biological effects of increasing deuterium content in water have been studied in microorganisms and plants, demonstrating that remarkably high levels of chronic substitution of D_2O in place of H_2O are tolerated across species.[4] It is safe to say that at this time the biological effects of deuterium, either at natural abundance or elevated levels, are not fully appreciated.

Medicinal chemists have long been familiar with the use of isotopically labeled compounds for mechanistic studies and as tools in biomedical research. Examples are the use of ^{18}F and ^{11}C in PET research, ^{14}C, ^{32}P, and ^{3}H in radioligands, and ^{2}H in internal standards for mass spectrometry detection during bioanalyses. However, the introduction of deuterium in active pharmaceutical ingredients has only recently been recognized as a potential path to better drugs, leading to an increase in patent applications containing examples of analogs bearing deuterium and other isotopic variants across a variety of therapeutic indications by companies both large and small. The topic of deuterium in medicinal chemistry has previously been reviewed.[5–7] In this perspective, an update will be provided focusing on the diversity of ways in which medicinal chemists are using deuteration in their efforts to deliver better drugs.

2. DEUTERATION AS A STRATEGY TO INCREASE DRUG EXPOSURE

Today, a number of commercial sources of chemically diverse deuterium-containing raw materials exist, generally providing materials of high isotopic purity. However, deuterium was only discovered in 1931, and it took decades to develop industrial methods to produce relatively high isotopic enrichment in large quantities, mainly D_2O as a modulator in certain nuclear reactors.[8] One of the earliest documented examples of assessing the enzymatic effects of deuteration in a drug used morphine (**1**, Fig. 1) as a substrate for rat liver microsomal enzymes, in 1961.[9] The first deuterated drug candidate we are aware of is fludalanine (**2**), an antimicrobial agent disclosed in the 1970s.[10]

The broad proposition for deuterium substitution in drug discovery is that if a drug is quickly metabolized via enzymatic cleavage of a C–H bond, and if this cleavage is at least partially rate-limiting, a deuterium isotope effect (DIE) may potentially slow the reaction kinetics and increase the metabolic stability of an appropriately deuterated drug. In turn, this could lead to lower efficacious doses and/or reduced frequency of administration. For most lipophilic drugs, turnover is carried out by enzymes in the cytochrome P-450 (CYP) superfamily. These CYP enzymes conduct oxidations in which a C–H bond is usually broken. These metabolizing reactions may potentially enable beneficial DIEs, depending on the enzymatic mechanism and the contribution of bond breaking to the overall enzymatic turnover.[11] The effects of deuteration on drug stability toward CYP metabolism are complex and unpredictable.[12] CYP catalysis is a multistep process wherein a number of steps could be partially or wholly rate limiting, thereby "masking" the isotope effect.[13] Additionally, as the system increases in complexity (Supersomes™ to microsomes to hepatocytes, etc.), other non-CYP clearance mechanisms may predominate and blunt the observed DIE. Furthermore, reducing CYP metabolism with deuterium at a known site of metabolism can "switch" metabolism to another site, resulting in no observed DIE (no change in clearance) or even an inverse DIE (faster clearance).[14–16] For example, maraviroc (**3**, Fig. 2) is a negative allosteric modulator of the chemokine receptor CCR5 marketed as Selzentry® for the treatment of human immunodeficiency virus infections. Oxidation at the pseudobenzylic methyl group to form "Metabolite 8" (**4**) and *N*-dealkylation adjacent to the tropane ring are two major metabolic pathways in humans.[17] Stability studies in human microsomal incubations showed that deuteration at both

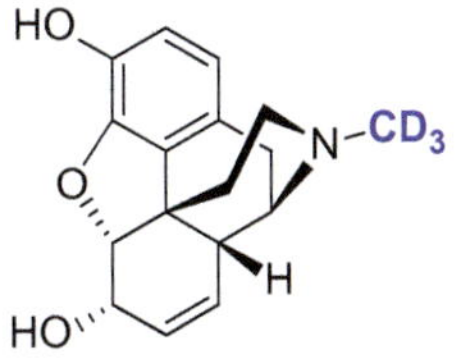

1

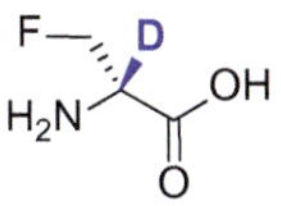

2

Fig. 1 Deuterated morphine and fludalanine.

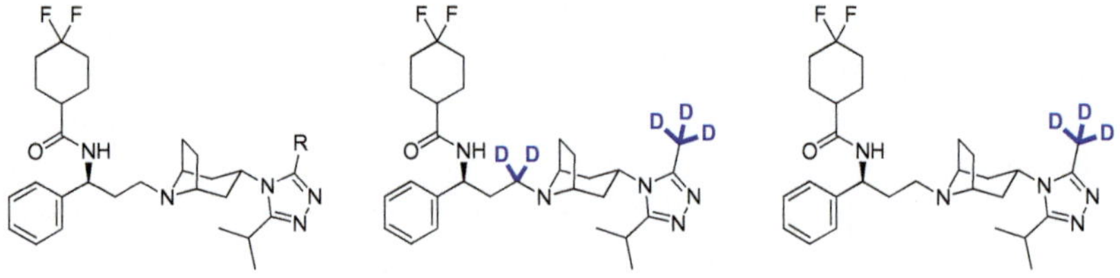

3 (R = CH_3; HLM $t_{1/2}$ = 97.2 min) **5** (HLM $t_{1/2}$ = 145.5 min) **6** (HLM $t_{1/2}$ = 45.5 min)
4 (R = CH_2OH)

Fig. 2 Maraviroc, deuterated analogs of maraviroc, and metabolite.

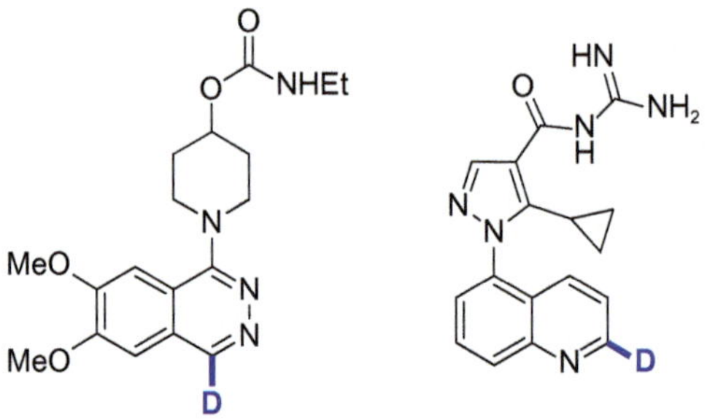

7 **8**

Fig. 3 Deuterated carbazeran and deuterated zoniporide.

of these positions (**5**) is required to reduce in vitro microsomal clearance. Unexpectedly, deuteration only at the methyl group (**6**) actually accelerated microsomal turnover compared with the protio drug **3**.[18]

Over time, medicinal chemists and drug metabolism scientists have succeeded in developing strategies to mitigate the risk of clinical failure due to poor pharmacokinetics when clearance is driven by CYP enzymes. However, other enzymatic mechanisms have also been recognized as causing rapid drug metabolism. Among these, aldehyde oxidase (AO) has gained notoriety for its ability to oxidize aromatic nitrogen heterocycles and for the difficulty scientists experience in attempting to translate in vitro AO intrinsic clearance to in vivo clearance.[19,20] DIEs have been experimentally determined for a number of AO substrates and have been correlated with in vitro and in vivo pharmacokinetic parameters for deuterated versions of two clinical compounds, 1-[^{2}H]-carbazeran (**7**) and 2-[^{2}H]-zoniporide (**8**) (Fig. 3).[21] While in vitro DIEs for the substrates investigated were significant (as high as 5), clearance mechanisms are complex, and thus it is difficult in general to predict whether deuteration of an AO substrate may lead to improved pharmacokinetic parameters in vivo.

3. A DEUTERATED DRUG GAINS REGULATORY APPROVAL

Deutetrabenazine (Austedo®, **9**, Fig. 4) is the first, and to date the only, deuterated drug approved by the FDA. Following 8 years during which only homochiral drugs gained FDA approval, this racemic drug was approved in April 2017 for the treatment of chorea associated with Huntington's disease. In addition, it is currently undergoing Phase 3 clinical trials for the treatment of tardive dyskinesia and is also in early clinical development for the treatment of Tourette's syndrome.[22]

Tetrabenazine (Xenazine®, **10**), the protio drug, has been known since the early 1950s. It gained FDA approval in 2008 and became the first treatment for Huntington's disease in the United States.[23] The development

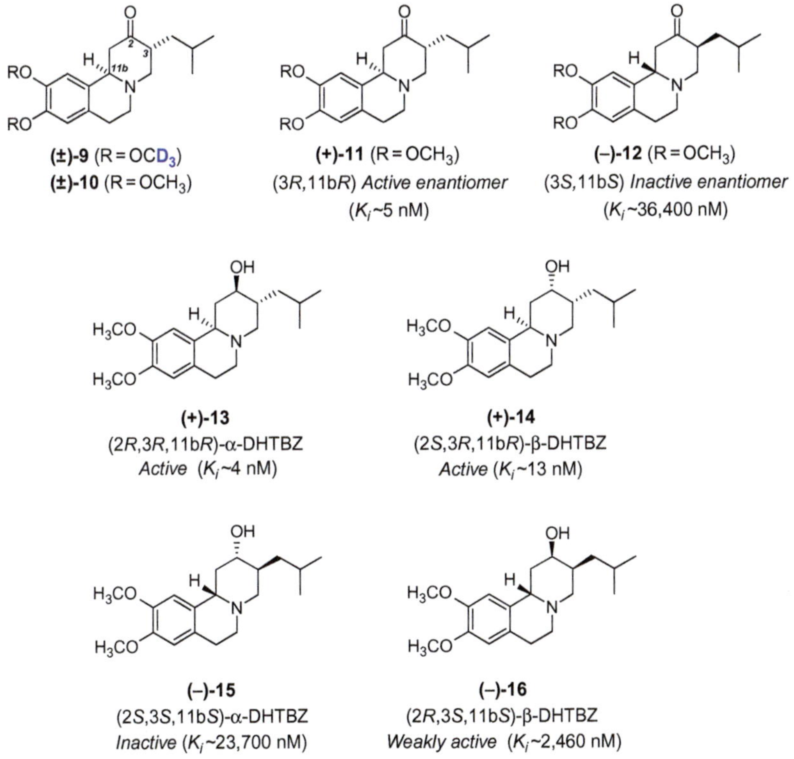

(DHTBZ = 3-isobutyl-9,10-dimethoxy-1,3,4,6,7,11b-hexahydro-2*H*-pyrido[2,1-*a*]isoquinolin-2-ol)

Fig. 4 Deutetrabenazine, tetrabenazine, and metabolites.

of deutetrabenazine avoided unnecessary duplication of studies by following a 505(b)(2) regulatory pathway, a development process that enabled referencing of certain nonclinical and safety findings previously reported in the tetrabenazine new drug application.[24,25] The mechanism of biological action of these drugs is thought to be the blockade of the vesicular monoamine transporter-2 (VMAT-2) in the brain. This leads to a decreased uptake of monoamines (e.g., dopamine, serotonin, norepinephrine, histamine) into synaptic vesicles and depletion of monoamine stores from nerve terminals.

In vitro, the binding affinities at VMAT-2 of the two enantiomers **11** and **12** differ significantly, and the reported values for (±)-**10**, (+)-**11**, and (−)-**12** are K_i of 8, 5, and 36,400 nM, respectively.[26] Following oral administration, the carbonyl group in the parent racemic drug is rapidly and extensively metabolized by carbonyl reductase to the diastereomeric alcohols (**13–16**) due to first pass metabolism, which results in low (5%–10%) drug oral bioavailability in rat and human.[27] Thus, given the low exposure of parent drug **9** or **10**, it is unlikely these two species have a significant contribution to the efficacy of the drug in vivo, whether in deuterated form or not.

The syntheses and biological properties of all possible diastereomers of the metabolite dihydrotetrabenazine have been reported.[26] For some of these isomers, absolute configuration was established by X-ray crystallography.[28,29] Characterization of the four metabolites **13–16** showed that for both α- and β-dihydrotetrabenazine metabolites, VMAT-2 affinity is highly enantioselective, with the 3*R*,11b*R* configuration (as in **13** and **14**) being highly preferred. The activity of the diastereomeric β-alcohol is similar to the α-alcohol. Therefore, not all circulating metabolites contribute to the biological activity of tetrabenazine. Deuteration does not change the intrinsic pharmacologic activity of a molecule, and thus a very similar pattern of VMAT-2 in vitro affinity was found for the $-OCD_3$ analogs, suggesting the active, circulating, brain-penetrant metabolites are the active species in deutetrabenazine as well.[30]

These active metabolites undergo further biotransformations. As a representative example, the pharmacologically active alcohol **13** (Fig. 5) is demethylated mainly by CYP2D6, leading to 9- or 10-*O*-demethylated species **17a,b**, which are both less active against VMAT-2.[30,31] Deuterium-mediated attenuation of the breakdown of active metabolites, combined with the use of controlled-release technology, leads to a differentiated

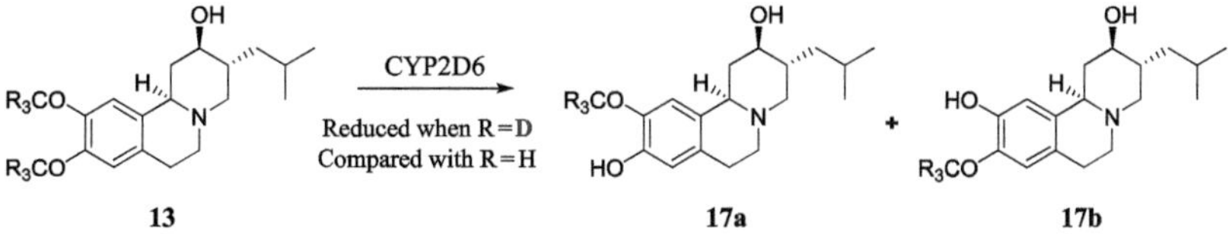

Fig. 5 CYP2D6 metabolism affords *O*-demethylated species.

pharmacokinetic profile for Austedo, enabling less frequent dosing, improving tolerability, and reducing the need for CYP2D6 genotyping compared with Xenazine.[31]

Results from clinical efficacy and safety studies compared with placebo were reported.[32,33] Although the deuterium impact is difficult to independently assess in terms of clinical effect since Austedo uses a controlled-release formulation while Xenazine does not, the deuterated methoxy groups in deutetrabenazine clearly attenuated the CYP2D6-driven metabolism and increased the half-lives of the active metabolites. Overall, Austedo provided more stable systemic exposure while preserving pharmacological activity. A comparison of the dosing recommendations for the protio drug Xenazine (three times daily) and the deuterated drug Austedo (twice daily) is shown in Table 1. While a clinical head-to-head comparison was not conducted, an indirect treatment comparison showed a substantially lower risk of moderate to severe adverse effects for the deuterated agent, indicating that for the treatment of Huntington's disease chorea Austedo appears to have a favorable tolerability profile compared to Xenazine.[37]

In summary, while translational aspects of the pharmacology of Austedo and Xenazine are complex, the beneficial impact of deuteration is driven by a reduction of the CYP2D6-catalyzed *O*-demethylation of the active, brain-penetrant, circulating metabolites.

4. CLINICAL COMPOUNDS

A number of deuterated compounds are progressing through the clinical pipelines of pharmaceutical companies. A select group of these compounds (Fig. 6) can be divided into three categories:

a. Deuterated versions of previously known drugs
b. Deuterated novel drugs
c. Deuterated forms of endogenous metabolites.

Table 1 Comparison of Some Properties of Xenazine and Austedo

	Xenazine (10)[34]		**Austedo** (9)[35,36]
Recommended maximum daily dose	50 mg in three divided doses per day; single dose not to exceed 25 mg		24 mg twice daily
Recommended starting dose	12.5 mg once daily		6 mg once daily
Dosage strengths	12.5 and 25 mg		6, 9, and 12 mg
Maximum dosing frequency	Three times a day		Two times a day
CYP2D6 phenotype effect	Intermediate and extensive metabolizers: dose may be increased to a maximum of 100 mg daily		Poor metabolizers: maximum dose should not exceed 36 mg
Food effect	Can be administered without regard to food. Food had no effect on mean plasma concentrations, C_{max}, or AUC of α-HTBZ or β-HTBZ		Administer with food. Food had no effect on the AUC of active metabolites; however, C_{max} increased ca. 50% in the presence of food
Pharmacokinetics (human, fasted)	Fasted		Fasted, extended release
	Dose = 25 mg		Dose = 15 mg
	C_{max} = 65.1 ng/mL		C_{max} = 22.5 ng/mL
	$t_{1/2}$ = 4.5 h		$t_{1/2}$ = 9.4 h
	T_{max} = 1.0 h		T_{max} = 2.3 h
	$AUC_{0\text{-}inf}$ = 257 ng h/mL		$AUC_{0\text{-}inf}$ = 273 ng h/mL
Metabolite properties	α-HTBZ, β-HTBZ, and 9-desmethyl-β-DHTBZ have half-lives of 7, 5, and 12 h, respectively		The half-life of total (α + β)-HTBZ from deutetrabenazine is approximately 9–10 h

All mean values except T_{max} (median).

4.1 Deuterated Versions of Previously Known Drugs

AVP-786, a combination of deuterium-substituted dextromethorphan **18** and a low dose of the CYP2D6 inhibitor quinidine, is being investigated in clinical trials for the treatment of neurologic and psychiatric disorders. The most advanced studies are Phase 3 clinical trials for the treatment of

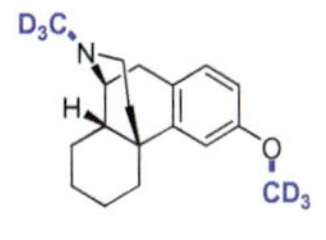

18

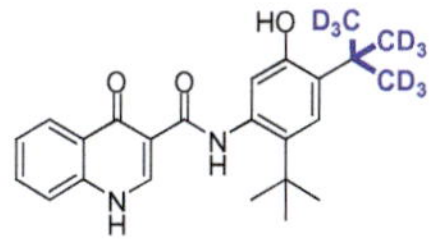

22

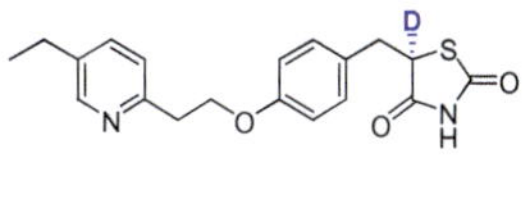

25

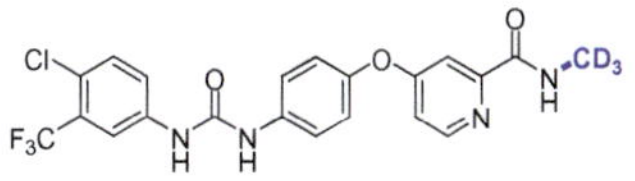

28

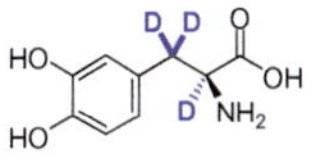

29
(Putative)

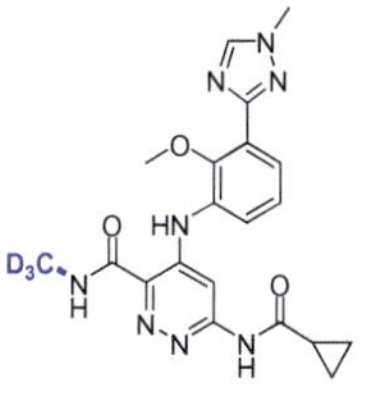

34

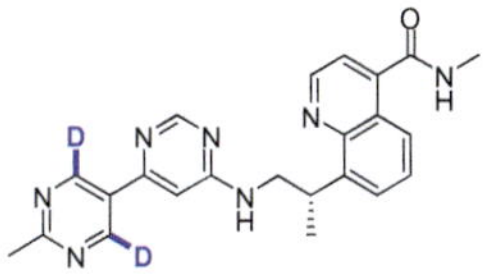

35

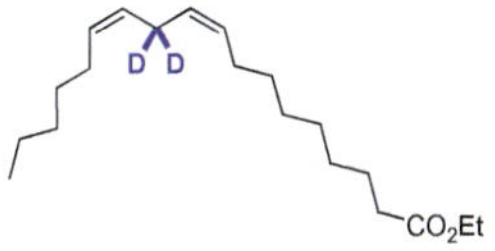

36

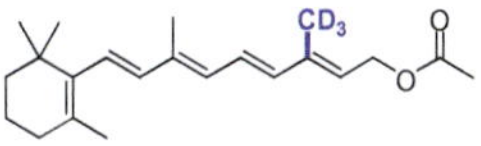

37

Fig. 6 Selected deuterated drugs studied in the clinic.

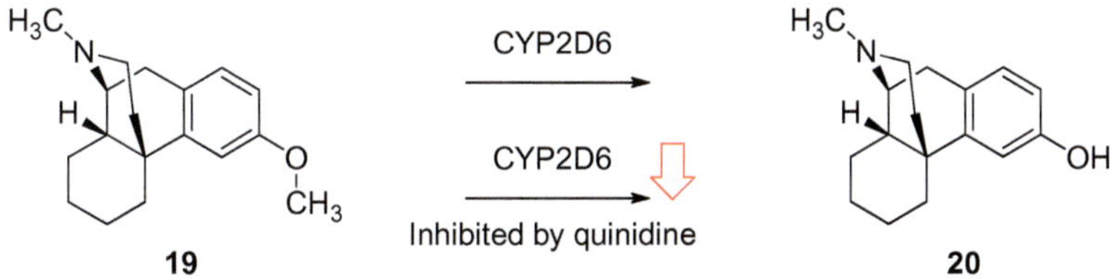

19 20

Fig. 7 Quinidine inhibits CYP2D6 turnover of **19** and reduces formation of phenol **20**.

agitation associated with Alzheimer's disease. A combination of 20 mg of non-deuterated dextromethorphan and a higher 10 mg dose of quinidine was previously approved by the FDA under the brand name Nuedexta®. It was approved for the treatment of pseudobulbar affect, a condition characterized by involuntary, sudden, and frequent episodes of laughing and/or crying.

In vivo, dextromethorphan (**19**) is rapidly *O*-demethylated by CYP2D6, a pharmacogenomically highly variable enzyme, which leads to variable and often very low oral bioavailability for dextromethorphan accompanied by the formation of the undesirable *O*-desmethyl metabolite, dextrorphan (**20**, Fig. 7). This metabolite is an NMDA receptor antagonist that causes psychoactive effects. In Nuedexta, quinidine is used to inhibit CYP2D6 and slow the formation of dextrorphan, while enhancing exposure to the parent drug. A further reduction in the required dose of quinidine can be accomplished by deuteration of the methyl groups as in **18**. Indeed, in a Phase 1 trial AVP-786 provided equivalent plasma exposure of **18** compared to nondeuterated dextromethorphan **19** at a substantially lower dose of quinidine.[38]

Cystic fibrosis (*CF*) is a genetic disease that results from mutations in the gene that encodes for the cystic fibrosis transmembrane conductance regulator (CFTR) protein. CFTR is a chloride channel that maintains anion homeostasis and mucus hydration necessary for epithelial tissue function. Defects in CFTR function affect multiple organs, including sweat glands, pancreas, colon, and genitourinary tract. Historically, the inability to digest food was a main cause of *CF* morbidity and mortality. Currently, due to the widespread use of oral pancreatic enzymes which enable *CF* patients to digest fats, the lung is the most critical organ with dysfunction resulting in morbidity and mortality. Ivacaftor (Kalydeco®, **21**) was the first drug approved for the treatment of *CF* in patients with specific mutations in CFTR. **CTP-656** (**22**), a deuterated version of ivacaftor, is a novel CFTR potentiator under clinical development for the treatment of cystic fibrosis in the approved mutations. In vivo, **21** is oxidized by CYP3A enzymes at one of its *tert*-butyl groups, leading to the formation of the primary alcohol (**23**) and the carboxylic acid (**24**) as major circulating metabolites (Fig. 8). Compared with the parent, both of these metabolites are less efficacious in enhancing functional response in vitro for short-circuit current (Isc) increase in the Ussing assay[39] in G551D/F508del HBE cells. In humans, deuteration of the targeted *tert*-butyl group leads to a ≥3-fold increase in exposure to CTP-656 vs ivacaftor and an increase in half-life from 11 to 15 h, while maintaining a similar exposure to the deuterated metabolites corresponding to **23** and **24** (Fig. 9).[40,41]

In a Phase 1 crossover study comparing CTP-656 and ivacaftor, the deuterated drug demonstrated reduced clearance, longer half-life, substantially increased exposure, and greater plasma levels at 24 h, all factors that may enable once-daily dosing for CTP-656.[40,41]

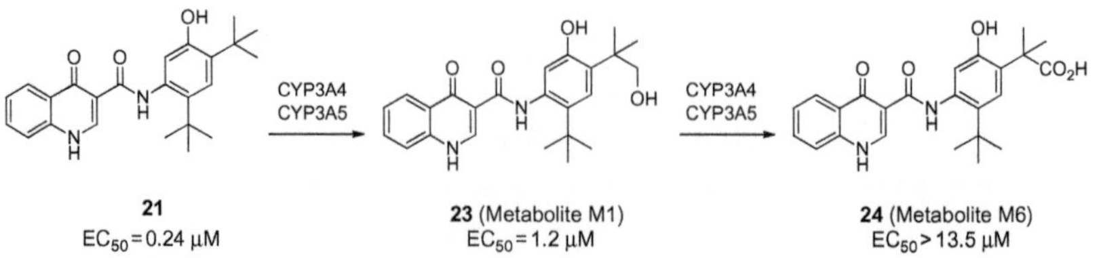

Fig. 8 Metabolism of ivacaftor in human.

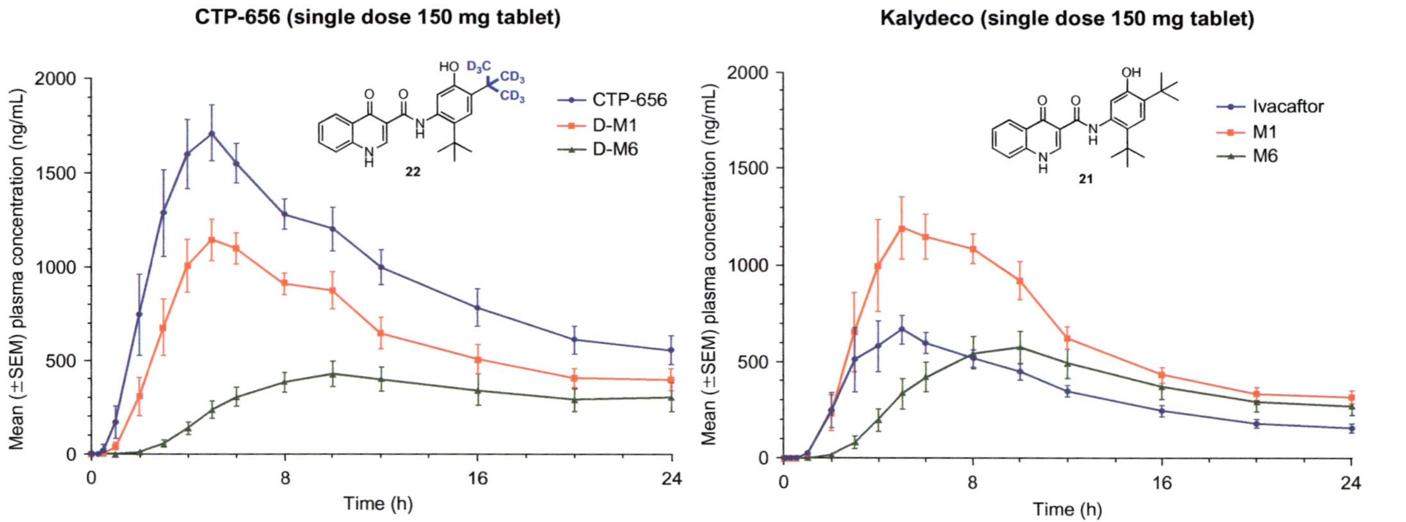

Fig. 9 Comparison of the oral exposure profiles for CTP-656 and Kalydeco (150 mg), including the alcohol (M1) and carboxylic acid (M6) circulating metabolites.

DRX-065 (**25**) is a deuterium-stabilized (*R*)-enantiomer of racemic pioglitazone ((±)-**26**), currently in Phase 1 studies for the potential treatment of adrenomyeloneuropathy and nonalcoholic steatohepatitis (NASH). Pioglitazone was approved with the brand name Actos® for the treatment of diabetes. Introduction of deuterium slows down the rate of racemization at C-5, which enables the isolation and characterization of individual enantiomers (Fig. 10). This strategy has previously been reported for other drug molecules[42,43] and is sometimes called "deuterium-enabled chiral switching" (DECS). The two pioglitazone enantiomers have different pharmacological properties. The (*S*)-enantiomer is a PPARγ agonist, a mechanism associated with side effects like weight gain and edema. Preclinical studies have shown that the (*R*)-enantiomer does not produce weight-gain effects, but shows modulation of mitochondrial function and antiinflammatory effects thought to derive from inhibition of the mitochondrial pyruvate carrier. Therefore, DRX-065 is expected to be devoid of the PPARγ-related side effects for NASH patients and to show a superior therapeutic index compared to pioglitazone. Clinical development follows a 505(b)(2) regulatory path, with initial doses tested in a Phase 1 open-label study of 22.5 mg for DRX-065 and 45 mg for Actos.[44]

Donafenib (**28**) is a protein kinase and angiogenesis inhibitor undergoing a number of clinical trials for oncology indications, including the treatment of metastatic colorectal cancer,[45,46] metastatic gastric cancer,[47] advanced esophageal cancer,[48] advanced hepatocellular cancer,[49,50] and metastatic nasopharyngeal cancer.[51] Tested oral doses are 200 and 300 mg twice daily. This compound is the deuterated version of sorafenib (Nexavar®), a drug approved for the treatment of primary kidney cancer (advanced renal cell

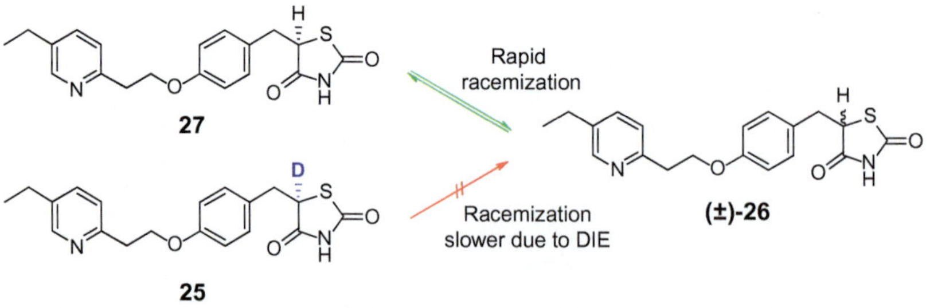

Fig. 10 Under physiologically relevant conditions, protio enantiomer **27** racemizes at a faster rate than deuterated enantiomer **25**. The rate difference renders the deuterated enantiomer configurationally stable in vivo, for practical purposes.

carcinoma), advanced primary liver cancer (hepatocellular carcinoma), and radioactive iodine-resistant advanced thyroid carcinoma.[52] Efficacious doses of sorafenib are up to 400 mg taken twice a day.[53]

Deulevodopa or "*deuldopa*," a deuterated version of levodopa (L-DOPA) of undisclosed chemical structure also known as SD-1077, is reportedly undergoing clinical Phase 1 studies.[54] The putative structure of this compound is thought to be **29**. In preclinical models of Parkinson's disease, deulevodopa was 1.5-fold more potent than L-DOPA with respect to immediate disability reduction and induction of dyskinesia. In addition, a significant prolongation of the reduction in disability and "ON time" was seen for deulevodopa compared to L-DOPA.[55] The effect of deuteration was attributed to decreased postsynaptic dopamine metabolism at MAO-B-containing sites. In this case, deuteration afforded equivalent in vivo efficacy at 60% of the dose of L-DOPA.[56] As to the mechanism of this effect, in vitro studies show that these enzymatic reactions occur with a very high degree of stereoselectivity, as shown in Fig. 11. Kinetic isotope effects for the decarboxylation of **29** to **30** by L-DOPA decarboxylase are close to unity,[57] and no significant contributions are envisioned from that step or the catechol methylation by COMT. The DIEs for MAO-catalyzed reactions on this substrate are relatively low (1.0–1.4).[58] On the other hand, the DBH oxidation of dopamine (**30**) to epinephrine (**33**) occurs with a large DIE (9.4–10.9),[59] suggesting this pathway may be relevant to the observed in vivo effects with deulevodopa.

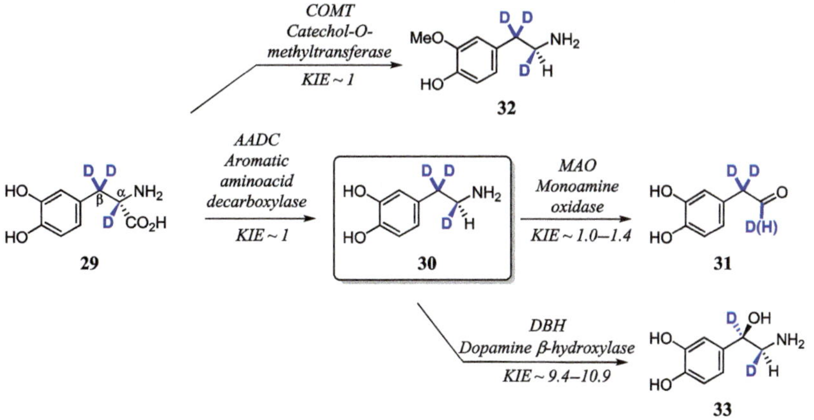

Fig. 11 Mechanism of the enzymatic transformation of deuterated 2,3,3-$^{2}H_3$-L-DOPA, indicating the participation of COMT, AADC, DBH, and MAO.

4.2 Deuterated Novel Drug Candidates

Medicinal chemists are increasingly employing deuterium substitution during the lead optimization phase of drug discovery projects. Deuteration has the potential to address a variety of issues beyond high CYP-driven metabolism. Two of these deuterated drug candidates are discussed in this section.

Compounds able to inhibit tyrosine kinase TYK2-mediated signal transduction may have utility in the treatment of autoimmune diseases and inflammation. **BMS-986165** (**34**) is an oral allosteric inhibitor of TYK2 in Phase 2 clinical trials for the treatment of psoriasis (NCT02534636 and NCT03004768) and in Phase 1 for other autoimmune disorders (NCT03044873).[60] Oral half-life was longer than 2 h after a 30 mg oral dose in human.[61] This compound is highly potent and selective toward TYK2, blocks IL-12, IL-23, and type I interferon signaling, and provides robust efficacy in preclinical models of systemic lupus erythematosus and inflammatory bowel disease.[62]

VX-984 (**35**) is an inhibitor of DNA-dependent protein kinase that also targets the DNA damage repair system. Deuteration was used as a strategy to mitigate AO-driven metabolism.[63] The drug has in vitro K_i values of 2 nM, 5.2 μM, >4 μM, 160 nM, 2.6 μM, and 0.3 μM for DNA-PK, serine-protein kinase ATM, serine/threonine-protein kinase ATR, phosphatidylinositol 3-kinase PI3Kα, PI3Kβ, and PI3Kγ, respectively. It shows good metabolic stability in human hepatocytes. It is currently undergoing Phase 1 clinical trials for the treatment of solid tumor and for the treatment of lymphoma. The solid tumor study is evaluating escalating doses of VX-984 alone and in combination with pegylated liposomal doxorubicin.[64]

4.3 Deuterated Endogenous Metabolites

When a disease is linked to a deficit or excess of an endogenous metabolite, deuteration of endogenous species at one or more key positions is being employed in certain cases as a tactic to correct this defect and provide a disease-modifying therapy. A few examples in this area are discussed in this section.

Cell survival processes require an optimal function of cellular membranes. Modification of the composition and properties of membranes can correct human cellular dysfunction.[65] Long-chain fatty acids, such as linoleic acid, are components of cell membranes. Mitochondrial membranes are particularly enriched in fatty acids, present in 80% of the total phospholipids and degraded by enzymatic and nonenzymatic mechanisms. Under pathophysiological conditions leading to excessive production of reactive oxygen species, membrane breakage may occur through a radical peroxidation mechanism, the limiting rate of which is hydrogen abstraction at the reactive *bis*-allylic position. In the absence of antioxidants, the chain reaction propagates by hydrogen atom transfer followed by diffusion-controlled oxygen addition to the resulting carbon radical. The consequences are deteriorating membrane fluidity and oxidative damage to biomolecules such as proteins and DNA through cross-linking of highly reactive carbonyl compounds (Fig. 12). Deuteration of this *bis*-allylic position, even partially, was shown to dramatically reduce the rate of peroxidation—as much as 36-fold vs the nondeuterated counterpart.[66] In turn, cell membranes are stabilized against such free radical processes.[66]

Friedreich's ataxia (FA) is an autosomal recessive disorder caused by reduced levels of the protein frataxin. This protein is located in mitochondria, where it has been linked to the biogenesis of iron–sulfur clusters relevant to the mitochondrial respiratory chain complexes. Disruption in iron biogenesis may lead to oxidative stress, which may cause mitochondrial energy imbalance and eventual cell death by the free radical peroxidation mechanism previously discussed.[67] **RT001** (**36**) is a deuterated derivative of linoleic acid undergoing clinical testing for the oral treatment of FA, having received orphan drug designation in 2016. Early results from studies dosing either 1.8 g once daily or 4.5 g twice daily, in 18 FA patients for 28 days, met primary safety, tolerability, and pharmacodynamics goals.[68] A number of activity measures were correlated to disease severity scales, and the drug showed efficacy at one or more doses. The maximum tolerated dose of

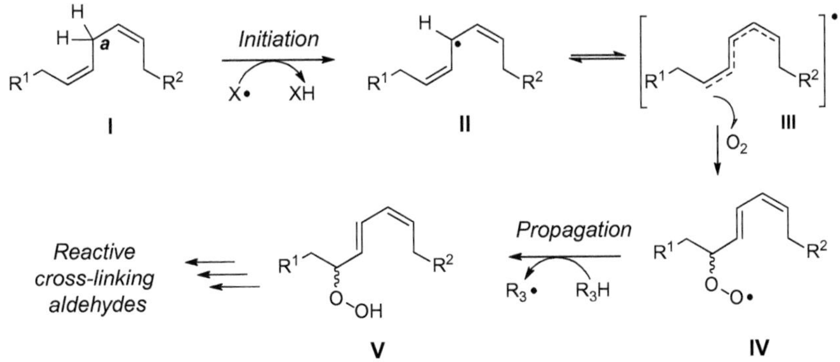

Fig. 12 Mechanism of radical chain reaction leading to cellular damage. Rate of H abstraction to form intermediate II is reduced when the *bis*-allylic position *"a"* is deuterated.

RT001 is 9 g/day due to uncontrolled diarrhea, a common complication of high fish oil doses in hypercholesterolemia. Metabolites of RT001 were also detected, consistent with normal fatty acid processing.[69]

Another example of a deuterated endogenous metabolite with therapeutic potential for disease modification is **ALK-001** (**37**). Retinal pigment epithelial (RPE) lipofuscin is a pigment implicated in the etiology of degenerative eye diseases, such as Stargardt disease, Best disease and age-related macular degeneration—some of the most common causes of blindness. More than 90% of the lipofuscin accumulated in RPE cells originates from conjugates formed by visual cycle retinoids in photoreceptor cells, including the polyenes *N*-retinylidene-*N*-retinylethanolamine (A2E, **38**, Fig. 13) and all-*trans*-retinaldehyde dimer (ATR dimer, **39**). These dimeric compounds containing the vitamin A scaffold cannot be enzymatically degraded, leading to accumulation and the formation of deposits in RPE cells.[70] Aging, environmental factors, drugs, or genetic mutations can all compromise the vitamin A cycle in the retina, resulting in the accelerated formation of vitamin A dimers.

Attempts to reduce vitamin A dimerization and prevent vision loss by this mechanism led to the discovery of **37**, a deuterated vitamin A analog in Phase 2 clinical trials for the oral treatment of Stargardt's macular dystrophy. The hypothesis is that in the eye, two molecules of all-*trans*-retinaldehyde (**40**, or **41** in deuterated form) combine to form A2E and ATR dimer following a multistep reaction sequence (Fig. 14). In the first

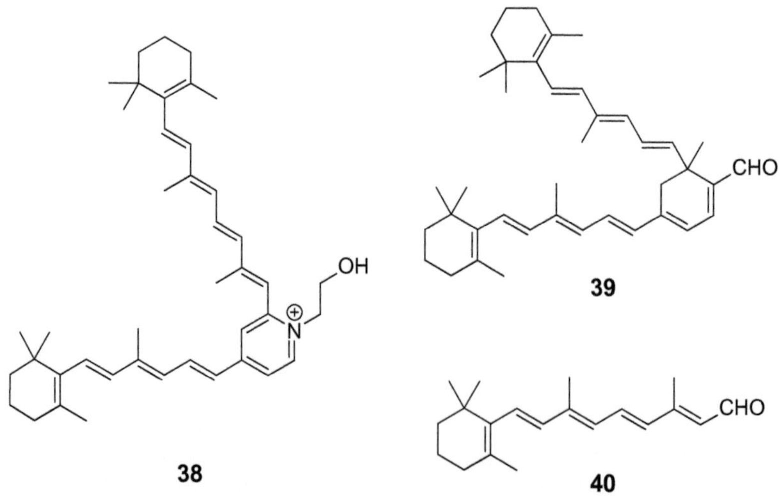

Fig. 13 Vitamin A derivatives implicated in damage to the eye.

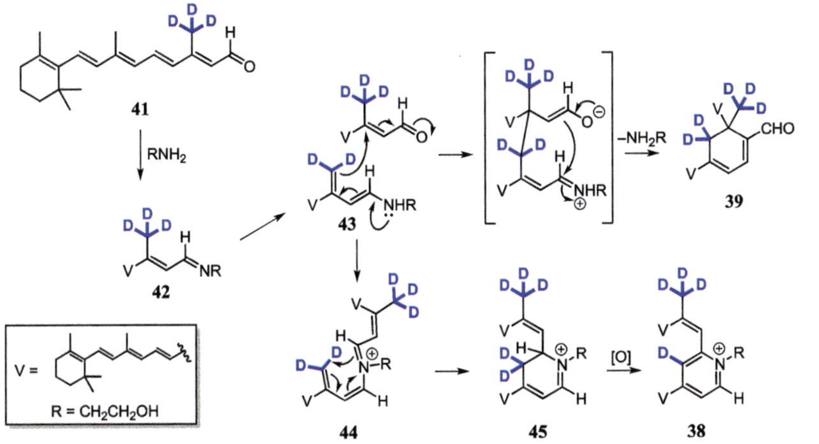

Fig. 14 Proposed biosynthesis of A2E and ATR dimer from deuterated all-*trans*-retinaldehyde.

step, retinaldehyde reacts with phosphatidylethanolamine, abundant in the lipid-rich environment of the disk membrane, to form a Schiff base (**42**), or either of the *trans* (**43**) or *cis* (**44**) isomeric enamines. Further condensation with retinaldehyde, either through amine condensation or 1,4-addition followed by cyclization (or direct Diels–Alder cycloaddition), yields the reduced forms of A2E (**45**) and deuterated ATR dimer (**39**), respectively. A2E is then formed through aromatization of the dihydropyridine, and ATR dimer is formed through deamination. Deuteration of the C-20 methyl group leads to a reduction in the rates of these cyclization reactions, thus potentially providing a disease-modifying treatment.[71]

5. CAN DEUTERIUM-CONTAINING DRUGS BE MANUFACTURED WITH ACCEPTABLE COSTS?

For the particular clinical compounds discussed earlier, the introduction of deuterium uses D_2O, CD_3OD, or other relatively inexpensive raw materials which are commercially available in large quantities and high chemical and isotopic purity (Table 2). Furthermore, as with every drug candidate that becomes commercialized, it is the ingenuity and dedication of process chemists that makes possible the delivery of a final synthetic route that meets cost and regulatory requirements. Thus, cost of goods or access to deuterated raw materials have not generally proven to be impediments to the successful use of a deuteration strategy in drug development.

Table 2 Comparison of Deuteration Effects for Selected Drugs Studied in the Clinic

Compound	Mechanism of Beneficial Deuterium Effect	Source of Deuterium in Synthesis[a]
Deutetrabenazine	Reduce rate of active metabolite demethylation driven by CYP2D6	CD_3I[72]
AVP-786	Reduce formation of toxic metabolite by CYP2D6	CD_3I, $LiAlD_4$[73]
CTP-656	Reduce rate of *t*-Bu group oxidation and in vivo clearance by CYP3A4	$(CD_3)_3COD$[74]
BMS-986165	Reduce formation of metabolites that impact kinase selectivity	$CD_3NH_2{\cdot}HCl$[75]
VX-984	Reduce aldehyde oxidase-driven metabolism	DCO_2D/ CD_3OD[76]
RT001	Minimize autooxidation at the *bis*-allylic position of linoleic acid esters, component of cell membranes	$(CD_2O)_n$[77]
ALK-001	Reduce rate of formation of all-*trans*-retinal dimers linked to detrimental deposits in the eye	D_2O[78]
DRX-065	Reduce rate of in vivo racemization, allowing characterization of individual enantiomers	$CD_3S(O)CD_3$, CD_3OD[79]
Donafenib	Reduce *N*-hydroxylation and demethylation by CYP3A4/2D6	CD_3OD[80]
Deulevodopa	Decrease dopamine metabolism	DCl/D_2O, D_2O/ CD_3OD[81]

[a]The synthetic pathways cited are generally those used to access initial quantities of compound as cited in patent literature. The commercial manufacturing routes may be different and may use alternative deuterium-containing reagents.

6. CHEMICAL BIOLOGY TOOLS

(±)-Ketamine (**46**, Fig. 15) is a marketed drug, known for decades, that was initially developed as a short-acting anesthetic. Common side effects include agitation, confusion, or hallucinations.[82] Recent clinical studies demonstrated that low, subhallucinogenic doses of **46** are efficacious as a fast-acting antidepressant treatment, with effects lasting for weeks after a

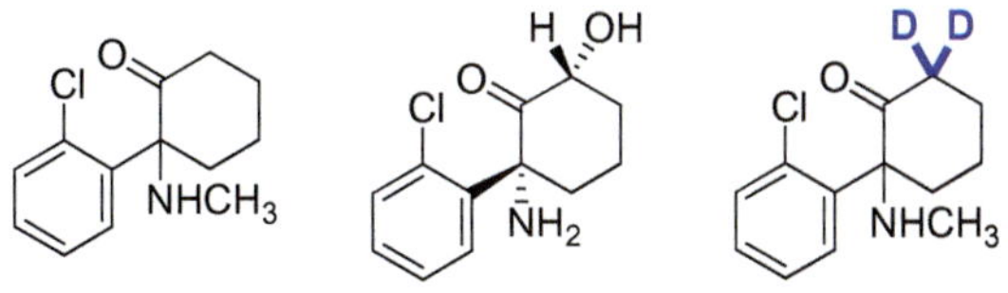

46 **47** **48**

Fig. 15 Ketamine, (2*R*,6*R*)-hydroxynorketamine, and deuterated probe compound.

single intravenous dosing.[83] Given the length of these effects, in contrast with the extensive metabolism and short pharmacokinetic half-life of the parent drug, major efforts are underway to understand the mechanism of action. A large number of metabolites were synthesized as single enantiomers, and their biological properties were evaluated in a number of tests. These included classical in vivo preclinical models of antidepressant-like effects, such as the forced-swim test and the learned helplessness paradigms. In these tests, the metabolite (2*R*,6*R*)-hydroxynorketamine (**47**) showed greater efficacy at a lower dose and longer-lasting effects than ketamine itself. Deuteration of position 6 to create probe compound (±)-**48** led to lower in vivo plasma concentrations of the active 6-hydroxy-metabolites, and provided correspondingly less efficacy. These results support the hypothesis that metabolism of ketamine to a 6-hydroxy derivative is a necessary step to produce antidepressant-like effects.[84]

As for potential clinical uses of deuterated versions of ketamine, an Auspex patent application directed toward deuterated ketamine was published in 2009.[85] Teva acquired Auspex in 2015 and reportedly is evaluating deuterium-labeled ketamine preclinically for an undisclosed indication.[86]

7. DEUTERATED PET LIGANDS

PET ligands are important translational tools. They are designed to have reversible binding at the target and fast clearance, but not too fast. Several strategies have been utilized which benefit from DIEs in developing improved PET ligands with optimal in vivo stability. In one case, the quantification of MAO-B using ^{18}F-fluorodeprenyl was made difficult by the MAO-B-catalyzed irreversible formation of a covalent bond to the enzyme. This bond formation caused the radioligand distribution in tissue to be controlled by blood flow rather than by the MAO-B enzyme concentration in regions with high activity. It was hypothesized that deuteration would reduce the irreversible binding to MAO-B through reduced formation of

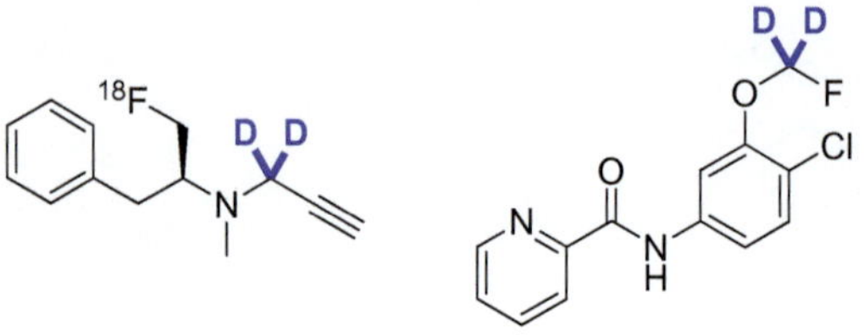

49 **50**

Fig. 16 Deuterated PET ligands of interest.

the reactive moiety. Indeed, the deuterated analog **49** (Fig. 16) was more stable in monkey blood plasma than the protio precursor, and deuteration led to a reduction in irreversible binding to the enzyme in vivo. The resulting improved sensitivity makes ^{18}F-[$^{2}H_2$]-fluorodeprenyl an improved PET tracer for MAO-B activity in the human brain.[87]

In another example, a series of *N*-aryl-picolinamide derivatives were studied as mGluR4 PET ligands. The dideuterofluoromethoxy analog **50** was identified as a promising high-affinity mGluR4 ligand, with improved in vitro microsomal stability over the protio counterpart, and was explored as PET ligand.[88]

8. CONCLUSIONS

The use of deuterium to drive DIEs and favorably alter pharmacokinetics and metabolism in drug discovery has accelerated in the last decade. The first major wave of patent applications came from small biotechnology organizations and were aimed at improving pharmacokinetic properties of marketed drugs or enhancing properties of failed drug candidates. A substantial number of patent applications were published beginning in 2006, and the deuterium chemistry patent estate continues to expand. Deuterated drug candidates have progressed from early patent filings to successful proof-of-concept human studies, including the first FDA-approved deuterated drug. The use of DIEs in drug design has expanded, and now deuteration of drug candidates and tool compounds, including imaging ligands, has become a welcome instrument in the arsenal of medicinal chemists. It is expected that experimentation with strategic introduction of deuterium into the structures of organic compounds will continue to increase and may enable the future development of a variety of novel drugs and chemical tools.

REFERENCES

1. As established by the internationally accepted Vienna Standard Mean Ocean Water (VSMOW). *See* Rosman, K. J. R.; Taylor, P. D. P. *Pure Appl. Chem.* **1998**, *70*(1), 217–235.
2. Cooper, G. J. T.; Surman, A. J.; McIver, J.; Colón-Santos, S. M.; Gromski, P. S.; Buchwald, S.; Suárez Marina, I.; Cronin, L. *Angew. Chem. Int. Ed. Engl.* **2017**, *56*, 8079–8082.
3. ClinicalTrials.gov NCT00990379, last updated: March 30, 2015, Accessed on May 8, 2017.
4. Katz, J. J.; Crespi, H. L. *Science* **1966**, *151*(3715), 1187–1194.
5. Harbeson, S. L.; Tung, R. D. *Annu. Rep. Med. Chem.* **2011**, *46*, 403–417.
6. Meanwell, N. A. *J. Med. Chem.* **2011**, *54*, 2529–2591.
7. Gant, T. G. *J. Med. Chem.* **2014**, *57*, 3595–3611.
8. Yang, J. *Deuterium, Discovery and Applications in Organic Chemistry*, 1st ed.; Elsevier: Cambridge, MA, 2016.
9. Elison, C.; Rapoport, H.; Laursen, R.; Elliott, H. W. *Science* **1961**, *134*(3485), 1078–1079.
10. Reinhold, D. F., US Patent 3,976,689, 1976.
11. Guengerich, F. P. *J. Label. Compd. Radiopharm.* **2013**, *56*(9–10), 428–431.
12. Fisher, M. B.; Henne, K. R.; Boer, J. *Curr. Opin. Drug Discov. Devel.* **2006**, *9*(1), 101–109.
13. Miwa, G. T.; Garland, W. A.; Hodshon, B. J.; Lu, A. Y. H.; Northrop, D. B. *J. Biol. Chem.* **1980**, *255*(13), 6049–6054.
14. Korzekwa, K. R.; Trager, W. F.; Gillette, J. R. *Biochem.* **1989**, *28*(23), 9012–9018.
15. Ling, K.-H. J.; Hanzlik, R. P. *Biochem. Biophys. Res. Commun.* **1989**, *160*(2), 844–849.
16. Nelson, S. D.; Trager, W. F. *Drug Metab. Dispos.* **2003**, *31*(2), 1481–1498.
17. Walker, D. K.; Abel, S.; Comby, P.; Muirhead, G. J.; Nedderman, A. N.; Smith, D. A. *Drug Metab. Dispos.* **2005**, *33*(4), 587–595.
18. Declaration filed on Nov. 16, 2010 in the file history of Tung, R., US Patent 7, 932,235, 2011.
19. Rashidi, M. R.; Soltani, S. *Expert Opin. Drug Discov.* **2017**, *12*(3), 305–316.
20. Pryde, D. C.; Tran, T. D.; Jones, P.; Duckworth, J.; Howard, M.; Gardner, I.; Hyland, R.; Webster, R.; Wenham, T.; Bagal, S.; Omoto, K.; Schneider, R. P.; Lin, J. *Bioorg. Med. Chem. Lett.* **2012**, *22*(8), 2856–2860.
21. Sharma, R.; Strelevitz, T. J.; Gao, H.; Clark, A. J.; Schildknegt, K.; Obach, R. S.; Ripp, S. L.; Spracklin, D. K.; Tremaine, L. M.; Vaz, A. D. *Drug Metab. Dispos.* **2012**, *40*(3), 625–634.
22. Teva Fact Sheet Q4 2016. www.TevaPharm.com. Accessed on April 13, 2017.
23. Yero, T.; Rey, J. A. *P. & T* **2008**, *33*(12), 690–694.
24. https://www.accessdata.fda.gov/drugsatfda_docs/appletter/2017/208082Orig1s000ltr.pdf. Accessed 19 May 2017.
25. https://www.fda.gov/downloads/Drugs/Guidances/ucm079345.pdf. Accessed 27 June 2017.
26. Yao, Z.; Wei, X.; Wu, X.; Katz, J. L.; Kopajtic, T.; Greig, N. H.; Sun, H. *Eur. J. Med. Chem.* **2011**, *46*(5), 1841–1848.
27. Mehvar, R.; Jamali, F.; Watson, M. W.; Skelton, D. *Drug Metab. Dispos.* **1987**, *15*(2), 250–255.
28. Kilbourn, M. R.; Lee, L. C.; Heeg, M. J.; Jewett, D. M. *Chirality* **1997**, *9*(1), 59–62.
29. Yu, Q. S.; Luo, W.; Deschamps, J.; Holloway, H. W.; Kopajtic, T.; Katz, J. L.; Brossi, A.; Greig, N. H. *ACS Med. Chem. Lett.* **2010**, *1*(3), 105–109.
30. https://www.accessdata.fda.gov/drugsatfda_docs/nda/2017/208082orig1s000toc.cfm. Accessed 19 May 2017.

31. US Securities and Exchange Commission, Form S-1, Auspex Pharmaceuticals, Inc. https://www.sec.gov/Archives/edgar/data/1454189/000119312513481239/d627086ds1.htm.
32. Huntington Study Group; Frank, S.; Testa, C. M.; Stamler, D.; Kayson, E.; Davis, C.; Edmondson, M. C.; Kinel, S.; Leavitt, B.; Oakes, D.; O'Neill, C.; Vaughan, C.; Goldstein, J.; Herzog, M.; Snively, V.; Whaley, J.; Wong, C.; Suter, G.; Jankovic, J.; Jimenez-Shahed, J.; Hunter, C.; Claassen, D. O.; Roman, O. C.; Sung, V.; Smith, J.; Janicki, S.; Clouse, R.; Saint-Hilaire, M.; Hohler, A.; Turpin, D.; James, R. C.; Rodriguez, R.; Rizer, K.; Anderson, K. E.; Heller, H.; Carlson, A.; Criswell, S.; Racette, B. A.; Revilla, F. J.; Nucifora, F., Jr.; Margolis, R. L.; Ong, M.; Mendis, T.; Mendis, N.; Singer, C.; Quesada, M.; Paulsen, J. S.; Brashers-Krug, T.; Miller, A.; Kerr, J.; Dubinsky, R. M.; Gray, C.; Factor, S. A.; Sperin, E.; Molho, E.; Eglow, M.; Evans, S.; Kumar, R.; Reeves, C.; Samii, A.; Chouinard, S.; Beland, M.; Scott, B. L.; Hickey, P. T.; Esmail, S.; Fung, W. L.; Gibbons, C.; Qi, L.; Colcher, A.; Hackmyer, C.; McGarry, A.; Klos, K.; Gudesblatt, M.; Fafard, L.; Graffitti, L.; Schneider, D. P.; Dhall, R.; Wojcieszek, J. M.; LaFaver, K.; Duker, A.; Neefus, E.; Wilson-Perez, H.; Shprecher, D.; Wall, P.; Blindauer, K. A.; Wheeler, L.; Boyd, J. T.; Houston, E.; Farbman, E. S.; Agarwal, P.; Eberly, S. W.; Watts, A.; Tariot, P. N.; Feigin, A.; Evans, S.; Beck, C.; Orme, C.; Edicola, J.; Christopher, E. *JAMA* **2016**, *316*(1), 40–50.
33. Geschwind, M. D.; Paras, N. *JAMA* **2016**, *316*(1), 33–35.
34. http://www.lundbeck.com/upload/us/files/pdf/Products/Xenazine_PI_US_EN.pdf. Accessed 4 July 2017.
35. https://www.accessdata.fda.gov/drugsatfda_docs/label/2017/208082s000lbl.pdf. Accessed 4 July 2017.
36. Stamler, D. A.; Brown, F.; Bradbury, M. *Mov. Disord.* **2013**, *28*(Suppl. 1), 765.
37. Claassen, D. O.; Carroll, B.; De Boer, L. M.; Wu, E.; Ayyagari, R.; Gandhi, S.; Stamler, D. *J. Clin. Mov. Disord.* **2017**, *4*(3), 1–11.
38. *Avanir Pharmaceuticals Announces Accelerated Development Path for AVP-786 Following Successful Pre-IND Meeting with FDA*. Press Release. June 4, 2013. www.avanir.com. Accessed 19 April 2017.
39. Devor, D. C.; Bridges, R. J.; Pilewski, J. M. *Am. J. Physiol. Cell Physiol.* **2000**, *279*(2), C461–C479.
40. Harbeson, S. L.; Nguyen, S.; Bridson, G.; Uttamsingh, V.; Wu, L.; Morgan, A. J.; Aslanian, A.; Braman, V.; Pilja, L. In: *Pharmacokinetic studies of deuterated analogs of ivacaftor in preclinical models and healthy volunteers; North American Cystic Fibrosis Conference (NACFC) Abst 29, October 8–10*, Phoenix, AZ; 2015. http://www.concertpharma.com/wp-content/uploads/2015/10/CTP656-Ph1SAD-NACFC-08OCT2015.pdf. Accessed 26 May 2017.
41. Uttamsingh, V.; Pilja, L.; Brummel, C. L.; Grotbeck, B.; Cassella, J. V.; Braman, G. In: *CTP-656 multiple dose pharmacokinetic profile continues to support a once-daily potentiator for cystic fibrosis patients with gating mutations; 30th Annual North American Cystic Fibrosis Conference (NACFC) Abst 224, October 27–29,* Orlando, FL; 2016. http://www.concertpharma.com/wp-content/uploads/2014/12/CTP-656-NACFC-POSTER-2016-27OCT16.pdf. Accessed 26 May 2017.
42. Uttamsingh, V.; Gallegos, R.; Cheng, C.; Aslanian, A.; Liu, J. F.; Tung, R.; Wu, L. In: *CTP-221, a deuterated S-enantiomer of lenalidomide, is greatly stabilized to epimerization and results in a more desirable pharmacokinetic profile than lenalidomide; American Association for Cancer Research (AACR) Annual Meeting 2013, April 6–10*; 2013.
43. Jacques, V.; Czarnik, A. W.; Judge, T. M.; Van der Ploeg, L. H. T.; DeWitt, S. H. *PNAS* **2015**, *112*(12), E1471–E1479.

44. Czarnik, A.; Dewitt, S.; Jacques, V.; Van der Ploeg, L. In: *Discovery of DRX-065: Characterizing the non-PPARγ, mitochondrial function modulation and anti-inflammatory activity of thiazolidinedione (TZD) enantiomers using deuterium; 252nd American Chemical Society (ACS) National Meeting (August 21–25, Philadelphia)*; 2016. Abst MEDI 274.
45. NCT02489916, last updated on January 3, 2016. www.clinicaltrials.gov, Accessed on April 11, 2017.
46. NCT02870582, last updated on March 20, 2017. www.clinicaltrials.gov, Accessed on April 11, 2017.
47. NCT02489214, last updated on August 12, 2016. www.clinicaltrials.gov, Accessed on April 11, 2017.
48. NCT02489201, last updated on August 12, 2016. www.clinicaltrials.gov, Accessed on April 11, 2017.
49. NCT02229071, last updated on January 23, 2017. www.clinicaltrials.gov, Accessed on April 11, 2017.
50. NCT02645981, last updated on March 21, 2017. www.clinicaltrials.gov, Accessed on April 11, 2017.
51. NCT02698111, last updated on October 12, 2016. www.clinicaltrials.gov, Accessed on April 11, 2017.
52. https://www.accessdata.fda.gov/scripts/cder/ob/search_product.cfm. Accessed 4 November 2017.
53. http://labeling.bayerhealthcare.com/html/products/pi/Nexavar_PI.pdf.
54. Reference to TEVA website. Accessed on April 13, 2017.
55. https://www.michaeljfox.org/foundation/grant-detail.php?grant_id=991. Accessed 13 April 2017.
56. Malmlöf, T.; Rylander, D.; Alken, R. G.; Schneider, F.; Svensson, T. H.; Cenci, M. A.; Schilström, B. *Exp. Neurol.* **2010**, *225*(2), 408–415.
57. Lin, Y. L.; Gao, J. *J. Am. Chem. Soc.* **2011**, *133*(12), 4398–4403.
58. Yu, P. H.; Bailey, B. A.; Durden, D. A.; Boulton, A. A. *Biochem. Pharmacol.* **1986**, *35*(6), 1027–1036.
59. Miller, S. M.; Klinman, J. P. *Biochemistry* **1983**, *22*(13), 3091–3096.
60. https://www.clinicaltrials.gov/ct2/results?term=BMS-986165+&Search=Search. Accessed 19 May 2017.
61. Weinstein, D.; Wrobleski, S.; Moslin, R.; et al. In: *Discovery of a pseudokinase domain ligand as an allosteric inhibitor of TYK2 for the treatment of autoimmune diseases; 252nd American Chemical Society (ACS) National Meeting (August 21–25, Philadelphia)*; 2016. Abst MEDI 272.
62. Gillooly, K.; Zhang, Y.; Yang, X.; et al. 80th Annual Scientific Meeting American College of Rheumatology (November 11–16, Washington D.C.) 2016, Abst 11L.
63. Cottrell, K.; Boucher, B.; Arimoto, R.; et al. In: *Discovery of VX-984: Mitigation of aldehyde oxidase metabolism through the use of targeted deuteration; 251st American Chemical Society (ACS) National Meeting (March 13–17, San Diego)*; 2016. Abst MEDI 283.
64. An Open-Label, Phase 1, First-in-Human Study of the Safety, Tolerability, and Pharmacokinetic/Pharmacodynamic Profile of VX-984 in Combination With Chemotherapy in Subjects With Advanced Solid Tumors. ClinicalTrials.gov Identifier: NCT02644278. Last updated: March 22, 2017. Accessed on March 27, 2017.
65. Wang, T. Y.; Libardo, M. D. J.; Angeles-Boza, A. M.; Pellois, J. P. *ACS Chem. Biol.* **2017**, *12*(5), 1170–1182.
66. Lamberson, C. R.; Xu, L.; Muchalski, H.; Montenegro-Burke, J. R.; Shmanai, V. V.; Bekish, A. V.; McLean, J. A.; Clarke, C. F.; Shchepinov, M. S.; Porter, N. A. *J. Am. Chem. Soc.* **2014**, *136*(3), 838–841.

67. Abeti, R.; Uzun, E.; Renganathan, I.; Honda, T.; Pook, M. A.; Giunti, P. *Pharmacol. Res.* **2015**, *99*, 344–350.
68. NCT02445794, last updated on October 3, 2016. www.clinicaltrials.gov, Accessed on April 12, 2017.
69. https://www.retrotope.com/news. Accessed 30 January 2017.
70. Sparrow, J. R.; Fishkin, N.; Zhou, J.; Cai, B.; Jang, Y. P.; Krane, S.; Itagaki, Y.; Nakanishi, K. *Vision Res.* **2003**, *43*(28), 2983–2990.
71. Kaufman, Y.; Ma, L.; Washington, I. *J. Biol. Chem.* **2011**, *286*(10), 7958–7965.
72. Zhang, C. Patent Application WO 2015/084622, 2015.
73. Tung, R. US Patent 9,314,440, 2016.
74. Morgan, A. J. US Patent 8,865,902, 2014.
75. Moslin, R. M.; Weinstein, D. S.; Wrobleski, S. T.; Tokarski, J. S.; Kumar, A.; Batt, D. G.; Lin, S.; Liu, C.; Spergel, S. H.; Zhang, Y.; Liu, O. Patent Application US 2017/0022192, 2017.
76. Charifson, P. S.; Cottrell, K. M.; Deng, H.; Duffy, J. P.; Gao, H.; Giroux, S.; Green, J.; Jackson, K. L.; Kennedy, J. M.; Lauffer, D. J.; Ledeboer, M. W.; Li, P.; Maxwell, J. P.; Morris, M. A.; Pierce, A. C.; Waal, N. D.; Xu, J. Patent Application US 2014/0045869, 2014.
77. Shchepmov, M. S. Patent Application US 2014/0044693, 2014.
78. Bergen, H. R.; Furr, H. C.; Olson, J. A. *J. Label. Compd. Radiopharm.* **1998**, *25*, 11–21.
79. Sheila DeWitt, S.; Jacques, V.; van der Ploeg, L. H. T. Patent Application US 2016/331737, 2016.
80. Feng, W., Gao, X., Dai, X, US Patent 8,618,306, 2013.
81. Gant, T. G.; Hodulik, C.; Woo, S. Patent Application US 2010/0172916, 2010.
82. Strayer, R. J.; Nelson, L. S. *Am. J. Emerg. Med.* **2008**, *26*(9), 985–1028.
83. Abdallah, C. G.; Adams, T. G.; Kelmendi, B.; Esterlis, I.; Sanacora, G.; Krystal, J. H. *Depress. Anxiety* **2016**, *33*(8), 689–697.
84. Zanos, P.; Moaddel, R.; Morris, P. J.; Georgiou, P.; Fischell, J.; Elmer, G. I.; Alkondon, M.; Yuan, P.; Pribut, H. J.; Singh, N. S.; Dossou, K. S.; Fang, Y.; Huang, X. P.; Mayo, C. L.; Wainer, I. W.; Albuquerque, E. X.; Thompson, S. M.; Thomas, C. J.; Zarate, C. A., Jr.; Gould, T. D. *Nature* **2016**, *533*(7604), 481–486.
85. Gant, T. G., Sarshar, S, US Patent 7,638,651, 2009.
86. https://pharma.globaldata.com/ProductsView.aspx?ProductType=0,1&ProductID=313573. Accessed 30 January 2017.
87. Nag, S.; Fazio, P.; Lehmann, L.; Kettschau, G.; Heinrich, T.; Thiele, A.; Svedberg, M.; Amini, N.; Leesch, S.; Catafau, A. M.; Hannestad, J.; Varrone, A.; Halldin, C. *J. Nucl. Med.* **2016**, *57*(2), 315–320.
88. Zhang, Z.; Kil, K. E.; Poutiainen, P.; Choi, J. K.; Kang, H. J.; Huang, X. P.; Roth, B. L.; Brownell, A.-L. *Bioorg. Med. Chem. Lett.* **2015**, *25*(18), 3956–3960.

CHAPTER FIFTEEN

From Natural Peptides to Market

Claudio D. Schteingart[1], Jolene L. Lau
Ferring Research Institute Inc., San Diego, CA, United States
[1]Corresponding author: e-mail address: claudio.schteingart@ferring.com

Contents

1. INTRODUCTION

The purpose of this chapter is to summarize for the small molecule medicinal chemist the scope and potential of peptides as drugs. In addition to a multitude of linear and cyclic peptides, nature offers an enormous variety of compounds based on amino acids with complex structures and atom connectivities. We will focus this review on the therapeutic peptides that have entered the market or are in advanced clinical development, which have structures that can be produced by conventional peptide chemical synthesis or recombinant methods. These peptides are listed on the

Annual Reports in Medicinal Chemistry, Volume 50
ISSN 0065-7743
https://doi.org/10.1016/bs.armc.2017.08.003

database of therapeutic peptides maintained by Ferring Research Institute Inc. It excludes insulin and analogues as their development has followed a pathway closer to that of recombinant proteins,[1] fermentation products (e.g., cyclosporine, vancomycin), peptide-based vaccines, peptides shorter than 4 amino acids, and peptides longer than 50 amino acids if they are prepared by recombinant methods. The boundary between peptides and proteins is arbitrary and has increased over time (in the minds of practitioners) as synthetic methods have improved. FDA categorizes "any alpha amino acid polymer" 40 amino acids or less in size as a peptide, and as a protein if larger. When made entirely by chemical synthesis, FDA labels it as a "chemically synthesized polypeptide" if less than 100 amino acids in size.

2. CHARACTERISTICS OF PEPTIDES IN NATURE

Natural peptides have a few key properties that determine the way they function in the body:

- Relatively large molecular weight
- Hydrophilicity
- Sensitivity to proteases
- Low or no plasma protein binding

2.1 Distribution

Their large molecular weight (compared to small molecules) and hydrophilicity prevent peptides from crossing cell membranes, thus the molecular targets that peptides address are extracellular, either exposed on the extracellular aspect of cell membranes or soluble in plasma or extracellular fluids.

2.2 Clearance and Half-Life

Most peptides function as messengers in the body. After secretion (be it endocrine, paracrine, or as a neurotransmitter) their action needs to be terminated to maintain fine regulation of the biological system they control. Paracrine or neurotransmitter actions can be terminated by diffusion out of the effect compartment and/or proteolysis. Endocrine actions require the elimination of the peptide from the circulation. Because most peptides have no or little plasma protein binding, they are always subject to passive glomerular filtration, which is not a very fast mechanism. The glomerular filtration rate in healthy young individuals is about 1.8 mL/kg/min and it decreases significantly with age. In addition, peptides distribute to extracellular water, with a volume of distribution at steady state of about 0.2 L/kg. Although this is a very small volume compared to small molecules, it contributes significantly to lengthening the half-life of peptides. A typical half-life in humans for a peptide subject only to passive renal filtration is 2–3 h, as in the case of the selective V_2 agonist desmopressin (*Minirin*™) **1**[2] or eptifibatide (*Integrilin*™) **2**.[3] Faster elimination of peptides from the circulation occurs by proteolysis by (mostly) tissue and/or plasma proteases, and other unknown mechanisms. In many cases, the half-life is difficult to measure as it is of the order of minutes. Although little is known, small peptides can also be taken up by tissues. In rats, the cyclic somatostatin agonist octreotide (*Sandostatin*™) **9** (MW 1638) is taken up by hepatocytes and secreted into bile via carrier mediated transport.[4]

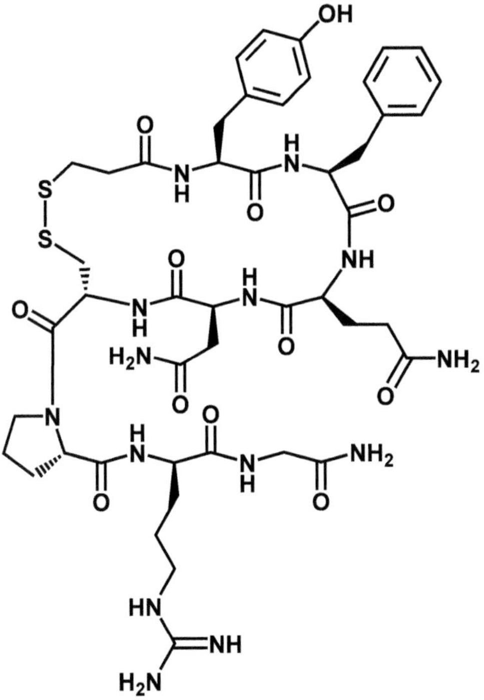

1

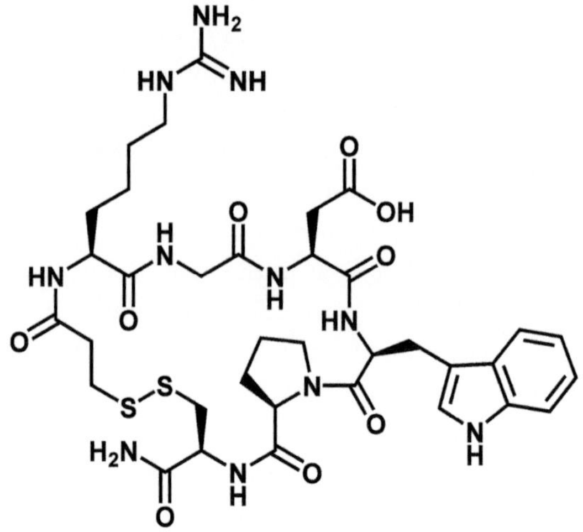

2

2.3 Potency and Selectivity

The structural variety of the 20 natural amino acids and the relatively large number of amino acids in a peptide chain makes them rich in structural information. In spite of being linear and containing a large number of rotatable bonds, the affinity of peptides for their GPCR receptors is high, with K_d values from pM to nM. In many cases, peptides achieve a certain level of ligand preorganization by cyclization via disulfide bridges between cysteines, for example, in endothelin-1 **3** (Fig. 1.[a])

---

[a] In peptide sequences, the one or three letter code of an amino acid represents the substructure -NH-CH(R)-CO-. Sequences are represented with the N-terminus on the left and the C-terminus on the right. Counting is from the N-terminus (not always starting with number 1). It is good practice to indicate the functionality of the termini: an "H'" at the N-terminus indicates a free amino terminus, and an "-OH" at the C-terminus indicates a free carboxylic acid. "-NH2" at the C-terminus indicates a primary amide. When the termini are not indicated, it is assumed that the amino terminus has an amino group and the C-terminus a carboxylic acid. Other modifications use standard abbreviations (e.g., Ac-, -OMe, -NHEt, etc.). Cycles with linkage by disulfide bonds between cysteines can be indicated by underlining the cysteines, or if there is more than one cycle by drawing lines between them or with a textual indication like cyclo(X–Y, W–Z) where X, Y, W, and Z are the numeric position of the bridged cysteines in the sequence.

It could be expected that nature would use these features to construct ligand-receptor systems that operate independently of each other with superb selectivity. However, this is not always the case as many natural peptides exhibit cross reactivities with several receptors in a family. For example, the cyclic nonapeptide arginine vasopressin (AVP) **4** secreted from the posterior pituitary is an agonist at the V_{1a}, V_{1b}, and V_2 receptors to cause vasoconstriction, ACTH release (together with corticotropin-releasing factor, CRF), and urine concentration, respectively. The tridecapeptide α-melanocyte-stimulating hormone, α-MSH, **5** is an agonist at the MC_1 (causing pigmentation), MC_3, MC_4, and MC_5 melanocortin receptors, but not at the MC_2 receptor.

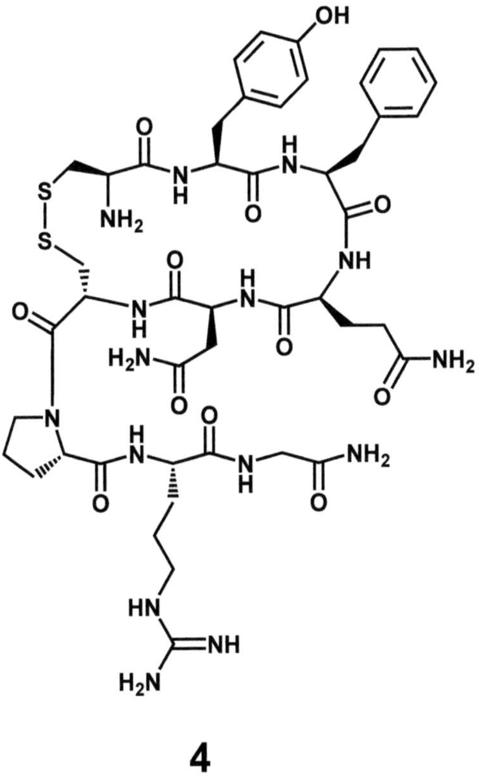

4

Ac-Ser-Tyr-Ser-Met-Glu-His-Phe-Arg-Trp-Gly-Lys-Pro-Val-NH_2

5

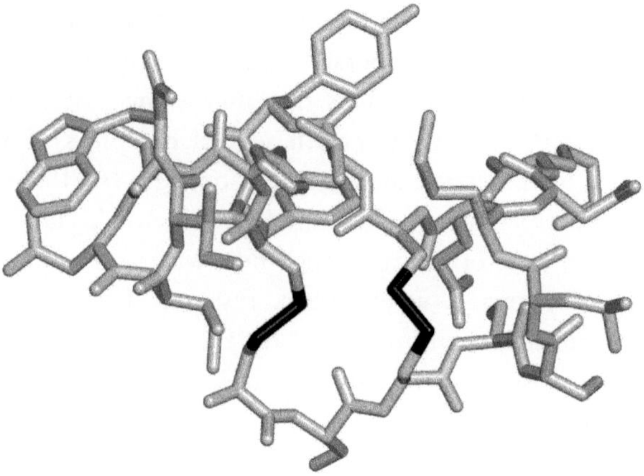

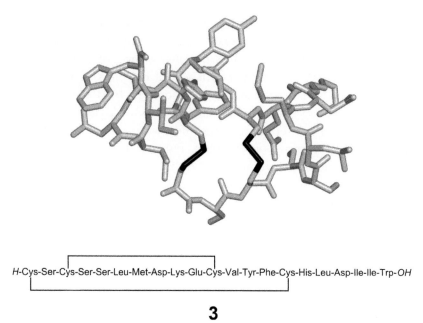

H-Cys-Ser-Cys-Ser-Ser-Leu-Met-Asp-Lys-Glu-Cys-Val-Tyr-Phe-Cys-His-Leu-Asp-Ile-Ile-Trp-*OH*

3

Fig. 1 Sequence of endothelin-1 **3** and stick model of its crystal structure showing the disulfide bridges in *black. Redrawn with PyMol from Protein Data Bank entry 1EDN, without hydrogens.*

Nature uses a variety of mechanisms to maintain functional selectivity, for example, distribution (CNS vs periphery), different circulating concentrations, costimulants, dilution (endocrine vs paracrine messaging), compartmentalization (neurotransmission in synapses), etc. An excellent review on the families of peptides, their receptors, and functions can be found in the *Handbook of Biologically Active Peptides* edited by Abba J. Kastin.[5]

3. FROM NATURAL PEPTIDES TO DRUGS

Most of the therapeutic peptides on the market today are derived from the corresponding natural ligand. In this case, the broad goals of the medicinal chemist are to provide a peptide drug product which is

- Safe
- Clinically effective
- With a mode of administration suitable for the indication
- With a frequency of administration suitable for the indication

- In an acceptable delivery device
- Stable in the supply chain
- With reasonable dose and cost of goods

3.1 Safety

It is generally found that peptides do not exhibit idiosyncratic toxicities and most adverse effects are due to exaggerated pharmacology. This is normally attributed to the fact that peptides cannot penetrate into cells and do not have an opportunity to interact with large numbers of intracellular molecular targets with unpredictable consequences. In addition, as most peptide drugs are very potent, the pharmacologically active plasma concentrations tend to be well below 100 nM and do not seem to unduly tax the elimination organs (liver, kidneys). Their structural complexity and size give the medicinal chemist many opportunities to dial the required selectivity, which together with the low plasma concentrations reduce the possibility of off-target effects and drug–drug interactions via transporters or metabolizing enzymes (cytochromes or proteases).

An adverse effect which has to be addressed early in any SAR program and during early clinical trials is the appearance of injection site reactions after subcutaneous administration. Many peptides and drugs have the capacity to cause mast cell degranulation and histamine release, especially considering that the solutions injected can have concentrations in the mM range. The MRGPRX2 receptor in mast cells has recently been identified as a contributor to histamine release by basic secretagogues.[6] It is not known whether screening at this receptor might be predictive of adverse effects in humans. It may be possible that histamine release is only one of several mechanisms contributing to the injection site reactions. Fortunately, in most cases these reactions are transient and not severe enough to stop treatment. A notable exception was the gonadotropin-releasing hormone (GnRH **31**) antagonist abarelix (*Plenaxis*™) **6**, which was withdrawn from the US market by the sponsor 2 years after its introduction due to immediate onset systemic allergic reactions, even after the first dose (which might rule out immune sensitization). It is well known that the first generations of peptidic GnRH antagonists had a tendency to produce histamine release, but there are no special warning structural features differentiating the structure of abarelix from other GnRH antagonists (Compounds **37–39**). The mechanism of these systemic allergic reactions was never elucidated and it is possible that the formulation used may have been a contributor.

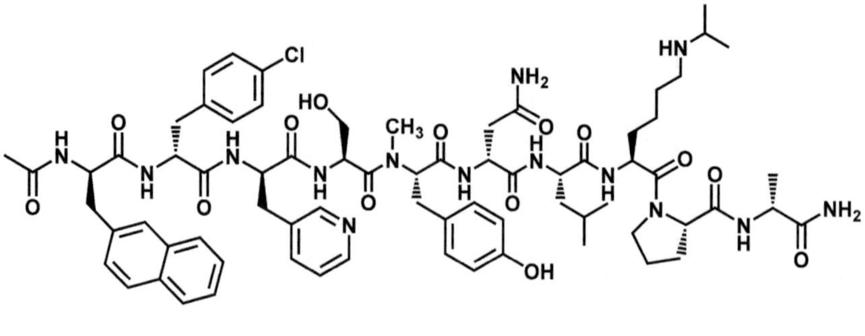

6

Administration of therapeutic peptides results in the appearance of anti-drug antibodies (ADAs) in a certain proportion of patients as the duration of treatment increases, but sometimes they disappear with continued exposure. In general, these ADAs do not seem to be "neutralizing" in the sense that they render the therapy ineffective. One notable exception is the peptidic GLP-1 agonist taspoglutide **29**; its development was terminated during Phase 3 trials due to mechanistic adverse effects and hypersensitivity reactions associated with ADAs.[7] In general, data on immunogenicity of peptides are not normally published in the open literature and have to be traced from regulatory documentation. A comparative review on approved GLP-1 agonists including immunogenicity has been published.[8]

3.2 Clinical Efficacy: Agonists and Antagonists

Peptides are subject to the same considerations and uncertainties as small molecules or biologicals with respect to clinical efficacy: the level of validation of the molecular target for the chosen indication is the most important factor. Most peptides approved to date (Table 1) address well-characterized molecular targets, which have been studied for many years as essential components of biology. However, many of the peptides currently in clinical development interact with less well-characterized targets, and it is likely that their success rate to demonstrate clinical proof of concept in Phase II studies will be lower than for established peptides.

For agonists, the natural peptide ligand generally provides a potent lead. The task of the medicinal chemist is to optimize the other necessary properties of the eventual NCE (discussed later) without excessive loss of potency, although in some cases peptides can be too potent. For example, the V_{1a} agonist selepressin **7**, currently in clinical trials to treat hypotension in septic shock, was designed to be less potent than the natural ligand AVP to allow measurement of its plasma concentrations by LC–MS during clinical trials.[9]

Table 1 List of Approved Peptides in the United States, European Union, and Japan

Compound Name	First Approval	Approved Indications	Approved RoA	Molecular Target	Action	Molecular Target Class	Site of Action	No. AA
Felypressin	1970s	Dental anesthesia adjunct	s.c.	AVP1 receptor	Agonist	Class A GPCR	Extracellular membrane	9
Ornipressin	1971	Bleeding esophageal varices	i.v.	AVP1 receptor	Agonist	Class A GPCR	Extracellular membrane	9
Terlipressin	1978	Bleeding esophageal varices	i.v.	AVP1 receptor	Agonist	Class A GPCR	Extracellular membrane	12
Vasopressin	1962	Vasodilatory shock; bleeding esophageal varices	i.v.	AVP1 receptor	Agonist	Class A GPCR	Extracellular membrane	9
Icatibant	2008	Hereditary angioedema in C1-esterase-inhibitor deficiency	s.c.	B2 receptor	Antagonist	Class A GPCR	Extracellular membrane	10
Calcitonin (human)	1986	Hypercalcemia; Paget's disease; osteoporosis	i.m.; i.n.; s.c.	Calcitonin receptor	Agonist	Class B GPCR	Extracellular membrane	32
Calcitonin (salmon)	1971	Hypercalcemia; Paget's disease; osteoporosis	i.n.; s.c.	Calcitonin receptor	Agonist	Class B GPCR	Extracellular membrane	32
Elcatonin	1981	Osteoporosis	i.m.	Calcitonin receptor	Agonist	Class B GPCR	Extracellular membrane	31
Pramlintide	2005	Type 1 and type 2 diabetes	s.c.	Calcitonin receptor	Agonist	Class B GPCR	Extracellular membrane	37
Etelcalcetide	2016	Secondary hyperparathyroidism	i.v.	CaSR	Agonist	Class C GPCR	Extracellular membrane	8

Continued

Table 1 List of Approved Peptides in the United States, European Union, and Japan—cont'd

Compound Name	First Approval	Approved Indications	Approved RoA	Molecular Target	Action	Molecular Target Class	Site of Action	No. AA
Carfilzomib	2012	Multiple myeloma	i.v.	20S proteasome	Inhibitor	Enzyme inhibitor	Intracellular	4
Linaclotide	2012	IBS-C; chronic idiopathic constipation	p.o.	GC-C receptor	Agonist	Guanylyl cyclase C	Extracellular membrane	14
Plecanatide	2017	Chronic idiopathic constipation	p.o.	GC-C receptor	Agonist	Guanylyl cyclase C	Extracellular membrane	16
Tesamorelin	2010	HIV lipodystrophy	s.c.	GHRH receptor	Agonist	Class B GPCR	Extracellular membrane	44
Albiglutide	2014	Type 2 diabetes	s.c.	GLP-1 receptor	Agonist	Class B GPCR	Extracellular membrane	30
Dulaglutide	2014	Type 2 diabetes	s.c.	GLP-1 receptor	Agonist	Class B GPCR	Extracellular membrane	31
Exenatide	2005	Type 2 diabetes	s.c.	GLP-1 receptor	Agonist	Class B GPCR	Extracellular membrane	39
Liraglutide	2009	Type 2 diabetes	s.c.	GLP-1 receptor	Agonist	Class B GPCR	Extracellular membrane	31
Lixisenatide	2013	Type 2 diabetes	s.c.	GLP-1 receptor	Agonist	Class B GPCR	Extracellular membrane	44
Teduglutide	2012	Short bowel syndrome	s.c.	GLP-2 receptor	Agonist	Class B GPCR	Extracellular membrane	33

Glucagon	1989	Hypoglycemia	s.c.	Glucagon receptor	Agonist	Class B GPCR	Extracellular membrane	29
Buserelin	1984	Prostate cancer; endometriosis; assisted reproduction	i.n.; s.c.	GnRH receptor	Agonist	Class A GPCR	Extracellular membrane	9
Gonadorelin	1989	Assisted reproduction; primary amenorrhea	s.c.	GnRH receptor	Agonist	Class A GPCR	Extracellular membrane	10
Goserelin	1987	Breast cancer; endometriosis; prostate cancer	s.c.	GnRH receptor	Agonist	Class A GPCR	Extracellular membrane	10
Histrelin	1991	Prostate cancer; precocious puberty	s.c.; i.m.	GnRH receptor	Agonist	Class A GPCR	Extracellular membrane	9
Leuprorelin	1984	Prostate cancer; endometriosis; uterine fibroids; precocious puberty	s.c.; i.m.	GnRH receptor	Agonist	Class A GPCR	Extracellular membrane	9
Nafarelin	1990	Precocious puberty; endometriosis	i.n.	GnRH receptor	Agonist	Class A GPCR	Extracellular membrane	9
Triptorelin	1986	Prostate cancer; endometriosis; breast cancer; precocious puberty	i.m.	GnRH receptor	Agonist	Class A GPCR	Extracellular membrane	10
Cetrorelix	1999	Assisted reproduction	s.c.	GnRH receptor	Antagonist	Class A GPCR	Extracellular membrane	10

Continued

Table 1 List of Approved Peptides in the United States, European Union, and Japan—cont'd

Compound Name	First Approval	Approved Indications	Approved RoA	Molecular Target	Action	Molecular Target Class	Site of Action	No. AA
Degarelix	2008	Prostate cancer	s.c.	GnRH receptor	Antagonist	Class A GPCR	Extracellular membrane	10
Ganirelix	1999	Assisted reproduction	s.c.	GnRH receptor	Antagonist	Class A GPCR	Extracellular membrane	10
Eptifibatide	1998	Acute coronary syndrome managed medically or with percutaneous coronary intervention	i.v.	GP IIb IIIa	Antagonist	Integrin	Extracellular membrane	7
Enfuvirtide	2003	HIV infection	s.c.	gp41	Inhibitor	Protein	Soluble EC fluid	36
Thymopentin	1985	Various autoimmune and infectious diseases	s.c.; i.m.	HLA-DR	Immunomodulator	Unknown	Extracellular membrane	5
Corticotropin	1952	Multiple inflammatory diseases; West syndrome	s.c.	MC receptors	Agonist	Class A GPCR	Extracellular membrane	39
Tetracosactide	1980	Multiple inflammatory diseases	s.c.	MC receptors	Agonist	Class A GPCR	Extracellular membrane	24
Afamelanotide	2014	Erythropoietic protoporphyria	s.c.	MC1 receptor	Agonist	Class A GPCR	Extracellular membrane	13
Glatiramer	1996	Multiple sclerosis	s.c.	MHC II	Decoy	Major histocompatibility complex	Extracellular membrane	Polymer

Carperitide	1995	Acute decompensated heart failure	i.v.	NPR-A	Agonist	Guanylyl cyclase A	Extracellular membrane	28
Nesiritide	2001	Acute decompensated heart failure	i.v.	NPR-A	Agonist	Guanylyl cyclase A	Extracellular membrane	32
Ziconotide	2004	Pain	Intrathecal	N-type calcium channels	Inhibitor	Ion channel	Extracellular membrane	25
Carbetocin	2001	Postpartum hemorrhage	i.v.	OT receptor	Agonist	Class A GPCR	Extracellular membrane	9
Oxytocin	1962	Labor induction; postpartum hemorrhage; lactation	i.m.; i.v.; i.n.	OT receptor	Agonist	Class A GPCR	Extracellular membrane	9
Atosiban	2000	Preterm labor	i.v.	OT receptor	Antagonist	Class A GPCR	Extracellular membrane	9
Abaloparatide	2017	Osteoporosis	s.c.	PTH1 receptor	Agonist	Class B GPCR	Extracellular membrane	34
Teriparatide	2002	Osteoporosis	s.c.	PTH1 receptor	Agonist	Class B GPCR	Extracellular membrane	34
Octreotide	1988	Acromegaly; neuroendocrine tumors	s.c.; i.m.; i.v.	SST2 receptor	Agonist	Class A GPCR	Extracellular membrane	8
Lanreotide	1994	Acromegaly; neuroendocrine tumors	s.c.	SST receptors	Agonist	Class A GPCR	Extracellular membrane	8
Pasireotide	2012	Acromegaly; Cushing's disease	s.c.	SST receptors	Agonist	Class A GPCR	Extracellular membrane	6

Continued

Table 1 List of Approved Peptides in the United States, European Union, and Japan—cont'd

Compound Name	First Approval	Approved Indications	Approved RoA	Molecular Target	Action	Molecular Target Class	Site of Action	No. AA
Somatostatin	1970s	Bleeding esophageal varices	i.v.	SST receptors	Agonist	Class A GPCR	Extracellular membrane	14
Lucinactant	2012	Neonatal respiratory distress syndrome	Inhaled	Surfactant replacement	Surfactant	Physicochemical action	Soluble EC fluid	21
Bivalirudin	1999	Acute coronary syndrome; heparin-induced thrombocytopenia; thrombosis	i.v.	Thrombin	Inhibitor	Enzyme inhibitor	Soluble EC fluid	20
Romiplostim	2008	Immune thrombocytopenic purpura	s.c.	Thrombopoietin receptor	Agonist	Cytokine receptor	Extracellular membrane	41
Desmopressin	1972	Diabetes insipidus; primary nocturnal enuresis, nocturia	p.o.; i.n.	V_2 receptor	Agonist	Class A GPCR	Extracellular membrane	9
Aviptadil	2000	Erectile dysfunction	Intracavernous	VIP1 receptor	Agonist	Class B GPCR	Extracellular membrane	28
P-15	1999	Bone grafting	Implanted	α2β1 Integrin	Agonist	Integrin	Extracellular membrane	15

i.m.: intramuscular; i.n.: intranasal; i.v.: intravenous; s.c.: subcutaneous.

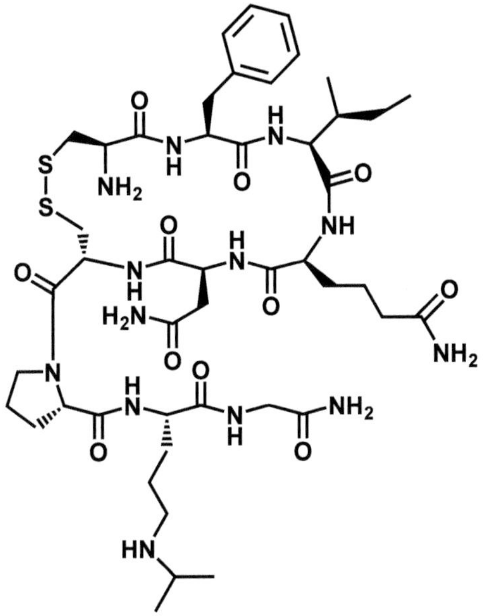

7

As mentioned earlier, the natural peptide often binds a family of related receptors and the selectivity for the desired one must be introduced as part of the medicinal chemistry program. Occasionally, selectivity for only one receptor is not desired, as in the case of the somatostatin agonist pasireotide (*Signifor*™) **8** approved for the treatment of Cushing's disease and acromegaly, which has high affinity for the SST1, SST2, SST3, and SST5 receptors.

It is beyond the scope of this short chapter to review the large number of chemical strategies used in medicinal chemistry of peptides. However, it is interesting to note that none of the peptides approved to date (Table 1) carry an additional cycle with the intention of preorganizing the structure to achieve higher affinity. The cycles, if present, are the same ones in the natural leads, independently of their origin. The medicinal chemist attempting to introduce a cycle on a linear peptide targeting a class A GPCR will be confronted with the task of mimicking the geometry of the main and side chains of the natural peptide in its active bound conformation. In the absence of additional information from crystal structures or from analogous peptides, an astronomical number of possible sizes and locations for the new ring, and of modifications to the amino acid side chains within it, may need to be tried. On the other hand, the somatostatin analogues octreotide (*Sandostatin*™) **9**, lanreotide (*Somatulin*™) **10**, and pasireotide (*Signifor*™) **8** are examples of contraction of the cycle in the natural peptide somatostatin-14

11 which successfully maintain the correct display conformation of the side chains of the amino acids mimicking those in the pharmacophore segment (Phe-Trp-Lys-Thr in somatostatin-14, in darker color in the structures).

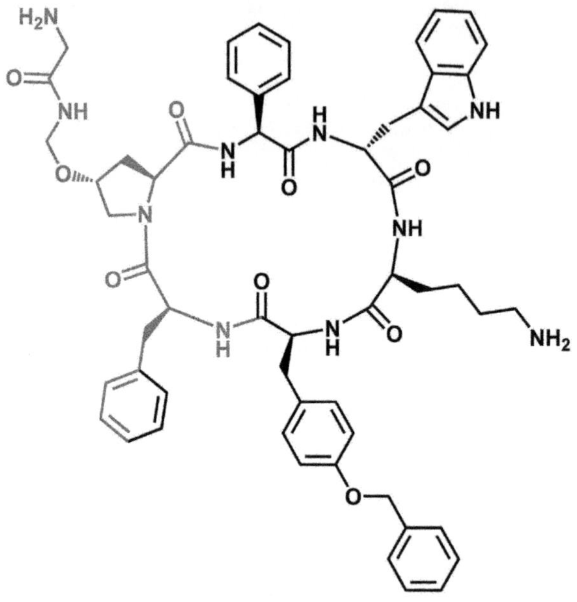

8

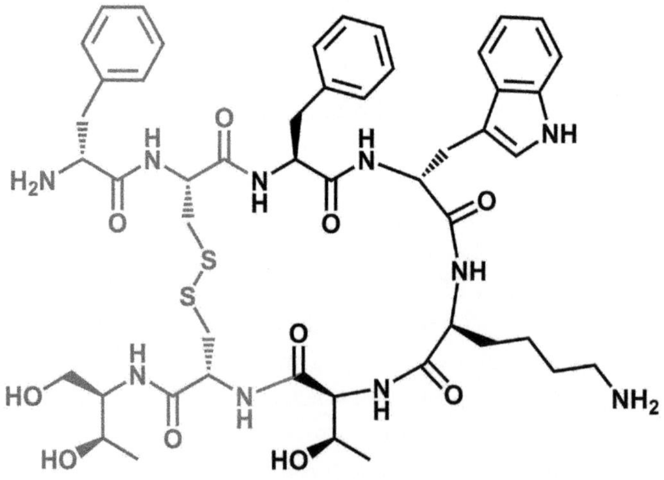

9

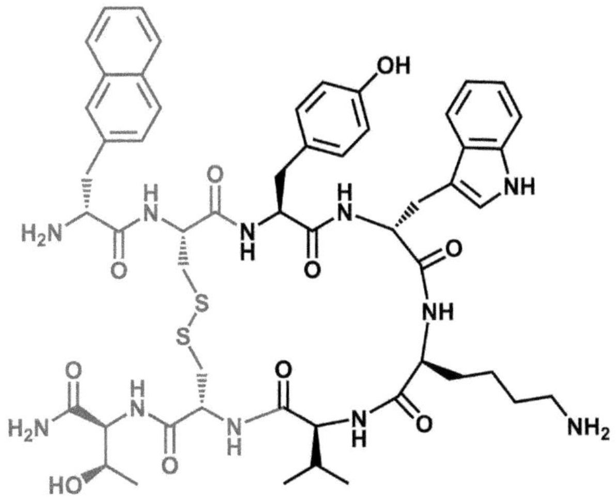

10

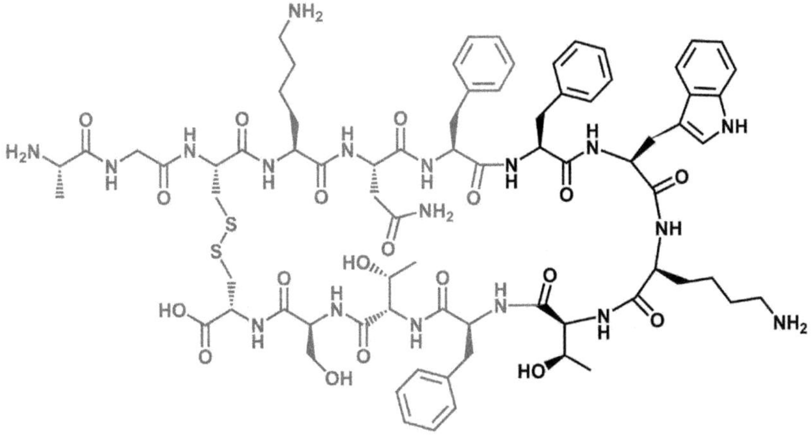

11

The majority of the approved peptides to date are agonists at their receptors (Table 1). Their high potency typically results in low doses and/or long intervals between doses, even when their clearance may not be especially low. Therapeutic peptidic antagonists are possible, but only a few have been approved so far, and require higher plasma concentrations than agonists, which translates into higher doses. Icatibant (*Firazyr*™) **12**, a B2 receptor antagonist for angioedema; eptifibatide

(*Integrilin*™) **2**, a GP IIb IIIa integrin antagonist for acute coronary syndrome;[3] and atosiban (*Tractocile*™) **14**, an oxytocin antagonist for preterm labor[10] are used for acute treatment in a hospital setting where i.v. bolus of high doses and infusions are normal procedures. The GnRH **31** antagonists cetrorelix (*Cetrotide*) **37** and ganirelix **38** are used by daily s.c. injections in short in vitro fertilization protocols. The GnRH antagonist degarelix (*Firmagon*™) **39** will be discussed later.

The switch from agonist to antagonist on peptides binding class A GPCRs can be accomplished by judicious changes in structure. Typically these are the introduction of D- or unnatural amino acids, local conformational restriction, addition or subtraction of amino acids, addition or subtraction of charge, etc. A typical example is the B2 antagonist icatibant (*Firazyr*™) **12**, which was derived from the natural agonist bradykinin **13**:

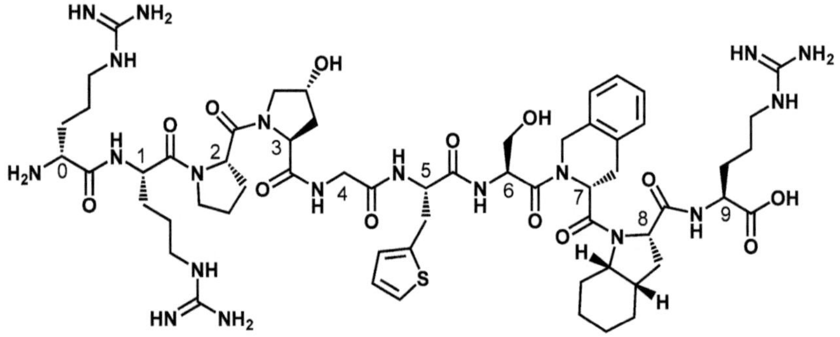

12

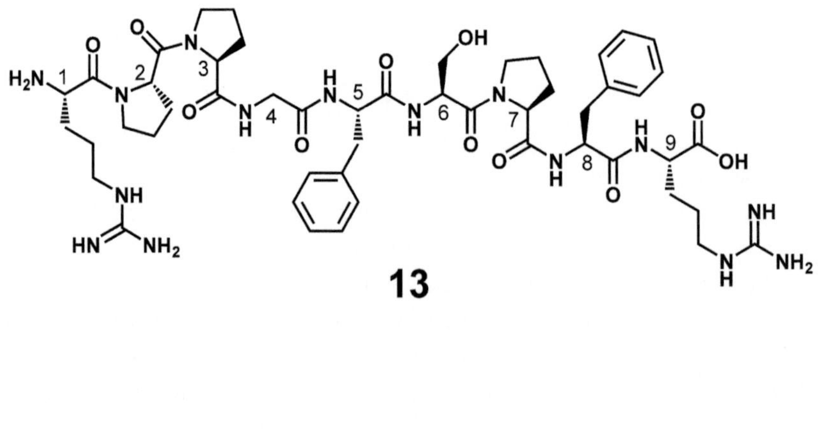

13

Five modifications of bradykinin resulted in the antagonist icatibant: introduction of an additional D-Arg at the N-terminus, hydroxyproline for Pro^3, a thienylalanine instead of Phe^5, and replacement of Pro^7 and Phe^8 by the conformationally restricting amino acids D-Tic (tetrahydoisoquinoline-3-carboxylic acid) and Oic (octahydroindole-2-carboxylic acid), respectively. These modifications provide antagonism, potency, selectivity, and metabolic stability against proteolysis.[11] The oxytocin antagonist atosiban (*Tractocile*™) **14**[12] was obtained by epimerization of Tyr^2 to a D-Tyr^2 *O*-ethyl analog, replacement of a leucine by ornithine and glutamine by threonine in oxytocin **15**, and elimination of the terminal amino group, which introduces selectivity. The *O*-alkylation of the D-tyrosine side chain introduces bulkiness, which prevents degradation by chymotrypsin.

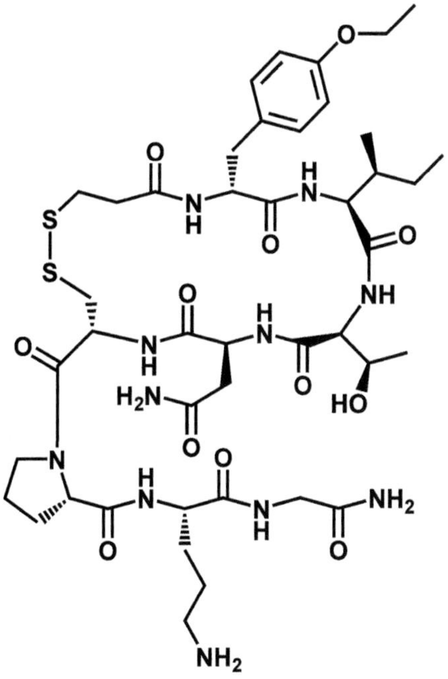

14

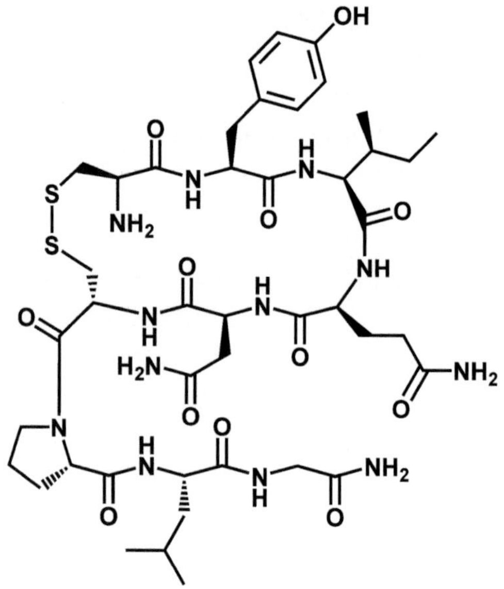

15

The availability of crystal structures of GPCRs has advanced the understanding of the molecular mechanisms of receptor activation, although many details are still missing.[13,14] It is tempting to hypothesize that the traditional approach to generating peptidic antagonists by modifications of the natural ligand works because it produces a peptide with high affinity for the orthosteric binding side (or a good portion of it), capable of displacing the agonist, while steering the receptor into a nonproductive conformation, which may require tilting some transmembrane domains by only a few degrees in the "wrong" direction.

Class B GPCRs bind their cognate peptidic ligands in a helical conformation, which interacts both with the transmembrane and extracellular domains of the receptors.[15] So far only peptidic agonists to this class have reached the market, but potent peptidic antagonists to the CRF receptor have been discovered.[16] The helical conformation of these peptides can be further stabilized by cross linking the side chain of an amino acid with that of the residue located one or more full turns away (e.g., i, i+3 or i, i+7 cycles) using a variety of chemistries (e.g., disulfide or amide bridges). This results not only in potentially an increase in affinity for the receptor, but also prevents proteolytic cleavage as the binding site of proteases generally requires the peptide to be in an extended conformation. This approach has been routinely used by peptide chemists decades[17] before it became popularized as "stapling" using ring-closing metathesis.[18]

3.3 Mode of Administration

The oral route of administration is not suitable for peptides; they are quickly degraded by digestive proteases in the stomach, duodenum, jejunum, and possibly in the colon. Even if they survive proteolysis they must then cross the epithelial barrier of the small intestine, an even more formidable barrier to surmount. Most of the peptides on the market (Table 1) are administered by the intravenous, intramuscular, or subcutaneous routes, with the latter preferred for the treatment of chronic conditions where the patient can self-administer the drug at home.

However, there are interesting cases where the peptide is well suited for an alternative route of administration. The antidiuretic desmopressin (*Minirin*™) **1** is the only truly orally dosed peptide on the market. Although its bioavailability is only 0.08%, the peptide is so potent ($EC_{50} \sim 2$ pM)[19] that its effect lasts overnight after a single dose.[2] This is an extraordinary case, driven by the very high sensitivity of the principal cells in the collecting ducts of the kidney to V_2 agonism.

The peptidic guanylyl cyclase C (GC-C) agonists linaclotide (*Linzess*™) **16** and plecanatide (*Trulance*™) **17** are also given by the oral route. Linaclotide is approved to treat chronic constipation and irritable bowel syndrome with constipation (IBS-C), whereas plecanatide is approved for chronic constipation. They act on their molecular target (GC-C) displayed on the luminal side of the enterocytes in the intestine to cause fluid secretion driven by chloride export via the CFTR channel. These peptides are not absorbed into the circulation; their effect is restricted to the lumen of the intestine with minimal systemic exposure and can therefore be seen as "topical." The active cores of these peptides survive the digestive enzymes because they are short (14 and 16 amino acids) and cross-linked via two or three disulfide bridges, which constrain the intervening loops to adopt a turn conformation probably inaccessible to proteases. Linaclotide is an analogue of the *Escherichia coli* heat stable toxin **18** (which is a stabilized version of the mammalian hormones guanylin **19** and uroguanylin **20**) and plecanatide is an analogue of uroguanylin **20**.

Linaclotide **16**
Plecanatide **17**
E. coli STh **18**
Human guanylin **19**
Human uroguanylin **20**

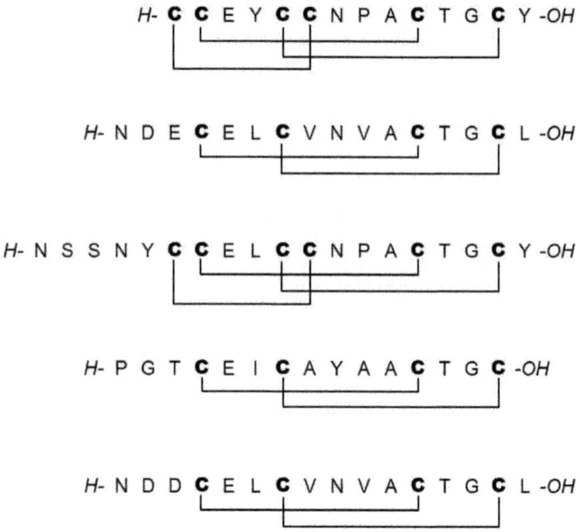

Over the years, there have been many attempts to dose peptides by the oral route employing various formulations containing protease inhibitors and penetration enhancers. In spite of multiple failures (none have reached the market), many companies continue investing in this route.[20]

The product lucinactant (*Surfaxin*™) is a suspension of the 21-amino acid peptide sinapultide, with sequence $(KL_4)_4$, with phospholipids and a fatty acid. It is administered by the intratracheal route to prevent respiratory distress syndrome (RDS) in premature babies. The drug combination functions as a surfactant that compensates for a deficiency of the natural pulmonary surfactant and helps stabilizing the alveoli against collapse.

The nasal epithelium is somewhat permeable to hydrophilic molecules, especially to those with molecular weight below 1000.[21] This allows systemic administration of peptides by the nasal route, but it is of low capacity. Only small volumes of reasonable concentration can be sprayed and retained in the nasal cavity (~100 μL/nostril) and the residence time of the solution in contact with the nasal turbinates is short due to ciliary movement in the direction of the throat, which results in absolute bioavailabilities below 5%. Currently, there are nasal products containing oxytocin (*Syntocinon*™) **15** (MW 1007), desmopressin (*Octostim*™) **1** (MW 1069), buserelin (*Suprefact*™) **32** (MW 1239), nafarelin (*Synarel*™) **33** (MW 1322), and somewhat surprisingly salmon calcitonin (*Miacalcin*™) **21** (MW 3432) although the last one seems to have high absorption variability. Penetration enhancers to increase intranasal bioavailability have been proposed for many years, but none are currently used on the market. Their potential for damaging the nasal mucosa, especially with continuous use, is the main stumbling block for their incorporation into nasal formulations.[22]

H-CSNLSTCVLGKLSQELHKLQTYPRTNTGSGTP-NH_2

21

3.4 Frequency of Administration

Most peptides must be administered by injection, which has low patient acceptance and brings up logistic issues, for example, intravenous infusions or boluses and intramuscular injections must be administered in a physician office or hospital setting. Patients can self-administer by the subcutaneous route some, but not all peptides. Therefore, any program to discover a new therapeutic peptide has to consider up front the frequency of administration of the eventual drug product. In general, the frequency of administration must be in accordance to the severity of the disease being treated.

The clearance and half-life of starting peptidic leads (natural or synthetic) can span a broad range of values (minutes to many hours) and can be modified by alterations to the peptide structure and/or formulations to certain extent. After safety, potency, and selectivity, the clearance and half-life of the drug candidate is one of the most important parameters to optimize.

For continuous intravenous administration in a hospital setting, the clearance is the parameter of interest. It should be sufficiently low to provide the required plasma concentration of the therapeutic peptide, given by its potency, using a reasonable dose of peptide for the duration of the infusion. The dose is typically constrained by its solubility and cost of goods. A well-known example is oxytocin (*Pitocin*™) **15** for labor induction, which is titrated from a low infusion rate of 1 mU/min (1.7 ng/min) up to 20 mU/min (34 ng/min).[23] Although the clearance of oxytocin is very high (~90 mL/kg/min in pregnant women at term)[24] the peptide is so potent that very small doses are needed over a typical induction period (~10 h). At the other end of the spectrum, the platelet aggregation inhibitor eptifibatide (*Integrilin*™) **2** is administered for acute coronary syndrome as a loading bolus of 180 μg/kg (~12 mg) followed by a continuous infusion at 2 μg/kg/min (~8 mg/h) for up to 72 h. The clearance of eptifibatide, ~0.8 mL/kg/min[3], is typical of a peptide with low plasma protein binding and elimination by glomerular filtration, but the effective plasma concentrations are in the μg/mL range,[25] which is high for a peptide. The resulting total dose of a few hundred mg is high, but acceptable given its straightforward synthesis and reasonable solubility.

Contrary to what can be assumed, there are instances where a short half-life for the peptide is desired. The most common situation is where rapid up and down titration of an intravenous infusion of the drug is desired, as in the case of oxytocin **15** in the above example or the vasopressor selepressin **7**. Sometimes the biological system may require only a short "pulse" of activity. The 84-amino acid natural parathyroid hormone (PTH) is secreted in discrete pulses by the parathyroid glands to regulate bone remodeling and maintain calcium homeostasis. The PTH (1–34)-OH fragment, teriparatide (*Forteo*™) **22** retains all its biological activity and it is used to reduce the risk of fractures in postmenopausal women with osteoporosis at high risk for fracture and to increase bone mineral density in men with primary or hypogonadal osteroporosis at high risk for fractures. Crucial to its anabolic effect is intermittent therapy by daily subcutaneous administration with limited duration of action due to a short terminal half-life (~1 h after subcutaneous administration).[26] By contrast, the continuous presence of PTH in the circulation (as in hyperparathyroidism) results in bone resorption.[27] Daily self-administration is acceptable given the benefit of fracture prevention and the limited duration of therapy (less than 2 years).

H-**SVSEIQLMHNLGKHLNSMERVEWLRKKLQDVHNF**-*OH*

22

Bolus administration of peptidic drugs to engage their molecular target continuously for 1 whole day, week, or months is much more challenging. The family of peptidic GLP-1 agonists approved to improve glycemic control in type 2 diabetes patients is an illustrative example. The endogenous GLP-1 peptide **23** is produced by L cells in the intestine and act on GLP-1 receptors in the pancreas to enhance glucose-dependent insulin secretion.

The natural GLP-1 peptide **23** has too short half-life in humans to be used as drug.[28] Its main metabolic liability is proteolysis of the two N-terminal amino acids *H*-His-Ala by the very abundant dipeptidyl peptidase DPP-IV. The first successful approach was the development of exenatide (*Byetta*™) **24**, a potent GLP-1 agonist from the saliva of the Gila monster, which is naturally stabilized against degradation by DPP-IV and appears to be eliminated only by glomerular filtration. Although its half-life is about 2.4 h, it is still not sufficiently long and must be dosed by subcutaneous injection twice a day.[29] Once daily subcutaneous dosing was finally achieved with liraglutide (*Victoza*™) **25**, a lipidated analogue of GLP-1 designed to bind albumin in plasma (>98% bound to plasma proteins). Its half-life is 13 h after subcutaneous administration, which is thought to be due to albumin binding in plasma and aggregation in the subcutaneous site of injection.[30]

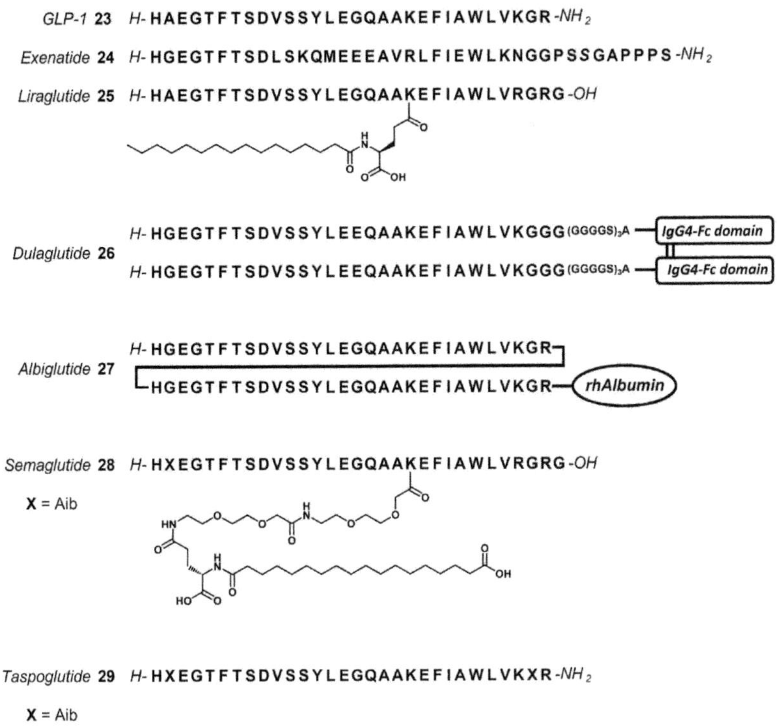

For weekly administration, exenatide **24** has been formulated by incorporation into poly(lactic-co-glycolic acid) copolymer (PLGA) microspheres (*Bydureon*™).[31] The two other GLP-1 analogues on the market for weekly dosing are not incorporated into microspheres, but required conjugation of the peptide to carrier proteins with intrinsically long half-life. Dulaglutide (*Trulicity*™) **26** contains two GLP-1 analogue chains fused to a modified IgG4 Fc moiety. After subcutaneous administration, its terminal half-life is about 5 days.[32] Albiglutide (*Tanzeum*™) **27** is composed of two tandem copies of a GLP-1 analogue fused to human albumin, one of which is believed to act as a linker. Its elimination half-life is also approximately 5 days.[33] One could argue whether this kind of fusion proteins should be qualified as "peptides" as they have high-molecular weight and must be manufactured exclusively by recombinant methods. It is remarkable that the half-lives of dulaglutide and albiglutide are notoriously shorter than the half-lives of albumin or IgG antibodies, both around 20 days in humans. It is likely that these fused peptides are still subject to degradation by plasma and tissue proteases and they must be modified to make them somewhat more resistant to them. In this sense, these fusion proteins behave like traditional peptides that are not subject to elimination by glomerular filtration.

Covalent attachment to other large molecular weight moieties can also extend the half-life of peptides in the circulation, for example, pegylation, hesylation, conjugation to heparosan, or conjugation to a recombinant well-defined artificial polypeptide (XTEN) among many others proposed.[34] The resulting increase in molecular weight will generally prevent elimination by glomerular filtration, but the peptide may still be subject to elimination by proteolysis. The moderate gain in half-life obtained comes at the cost of a significant increase in dose by mass and added complexity of the new chemical entity which will result in more elaborate analytical development and higher cost of goods. Only one pegylated peptide has reached the market, peginesatide (*Omontys*™) **30**, an erythropoiesis-stimulating agent approved in 2012 for the treatment of anemia in chronic kidney disease. Peginesatide is a pegylated dimeric peptide (21 amino acids long) with MW 45,000 and a half-life of 25 h by i.v. route in healthy subjects. Unfortunately, the product had to be withdrawn from the market in 2013 due to anaphylactoid reactions after the first exposure to the drug. The reasons for these reactions have not been elucidated although it has been suggested that phenol, a common antimicrobial preservative used in many injectable products, may be responsible.[35]

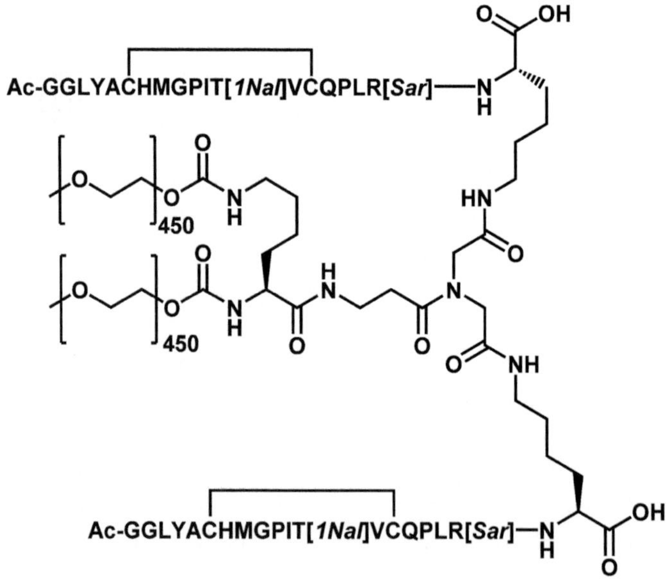

30

Fusion to a large protein or polymer may not always be necessary however. Although not yet approved, the GLP-1 analogue semaglutide **28** has sufficiently long half-life (168 h) by subcutaneous administration in humans to allow weekly dosing.[36] This was achieved by introduction of an albumin binding fatty acid mimicking moiety containing an additional distal negative charge, and by stabilizing the peptide against DPP-IV by replacement of alanine by the unnatural amino acid 2-aminoisobutyric (Aib) in position 2.[37]

Another way to lengthen the duration of action of peptides is by incorporation into a parenteral formulation. Typical examples are GnRH (**31**) agonists like leuprorelin **34** or goserelin **35**, used for the treatment of androgen-dependent prostate cancer by desensitization of pituitary gonadotrophs after 2 weeks of treatment. Goserelin (*Zoladex*™) is administered as a s.c. implant (which requires a minor office procedure) of the peptide dispersed in poly(lactic-co-glycolic acid) copolymer (PLGA). Leuprorelin (*Lupron depot*™) is formulated in PLGA microspheres that are injected as a s.c. suspension. These products are available for injection every 1 or 3 months. The GnRH agonist triptorelin **36** when formulated as the pamoate salt incorporated in PLGA (*Trelstar*™) can be injected by the i.m. route every, 1, 3, and 6 months, depending on the dose. The requirements for a successful formulation product are that the peptide should be very potent, typically an agonist, and of low clearance so that only small amounts of peptide need to be incorporated in the formulation. As an example, the 6 month *Trelstar* product contains 22.5 mg of triptorelin in 183 mg of PLGA.[38] Although the release from these type of formulations is not uniform, the equivalent notional daily dose of *Trelstar* is just 0.14 mg/day. However, the development of a long-term release formulation of a peptide is a complex, expensive, and time-consuming task, which should not be undertaken lightly.

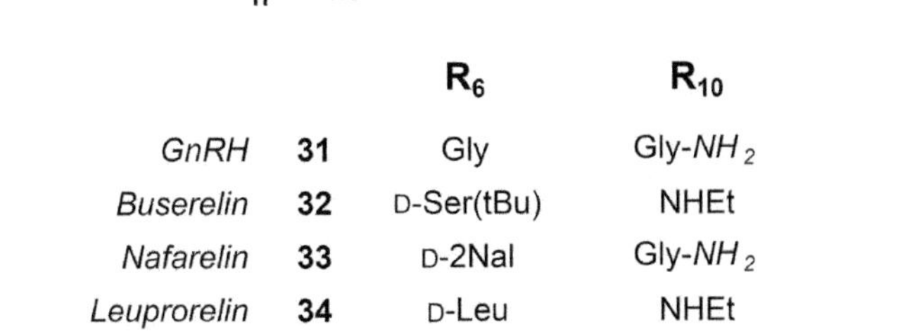

		R_6	R_{10}
GnRH	**31**	Gly	Gly-*NH*$_2$
Buserelin	**32**	D-Ser(tBu)	NHEt
Nafarelin	**33**	D-2Nal	Gly-*NH*$_2$
Leuprorelin	**34**	D-Leu	NHEt
Goserelin	**35**	D-Ser(tBu)	Azagly-*NH*$_2$
Triptorelin	**36**	D-Trp	Gly-*NH*$_2$

A more unusual way to achieve slow release from the injection site is to rely on the physicochemical properties of the peptide itself. The GnRH antagonist degarelix (*Firmagon*™) **39** forms a depot after s.c. injection of a solution in 5% mannitol. The change to the pH and ionic strength of the subcutaneous space causes the peptide to aggregate into fibers that release the peptide slowly to allow monthly dosing for the treatment of androgen-dependent prostate cancer by direct antagonism of the receptor.[39] This careful balance between depoting and redissolution was achieved in an SAR program biased to increase the number of hydrogen bond donors and acceptors in the molecule and screening for duration of pharmacodynamic action in a rat model using the subcutaneous route.[40]

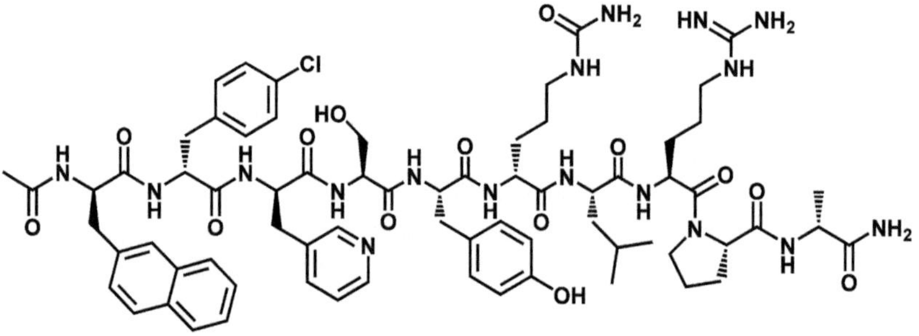

37

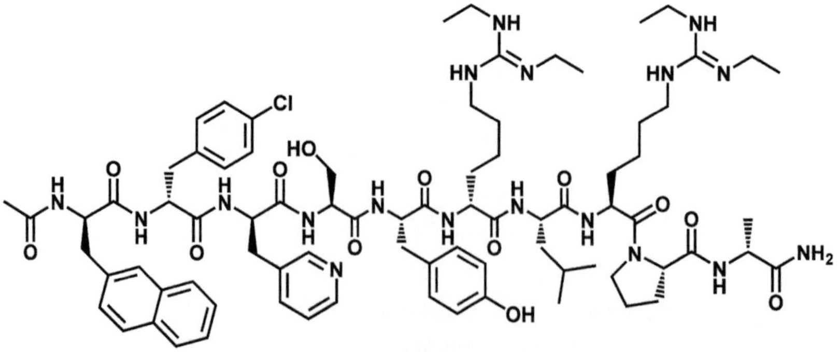

38

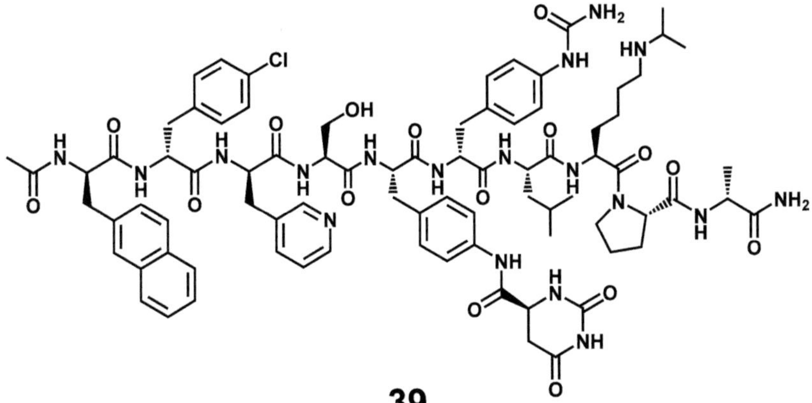

Another example of a self-depoting peptide is the somatostatin agonist lanreotide **10**, for the treatment of acromegaly. One of its presentations, *Somatuline Depot*™/*Somatuline Autogel*™, consists of a concentrated, stable, premade suspension of nanotubes packed into an hexagonal phase.[41] It also forms a depot after deep subcutaneous administration which allows monthly dosing. With so few examples, we cannot venture a guess as to whether introducing self-depoting properties will be possible for even some peptides in the future. Most likely it will depend on myriad structural factors of the sequence of the starting lead peptide like solubility, p*K*a value of charged groups, arrangement of the lipophilic and hydrophilic patches, secondary structure, spatial orientation of hydrogen bond donors and acceptors, etc.

3.5 Delivery Device, Dose, and Stability

Since most peptides are given by a parenteral route, the expected dose, delivery device, volume administered, product presentation, etc. have to be considered up front in any drug discovery program. The potency, clearance, plasma protein binding, and solubility should be tailored to result in a dose compatible with the mode and frequency of administration expected. Whereas there are no major constraints on solubility and volume of administration by the intravenous route, the volume for a self-administered s.c. injection should be about 1 mL or less. This immediately restricts the possible doses to the peptide solubility at pH values tolerable for a s.c. injection site (pH 4–8). Too high concentrations may result in injection site reactions, second order degradation reactions in solution, and sometimes unacceptable high viscosity. An example of a low dose product is teriparatide (*Forteo*™) **22** which is injected daily at a dose of 20 μg in 80 μL using a pen injector prefilled with a solution of the peptide at 250 μg/mL at pH 4.[26] At the other extreme, the 36-amino acid long inhibitor of the fusion of HIV with $CD4^+$ cells enfuvirtide (*Fuzeon*™) **40** is injected s.c. twice daily at a dose of 90 mg in 1 mL of vehicle immediately following reconstitution from the lyophilized powder. A significant proportion of patients experience injection site reactions like pain, induration, erythema, and the formation of nodules.[42]

Ac-**YTSLIHSLIEESQNQQEKNEQELLEDKWASLWNWF**-*NH*$_2$

40

For injectables, two practical qualities of the product are remarkably important for acceptability by physicians and patients: the ease of preparation of the product and the size of the injection needle. When several products with similar clinical efficacy and side effect profile are present on the market, these factors may drive a strong preference for the most convenient or acceptable one.

The most desirable presentation is a prefilled syringe/cartridge with a solution ready to inject, as in *Forteo* or *Victoza*. Next in convenience are premade solutions in vials, to load into a syringe. Presentation in solution requires that the peptide is chemically stable in solution during its shelf life. Peptides are typically more stable in slightly acidic solutions (pH 4–5), which minimize deamidation reactions, and at low temperature, requiring refrigeration for long-term storage. One approach to enhance stability in solution is to replace labile asparagines or glutamines, which can be hydrolyzed to the corresponding acids, by other polar amino acids, e.g., serine, homoserine, or unnatural amino acids. In synthetic peptides, methionines can be replaced by norleucine or other lipophilic amino acids to avoid sulfur oxidation during storage. Depending on the details of its structure, a peptide family may exhibit unpredictable "idiosyncratic" weak spots subject to degradation in solution, e.g., oxytocin **15** unexpectedly forms dimers via an aldol condensation at neutral pH.[43] These weak spots can be identified early in a drug discovery program and appropriate chemical modifications introduced to prevent them.

When the peptide is not stable in solution or the pH of better stability does not match the pH providing adequate solubility, it must be presented as a lyophilized powder for reconstitution (in a vial or dual chamber syringe). This adds more time to the preparation and introduces the possibility of incomplete dissolution and underdosing. Peptides formulated into solid polymeric microparticles require resuspension in a slightly viscous vehicle, which may take the physician staff or the patient considerable time to prepare.

Small volumes of nonviscous solutions can be easily injected using 31G "insulin needles" of very small outer diameter (OD 0.254 mm) which are almost painless. Suspensions of microparticles require larger diameter needles, e.g., the *Trelstar*™ 22.5 mg suspension of triptorelin pamoate in PLGA requires a 21G needle (OD 0.812 mm). The highest strength of *Somatuline Depot*™, 120 mg lanreotide stable solution, requires an 18G needle with OD 1.27 mm and a relatively thin syringe body to facilitate dosing of the viscous solution. In general, the inconvenience of these injections for the patient is balanced by the clinical benefit obtained and the need for less frequent administrations.

Although not much is known about the stability of peptides in the subcutaneous space,[44] the peptides that have moved forward to approval appear to be reasonably stable for the duration of the absorption phase and have bioavailability above 50%. A notable exception is the growth hormone releasing hormone analogue tesamorelin (*Egrifta*™) **41** for the treatment of lipodystrophy in HIV-infected patients. It is the natural GHRH hormone acylated at the N-terminus with a *trans*-3-hexenoyl moiety.

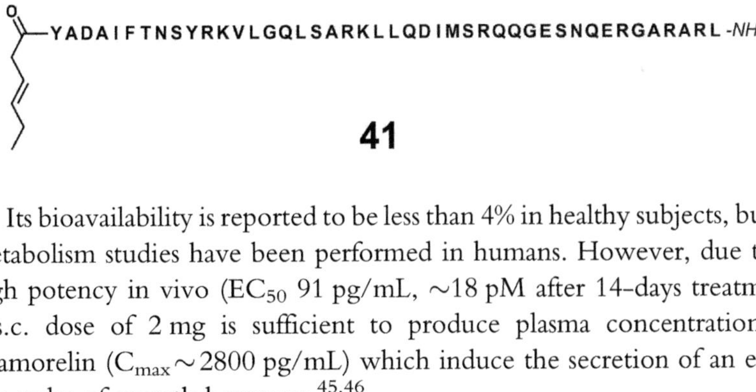

41

Its bioavailability is reported to be less than 4% in healthy subjects, but no metabolism studies have been performed in humans. However, due to its high potency in vivo (EC_{50} 91 pg/mL, ~18 pM after 14-days treatment) a s.c. dose of 2 mg is sufficient to produce plasma concentrations of tesamorelin (C_{max} ~2800 pg/mL) which induce the secretion of an effective pulse of growth hormone.[45,46]

3.6 Cost of Goods

Many legacy peptides are still manufactured at commercial scale by liquid phase synthesis. However, these days the large scale synthesis of new peptides entering clinical trials is performed mostly by solid phase peptide synthesis using Fmoc chemistry. It is the most expedient method, with the shortest development time, and the cost of Fmoc protected amino acids has decreased substantially.[47] The maximum length synthesis that can be undertaken depends on the experience of the contract manufacturing organization and the details of the peptide sequence, nowadays about 50 to 60 amino acids. Peptides containing only natural amino acids can be made by recombinant methods, e.g., teriparatide (*Forteo*™) **22**, or teduglutide (*Gattex*™) **42** for short bowel syndrome. Liraglutide (*Victoza*™) **25** is made by chemical acylation of a recombinant peptide modified to contain only one reactive lysine side chain. The complexity of the synthesis increases when multiple disulfide bridges need to be introduced with controlled regioselectivity, as in the case of linaclotide (*Linzess*™) **16** or ziconotide (*Prialt*™, for severe chronic pain) **43**.

H-**HGDGSFSDEMNTILDNLAARDFINWLIQTKITD**-*OH*

42

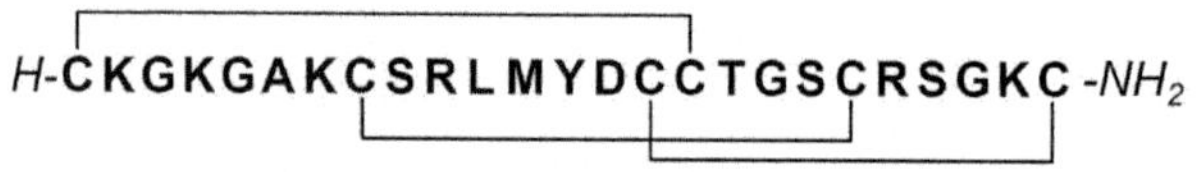

43

Although peptides are more expensive than small molecule drugs on a per gram basis, the doses of peptides are generally much lower, resulting in comparable cost of goods per dose. As mentioned earlier, the limitations of solubility and volume of administration for s.c., i.m., intranasal, or any other parenteral administration impose an explicit maximum dose target to the teams working on peptide drug discovery programs. These in turn must bias their SAR to achieve sufficiently high potency and low clearance to make such low doses possible, and cost of goods does not become a serious issue.

4. A BRIEF LOOK AT THE PEPTIDES ON THE MARKET

Table 1 lists the peptides that have been approved in the United States, the European Union, or Japan, and that remain on the market by the date of this review.

The majority of peptides address a membrane target exposed to the extracellular milieu with a few interacting with soluble proteins. Only one, carfilzomib (*Kyprolis*™) **44**, acts inside the cell to inhibit the 20S proteasome, but it is a tetrapeptide with MW 720, which could be predicted to cross cell membranes based on the Lipinski rules.

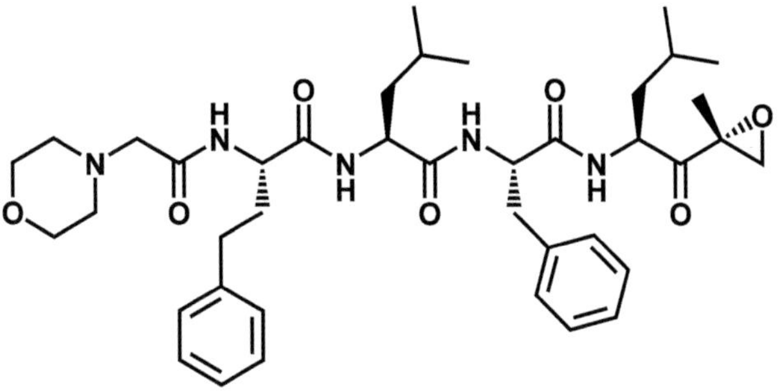

44

Most peptides are agonists at class A or class B GPCR receptors and either the natural peptide or an optimized analogue. As mentioned earlier, only a few antagonists at class A GPCRs are approved, and none at class B receptors. In general, it has been more difficult to find potent small molecule or peptidic antagonists at class B GPCRs, but this may change with the availability of detailed structures of these receptors, e.g., GLP-1,[48] glucagon,[49] and calcitonin.[50] Also, it may be a historical accident that the desired clinical effect for the indications of interest was achieved by agonism. Currently, CGRP antagonists (small molecules and antibodies) are being developed for the treatment of migraine.[51]

There is only one class C GPCR allosteric agonist, the calcimimetic etelcalcetide (*Parsabiv*™) **45** for the treatment of secondary hyperparathyroidism in patients with chronic kidney disease on hemodialysis. It is an *N*-acetylated all D heptapeptide with an L-cysteine linked by a disulfide bridge to its D-Cys^{1}. It is administered as an i.v. bolus at the end of hemodialysis treatments and is in dynamic equilibrium via disulfide bond exchange with the free Cys^{34} in albumin, free cysteine, and other thiol-containing compounds in serum.[52] It acts as an allosteric modulator of the calcium-sensing receptor when it binds via disulfide exchange with free Cys^{482} in the receptor.[53]

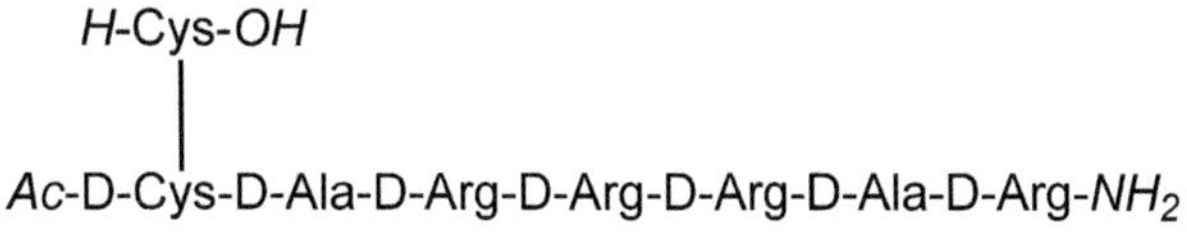

45

The utility of peptides as therapeutics is not restricted to GPCRs; peptides have been approved that target a variety of other molecular targets: guanylyl cyclases, enzymes, cytokine receptors, ion channels, and integrins.

It is interesting to note that although nature provides an enormous library of ion channel blockers in the venoms from many insects and animals (spiders, snakes, snails, scorpions, etc.)[54] only one has reached the market, ziconotide (*Prialt*™) **43**, to manage severe chronic pain.

Ziconotide is the synthetic version of a conopeptide in the marine snail *Conus magus*. It is administered by intrathecal infusion with an external or an implanted infusion pump to block N-type calcium channels in afferent nerves in the dorsal horn of the spinal cord. Because its effects are local the doses are very low: it is titrated from a starting infusion rate of 2.4 μg/day up to a maximum of 19.2 μg/day, as required by the patient.[55]

However, natural products, including venoms, have been a source of inspiration for a significant number of the therapeutic peptides on the market (Table 2).

It is interesting to note that some families of peptides do not appear in Table 1, in spite of considerable research. No peptidic antibiotics based on cationic antimicrobial peptides or host defense peptides, which act by disrupting bacterial cell membranes, have reached the market. The causes may be multiple but a small therapeutic window and toxicity may be important factors. Also, no cell-penetrating peptides (CPPs) or conjugates of bioactive peptides with a CPP have reached the market yet. The most likely main obstacle may be that a high local concentration (μM) seems necessary for

Table 2 Source of Inspiration of Some Therapeutic Peptides on the Market

INN Name	Activity	Inspiration or Source
Bivalirudin	Thrombin inhibitor	Hirudin from leeches *Hirudo medicinalis*
Carfilzomib	Proteasome inhibitor	Epoxomicin from an *actinomycete* strain
Eptifibatide	Integrin antagonist	Barbourin toxin from the dusky pigmy rattlesnake *Sistrurus m. barbouri*
Exenatide	GLP-1 agonist	In saliva of the Gila monster *Heloderma suspectum*
Linaclotide	GC-C agonist	*E. coli* heat stable toxin
Ziconotide	N-type calcium channel blocker	*Conus magus* conotoxin MVIIA

them to penetrate cell membranes. Endosomal localization once inside the cell, and preferential distribution to elimination organs (kidney, liver) after systemic administration may also be contributing factors. Hopefully this situation will change in the future as the understanding of these problematic issues advances and solutions are found. Although many therapeutic peptides address class B GPCRs and are expected to bind in a helical conformation, none of the peptides in Table 1 have been "stapled" so far.

Table 1 shows that some molecular targets have attracted the development of multiple peptides, typically for the same indications. In addition to the possibilities afforded (or denied) by nature and the state of the art at the time of their invention, the commercial attractiveness of the indication may have played a significant role. Fig. 2 shows the revenue generated by the top peptide brand sellers in 2015.[56]

The top seller is glatiramer (*Copaxone*), for the treatment of relapsing–remitting multiple sclerosis. It is a synthetic random copolymer containing the L-amino acids glutamic acid, alanine, tyrosine, and lysine in defined proportions mimicking myelin basic protein, with average MW 5000–9000. Its mechanism of action is not fully understood; it is believed that after s.c. injection it gets hydrolyzed locally to smaller fragments that can bind MHC II receptors in antigen-presenting cells which interact with naïve T-cells to generate regulatory T_{reg} and Th_2 helper cells.[57]

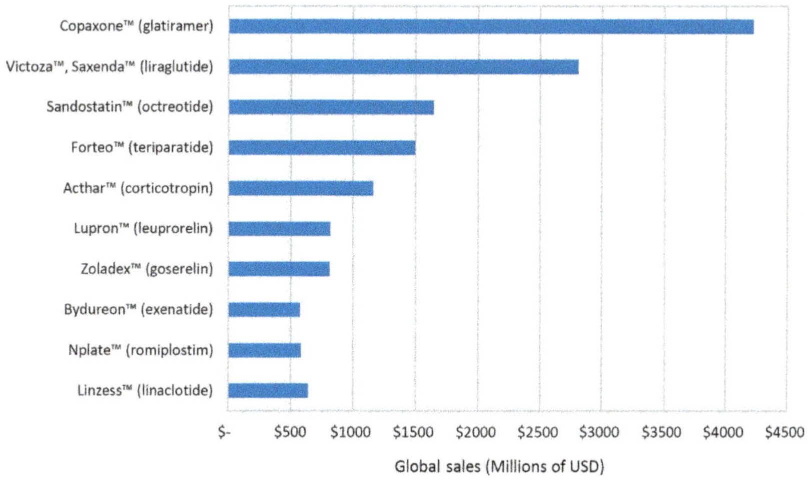

Fig. 2 Global revenue for top seller peptides in 2015.

The next big sellers were the group of GLP-1 agonists for the treatment of type 2 diabetes: liraglutide (*Victoza*™, *Saxenda*™) and exenatide CR (*Bydureon*™). Since then three new GLP-1 agonists have been approved, one for daily and two for weekly dosing, which has increased the market size for this group significantly.

The somatostatin agonist family to treat acromegaly is represented by octreotide (*Sandostatin*™) and lanreotide (*Somatulin*™), with pasireotide (*Signifor*™) approved in 2012.

The large group of GnRH agonists used to treat hormone-sensitive prostate cancer is represented by leuprorelin acetate (*Lupron depot*™) and goserelin (*Zoladex*™), but there are a total of 10 agonists and antagonists on the market. Teriparatide (*Forteo*™) was the only product approved in 2015 for the treatment of osteoporosis, and in 2017 abaloparatide (*Tymlos*™) was added. Natural corticotropin, ACTH (1-39) (*H.P. Acthar Gel*™), is used to treat infantile spasms and for multiple sclerosis; it is priced at a high level in the United States contributing to considerable total sales. Romiplostim is a fusion protein of an Fc domain with two peptides, each one in turn a dimer of a 14-amino acid peptide; it is a thrombopoietin mimetic for the treatment of thrombocytopenia in patients with chronic immune thrombocytopenia.

5. A PROMISING FUTURE

It is beyond the scope of this chapter to review all the traditional and novel chemical approaches to the discovery of new therapeutic peptides.[58] In addition to the natural peptide ligands many other sources of leads are possible, e.g., phage display, combinatorial approaches, designer cyclic peptides, libraries of natural peptides, cyclotides, etc.[59] These approaches are valuable when it is desired to address a target that does not have a natural peptide partner that can be optimized or the natural peptide lead is not suitable, e.g., due to complexity, cost, intractable lack of selectivity. Examples of these situations are interfering with protein–protein interactions, antagonizing a membrane receptor with a large protein partner or a nonproteic ligand (sugars, lipids, ions), inhibiting enzymes, transporters, or ion channels, etc. Independently of the origin of the peptide lead, the optimization process and considerations to convert it into a therapeutic are the same as when starting from known peptide ligands.

6. SCOPE AND OPPORTUNITY FOR PEPTIDE THERAPEUTICS

There are several scenarios where the discovery and development of a therapeutic peptide may be preferable or at least competitive with that of a small molecule or a biological.

Screening of small molecule libraries has not yielded suitable leads for many molecular targets, or their elaboration into drug candidates may take a long time, and often ultimately fail. In those situations developing a therapeutic peptide (or a biological) may be the most expedient way to reach the market. The examples are many, including the class A receptor families in Table 1. Only peptidic somatostatin agonists are on the market, after decades of intense research in small molecules. The first peptidic agonist at the GnRH receptor was approved in 1984 and the first antagonist in 1999. The oral small molecules elagolix (Neurocrine Biosciences) **46**, relugolix (Takeda Pharmaceuticals) **47**, and KLH-2109 (Kissei Pharmaceutical Co. Ltd. and ObsEva SA) are only now undergoing Phase 3 clinical trials. A similar situation seems to have occurred with the oxytocin and vasopressin family of receptors. Peptidic agonists at the V_{1a} receptor (terlipressin, ornipressin, felypressin) and the V_2 receptor (desmopressin), and an antagonist at the oxytocin receptor (atosiban) were approved in the 1970s. Sufficient commercial interest and progress in drug discovery of small molecule leads resulted in the approval of the mixed V_{1a}/V_2 antagonist conivaptan (*Vaprisol*™) **48** only in 2005, and the V_2 antagonist tolvaptan (*Samsca*™) **49** in 2009, both for the correction of hyponatremia. Small molecule oxytocin antagonists for the treatment of preterm labor are still in clinical trials, e.g., retosiban **50** and nolasiban **51**.

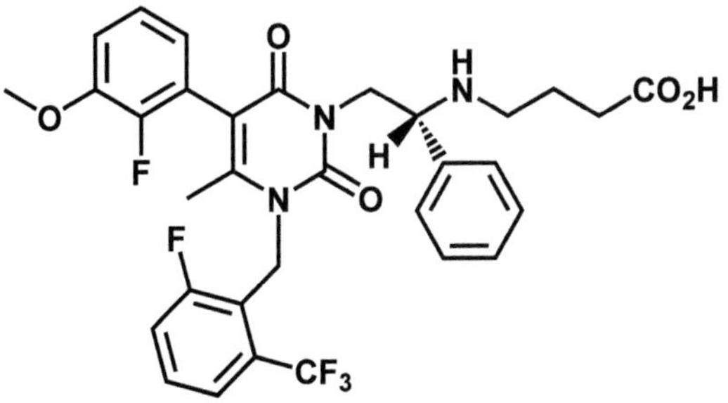

46

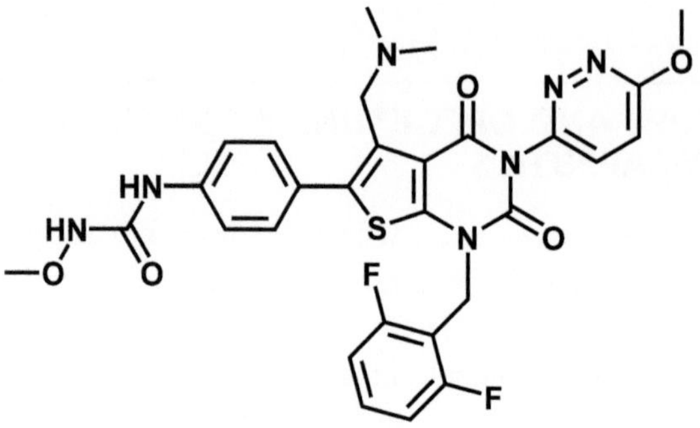

47

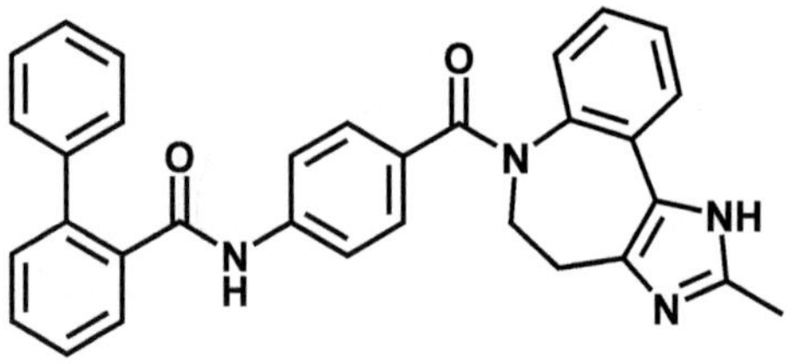

48

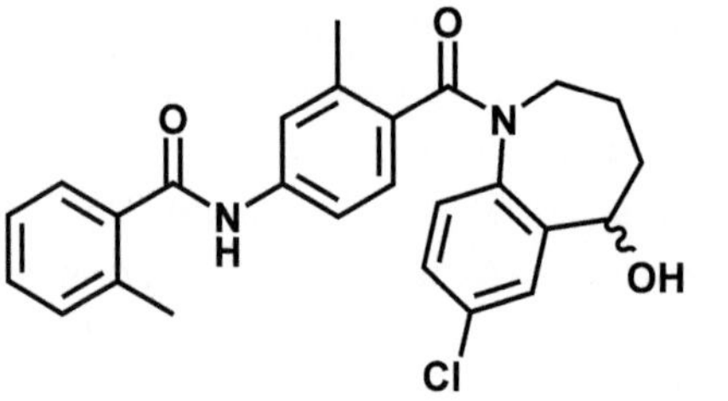

49

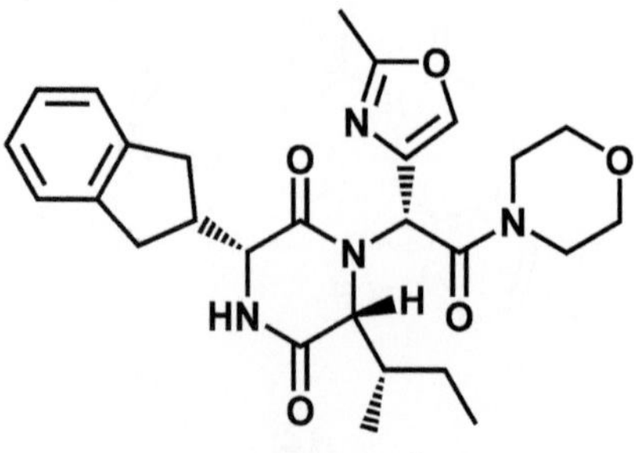

50

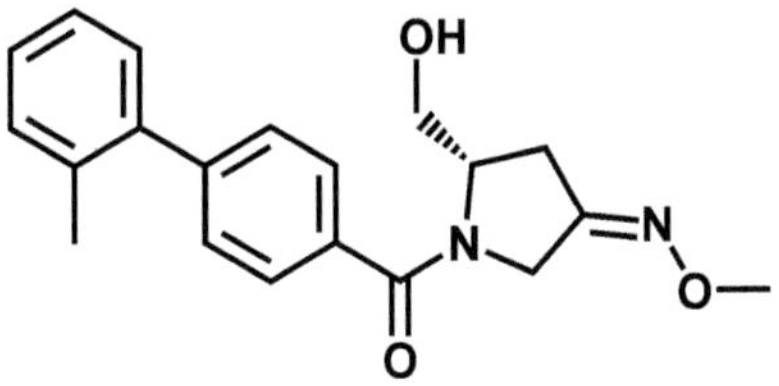

51

Based on the above examples, it would appear that small molecule drugs could be found for all class A GPCRs with natural peptide ligands. However, in some cases the obstacles thrown up by nature might require considerable time and effort to conquer them. The need for large MW or very hydrophobic molecules can make progress difficult, as was the case of the GnRH antagonists. In the case of the melanocortin or somatostatin receptors, obtaining selectivity for the desired receptor subtype may be a significant barrier.

The class B GPCRs in Table 1 address very attractive markets, but potent and selective small molecule agonist drugs for these receptors are very difficult to find. Instead, a series of analogues of the natural peptide hormones have very successfully occupied this competitive space.

The disruption of protein–protein interactions (referred to "undruggable targets" in some cases) is difficult to achieve with small molecules and there is currently interest in exploring peptides for this purpose.[60] However, to the extent that the proteins are intracellular, the cell membrane will continue being a formidable obstacle to the diffusion of the drug from the extracellular fluid to the interior of the cell.

In indications where the patient already has an established intravenous access (e.g., in a hospital setting or nursing home), a peptide may be desirable if the drug needs to be administered by constant infusion with the possibility of rapid up and down titration of the infusion rate as the condition of the patient changes, e.g., oxytocin (*Pitocin*™) **15** for labor induction or eptifibatide (*Integrilin*™) **2** to prevent platelet aggregation during percutaneous coronary intervention.[3]

For treatment of acute situations an injectable is perfectly acceptable, e.g., icatibant (*Firazyr*™) **12** for acute attacks of hereditary angioedema.[61] This is also the case for treatments of limited duration and significant clinical benefit, like teriparatide (*Forteo*™) **22** for bone mass restoration in osteoporosis. Injectables are also tolerable for chronic diseases with indefinite duration of treatment if the delivery device is convenient and the benefit for the patient abundantly clear, to induce good compliance, as in the case of the GLP-1 agonists for type 2 diabetes.

Peptides generally do not cross the blood–brain barrier and might be preferable when the need is to interact with a molecular target only in the periphery without affecting the same target inside the brain. An example is the peripheral kappa agonist CR845, difelikefalin **52**, in clinical trials for pruritus and postoperative and chronic pain, which does not cause central opioid effects.[62]

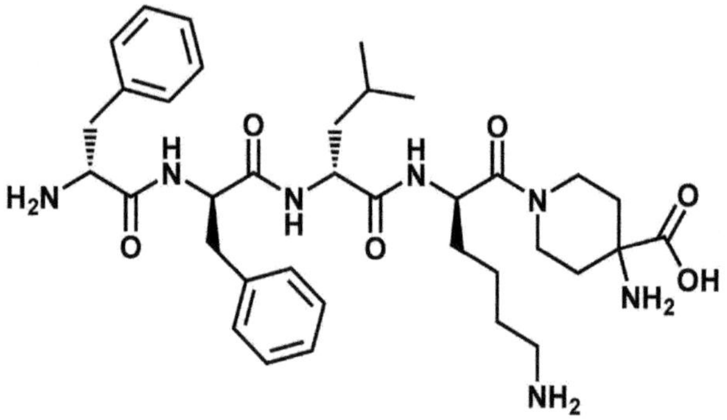

52

In general, peptides have a lower probability of encountering unexpected safety issues, other than mechanistic, during clinical development. Although the median time from initiation of clinical development to approval is not very different from small molecules, about 10 years, they have a higher probability of success from any development stage to regulatory approval than small molecules.[63] For specific niche indications with a significant but unsatisfied medical need, or orphan diseases, a therapeutic peptide may be strategically the faster choice to market, especially if the development of a competing small molecule NCE or biological is not expected due to their longer timelines and the higher investment necessary.

REFERENCES

1. Zaykov, A. N.; Mayer, J. P.; DiMarchi, R. D. Pursuit of a Perfect Insulin. *Nat. Rev. Drug Discov.* **2016**, *15*(6), 425–439.
2. Rembratt, A.; Graugaard-Jensen, C.; Senderovitz, T.; Norgaard, J. P.; Djurhuus, J. C. Pharmacokinetics and Pharmacodynamics of Desmopressin Administered Orally Versus Intravenously at Daytime Versus Night-Time in Healthy Men Aged 55–70 Years. *Eur. J. Clin. Pharmacol.* **2004**, *60*(6), 397–402.
3. Integrilin [Prescribing Information]. Merck & Co., Inc., Whitehouse Station, NJ; 2013.

4. Yamada, T.; Niinuma, K.; Lemaire, M.; Terasaki, T.; Sugiyama, Y. Carrier-Mediated Hepatic Uptake of the Cationic Cyclopeptide, Octreotide, in Rats. Comparison Between In Vivo and In Vitro. *Drug Metab. Dispos.* **1997**, *25*(5), 536–543.
5. Kastin, A. J. *Handbook of Biologically Active Peptides*, 2nd Ed.; Academic Press: New York; 2013.
6. McNeil, B. D.; Pundir, P.; Meeker, S.; Han, L.; Undem, B. J.; Kulka, M.; Dong, X. Identification of a Mast-Cell-Specific Receptor Crucial for Pseudo-Allergic Drug Reactions. *Nature* **2015**, *519*(7542), 237–241.
7. Rosenstock, J.; Balas, B.; Charbonnel, B.; Bolli, G. B.; Boldrin, M.; Ratner, R.; Balena, R. The Fate of Taspoglutide, a Weekly GLP-1 Receptor Agonist, Versus Twice-Daily Exenatide for Type 2 Diabetes: The T-Emerge 2 Trial. *Diabetes Care* **2013**, *36*(3), 498–504.
8. Madsbad, S. Review of Head-to-Head Comparisons of Glucagon-Like Peptide-1 Receptor Agonists. *Diabetes Obes. Metab.* **2016**, *18*(4), 317–332.
9. Laporte, R.; Kohan, A.; Heitzmann, J.; Wisniewska, H.; Toy, J.; La, E.; Tariga, H.; Alagarsamy, S.; Ly, B.; Dykert, J.; Qi, S.; Wisniewski, K.; Galyean, R.; Croston, G.; Schteingart, C. D.; Riviere, P. J. Pharmacological Characterization of FE 202158, a Novel, Potent, Selective, and Short-Acting Peptidic Vasopressin V1a Receptor Full Agonist for the Treatment of Vasodilatory Hypotension. *J. Pharmacol. Exp. Ther.* **2011**, *337*(3), 786–796.
10. Husslein, P.; Roura, L. C.; Dudenhausen, J.; Helmer, H.; Frydman, R.; Rizzo, N.; Schneider, D. Clinical Practice Evaluation of Atosiban in Preterm Labour Management in Six European Countries. *BJOG* **2006**, *113*(Suppl. 3), 105–110.
11. Marceau, F.; Regoli, D. Bradykinin Receptor Ligands: Therapeutic Perspectives. *Nat. Rev. Drug Discov.* **2004**, *3*(10), 845–852.
12. Melin, P.; Vilhardt, H.; Lindeberg, G.; Larsson, L. E.; Akerlund, M. Inhibitory Effect of O-Alkylated Analogues of Oxytocin and Vasopressin on Human and Rat Myometrial Activity. *J. Endocrinol.* **1981**, *88*(2), 173–180.
13. Dalton, J. A.; Lans, I.; Giraldo, J. Quantifying Conformational Changes in GPCRs: Glimpse of a Common Functional Mechanism. *BMC Bioinformatics* **2015**, *16*, 124.
14. Katritch, V.; Cherezov, V.; Stevens, R. C. Structure-Function of the G Protein-Coupled Receptor Superfamily. *Annu. Rev. Pharmacol. Toxicol.* **2013**, *53*, 531–556.
15. Bortolato, A.; Dore, A. S.; Hollenstein, K.; Tehan, B. G.; Mason, J. S.; Marshall, F. H. Structure of Class B GPCRs: New Horizons for Drug Discovery. *Br. J. Pharmacol.* **2014**, *171*(13), 3132–3145.
16. Rivier, J. E.; Rivier, C. L. Corticotropin-Releasing Factor Peptide Antagonists: Design, Characterization and Potential Clinical Relevance. *Front. Neuroendocrinol.* **2014**, *35*(2), 161–170.
17. Ravi, A.; Prasad, B. V.; Balaram, P. Cyclic Peptide Disulfides. Solution and Solid-State Conformation of Boc-Cys-Pro-Aib-Cys-NHMe with a Disulfide Bridge from Cys to Cys, a Disulfide-Bridged Peptide Helix. *J. Am. Chem. Soc.* **1983**, *105*(1), 105–109.
18. Blackwell, H. E.; Grubbs, R. H. Highly Efficient Synthesis of Covalently Cross-Linked Peptide Helices by Ring-Closing Metathesis. *Angew. Chem. Int. Ed.* **1998**, *37*(23), 3281–3284.
19. Juul, K. V.; Erichsen, L.; Robertson, G. L. Temporal Delays and Individual Variation in Antidiuretic Response to Desmopressin. *Am. J. Physiol. Renal Physiol.* **2013**, *304*(3), F268–F278.
20. Thwala, L. N.; Preat, V.; Csaba, N. S. Emerging Delivery Platforms for Mucosal Administration of Biopharmaceuticals: A Critical Update on Nasal, Pulmonary and Oral Routes. *Expert Opin. Drug Deliv.* **2017**, *14*(1), 23–36.
21. McMartin, C.; Hutchinson, L. E.; Hyde, R.; Peters, G. E. Analysis of Structural Requirements for the Absorption of Drugs and Macromolecules from the Nasal Cavity. *J. Pharm. Sci.* **1987**, *76*(7), 535–540.

22. Lemmer, H. J.; Hamman, J. H. Paracellular Drug Absorption Enhancement through Tight Junction Modulation. *Expert Opin. Drug Deliv.* **2013**, *10*(1), 103–114.
23. Zhang, J.; Branch, D. W.; Ramirez, M. M.; Laughon, S. K.; Reddy, U.; Hoffman, M.; Bailit, J.; Kominiarek, M.; Chen, Z.; Hibbard, J. U. Oxytocin Regimen for Labor Augmentation, Labor Progression, and Perinatal Outcomes. *Obstet. Gynecol.* **2011**, *118*(2 Pt. 1), 249–256.
24. Thornton, S.; Davison, J. M.; Baylis, P. H. Effect of Human Pregnancy on Metabolic Clearance Rate of Oxytocin. *Am. J. Physiol.* **1990**, *259*(1 Pt. 2), R21–R24.
25. Gretler, D. D.; Guerciolini, R.; Williams, P. J. Pharmacokinetic and Pharmacodynamic Properties of Eptifibatide in Subjects with Normal or Impaired Renal Function. *Clin. Ther.* **2004**, *26*(3), 390–398.
26. Forteo [Prescribing Information]. Eli Lilly and Company, Indianapolis, IN; 2017.
27. Rubin, M. R.; Cosman, F.; Lindsay, R.; Bilezikian, J. P. The Anabolic Effects of Parathyroid Hormone. *Osteoporos. Int.* **2002**, *13*(4), 267–277.
28. Deacon, C. F.; Nauck, M. A.; Toft-Nielsen, M.; Pridal, L.; Willms, B.; Holst, J. J. Both Subcutaneously and Intravenously Administered Glucagon-like Peptide I are Rapidly Degraded from the NH2-Terminus in Type II Diabetic Patients and in Healthy Subjects. *Diabetes* **1995**, *44*(9), 1126–1131.
29. Byetta [Prescribing Information]. AstraZeneca Pharmaceuticals LP, Wilmington, DE; 2015.
30. Victoza [Prescribing Information]. Novo Nordisk A/S, Bagsvaerd, Denmark; 2016.
31. Bydureon [Prescribing Information]. AstraZeneca Pharmaceuticals LP, Wilmington, DE; 2017.
32. Trulicity [Prescribing Information]. Eli Lilly and Company, Indianapolis, IN; 2016.
33. Tanzeum [Prescribing Information]. GlaxoSmithKline LLC, Wilmington, DE; 2016.
34. Kontermann, R. E. Half-Life Extended Biotherapeutics. *Expert Opin. Biol. Ther.* **2016**, *16*(7), 903–915.
35. Weaver, J. L.; Boyne, M.; Pang, E.; Chimalakonda, K.; Howard, K. E. Nonclinical Evaluation of the Potential for Mast Cell Activation by an Erythropoietin Analog. *Toxicol. Appl. Pharmacol.* **2015**, *287*(3), 246–252.
36. Jensen, L.; Helleberg, H.; Roffel, A.; van Lier, J. J.; Bjornsdottir, I.; Pedersen, P. J.; Rowe, E.; Derving, K. J.; Pedersen, M. L. Absorption, Metabolism and Excretion of the GLP-1 Analogue Semaglutide in Humans and Nonclinical Species. *Eur. J. Pharm. Sci.* **2017**, *104*, 31–41.
37. Lau, J.; Bloch, P.; Schaffer, L.; Pettersson, I.; Spetzler, J.; Kofoed, J.; Madsen, K.; Knudsen, L. B.; McGuire, J.; Steensgaard, D. B.; Strauss, H. M.; Gram, D. X.; Knudsen, S. M.; Nielsen, F. S.; Thygesen, P.; Reedtz-Runge, S.; Kruse, T. Discovery of the Once-Weekly Glucagon-like Peptide-1 (GLP-1) Analogue Semaglutide. *J. Med. Chem.* **2015**, *58*(18), 7370–7380.
38. Trelstar [Prescribing Information]. Allergan USA, Inc., Irvine, CA; 2016.
39. Shore, N. D. Experience with Degarelix in the Treatment of Prostate Cancer. *Ther. Adv. Urol.* **2013**, *5*(1), 11–24.
40. Jiang, G.; Stalewski, J.; Galyean, R.; Dykert, J.; Schteingart, C.; Broqua, P.; Aebi, A.; Aubert, M. L.; Semple, G.; Robson, P.; Akinsanya, K.; Haigh, R.; Riviere, P.; Trojnar, J.; Junien, J. L.; Rivier, J. E. GnRH Antagonists: A new Generation of Long Acting Analogues Incorporating p-Ureido-Phenylalanines at Positions 5 and 6. *J. Med. Chem.* **2001**, *44*(3), 453–467.
41. Valery, C.; Paternostre, M.; Robert, B.; Gulik-Krzywicki, T.; Narayanan, T.; Dedieu, J. C.; Keller, G.; Torres, M. L.; Cherif-Cheikh, R.; Calvo, P.; Artzner, F. Biomimetic Organization: Octapeptide Self-Assembly into Nanotubes of Viral Capsid-like Dimension. *Proc. Natl. Acad. Sci. U. S. A.* **2003**, *100*(18), 10258–10262.
42. Fuzeon [Prescribing Information]. Genentech USA, Inc., South San Francisco, CA; 2017.

43. Wisniewski, K.; Finnman, J.; Flipo, M.; Galyean, R.; Schteingart, C. D. On the Mechanism of Degradation of Oxytocin and Its Analogues in Aqueous Solution. *Biopolymers* **2013**, *100*(4), 408–421.
44. Richter, W. F.; Bhansali, S. G.; Morris, M. E. Mechanistic Determinants ofBiotherapeutics Absorption Following SC Administration. *AAPS J.* **2012**, *14*(3), 559–570.
45. Egrifta [Prescribing Information]. Theratechnologies Inc., Montreal, Quebec, Canada.
46. Gonzalez-Sales, M.; Barriere, O.; Tremblay, P. O.; Nekka, F.; Mamputu, J. C.; Boudreault, S.; Tanguay, M. Population Pharmacokinetic and Pharmacodynamic Analysis of Tesamorelin in HIV-Infected Patients and Healthy Subjects. *J. Pharmacokinet. Pharmacodyn.* **2015**, *42*(3), 287–299.
47. Bray, B. L. Large-Scale Manufacture of Peptide Therapeutics by Chemical Synthesis. *Nat. Rev. Drug Discov.* **2003**, *2*(7), 587–593.
48. Jazayeri, A.; Rappas, M.; Brown, A. J. H.; Kean, J.; Errey, J. C.; Robertson, N. J.; Fiez-Vandal, C.; Andrews, S. P.; Congreve, M.; Bortolato, A.; Mason, J. S.; Baig, A. H.; Teobald, I.; Dore, A. S.; Weir, M.; Cooke, R. M.; Marshall, F. H. Crystal Structure of the GLP-1 Receptor Bound to a Peptide Agonist. *Nature* **2017**, *546*(7657), 254–258.
49. Zhang, H.; Qiao, A.; Yang, D.; Yang, L.; Dai, A.; de, G. C.; Reedtz-Runge, S.; Dharmarajan, V.; Zhang, H.; Han, G. W.; Grant, T. D.; Sierra, R. G.; Weierstall, U.; Nelson, G.; Liu, W.; Wu, Y.; Ma, L.; Cai, X.; Lin, G.; Wu, X.; Geng, Z.; Dong, Y.; Song, G.; Griffin, P. R.; Lau, J.; Cherezov, V.; Yang, H.; Hanson, M. A.; Stevens, R. C.; Zhao, Q.; Jiang, H.; Wang, M. W.; Wu, B. Structure of the Full-Length Glucagon Class B G-Protein-Coupled Receptor. *Nature* **2017**, *546*(7657), 259–264.
50. Liang, Y. L.; Khoshouei, M.; Radjainia, M.; Zhang, Y.; Glukhova, A.; Tarrasch, J.; Thal, D. M.; Furness, S. G. B.; Christopoulos, G.; Coudrat, T.; Danev, R.; Baumeister, W.; Miller, L. J.; Christopoulos, A.; Kobilka, B. K.; Wootten, D.; Skiniotis, G.; Sexton, P. M. Phase-Plate Cryo-EM Structure of a Class B GPCR-G-Protein Complex. *Nature* **2017**, *546*(7656), 118–123.
51. Schuster, N. M.; Rapoport, A. M. New Strategies for the Treatment and Prevention of Primary Headache Disorders. *Nat. Rev. Neurol.* **2016**, *12*(11), 635–650.
52. Subramanian, R.; Zhu, X.; Hock, M. B.; Sloey, B. J.; Wu, B.; Wilson, S. F.; Egbuna, O.; Slatter, J. G.; Xiao, J.; Skiles, G. L. Pharmacokinetics, Biotransformation, and Excretion of [14C]Etelcalcetide (AMG 416) Following a Single Microtracer Intravenous Dose in Patients with Chronic Kidney Disease on Hemodialysis. *Clin. Pharmacokinet.* **2017**, *56*(2), 179–192.
53. Alexander, S. T.; Hunter, T.; Walter, S.; Dong, J.; Maclean, D.; Baruch, A.; Subramanian, R.; Tomlinson, J. E. Critical Cysteine Residues in Both the Calcium-Sensing Receptor and the Allosteric Activator AMG 416 Underlie the Mechanism of Action. *Mol. Pharmacol.* **2015**, *88*(5), 853–865.
54. Mobli, M.; Undheim, E. A. B.; Rash, L. D. Modulation of Ion Channels by Cysteine-Rich Peptides: From Sequence to Structure. *Adv. Pharmacol.* **2017**, *79*, 199–223.
55. Prialt [Prescribing Information]. Jazz Pharmaceuticals, Inc.,Palo Alto, CA; 2013.
56. Lau, J. L.; Dunn, M. K. *Development Trends for Peptide Therapeutics: Status in 2016*, 11th Annual Peptide Therapeutics Symposium La Jolla, CA, October 27–28, 2016 http://www.peptidetherapeutics.org/wp-content/uploads/2017/02/2016-Peptide-Therapeutics-Poster-Ferring-Research-Institute.pdf.
57. Hasson, T.; Kolitz, S.; Towfic, F.; Laifenfeld, D.; Bakshi, S.; Beriozkin, O.; Shacham-Abramson, M.; Timan, B.; Fowler, K. D.; Birnberg, T.; Konya, A.; Komlosh, A.; Ladkani, D.; Hayden, M. R.; Zeskind, B.; Grossman, I. Functional Effects of the Antigen Glatiramer Acetate are Complex and Tightly Associated with Its Composition. *J. Neuroimmunol.* **2016**, *290*, 84–95.
58. Fosgerau, K.; Hoffmann, T. Peptide Therapeutics: Current Status and Future Directions. *Drug Discov. Today* **2015**, *20*(1), 122–128.

59. Henninot, A.; Collins, J. C.; Nuss, J. M. The Current State of Peptide Drug Discovery: Back to the Future? *J. Med. Chem.* **2017**, https://doi.org/10.1021/acs.jmedchem.7b00318.
60. Cardote, T. A.; Ciulli, A. Cyclic and Macrocyclic Peptides as Chemical Tools to Recognise Protein Surfaces and Probe Protein-Protein Interactions. *ChemMedChem* **2016**, *11*(8), 787–794.
61. Firazyr [Prescribing Information]. Shire Orphan Therapies LLC, Lexington, MA; 2015.
62. Cara Therapeutics Website http://www.caratherapeutics.com/; 2017.
63. Lau, J. L.; Dunn, M. K. *Therapeutic Peptides: Historical Perspectives, Current Development Trends, and Future Directions. Bioorg. Med. Chem.* **2017**, https://doi.org/10.1016/j.bmc.2017.06.052

KEYWORD INDEX, VOLUME 50

Note: Page numbers followed by "*f*" indicate figures, "*t*" indicate tables, and "*s*" indicate schemes.

A

B

C

O

P

R

S

CUMULATIVE CHAPTER TITLES KEYWORD INDEX, VOLUME 1 – 50

CUMULATIVE NCE INTRODUCTION INDEX, 1983–2017

GENERIC NAME	INDICATION	YEAR INTRODUCED	ARMC VOL., (PAGE)
abacavir sulfate	antiviral	1999	35 (333)
abaloparatide	osteoporosis	2017	50 (555)
abarelix	anticancer	2004	40 (446)
abatacept	antiarthritic	2006	42 (509)
abiraterone acetate	anticancer	2011	47 (505)
acarbose	antidiabetic	1990	26 (297)
aceclofenac	antiinflammatory	1992	28 (325)
acemannan	wound healing agent	2001	37 (259)
acetohydroxamic acid	urinary tract/bladder disorders	1983	19 (313)
acetorphan	antidiarrheal	1993	29 (332)
acipimox	antihypercholesterolemic	1985	21 (323)
acitretin	antipsoriasis	1989	25 (309)
aclidinium bromide	chronic obstructive pulmonary disorder	2012	48 (481)
acotiamide	dyspepsia	2013	49 (447)
acrivastine	antiallergy	1988	24 (295)
actarit	antiinflammatory	1994	30 (296)
adalimumab	antiarthritic	2003	39 (267)
adamantanium bromide	antibacterial	1984	20 (315)
adefovir dipivoxil	antiviral	2002	38 (348)
ado-trastuzumab emtansine	anticancer	2013	49 (449)
adrafinil	sleep disorders	1986	22 (315)
AF-2259	antiinflammatory	1987	23 (325)
afamelanotide	erythropoietic protoporphyria	2014	50 (554)
afatinib	anticancer	2013	49 (451)
aflibercept	ophthalmologic, macular degeneration	2011	47 (507)
afloqualone	muscle relaxant	1983	19 (313)
agalsidase alfa	Fabry's disease	2001	37 (259)
alacepril	antihypertensive	1988	24 (296)
albiglutide	antidiabetic	2014	50 (552)
alcaftadine	ophthalmologic (allergic conjunctivitis)	2010	46 (444)
alclometasone dipropionate	antiinflammatory	1985	21 (323)
alefacept	antipsoriasis	2003	39 (267)
alemtuzumab	anticancer	2001	37 (260)
alendronate sodium	osteoporosis	1993	29 (332)
alfentanil hydrochloride	analgesic	1983	19 (314)
alfuzosin hydrochloride	antihypertensive	1988	24 (296)
alglucerase	Gaucher's disease	1991	27 (321)

GENERIC NAME	INDICATION	YEAR INTRODUCED	ARMC VOL., (PAGE)
alglucosidase alfa	Pompe disease	2006	42 (511)
aliskiren	antihypertensive	2007	43 (461)
alitretinoin	anticancer	1999	35 (333)
alminoprofen	analgesic	1983	19 (314)
almotriptan	antimigraine	2000	36 (295)
alogliptin	antidiabetic	2010	46 (446)
alosetron hydrochloride	irritable bowel syndrome	2000	36 (295)
alpha-1 antitrypsin	emphysema	1988	24 (297)
alpidem	anxiolytic	1991	27 (322)
alpiropride	antimigraine	1988	24 (296)
alteplase	antithrombotic	1987	23 (326)
alvimopan	post-operative ileus	2008	44 (584)
ambrisentan	pulmonary hypertension	2007	43 (463)
amfenac sodium	antiinflammatory	1986	22 (315)
amifostine	cytoprotective	1995	31 (338)
aminoprofen	antiinflammatory	1990	26 (298)
amisulpride	antipsychotic	1986	22 (316)
amlexanox	antiasthma	1987	23 (327)
amlodipine besylate	antihypertensive	1990	26 (298)
amorolfine hydrochloride	antifungal	1991	27 (322)
amosulalol	antihypertensive	1988	24 (297)
ampiroxicam	antiinflammatory	1994	30 (296)
amprenavir	antiviral	1999	35 (334)
amrinone	congestive heart failure	1983	19 (314)
amrubicin hydrochloride	anticancer	2002	38 (349)
amsacrine	anticancer	1987	23 (327)
amtolmetin guacil	antiinflammatory	1993	29 (332)
anagliptin	antidiabetic	2012	48 (483)
anagrelide hydrochloride	antithrombotic	1997	33 (328)
anakinra	antiarthritic	2001	37 (261)
anastrozole	anticancer	1995	31 (338)
angiotensin II	anticancer adjuvant	1994	30 (296)
anidulafungin	antifungal	2006	42 (512)
aniracetam	cognition enhancer	1993	29 (333)
anti-digoxin polyclonal antibody	antidote, digoxin poisoning	2002	38 (350)
APD	osteoporosis	1987	23 (326)
apixaban	antithrombotic	2011	47 (509)
apraclonidine hydrochloride	antiglaucoma	1988	24 (297)
aprepitant	antiemetic	2003	39 (268)
APSAC	antithrombotic	1987	23 (326)
aranidipine	antihypertensive	1996	32 (306)
arbekacin	antibacterial	1990	26 (298)
arformoterol	antiasthma	2007	43 (465)

GENERIC NAME	INDICATION	YEAR INTRODUCED	ARMC VOL., (PAGE)
argatroban	antithrombotic	1990	26 (299)
arglabin	anticancer	1999	35 (335)
aripiprazole	antipsychotic	2002	38 (350)
armodafinil	sleep disorders	2009	45 (478)
arotinolol hydrochloride	antihypertensive	1986	22 (316)
arteether	antimalarial	2000	36 (296)
artemisinin	antimalarial	1987	23 (327)
asenapine	antipsychotic	2009	45 (479)
aspoxicillin	antibacterial	1987	23 (328)
astemizole	antiallergy	1983	19 (314)
astromycin sulfate	antibacterial	1985	21 (324)
atazanavir	antiviral	2003	39 (269)
atomoxetine	attention deficit hyperactivity disorder	2003	39 (270)
atorvastatin calcium	antihypercholesterolemic	1997	33 (328)
atosiban	premature labor	2000	36 (297), 50 (555)
atovaquone	antiparasitic	1992	28 (326)
auranofin	antiarthritic	1983	19 (314)
avanafil	male sexual dysfunction	2011	47 (512)
aviptadil	erectile dysfunction	2000	50 (556)
axitinib	anticancer	2012	48 (485)
azacitidine	anticancer	2004	40 (447)
azelaic acid	acne	1989	25 (310)
azelastine hydrochloride	antiallergy	1986	22 (316)
azelnidipine	antihypertensive	2003	39 (270)
azilsartan	antihypertensive	2011	47 (514)
azithromycin	antibacterial	1988	24 (298)
azosemide	diuretic	1986	22 (316)
aztreonam	antibacterial	1984	20 (315)
balofloxacin	antibacterial	2002	38 (351)
balsalazide disodium	ulcerative colitis	1997	33 (329)
bambuterol	antiasthma	1990	26 (299)
barnidipine hydrochloride	antihypertensive	1992	28 (326)
beclobrate	antihypercholesterolemic	1986	22 (317)
bedaquiline	antibacterial	2012	48 (487)
befunolol hydrochloride	antiglaucoma	1983	19 (315)
belatacept	immunosuppressant	2011	47 (516)
belimumab	lupus	2011	47 (519)
belotecan	anticancer	2004	40 (449)
benazepril hydrochloride	antihypertensive	1990	26 (299)
benexate hydrochloride	antiulcer	1987	23 (328)
benidipine hydrochloride	antihypertensive	1991	27 (322)
beraprost sodium	antiplatelet	1992	28 (326)
besifloxacin	antibacterial	2009	45 (482)

GENERIC NAME	INDICATION	YEAR INTRODUCED	ARMC VOL., (PAGE)
betamethasone butyrate propionate	antiinflammatory	1994	30 (297)
betaxolol hydrochloride	antihypertensive	1983	19 (315
betotastine besilate	antiallergy	2000	36 (297)
bevacizumab	anticancer	2004	40 (450)
bevantolol hydrochloride	antihypertensive	1987	23 (328)
bexarotene	anticancer	2000	36 (298)
biapenem	antibacterial	2002	38 (351)
bicalutamide	anticancer	1995	31 (338)
bifemelane hydrochloride	nootropic	1987	23 (329)
bilastine	antiallergy	2010	46 (449)
bimatoprost	antiglaucoma	2001	37 (261)
binfonazole	sleep disorders	1983	19 (315)
binifibrate	antihypercholesterolemic	1986	22 (317)
biolimus drug-eluting stent	coronary artery disease, antirestenotic	2008	44 (586)
bisantrene hydrochloride	anticancer	1990	26 (300)
bisoprolol fumarate	antihypertensive	1986	22 (317)
bivalirudin	acute coronary syndrome	1999	50 (556)
bivalirudin	antithrombotic	1999	50 (556)
bivalirudin	thrombosis	1999	50 (556)
bivalirudin	antithrombotic	2000	36 (298)
blonanserin	antipsychotic	2008	44 (587)
boceprevir	antiviral	2011	47 (521)
bopindolol	antihypertensive	1985	21 (324)
bortezomib	anticancer	2003	39 (271)
bosentan	antihypertensive	2001	37 (262)
bosutinib	anticancer	2012	48 (489)
brentuximab	anticancer	2011	47 (523)
brimonidine	antiglaucoma	1996	32 (306)
brinzolamide	antiglaucoma	1998	34 (318)
brodimoprin	antibacterial	1993	29 (333)
bromfenac sodium	antiinflammatory	1997	33 (329)
brotizolam	sleep disorders	1983	19 (315)
brovincamine fumarate	cerebral vasodilator	1986	22 (317)
bucillamine	immunomodulator	1987	23 (329)
bucladesine sodium	congestive heart failure	1984	20 (316)
budipine	Parkinson's disease	1997	33 (330)
budralazine	antihypertensive	1983	19 (315
bulaquine	antimalarial	2000	36 (299)
bunazosin hydrochloride	antihypertensive	1985	21 (324)
bupropion hydrochloride	antidepressant	1989	25 (310)
buserelin	infertility	1984	50 (553)
buserelin	endometriosis	1984	50 (553)
buserelin	anticancer	1984	50 (553)

GENERIC NAME	INDICATION	YEAR INTRODUCED	ARMC VOL., (PAGE)
buserelin acetate	hormone therapy	1984	20 (316)
buspirone hydrochloride	anxiolytic	1985	21 (324)
butenafine hydrochloride	antifungal	1992	28 (327)
butibufen	antiinflammatory	1992	28 (327)
butoconazole	antifungal	1986	22 (318)
butoctamide	sleep disorders	1984	20 (316)
butyl flufenamate	antiinflammatory	1983	19 (316)
cabazitaxel	anticancer	2010	46 (451)
cabergoline	antiprolactin	1993	29 (334)
cabozantinib	anticancer	2012	48 (491)
cadexomer iodine	wound healing agent	1983	19 (316)
cadralazine	antihypertensive	1988	24 (298)
calcipotriol	antipsoriasis	1991	27 (323)
calcitonin (human)	hypercalcemia	1986	50 (551)
calcitonin (human)	Paget's disease	1986	50 (551)
calcitonin (human)	osteoporosis	1986	50 (551)
calcitonin (salmon)	hypercalcemia	1971	50 (551)
calcitonin (salmon)	Paget's disease	1971	50 (551)
calcitonin (salmon)	osteoporosis	1971	50 (551)
camostat mesylate	anticancer	1985	21 (325)
canagliflozin	antidiabetic	2013	49 (453)
canakinumab	antiinflammatory	2009	45 (484)
candesartan cilexetil	antihypertensive	1997	33 (330)
capecitabine	anticancer	1998	34 (319)
captopril	antihypertensive	1982	13 (086)
carbetocin	postpartum hemorrhage	2001	50 (555)
carboplatin	antibacterial	1986	22 (318)
carfilzomib	anticancer	2012	48 (492), 50 (552)
carperitide	congestive heart failure	1995	31 (339), 50 (555)
carumonam	antibacterial	1988	24 (298)
carvedilol	antihypertensive	1991	27 (323)
caspofungin acetate	antifungal	2001	37 (263)
catumaxomab	anticancer	2009	45 (486)
cefbuperazone sodium	antibacterial	1985	21 (325)
cefcapene pivoxil	antibacterial	1997	33 (330)
cefdinir	antibacterial	1991	27 (323)
cefditoren pivoxil	antibacterial	1994	30 (297)
cefepime	antibacterial	1993	29 (334)
cefetamet pivoxil hydrochloride	antibacterial	1992	28 (327)
cefixime	antibacterial	1987	23 (329)
cefmenoxime hydrochloride	antibacterial	1983	19 (316)

GENERIC NAME	INDICATION	YEAR INTRODUCED	ARMC VOL., (PAGE)
cefminox sodium	antibacterial	1987	23 (330)
cefodizime sodium	antibacterial	1990	26 (300)
cefonicid sodium	antibacterial	1984	20 (316)
ceforanide	antibacterial	1984	20 (317)
cefoselis	antibacterial	1998	34 (319)
cefotetan disodium	antibacterial	1984	20 (317)
cefotiam hexetil hydrochloride	antibacterial	1991	27 (324)
cefozopran hydrochloride	antibacterial	1995	31 (339)
cefpimizole	antibacterial	1987	23 (330)
cefpiramide sodium	antibacterial	1985	21 (325)
cefpirome sulfate	antibacterial	1992	28 (328)
cefpodoxime proxetil	antibacterial	1989	25 (310)
cefprozil	antibacterial	1992	28 (328)
ceftaroline fosamil	antibacterial	2010	46 (453)
ceftazidime	antibacterial	1983	19 (316)
cefteram pivoxil	antibacterial	1987	23 (330)
ceftibuten	antibacterial	1992	28 (329)
ceftobiprole medocaril	antibacterial	2008	44 (589)
cefuroxime axetil	antibacterial	1987	23 (331)
cefuzonam sodium	antibacterial	1987	23 (331)
celecoxib	antiarthritic	1999	35 (335)
celiprolol hydrochloride	antihypertensive	1983	19 (317
centchroman	contraception	1991	27 (324)
centoxin	immunomodulator	1991	27 (325)
cerivastatin	antihypercholesterolemic	1997	33 (331)
certolizumab pegol	irritable bowel syndrome	2008	44 (592)
cetilistat	antiobesity	2013	49 (454)
cetirizine hydrochloride	antiallergy	1987	23 (331)
cetrorelix	infertility	1999	35 (336), 50 (553)
cetuximab	anticancer	2003	39 (272)
cevimeline hydrochloride	antixerostomia	2000	36 (299)
chenodiol	gallstones	1983	19 (317)
CHF-1301	Parkinson's disease	1999	35 (336)
choline alfoscerate	cognition enhancer	1990	26 (300)
choline fenofibrate	antihypercholesterolemic	2008	44 (594)
cibenzoline	antiarrhythmic	1985	21 (325)
ciclesonide	antiasthma	2005	41 (443)
cicletanine	antihypertensive	1988	24 (299)
cidofovir	antiviral	1996	32 (306)
cilazapril	antihypertensive	1990	26 (301)
cilostazol	antithrombotic	1988	24 (299)
cimetropium bromide	antispasmodic	1985	21 (326)
cinacalcet	hyperparathyroidism	2004	40 (451)

GENERIC NAME	INDICATION	YEAR INTRODUCED	ARMC VOL., (PAGE)
cinildipine	antihypertensive	1995	31 (339)
cinitapride	gastroprokinetic	1990	26 (301)
cinolazepam	anxiolytic	1993	29 (334)
ciprofibrate	antihypercholesterolemic	1985	21 (326)
ciprofloxacin	antibacterial	1986	22 (318)
cisapride	gastroprokinetic	1988	24 (299)
cisatracurium besilate	muscle relaxant	1995	31 (340)
citalopram	antidepressant	1989	25 (311)
cladribine	anticancer	1993	29 (335)
clarithromycin	antibacterial	1990	26 (302)
clevidipine	antihypertensive	2008	44 (596)
clevudine	antiviral	2007	43 (466)
clobenoside	antiinflammatory	1988	24 (300)
cloconazole hydrochloride	antifungal	1986	22 (318)
clodronate disodium	calcium regulation	1986	22 (319)
clofarabine	anticancer	2005	41 (444)
clopidogrel hydrogensulfate	antithrombotic	1998	34 (320)
cloricromen	antithrombotic	1991	27 (325)
clospipramine hydrochloride	antipsychotic	1991	27 (325)
cobicistat	antiviral pharmacokinetic enhancer)	2013	49 (456)
colesevelam hydrochloride	antihypercholesterolemic	2000	36 (300)
colestimide	antihypercholesterolemic	1999	35 (337)
colforsin daropate hydrochloride	congestive heart failure	1999	35 (337)
conivaptan	hyponatremia	2006	42 (514)
corifollitropin alfa	infertility	2010	46 (455)
corticotropin	multiple inflammatory diseases	1952	50 (554)
corticotropin	West syndrome	1952	50 (554)
crizotinib	anticancer	2011	47 (525)
crofelemer	antidiarrheal	2012	48 (494)
crotelidae polyvalent immune fab	antidote, snake venom poisoning	2001	37 (263)
cyclosporine	immunosuppressant	1983	19 (317)
cytarabine ocfosfate	anticancer	1993	29 (335)
dabigatran etexilate	anticoagulant	2008	44 (598)
dabrafenib	anticancer	2013	49 (458)
dalfampridine	multiple sclerosis	2010	46 (458)
dalfopristin	antibacterial	1999	35 (338)
dapagliflozin	antidiabetic	2012	48 (495)
dapiprazole hydrochloride	antiglaucoma	1987	23 (332)
dapoxetine	premature ejaculation	2009	45 (488)
daptomycin	antibacterial	2003	39 (272)

GENERIC NAME	INDICATION	YEAR INTRODUCED	ARMC VOL., (PAGE)
darifenacin	urinary tract/bladder disorders	2005	41 (445)
darunavir	antiviral	2006	42 (515)
dasatinib	anticancer	2006	42 (517)
decitabine	myelodysplastic syndromes	2006	42 (519)
defeiprone	iron chelation therapy	1995	31 (340)
deferasirox	iron chelation therapy	2005	41 (446)
defibrotide	antithrombotic	1986	22 (319)
deflazacort	antiinflammatory	1986	22 (319)
degarelix	anticancer	2008	50 (554)
degarelix acetate	anticancer	2009	45 (490)
delapril	antihypertensive	1989	25 (311)
delavirdine mesylate	antiviral	1997	33 (331)
denileukin diftitox	anticancer	1999	35 (338)
denopamine	congestive heart failure	1988	24 (300)
denosumab	osteoporosis	2010	46 (459)
deprodone propionate	antiinflammatory	1992	28 (329)
desflurane	anesthetic	1992	28 (329)
desloratadine	antiallergy	2001	37 (264)
desmopressin	diabetes insipidus	1972	50 (556)
desmopressin	primary nocturnal enuresis, nocturia	1972	50 (556)
desvenlafaxine	antidepressant	2008	44 (600)
dexfenfluramine	antiobesity	1997	33 (332)
dexibuprofen	antiinflammatory	1994	30 (298)
dexlansoprazole	antiulcer	2009	45 (492)
dexmedetomidine hydrochloride	sleep disorders	2000	36 (301)
dexmethylphenidate hydrochloride	attention deficit hyperactivity disorder	2002	38 (352)
dexrazoxane	cardioprotective	1992	28 (330)
dezocine	analgesic	1991	27 (326)
diacerein	antiinflammatory	1985	21 (326)
didanosine	antiviral	1991	27 (326)
dilevalol	antihypertensive	1989	25 (311)
dimethyl fumarate	multiple sclerosis	2013	49 (460)
diquafosol tetrasodium	ophthalmologic (dry eye)	2010	46 (462)
dirithromycin	antibacterial	1993	29 (336)
disodium pamidronate	osteoporosis	1989	25 (312)
divistyramine	antihypercholesterolemic	1984	20 (317)
docarpamine	congestive heart failure	1994	30 (298)
docetaxel	anticancer	1995	31 (341)
dofetilide	antiarrhythmic	2000	36 (301)
dolasetron mesylate	antiemetic	1998	34 (321)
dolutegravir	antiviral	2013	49 (461)
donepezil hydrochloride	Alzheimer's disease	1997	33 (332)
dopexamine	congestive heart failure	1989	25 (312)

GENERIC NAME	INDICATION	YEAR INTRODUCED	ARMC VOL., (PAGE)
doripenem	antibacterial	2005	41 (448)
dornase alfa	cystic fibrosis	1994	30 (298)
dorzolamide hydrochloride	antiglaucoma	1995	31 (341)
dosmalfate	antiulcer	2000	36 (302)
doxacurium chloride	muscle relaxant	1991	27 (326)
doxazosin mesylate	antihypertensive	1988	24 (300)
doxefazepam	anxiolytic	1985	21 (326)
doxercalciferol	hyperparathyroidism	1999	35 (339)
doxifluridine	anticancer	1987	23 (332)
doxofylline	antiasthma	1985	21 (327)
dronabinol	antiemetic	1986	22 (319)
dronedarone	antiarrhythmic	2009	45 (495)
drospirenone	contraception	2000	36 (302)
drotrecogin alfa	antisepsis	2001	37 (265)
droxicam	antiinflammatory	1990	26 (302)
droxidopa	Parkinson's disease	1989	25 (312)
dulaglutide	antidiabetic	2014	50 (552)
duloxetine	antidepressant	2004	40 (452)
dutasteride	benign prostatic hyperplasia	2002	38 (353)
duteplase	anticoagulant	1995	31 (342)
ebastine	antiallergy	1990	26 (302)
eberconazole	antifungal	2005	41 (449)
ebrotidine	antiulcer	1997	33 (333)
ecabet sodium	antiulcer	1993	29 (336)
ecallantide	angioedema, hereditary	2009	46 (464)
eculizumab	hemoglobinuria	2007	43 (468)
edaravone	neuroprotective	2001	37 (265)
edoxaban	antithrombotic	2011	47 (527)
efalizumab	antipsoriasis	2003	39 (274)
efavirenz	antiviral	1998	34 (321)
efinaconazole	antifungal	2013	49 (463)
efonidipine	antihypertensive	1994	30 (299)
egualen sodium	antiulcer	2000	36 (303)
elcatonin	osteoporosis	1981	50 (551)
eldecalcitol	osteoporosis	2011	47 (529)
eletriptan	antimigraine	2001	37 (266)
eltrombopag	antithrombocytopenic	2009	45 (497)
elvitegravir	antiviral	2013	49 (465)
emedastine difumarate	antiallergy	1993	29 (336)
emorfazone	analgesic	1984	20 (317)
emtricitabine	antiviral	2003	39 (274)
enalapril maleate	antihypertensive	1984	20 (317)
enalaprilat	antihypertensive	1987	23 (332)
encainide hydrochloride	antiarrhythmic	1987	23 (333)

GENERIC NAME	INDICATION	YEAR INTRODUCED	ARMC VOL., (PAGE)
enfuvirtide	antiviral	2003	39 (275), 50 (554)
enocitabine	anticancer	1983	19 (318)
enoxacin	antibacterial	1986	22 (320)
enoxaparin	anticoagulant	1987	23 (333)
enoximone	congestive heart failure	1988	24 (301)
enprostil	antiulcer	1985	21 (327)
entacapone	Parkinson's disease	1998	34 (322)
entecavir	antiviral	2005	41 (450)
enzalutamide	anticancer	2012	48 (497)
epalrestat	antidiabetic	1992	28 (330)
eperisone hydrochloride	muscle relaxant	1983	19 (318)
epidermal growth factor	wound healing agent	1987	23 (333)
epinastine	antiallergy	1994	30 (299)
epirubicin hydrochloride	anticancer	1984	20 (318)
eplerenone	antihypertensive	2003	39 (276)
epoprostenol sodium	antiplatelet	1983	19 (318)
eprosartan	antihypertensive	1997	33 (333)
eptazocine hydrobromide	analgesic	1987	23 (334)
eptifibatide	acute coronary syndrome	1998	50 (554)
eptilfibatide	antithrombotic	1999	35 (340)
erdosteine	expectorant	1995	31 (342)
eribulin mesylate	anticancer	2010	46 (465)
erlotinib	anticancer	2004	40 (454)
ertapenem sodium	antibacterial	2002	38 (353)
erythromycin acistrate	antibacterial	1988	24 (301)
erythropoietin	hematopoietic	1988	24 (301)
escitalopram oxolate	antidepressant	2002	38 (354)
eslicarbazepine acetate	anticonvulsant	2009	45 (498)
esmolol hydrochloride	antiarrhythmic	1987	23 (334)
esomeprazole magnesium	antiulcer	2000	36 (303)
eszopiclone	sleep disorders	2005	41 (451)
etelcalcetide	secondary hyperparathyroidism	2016	50 (551)
ethyl icosapentate	antithrombotic	1990	26 (303)
etizolam	anxiolytic	1984	20 (318)
etodolac	antiinflammatory	1985	21 (327)
etoricoxibe	antiarthritic	2002	38 (355)
etravirine	antiviral	2008	44 (602)
everolimus	immunosuppressant	2004	40 (455)
exemestane	anticancer	2000	36 (304)
exenatide	antidiabetic	2005	41 (452), 50 (552)
exifone	cognition enhancer	1988	24 (302)
ezetimibe	antihypercholesterolemic	2002	38 (355)
factor VIIa	haemophilia	1996	32 (307)

GENERIC NAME	INDICATION	YEAR INTRODUCED	ARMC VOL., (PAGE)
factor VIII	hemostatic	1992	28 (330)
fadrozole hydrochloride	anticancer	1995	31 (342)
falecalcitriol	hyperparathyroidism	2001	37 (266)
famciclovir	antiviral	1994	30 (300)
famotidine	antiulcer	1985	21 (327)
fasudil hydrochloride	amyotrophic lateral sclerosis	1995	31 (343)
febuxostat	gout	2009	45 (501)
felbamate	anticonvulsant	1993	29 (337)
felbinac	antiinflammatory	1986	22 (320)
felypressin	dental anesthesia	1970	50 (551)
felodipine	antihypertensive	1988	24 (302)
fenbuprol	biliary tract dysfunction	1983	19 (318)
fenoldopam mesylate	antihypertensive	1998	34 (322)
fenticonazole nitrate	antifungal	1987	23 (334)
fesoterodine	urinary tract/bladder disorders	2008	44 (604)
fexofenadine	antiallergy	1996	32 (307)
fidaxomicin	antibacterial	2011	47 (531)
filgrastim	immunostimulant	1991	27 (327)
finasteride	benign prostatic hyperplasia	1992	28 (331)
fingolimod	multiple sclerosis	2010	46 (468)
fisalamine	antiinflammatory	1984	20 (318)
fleroxacin	antibacterial	1992	28 (331)
flomoxef sodium	antibacterial	1988	24 (302)
flosequinan	congestive heart failure	1992	28 (331)
fluconazole	antifungal	1988	24 (303)
fludarabine phosphate	anticancer	1991	27 (327)
flumazenil	antidote, benzodiazepine overdose	1987	23 (335)
flunoxaprofen	antiinflammatory	1987	23 (335)
fluoxetine hydrochloride	antidepressant	1986	22 (320)
flupirtine maleate	analgesic	1985	21 (328)
flurithromycin ethylsuccinate	antibacterial	1997	33 (333)
flutamide	anticancer	1983	19 (318)
flutazolam	anxiolytic	1984	20 (318)
fluticasone furoate	antiallergy	2007	43 (469)
fluticasone propionate	antiinflammatory	1990	26 (303)
flutoprazepam	anxiolytic	1986	22 (320)
flutrimazole	antifungal	1995	31 (343)
flutropium bromide	antiasthma	1988	24 (303)
fluvastatin	antihypercholesterolemic	1994	30 (300)
fluvoxamine maleate	antidepressant	1983	19 (319)
follitropin alfa	infertility	1996	32 (307)
follitropin beta	infertility	1996	32 (308)

GENERIC NAME	INDICATION	YEAR INTRODUCED	ARMC VOL., (PAGE)
fomepizole	antidote, ethylene glycol poisoning	1998	34 (323)
fomivirsen sodium	antiviral	1998	34 (323)
fondaparinux sodium	antithrombotic	2002	38 (356)
formestane	anticancer	1993	29 (337)
formoterol fumarate	chronic obstructive pulmonary disorder	1986	22 (321)
fosamprenavir	antiviral	2003	39 (277)
fosaprepitant dimeglumine	antiemetic	2008	44 (606)
foscarnet sodium	antiviral	1989	25 (313)
fosfluconazole	antifungal	2004	40 (457)
fosfosal	analgesic	1984	20 (319)
fosinopril sodium	antihypertensive	1991	27 (328)
fosphenytoin sodium	anticonvulsant	1996	32 (308)
fotemustine	anticancer	1989	25 (313)
fropenam	antibacterial	1997	33 (334)
frovatriptan	antimigraine	2002	38 (357)
fudosteine	expectorant	2001	37 (267)
fulveristrant	anticancer	2002	38 (357)
gabapentin	anticonvulsant	1993	29 (338)
gabapentin Enacarbil	restless leg syndrome	2011	47 (533)
gadoversetamide	diagnostic	2000	36 (304)
gallium nitrate	calcium regulation	1991	27 (328)
gallopamil hydrochloride	antianginal	1983	19 (3190)
galsulfase	mucopolysaccharidosis VI	2005	41 (453)
ganciclovir	antiviral	1988	24 (303)
ganirelix	infertility	1999	50 (554)
ganirelix acetate	infertility	2000	36 (305)
garenoxacin	antibacterial	2007	43 (471)
gatilfloxacin	antibacterial	1999	35 (340)
gefitinib	anticancer	2002	38 (358)
gemcitabine hydrochloride	anticancer	1995	31 (344)
gemeprost	abortifacient	1983	19 (319)
gemifloxacin	antibacterial	2004	40 (458)
gemtuzumab ozogamicin	anticancer	2000	36 (306)
gestodene	contraception	1987	23 (335)
gestrinone	contraception	1986	22 (321)
glatiramer	multiple sclerosis	1996	50 (554)
glatiramer acetate	multiple sclerosis	1997	33 (334)
glimepiride	antidiabetic	1995	31 (344)
glucagon	hypoglycemia	1989	50 (553)
glucagon, rDNA	antidiabetic	1993	29 (338)
GMDP	immunostimulant	1996	32 (308)
golimumab	antiinflammatory	2009	45 (503)
gonadorelin	infertility	1989	50 (553)

GENERIC NAME	INDICATION	YEAR INTRODUCED	ARMC VOL., (PAGE)
gonadorelin	primary amenorrhea	1989	50 (553)
goserelin	hormone therapy	1987	23 (336), 50 (553)
granisetron hydrochloride	antiemetic	1991	27 (329)
guanadrel sulfate	antihypertensive	1983	19 (319)
gusperimus	immunosuppressant	1994	30 (300)
halobetasol propionate	antiinflammatory	1991	27 (329)
halofantrine	antimalarial	1988	24 (304)
halometasone	antiinflammatory	1983	19 (320)
histrelin	anticancer	1991	50 (553)
histrelin	precocious puberty	1991	50 (553)
histrelin	precocious puberty	1993	29 (338)
hydrocortisone aceponate	antiinflammatory	1988	24 (304)
hydrocortisone butyrate	antiinflammatory	1983	19 (320)
ibandronic acid	osteoporosis	1996	32 (309)
ibopamine hydrochloride	congestive heart failure	1984	20 (319)
ibritunomab tiuxetan	anticancer	2002	38 (359)
ibrutinib	anticancer	2013	49 (466)
ibudilast	antiasthma	1989	25 (313)
ibutilide fumarate	antiarrhythmic	1996	32 (309)
icatibant	angioedema, hereditary	2008	44 (608), 50 (551)
idarubicin hydrochloride	anticancer	1990	26 (303)
idebenone	nootropic	1986	22 (321)
idursulfase	mucopolysaccharidosis II (Hunter syndrome)	2006	42 (520)
iguratimod	antiarthritic	2011	47 (535)
iloprost	antiplatelet	1992	28 (332)
imatinib mesylate	anticancer	2001	37 (267)
imidafenacin	urinary tract/bladder disorders	2007	43 (472)
imidapril hydrochloride	antihypertensive	1993	29 (339)
imiglucerase	Gaucher's disease	1994	30 (301)
imipenem/cilastatin	antibacterial	1985	21 (328)
imiquimod	antiviral	1997	33 (335)
incadronic acid	osteoporosis	1997	33 (335)
indacaterol	chronic obstructive pulmonary disease	2009	45 (505)
indalpine	antidepressant	1983	19 (320)
indeloxazine hydrochloride	nootropic	1988	24 (304)
indinavir sulfate	antiviral	1996	32 (310)
indisetron	antiemetic	2004	40 (459)
indobufen	antithrombotic	1984	20 (319)
influenza virus (live)	antiviral	2003	39 (277)
ingenol mebutate	anticancer	2012	48 (499)

GENERIC NAME	INDICATION	YEAR INTRODUCED	ARMC VOL., (PAGE)
insulin lispro	antidiabetic	1996	32 (310)
interferon alfacon-1	antiviral	1997	33 (336)
interferon gamma-1b	immunostimulant	1991	27 (329)
interferon, b-1a	multiple sclerosis	1996	32 (311)
interferon, b-1b	multiple sclerosis	1993	29 (339)
interferon, gamma	antiinflammatory	1989	25 (314)
interferon, gamma-1	anticancer	1992	28 (332)
interleukin-2	anticancer	1989	25 (314)
ioflupane	diagnostic	2000	36 (306)
ipilimumab	anticancer	2011	47 (537)
ipriflavone	osteoporosis	1989	25 (314)
irbesartan	antihypertensive	1997	33 (336)
irinotecan	anticancer	1994	30 (301)
irsogladine	antiulcer	1989	25 (315)
isepamicin	antibacterial	1988	24 (305)
isofezolac	antiinflammatory	1984	20 (319)
isoxicam	antiinflammatory	1983	19 (320)
isradipine	antihypertensive	1989	25 (315)
istradefylline	Parkinson's disease	2013	49 (468)
itopride hydrochloride	gastroprokinetic	1995	31 (344)
itraconazole	antifungal	1988	24 (305)
ivabradine	antianginal	2006	42 (522)
ivacaftor	cystic fibrosis	2012	48 (501)
ivermectin	antiparasitic	1987	23 (336)
ixabepilone	anticancer	2007	43 (473)
ketanserin	antihypertensive	1985	21 (328)
ketorolac tromethamine	analgesic	1990	26 (304)
kinetin	dermatologic, skin photodamage	1999	35 (341)
lacidipine	antihypertensive	1991	27 (330)
lacosamide	anticonvulsant	2008	44 (610)
lafutidine	antiulcer	2000	36 (307)
lamivudine	antiviral	1995	31 (345)
lamotrigine	anticonvulsant	1990	26 (304)
landiolol	antiarrhythmic	2002	38 (360)
laninamivir octanoate	antiviral	2010	46 (470)
lanoconazole	antifungal	1994	30 (302)
lanreotide	growth disorders	1994	50 (555)
lanreotide	neuroendocrine tumors	1994	50 (555)
lanreotide acetate	growth disorders	1995	31 (345)
lansoprazole	antiulcer	1992	28 (332)
lapatinib	anticancer	2007	43 (475)
laronidase	mucopolysaccharidosis I	2003	39 (278)
latanoprost	antiglaucoma	1996	32 (311)
lefunomide	antiarthritic	1998	34 (324)

GENERIC NAME	INDICATION	YEAR INTRODUCED	ARMC VOL., (PAGE)
lenalidomide	myelodysplastic syndromes, multiple myeloma	2006	42 (523)
lenampicillin hydrochloride	antibacterial	1987	23 (336)
lentinan	immunostimulant	1986	22 (322)
lepirudin	anticoagulant	1997	33 (336)
lercanidipine	antihypertensive	1997	33 (337)
letrazole	anticancer	1996	32 (311)
leuprolide acetate	hormone therapy	1984	20 (319)
leuprorelin	anticancer	1984	50 (553)
leuprorelin	endometriosis	1984	50 (553)
leuprorelin	uterine fibroids	1984	50 (553)
leuprorelin	precocious puberty	1984	50 (553)
levacecarnine hydrochloride	cognition enhancer	1986	22 (322)
levalbuterol hydrochloride	antiasthma	1999	35 (341)
levetiracetam	anticonvulsant	2000	36 (307)
levobunolol hydrochloride	antiglaucoma	1985	21 (328)
levobupivacaine hydrochloride	anesthetic	2000	36 (308)
levocabastine hydrochloride	antiallergy	1991	27 (330)
levocetirizine	antiallergy	2001	37 (268)
levodropropizine	antitussive	1988	24 (305)
levofloxacin	antibacterial	1993	29 (340)
levosimendan	congestive heart failure	2000	36 (308)
lidamidine hydrochloride	antidiarrheal	1984	20 (320)
limaprost	antithrombotic	1988	24 (306)
linaclotide	chronic idiopathic constipation	2012	50 (552)
linaclotide	irritable bowel syndrome	2012	48 (502), 50 (552)
linagliptin	antidiabetic	2011	47 (540)
linezolid	antibacterial	2000	36 (309)
liraglutide	antidiabetic	2009	45 (507) 50 (552)
liranaftate	antifungal	2000	36 (309)
lisdexamfetamine	attention deficit hyperactivity disorder	2007	43 (477)
lisinopril	antihypertensive	1987	23 (337)
lixisenatide	antidiabetic	2013	49 (470) 50 (552)
lobenzarit sodium	antiinflammatory	1986	22 (322)
lodoxamide tromethamine	antiallergy	1992	28 (333)
lomefloxacin	antibacterial	1989	25 (315)
lomerizine hydrochloride	antimigraine	1999	35 (342)

GENERIC NAME	INDICATION	YEAR INTRODUCED	ARMC VOL., (PAGE)
lomitapide	antihypercholersteremic	2012	48 (504)
lonidamine	anticancer	1987	23 (337)
lopinavir	antiviral	2000	36 (310)
loprazolam mesylate	sleep disorders	1983	19 (321)
loprinone hydrochloride	congestive heart failure	1996	32 (312)
loracarbef	antibacterial	1992	28 (333)
loratadine	antiallergy	1988	24 (306)
lorcaserin hydrochloride	antiobesity	2012	48 (506)
lornoxicam	antiinflammatory	1997	33 (337)
losartan	antihypertensive	1994	30 (302)
loteprednol etabonate	antiallergy	1998	34 (324)
lovastatin	antihypercholesterolemic	1987	23 (337)
loxoprofen sodium	antiinflammatory	1986	22 (322)
lucinactant	neonatal respiratory distress syndrome	2012	50 (556)
lulbiprostone	constipation	2006	42 (525)
luliconazole	antifungal	2005	41 (454)
lumiracoxib	antiinflammatory	2005	41 (455)
lurasidone hydrochloride	antipsychotic	2010	46 (473)
Lyme disease vaccine	Lyme disease	1999	35 (342)
mabuterol hydrochloride	antiasthma	1986	22 (323)
macitentan	pulmonary hypertension	2013	49 (472)
malotilate	hepatoprotective	1985	21 (329)
manidipine hydrochloride	antihypertensive	1990	26 (304)
maraviroc	antiviral	2007	43 (478)
masoprocol	anticancer	1992	28 (333)
maxacalcitol	hyperparathyroidism	2000	36 (310)
mebefradil hydrochloride	antihypertensive	1997	33 (338)
medifoxamine fumarate	antidepressant	1986	22 (323)
mefloquine hydrochloride	antimalarial	1985	21 (329)
meglutol	antihypercholesterolemic	1983	19 (321)
melinamide	antihypercholesterolemic	1984	20 (320)
meloxicam	antiarthritic	1996	32 (312)
mepixanox	respiratory stimulant	1984	20 (320)
meptazinol hydrochloride	analgesic	1983	19 (321)
meropenem	antibacterial	1994	30 (303)
metaclazepam	anxiolytic	1987	23 (338)
metapramine	antidepressant	1984	20 (320)
methylnaltrexone bromide	constipation	2008	44 (612)
metreleptin	lipodystrophy	2013	49 (474)
mexazolam	anxiolytic	1984	20 (321)
micafungin	antifungal	2002	38 (360)
mifamurtide	anticancer	2009	46 (476)
mifepristone	abortifacient	1988	24 (306)
miglitol	antidiabetic	1998	34 (325)

GENERIC NAME	INDICATION	YEAR INTRODUCED	ARMC VOL., (PAGE)
miglustat	Gaucher's disease	2003	39 (279)
milnacipran	antidepressant	1997	33 (338)
milrinone	congestive heart failure	1989	25 (316)
miltefosine	anticancer	1993	29 (340)
minodronic acid	osteoporosis	2009	45 (509)
miokamycin	antibacterial	1985	21 (329)
mipomersen	antihypercholesterolemic	2013	49 (476)
mirabegron	urinary tract/bladder disorders	2011	47 (542)
mirtazapine	antidepressant	1994	30 (303)
misoprostol	antiulcer	1985	21 (329)
mitiglinide	antidiabetic	2004	40 (460)
mitoxantrone hydrochloride	anticancer	1984	20 (321)
mivacurium chloride	muscle relaxant	1992	28 (334)
mivotilate	hepatoprotective	1999	35 (343)
mizolastine	antiallergy	1998	34 (325)
mizoribine	immunosuppressant	1984	20 (321)
moclobemide	antidepressant	1990	26 (305)
modafinil	sleep disorders	1994	30 (303)
moexipril hydrochloride	antihypertensive	1995	31 (346)
mofezolac	analgesic	1994	30 (304)
mogamulizumab	anticancer	2012	48 (507)
mometasone furoate	antiinflammatory	1987	23 (338)
montelukast sodium	antiasthma	1998	34 (326)
moricizine hydrochloride	antiarrhythmic	1990	26 (305)
mosapride citrate	gastroprokinetic	1998	34 (326)
moxifloxacin hydrochloride	antibacterial	1999	35 (343)
moxonidine	antihypertensive	1991	27 (330)
mozavaptan	hyponatremia	2006	42 (527)
mupirocin	antibacterial	1985	21 (330)
muromonab-CD3	immunosuppressant	1986	22 (323)
muzolimine	diuretic	1983	19 (321)
mycophenolate mofetil	immunosuppressant	1995	31 (346)
mycophenolate sodium	immunosuppressant	2003	39 (279)
nabumetone	antiinflammatory	1985	21 (330)
nadifloxacin	antibacterial	1993	29 (340)
nafamostat mesylate	pancreatitis	1986	22 (323)
nafarelin	precocious puberty	1990	50 (553)
nafarelin	endometriosis	1990	50 (553)
nafarelin acetate	hormone therapy	1990	26 (306)
naftifine hydrochloride	antifungal	1984	20 (321)
naftopidil	urinary tract/bladder disorders	1999	35 (344)
nalfurafine hydrochloride	pruritus	2009	45 (510)
nalmefene hydrochloride	addiction, opioids	1995	31 (347)

GENERIC NAME	INDICATION	YEAR INTRODUCED	ARMC VOL., (PAGE)
naltrexone hydrochloride	addiction, opioids	1984	20 (322)
naratriptan hydrochloride	antimigraine	1997	33 (339)
nartograstim	leukopenia	1994	30 (304)
natalizumab	multiple sclerosis	2004	40 (462)
nateglinide	antidiabetic	1999	35 (344)
nazasetron	antiemetic	1994	30 (305)
nebivolol	antihypertensive	1997	33 (339)
nedaplatin	anticancer	1995	31 (347)
nedocromil sodium	antiallergy	1986	22 (324)
nefazodone	antidepressant	1994	30 (305)
nelarabine	anticancer	2006	42 (528)
nelfinavir mesylate	antiviral	1997	33 (340)
neltenexine	cystic fibrosis	1993	29 (341)
nemonapride	antipsychotic	1991	27 (331)
nepafenac	antiinflammatory	2005	41 (456)
neridronic acide	calcium regulation	2002	38 (361)
nesiritide	congestive heart failure	2001	37 (269), 50 (555)
neticonazole hydrochloride	antifungal	1993	29 (341)
nevirapine	antiviral	1996	32 (313)
nicorandil	antianginal	1984	20 (322)
nif ekalant hydrochloride	antiarrhythmic	1999	35 (344)
nilotinib	anticancer	2007	43 (480)
nilutamide	anticancer	1987	23 (338)
nilvadipine	antihypertensive	1989	25 (316)
nimesulide	antiinflammatory	1985	21 (330)
nimodipine	cerebral vasodilator	1985	21 (330)
nimotuzumab	anticancer	2006	42 (529)
nipradilol	antihypertensive	1988	24 (307)
nisoldipine	antihypertensive	1990	26 (306)
nitisinone	antityrosinaemia	2002	38 (361)
nitrefazole	addiction, alcohol	1983	19 (322)
nitrendipine	antihypertensive	1985	21 (331)
nizatidine	antiulcer	1987	23 (339)
nizofenzone	nootropic	1988	24 (307)
nomegestrol acetate	contraception	1986	22 (324)
norelgestromin	contraception	2002	38 (362)
norfloxacin	antibacterial	1983	19 (322)
norgestimate	contraception	1986	22 (324)
obinutuzumab	anticancer	2013	49 (478)
OCT-43	anticancer	1999	35 (345)
octreotide	neuroendocrine tumors	1988	50 (555)
octreotide	growth disorders	1988	24 (307), 50 (555)
ofatumumab	anticancer	2009	45 (512)

GENERIC NAME	INDICATION	YEAR INTRODUCED	ARMC VOL., (PAGE)
ofloxacin	antibacterial	1985	21 (331)
olanzapine	antipsychotic	1996	32 (313)
olimesartan Medoxomil	antihypertensive	2002	38 (363)
olodaterol	chronic obstructive pulmonary disorder	2013	49 (480)
olopatadine hydrochloride	antiallergy	1997	33 (340)
omacetaxine mepesuccinate	anticancer	2012	48 (510)
omalizumab allergic	antiasthma	2003	39 (280)
omeprazole	antiulcer	1988	24 (308)
ondansetron hydrochloride	antiemetic	1990	26 (306)
OP-1	osteoinductor	2001	37 (269)
orlistat	antiobesity	1998	34 (327)
ornipressin	bleeding esophageal varices	1971	50 (551)
ornoprostil	antiulcer	1987	23 (339)
osalazine sodium	antiinflammatory	1986	22 (324)
oseltamivir phosphate	antiviral	1999	35 (346)
ospemifene	dyspareunia	2013	49 (482)
oxaliplatin	anticancer	1996	32 (313)
oxaprozin	antiinflammatory	1983	19 (322)
oxcarbazepine	anticonvulsant	1990	26 (307)
oxiconazole nitrate	antifungal	1983	19 (322)
oxiracetam	cognition enhancer	1987	23 (339)
oxitropium bromide	antiasthma	1983	19 (323)
oxytocin	labor induction	1962	50 (555)
oxytocin	postpartum hemorrhage	1962	50 (555)
oxytocin	lactation	1962	50 (555)
ozagrel sodium	antithrombotic	1988	24 (308)
P-15	bone grafting	1999	50 (556)
paclitaxal	anticancer	1993	29 (342)
palifermin	mucositis	2005	41 (461)
paliperidone	antipsychotic	2007	43 (482)
palonosetron	antiemetic	2003	39 (281)
panipenem/ betamipron carbapenem	antibacterial	1994	30 (305)
panitumumab	anticancer	2006	42 (531)
pantoprazole sodium	antiulcer	1995	30 (306)
parecoxib sodium	analgesic	2002	38 (364)
paricalcitol	hyperparathyroidism	1998	34 (327)
parnaparin sodium	anticoagulant	1993	29 (342)
paroxetine	antidepressant	1991	27 (331)
pasireotide	growth disorders	2012	50 (555)
pasireotide	Cushing's Disease	2012	48 (512), 50 (555)
pazopanib	anticancer	2009	45 (514)

GENERIC NAME	INDICATION	YEAR INTRODUCED	ARMC VOL., (PAGE)
pazufloxacin	antibacterial	2002	38 (364)
pefloxacin mesylate	antibacterial	1985	21 (331)
pegademase bovine	immunostimulant	1990	26 (307)
pegaptanib	ophthalmologic (macular degeneration)	2005	41 (458)
pegaspargase	anticancer	1994	30 (306)
peginesatide acetate	hematopoietic	2012	48 (514)
pegvisomant	growth disorders	2003	39 (281)
pemetrexed	anticancer	2004	40 (463)
pemirolast potassium	antiasthma	1991	27 (331)
penciclovir	antiviral	1996	32 (314)
pentostatin	anticancer	1992	28 (334)
peramivir	antiviral	2010	46 (477)
perampanel	anticonvulsant	2012	48 (516)
pergolide mesylate	Parkinson's disease	1988	24 (308)
perindopril	antihypertensive	1988	24 (309)
perospirone hydrochloride	antipsychotic	2001	37 (270)
pertuzumab	anticancer	2012	48 (517)
picotamide	antithrombotic	1987	23 (340)
pidotimod	immunostimulant	1993	29 (343)
piketoprofen	antiinflammatory	1984	20 (322)
pilsicainide hydrochloride	antiarrhythmic	1991	27 (332)
pimaprofen	antiinflammatory	1984	20 (322)
pimecrolimus	immunosuppressant	2002	38 (365)
pimobendan	congestive heart failure	1994	30 (307)
pinacidil	antihypertensive	1987	23 (340)
pioglitazone hydrochloride	antidiabetic	1999	35 (346)
pirarubicin	anticancer	1988	24 (309)
pirfenidone	pulmonary fibrosis, idiopathic	2008	44 (614)
pirmenol	antiarrhythmic	1994	30 (307)
piroxicam cinnamate	antiinflammatory	1988	24 (309)
pitavastatin	antihypercholesterolemic	2003	39 (282)
pivagabine	antidepressant	1997	33 (341)
pixantrone dimaleate	anticancer	2012	48 (519)
plaunotol	antiulcer	1987	23 (340)
plecanatide	chronic idiopathic constipation	2017	50 (552)
plerixafor hydrochloride	stem cell mobilizer	2009	45 (515)
polaprezinc	antiulcer	1994	30 (307)
pomalidomide	anticancer	2013	49 (484)
ponatinib	anticancer	2012	48 (521)
porfimer sodium	anticancer	1993	29 (343)
posaconazole	antifungal	2006	42 (532)
pralatrexate	anticancer	2009	45 (517)
pramlintide	antidiabetic	2005	50 (551)
pramipexole hydrochloride	Parkinson's disease	1997	33 (341)

GENERIC NAME	INDICATION	YEAR INTRODUCED	ARMC VOL., (PAGE)
pramiracetam sulfate	cognition enhancer	1993	29 (343)
pramlintide	antidiabetic	2005	41 (460)
pranlukast	antiasthma	1995	31 (347)
prasugrel	antiplatelet	2009	45 (519)
pravastatin	antihypercholesterolemic	1989	25 (316)
prednicarbate	antiinflammatory	1986	22 (325)
pregabalin	anticonvulsant	2004	40 (464)
prezatide copper acetate	wound healing agent	1996	32 (314)
progabide	anticonvulsant	1985	21 (331)
promegestrone	contraception	1983	19 (323)
propacetamol hydrochloride	analgesic	1986	22 (325)
propagermanium	antiviral	1994	30 (308)
propentofylline propionate	cerebral vasodilator	1988	24 (310)
propiverine hydrochloride	urinary tract/bladder disorders	1992	28 (335)
propofol	anesthetic	1986	22 (325)
prulifloxacin	antibacterial	2002	38 (366)
pumactant	respiratory distress syndrome	1994	30 (308)
quazepam	sleep disorders	1985	21 (332)
quetiapine fumarate	antipsychotic	1997	33 (341)
quinagolide	hyperprolactinemia	1994	30 (309)
quinapril	antihypertensive	1989	25 (317)
quinf amideamebicide	antiparasitic	1984	20 (322)
quinupristin	antibacterial	1999	35 (338)
rabeprazole sodium	antiulcer	1998	34 (328)
radotinib	anticancer	2012	48 (523)
raloxifene hydrochloride	osteoporosis	1998	34 (328)
raltegravir	antiviral	2007	43 (484)
raltitrexed	anticancer	1996	32 (315)
ramatroban	antiallergy	2000	36 (311)
ramelteon	sleep disorders	2005	41 (462)
ramipril	antihypertensive	1989	25 (317)
ramosetron	antiemetic	1996	32 (315)
ranibizumab	ophthalmologic (macular degeneration)	2006	42 (534)
ranimustine	anticancer	1987	23 (341)
ranitidine bismuth citrate	antiulcer	1995	31 (348)
ranolazine	antianginal	2006	42 (535)
rapacuronium bromide	muscle relaxant	1999	35 (347)
rasagiline	Parkinson's disease	2005	41 (464)
rebamipide	antiulcer	1990	26 (308)
reboxetine	antidepressant	1997	33 (342)
regorafenib	anticancer	2012	48 (524)
remifentanil hydrochloride	analgesic	1996	32 (316)
remoxipride hydrochloride	antipsychotic	1990	26 (308)

GENERIC NAME	INDICATION	YEAR INTRODUCED	ARMC VOL., (PAGE)
repaglinide	antidiabetic	1998	34 (329)
repirinast	antiallergy	1987	23 (341)
retapamulin	antibacterial	2007	43 (486)
reteplase	antithrombotic	1996	32 (316)
retigabine	anticonvulsant	2011	47 (544)
reviparin sodium	anticoagulant	1993	29 (344)
rifabutin	antibacterial	1992	28 (335)
rifapentine	antibacterial	1988	24 (310)
rifaximin	antibacterial	1987	23 (341)
rifaximin	antibacterial	1985	21 (332)
rilmazafone	sleep disorders	1989	25 (317)
rilmenidine	antihypertensive	1988	24 (310)
rilonacept	genetic autoinflammatory syndromes	2008	44 (615)
rilpivirine	antiviral	2011	47 (546)
riluzole	amyotrophic lateral sclerosis	1996	32 (316)
rimantadine hydrochloride	antiviral	1987	23 (342)
rimexolone	antiinflammatory	1995	31 (348)
rimonabant	antiobesity	2006	42 (537)
riociguat	pulmonary hypertension	2013	49 (486)
risedronate sodium	osteoporosis	1998	34 (330)
risperidone	antipsychotic	1993	29 (344)
ritonavir	antiviral	1996	32 (317)
rivaroxaban	anticoagulant	2008	44 (617)
rivastigmin	Alzheimer's disease	1997	33 (342)
rizatriptan benzoate	antimigraine	1998	34 (330)
rocuronium bromide	muscle relaxant	1994	30 (309)
rofecoxib	antiarthritic	1999	35 (347)
roflumilast	chronic obstructive pulmonary disorder	2010	46 (480)
rokitamycin	antibacterial	1986	22 (325)
romidepsin	anticancer	2009	46 (482)
romiplostim	antithrombocytopenic	2008	44 (619), 50 (556)
romurtide	immunostimulant	1991	27 (332)
ronafibrate	antihypercholesterolemic	1986	22 (326)
ropinirole hydrochloride	Parkinson's disease	1996	32 (317)
ropivacaine	anesthetic	1996	32 (318)
rosaprostol	antiulcer	1985	21 (332)
rosiglitazone maleate	antidiabetic	1999	35 (348)
rosuvastatin	antihypercholesterolemic	2003	39 (283)
rotigotine	Parkinson's disease	2006	42 (538)
roxatidine acetate hydrochloride	antiulcer	1986	22 (326)
roxithromycin	antiulcer	1987	23 (342)

GENERIC NAME	INDICATION	YEAR INTRODUCED	ARMC VOL., (PAGE)
rufinamide	anticonvulsant	2007	43 (488)
rufloxacin hydrochloride	antibacterial	1992	28 (335)
rupatadine fumarate	antiallergy	2003	39 (284)
ruxolitinib	anticancer	2011	47 (548)
RV-11	antibacterial	1989	25 (318)
salmeterol hydroxynaphthoate	antiasthma	1990	26 (308)
sapropterin hydrochloride	phenylketouria	1992	28 (336)
saquinavir mesvlate	antiviral	1995	31 (349)
sargramostim	immunostimulant	1991	27 (332)
saroglitazar	antidiabetic	2013	49 (488)
sarpogrelate hydrochloride	antithrombotic	1993	29 (344)
saxagliptin	antidiabetic	2009	45 (521)
schizophyllan	immunostimulant	1985	22 (326)
seratrodast	antiasthma	1995	31 (349)
sertaconazole nitrate	antifungal	1992	28 (336)
sertindole	antipsychotic	1996	32 (318)
setastine hydrochloride	antiallergy	1987	23 (342)
setiptiline	antidepressant	1989	25 (318)
setraline hydrochloride	antidepressant	1990	26 (309)
sevoflurane	anesthetic	1990	26 (309)
sibutramine	antiobesity	1998	34 (331)
sildenafil citrate	male sexual dysfunction	1998	34 (331)
silodosin	urinary tract/bladder disorders	2006	42 (540)
simeprevir	antiviral	2013	49 (489)
simvastatin	antihypercholesterolemic	1988	24 (311)
sipuleucel-t	anticancer	2010	46 (484)
sitafloxacin hydrate	antibacterial	2008	44 (621)
sitagliptin	antidiabetic	2006	42 (541)
sitaxsentan	pulmonary hypertension	2006	42 (543)
sivelestat	antiinflammatory	2002	38 (366)
SKI-2053R	anticancer	1999	35 (348)
sobuzoxane	anticancer	1994	30 (310)
sodium cellulose phosphate	urinary tract/bladder disorders	1983	19 (323)
sofalcone	antiulcer	1984	20 (323)
sofosbuvir	antiviral	2013	49 (492)
solifenacin	urinary tract/bladder disorders	2004	40 (466)
somatomedin-1	growth disorders	1994	30 (310)
somatostatin	bleeding esophageal varices	1970	50 (556)
somatotropin	growth disorders	1994	30 (310)
somatropin	growth disorders	1987	23 (343)
sorafenib	anticancer	2005	41 (466)
sorivudine	antiviral	1993	29 (345)
sparfloxacin	antibacterial	1993	29 (345)
spirapril hydrochloride	antihypertensive	1995	31 (349)

GENERIC NAME	INDICATION	YEAR INTRODUCED	ARMC VOL., (PAGE)
spizofurone	antiulcer	1987	23 (343)
stavudine	antiviral	1994	30 (311)
strontium ranelate	osteoporosis	2004	40 (466)
succimer	antidote, lead poisoning	1991	27 (333)
sufentanil	analgesic	1983	19 (323)
sugammadex	neuromuscular blockade, reversal	2008	44 (623)
sulbactam sodium	antibacterial	1986	22 (326)
sulconizole nitrate	antifungal	1985	21 (332)
sultamycillin tosylate	antibacterial	1987	23 (343)
sumatriptan succinate	antimigraine	1991	27 (333)
sunitinib	anticancer	2006	42 (544)
suplatast tosilate	antiallergy	1995	31 (350)
suprofen	analgesic	1983	19 (324)
surfactant TA	respiratory surfactant	1987	23 (344)
tacalcitol	antipsoriasis	1993	29 (346)
tacrine hydrochloride	Alzheimer's disease	1993	29 (346)
tacrolimus	immunosuppressant	1993	29 (347)
tadalafil	male sexual dysfunction	2003	39 (284)
tafamidis	neurodegeneration	2011	47 (550)
tafluprost	antiglaucoma	2008	44 (625)
talaporfin sodium	anticancer	2004	40 (469)
talipexole	Parkinson's disease	1996	32 (318)
taltirelin	neurodegeneration	2000	36 (311)
tamibarotene	anticancer	2005	41 (467)
tamsulosin hydrochloride	benign prostatic hyperplasia	1993	29 (347)
tandospirone	anxiolytic	1996	32 (319)
tapentadol hydrochloride	analgesic	2009	45 (523)
tasonermin	anticancer	1999	35 (349)
tazanolast	antiallergy	1990	26 (309)
tazarotene	antipsoriasis	1997	33 (343)
tazobactam sodium	antibacterial	1992	28 (336)
teduglutide	short bowel syndrome	2012	48 (526), 50 (552)
tegaserod maleate	irritable bowel syndrome	2001	37 (270)
teicoplanin	antibacterial	1988	24 (311)
telaprevir	antiviral	2011	47 (552)
telavancin	antibacterial	2009	45 (525)
telbivudine	antiviral	2006	42 (546)
telithromycin	antibacterial	2001	37 (271)
telmesteine	expectorant	1992	28 (337)
telmisartan	antihypertensive	1999	35 (349)
temafloxacin hydrochloride	antibacterial	1991	27 (334)
temocapril	antihypertensive	1994	30 (311)

GENERIC NAME	INDICATION	YEAR INTRODUCED	ARMC VOL., (PAGE)
temocillin disodium	antibacterial	1984	20 (323)
temoporphin	anticancer	2002	38 (367)
temozolomide	anticancer	1999	35 (349)
temsirolimus	anticancer	2007	43 (490)
teneligliptin	antidiabetic	2012	48 (528)
tenofovir disoproxil fumarate	antiviral	2001	37 (271)
tenoxicam	antiinflammatory	1987	23 (344)
teprenone	antiulcer	1984	20 (323)
terazosin hydrochloride	antihypertensive	1984	20 (323)
terbinafine hydrochloride	antifungal	1991	27 (334)
terconazole	antifungal	1983	19 (324)
teriflunomide	multiple sclerosis	2012	48 (530)
teriparatide	osteoporosis	2002	50 (555)
terlipressin	bleeding esophageal varices	1978	50 (551)
tertatolol hydrochloride	antihypertensive	1987	23 (344)
tesamorelin acetate	lipodystrophy	2010	46 (486), 50 (552)
tetracosactide	multiple inflammatory diseases	1980	50 (554)
thrombin alfa	hemostatic	2008	44 (627)
thrombomodulin (recombinant)	anticoagulant	2008	44 (628)
thymopentin	immunomodulator	1985	21 (333), 50 (554)
tiagabine	anticonvulsant	1996	32 (319)
tiamenidine hydrochloride	antihypertensive	1988	24 (311)
tianeptine sodium	antidepressant	1983	19 (324)
tibolone	hormone therapy	1988	24 (312)
ticagrelor	antithrombotic	2010	46 (488)
tigecycline	antibacterial	2005	41 (468)
tilisolol hydrochloride	antihypertensive	1992	28 (337)
tiludronate disodium	Paget's disease	1995	31 (350)
timiperone	antipsychotic	1984	20 (323)
tinazoline	nasal decongestant	1988	24 (312)
tioconazole	antifungal	1983	19 (324)
tiopronin	urolithiasis	1989	25 (318)
tiotropium bromide	chronic obstructive pulmonary disorder	2002	38 (368)
tipranavir	antiviral	2005	41 (470)
tiquizium bromide	antispasmodic	1984	20 (324)
tiracizine hydrochloride	antiarrhythmic	1990	26 (310)
tirilazad mesylate	subarachnoid hemorrhage	1995	31 (351)
tirofiban hydrochloride	antithrombotic	1998	34 (332)
tiropramide hydrochloride	muscle relaxant	1983	19 (324)
tizanidine	muscle relaxant	1984	20 (324)

GENERIC NAME	INDICATION	YEAR INTRODUCED	ARMC VOL., (PAGE)
tofacitinib	antiarthritic	2012	48 (532)
tolcapone	Parkinson's disease	1997	33 (343)
toloxatone	antidepressant	1984	20 (324)
tolrestat	antidiabetic	1989	25 (319)
tolvaptan	hyponatremia	2009	45 (528)
topiramate	anticonvulsant	1995	31 (351)
topotecan hydrochloride	anticancer	1996	32 (320)
torasemide	diuretic	1993	29 (348)
toremifene	anticancer	1989	25 (319)
tositumomab	anticancer	2003	39 (285)
tosufloxacin tosylate	antibacterial	1990	26 (310)
trabectedin	anticancer	2007	43 (492)
trametinib	anticancer	2013	49 (494)
trandolapril	antihypertensive	1993	29 (348)
travoprost	antiglaucoma	2001	37 (272)
treprostinil sodium	antihypertensive	2002	38 (368)
tretinoin tocoferil	antiulcer	1993	29 (348)
trientine hydrochloride	antidote, copper poisoning	1986	22 (327)
trimazosin hydrochloride	antihypertensive	1985	21 (333)
trimegestone	contraception	2001	37 (273)
trimetrexate glucuronate	antifungal	1994	30 (312)
triptorelin	precocious puberty	1986	50 (553)
triptorelin	anticancer	1986	50 (553)
triptorelin	endometriosis	1986	50 (553)
triptorelin	breast cancer	1986	50 (553)
troglitazone	antidiabetic	1997	33 (344)
tropisetron	antiemetic	1992	28 (337)
trovafloxacin mesylate	antibacterial	1998	34 (332)
troxipide	antiulcer	1986	22 (327)
ubenimex	immunostimulant	1987	23 (345)
udenafil	male sexual dysfunction	2005	41 (472)
ulipristal acetate	contraception	2009	45 (530)
unoprostone isopropyl ester	antiglaucoma	1994	30 (312)
ustekinumab	antipsoriasis	2009	45 (532)
vadecoxib	antiarthritic	2002	38 (369)
vaglancirclovir hydrochloride	antiviral	2001	37 (273)
valaciclovir hydrochloride	antiviral	1995	31 (352)
valrubicin	anticancer	1999	35 (350)
valsartan	antihypertensive	1996	32 (320)
vandetanib	anticancer	2011	47 (555)
vardenafil	male sexual dysfunction	2003	39 (286)
varenicline	addiction, nicotine	2006	42 (547)
vasopressin	bleeding esophageal varices	1962	50 (551)

GENERIC NAME	INDICATION	YEAR INTRODUCED	ARMC VOL., (PAGE)
vasopressin	vasodilatory shock	1962	50 (551)
vemurafenib	anticancer	2011	47 (556)
venlafaxine	antidepressant	1994	30 (312)
vernakalant	antiarrhythmic	2010	46 (491)
verteporfin	ophthalmologic (macular degeneration)	2000	36 (312)
vesnarinone	congestive heart failure	1990	26 (310)
vigabatrin	anticonvulsant	1989	25 (319)
vilazodone	antidepressant	2011	47 (558)
vildagliptin	antidiabetic	2007	43 (494)
vinflunine	anticancer	2009	46 (493)
vinorelbine	anticancer	1989	25 (320)
vismodegib	anticancer	2012	48 (534)
voglibose	antidiabetic	1994	30 (313)
voriconazole	antifungal	2002	38 (370)
vorinostat	anticancer	2006	42 (549)
vortioxetine	antidepressant	2013	49 (496)
xamoterol fumarate	congestive heart failure	1988	24 (312)
ximelagatran	anticoagulant	2004	40 (470)
zafirlukast	antiasthma	1996	32 (321)
zalcitabine	antiviral	1992	28 (338)
zaleplon	sleep disorders	1999	35 (351)
zaltoprofen	antiinflammatory	1993	29 (349)
zanamivir	antiviral	1999	35 (352)
ziconotide	analgesic	2004	50 (555)
ziconotide	analgesic	2005	41 (473)
zidovudine	antiviral	1987	23 (345)
zileuton	antiasthma	1997	33 (344)
zinostatin stimalamer	anticancer	1994	30 (313)
ziprasidone hydrochloride	antipsychotic	2000	36 (312)
zofenopril calcium	antihypertensive	2000	36 (313)
zoledronate disodium	osteoporosis	2000	36 (314)
zolpidem hemitartrate	sleep disorders	1988	24 (313)
zomitriptan	antimigraine	1997	33 (345)
zonisamide	anticonvulsant	1989	25 (320)
zopiclone	sleep disorders	1986	22 (327)
zucapsaicin	analgesic	2010	46 (495)
zuclopenthixol acetate	antipsychotic	1987	23 (345)

CUMULATIVE NCE INTRODUCTION INDEX, 1983–2017 (BY INDICATION)

GENERIC NAME	INDICATION	YEAR INTRODUCED	ARMC VOL., (PAGE)
gemeprost	abortifacient	1983	19 (319)
mifepristone	abortifacient	1988	24 (306)
azelaic acid	acne	1989	25 (310)
bivalirudin	acute coronary syndrome	1999	50 (556)
eptifibatide	acute coronary syndrome	1998	50 (554)
nitrefazole	addiction, alcohol	1983	19 (322)
varenicline	addiction, nicotine	2006	42 (547)
nalmefene hydrochloride	addiction, opioids	1995	31 (347)
naltrexone hydrochloride	addiction, opioids	1984	20 (322)
donepezil hydrochloride	Alzheimer's disease	1997	33 (332)
rivastigmin	Alzheimer's disease	1997	33 (342)
tacrine hydrochloride	Alzheimer's disease	1993	29 (346)
fasudil hydrochloride	amyotrophic lateral sclerosis	1995	31 (343)
riluzole	amyotrophic lateral sclerosis	1996	32 (316)
alfentanil hydrochloride	analgesic	1983	19 (314)
alminoprofen	analgesic	1983	19 (314)
dezocine	analgesic	1991	27 (326)
emorfazone	analgesic	1984	20 (317)
eptazocine hydrobromide	analgesic	1987	23 (334)
flupirtine maleate	analgesic	1985	21 (328)
fosfosal	analgesic	1984	20 (319)
ketorolac tromethamine	analgesic	1990	26 (304)
meptazinol hydrochloride	analgesic	1983	19 (321)
mofezolac	analgesic	1994	30 (304)
parecoxib sodium	analgesic	2002	38 (364)
propacetamol hydrochloride	analgesic	1986	22 (325)
remifentanil hydrochloride	analgesic	1996	32 (316)
sufentanil	analgesic	1983	19 (323)
suprofen	analgesic	1983	19 (324)
tapentadol hydrochloride	analgesic	2009	45 (523)
ziconotide	analgesic	2005	41 (473)
zucapsaicin	analgesic	2010	46 (495)
desflurane	anesthetic	1992	28 (329)
felypressin	anesthetic	1970	50 (551)
levobupivacaine hydrochloride	anesthetic	2000	36 (308)
propofol	anesthetic	1986	22 (325)
ropivacaine	anesthetic	1996	32 (318)
sevoflurane	anesthetic	1990	26 (309)
ecallantide	angioedema, hereditary	2009	46 (464)

GENERIC NAME	INDICATION	YEAR INTRODUCED	ARMC VOL., (PAGE)
icatibant	angioedema, hereditary	2008	44 (608), 50 (551)
acrivastine	antiallergy	1988	24 (295)
astemizole	antiallergy	1983	19 (314)
azelastine hydrochloride	antiallergy	1986	22 (316)
betotastine besilate	antiallergy	2000	36 (297)
bilastine	antiallergy	2010	46 (449)
cetirizine hydrochloride	antiallergy	1987	23 (331)
desloratadine	antiallergy	2001	37 (264)
ebastine	antiallergy	1990	26 (302)
emedastine difumarate	antiallergy	1993	29 (336)
epinastine	antiallergy	1994	30 (299)
fexofenadine	antiallergy	1996	32 (307)
fluticasone furoate	antiallergy	2007	43 (469)
levocabastine hydrochloride	antiallergy	1991	27 (330)
levocetirizine	antiallergy	2001	37 (268)
lodoxamide tromethamine	antiallergy	1992	28 (333)
loratadine	antiallergy	1988	24 (306)
loteprednol etabonate	antiallergy	1998	34 (324)
mizolastine	antiallergy	1998	34 (325)
nedocromil sodium	antiallergy	1986	22 (324)
olopatadine hydrochloride	antiallergy	1997	33 (340)
ramatroban	antiallergy	2000	36 (311)
repirinast	antiallergy	1987	23 (341)
rupatadine fumarate	antiallergy	2003	39 (284)
setastine hydrochloride	antiallergy	1987	23 (342)
suplatast tosilate	antiallergy	1995	31 (350)
tazanolast	antiallergy	1990	26 (309)
gallopamil hydrochloride	antianginal	1983	19 (319)
ivabradine	antianginal	2006	42 (522)
nicorandil	antianginal	1984	20 (322)
ranolazine	antianginal	2006	42 (535)
cibenzoline	antiarrhythmic	1985	21 (325)
dofetilide	antiarrhythmic	2000	36 (301)
dronedarone	antiarrhythmic	2009	45 (495)
encainide hydrochloride	antiarrhythmic	1987	23 (333)
esmolol hydrochloride	antiarrhythmic	1987	23 (334)
ibutilide fumarate	antiarrhythmic	1996	32 (309)
landiolol	antiarrhythmic	2002	38 (360)
moricizine hydrochloride	antiarrhythmic	1990	26 (305)
nif ekalant hydrochloride	antiarrhythmic	1999	35 (344)
pilsicainide hydrochloride	antiarrhythmic	1991	27 (332)
pirmenol	antiarrhythmic	1994	30 (307)
tiracizine hydrochloride	antiarrhythmic	1990	26 (310)
vernakalant	antiarrhythmic	2010	46 (491)
abatacept	antiarthritic	2006	42 (509)

GENERIC NAME	INDICATION	YEAR INTRODUCED	ARMC VOL., (PAGE)
adalimumab	antiarthritic	2003	39 (267)
anakinra	antiarthritic	2001	37 (261)
auranofin	antiarthritic	1983	19 (314)
celecoxib	antiarthritic	1999	35 (335)
etoricoxibe	antiarthritic	2002	38 (355)
iguratimod	antiarthritic	2011	47 (535)
lefunomide	antiarthritic	1998	34 (324)
meloxicam	antiarthritic	1996	32 (312)
rofecoxib	antiarthritic	1999	35 (347)
tofacitinib	antiarthritic	2012	48 (532)
vadecoxib	antiarthritic	2002	38 (369)
amlexanox	antiasthma	1987	23 (327)
arformoterol	antiasthma	2007	43 (465)
bambuterol	antiasthma	1990	26 (299)
ciclesonide	antiasthma	2005	41 (443)
doxofylline	antiasthma	1985	21 (327)
flutropium bromide	antiasthma	1988	24 (303)
ibudilast	antiasthma	1989	25 (313)
levalbuterol hydrochloride	antiasthma	1999	35 (341)
mabuterol hydrochloride	antiasthma	1986	22 (323)
montelukast sodium	antiasthma	1998	34 (326)
omalizumab allergic	antiasthma	2003	39 (280)
oxitropium bromide	antiasthma	1983	19 (323)
pemirolast potassium	antiasthma	1991	27 (331)
pranlukast	antiasthma	1995	31 (347)
salmeterol	antiasthma	1990	26 (308)
hydroxynaphthoate			
seratrodast	antiasthma	1995	31 (349)
zafirlukast	antiasthma	1996	32 (321)
zileuton	antiasthma	1997	33 (344)
adamantanium bromide	antibacterial	1984	20 (315)
arbekacin	antibacterial	1990	26 (298)
aspoxicillin	antibacterial	1987	23 (328)
astromycin sulfate	antibacterial	1985	21 (324)
azithromycin	antibacterial	1988	24 (298)
aztreonam	antibacterial	1984	20 (315)
balofloxacin	antibacterial	2002	38 (351)
bedaquiline	antibacterial	2012	48 (487)
besifloxacin	antibacterial	2009	45 (482)
biapenem	antibacterial	2002	38 (351)
brodimoprin	antibacterial	1993	29 (333)
carboplatin	antibacterial	1986	22 (318)
carumonam	antibacterial	1988	24 (298)
cefbuperazone sodium	antibacterial	1985	21 (325)
cefcapene pivoxil	antibacterial	1997	33 (330)
cefdinir	antibacterial	1991	27 (323)
cefditoren pivoxil	antibacterial	1994	30 (297)

GENERIC NAME	INDICATION	YEAR INTRODUCED	ARMC VOL., (PAGE)
cefepime	antibacterial	1993	29 (334)
cefetamet pivoxil hydrochloride	antibacterial	1992	28 (327)
cefixime	antibacterial	1987	23 (329)
cefmenoxime hydrochloride	antibacterial	1983	19 (316)
cefminox sodium	antibacterial	1987	23 (330)
cefodizime sodium	antibacterial	1990	26 (300)
cefonicid sodium	antibacterial	1984	20 (316)
ceforanide	antibacterial	1984	20 (317)
cefoselis	antibacterial	1998	34 (319)
cefotetan disodium	antibacterial	1984	20 (317)
cefotiam hexetil hydrochloride	antibacterial	1991	27 (324)
cefozopran hydrochloride	antibacterial	1995	31 (339)
cefpimizole	antibacterial	1987	23 (330)
cefpiramide sodium	antibacterial	1985	21 (325)
cefpirome sulfate	antibacterial	1992	28 (328)
cefpodoxime proxetil	antibacterial	1989	25 (310)
cefprozil	antibacterial	1992	28 (328)
ceftaroline fosamil	antibacterial	2010	46 (453)
ceftazidime	antibacterial	1983	19 (316)
cefteram pivoxil	antibacterial	1987	23 (330)
ceftibuten	antibacterial	1992	28 (329)
ceftobiprole medocaril	antibacterial	2008	44 (589)
cefuroxime axetil	antibacterial	1987	23 (331)
cefuzonam sodium	antibacterial	1987	23 (331)
ciprofloxacin	antibacterial	1986	22 (318)
clarithromycin	antibacterial	1990	26 (302)
dalfopristin	antibacterial	1999	35 (338)
daptomycin	antibacterial	2003	39 (272)
dirithromycin	antibacterial	1993	29 (336)
doripenem	antibacterial	2005	41 (448)
enoxacin	antibacterial	1986	22 (320)
ertapenem sodium	antibacterial	2002	38 (353)
erythromycin acistrate	antibacterial	1988	24 (301)
fidaxomicin	antibacterial	2011	47 (531)
fleroxacin	antibacterial	1992	28 (331)
flomoxef sodium	antibacterial	1988	24 (302)
flurithromycin ethylsuccinate	antibacterial	1997	33 (333)
fropenam	antibacterial	1997	33 (334)
garenoxacin	antibacterial	2007	43 (471)
gatilfloxacin	antibacterial	1999	35 (340)
gemifloxacin	antibacterial	2004	40 (458)
imipenem/cilastatin	antibacterial	1985	21 (328)
isepamicin	antibacterial	1988	24 (305)

GENERIC NAME	INDICATION	YEAR INTRODUCED	ARMC VOL., (PAGE)
lenampicillin hydrochloride	antibacterial	1987	23 (336)
levofloxacin	antibacterial	1993	29 (340)
linezolid	antibacterial	2000	36 (309)
lomefloxacin	antibacterial	1989	25 (315)
loracarbef	antibacterial	1992	28 (333)
meropenem	antibacterial	1994	30 (303)
miokamycin	antibacterial	1985	21 (329)
moxifloxacin hydrochloride	antibacterial	1999	35 (343)
mupirocin	antibacterial	1985	21 (330)
nadifloxacin	antibacterial	1993	29 (340)
norfloxacin	antibacterial	1983	19 (322)
ofloxacin	antibacterial	1985	21 (331)
panipenem/ betamipron carbapenem	antibacterial	1994	30 (305)
pazufloxacin	antibacterial	2002	38 (364)
pefloxacin mesylate	antibacterial	1985	21 (331)
prulifloxacin	antibacterial	2002	38 (366)
quinupristin	antibacterial	1999	35 (338)
retapamulin	antibacterial	2007	43 (486)
rifabutin	antibacterial	1992	28 (335)
rifapentine	antibacterial	1988	24 (310)
rifaximin	antibacterial	1987	23 (341)
rifaximin	antibacterial	1985	21 (332)
rokitamycin	antibacterial	1986	22 (325)
rufloxacin hydrochloride	antibacterial	1992	28 (335)
RV-11	antibacterial	1989	25 (318)
sitafloxacin hydrate	antibacterial	2008	44 (621)
sparfloxacin	antibacterial	1993	29 (345)
sulbactam sodium	antibacterial	1986	22 (326)
sultamycillin tosylate	antibacterial	1987	23 (343)
tazobactam sodium	antibacterial	1992	28 (336)
teicoplanin	antibacterial	1988	24 (311)
telavancin	antibacterial	2009	45 (525)
telithromycin	antibacterial	2001	37 (271)
temafloxacin hydrochloride	antibacterial	1991	27 (334)
temocillin disodium	antibacterial	1984	20 (323)
tigecycline	antibacterial	2005	41 (468)
tosufloxacin tosylate	antibacterial	1990	26 (310)
trovafloxacin mesylate	antibacterial	1998	34 (332)
abarelix	anticancer	2004	40 (446)
abiraterone acetate	anticancer	2011	47 (505)
ado-trastuzumab emtansine	anticancer	2013	49 (449)
afatinib	anticancer	2013	49 (451)

GENERIC NAME	INDICATION	YEAR INTRODUCED	ARMC VOL., (PAGE)
alemtuzumab	anticancer	2001	37 (260)
alitretinoin	anticancer	1999	35 (333)
amrubicin hydrochloride	anticancer	2002	38 (349)
amsacrine	anticancer	1987	23 (327)
anastrozole	anticancer	1995	31 (338)
arglabin	anticancer	1999	35 (335)
axitinib	anticancer	2012	48 (485)
azacitidine	anticancer	2004	40 (447)
belotecan	anticancer	2004	40 (449)
bevacizumab	anticancer	2004	40 (450)
bexarotene	anticancer	2000	36 (298)
bicalutamide	anticancer	1995	31 (338)
bisantrene hydrochloride	anticancer	1990	26 (300)
bortezomib	anticancer	2003	39 (271)
bosutinib	anticancer	2012	48 (489)
brentuximab	anticancer	2011	47 (523)
buserelin	anticancer	1984	50 (553)
cabozantinib	anticancer	2012	48 (491)
camostat mesylate	anticancer	1985	21 (325)
capecitabine	anticancer	1998	34 (319)
carfilzomib	anticancer	2012	48 (492)
catumaxomab	anticancer	2009	45 (486)
cetuximab	anticancer	2003	39 (272)
cladribine	anticancer	1993	29 (335)
clofarabine	anticancer	2005	41 (444)
crizotinib	anticancer	2011	47 (525)
cytarabine ocfosfate	anticancer	1993	29 (335)
dabrafenib	anticancer	2013	49 (458)
dasatinib	anticancer	2006	42 (517)
degarelix	anticancer	2008	50 (554)
degarelix acetate	anticancer	2009	45 (490)
denileukin diftitox	anticancer	1999	35 (338)
docetaxel	anticancer	1995	31 (341)
doxifluridine	anticancer	1987	23 (332)
enocitabine	anticancer	1983	19 (318)
enzalutamide	anticancer	2012	48 (497)
epirubicin hydrochloride	anticancer	1984	20 (318)
erlotinib	anticancer	2004	40 (454)
exemestane	anticancer	2000	36 (304)
fadrozole hydrochloride	anticancer	1995	31 (342)
fludarabine phosphate	anticancer	1991	27 (327)
flutamide	anticancer	1983	19 (318)
formestane	anticancer	1993	29 (337)
fotemustine	anticancer	1989	25 (313)
fulveristrant	anticancer	2002	38 (357)
gefitinib	anticancer	2002	38 (358)
gemcitabine hydrochloride	anticancer	1995	31 (344)

GENERIC NAME	INDICATION	YEAR INTRODUCED	ARMC VOL., (PAGE)
gemtuzumab ozogamicin	anticancer	2000	36 (306)
histrelin	anticancer	1991	50 (553)
ibritunomab tiuxetan	anticancer	2002	38 (359)
ibrutinib	anticancer	2013	49 (466)
idarubicin hydrochloride	anticancer	1990	26 (303)
imatinib mesylate	anticancer	2001	37 (267)
ingenol mebutate	anticancer	2012	48 (499)
interferon, gamma-1	anticancer	1992	28 (332)
interleukin-2	anticancer	1989	25 (314)
ipilimumab	anticancer	2011	47 (537)
irinotecan	anticancer	1994	30 (301)
ixabepilone	anticancer	2007	43 (473)
lapatinib	anticancer	2007	43 (475)
letrazole	anticancer	1996	32 (311)
leuprorelin	anticancer	1984	50 (553)
lonidamine	anticancer	1987	23 (337)
masoprocol	anticancer	1992	28 (333)
miltefosine	anticancer	1993	29 (340)
mitoxantrone hydrochloride	anticancer	1984	20 (321)
mogamulizumab	anticancer	2012	48 (507)
nedaplatin	anticancer	1995	31 (347)
nelarabine	anticancer	2006	42 (528)
nilutamide	anticancer	1987	23 (338)
nimotuzumab	anticancer	2006	42 (529)
obinutuzumab	anticancer	2013	49 (478)
OCT-43	anticancer	1999	35 (345)
ofatumumab	anticancer	2009	45 (512)
omacetaxine mepesuccinate	anticancer	2012	48 (510)
oxaliplatin	anticancer	1996	32 (313)
paclitaxal	anticancer	1993	29 (342)
panitumumab	anticancer	2006	42 (531)
pazopanib	anticancer	2009	45 (514)
pegaspargase	anticancer	1994	30 (306)
pemetrexed	anticancer	2004	40 (463)
pentostatin	anticancer	1992	28 (334)
pertuzumab	anticancer	2012	48 (517)
pirarubicin	anticancer	1988	24 (309)
pixantrone dimaleate	anticancer	2012	48 (519)
pomalidomide	anticancer	2013	49 (484)
ponatinib	anticancer	2012	48 (521)
porfimer sodium	anticancer	1993	29 (343)
pralatrexate	anticancer	2009	45 (517)
radotinib	anticancer	2012	48 (523)
raltitrexed	anticancer	1996	32 (315)
ranimustine	anticancer	1987	23 (341)

GENERIC NAME	INDICATION	YEAR INTRODUCED	ARMC VOL., (PAGE)
regorafenib	anticancer	2012	48 (524)
ruxolitinib	anticancer	2011	47 (548)
SKI-2053R	anticancer	1999	35 (348)
sobuzoxane	anticancer	1994	30 (310)
sorafenib	anticancer	2005	41 (466)
sunitinib	anticancer	2006	42 (544)
talaporfin sodium	anticancer	2004	40 (469)
tamibarotene	anticancer	2005	41 (467)
tasonermin	anticancer	1999	35 (349)
temoporphin	anticancer	2002	38 (367)
temozolomide	anticancer	1999	35 (349)
temsirolimus	anticancer	2007	43 (490)
topotecan hydrochloride	anticancer	1996	32 (320)
toremifene	anticancer	1989	25 (319)
tositumomab	anticancer	2003	39 (285)
trabectedin	anticancer	2007	43 (492)
trametinib	anticancer	2013	49 (494)
triptorelin	anticancer	1986	50 (553)
valrubicin	anticancer	1999	35 (350)
vandetanib	anticancer	2011	47 (555)
vemurafenib	anticancer	2011	47 (556)
vinorelbine	anticancer	1989	25 (320)
vismodegib	anticancer	2012	48 (534)
vorinostat	anticancer	2006	42 (549)
zinostatin stimalamer	anticancer	1994	30 (313)
cabazitaxel	anticancer	2010	46 (451)
eribulin mesylate	anticancer	2010	46 (465)
mifamurtide	anticancer	2009	46 (476)
nilotinib	anticancer	2007	43 (480)
romidepsin	anticancer	2009	46 (482)
sipuleucel-t	anticancer	2010	46 (484)
vinflunine	anticancer	2009	46 (493)
angiotensin II	anticancer adjuvant	1994	30 (296)
dabigatran etexilate	anticoagulant	2008	44 (598)
duteplase	anticoagulant	1995	31 (342)
enoxaparin	anticoagulant	1987	23 (333)
lepirudin	anticoagulant	1997	33 (336)
parnaparin sodium	anticoagulant	1993	29 (342)
reviparin sodium	anticoagulant	1993	29 (344)
rivaroxaban	anticoagulant	2008	44 (617)
thrombomodulin (recombinant)	anticoagulant	2008	44 (628)
ximelagatran	anticoagulant	2004	40 (470)
eslicarbazepine acetate	anticonvulsant	2009	45 (498)
felbamate	anticonvulsant	1993	29 (337)
fosphenytoin sodium	anticonvulsant	1996	32 (308)

GENERIC NAME	INDICATION	YEAR INTRODUCED	ARMC VOL., (PAGE)
gabapentin	anticonvulsant	1993	29 (338)
lacosamide	anticonvulsant	2008	44 (610)
lamotrigine	anticonvulsant	1990	26 (304)
levetiracetam	anticonvulsant	2000	36 (307)
oxcarbazepine	anticonvulsant	1990	26 (307)
perampanel	anticonvulsant	2012	48 (516)
pregabalin	anticonvulsant	2004	40 (464)
progabide	anticonvulsant	1985	21 (331)
retigabine	anticonvulsant	2011	47 (544)
rufinamide	anticonvulsant	2007	43 (488)
tiagabine	anticonvulsant	1996	32 (319)
topiramate	anticonvulsant	1995	31 (351)
vigabatrin	anticonvulsant	1989	25 (319)
zonisamide	anticonvulsant	1989	25 (320)
bupropion hydrochloride	antidepressant	1989	25 (310)
citalopram	antidepressant	1989	25 (311)
desvenlafaxine	antidepressant	2008	44 (600)
duloxetine	antidepressant	2004	40 (452)
escitalopram oxolate	antidepressant	2002	38 (354)
fluoxetine hydrochloride	antidepressant	1986	22 (320)
fluvoxamine maleate	antidepressant	1983	19 (319)
indalpine	antidepressant	1983	19 (320)
medifoxamine fumarate	antidepressant	1986	22 (323)
metapramine	antidepressant	1984	20 (320)
milnacipran	antidepressant	1997	33 (338)
mirtazapine	antidepressant	1994	30 (303)
moclobemide	antidepressant	1990	26 (305)
nefazodone	antidepressant	1994	30 (305)
paroxetine	antidepressant	1991	27 (331)
pivagabine	antidepressant	1997	33 (341)
reboxetine	antidepressant	1997	33 (342)
setiptiline	antidepressant	1989	25 (318)
setraline hydrochloride	antidepressant	1990	26 (309)
tianeptine sodium	antidepressant	1983	19 (324)
toloxatone	antidepressant	1984	20 (324)
venlafaxine	antidepressant	1994	30 (312)
vilazodone	antidepressant	2011	47 (558)
vortioxetine	antidepressant	2013	49 (496)
acarbose	antidiabetic	1990	26 (297)
albiglutide	antidiabetic	2014	50 (552)
alogliptin	antidiabetic	2010	46 (446)
anagliptin	antidiabetic	2012	48 (483)
canagliflozin	antidiabetic	2013	49 (453)
dapagliflozin	antidiabetic	2012	48 (495)
dulaglutide	antidiabetic	2014	50 (552)
epalrestat	antidiabetic	1992	28 (330)

GENERIC NAME	INDICATION	YEAR INTRODUCED	ARMC VOL., (PAGE)
exenatide	antidiabetic	2005	41 (452), 50 (552)
glimepiride	antidiabetic	1995	31 (344)
glucagon, rDNA	antidiabetic	1993	29 (338)
insulin lispro	antidiabetic	1996	32 (310)
linagliptin	antidiabetic	2011	47 (540)
liraglutide	antidiabetic	2009	45 (507), 50 (552)
lixisenatide	antidiabetic	2013	49 (470), 50 (552)
miglitol	antidiabetic	1998	34 (325)
mitiglinide	antidiabetic	2004	40 (460)
nateglinide	antidiabetic	1999	35 (344)
pioglitazone hydrochloride	antidiabetic	1999	35 (346)
pramlintide	antidiabetic	2005	41 (460), 50 (551)
repaglinide	antidiabetic	1998	34 (329)
rosiglitazone maleate	antidiabetic	1999	35 (348)
saroglitazar	antidiabetic	2013	49 (488)
saxagliptin	antidiabetic	2009	45 (521)
sitagliptin	antidiabetic	2006	42 (541)
teneligliptin	antidiabetic	2012	48 (528)
tolrestat	antidiabetic	1989	25 (319)
troglitazone	antidiabetic	1997	33 (344)
vildagliptin	antidiabetic	2007	43 (494)
voglibose	antidiabetic	1994	30 (313)
acetorphan	antidiarrheal	1993	29 (332)
crofelemer	antidiarrheal	2012	48 (494)
lidamidine hydrochloride	antidiarrheal	1984	20 (320)
flumazenil	antidote, benzodiazepine overdose	1987	23 (335)
trientine hydrochloride	antidote, copper poisoning	1986	22 (327)
anti-digoxin polyclonal antibody	antidote, digoxin poisoning	2002	38 (350)
fomepizole	antidote, ethylene glycol poisoning	1998	34 (323)
succimer	antidote, lead poisoning	1991	27 (333)
crotelidae polyvalent immune fab	antidote, snake venom poisoning	2001	37 (263)
aprepitant	antiemetic	2003	39 (268)
dolasetron mesylate	antiemetic	1998	34 (321)
dronabinol	antiemetic	1986	22 (319)
fosaprepitant dimeglumine	antiemetic	2008	44 (606)
granisetron hydrochloride	antiemetic	1991	27 (329)
indisetron	antiemetic	2004	40 (459)
nazasetron	antiemetic	1994	30 (305)

GENERIC NAME	INDICATION	YEAR INTRODUCED	ARMC VOL., (PAGE)
ondansetron hydrochloride	antiemetic	1990	26 (306)
palonosetron	antiemetic	2003	39 (281)
ramosetron	antiemetic	1996	32 (315)
tropisetron	antiemetic	1992	28 (337)
amorolfine hydrochloride	antifungal	1991	27 (322)
anidulafungin	antifungal	2006	42 (512)
butenafine hydrochloride	antifungal	1992	28 (327)
butoconazole	antifungal	1986	22 (318)
caspofungin acetate	antifungal	2001	37 (263)
cloconazole hydrochloride	antifungal	1986	22 (318)
eberconazole	antifungal	2005	41 (449)
efinaconazole	antifungal	2013	49 (463)
fenticonazole nitrate	antifungal	1987	23 (334)
fluconazole	antifungal	1988	24 (303)
flutrimazole	antifungal	1995	31 (343)
fosfluconazole	antifungal	2004	40 (457)
itraconazole	antifungal	1988	24 (305)
lanoconazole	antifungal	1994	30 (302)
liranaftate	antifungal	2000	36 (309)
luliconazole	antifungal	2005	41 (454)
micafungin	antifungal	2002	38 (360)
naftifine hydrochloride	antifungal	1984	20 (321)
neticonazole hydrochloride	antifungal	1993	29 (341)
oxiconazole nitrate	antifungal	1983	19 (322)
posaconazole	antifungal	2006	42 (532)
sertaconazole nitrate	antifungal	1992	28 (336)
sulconizole nitrate	antifungal	1985	21 (332)
terbinafine hydrochloride	antifungal	1991	27 (334)
terconazole	antifungal	1983	19 (324)
tioconazole	antifungal	1983	19 (324)
trimetrexate glucuronate	antifungal	1994	30 (312)
voriconazole	antifungal	2002	38 (370)
apraclonidine hydrochloride	antiglaucoma	1988	24 (297)
befunolol hydrochloride	antiglaucoma	1983	19 (315)
bimatoprost	antiglaucoma	2001	37 (261)
brimonidine	antiglaucoma	1996	32 (306)
brinzolamide	antiglaucoma	1998	34 (318)
dapiprazole hydrochloride	antiglaucoma	1987	23 (332)
dorzolamide hydrochloride	antiglaucoma	1995	31 (341)
latanoprost	antiglaucoma	1996	32 (311)
levobunolol hydrochloride	antiglaucoma	1985	21 (328)
tafluprost	antiglaucoma	2008	44 (625)
travoprost	antiglaucoma	2001	37 (272)

GENERIC NAME	INDICATION	YEAR INTRODUCED	ARMC VOL., (PAGE)
unoprostone isopropyl ester	antiglaucoma	1994	30 (312)
lomitapide	antihypercholersteremic	2012	48 (504)
acipimox	antihypercholesterolemic	1985	21 (323)
atorvastatin calcium	antihypercholesterolemic	1997	33 (328)
beclobrate	antihypercholesterolemic	1986	22 (317)
binifibrate	antihypercholesterolemic	1986	22 (317)
cerivastatin	antihypercholesterolemic	1997	33 (331)
choline fenofibrate	antihypercholesterolemic	2008	44 (594)
ciprofibrate	antihypercholesterolemic	1985	21 (326)
colesevelam hydrochloride	antihypercholesterolemic	2000	36 (300)
colestimide	antihypercholesterolemic	1999	35 (337)
divistyramine	antihypercholesterolemic	1984	20 (317)
ezetimibe	antihypercholesterolemic	2002	38 (355)
fluvastatin	antihypercholesterolemic	1994	30 (300)
lovastatin	antihypercholesterolemic	1987	23 (337)
meglutol	antihypercholesterolemic	1983	19 (321)
melinamide	antihypercholesterolemic	1984	20 (320)
mipomersen	antihypercholesterolemic	2013	49 (476)
pitavastatin	antihypercholesterolemic	2003	39 (282)
pravastatin	antihypercholesterolemic	1989	25 (316)
ronafibrate	antihypercholesterolemic	1986	22 (326)
rosuvastatin	antihypercholesterolemic	2003	39 (283)
simvastatin	antihypercholesterolemic	1988	24 (311)
alacepril	antihypertensive	1988	24 (296)
alfuzosin hydrochloride	antihypertensive	1988	24 (296)
aliskiren	antihypertensive	2007	43 (461)
amlodipine besylate	antihypertensive	1990	26 (298)
amosulalol	antihypertensive	1988	24 (297)
aranidipine	antihypertensive	1996	32 (306)
arotinolol hydrochloride	antihypertensive	1986	22 (316)
azelnidipine	antihypertensive	2003	39 (270)
azilsartan	antihypertensive	2011	47 (514)
barnidipine hydrochloride	antihypertensive	1992	28 (326)
benazepril hydrochloride	antihypertensive	1990	26 (299)
benidipine hydrochloride	antihypertensive	1991	27 (322)
betaxolol hydrochloride	antihypertensive	1983	19 (315)
bevantolol hydrochloride	antihypertensive	1987	23 (328)
bisoprolol fumarate	antihypertensive	1986	22 (317)
bopindolol	antihypertensive	1985	21 (324)
bosentan	antihypertensive	2001	37 (262)
budralazine	antihypertensive	1983	19 (315)
bunazosin hydrochloride	antihypertensive	1985	21 (324)
cadralazine	antihypertensive	1988	24 (298)
candesartan cilexetil	antihypertensive	1997	33 (330)
captopril	antihypertensive	1982	13 (086)

GENERIC NAME	INDICATION	YEAR INTRODUCED	ARMC VOL., (PAGE)
carvedilol	antihypertensive	1991	27 (323)
celiprolol hydrochloride	antihypertensive	1983	19 (317)
cicletanine	antihypertensive	1988	24 (299)
cilazapril	antihypertensive	1990	26 (301)
cinildipine	antihypertensive	1995	31 (339)
clevidipine	antihypertensive	2008	44 (596)
delapril	antihypertensive	1989	25 (311)
dilevalol	antihypertensive	1989	25 (311)
doxazosin mesylate	antihypertensive	1988	24 (300)
efonidipine	antihypertensive	1994	30 (299)
enalapril maleate	antihypertensive	1984	20 (317)
enalaprilat	antihypertensive	1987	23 (332)
eplerenone	antihypertensive	2003	39 (276)
eprosartan	antihypertensive	1997	33 (333)
felodipine	antihypertensive	1988	24 (302)
fenoldopam mesylate	antihypertensive	1998	34 (322)
fosinopril sodium	antihypertensive	1991	27 (328)
guanadrel sulfate	antihypertensive	1983	19 (319)
imidapril hydrochloride	antihypertensive	1993	29 (339)
irbesartan	antihypertensive	1997	33 (336)
isradipine	antihypertensive	1989	25 (315)
ketanserin	antihypertensive	1985	21 (328)
lacidipine	antihypertensive	1991	27 (330)
lercanidipine	antihypertensive	1997	33 (337)
lisinopril	antihypertensive	1987	23 (337)
losartan	antihypertensive	1994	30 (302)
manidipine hydrochloride	antihypertensive	1990	26 (304)
mebefradil hydrochloride	antihypertensive	1997	33 (338)
moexipril hydrochloride	antihypertensive	1995	31 (346)
moxonidine	antihypertensive	1991	27 (330)
nebivolol	antihypertensive	1997	33 (339)
nilvadipine	antihypertensive	1989	25 (316)
nipradilol	antihypertensive	1988	24 (307)
nisoldipine	antihypertensive	1990	26 (306)
nitrendipine	antihypertensive	1985	21 (331)
olimesartan Medoxomil	antihypertensive	2002	38 (363)
perindopril	antihypertensive	1988	24 (309)
pinacidil	antihypertensive	1987	23 (340)
quinapril	antihypertensive	1989	25 (317)
ramipril	antihypertensive	1989	25 (317)
rilmenidine	antihypertensive	1988	24 (310)
spirapril hydrochloride	antihypertensive	1995	31 (349)
telmisartan	antihypertensive	1999	35 (349)
temocapril	antihypertensive	1994	30 (311)
terazosin hydrochloride	antihypertensive	1984	20 (323)
tertatolol hydrochloride	antihypertensive	1987	23 (344)

GENERIC NAME	INDICATION	YEAR INTRODUCED	ARMC VOL., (PAGE)
tiamenidine hydrochloride	antihypertensive	1988	24 (311)
tilisolol hydrochloride	antihypertensive	1992	28 (337)
trandolapril	antihypertensive	1993	29 (348)
treprostinil sodium	antihypertensive	2002	38 (368)
trimazosin hydrochloride	antihypertensive	1985	21 (333)
valsartan	antihypertensive	1996	32 (320)
zofenopril calcium	antihypertensive	2000	36 (313)
aceclofenac	antiinflammatory	1992	28 (325)
actarit	antiinflammatory	1994	30 (296)
AF-2259	antiinflammatory	1987	23 (325)
alclometasone dipropionate	antiinflammatory	1985	21 (323)
amfenac sodium	antiinflammatory	1986	22 (315)
aminoprofen	antiinflammatory	1990	26 (298)
ampiroxicam	antiinflammatory	1994	30 (296)
amtolmetin guacil	antiinflammatory	1993	29 (332)
betamethasone butyrate propionate	antiinflammatory	1994	30 (297)
bromfenac sodium	antiinflammatory	1997	33 (329)
butibufen	antiinflammatory	1992	28 (327)
butyl flufenamate	antiinflammatory	1983	19 (316)
canakinumab	antiinflammatory	2009	45 (484)
clobenoside	antiinflammatory	1988	24 (300)
deflazacort	antiinflammatory	1986	22 (319)
deprodone propionate	antiinflammatory	1992	28 (329)
dexibuprofen	antiinflammatory	1994	30 (298)
diacerein	antiinflammatory	1985	21 (326)
droxicam	antiinflammatory	1990	26 (302)
etodolac	antiinflammatory	1985	21 (327)
felbinac	antiinflammatory	1986	22 (320)
fisalamine	antiinflammatory	1984	20 (318)
flunoxaprofen	antiinflammatory	1987	23 (335)
fluticasone propionate	antiinflammatory	1990	26 (303)
golimumab	antiinflammatory	2009	45 (503)
halobetasol propionate	antiinflammatory	1991	27 (329)
halometasone	antiinflammatory	1983	19 (320)
hydrocortisone aceponate	antiinflammatory	1988	24 (304)
hydrocortisone butyrate	antiinflammatory	1983	19 (320)
interferon, gamma	antiinflammatory	1989	25 (314)
isofezolac	antiinflammatory	1984	20 (319)
isoxicam	antiinflammatory	1983	19 (320)
lobenzarit sodium	antiinflammatory	1986	22 (322)
lornoxicam	antiinflammatory	1997	33 (337)
loxoprofen sodium	antiinflammatory	1986	22 (322)
lumiracoxib	antiinflammatory	2005	41 (455)
mometasone furoate	antiinflammatory	1987	23 (338)

GENERIC NAME	INDICATION	YEAR INTRODUCED	ARMC VOL., (PAGE)
nabumetone	antiinflammatory	1985	21 (330)
nepafenac	antiinflammatory	2005	41 (456)
nimesulide	antiinflammatory	1985	21 (330)
osalazine sodium	antiinflammatory	1986	22 (324)
oxaprozin	antiinflammatory	1983	19 (322)
piketoprofen	antiinflammatory	1984	20 (322)
pimaprofen	antiinflammatory	1984	20 (322)
piroxicam cinnamate	antiinflammatory	1988	24 (309)
prednicarbate	antiinflammatory	1986	22 (325)
rimexolone	antiinflammatory	1995	31 (348)
sivelestat	antiinflammatory	2002	38 (366)
tenoxicam	antiinflammatory	1987	23 (344)
zaltoprofen	antiinflammatory	1993	29 (349)
arteether	antimalarial	2000	36 (296)
artemisinin	antimalarial	1987	23 (327)
bulaquine	antimalarial	2000	36 (299)
halofantrine	antimalarial	1988	24 (304)
mefloquine hydrochloride	antimalarial	1985	21 (329)
almotriptan	antimigraine	2000	36 (295)
alpiropride	antimigraine	1988	24 (296)
eletriptan	antimigraine	2001	37 (266)
frovatriptan	antimigraine	2002	38 (357)
lomerizine hydrochloride	antimigraine	1999	35 (342)
naratriptan hydrochloride	antimigraine	1997	33 (339)
rizatriptan benzoate	antimigraine	1998	34 (330)
sumatriptan succinate	antimigraine	1991	27 (333)
zomitriptan	antimigraine	1997	33 (345)
cetilistat	antiobesity	2013	49 (454)
dexfenfluramine	antiobesity	1997	33 (332)
lorcaserin hydrochloride	antiobesity	2012	48 (506)
orlistat	antiobesity	1998	34 (327)
rimonabant	antiobesity	2006	42 (537)
sibutramine	antiobesity	1998	34 (331)
atovaquone	antiparasitic	1992	28 (326)
ivermectin	antiparasitic	1987	23 (336)
quinf amideamebicide	antiparasitic	1984	20 (322)
beraprost sodium	antiplatelet	1992	28 (326)
epoprostenol sodium	antiplatelet	1983	19 (318)
iloprost	antiplatelet	1992	28 (332)
prasugrel	antiplatelet	2009	45 (519)
cabergoline	antiprolactin	1993	29 (334)
acitretin	antipsoriasis	1989	25 (309)
alefacept	antipsoriasis	2003	39 (267)
calcipotriol	antipsoriasis	1991	27 (323)
efalizumab	antipsoriasis	2003	39 (274)
tacalcitol	antipsoriasis	1993	29 (346)

GENERIC NAME	INDICATION	YEAR INTRODUCED	ARMC VOL., (PAGE)
tazarotene	antipsoriasis	1997	33 (343)
ustekinumab	antipsoriasis	2009	45 (532)
amisulpride	antipsychotic	1986	22 (316)
aripiprazole	antipsychotic	2002	38 (350)
asenapine	antipsychotic	2009	45 (479)
blonanserin	antipsychotic	2008	44 (587)
clospipramine hydrochloride	antipsychotic	1991	27 (325)
lurasidone hydrochloride	antipsychotic	2010	46 (473)
nemonapride	antipsychotic	1991	27 (331)
olanzapine	antipsychotic	1996	32 (313)
paliperidone	antipsychotic	2007	43 (482)
perospirone hydrochloride	antipsychotic	2001	37 (270)
quetiapine fumarate	antipsychotic	1997	33 (341)
remoxipride hydrochloride	antipsychotic	1990	26 (308)
risperidone	antipsychotic	1993	29 (344)
sertindole	antipsychotic	1996	32 (318)
timiperone	antipsychotic	1984	20 (323)
ziprasidone hydrochloride	antipsychotic	2000	36 (312)
zuclopenthixol acetate	antipsychotic	1987	23 (345)
drotrecogin alfa	antisepsis	2001	37 (265)
cimetropium bromide	antispasmodic	1985	21 (326)
tiquizium bromide	antispasmodic	1984	20 (324)
eltrombopag	antithrombocytopenic	2009	45 (497)
romiplostim	antithrombocytopenic	2008	44 (619), 50 (556)
alteplase	antithrombotic	1987	23 (326)
anagrelide hydrochloride	antithrombotic	1997	33 (328)
apixaban	antithrombotic	2011	47 (509)
APSAC	antithrombotic	1987	23 (326)
argatroban	antithrombotic	1990	26 (299)
bivalirudin	antithrombotic	1999	50 (556)
bivalirudin	antithrombotic	2000	36 (298)
cilostazol	antithrombotic	1988	24 (299)
clopidogrel hydrogensulfate	antithrombotic	1998	34 (320)
cloricromen	antithrombotic	1991	27 (325)
defibrotide	antithrombotic	1986	22 (319)
edoxaban	antithrombotic	2011	47 (527)
eptilfibatide	antithrombotic	1999	35 (340)
ethyl icosapentate	antithrombotic	1990	26 (303)
fondaparinux sodium	antithrombotic	2002	38 (356)
indobufen	antithrombotic	1984	20 (319)
limaprost	antithrombotic	1988	24 (306)
ozagrel sodium	antithrombotic	1988	24 (308)

GENERIC NAME	INDICATION	YEAR INTRODUCED	ARMC VOL., (PAGE)
picotamide	antithrombotic	1987	23 (340)
reteplase	antithrombotic	1996	32 (316)
sarpogrelate hydrochloride	antithrombotic	1993	29 (344)
ticagrelor	antithrombotic	2010	46 (488)
tirofiban hydrochloride	antithrombotic	1998	34 (332)
levodropropizine	antitussive	1988	24 (305)
nitisinone	antityrosinaemia	2002	38 (361)
benexate hydrochloride	antiulcer	1987	23 (328)
dexlansoprazole	antiulcer	2009	45 (492)
dosmalfate	antiulcer	2000	36 (302)
ebrotidine	antiulcer	1997	33 (333)
ecabet sodium	antiulcer	1993	29 (336)
egualen sodium	antiulcer	2000	36 (303)
enprostil	antiulcer	1985	21 (327)
esomeprazole magnesium	antiulcer	2000	36 (303)
famotidine	antiulcer	1985	21 (327)
irsogladine	antiulcer	1989	25 (315)
lafutidine	antiulcer	2000	36 (307)
lansoprazole	antiulcer	1992	28 (332)
misoprostol	antiulcer	1985	21 (329)
nizatidine	antiulcer	1987	23 (339)
omeprazole	antiulcer	1988	24 (308)
ornoprostil	antiulcer	1987	23 (339)
pantoprazole sodium	antiulcer	1995	30 (306)
plaunotol	antiulcer	1987	23 (340)
polaprezinc	antiulcer	1994	30 (307)
rabeprazole sodium	antiulcer	1998	34 (328)
ranitidine bismuth citrate	antiulcer	1995	31 (348)
rebamipide	antiulcer	1990	26 (308)
rosaprostol	antiulcer	1985	21 (332)
roxatidine acetate hydrochloride	antiulcer	1986	22 (326)
roxithromycin	antiulcer	1987	23 (342)
sofalcone	antiulcer	1984	20 (323)
spizofurone	antiulcer	1987	23 (343)
teprenone	antiulcer	1984	20 (323)
tretinoin tocoferil	antiulcer	1993	29 (348)
troxipide	antiulcer	1986	22 (327)
abacavir sulfate	antiviral	1999	35 (333)
adefovir dipivoxil	antiviral	2002	38 (348)
amprenavir	antiviral	1999	35 (334)
atazanavir	antiviral	2003	39 (269)
boceprevir	antiviral	2011	47 (521)
cidofovir	antiviral	1996	32 (306)
clevudine	antiviral	2007	43 (466)
darunavir	antiviral	2006	42 (515)

GENERIC NAME	INDICATION	YEAR INTRODUCED	ARMC VOL., (PAGE)
delavirdine mesylate	antiviral	1997	33 (331)
didanosine	antiviral	1991	27 (326)
dolutegravir	antiviral	2013	49 (461)
efavirenz	antiviral	1998	34 (321)
elvitegravir	antiviral	2013	49 (465)
emtricitabine	antiviral	2003	39 (274)
enfuvirtide	antiviral	2003	39 (275)
entecavir	antiviral	2005	41 (450)
etravirine	antiviral	2008	44 (602)
famciclovir	antiviral	1994	30 (300)
fomivirsen sodium	antiviral	1998	34 (323)
fosamprenavir	antiviral	2003	39 (277)
foscarnet sodium	antiviral	1989	25 (313)
ganciclovir	antiviral	1988	24 (303)
imiquimod	antiviral	1997	33 (335)
indinavir sulfate	antiviral	1996	32 (310)
influenza virus (live)	antiviral	2003	39 (277)
interferon alfacon-1	antiviral	1997	33 (336)
lamivudine	antiviral	1995	31 (345)
laninamivir octanoate	antiviral	2010	46 (470)
lopinavir	antiviral	2000	36 (310)
maraviroc	antiviral	2007	43 (478)
nelfinavir mesylate	antiviral	1997	33 (340)
nevirapine	antiviral	1996	32 (313)
oseltamivir phosphate	antiviral	1999	35 (346)
penciclovir	antiviral	1996	32 (314)
peramivir	antiviral	2010	46 (477)
propagermanium	antiviral	1994	30 (308)
raltegravir	antiviral	2007	43 (484)
rilpivirine	antiviral	2011	47 (546)
rimantadine hydrochloride	antiviral	1987	23 (342)
ritonavir	antiviral	1996	32 (317)
saquinavir mesvlate	antiviral	1995	31 (349)
simeprevir	antiviral	2013	49 (489)
sofosbuvir	antiviral	2013	49 (492)
sorivudine	antiviral	1993	29 (345)
stavudine	antiviral	1994	30 (311)
telaprevir	antiviral	2011	47 (552)
telbivudine	antiviral	2006	42 (546)
tenofovir disoproxil fumarate	antiviral	2001	37 (271)
tipranavir	antiviral	2005	41 (470)
vaglancirclovir hydrochloride	antiviral	2001	37 (273)
valaciclovir hydrochloride	antiviral	1995	31 (352)

GENERIC NAME	INDICATION	YEAR INTRODUCED	ARMC VOL., (PAGE)
zalcitabine	antiviral	1992	28 (338)
zanamivir	antiviral	1999	35 (352)
zidovudine	antiviral	1987	23 (345)
cobicistat	antiviral pharmacokinetic enhancer)	2013	49 (456)
cevimeline hydrochloride	antixerostomia	2000	36 (299)
alpidem	anxiolytic	1991	27 (322)
buspirone hydrochloride	anxiolytic	1985	21 (324)
cinolazepam	anxiolytic	1993	29 (334)
doxefazepam	anxiolytic	1985	21 (326)
etizolam	anxiolytic	1984	20 (318)
flutazolam	anxiolytic	1984	20 (318)
flutoprazepam	anxiolytic	1986	22 (320)
metaclazepam	anxiolytic	1987	23 (338)
mexazolam	anxiolytic	1984	20 (321)
tandospirone	anxiolytic	1996	32 (319)
atomoxetine	attention deficit hyperactivity disorder	2003	39 (270)
dexmethylphenidate hydrochloride	attention deficit hyperactivity disorder	2002	38 (352)
lisdexamfetamine	attention deficit hyperactivity disorder	2007	43 (477)
dutasteride	benign prostatic hyperplasia	2002	38 (353)
finasteride	benign prostatic hyperplasia	1992	28 (331)
tamsulosin hydrochloride	benign prostatic hyperplasia	1993	29 (347)
fenbuprol	biliary tract dysfunction	1983	19 (318)
ornipressin	bleeding esophageal varices	1971	50 (551)
somatostatin	bleeding esophageal varices	1970	50 (556)
terlipressin	bleeding esophageal varices	1978	50 (551)
vasopressin	bleeding esophageal varices	1962	50 (551)
P-15	bone grafting	1999	50 (556)
triptorelin	breast cancer	1986	50 (553)
clodronate disodium	calcium regulation	1986	22 (319)
gallium nitrate	calcium regulation	1991	27 (328)
neridronic acide	calcium regulation	2002	38 (361)
dexrazoxane	cardioprotective	1992	28 (330)
brovincamine fumarate	cerebral vasodilator	1986	22 (317)
nimodipine	cerebral vasodilator	1985	21 (330)
propentofylline propionate	cerebral vasodilator	1988	24 (310)
linaclotide	chronic idiopathic constipation	2012	50 (552)
plecanatide	chronic idiopathic constipation	2017	50 (552)
indacaterol	chronic obstructive pulmonary disease	2009	45 (505)
aclidinium bromide	chronic obstructive pulmonary disorder	2012	48 (481)

GENERIC NAME	INDICATION	YEAR INTRODUCED	ARMC VOL., (PAGE)
formoterol fumarate	chronic obstructive pulmonary disorder	1986	22 (321)
olodaterol	chronic obstructive pulmonary disorder	2013	49 (480)
roflumilast	chronic obstructive pulmonary disorder	2010	46 (480)
tiotropium bromide	chronic obstructive pulmonary disorder	2002	38 (368)
aniracetam	cognition enhancer	1993	29 (333)
choline alfoscerate	cognition enhancer	1990	26 (300)
exifone	cognition enhancer	1988	24 (302)
levacecarnine hydrochloride	cognition enhancer	1986	22 (322)
oxiracetam	cognition enhancer	1987	23 (339)
pramiracetam sulfate	cognition enhancer	1993	29 (343)
amrinone	congestive heart failure	1983	19 (314)
bucladesine sodium	congestive heart failure	1984	20 (316)
carperitide	congestive heart failure	1995	31 (339)
colforsin daropate hydrochloride	congestive heart failure	1999	35 (337)
denopamine	congestive heart failure	1988	24 (300)
docarpamine	congestive heart failure	1994	30 (298)
dopexamine	congestive heart failure	1989	25 (312)
enoximone	congestive heart failure	1988	24 (301)
flosequinan	congestive heart failure	1992	28 (331)
ibopamine hydrochloride	congestive heart failure	1984	20 (319)
levosimendan	congestive heart failure	2000	36 (308)
loprinone hydrochloride	congestive heart failure	1996	32 (312)
milrinone	congestive heart failure	1989	25 (316)
nesiritide	congestive heart failure	2001	37 (269), 50 (555)
pimobendan	congestive heart failure	1994	30 (307)
vesnarinone	congestive heart failure	1990	26 (310)
xamoterol fumarate	congestive heart failure	1988	24 (312)
lulbiprostone	constipation	2006	42 (525)
methylnaltrexone bromide	constipation	2008	44 (612)
centchroman	contraception	1991	27 (324)
drospirenone	contraception	2000	36 (302)
gestodene	contraception	1987	23 (335)
gestrinone	contraception	1986	22 (321)
nomegestrol acetate	contraception	1986	22 (324)
norelgestromin	contraception	2002	38 (362)
norgestimate	contraception	1986	22 (324)
promegestrone	contraception	1983	19 (323)

GENERIC NAME	INDICATION	YEAR INTRODUCED	ARMC VOL., (PAGE)
trimegestone	contraception	2001	37 (273)
ulipristal acetate	contraception	2009	45 (530)
biolimus drug-eluting stent	coronary artery disease, antirestenotic	2008	44 (586)
pasireotide	Cushing's Disease	2012	48 (512), 50 (555)
dornase alfa	cystic fibrosis	1994	30 (298)
ivacaftor	cystic fibrosis	2012	48 (501)
neltenexine	cystic fibrosis	1993	29 (341)
amifostine	cytoprotective	1995	31 (338)
kinetin	dermatologic, skin photodamage	1999	35 (341)
desmopressin	diabetes insipidus	1972	50 (556)
gadoversetamide	diagnostic	2000	36 (304)
ioflupane	diagnostic	2000	36 (306)
azosemide	diuretic	1986	22 (316)
muzolimine	diuretic	1983	19 (321)
torasemide	diuretic	1993	29 (348)
ospemifene	dyspareunia	2013	49 (482)
acotiamide	dyspepsia	2013	49 (447)
alpha-1 antitrypsin	emphysema	1988	24 (297)
buserelin	endometriosis	1984	50 (553)
leuprorelin	endometriosis	1984	50 (553)
nafarelin	endometriosis	1990	50 (553)
triptorelin	endometriosis	1986	50 (553)
afamelanotide	erythropoietic protoporphyria	2014	50 (554)
erdosteine	expectorant	1995	31 (342)
fudosteine	expectorant	2001	37 (267)
telmesteine	expectorant	1992	28 (337)
agalsidase alfa	Fabry's disease	2001	37 (259)
chenodiol	gallstones	1983	19 (317)
cinitapride	gastroprokinetic	1990	26 (301)
cisapride	gastroprokinetic	1988	24 (299)
itopride hydrochloride	gastroprokinetic	1995	31 (344)
mosapride citrate	gastroprokinetic	1998	34 (326)
alglucerase	Gaucher's disease	1991	27 (321)
imiglucerase	Gaucher's disease	1994	30 (301)
miglustat	Gaucher's disease	2003	39 (279)
rilonacept	genetic autoinflammatory syndromes	2008	44 (615)
febuxostat	gout	2009	45 (501)
lanreotide	growth disorders	1994	50 (555)
lanreotide acetate	growth disorders	1995	31 (345)
octreotide	growth disorders	1988	24 (307), 50 (555)

GENERIC NAME	INDICATION	YEAR INTRODUCED	ARMC VOL., (PAGE)
pasireotide	growth disorders	2012	50 (555)
pegvisomant	growth disorders	2003	39 (281)
somatomedin-1	growth disorders	1994	30 (310)
somatotropin	growth disorders	1994	30 (310)
somatropin	growth disorders	1987	23 (343)
factor VIIa	haemophilia	1996	32 (307)
erythropoietin	hematopoietic	1988	24 (301)
peginesatide acetate	hematopoietic	2012	48 (514)
eculizumab	hemoglobinuria	2007	43 (468)
factor VIII	hemostatic	1992	28 (330)
thrombin alfa	hemostatic	2008	44 (627)
malotilate	hepatoprotective	1985	21 (329)
mivotilate	hepatoprotective	1999	35 (343)
buserelin acetate	hormone therapy	1984	20 (316)
goserelin	hormone therapy	1987	23 (336), 50 (553)
leuprolide acetate	hormone therapy	1984	20 (319)
nafarelin acetate	hormone therapy	1990	26 (306)
tibolone	hormone therapy	1988	24 (312)
calcitonin (human)	hypercalcemia	1986	50 (551)
cinacalcet	hyperparathyroidism	2004	40 (451)
doxercalciferol	hyperparathyroidism	1999	35 (339)
falecalcitriol	hyperparathyroidism	2001	37 (266)
maxacalcitol	hyperparathyroidism	2000	36 (310)
paricalcitol	hyperparathyroidism	1998	34 (327)
quinagolide	hyperprolactinemia	1994	30 (309)
glucagon	hypoglycemia	1989	50 (553)
conivaptan	hyponatremia	2006	42 (514)
mozavaptan	hyponatremia	2006	42 (527)
tolvaptan	hyponatremia	2009	45 (528)
bucillamine	immunomodulator	1987	23 (329)
centoxin	immunomodulator	1991	27 (325)
thymopentin	immunomodulator	1985	21 (333), 50 (554)
filgrastim	immunostimulant	1991	27 (327)
GMDP	immunostimulant	1996	32 (308)
interferon gamma-1b	immunostimulant	1991	27 (329)
lentinan	immunostimulant	1986	22 (322)
pegademase bovine	immunostimulant	1990	26 (307)
pidotimod	immunostimulant	1993	29 (343)
romurtide	immunostimulant	1991	27 (332)
sargramostim	immunostimulant	1991	27 (332)
schizophyllan	immunostimulant	1985	22 (326)
ubenimex	immunostimulant	1987	23 (345)
belatacept	immunosuppressant	2011	47 (516)
cyclosporine	immunosuppressant	1983	19 (317)

GENERIC NAME	INDICATION	YEAR INTRODUCED	ARMC VOL., (PAGE)
everolimus	immunosuppressant	2004	40 (455)
gusperimus	immunosuppressant	1994	30 (300)
mizoribine	immunosuppressant	1984	20 (321)
muromonab-CD3	immunosuppressant	1986	22 (323)
mycophenolate mofetil	immunosuppressant	1995	31 (346)
mycophenolate sodium	immunosuppressant	2003	39 (279)
pimecrolimus	immunosuppressant	2002	38 (365)
tacrolimus	immunosuppressant	1993	29 (347)
buserelin	infertility	1984	50 (553)
cetrorelix	infertility	1999	35 (336)
corifollitropin alfa	infertility	2010	46 (455)
follitropin alfa	infertility	1996	32 (307)
follitropin beta	infertility	1996	32 (308)
ganirelix	infertility	1999	50 (554)
ganirelix acetate	infertility	2000	36 (305)
gonadorelin	infertility	1989	50 (553)
defeiprone	iron chelation therapy	1995	31 (340)
deferasirox	iron chelation therapy	2005	41 (446)
alosetron hydrochloride	irritable bowel syndrome	2000	36 (295)
certolizumab pegol	irritable bowel syndrome	2008	44 (592)
linaclotide	irritable bowel syndrome	2012	48 (502), 50 (552)
tegaserod maleate	irritable bowel syndrome	2001	37 (270)
oxytocin	labor induction	1962	50 (555)
oxytocin	lactation	1962	50 (555)
nartograstim	leukopenia	1994	30 (304)
metreleptin	lipodystrophy	2013	49 (474)
tesamorelin acetate	lipodystrophy	2010	46 (486), 50 (552)
belimumab	lupus	2011	47 (519)
Lyme disease vaccine	Lyme disease	1999	35 (342)
avanafil	male sexual dysfunction	2011	47 (512)
sildenafil citrate	male sexual dysfunction	1998	34 (331)
tadalafil	male sexual dysfunction	2003	39 (284)
udenafil	male sexual dysfunction	2005	41 (472)
vardenafil	male sexual dysfunction	2003	39 (286)
laronidase	mucopolysaccharidosis I	2003	39 (278)
idursulfase	mucopolysaccharidosis II (Hunter syndrome)	2006	42 (520)
galsulfase	mucopolysaccharidosis VI	2005	41 (453)
palifermin	mucositis	2005	41 (461)
corticotropin	multiple inflammatory diseases	1952	50 (554)
tetracosactide	multiple inflammatory diseases	1980	50 (554)
dalfampridine	multiple sclerosis	2010	46 (458)
dimethyl fumarate	multiple sclerosis	2013	49 (460)
fingolimod	multiple sclerosis	2010	46 (468)

GENERIC NAME	INDICATION	YEAR INTRODUCED	ARMC VOL., (PAGE)
glatiramer	multiple sclerosis	1996	50 (554)
glatiramer acetate	multiple sclerosis	1997	33 (334)
interferon, b-1a	multiple sclerosis	1996	32 (311)
interferon, b-1b	multiple sclerosis	1993	29 (339)
natalizumab	multiple sclerosis	2004	40 (462)
teriflunomide	multiple sclerosis	2012	48 (530)
afloqualone	muscle relaxant	1983	19 (313)
cisatracurium besilate	muscle relaxant	1995	31 (340)
doxacurium chloride	muscle relaxant	1991	27 (326)
eperisone hydrochloride	muscle relaxant	1983	19 (318)
mivacurium chloride	muscle relaxant	1992	28 (334)
rapacuronium bromide	muscle relaxant	1999	35 (347)
rocuronium bromide	muscle relaxant	1994	30 (309)
tiropramide hydrochloride	muscle relaxant	1983	19 (324)
tizanidine	muscle relaxant	1984	20 (324)
decitabine	myelodysplastic syndromes	2006	42 (519)
lenalidomide	myelodysplastic syndromes, multiple myeloma	2006	42 (523)
tinazoline	nasal decongestant	1988	24 (312)
lucinactant	neonatal respiratory distress syndrome	2012	50 (556)
tafamidis	neurodegeneration	2011	47 (550)
taltirelin	neurodegeneration	2000	36 (311)
lanreotide	neuroendocrine tumors	1994	50 (555)
sugammadex	neuromuscular blockade, reversal	2008	44 (623)
edaravone	neuroprotective	2001	37 (265)
bifemelane hydrochloride	nootropic	1987	23 (329)
idebenone	nootropic	1986	22 (321)
indeloxazine hydrochloride	nootropic	1988	24 (304)
nizofenzone	nootropic	1988	24 (307)
alcaftadine	ophthalmologic (allergic conjunctivitis)	2010	46 (444)
diquafosol tetrasodium	ophthalmologic (dry eye)	2010	46 (462)
pegaptanib	ophthalmologic (macular degeneration)	2005	41 (458)
ranibizumab	ophthalmologic (macular degeneration)	2006	42 (534)
verteporfin	ophthalmologic (macular degeneration)	2000	36 (312)
aflibercept	ophthalmologic, macular degeneration	2011	47 (507)
OP-1	osteoinductor	2001	37 (269)

GENERIC NAME	INDICATION	YEAR INTRODUCED	ARMC VOL., (PAGE)
abaloparatide	osteoporosis	2017	50 (555)
alendronate sodium	osteoporosis	1993	29 (332)
APD	osteoporosis	1987	23 (326)
calcitonin (human)	osteoporosis	1986	50 (551)
denosumab	osteoporosis	2010	46 (459)
disodium pamidronate	osteoporosis	1989	25 (312)
elcatonin	osteoporosis	1981	50 (551)
eldecalcitol	osteoporosis	2011	47 (529)
ibandronic acid	osteoporosis	1996	32 (309)
incadronic acid	osteoporosis	1997	33 (335)
ipriflavone	osteoporosis	1989	25 (314)
minodronic acid	osteoporosis	2009	45 (509)
raloxifene hydrochloride	osteoporosis	1998	34 (328)
risedronate sodium	osteoporosis	1998	34 (330)
strontium ranelate	osteoporosis	2004	40 (466)
teriparatide	osteoporosis	2002	50 (555)
zoledronate disodium	osteoporosis	2000	36 (314)
calcitonin (human)	Paget's disease	1986	50 (551)
tiludronate disodium	Paget's disease	1995	31 (350)
nafamostat mesylate	pancreatitis	1986	22 (323)
budipine	Parkinson's disease	1997	33 (330)
CHF-1301	Parkinson's disease	1999	35 (336)
droxidopa	Parkinson's disease	1989	25 (312)
entacapone	Parkinson's disease	1998	34 (322)
istradefylline	Parkinson's disease	2013	49 (468)
pergolide mesylate	Parkinson's disease	1988	24 (308)
pramipexole hydrochloride	Parkinson's disease	1997	33 (341)
rasagiline	Parkinson's disease	2005	41 (464)
ropinirole hydrochloride	Parkinson's disease	1996	32 (317)
rotigotine	Parkinson's disease	2006	42 (538)
talipexole	Parkinson's disease	1996	32 (318)
tolcapone	Parkinson's disease	1997	33 (343)
sapropterin hydrochloride	phenylketouria	1992	28 (336)
alglucosidase alfa	Pompe disease	2006	42 (511)
alvimopan	post-operative ileus	2008	44 (584)
carbetocin	postpartum hemorrhage	2001	50 (555)
oxytocin	postpartum hemorrhage	1962	50 (555)
histrelin	precocious puberty	1993	29 (338)
histrelin	precocious puberty	1991	50 (553)
leuprorelin	precocious puberty	1984	50 (553)
nafarelin	precocious puberty	1990	50 (553)
triptorelin	precocious puberty	1986	50 (553)
dapoxetine	premature ejaculation	2009	45 (488)
atosiban	premature labor	2000	36 (297)

GENERIC NAME	INDICATION	YEAR INTRODUCED	ARMC VOL., (PAGE)
gonadorelin	primary amenorrhea	1989	50 (553)
desmopressin	primary nocturnal enuresis, nocturia	1972	50 (556)
nalfurafine hydrochloride	pruritus	2009	45 (510)
pirfenidone	pulmonary fibrosis, idiopathic	2008	44 (614)
ambrisentan	pulmonary hypertension	2007	43 (463)
macitentan	pulmonary hypertension	2013	49 (472)
riociguat	pulmonary hypertension	2013	49 (486)
sitaxsentan	pulmonary hypertension	2006	42 (543)
pumactant	respiratory distress syndrome	1994	30 (308)
mepixanox	respiratory stimulant	1984	20 (320)
surfactant TA	respiratory surfactant	1987	23 (344)
gabapentin Enacarbil	restless leg syndrome	2011	47 (533)
etelcalcetide	secondary hyperparathyroidism	2016	50 (551)
teduglutide	short bowel syndrome	2012	48 (526), 50 (552)
adrafinil	sleep disorders	1986	22 (315)
armodafinil	sleep disorders	2009	45 (478)
binfonazole	sleep disorders	1983	19 (315)
brotizolam	sleep disorders	1983	19 (315)
butoctamide	sleep disorders	1984	20 (316)
dexmedetomidine hydrochloride	sleep disorders	2000	36 (301)
eszopiclone	sleep disorders	2005	41 (451)
loprazolam mesylate	sleep disorders	1983	19 (321)
modafinil	sleep disorders	1994	30 (303)
quazepam	sleep disorders	1985	21 (332)
ramelteon	sleep disorders	2005	41 (462)
rilmazafone	sleep disorders	1989	25 (317)
zaleplon	sleep disorders	1999	35 (351)
zolpidem hemitartrate	sleep disorders	1988	24 (313)
zopiclone	sleep disorders	1986	22 (327)
plerixafor hydrochloride	stem cell mobilizer	2009	45 (515)
tirilazad mesylate	subarachnoid hemorrhage	1995	31 (351)
bivalirudin	thrombosis	1999	50 (556)
balsalazide disodium	ulcerative colitis	1997	33 (329)
acetohydroxamic acid	urinary tract/bladder disorders	1983	19 (313)
darifenacin	urinary tract/bladder disorders	2005	41 (445)
fesoterodine	urinary tract/bladder disorders	2008	44 (604)
imidafenacin	urinary tract/bladder disorders	2007	43 (472)
mirabegron	urinary tract/bladder disorders	2011	47 (542)
naftopidil	urinary tract/bladder disorders	1999	35 (344)
propiverine hydrochloride	urinary tract/bladder disorders	1992	28 (335)
silodosin	urinary tract/bladder disorders	2006	42 (540)

GENERIC NAME	INDICATION	YEAR INTRODUCED	ARMC VOL., (PAGE)
sodium cellulose phosphate	urinary tract/bladder disorders	1983	19 (323)
solifenacin	urinary tract/bladder disorders	2004	40 (466)
tiopronin	urolithiasis	1989	25 (318)
leuprorelin	uterine fibroids	1984	50 (553)
vasopressin	vasodilatory shock	1962	50 (551)
corticotropin	West syndrome	1952	50 (554)
acemannan	wound healing agent	2001	37 (259)
cadexomer iodine	wound healing agent	1983	19 (316)
epidermal growth factor	wound healing agent	1987	23 (333)
prezatide copper acetate	wound healing agent	1996	32 (314)

CPI Antony Rowe
Chippenham, UK
2017-11-24 11:21